Lecture Notes in Mathematics

Volume 2290

More information about this series at http://www.springer.com/series/304

Chaire Jean-Morlet

The CIRM Jean-Morlet Series is a collection of scientific publications centering on the themes developed by successive holders of the Jean Morlet Chair.

This chair has been hosted by the *Centre International de Rencontres Mathématiques* (CIRM, Luminy, France) since its creation in 2013. The Chair is named in honour of Jean Morlet (1931–2007). He was an engineer at the French oil company Elf (now Total) and, together with the physicist Alex Grossman, conducted pioneering work in wavelet analysis. This theory has since become a building block of modern mathematics. It was at CIRM that they met on several occasions, and the center then played host to some of the key conferences in this field.

Appointments to the *Jean-Morlet* Chair are made to world-class researchers based outside France and who work in collaboration with local project leaders in order to conduct original and ambitious scientific programs. The Chair is supported financially by CIRM, Aix-Marseille Université and the City of Marseille.

A key feature of the Chair is that it does not focus solely on the research themes developed by Jean Morlet. The idea is to support the freedom of pioneers in mathematical sciences and to nurture the enthusiasm that comes from opening new avenues of research.

CIRM: a beacon for international cooperation

Situated at the heart of the *Parc des Calanques*, an area of outstanding natural beauty, CIRM is one of the largest conference centers dedicated to mathematical and related sciences in the world, with close to 3500 visitors per year. Jointly supervised by SMF (the French Mathematical Society) and CNRS (French National Center for Scientific Research), CIRM has been a hub for international research in mathematics since 1981. CIRM's *raison d'être* is to be a venue that fosters exchanges, pioneering research in mathematics in interaction with other sciences and the dissemination of knowledge to the younger scientific community.

www.chairejeanmorlet.com
www.cirm-math.fr

Mark Pollicott • Sandro Vaienti

Editors

Thermodynamic Formalism

CIRM Jean-Morlet Chair, Fall 2019

 Springer

Editors
Mark Pollicott
Department of Mathematics
Warwick University
Coventry, UK

Sandro Vaienti
Institute de Mathématiques de Toulon
Toulon, France
Centre de Physique Théorique de Luminy
Marseille, France

ISSN 0075-8434   ISSN 1617-9692   (electronic)
Lecture Notes in Mathematics
ISBN 978-3-030-74862-3   ISBN 978-3-030-74863-0   (eBook)
https://doi.org/10.1007/978-3-030-74863-0

Mathematics Subject Classification: 37D35, 37C35

This Springer imprint is published by the registered company Springer Nature Switzerland AG.
The registered company address is: Gewerbestrasse 11, 6330 Cham, Switzerland

Il semble que la perfection soit atteinte non quand il n'y a plus rien à ajouter, mais quand il n'y a plus rien à retrancher
  –Antoine de Saint-Exupéry - (Terre des Hommes, 1939)

Foreword

Bulk matter usually appears to us as solid, liquid, or gas. And some states of bulk matter can be characterized as equilibrium states. These states have features—like temperature—which have no obvious interpretation in terms of classical mechanics. A macroscopic theory of equilibrium states has been developed, involving somewhat obscure new quantities like entropy. This macroscopic theory of equilibrium states is called *thermodynamics*.

At the end of the nineteenth century, the underlying microscopic mechanical structure of thermodynamics was revealed by Maxwell, Boltzmann, and Gibbs: *statistical mechanics* was created. It turned out that the microscopic definition of equilibrium states involves a statistical superposition of many configurations of particles (the volume of these configurations being related to entropy).

While the physics of statistical mechanics was clear to its founding fathers, it must be realized that the mathematics available to them was extremely deficient compared to what is available to us. They lacked measure theory, and the basic ergodic theory necessary for the understanding of dynamical systems. When these became available, the extreme mathematical richness underlying statistical mechanics became progressively visible: the *thermodynamic formalism* was born.

An important element in the thermodynamic formalism is the concept of Gibbs state. This is a mathematically local version of equilibrium states. As seen by Sinai, Gibbs states on a one-dimensional lattice correspond to probability measures of great interest for an important class of dynamical systems (those which are uniformly hyperbolic). There is thus an unexpected relation between statistical mechanics and smooth dynamics: this has many consequences. The present set of lectures will present some aspects of the unification brought by the thermodynamic formalism to different domains of mathematics and physics.

Bures-sur-Yvette, France
January 2021

David Ruelle

Preface

During the latter half of 2019, CIRM hosted a semester on *Thermodynamic Formalism: Applications to Probability, Geometry and Fractals (Formalisme thermodynamique : applications aux probabilités, à la géométrie et aux fractales)*. This was under the auspices of the Jean Morlet chair programme, where the Jean Morlet chair holder was Mark Pollicott (Warwick University) and the local coordinator was Sandro Vaienti (University of Toulon and CPT Marseille). Luminy provided a backdrop of great natural beauty for the diverse scientific activities. This volume arose from minicourse notes, surveys and research articles that were a consequence of the research workshop, summer school and conference and other scientific activities that took place between 1st of July 2019 and the 31st of December 2019.

The name of the semester was inspired by the title of the highly influential book *Thermodynamic Formalism*, by David Ruelle. The programme began with a summer school on *Thermodynamic Formalism: Modern Techniques in Smooth Ergodic Theory (Formalisme thermodynamique : Techniques modernes en théorie ergodique)* organized by Mark Pollicott and Sandro Vaienti, which had a stimulating mixture of short lecture courses and individual keynote speakers. The aim of the school was to introduce participants to some of the current themes and ideas within the broad panorama of thermodynamic formalism and, in particular, its applications to geometry, probability theory and ergodic theory, dynamical systems, fractals, and number theory. At the end of the semester, there was a large conference on *Thermodynamic Formalism: Dynamical Systems, Statistical Properties and their Applications Formalisme thermodynamique : Systèmes dynamiques, propriétés statistiques et leur applications*. The scientific and organizing committee consisted of: Matthew Nicol (University of Houston), Mark Pollicott, Serge Troubetzkoy (Aix-Marseille University), and Sandro Vaienti. This conference brought together experts in thermodynamic formalism and specialists in related areas to review the current state of the subject and its myriad applications. The conference was enlivened by coinciding with a national rail strike. In addition, there was a more focused small research workshop *Thermodynamic Formalism: Ergodic Theory and Validated Numerics Formalisme thermodynamique : Théorie ergodique et validation numérique*, which further added to the diversity of the topics.

We now briefly describe the contents of this volume. We have grouped the chapters according to sub-areas of the subject even though they share a common use of "thermodynamic ideas" (including specification, Gibbs measures or equilibrium states and transfer operators).

Part I deals with two of the basic tools in the area: specification and expansiveness. The substantial survey by Climenhaga and Thompson on Bowen's specification property and its generalizations illustrates the role of this classical technique to show uniqueness of equilibrium measures. The chapter by Troubetzkoy and Varandas further explores the interrelationship between expansiveness and other familiar dynamical properties.

Part II contains a chapter dedicated to thermodynamical formalism in the context of low dimensional systems, an area where this approach has proved fruitful. The chapter by Mayer and Urbanski gives an exposition of these ideas in the context of complex holomorphic maps.

Part III contains chapters on probability and ergodic theory. Conze's survey illustrates the connections with classical ergodic theory, cocycles and recurrence. In a similar spirit, the chapter by Pène and Saussol describes with visits to small sets. On the other hand, the chapter by Dragicevic and Hafouta deals with stronger statistical properties and invariance principles, and that of Ngo and Peigne deals with random walks.

The theme of Part IV is the application of thermodynamic ideas to geometry. The chapter by Broise-Alamichel, Parkkonen and Paulin on the rate of mixing for equilibrium states in negative curvature and trees corresponds to a lecture course given by the third author. The chapter by Aimino and Pollicott deals with translation surfaces, and the chapters by Kao and Pollicott and Sharp deal with the pressure metric on moduli spaces.

Finally, Part V contains chapters on fractal geometry. The first is illustrated by the chapters by Falconer, Fraser and Simon. Falconer's chapter addresses different notions of dimension (box dimension, Hausdorff dimension and, now, intermediate dimension). The chapter by Feng and Simon deals with iterated function schemes and dimension estimates. Finally, the chapter of Fraser focuses on one of the motivating examples in this area: The Bedford-McMullen carpet. Finally, Matheus' chapter explores visible parts of fractal sets.

Acknowledgments We are grateful to the director and staff at CIRM for their help and support throughout the programme. The semester received generous core funding by CIRM and the University of Aix-Marseilles. There was additional funding from the city of Marseille, the two Marseille laboratories I2M and CPT, CNRS-Peps, Labex Archimede, Labex Carmin, FRUMAM, and ANR grants. Specific activities benefited from gratefully received targeted support from the Clay Institute, the US National Science Foundation, the UK Engineering and Physical Sciences Research Council, and the European Mathematical Society.

Kenilworth, UK   Mark Pollicott
Marseille, France   Sandro Vaienti
January 2021

Contents

Contributors

Romain Aimino Centro de Matemática da Universidade do Porto, Porto, Portugal

Anne Broise-Alamichel Laboratoire de mathématique d'Orsay, UMR 8628 CNRS, Université Paris-Saclay, Orsay, France

Vaughn Climenhaga Department of Mathematics, University of Houston, Houston, TX, USA

Jean-Pierre Conze Univ Rennes, CNRS, IRMAR - UMR 6625, Rennes, France

Davor Dragičević Department of Mathematics, University of Rijeka, Rijeka, Croatia

Kenneth J. Falconer Mathematical Institute, University of St Andrews, St Andrews, Fife, UK

De-Jun Feng Department of Mathematics, The Chinese University of Hong Kong, Shatin, Hong Kong

Jonathan M. Fraser University of St Andrews, St Andrews, Scotland

Yeor Hafouta Department of Mathematics, The Ohio State University, Columbus, OH, USA

Lien-Yung Kao Department of Mathematics, The George Washington University, Washington, DC, USA

Yuri Kifer Institute of Mathematics, The Hebrew University, Jerusalem, Israel

Carlos Matheus CMLS, CNRS, École polytechnique, Institut Polytechnique de Paris, Palaiseau Cedex, France

Volker Mayer Université de Lille, Département de Mathématiques, UMR 8524 du CNRS, Villeneuve d'Ascq Cedex, France

Hoang-Long Ngo Hanoi National University of Education, Cau Giay, Vietnam

Jouni Parkkonen Department of Mathematics and Statistics, University of Jyväskylä, Jyväskylä, Finland

Frédéric Paulin Laboratoire de mathématique d'Orsay, UMR 8628 CNRS, Université Paris-Saclay, Orsay Cedex, France

Marc Peigné Laboratoire de Mathématiques et Physique Théorique, CNRS UMR 7350, Faculté des Sciences et Techniques, Université François Rabelais, Parc de Grandmont, Tours, France

Françoise Pène Univ Brest, Université de Brest, LMBA, Laboratoire de Mathématiques de Bretagne Atlantique, CNRS UMR 6205, Brest, France

Mark Pollicott Mathematics Institute, University of Warwick, Coventry, UK

Benoît Saussol Univ Brest, Université de Brest, LMBA, Laboratoire de Mathématiques de Bretagne Atlantique, CNRS UMR 6205, Brest, France

Richard Sharp Mathematics Institute, University of Warwick, Coventry, UK

Károly Simon Budapest University of Technology and Economics, Department of Stochastics, Institute of Mathematics and MTA-BME Stochastics Research Group, Budapest, Hungary

Dan Thompson Department of Mathematics, Ohio State University, Columbus, OH, USA

Serge Troubetzkoy Aix Marseille Univ, CNRS, Centrale Marseille, I2M, Marseille, France

Mariusz Urbański Department of Mathematics, University of North Texas, Denton, TX, USA

Paulo Varandas CMUP and Departamento de Matemática, Universidade Federal da Bahia Av. Ademar de Barros s/n, Salvador, Brazil

Part I
Specifications and Expansiveness

Chapter 1
Beyond Bowen's Specification Property

Vaughn Climenhaga and Daniel J. Thompson

Abstract A classical result in thermodynamic formalism is that for uniformly hyperbolic systems, every Hölder continuous potential has a unique equilibrium state. One proof of this fact is due to Rufus Bowen and uses the fact that such systems satisfy expansivity and specification properties. In these notes, we survey recent progress that uses generalizations of these properties to extend Bowen's arguments beyond uniform hyperbolicity, including applications to partially hyperbolic systems and geodesic flows beyond negative curvature. We include a new criterion for uniqueness of equilibrium states for partially hyperbolic systems with 1-dimensional center.

1.1 Introduction

We survey recent progress in the study of existence and uniqueness of measures of maximal entropy and equilibrium states in settings beyond uniform hyperbolicity using weakened versions of specification and expansivity. Our focus is a long-running joint project initiated by the authors in [1], and extended in a series of papers including [2, 3]. This approach is based on the fundamental insights of Rufus Bowen in the 1970s [4, 5], who identified and formalized three properties enjoyed by uniformly hyperbolic systems that serve as foundations for the equilibrium state theory: these properties are specification, expansivity, and a regularity condition now known as the Bowen property. We relax all three of these properties in order to study systems exhibiting various types of non-uniform structure. These notes start by recalling the basic mechanisms of Bowen, and then gradually build up in generality, introducing the ideas needed to move to non-uniform versions of

V. Climenhaga
Department of Mathematics, University of Houston, Houston, TX, USA
e-mail: climenha@math.uh.edu

D. J. Thompson (✉)
Department of Mathematics, Ohio State University, Columbus, OH, USA
e-mail: thompson@math.osu.edu

© The Author(s), under exclusive license to Springer Nature Switzerland AG 2021
M. Pollicott, S. Vaienti (eds.), *Thermodynamic Formalism*, Lecture Notes
in Mathematics 2290, https://doi.org/10.1007/978-3-030-74863-0_1

Bowen's hypotheses. The generality is motivated by, and illustrated by, examples: we discuss applications in symbolic dynamics, to certain partially hyperbolic systems, and to wide classes of geodesic flows with non-uniform hyperbolicity. This survey has its roots in the authors' 6-part minicourse at the *Dynamics Beyond Uniform Hyperbolicity* conference at CIRM in May 2019.

Section 1.2 describes Bowen's result for MMEs and the simplest case of our generalization. It begins by recalling the basic ideas of thermodynamic formalism (Sect. 1.2.1) and outlining Bowen's original argument in the simplest case: the measure of maximal entropy (MME) for a shift space with specification (Sect. 1.2.2). In Sect. 1.2.3, we introduce the main idea of our approach, the use of *decompositions* to quantify the idea of "obstructions to specification", and we give an application to β-shifts. Moving beyond the symbolic case requires the notion of expansivity, and in Sect. 1.2.4 we discuss the role this plays in Bowen's argument.

Section 1.3 develops our general results for discrete-time systems. The notion of "obstructions to expansivity" is introduced in Sect. 1.3.1, and an application to partial hyperbolicity (the Mañé example) is described in Sect. 1.3.2. Combining the notions of obstructions to specification and expansivity leads to the general result for MMEs in discrete-time in Sect. 1.3.3, which is applied in Sect. 1.3.4 to the broader class of partially hyperbolic diffeomorphisms with one-dimensional center. The extension to equilibrium states for nonzero potential functions is given in Sect. 1.3.5.

Section 1.4 is devoted to equilibrium states for geodesic flows, with particular emphasis on the case of non-positive curvature, which is one of the most widely studied examples of a non-uniformly hyperbolic flow. After recalling some geometric background in Sect. 1.4.1, we give an introduction in Sect. 1.4.2 to the ideas in the paper [3], including the main "pressure gap" criterion for uniqueness, and how to decompose the space of orbit segments using a function λ that measures curvature of horospheres. We also outline recent results for manifolds without conjugate points and CAT(-1) spaces. In Sect. 1.4.3, we discuss how to improve ergodicity of the equilibrium states in non-positive curvature to the much stronger Kolmogorov K-property. Finally, in Sect. 1.4.4, we describe our proof of Knieper's "entropy gap" for geodesic flow on a rank 1 non-positive curvature manifold.

To illustrate the broad utility of the specification-based approach to uniqueness, we mention the following applications of the machinery we describe, which go well beyond what we are able to discuss in detail in this survey.

- Measures of maximal entropy for symbolic examples: β-shifts, S-gap shifts, and their factors [1]; certain shifts of quasi-finite type [6]; S-limited shifts [7]; shifts with "one-sided almost specification" [8]; $(-\beta)$-shifts [9];
- Equilibrium states for symbolic examples: β-shifts in [10], their factors in [6, 11] (in particular, [11] studies general conditions under which the "pressure gap" condition holds); S-gap shifts in [12]; certain α-β shifts [13]; applications to Manneville–Pomeau and related interval maps [10].

- Diffeomorphisms beyond uniform hyperbolicity: Bonatti–Viana examples [14]; Mañé examples [15]; Katok examples [16]; certain partially hyperbolic attractors [17].
- Geodesic flows: non-positive curvature [3]; no focal points [18, 19]; no conjugate points [20]; CAT(-1) geodesic flows [21].

We also mention two related results: the machinery we describe has recently been used to prove "denseness of intermediate pressures" [22]; an approach to uniqueness (and non-uniqueness) for equilibrium states using various weak specification properties has been developed by Pavlov [23, 24] for symbolic and expansive systems.

The current literature in the field is vibrant and continually growing. The scope of this article is restricted to the specification approach to equilibrium states, and we largely do not address the literature beyond that. Other uses for the specification property that we do not discuss include large deviations properties, multifractal analysis, and universality constructions; see e.g. [25–31] (among many others). Different variants of the specification property are sometimes more appropriate for these arguments; various definitions are surveyed in [32, 33].

We stress that we do not address the use of other techniques to study existence and uniqueness of equilibrium states. These approaches include transfer operator techniques, Margulis-type constructions, symbolic dynamics, and the Patterson-Sullivan approach. We suggest the following recent references as a starting point to delve into the literature: [34–39]. Classic references include [40–42].

We also do not discuss the large and important area of statistical properties for equilibrium states. If f is a $C^{1+\alpha}$ Anosov diffeomorphism (or if X is an Axiom A attractor) then the unique equilibrium state for the *geometric potential* $\varphi(x) = -\log|\det Df|_{E^u(x)}|$ is the physically relevant Sinai–Ruelle–Bowen (SRB) measure. This provides important motivation and application for thermodynamic formalism, and this general setting is one of the major approaches to studying the statistical properties of the SRB measure. References include [40, 41, 43–48].

We sometimes adopt a conversational writing style. We hope that the informal style will be helpful for current purposes; we invite the reader to look at our original papers, particularly [1–3] for a more precise account.

1.2   Main Ideas: Uniqueness of the Measure of Maximal Entropy

We introduce our main ideas in the case of a discrete-time dynamical system (X, f). In this section, we often consider the case when (X, f) is a shift space. We also consider the general topological dynamics setting where X is a compact metric space and $f\colon X \to X$ is continuous. In many of our examples of interest, X is a smooth manifold and f is a diffeomorphism.

1.2.1 Entropy and Thermodynamic Formalism

For a *probability vector* $\mathbf{p} = (p_1, \ldots, p_N) \in [0, 1]^N$, where $\sum p_i = 1$, the *entropy* of $\mathbf{p}$ is $H(\mathbf{p}) = \sum_i -p_i \log p_i$. The following is an elementary exercise:

- $\max_{\mathbf{p}} H(\mathbf{p}) = \log N$;
- $H(\mathbf{p}) = \log N \quad \Leftrightarrow \quad p_i = \frac{1}{N}$ for all i $\quad \Leftrightarrow \quad p_i = p_j$ for all i, j.

These general principles lie at the heart of thermodynamic formalism for uniformly hyperbolic dynamical systems, with 'probability vector' replaced by 'invariant probability measure':

- there is a function called 'entropy' that we wish to maximize;
- it is maximized at a unique measure (variational principle and uniqueness);
- that measure is characterized by an equidistribution (Gibbs) property.

Now we recall the formal definitions, referring to [49–52] for further details and properties.

Let X be a compact metric space and $f : X \to X$ a continuous map. This gives a discrete-time topological dynamical system (X, f). Let $\mathcal{M}_f(X)$ denote the space of Borel f-invariant probability measures on X.

When f exhibits some hyperbolic behavior, $\mathcal{M}_f(X)$ is typically extremely large—an infinite-dimensional simplex—and it becomes important to identify certain "distinguished measures" in $\mathcal{M}_f(X)$. This includes SRB measures, measures of maximal entropy, and more generally, equilibrium measures.

Definition 1.2.1.1 (Measure-Theoretic Kolmogorov–Sinai Entropy) Fix $\mu \in \mathcal{M}_f(X)$. Given a countable partition α of X into Borel sets, write

$$H_\mu(\alpha) := \sum_{A \in \alpha} -\mu(A) \log \mu(A) = \int -\log \mu(\alpha(x)) \, d\mu(x)$$

for the *static entropy* of α, where we write $\alpha(x)$ for the element of α containing x. One can interpret $H_\mu(\alpha)$ as the expected amount of information gained by observing which partition element a point $x \in X$ lies in. Given $j \leq k$, the corresponding *dynamical refinement* of α records which elements of α the iterates $f^j x, \ldots, f^k x$ lie in:

$$\alpha_j^k = \bigvee_{i=j}^{k} f^{-i} \alpha \quad \Leftrightarrow \quad \alpha_j^k(x) = \bigcap_{i=j}^{k} f^{-i}(\alpha(f^i x)).$$

A standard short argument shows that

$$H_\mu(\alpha_0^{n+m-1}) \leq H_\mu(\alpha_0^{n-1}) + H_\mu(\alpha_n^{n+m-1}) = H_\mu(\alpha_0^{n-1}) + H_\mu(\alpha_0^{0+m-1}),$$

so that the sequence $c_n = H_\mu(\alpha_0^{n-1})$ is subadditive: $c_{n+m} \leq c_n + c_m$. Thus, by Fekete's lemma [53], $\lim \frac{c_n}{n}$ exists, and equals $\inf \frac{c_n}{n}$. We can therefore define the *dynamical entropy* of α with respect to f to be

$$h_\mu(f, \alpha) := \lim_{n \to \infty} \frac{1}{n} H_\mu(\alpha_0^{n-1}) = \inf_{n \in \mathbb{N}} \frac{1}{n} H_\mu(\alpha_0^{n-1}).$$

The *measure-theoretic (Kolmogorov–Sinai) entropy* of (X, f, μ) is

$$h_\mu(f) = \sup_\alpha h_\mu(f, \alpha),$$

where the supremum is taken over all partitions α as above for which $H_\mu(\alpha) < \infty$.

The *variational principle* [50, Theorem 8.6] states that

$$\sup_{\mu \in \mathcal{M}_f(X)} h_\mu(f) = h_{\text{top}}(X, f),$$

where $h_{\text{top}}(X, f)$ is the *topological entropy* of $f : X \to X$, which we will define more carefully below (Definition 1.2.4.2). Now we define a central object in our study.

Definition 1.2.1.2 (MMEs) A measure $\mu \in \mathcal{M}_f(X)$ is a *measure of maximal entropy (MME)* for (X, f) if $h_\mu(f) = h_{\text{top}}(X, f)$; equivalently, if $h_\nu(f) \leq h_\mu(f)$ for every $\nu \in \mathcal{M}_f(X)$.

The following theorem on uniformly hyperbolic systems is classical.

Theorem 1.2.1.1 (Existence and Uniqueness) *Suppose one of the following is true.*

1. *$(X, f = \sigma)$ is a transitive shift of finite type (SFT).*
2. *$f : M \to M$ is a C^1 diffeomorphism and $X \subset M$ is a compact f-invariant topologically transitive locally maximal hyperbolic set.[1]*

Then there exists a unique measure of maximal entropy μ for (X, f).

Remark 1.1 The unique MME can be thought of as the 'most complex' invariant measure for a system, and often encodes dynamically relevant information such as the distribution and asymptotic behavior of the set of periodic points.

---

[1] In particular, this holds if $X = M$ is compact and f is a transitive Anosov diffeomorphism.

1.2.2 Bowen's Original Argument: The Symbolic Case

1.2.2.1 The Specification Property in a Shift Space

Following Bowen [5], we outline a proof of Theorem 1.2.1.1 in the first case, when (X, σ) is a transitive SFT. The original construction of the MME in this setting is due to Parry and uses the transition matrix. Bowen's proof works for a broader class of systems, which we now describe.

Fix a finite set A (the *alphabet*), let $\sigma : A^{\mathbb{N}} \to A^{\mathbb{N}}$ be the shift map $\sigma(x_1 x_2 \ldots) = x_2 x_3 \ldots$, and let $X \subset A^{\mathbb{N}}$ be closed and σ-invariant: $\sigma(X) = X$. Here $A^{\mathbb{N}}$ (and hence X) is equipped with the metric $d(x, y) = 2^{-\min\{n : x_n \neq y_n\}}$. We refer to X as a *one-sided shift space*. One could just as well consider two-sided shift spaces by replacing $\mathbb{N}$ with $\mathbb{Z}$ (and using $|n|$ in the definition of d); all the results below would be the same, with natural modifications to the proofs. Note that so far we do not assume that X is an SFT or anything of the sort.

Given $x \in A^{\mathbb{N}}$ and $i < j$, we write $x_{[i,j]} = x_i x_{i+1} \cdots x_j$ for the *word* that appears in positions i through j. We use similar notation to denote subwords of a word $w \in A^* := \bigcup_n A^n$. Given $w \in A^n$, we write $|w| = n$ for the *length* of the word, and $[w] = \{x \in X : x_{[1,n]} = w\}$ for the *cylinder* it determines in X. We write

$$\mathcal{L}_n := \{w \in A^n : [w] \neq \emptyset\}, \qquad \mathcal{L} := \bigcup_{n \geq 0} \mathcal{L}_n,$$

and refer to $\mathcal{L}$ as the *language* of X.

Definition 1.2.2.1 The *topological entropy* of X is $h_{\text{top}}(X) = \lim_{n \to \infty} \frac{1}{n} \log \#\mathcal{L}_n$. We often write $h(X)$ for brevity. The limit exists by Fekete's lemma using the fact that $\log \#\mathcal{L}_n$ is subadditive, which we prove in Lemma 1.2.2.1 below.

It is a simple exercise to verify that every transitive SFT has the following property: there is $\tau \in \mathbb{N}$ such that for every $v, w \in \mathcal{L}$ there is $u \in \mathcal{L}$ with $|u| \leq \tau$ such that $vuw \in \mathcal{L}$. Iterating this, we see that

for every $w^1, \ldots, w^k \in \mathcal{L}$ there are $u^1, \ldots, u^{k-1} \in \mathcal{L}$

such that $|u^i| \leq \tau$ for all i, and $w^1 u^1 w^2 u^2 \cdots u^{k-1} w^k \in \mathcal{L}$.   (1.2.2.1)

We say that a shift space whose language satisfies (1.2.2.1) has the *specification property*. There are a number of different variants of specification in the literature:[2] for example, one might ask that the connecting words $u^i \in \mathcal{L}$ satisfy $|u^i| = \tau$,

---

[2] The terminology in the literature for these different variants (weak specification, almost specification, almost weak specification, transitive orbit gluing, etc.) is not always consistent, and we make no attempt to survey or standardize it here. To keep our terminology as simple as possible, we just use the word *specification* for the version of the definition which is our main focus. In places where a different variant is considered, we take care to emphasize this.

which implies topological mixing, not just transitivity (this stronger property holds for mixing SFTs). The version in (1.2.2.1) is sufficient for the uniqueness argument, which is the main goal of these notes.[3]

Theorem 1.2.2.1 (Shift Spaces with Specification) *Let (X, σ) be a shift space with the specification property. Then there is a unique measure of maximal entropy on X.*

In the remainder of this section, we outline the two main steps in the proof of Theorem 1.2.2.1: proving uniqueness using a Gibbs property (Sect. 1.2.2.2), and building a measure with the Gibbs property using specification (Sect. 1.2.2.3).[4]

Remark 1.2 As mentioned above, the original proof that a transitive SFT has a unique MME is due to Parry [56]. Parry constructed the MME using eigendata of the transition matrix for the SFT, and proved uniqueness by showing that any MME must be a Markov measure, then showing that there is only one MME among Markov measures.

A different proof of uniqueness in the SFT case was given by Adler and Weiss, who gave a more flexible argument based on showing that if μ is the Parry measure, then every $\nu \perp \mu$ must have smaller entropy. The argument is described in [57], with full details in [58]. A key step in the proof is to consider an arbitrary set $E \subset X$ and relate $\mu(E)$ to the number of n-cylinders intersecting E. In extending the uniqueness result to sofic shifts (factors of SFTs), Weiss [59] clarified the crucial role of what we refer to below as the "lower Gibbs bound" in carrying out this step. This is essentially the proof of uniqueness that we use in all the results in this survey.

The crucial difference between Theorem 1.2.2.1 and the results of Parry, Adler, and Weiss is the construction of the MME using the specification property rather than eigendata of a matrix. This is due to Bowen, as is the further generalization to non-symbolic systems and equilibrium states for non-zero potentials [5]. Thus we often refer informally to the proof below as "Bowen's argument".

1.2.2.2   The Lower Gibbs Bound as the Mechanism for Uniqueness

It follows from the Shannon–McMillan–Breiman theorem that if μ is an ergodic shift-invariant measure, then for μ-a.e. x we have

$$-\frac{1}{n} \log \mu[x_{[1,n]}] \to h_\mu(\sigma) \text{ as } n \to \infty.$$

---

[3]For other purposes, and especially in the absence of any expansivity property, the difference between $\leq \tau$ and $= \tau$ can be quite substantial, see for example [54, 55].

[4]The notes at https://vaughnclimenhaga.wordpress.com/2020/06/23/specification-and-the-measure-of-maximal-entropy/ give a slightly more detailed version of this proof.

This can be rewritten as

$$\frac{1}{n} \log \left(\frac{\mu[x_{[1,n]}]}{e^{-nh_\mu(\sigma)}} \right) \to 0 \quad \text{for } \mu\text{-a.e. } x.$$

In other words, for μ-typical x, the measure $\mu[x_{[1,n]}]$ decays like $e^{-nh_\mu(\sigma)}$ in the sense that $\mu[x_{[1,n]}]/e^{-nh_\mu(\sigma)}$ is "subexponential in n". The mechanism for uniqueness in the Parry–Adler–Weiss–Bowen argument is to produce an ergodic measure for which this subexponential growth is strengthened to uniform boundedness[5] and applies for all x.

The next proposition makes this *Gibbs property* precise and explain how uniqueness follows; then in Sect. 1.2.2.3 we describe how to construct such a measure. The following argument appears in [59, Lemma 2] (see also [57, 58]); see [60] for a version that works in the nonsymbolic setting, which we will describe in Sect. 1.2.4.4 below.

Proposition 1.1 *Let* $X \subset A^{\mathbb{N}}$ *be a shift space and* μ *an ergodic* σ-*invariant measure on* X. *Suppose that there are* $K, h > 0$ *such that for every* $x \in X$ *and* $n \in \mathbb{N}$, *we have the* Gibbs bounds

$$K^{-1} e^{-nh} \leq \mu[x_{[1,n]}] \leq K e^{-nh}. \tag{1.2.2.2}$$

Then $h = h_\mu(\sigma) = h_{\text{top}}(X, \sigma)$, *and* μ *is the unique MME for* (X, σ).

Proof First observe that by the Shannon–McMillan–Breiman theorem, the upper bound in (1.2.2.2) gives $h_\mu(\sigma) \geq h$, while the lower bound gives $h_\mu(\sigma) \leq h$.[6] Moreover, summing (1.2.2.2) over all words in $\mathcal{L}_n$ gives $K^{-1} e^{nh} \leq \#\mathcal{L}_n \leq K e^{nh}$, so $h_{\text{top}}(X, \sigma) = h$.

The remainder of the proof is devoted to using the lower bound to show that

$$h_\nu(\sigma) < h = h_\mu(\sigma) \text{ for all } \nu \in \mathcal{M}_\sigma(X) \text{ with } \nu \neq \mu. \tag{1.2.2.3}$$

This will show that μ is the unique MME.

Given $\nu \in \mathcal{M}_\sigma(X)$, the Lebesgue decomposition theorem gives $\nu = t\nu_1 + (1 - t)\nu_2$ for some $t \in [0, 1]$ and $\nu_1, \nu_2 \in \mathcal{M}_f(X)$ with $\nu_1 \perp \mu$ and $\nu_2 \ll \mu$. By ergodicity, $\nu_2 = \mu$, and thus if $\nu \neq \mu$ we must have $t > 0$. Since $h_\nu(\sigma) = th_{\nu_1}(\sigma) + (1 - t)h_{\nu_2}(\sigma)$ and $h_{\nu_2}(\sigma) = h_\mu(\sigma) \leq h$, we see that to prove (1.2.2.3), it suffices to prove that $h_\nu(\sigma) < h$ whenever $\nu \perp \mu$.

---

[5] We will encounter this general principle multiple times: many of our proofs rely on obtaining uniform bounds (away from 0 and ∞) for quantities that *a priori* can grow or decay subexponentially.

[6] This requires ergodicity of μ; one can also give a short argument directly from the definition of $h_\mu(\sigma)$ that does not need ergodicity.

Writing α for the (generating) partition into 1-cylinders, we see that for any $\nu \in \mathcal{M}_\sigma(X)$ we have

$$nh_\nu(\sigma) = h_\nu(\sigma^n) = h_\nu(\sigma^n, \alpha_0^{n-1}) \le H_\nu(\alpha_0^{n-1})$$

$$= \sum_{w \in \mathcal{L}_n} -\nu[w] \log \nu[w]. \qquad (1.2.2.4)$$

When $\nu \perp \mu$, there is a Borel set $D \subset X$ such that $\mu(D) = 1$ and $\nu(D) = 0$. Since cylinders generate the σ-algebra, there is $\mathcal{D} \subset \mathcal{L}(X)$ such that $\mu(\mathcal{D}_n) \to 1$ and $\nu(\mathcal{D}_n) \to 0$, where $\mu(\mathcal{D}_n) := \mu\left(\bigcup_{w \in \mathcal{D}_n}[w]\right)$. We break the sum in (1.2.2.4) into two pieces, one over $\mathcal{D}_n$ and one over $\mathcal{D}_n^c = \mathcal{L}_n \setminus \mathcal{D}_n$. Observe that

$$\sum_{w \in \mathcal{D}_n} -\nu[w] \log \nu[w] = \sum_{w \in \mathcal{D}_n} -\nu[w]\left(\log \frac{\nu[w]}{\nu(\mathcal{D}_n)} + \log \nu(\mathcal{D}_n)\right)$$

$$= \left(\nu(\mathcal{D}_n) \sum_{w \in \mathcal{D}_n} -\frac{\nu[w]}{\nu(\mathcal{D}_n)} \log \frac{\nu[w]}{\nu(\mathcal{D}_n)}\right) - \nu(\mathcal{D}_n) \log \nu(\mathcal{D}_n)$$

$$\le (\nu(\mathcal{D}_n) \log \#\mathcal{D}_n) + 1,$$

where the last line uses the fact that $\sum_{i=1}^{k} p_i \log p_i \le \log k$ whenever $p_i \ge 0$, $\sum p_i = 1$, as well as the fact that $-t \log t \le 1$ for all $t \in [0, 1]$. A similar computation holds for $\mathcal{D}_n^c$, and together with (1.2.2.4) this gives

$$nh_\nu(\sigma) \le 2 + \nu(\mathcal{D}_n) \log \#\mathcal{D}_n + \nu(\mathcal{D}_n^c) \log \#\mathcal{D}_n^c. \qquad (1.2.2.5)$$

Using (1.2.2.2) and summing over $\mathcal{D}_n$ gives

$$\mu(\mathcal{D}_n) = \sum_{w \in \mathcal{D}_n} \mu[w] \ge K^{-1} e^{-nh} \#\mathcal{D}_n \quad \Rightarrow \quad \#\mathcal{D}_n \le K e^{nh} \mu(\mathcal{D}_n),$$

and similarly for $\mathcal{D}_n^c$, so (1.2.2.5) gives

$$nh_\nu(\sigma) \le 2 + \nu(\mathcal{D}_n)\left(\log K + nh + \log \mu(\mathcal{D}_n)\right)$$

$$+ \nu(\mathcal{D}_n^c)\left(\log K + nh + \log \mu(\mathcal{D}_n^c)\right)$$

$$= 2 + \log K + nh + \nu(\mathcal{D}_n) \log \mu(\mathcal{D}_n) + \nu(\mathcal{D}_n^c) \log \mu(\mathcal{D}_n^c).$$

Rewriting this as

$$n(h_\nu(\sigma) - h) \le 2 + \log K + \nu(\mathcal{D}_n) \log \mu(\mathcal{D}_n) + \nu(\mathcal{D}_n^c) \log \mu(\mathcal{D}_n^c),$$

we see that the right-hand side goes to $-\infty$ as $n \to \infty$, since $\nu(\mathcal{D}_n) \to 0$ and $\mu(\mathcal{D}_n) \to 1$, so the left-hand side must be negative for large enough n, which implies that $h_\nu(\sigma) < h$ and completes the proof. $\qquad\square$

1.2.2.3 Building a Gibbs Measure

Now the question becomes how to build an ergodic measure satisfying the lower Gibbs bound. There is a standard construction of an MME for a shift space, which proceeds as follows: let ν_n be any measure on X such that $\nu_n[w] = 1/\#\mathcal{L}_n$ for every $w \in \mathcal{L}_n$, and then consider the measures

$$\mu_n := \frac{1}{n}\sum_{k=0}^{n-1}\sigma_*^k \nu_n = \frac{1}{n}\sum_{k=0}^{n-1}\nu_n \circ \sigma^{-k}. \qquad (1.2.2.6)$$

A general argument (which appears in the proof of the variational principle, see for example [50, Theorem 8.6]) shows that any weak* limit point of the sequence μ_n is an MME. If the shift space satisfies the specification property, one can prove more.

Proposition 1.2 *Let (X, σ) be a shift space with the specification property, let μ_n be given by (1.2.2.6), and suppose that $\mu_{n_j} \to \mu$ in the weak* topology. Then μ is σ-invariant, ergodic, and there is $K \geq 1$ such that μ satisfies the following* Gibbs *property:*

$$K^{-1}e^{-nh_{\text{top}}(X)} \leq \mu[w] \leq Ke^{-nh_{\text{top}}(X)} \text{ for all } w \in \mathcal{L}_n.$$

Combining Propositions 1.1 and 1.2 shows that there is a unique MME μ, which is the weak* limit of the sequence μ_n from (1.2.2.6). Thus to prove Theorem 1.2.2.1 it suffices to prove Proposition 1.2. We omit the full proof, and highlight only the most important part of the associated counting estimates.

Lemma 1.2.2.1 *Let (X, σ) be a shift space with the specification property, with gap size τ. Then for every $n \in \mathbb{N}$, we have*

$$e^{nh_{\text{top}}(X)} \leq \#\mathcal{L}_n \leq Qe^{nh_{\text{top}}(X)}, \quad \text{where } Q = (\tau+1)e^{\tau h_{\text{top}}(X)}. \qquad (1.2.2.7)$$

Proof For every $m, n \in \mathbb{N}$, there is an injective map $\mathcal{L}_{m+n} \to \mathcal{L}_m \times \mathcal{L}_n$ defined by $w \mapsto (w_{[1,m]}, w_{[m+1,m+n]})$, so $\#\mathcal{L}_{m+n} \leq \#\mathcal{L}_m\#\mathcal{L}_n$. Iterating this gives

$$\#\mathcal{L}_{kn} \leq (\#\mathcal{L}_n)^k \quad \Rightarrow \quad \frac{1}{kn}\log\#\mathcal{L}_{kn} \leq \frac{1}{n}\log\#\mathcal{L}_n,$$

and sending $k \to \infty$ we get $h_{\text{top}}(X) \leq \frac{1}{n}\log\#\mathcal{L}_n$ for all n, which proves the lower bound. For the upper bound we observe that specification gives a map $\mathcal{L}_m \times \mathcal{L}_n \to \mathcal{L}_{m+n+\tau}$ defined by mapping (v, w) to vuw', where $u = u(v, w) \in \mathcal{L}$ with $|u| \leq \tau$

Fig. 1.1 Estimating
$\nu_n(\sigma^{-k}[w])$

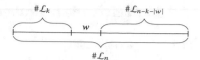

is the 'gluing word' provided by the specification property, and u' is *any* word of length $\tau - |u|$ that can legally follow vuw. This map may not be injective because w can appear in different positions, but each word in $\mathcal{L}_{m+n}$ can have at most $(\tau + 1)$ preimages, since v, w are completely determined by $vuwu'$ and the length of u. This shows that

$$\#\mathcal{L}_{m+n+\tau} \geq \frac{1}{\tau + 1}\#\mathcal{L}_m\#\mathcal{L}_n \quad \Rightarrow \quad \#\mathcal{L}_{k(n+\tau)} \geq \left(\frac{\#\mathcal{L}_n}{\tau + 1}\right)^k.$$

Taking logs and dividing by $k(n + \tau)$ gives

$$\frac{1}{k(n + \tau)}\#\mathcal{L}_{k(n+\tau)} \geq \frac{1}{n + \tau}\left(\log\#\mathcal{L}_n - \log(\tau + 1)\right).$$

Sending $k \to \infty$ and rearranging gives $\log\#\mathcal{L}_n \leq \log(\tau + 1) + (n + \tau)h_{\text{top}}(X)$. Taking an exponential proves the upper bound. $\qquad\square$

With Lemma 1.2.2.1 in hand, the idea of Proposition 1.2 is to first prove the bounds on $\mu[w]$ by estimating, for each $n \gg |w|$ and $k \in \{1, \ldots, n - |w|\}$, the number of words $u \in \mathcal{L}_n$ for which w appears in position k; see Fig. 1.1. By considering the subwords of u lying before and after w, one sees that there are at most $(\#\mathcal{L}_k)(\#\mathcal{L}_{n-k-|w|})$ such words, as in the proof of Lemma 1.2.2.1, and thus the bounds from that lemma give

$$\nu_n(\sigma^{-k}[w]) \leq \frac{(\#\mathcal{L}_k)(\#\mathcal{L}_{n-k-|w|})}{\#\mathcal{L}_n}$$

$$\leq \frac{Qe^{kh_{\text{top}}(X)}Qe^{(n-k-|w|)h_{\text{top}}(X)}}{e^{nh_{\text{top}}(X)}} = Q^2 e^{-|w|h_{\text{top}}(X,\sigma)};$$

averaging over k gives the upper Gibbs bound, and the lower Gibbs bound follows from a similar estimate that uses the specification property.

Next, one can use similar arguments to produce $c > 0$ such that, for each pair of words v, w, there are arbitrarily large $j \in \mathbb{N}$ such that $\mu([v] \cap \sigma^{-j}[w]) \geq c\mu[v]\mu[w]$; this is once again done by counting the number of long words that have v, w in the appropriate positions.

Since any measurable sets V and W can be approximated by unions of cylinders, one can use this to prove that $\overline{\lim}_n \mu(V \cap \sigma^{-n}W) \geq c\mu(V)\mu(W)$. Considering the case when $V = W$ is σ-invariant demonstrates that μ is ergodic.

1.2.3 Relaxing Specification: Decompositions of the Language

1.2.3.1 Decompositions

There are many shift spaces that can be shown to have a unique MME despite not having the specification property; see Sect. 1.2.3.2 below for the example that motivated the present work. We want to consider shift spaces for which the specification property holds if we restrict our attention to "good words", and will see that the uniqueness result in Theorem 1.2.2.1 can be extended to this setting provided the collection of "good words" is "large enough" in an appropriate sense.

To make this more precise, let X be a shift space on a finite alphabet, and $\mathcal{L}$ its language. We consider the following more general version of (1.2.2.1).

Definition 1.2.3.1 A collection of words $\mathcal{G} \subset \mathcal{L}$ has *specification* if there exists $\tau \in \mathbb{N}$ such that for every finite set of words $w^1, \ldots, w^k \in \mathcal{G}$, there are $u^1, \ldots, u^{k-1} \in \mathcal{L}$ with $|u^i| \leq \tau$ such that $w^1 u^1 w^2 u^2 \cdots u^{k-1} w^k \in \mathcal{L}$.

The only difference between this definition and (1.2.2.1) is that here we only require the gluing property to hold for words in $\mathcal{G}$, not for all words.

Remark 1.3 In particular, $\mathcal{G}$ has specification if there is $\tau \in \mathbb{N}$ such that for every $v, w \in \mathcal{G}$, there is $u \in \mathcal{L}$ with $|u| \leq \tau$ and $vuw \in \mathcal{G}$, because iterating this property gives the one stated above. The property above, which is sufficient for our uniqueness results, is a priori more general because the concatenated word is not required to lie in $\mathcal{G}$.

Now we need a way to say that a collection $\mathcal{G}$ on which specification holds is sufficiently large.

Definition 1.2.3.2 A *decomposition* of the language $\mathcal{L}$ consists of three collections of words $C^p, \mathcal{G}, C^s \subset \mathcal{L}$ with the property that

for every $w \in \mathcal{L}$, there are $u^p \in C^p$, $v \in \mathcal{G}$, $u^s \in C^s$ such that $w = u^p v u^s$.

Given a decomposition of $\mathcal{L}$, we also consider for each $M \in \mathbb{N}$ the collection of words

$$\mathcal{G}^M := \{u^p v u^s \in \mathcal{L} : u^p \in C^p, v \in \mathcal{G}, u^s \in C^s, |u^p|, |u^s| \leq M\}.$$

If each $\mathcal{G}^M$ has specification, then the set $C^p \cup C^s$ can be thought of as the set of *obstructions* to the specification property.

Definition 1.2.3.3 The *entropy* of a collection of words $C \subset \mathcal{L}$ is

$$h(C) = \varlimsup_{n \to \infty} \frac{1}{n} \log \#C_n.$$

Theorem 1.2.3.1 (Uniqueness Using a Decomposition [1]) *Let X be a shift space on a finite alphabet, and suppose that the language $\mathcal{L}$ of X admits a decomposition $C^p \mathcal{G} C^s$ such that*

(I) *every collection $\mathcal{G}^M$ has specification, and*
(II) $h(C^p \cup C^s) < h(X)$.

Then (X, σ) has a unique MME μ.

Remark 1.4 Note that $\mathcal{L} = \bigcup_{M \in \mathbb{N}} \mathcal{G}^M$; the sets $\mathcal{G}^M$ play a similar role to the regular level sets that appear in Pesin theory.[7] The gap size τ appearing in the specification property for $\mathcal{G}^M$ is allowed to depend on M, just as the constants appearing in the definition of hyperbolicity are allowed to depend on which regular level set a point lies in. Similarly, for the unique MME μ one can prove that $\lim_{M \to \infty} \mu(\mathcal{G}^M) = 1$, which mirrors a standard result for hyperbolic measures and Pesin sets.

Remark 1.5 In fact we do not quite need *every* $w \in \mathcal{L}$ to admit a decomposition as in definition 1.2.3.2. It is enough to have $C^p, \mathcal{G}, C^s \subset \mathcal{L}$ such that $h(\mathcal{L} \setminus (C^p \mathcal{G} C^s)) < h(X)$, in addition to the conditions above [6].

We outline the proof of Theorem 1.2.3.1. The idea is to mimic Bowen's proof using Propositions 1.1 and 1.2 by completing the following steps.

1. Prove uniform counting bounds as in Lemma 1.2.2.1.
2. Use these to establish the following *non-uniform* Gibbs property for any limit point μ of the sequence of measures in (1.2.2.6): there are constants $K, K_M \geq 1$ such that for all $M \in \mathbb{N}$ and $w \in \mathcal{G}^M$

$$K_M^{-1} e^{-|w| h_{\text{top}}(X)} \leq \mu[w] \leq K e^{-|w| h_{\text{top}}(X)}. \tag{1.2.3.1}$$

We emphasize that the Gibbs property is non-uniform in the sense that the lower Gibbs constant depends on M.[8] The upper bound that we will obtain from our hypotheses is uniform in M. On a fixed $\mathcal{G}^M$, we have uniform Gibbs estimates.
3. Give a similar argument for ergodicity, and then prove that the non-uniform lower Gibbs bound in (1.2.3.1) still gives uniqueness as in Proposition 1.1.

Once the uniform counting bounds are established, the proof of (1.2.3.1) follows the same approach as before. We do not discuss the third step at this level of generality except to emphasize that it follows the approach given in Proposition 1.1.

---

[7] Since $\mathcal{G}^M$ corresponds to a collection of orbit segments rather than a subset of the space, the most accurate analogy might be to think of $\mathcal{G}^M$ as corresponding to orbit segments that start and end in a given regular level set.

[8] The constant K_M increases exponentially with the transition time in the specification property for $\mathcal{G}^M$, so we do not expect any explicit relationship between M and K_M in general. Examples of S-gap shifts (see Remark 1.9) can be easily constructed to make the constants K_M^{-1} decay fast.

For the counting bounds in the first step, we start by observing that the bound $\#\mathcal{L}_n \geq e^{nh_{\text{top}}(X)}$ did not require any hypotheses on the symbolic space X and thus continues to hold. The argument for the upper bound in Lemma 1.2.2.1 can be easily adapted to show that there is a constant Q such that $\#\mathcal{G}_n \leq Q e^{nh_{\text{top}}(X)}$ for all n. Then the desired upper bound for $\#\mathcal{L}_n$ is a consequence of the following.

Lemma 1.2.3.1 *For any* $r \in (0, 1)$, *there is* M *such that* $\#\mathcal{G}_n^M \geq r\#\mathcal{L}_n$ *for all n.*

Proof Let $a_i = \#(C_i^p \cup C_i^s)e^{-ih_{\text{top}}(X)}$, so that in particular $\sum a_i < \infty$ by (II). Since any $w \in \mathcal{L}_n$ can be written as $w = u^p v u^s$ for some $u \in C_i^p$, $v \in \mathcal{G}_j$, and $w \in C_k^s$ with $i + j + k = n$, we have

$$\#\mathcal{L}_n \leq \#\mathcal{G}_n^M + \sum_{\substack{i+j+k=n \\ \max(i,k)>M}} (\#C_i^p)(\#\mathcal{G}_j)(\#C_k^s) \leq \#\mathcal{G}_n^M + \sum_{\substack{i+j+k=n \\ \max(i,k)>M}} a_i a_k Q e^{nh_{\text{top}}(X)},$$

where the second inequality uses the upper bound $\#\mathcal{G}_j \leq Q e^{jh_{\text{top}}(X)}$. Since $\sum a_i < \infty$, there is M such that

$$\sum_{\substack{i+j+k=n \\ \max(i,k)>M}} a_i a_k Q e^{nh_{\text{top}}(X)} < (1 - r)e^{nh_{\text{top}}(X)} \leq (1 - r)\#\mathcal{L}_n,$$

where the second inequality uses the lower bound $\#\mathcal{L}_n \geq e^{nh_{\text{top}}(X)}$. Combining these estimates gives $\#\mathcal{L}_n \leq \#\mathcal{G}_n^M + (1 - r)\#\mathcal{L}_n$, which proves the lemma. $\qquad\square$

The same specification argument that gives the upper bound on $\#\mathcal{G}_n$ gives a corresponding upper bound on $\mathcal{G}_n^M$ (with a different constant), and thus we deduce the following consequence of Lemma 1.2.3.1.

Corollary 1.2.3.1 *There are constants a, $A > 0$ and $M \in \mathbb{N}$ such that*

$$e^{nh_{\text{top}}(X)} \leq \#\mathcal{L}_n \leq A e^{nh_{\text{top}}(X)} \quad \text{and} \quad \#\mathcal{G}_n^M \geq a e^{nh_{\text{top}}(X)} \text{ for all } n \in \mathbb{N}.$$

Remark 1.6 In fact, the proof of Lemma 1.2.3.1 can easily be adapted to show a stronger result: given any $\gamma > 0$ and $r \in (0, 1)$, there is M such that if $\mathcal{D}_n \subset \mathcal{L}_n$ has $\#\mathcal{D}_n \geq \gamma e^{nh_{\text{top}}(X)}$, then $\#(\mathcal{D}_n \cap \mathcal{G}_n^M) \geq r\#\mathcal{D}_n$. These types of estimates are what lie behind the claim in Remark 1.4 that the (non-uniform) Gibbs property implies $\mu(\mathcal{G}^M) \to 1$ as $M \to \infty$.

1.2.3.2   An Example: Beta Shifts

Given a real number $\beta > 1$, the corresponding β-transformation $f: [0, 1) \to [0, 1)$ is $f(x) = \beta x \pmod 1$. Let $A = \{0, 1, \ldots, \lceil \beta \rceil - 1\}$; then every $x \in [0, 1)$ admits a coding $y = \pi(x) \in A^{\mathbb{N}}$ defined by $y_n = \lfloor \beta f^{n-1}(x) \rfloor$, and we have $\pi \circ f = \sigma \circ \pi$, where $\sigma: A^{\mathbb{N}} \to A^{\mathbb{N}}$ is the left shift. Observe that $\pi(x)_n = a$ if and only if

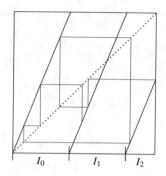 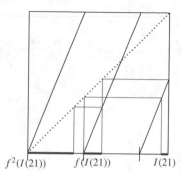

I_0 I_1 I_2 $f^2(I(21))$ $f(I(21))$ $I(21)$

Fig. 1.2 Coding a β-transformation

$f^{n-1}(x) \in I_a$, where the intervals I_a are as shown in Fig. 1.2.[9] Given $n \in \mathbb{N}$ and $w \in A^n$, let

$$I(w) := \bigcap_{k=1}^{n} f^{-(k-1)}(I_{w_k})$$

be the interval in $[0, 1)$ containing all points x for which the first n iterates are coded by w. The figure shows an example for which $f^n(I(w))$ is not the whole interval $[0, 1)$; it is worth checking some other examples and seeing if you can tell for which words $f^n(I(w))$ *is* equal to the whole interval. Observe that if β is an integer then this is true for every word.

Definition 1.2.3.4 The *β-shift* X_β is the closure of the image of π, and is σ-invariant. Equivalently, X_β is the shift space whose language $\mathcal{L}$ is the set of all $w \in A^*$ such that $I(w) \neq \emptyset$; thus $y \in A^{\mathbb{N}}$ is in X_β if and only if $I(y_1 \cdots y_n) \neq \emptyset$ for all $n \in \mathbb{N}$.

For further background on the β-shifts, see [61–63]. We summarize the properties relevant for our purposes.

Write $\preceq$ for the lexicographic order on $A^{\mathbb{N}}$ and observe that π is order-preserving. Let $\mathbf{z} = \lim_{x \nearrow 1} \pi(x)$ denote the supremum of X_β in this ordering. It will be convenient to extend $\preceq$ to A^*, writing $v \preceq w$ if for $n = \min(|v|, |w|)$ we have $v_{[1,n]} \preceq w_{[1,n]}$.

Remark 1.7 Observe that on $A^* \cup A^{\mathbb{N}}$, $\preceq$ is only a pre-order, because there are $v \neq w$ such that $v \preceq w$ and $w \preceq v$; this occurs whenever one of v, w is a prefix of the other.

---

[9]Formally, $I_a = \{x \in [0, 1) : \lfloor \beta x \rfloor = a\}$, so $I_a = [\frac{a}{\beta}, 1)$ if $a = \lceil \beta \rceil - 1$, and $[\frac{a}{\beta}, \frac{a+1}{\beta})$ otherwise.

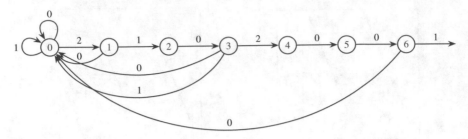

Fig. 1.3 A graph representation of X_β

The β-shift can be described in terms of the lexicographic ordering, or in terms of the following countable-state graph:

- the vertex set is $\mathbb{N}_0 = \{0, 1, 2, 3, \ldots\}$;
- the vertex n has $1 + z_{n+1}$ outgoing edges, labeled with $\{0, 1, \ldots, z_{n+1}\}$; the edge labeled z_{n+1} goes to $n + 1$, and the rest go to the 'base' vertex 0.

Figure 1.3 shows (part of) the graph when $\mathbf{z} = 2102001\ldots$, as in Fig. 1.2.

Proposition 1.3 *Given $n \in \mathbb{N}$ and $w \in A^n$, the following are equivalent.*

1. $I(w) \neq \emptyset$ (which is equivalent to $w \in \mathcal{L}(X_\beta)$ by definition).
2. $w_{[j,n]} \preceq \mathbf{z}$ for every $1 \leq j \leq n$.
3. w labels the edges of a path on the graph that starts at the base vertex 0.

Idea of Proof Using induction, check that the following are equivalent for every $n \in \mathbb{N}$, $0 \leq k \leq n$, and $w \in A^n$.

1. $f^n(I(w)) = f^k(I(\mathbf{z}_{[1,k]}))$, where we write $I(\mathbf{z}_{[1,0]}) := [0, 1)$.
2. $w_{[j,n]} \preceq \mathbf{z}$ for every $1 \leq j \leq n$, and k is maximal such that $w_{[n-k+1,n]} = \mathbf{z}_{[1,k]}$.
3. w labels the edges of a path on the graph that starts at the base vertex 0 and ends at the vertex k.

Corollary 1.2.3.2 *Given $x \in A^{\mathbb{N}}$, the following are equivalent.*

1. $x \in X_\beta$.
2. $\sigma^n(x) \preceq \mathbf{z}$ for every n.
3. x labels the edges of an infinite path of the graph starting at the vertex 0.

Exercise 1.1 Prove that X_β has the specification property if and only if $\mathbf{z}$ does not contain arbitrarily long strings of 0s.

In fact, Schmeling showed [64] that for Lebesgue-a.e. $\beta > 1$, the β-shift X_β does *not* have the specification property. Nevertheless, every β-shift has a unique MME. This was originally proved by Hofbauer [65] and Walters [66] using techniques not based on specification. Theorem 1.2.3.1 gives an alternate proof: writing $\mathcal{G}$ for the set of words that label a path starting *and ending* at the base vertex, and C^s for the set or words that label a path starting at the base vertex *and never returning to it*, one quickly deduces the following.

- $\mathcal{G}C^s$ is a decomposition of $\mathcal{L}$.
- $\mathcal{G}^M$ is the set of words labeling a path starting at the base vertex and ending somewhere in the first M vertices; writing τ for the maximum graph distance from such a vertex to the base vertex, $\mathcal{G}^M$ has specification with gap size τ.
- $\#C_n^s = 1$ for every n, and thus $h(C^s) = 0 < h_{\text{top}}(X_\beta) = \log \beta$.

This verifies the conditions of Theorem 1.2.3.1 and thus provides another proof of uniqueness of the MME.

Remark 1.8 Because the earlier proofs of uniqueness did not pass to subshift factors of β-shifts, it was for several years an open problem (posed by Klaus Thomsen) whether such factors still had a unique MME. The inclusion of this problem in Mike Boyle's article "Open problems in symbolic dynamics" [67] was our original motivation for studying uniqueness using non-uniform versions of the specification property, which led us to formulate the conditions in Theorem 1.2.3.1; these can be shown to pass to factors, providing a positive answer to Thomsen's question [1].

Remark 1.9 Theorem 1.2.3.1 can be applied to other symbolic examples as well, including S-gap shifts [1]. The S-gap shifts are a family of subshifts of $\{0, 1\}^{\mathbb{Z}}$ defined by the property that the number of 0's that appear between any two 1's is an element of a prescribed set $S \subset \mathbb{Z}$. A specific example is the prime gap shift, where S is taken to be the prime numbers. The theorem also admits an extension to equilibrium states for nonzero potential functions along the lines described in Sect. 1.3.5 below, which has been applied to β-shifts [10], S-gap shifts [12], shifts of quasi-finite type [6], and α-β shifts (which code $x \mapsto \alpha + \beta x \pmod 1$) [13].

1.2.3.3 Periodic Points

It is often the case that one can prove a stronger version of specification, for example, when X is a mixing SFT.

Definition 1.2.3.5 Say that $\mathcal{G} \subset \mathcal{L}$ has *periodic strong specification* if there exists $\tau \in \mathbb{N}$ such that for all $w^1, \ldots, w^k \in \mathcal{G}$, there are $u^1, \ldots, u^k \in \mathcal{L}_\tau$ such that $v := w^1 u^1 \cdots w^k u^k \in \mathcal{L}$, and moreover $x = vvvvv \cdots \in X$.

There are two strengthenings of specification, in the sense of (1.2.2.1), here: first, we assume that the gap size is equal to τ, not just $\leq \tau$, and second, we assume that the "glued word" can be extended periodically after adding τ more symbols.

If we replace specification in Theorem 1.2.3.1 with periodic strong specification for each $\mathcal{G}^M$, then the counting estimates in Lemma 1.2.2.1 immediately lead to the following estimates on the number of periodic points: writing $\text{Per}_n = \{x \in X : \sigma^n x = x\}$, we have

$$C^{-1} e^{n h_{\text{top}}(X)} \leq \#\text{Per}_n \leq C e^{n h_{\text{top}}(X)}.$$

Using this fact and the construction of the unique MME given just before Proposition 1.2, one can also conclude that the unique MME μ is the limiting distribution of periodic orbits in the following sense:

$$\frac{1}{\#\mathrm{Per}_n} \sum_{x \in \mathrm{Per}_n} \delta_x \xrightarrow{\text{weak*}} \mu \text{ as } n \to \infty.$$

This argument holds true in the classical Theorem 1.2.2.1, and for β-shifts. It also extends beyond the symbolic setting, and a natural analogue of the argument holds for regular closed geodesics on rank one non-positive curvature manifolds.

1.2.4 Beyond Shift Spaces: Expansivity in Bowen's Argument

Now we move to the non-symbolic setting and describe how Bowen's approach works for a continuous map on a compact metric space. In particular, his assumptions apply to and were inspired by the case when X is a transitive locally maximal hyperbolic set for a diffeomorphism f. First we recall some basic definitions.

1.2.4.1 Topological Entropy

Definition 1.2.4.1 Given $n \in \mathbb{N}$, the nth *dynamical metric* on X is

$$d_n(x, y) := \max\{d(f^k x, f^k y) : 0 \leq k < n\}. \tag{1.2.4.1}$$

The *Bowen ball of order n and radius $\epsilon > 0$ centered at $x \in X$* is

$$B_n(x, \epsilon) := \{y \in X : d_n(x, y) < \epsilon\}. \tag{1.2.4.2}$$

A set $E \subset X$ is called (n, ϵ)-*separated* if $d_n(x, y) > \epsilon$ for all $x, y \in E$ with $x \neq y$; equivalently, if $y \notin B_n(x, \epsilon)$ for all such x, y.

We define entropy in a more general way than is standard, reflecting our focus on the space of finite-length orbit segments $X \times \mathbb{N}$ as the relevant object of study; this replaces the language $\mathcal{L}$ that we used in the symbolic setting. We interpret $(x, n) \in X \times \mathbb{N}$ as representing the orbit segment $(x, fx, f^2x, \ldots, f^{n-1}x)$. Then the analogy is that a cylinder $[w]$ for a word in the language corresponds to a Bowen ball $B_n(x, \epsilon)$ associated to an orbit segment $(x, n) \in X \times \mathbb{N}$. Given a collection of orbit segments $\mathcal{D} \subset X \times \mathbb{N}$, for each $n \in \mathbb{N}$ we write

$$\mathcal{D}_n := \{x \in X : (x, n) \in \mathcal{D}\}$$

for the collection of points that begin a length-n orbit segment in $\mathcal{D}$.

Definition 1.2.4.2 (Topological Entropy) Given a collection of orbit segments $\mathcal{D} \subset X \times \mathbb{N}$, for each $\epsilon > 0$ and $n \in \mathbb{N}$ we write

$$\Lambda(\mathcal{D}, \epsilon, n) := \max\{\#E : E \subset \mathcal{D}_n \text{ is } (n, \epsilon)\text{-separated}\}.$$

The *entropy* of $\mathcal{D}$ at scale $\epsilon > 0$ is

$$h(\mathcal{D}, \epsilon) := \varlimsup_{n \to \infty} \frac{1}{n} \log \Lambda(\mathcal{D}, \epsilon, n),$$

and the entropy of $\mathcal{D}$ is

$$h(\mathcal{D}) := \lim_{\epsilon \to 0} h(\mathcal{D}, \epsilon).$$

When $\mathcal{D} = Y \times \mathbb{N}$ for some $Y \subset X$, we write $\Lambda(Y, \epsilon, n) = \Lambda(Y \times \mathbb{N}, \epsilon, n)$, $h_{\text{top}}(Y, \epsilon) = h(Y \times \mathbb{N}, \epsilon)$ and $h_{\text{top}}(Y) = \lim_{\epsilon \to 0} h_{\text{top}}(Y, \epsilon)$. In particular, when $\mathcal{D} = X \times \mathbb{N}$ we write $h_{\text{top}}(X, f) = h_{\text{top}}(X) = h(X \times \mathbb{N})$ for the *topological entropy* of $f \colon X \to X$.

When different orbit segments in $\mathcal{D}$ are given weights according to their ergodic sum w.r.t. a given potential φ, we obtain a notion of *topological pressure*, which we will discuss in Sect. 1.3.5.

Theorem 1.2.4.1 (Variational Principle) *Let X be a compact metric space and $f \colon X \to X$ a continuous map. Then*

$$h_{\text{top}}(X, f) = \sup_{\mu \in \mathcal{M}_f(X)} h_\mu(f).$$

The following construction forms one half of the proof of the variational principle.

Proposition 1.4 (Building a Measure of Almost Maximal Entropy) *With X, f as above, fix $\epsilon > 0$, and for each $n \in \mathbb{N}$, let $E_n \subset X$ be an (n, ϵ)-separated set. Consider the Borel probability measures*

$$\nu_n := \frac{1}{\#E_n} \sum_{x \in E_n} \delta_x, \quad \mu_n := \frac{1}{n} \sum_{k=0}^{n-1} f_*^k \nu_n = \frac{1}{n} \sum_{k=0}^{n-1} \nu_n \circ f^{-k}. \tag{1.2.4.3}$$

Let μ_{n_j} be any subsequence that converges in the weak*-topology to a limiting measure μ. Then $\mu \in \mathcal{M}_f(X)$ and

$$h_\mu(f) \geq \varlimsup_{j \to \infty} \frac{1}{n_j} \log \#E_{n_j}.$$

In particular, for every $\delta > 0$ there exists $\mu \in M_f(X)$ such that $h_\mu(f) \geq h_{\text{top}}(X, f, \delta)$.

Proof See [50, Theorem 8.6]. □

Corollary 1.2.4.1 *Let X, f be as above, and suppose that there is $\delta > 0$ such that $h_{\text{top}}(X, f, \delta) = h_{\text{top}}(X, f)$. Then there exists a measure of maximal entropy for (X, f). Indeed, given any sequence $\{E_n \subset X\}_{n=1}^\infty$ of maximal (n, δ)-separated sets, every weak*-limit point of the sequence μ_n from (1.2.4.3) is an MME.*

In our applications, it will often be relatively easy to verify that $h_{\text{top}}(X, f, \delta) = h_{\text{top}}(X, f)$ for some $\delta > 0$, and so Corollary 1.2.4.1 establishes existence of a measure of maximal entropy. Thus the real challenge is to prove uniqueness, and this will be our focus.

1.2.4.2 Expansivity

In Bowen's general result, the assumption that X is a shift space is replaced by the following condition.

Definition 1.2.4.3 (Expansivity) Given $x \in X$ and $\epsilon > 0$, let

$$\Gamma_\epsilon^+(x) := \{y \in X : d(f^n y, f^n x) < \epsilon \text{ for all } n \geq 0\} = \bigcap_{n \in \mathbb{N}} B_n(x, \epsilon)$$

be the forward infinite Bowen ball. If f is invertible, let

$$\Gamma_\epsilon^-(x) := \{y \in X : d(f^n y, f^n x) < \epsilon \text{ for all } n \geq 0\}$$

be the backward infinite Bowen ball, and let

$$\Gamma_\epsilon(x) := \Gamma_\epsilon^+(x) \cap \Gamma_\epsilon^-(x) = \{y \in X : d(f^n y, f^n x) < \epsilon \text{ for all } n \in \mathbb{Z}\}$$

be the bi-infinite Bowen ball. The system (X, f) is *positively expansive at scale* $\epsilon > 0$ if $\Gamma_\epsilon^+(x) = \{x\}$ for all $x \in X$, and *(two-sided) expansive at scale* $\epsilon > 0$ if $\Gamma_\epsilon(x) = \{x\}$. The system is *(positively) expansive* if there exists $\epsilon > 0$ such that it is (positively) expansive at scale ϵ.

It is an easy exercise to check that one-sided shift spaces are positively expansive. A system (X, f) is *uniformly expanding* if there are $\epsilon, \lambda > 0$ such that $d(fy, fx) \geq e^\lambda d(y, x)$ whenever $x, y \in X$ have $d(x, y) < \epsilon$. Iterating this property gives $\text{diam } B_n(x, \epsilon) \leq \epsilon e^{-\lambda n}$ for all n, and thus $\Gamma_\epsilon^+(x) = \{x\}$, so (X, f) is positively expansive.

Two-sided shift spaces can easily be checked to be (two-sided) expansive, and we also have the following.

Proposition 1.5 *If X is a hyperbolic set for a diffeomorphism f, then (X, f) is expansive.*

Sketch of Proof Choose $\epsilon > 0$ small enough that given any $x, y \in X$ with $d(x, y) < \epsilon$, the local leaves $W^s(x)$ and $W^u(y)$ intersect in a unique point $[x, y]$ (we do not require that this point is in X). Write

$$d^u(x, y) = d(x, [x, y]) \quad \text{and} \quad d^s(x, y) = d(y, [x, y]).$$

Passing to an adapted metric if necessary, hyperbolicity gives $\lambda > 0$ such that

$$d^u(f^n x, f^n y) \geq e^{\lambda n} d^u(x, y) \text{ if } d(f^k x, f^k y) < \epsilon \text{ for all } 0 \leq k \leq n, \qquad (1.2.4.4)$$

$$d^s(f^{-n} x, f^{-n} y) \geq e^{\lambda n} d^s(x, y) \text{ if } d(f^{-k} x, f^{-k} y) < \epsilon \text{ for all } 0 \leq k \leq n.$$
$$(1.2.4.5)$$

In particular, if $y \in \Gamma_\epsilon(x)$ then $d^u(f^n x, f^n y)$ is uniformly bounded for all n, so $d^u(x, y) = 0$, and similarly for d^s, which implies that $x = [x, y] = y$.

One important consequence of expansivity is the following.

Proposition 1.6 *If (X, f) is expansive at scale ϵ, then $h_{\text{top}}(X, f, \epsilon) = h_{\text{top}}(X, f)$.*

Two Proof Ideas We outline two proofs in the positively expansive case.

One argument uses a compactness argument to show that for every $0 < \delta < \epsilon$, there is $N \in \mathbb{N}$ such that $B_N(x, \epsilon) \subset B(x, \delta)$ for all $x \in X$. This implies that $B_{n+N}(x, \epsilon) \subset B_n(x, \delta)$ for all x, and then one can show that the definition of topological entropy via (n, ϵ)-separated sets gives the same value at δ as at ϵ.

Another method, which is better for our purposes, is to observe that since ϵ-expansivity gives $\bigcap_n B_n(x, \epsilon) = \{x\}$ for all x, one can easily show that for every $\nu \in \mathcal{M}_f(X)$, we have:

if β is a partition with d_n-diameter $< \epsilon$, then β is generating for (f^n, ν).

Given a maximal (n, ϵ)-separated set E_n, we can choose a partition β_n such that each element of β_n is contained in $B_n(x, \epsilon)$ for some $x \in E_n$, so β_n has exactly $\#E_n$ elements. Then we have

$$h_\mu(f) = \frac{1}{n} h_\mu(f^n) = \frac{1}{n} h_\mu(f^n, \beta_n) \leq \frac{1}{n} H_\mu(\beta_n) \leq \frac{1}{n} \log \#E_n. \qquad (1.2.4.6)$$

Sending $n \to \infty$ gives $h_\mu(f) \leq h_{\text{top}}(X, f, \epsilon)$, and taking a supremum over all $\mu \in \mathcal{M}_f(X)$ proves that $h_{\text{top}}(X, f, \epsilon) = h_{\text{top}}(X, f)$.

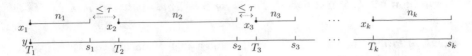

Fig. 1.4 Bookkeeping in the specification property

1.2.4.3 Specification

The following formulation of the specification property is given for a collection of orbit segments $\mathcal{D} \subset X \times \mathbb{N}$, and thus is not quite the classical one, but reduces to (a version of) the classical definition when we take $\mathcal{D} = X \times \mathbb{N}$. Observe that when X is a shift space and we associate to each (x, n) the word $x_{[1,n]} \in \mathcal{L}(X)$, the following agrees with the definition from (1.2.2.1).

Definition 1.2.4.4 (Specification) A collection of orbit segments $\mathcal{D} \subset X \times \mathbb{N}$ has the *specification property at scale* $\delta > 0$ if there exists $\tau \in \mathbb{N}$ (the *gap size* or *transition time*) such that for every $(x_1, n_1), \ldots, (x_k, n_k) \in \mathcal{D}$, there exist $0 = T_1 < T_2 < \cdots < T_k \in \mathbb{N}$ and $y \in X$ such that

$$f^{T_i}(y) \in B_{n_i}(x_i, \delta) \text{ and } T_i - (T_{i-1} + n_{i-1}) \in [0, \tau] \text{ for all } 1 \le i \le k,$$

see Fig. 1.4. That is, starting from time T_i the orbit of y shadows the orbit of x_i, and moreover, writing $s_i = T_i + n_i$ for the time at which this shadowing ends, we have

$$s_i \le T_{i+1} \le s_i + \tau \text{ for all } 1 \le i < k.$$

We say that $\mathcal{D}$ has the *specification property* if the above holds for every $\delta > 0$. We say that (X, f) has the specification property if $X \times \mathbb{N}$ does. We say that $\mathcal{D}$ has periodic specification if y can be chosen to be periodic with period in $[s_k, s_k + \tau]$.

First we explain how specification (for the whole system) is established in the uniformly hyperbolic case. Recall from (1.2.2.1) and the paragraph preceding it that in the symbolic case, one can establish specification by verifying it in the case $k = 2$ and then iterating. In the non-symbolic case, the proof of specification usually follows this same approach, but one needs to verify a mildly stronger property for $k = 2$ to allow the iteration step; one possible version of this property is formulated in the next lemma.

Lemma 1.2.4.1 *Given* $f \colon X \to X$, *suppose that* $\delta_1 > 0$, $\delta_2 \ge 0$, $\chi \in (0, 1)$, *and* $\tau \in \mathbb{N}$ *are such that for every* $(x_1, n_1), (x_2, n_2) \in X \times \mathbb{N}$, *there are* $t \in \{0, 1, \ldots, \tau\}$ *and* $y \in X$ *such that*

$$d(f^k y, f^k x_1) \le \delta_1 \chi^{n_1 - k} \text{ for all } 0 \le k < n_1 \text{ and } d_{n_2}(f^{n_1 + t} y, x_2) \le \delta_2.$$
(1.2.4.7)

Then (X, f) *has the specification property at scale* $\delta_2 + \delta_1/(1 - \chi)$ *with gap size* τ.

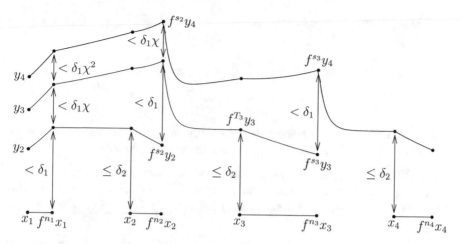

Fig. 1.5 Proving specification using a one-step property

Proof Given $(x_1, n_1), \ldots, (x_k, n_k) \in X \times \mathbb{N}$, we will apply (1.2.4.7) iteratively to produce $y_1, \ldots, y_k$ and $T_1, \ldots, T_k \in \mathbb{N}$ such that writing $\delta' = \delta_2 + \delta_1/(1 - \chi)$, we have

$$f^{T_i}(y_j) \in B_{n_i}(x_i, \delta') \text{ for all } 1 \leq i \leq j \leq k. \tag{1.2.4.8}$$

Once this is done, y_k is the desired shadowing point. See Fig. 1.5 for an illustration of the following procedure and estimates.

Along with y_i, T_i, we will produce $s_i = T_i + n_i$ and $t_i \in \{0, \ldots, \tau\}$ such that $T_{i+1} = s_i + t_i$. Start by putting $y_1 = x_1$, $T_1 = 0$, and $s_1 = n_1$. Then apply (1.2.4.7) to (y_1, s_1) and (x_2, n_2) to get $y_2 \in X$ and $t_1 \in \{0, 1, \ldots, \tau\}$ such that writing $T_2 = s_1 + t_1$, we have

$$d(f^k y_2, f^k y_1) \leq \delta_1 \chi^{s_1 - k} \text{ for all } 0 \leq k < s_1 \text{ and } d_{n_2}(f^{T_2} y_2, x_2) \leq \delta_2.$$

In general, once y_i, s_i are determined (with $T_i = s_i - n_i$), we apply (1.2.4.7) to (y_i, s_i) and (x_{i+1}, n_{i+1}) to get $t_i \in \{0, 1, \ldots, \tau\}$ and $y_{i+1} \in X$ such that writing $T_{i+1} = s_i + t_i$, we have

$$d(f^k y_{i+1}, f^k y_i) \leq \delta_1 \chi^{s_i - k} \text{ for all } 0 \leq k < s_i \text{ and } d_{n_{i+1}}(f^{T_{i+1}} y_{i+1}, x_{i+1}) \leq \delta_2.$$

Now we can verify (1.2.4.8) by observing that for all $1 \leq i \leq j \leq k$, we have

$$d_{n_i}(f^{T_i}(y_j), f^{T_i}(y_i)) \leq \sum_{\ell=i}^{j-1} d_{n_i}(f^{T_i}(y_{\ell+1}), f^{T_i}(y_\ell))$$

$$\leq \sum_{\ell=i}^{j-1} \delta_1 \chi^{s_\ell - s_i} < \frac{\delta_1}{1 - \chi}, \tag{1.2.4.9}$$

(the last inequality uses the fact that $s_\ell - s_i \geq \ell - i$), and also $d_{n_i}(f^{T_i}(y_i), x_i) \leq \delta_2$, so

$$d_{n_i}(x_i, f^{T_i}y_j) \leq d_{n_i}(x_i, f^{T_i}y_i) + d_{n_i}(f^{T_i}y_i, f^{T_i}y_j) \leq \delta_2 + \frac{\delta_1}{1 - \chi} = \delta'.$$

$\square$

Proposition 1.7 *If X is a topologically transitive locally maximal hyperbolic set for a diffeomorphism f, then (X, f) has the specification property.*

Proof By Lemma 1.2.4.1, it suffices to show that for every sufficiently small $\delta > 0$, there are $\chi \in (0, 1)$ and $\tau \in \mathbb{N}$ such that for every $(x_1, n_1), (x_2, n_2) \in X \times \mathbb{N}$, there are $t \in \{0, 1, \ldots, \tau\}$ and $y \in X$ such that (1.2.4.7) holds. To prove this, let $\delta, \rho > 0$ be such that

- every $x \in X$ has local stable and unstable leaves $W_\delta^s(x)$ and $W_\delta^u(x)$ with diameter $< \delta$, and
- for every $x, y \in X$ with $d(x, y) < \rho$, the intersection $W_\delta^s(x) \cap W_\delta^u(y)$ is a single point, which lies in X.

By topological transitivity and compactness, there is $\tau \in \mathbb{N}$ such that for every $x, y \in X$ there is $t \in \{0, 1, \ldots, \tau\}$ with $d(f^t x, y) < \rho$, and thus $f^t(W_\delta^u(x)) \cap W_\delta^s(y) \neq \emptyset$.

Using this fact, given $(x_1, n_1), (x_2, n_2) \in X \times \mathbb{N}$, we can let $t \in \{0, 1, \ldots, \tau\}$ be such that $f^t(W_\delta^u(f^{n_1}x_1))$ intersects $W_\delta^s(x_2)$. Choosing z in this intersection and putting $y = f^{-(t+n_1)}(z)$, we see that y satisfies (1.2.4.7) with $\delta_1 = \delta_2 = \delta$, and thus Lemma 1.2.4.1 proves the proposition. $\square$

Remark 1.10 Uniform contraction of f along W^s is not used; to prove specification at scale δ', it would suffice to know that if x, y lie on the same local stable leaf and $d(x, y) \leq \delta_2$, then the same is true of $f(x), f(y)$, which still gives the second half of (1.2.4.7). In particular, this follows as soon as $\|Df|_{E^s}\| \leq 1$. The same idea can also be applied to obtain specification on suitable collections $\mathcal{G} \subset X \times \mathbb{N}$, and can be extended naturally to the continuous-time case.

We also emphasize that the exponential contraction asked for in the first half of (1.2.4.7), which is obtained from uniform backwards contraction along W^u, can be significantly weakened. What is really essential for the argument is backwards contraction in the local unstables by a fixed amount in each of the orbit segments (not necessarily proportional to length), and this is enough to obtain a uniform distance estimate analogous to (1.2.4.9). We carried out the details of this argument in [3, §4] in the non-uniformly hyperbolic setting of rank one geodesic flow for the family of orbit segments $C(\eta)$, which are defined in this survey in Sect. 1.4.2.6.

The following gives the corresponding result in the non-invertible case.

Proposition 1.8 *Suppose that* $f: X \to X$ *is topologically transitive and has the following properties.*

- Uniformly expanding: $d(fx, fy) \geq e^\lambda d(x, y)$ *whenever* $d(x, y) < \delta$.
- Locally onto: *For every* $x \in X$, *we have* $f(B(x, \delta)) \supset B(fx, \delta)$.[10]

Then (X, f) *has the specification property at scale* $\delta/(1 - e^{-\lambda})$.

Proof It suffices to verify (1.2.4.7) with $\delta_1 = \delta$, $\delta_2 = 0$, and $\chi = e^{-\lambda}$; then we can apply Lemma 1.2.4.1. We need the following consequence of the locally onto property:

for every $x \in X$ and $n \in \mathbb{N}$, we have $f^n(B_n(x, \delta)) \supset B(f^n x, \delta)$. (1.2.4.10)

As in the previous proposition, we use the following consequence of topological transitivity and compactness: given $\delta > 0$, there is $\tau \in \mathbb{N}$ such that for every $x, y \in X$ there is $t \in \{0, 1, \ldots, \tau\}$ with $f^t(x) \in B(y, \delta)$. Now given $(x_1, n_1), (x_2, n_2) \in X \times \mathbb{N}$, there is $t \in \{0, 1, \ldots, \tau\}$ such that $f^t(f^{n_1}(x_1)) \in B(x_2, \delta)$, and thus (1.2.4.10) gives

$$f^t(f^{n_1} B_{n_1}(x_1, \delta)) \supset f^t B(f^{n_1} x_1, \delta) \supset B(f^{n_1+t} x_1, \delta) \ni x_2.$$

Thus there is $y \in B_{n_1}(x_1, \delta)$ such that $f^{n_1+t}(y) = x_2$, which verifies (1.2.4.7); Lemma 1.2.4.1 completes the proof. □

1.2.4.4 Bowen's Proof Revisited

Bowen's original uniqueness result [5], which we outlined in Sect. 1.2.2, was actually given not for shift spaces, but for more general expansive systems.

Theorem 1.2.4.2 (Expansivity and Specification (Bowen)) *Let* X *be a compact metric space and* $f: X \to X$ *a continuous map. Suppose that* $\epsilon > 40\delta > 0$ *are such that* f *has expansivity at scale* ϵ *and the specification property at scale* δ. *Then* (X, f) *has a unique measure of maximal entropy.*

Remark 1.11 Bowen's original paper assumed expansivity and periodic specification at all scales. We relax the proof mildly so that it does not use periodic orbits and only uses specification at a fixed scale, small relative to an expansivity constant.[11] We will see examples later where this additional generality is beneficial.

---

[10]In the symbolic setting, this corresponds to X being a subshift of finite type.

[11]The statements in [68] used $\epsilon \geq 28\delta$ but this must be corrected to $\epsilon > 40\delta$; see [2, §5.7].

The proof of Theorem 1.2.4.2 extends the strategy in the symbolic case:

1. establish uniform counting bounds;
2. show that the usual construction of an MME gives an ergodic Gibbs measure;
3. prove that an ergodic Gibbs measure must be the unique MME.

We must also examine the role played by expansivity.

In the symbolic setting, the first step was to prove the *counting bounds* on #$\mathcal{L}_n$ given in (1.2.2.7). In the general setting, #$\mathcal{L}_n$ is replaced with $\Lambda(X, \epsilon, n)$ from Definition 1.2.4.2, and mimicking the arguments in Lemma 1.2.2.1 leads to the estimates

$$e^{nh_{\text{top}}(X,f,6\delta)} \leq \Lambda(X, 3\delta, n) \leq Qe^{nh_{\text{top}}(X,f,\delta)}, \quad \text{where } Q = (\tau + 1)e^{\tau h_{\text{top}}(X,f,\delta)}.$$
$$(1.2.4.11)$$

Observe that the lower and upper bounds in (1.2.4.11) involve the entropy of f at different scales, a phenomenon which did not appear in (1.2.2.7). To see why this occurs, recall that in the proof of Lemma 1.2.2.1 we used an injective map

$$\mathcal{L}_{m+n} \to \mathcal{L}_m \times \mathcal{L}_n, \qquad w \mapsto (w_{[1,m]}, w_{[m+1,m+n]}), \tag{1.2.4.12}$$

as well as an at-most-$(\tau + 1)$-to-1 map given by specification:

$$\mathcal{L}_m \times \mathcal{L}_n \to \mathcal{L}_{m+n+\tau}, \qquad (v, w) \mapsto vuwu'. \tag{1.2.4.13}$$

In a general metric space, to generalize (1.2.4.12) one might first attempt the following:

- fixing $\rho > 0$, let $E_k^\rho \subset X$ be a maximal (k, ρ)-separated set for each $k \in \mathbb{N}$;
- by maximality, for every $x \in X$ and $k \in \mathbb{N}$ there is $\pi_k(x) \in E_k^\rho$ such that $x \in B_k(\pi_k(x), \rho)$;
- then consider the map $E_{m+n}^\rho \to E_m^\rho \times E_n^\rho$ given by $x \mapsto (\pi_m(x), \pi_{n-m}(f^m x))$.

The problem is that injectivity may fail: there could be $z \in E_m$ such that $B_m(z, \rho)$ contains two distinct points $x, y \in E_{m+n}$, even though $d_m(x, y) \geq d_{m+n}(x, y) \geq \rho$. This possibility can be ruled out by considering a map $E_{m+n}^{2\rho} \to E_m^\rho \times E_n^\rho$; note the use of two different scales. With $\rho = 3\delta$, this leads to the lower bound in (1.2.4.11). See [68, §3.1] for details.

For the upper bound in (1.2.4.11), one must play a similar game with (1.2.4.13). With E_k^ρ as above, specification (at scale δ) gives a "gluing map" $\pi : E_m^\rho \times E_n^\rho \to X$. As long as $\rho \geq \delta$, the multiplicity of this map is at most $\tau + 1$ for the same reasons as in Lemma 1.2.2.1. However, since the gluing process in specification can move orbit segments by up to δ, the image set $\pi(E_m^\rho \times E_n^\rho)$ can only be guaranteed to be

$(\rho - 2\delta, m + n + \tau)$-separated. Again, taking $\rho = 3\delta$ gives (1.2.4.11); see [68, §3.2] for details.

Remark 1.12 The reason that these issues do not arise in the symbolic setting is that there, if $\delta = \frac{1}{4}$ and $y \in B_n(x, \delta)$, then $B_n(y, \delta) = [y_{[1,n]}] = [x_{[1,n]}] = B_n(x, \delta)$. In other words, in a shift space, each d_n is an *ultrametric*, for which the triangle inequality is strengthened to $d_n(x, z) \leq \max\{d_n(x, y), d_n(y, z)\}$. In the non-symbolic setting, if $y \in B_n(x, \delta)$ then the most we can say is that $B_n(y, \delta) \subset B_n(x, 2\delta)$, and vice versa. This leads to the "changing scales" aspect of the arguments above, which appears at several other places in the general proofs.

With the counting bounds established as in (1.2.2.7) and (1.2.4.11), the next step in the symbolic proof was to consider measures ν_n giving equal weight to every n-cylinder, and prove a Gibbs property for any limit point of the measures $\mu_n = \frac{1}{n} \sum_{k=0}^{n-1} \sigma_*^k \nu_n$. For non-symbolic systems, one replaces the collection of n-cylinders with a maximal (n, δ)-separated set, and proves the following.

Proposition 1.9 *Let X be a compact metric space and $f: X \to X$ a continuous map with the specification property at scale $\delta > 0$ and expansivity at scale ϵ, with $\epsilon > 40\delta$, and let $\rho \in (5\delta, \epsilon/8]$. Let $E_n \subset X$ be a maximal $(n, \rho - \delta)$-separated set for each n, and consider the measures*

$$\mu_n := \frac{1}{\#E_n} \sum_{x \in E_n} \frac{1}{n} \sum_{k=0}^{n-1} \delta_{f^k x}. \tag{1.2.4.14}$$

Then there is $K \geq 1$ such that every weak* limit point μ of the sequence μ_n is f-invariant and satisfies the Gibbs property

$$K^{-1} e^{-n h_{\text{top}}(X,f)} \leq \mu(B_n(x, \rho)) \leq K e^{-n h_{\text{top}}(X,f)} \quad \text{for all } x \in X, n \in \mathbb{N}.$$
$$\tag{1.2.4.15}$$

This statement is a mild extension of the argument in [5], which is simplified by having periodic specification at all scales and constructing μ_n using periodic orbits. Proposition 1.9 is proven, with the same level of detail on the choice of scales in [2, §6]. For the purposes of this survey, the main point is simply that the expansivity scale is a suitably large multiple of the specification scale. However, we state the exact range of scales carefully for consistency with [2, §6]. See also [68] for a proof of the lower Gibbs bound. In that paper, many of the intermediate statements and bounds are given in terms of $h_{\text{top}}(X, f, c\rho)$, with $c \in \{1, 2, 3, 4\}$. It is thus crucial that $h_{\text{top}}(X, f, c\rho) = h_{\text{top}}(X, f)$, which is provided in this statement by the expansivity assumption. This is the only way in which expansivity is used in the above proposition.

Observe that we have not yet claimed anything about ergodicity of the Gibbs measure μ. In the symbolic case, the argument for the Gibbs property can be used to deduce that there is $c > 0$ and $k \in \mathbb{N}$ such that for every $v, w \in \mathcal{L}$ and $\ell \geq |v|$, there is $j \in [\ell, \ell + k)$ such that

$$\mu([v] \cap \sigma^{-j}[w]) \geq c\mu[v]\mu[w].$$

Since any Borel set can be approximated (w.r.t. μ) by unions of cylinders, this can be used to deduce that

$$\varliminf_{j \to \infty} \mu(V \cap \sigma^{-j} W) \geq \frac{c}{k}\mu(V)\mu(W)$$

for all $V, W \subset X$, which gives ergodicity. In the non-symbolic setting, one can still mimic the Gibbs argument to produce $c > 0$ and $k \in \mathbb{N}$ such that for every $(x, n), (y, m) \in X \times \mathbb{N}$ and any $\ell \geq n$, there is $j \in [\ell, \ell + k)$ such that

$$\mu(B_n(x, \rho) \cap f^{-j} B_m(y, \rho)) \geq c\mu(B_n(x, \rho))\mu(B_m(y, \rho)). \tag{1.2.4.16}$$

To establish ergodicity from this one needs to approximate arbitrary Borel sets by sets whose μ-measure we control; this can be done by using a sequence of partitions β_n, for which each element of β_n contains a Bowen ball $B_n(x, \rho)$ and is contained inside a Bowen ball $B_n(x, 2\rho)$. Expansivity implies that this sequence of partitions is generating w.r.t. μ, so the rest of the argument goes through as before, and establishes ergodicity. As we saw in the proof of Proposition 1.6, this is also enough to guarantee that $h_{\text{top}}(X, f, \epsilon) = h_{\text{top}}(X, f)$. We summarize our conclusions as follows.

Proposition 1.10 *Let* X, f, δ, μ *be as in Proposition 1.9. Suppose that* f *is expansive at a scale greater than* 40δ. *Then* μ *is ergodic and satisfies the Gibbs property* (1.2.4.15).

The proof that an ergodic Gibbs measure is the unique MME (Proposition 1.1) has the following generalization to the non-symbolic setting.

Proposition 1.11 *Let X be a compact metric space, $f \colon X \to X$ a continuous map, and μ an ergodic f-invariant measure on X. Suppose $\rho > 0$ is such that*

- f *is expansive (or positively expansive) at scale* 4ρ;
- *there are $K, h > 0$ such that μ satisfies the Gibbs bound*

$$K^{-1}e^{-nh} \leq \mu(B_n(x, \rho)) \leq Ke^{-nh} \text{ for every } x \in X \text{ and } n \in \mathbb{N}. \tag{1.2.4.17}$$

Then $h = h_\mu(f) = h_{\text{top}}(X, f)$, and μ is the unique MME for (X, f).

Outline of Proof As before, one starts by using general arguments to prove that $h = h_\mu(f) = h_{\text{top}}(X, f)$ and to reduce to the case of considering an invariant measure $\nu \perp \mu$, for which we must show $h_\nu(f) < h_\mu(f)$; this is unchanged from the symbolic case. The next step there was to choose $D \subset X$ with $\mu(D) = 1$ and $\nu(D) = 0$, and approximate D by a union of cylinders; then similar to (1.2.4.6), writing

$$nh_\nu(f) = h_\nu(f^n) = h_\nu(f^n, \alpha_0^{n-1}) \le H_\nu(\alpha_0^{n-1}) = \sum_{w \in \mathcal{L}_n} -\nu[w] \log \nu[w],$$

(1.2.4.18)

and splitting the sum between cylinders in $\mathcal{D}_n$ and those in $\mathcal{D}_n^c$, one eventually proves that $h_\nu(f) < h_\mu(f)$ by using the Gibbs bound $\mu[w] \ge K^{-1} e^{-|w| h_{\text{top}}(X)}$.

In the non-symbolic setting, the approximation of D follows just as in the paragraph after (1.2.4.16). Moreover, we can obtain an analogue of (1.2.4.18) by replacing α_0^{n-1} with a partition β_n such that every element of β_n is contained in $B_n(x, 2\rho)$ for some point x in a maximal $(n, 2\rho)$-separated set E_n. Finally, as long as we also arrange that each element of β_n contain $B_n(x, \rho)$, we can use the lower Gibbs bound to complete the proof just as in the symbolic case.

Remark 1.13 The partition β_n which appears in the above proofs is called an *adapted partition* for E_n. Adapted partitions exist for any $(n, 2\rho)$ separated set of maximal cardinality since the sets $B_n(x, \rho)$ are disjoint and the sets $\overline{B}_n(x, 2\rho)$ cover X.

Remark 1.14 In the two-sided expansive case, the same argument works, provided we replace d_n and B_n with their two-sided versions. That is, we consider balls in the metric $d_{[-n,n]}(x, y) = \max\{d(f^k x, f^k y) : -n \le k \le n\}$ in place of B_n. Then one uses adapted partitions and proceeds as in the positively expansive case.

1.3 Non-uniform Bowen Hypotheses and Equilibrium States

In Sect. 1.3.1, we recall the role played by expansivity in Bowen's proof of uniqueness, and formulate a uniqueness result using a weaker version of expansivity. Then in Sect. 1.3.2 we describe an explicit class of partially hyperbolic diffeomorphisms where expansivity fails but this result still applies. In Sect. 1.3.3 we combine the weakened versions of expansivity and specification to formulate our most general result on MMEs for discrete-time systems, which we apply in Sect. 1.3.4 to a more general partially hyperbolic setting. Finally, in Sect. 1.3.5 we describe how this theory extends to equilibrium states for nonzero potential functions.

1.3.1 Relaxing the Expansivity Hypothesis

In this section, we describe how we relax the expansivity property. Our motivating examples are diffeomorphisms for which expansivity fails, but for which the failure of expansivity is "invisible" to the MME. In these examples, the failure of expansivity is a lower entropy phenomenon, and this leaves room for us to develop a version of Bowen's argument for the MME.

As explained in the previous section, Bowen's proof of uniqueness uses expansivity to guarantee that certain sequences of partitions are generating with respect to every invariant ν. In fact, in every place where this property is used, it is enough to know that this holds for all ν with sufficiently large entropy.

More precisely, at the end of the proof, in (the analogue of) (1.2.4.18), it suffices to know that α_0^{n-1} is generating for (f^n, ν) when ν is an arbitrary MME, because if ν is not an MME then we already have $h_\nu < h_\mu$, which was the goal. This is also sufficient for the approximation of D by elements of the partitions β_n, and thus Proposition 1.11 remains true if we replace expansivity with the assumption that for every MME ν, we have $\Gamma_\epsilon(x) = \{x\}$ for ν-a.e. x.

In Proposition 1.9, the argument for ergodicity required a similar generating property. Finally, in Proposition 1.6, it suffices to have this generating property w.r.t. a family of measures ν over which $\sup_\nu h_\nu(f) = h_{\text{top}}(X, f)$.

With these observations in mind, we make the following definitions.

Definition 1.3.1.1 ([69]) An f-invariant measure μ is *almost expansive at scale ϵ* if $\Gamma_\epsilon(x) = \{x\}$ for μ-a.e. x; equivalently, if the *non-expansive set* $\text{NE}(\epsilon) = \{x \in X : \Gamma_\epsilon(x) \neq \{x\}\}$ has $\mu(\text{NE}(\epsilon)) = 0$. Replacing Γ_ϵ by Γ_ϵ^+ gives NE^+ and a notion of *almost positively expansive*.

Definition 1.3.1.2 ([68]) The *entropy of obstructions to expansivity at scale ϵ* is

$$h_{\exp}^\perp(X, f, \epsilon) := \sup\{h_\mu(f) : \mu \in \mathcal{M}_f^e(X) \text{ is not almost expansive at scale } \epsilon\}$$

$$= \sup\{h_\mu(f) : \mu \in \mathcal{M}_f^e(X) \text{ and } \mu(\text{NE}(\epsilon)) > 0\}.$$

We write $h_{\exp}^\perp(X, f) = \lim_{\epsilon \to 0} h_{\exp}^\perp(X, f, \epsilon)$ for the *entropy of obstructions to expansivity*, without reference to scale. The entropy of obstructions to positive expansivity $h_{\exp^+}^\perp$ is defined analogously.

From the discussion after Proposition 1.9, we see that we can replace the assumption of expansivity with the assumption that $h_{\exp}^\perp(X, f, \rho) < h_{\text{top}}(X, f)$, since then every ergodic ν with $h_\nu(f) > h_{\exp}^\perp(X, f, \rho)$ is almost expansive, so the Proposition goes through.[12] Similarly in Propositions 1.10 and 1.11, it suffices to assume that $h_{\exp}^\perp(X, f, 4\rho) < h_{\text{top}}(X, f)$.

---

[12]See [68, Proposition 2.7] for a detailed proof that $h_{\text{top}}(X, f, \rho) = h_{\text{top}}(X, f)$ in this case.

Now we have all the pieces for a uniqueness result using non-uniform expansivity.

Theorem 1.3.1.1 (Unique MME with Non-uniform Expansivity [68]) *Let X be a compact metric space and $f : X \to X$ a continuous map. Suppose that $\epsilon > 40\delta > 0$ are such that $h_{\exp}^{\perp}(X, f, \epsilon) < h_{\text{top}}(X, f)$, and that f has the specification property at scale δ. Then (X, f) has a unique measure of maximal entropy.*

1.3.2 Derived-from-Anosov Systems

We describe a class of smooth systems for which expansivity fails but the entropy of obstructions to expansivity is small. The following example is due to Mañé [70]; we primarily follow the discussion in [15], and refer to that paper for further details and references.

1.3.2.1 Construction of the Mañé Example

Fix a matrix $A \in SL(3, \mathbb{Z})$ with simple real eigenvalues $\lambda_u > 1 > \lambda_s > \lambda_{ss} > 0$, and corresponding eigenspaces $F^{u,s,ss} \subset \mathbb{R}^3$. Let $f_0 : \mathbb{T}^3 \to \mathbb{T}^3$ be the hyperbolic toral automorphism defined by A, and let $\mathcal{F}^{u,s,ss}$ be the corresponding foliations of $\mathbb{T}^3$. Define a perturbation f of f_0 as follows.

Fix $\rho > \rho' > 0$ such that f_0 is expansive at scale ρ. Let $q \in \mathbb{T}^3$ be a fixed point of f, and set $f = f_0$ outside of $B(q, \rho)$. Inside $B(q, \rho)$, perform a pitchfork bifurcation in the center direction as shown in Fig. 1.6, in such a way that

- the foliation $W^c := \mathcal{F}^s$ remains f-invariant, and we write $E^c = TW^c$;
- the cones around F^u and F^{ss} remain invariant and uniformly expanding for Df and Df^{-1}, respectively, so they contain Df-invariant distributions $E^{u,ss}$ that integrate to f-invariant foliations $W^{u,ss}$;

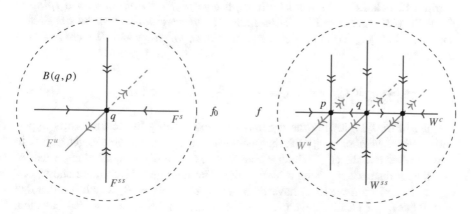

Fig. 1.6 Mañé's construction

- $E^{cs} = E^c \oplus E^{ss}$ integrates to a foliation W^{cs};
- outside of $B(q, \rho')$, we have $\|Df|_{E^{cs}}\| \le \lambda_s < 1$.

Thus f is partially hyperbolic with $T\mathbb{T}^3 = E^u \oplus E^c \oplus E^{ss} = E^u \oplus E^{cs}$. Observe that

$$\lambda_c(f) := \sup\{\|Df|_{E^{cs}(x)}\| : x \in \mathbb{T}^3\} > 1 \tag{1.3.2.1}$$

because the center direction is expanding at q.

Now consider a diffeomorphism $g \colon \mathbb{T}^3 \to \mathbb{T}^3$ that is C^1-close to f. Such a g remains partially hyperbolic, with

$$\lambda_c(g) > 1 > \lambda_s(g) := \sup\{\|Df|_{E^{cs}(x)}\| : x \in \mathbb{T}^3 \setminus B(q, \rho')\}. \tag{1.3.2.2}$$

Existence of a unique MME was proved for such g by Ures [71] and by Buzzi et al. [72], using the fact that there is a semiconjugacy from g back to the hyperbolic toral automorphism f_0. We outline an alternate proof using Theorem 1.3.1.1, which has the benefit of extending to a class of nonzero Hölder continuous potential functions [15].

1.3.2.2   Estimating the Entropy of Obstructions

Although the map g behaves as if it is uniformly hyperbolic outside of $B(q, \rho)$, the presence of fixed points with different indices inside this ball causes expansivity to fail. Indeed, let p denote one of the two fixed points created via the pitchfork bifurcation, and let x be any point on the leaf of W^c that connects p to q. Then for every $\epsilon > 0$, the bi-infinite Bowen ball $\Gamma_\epsilon(x)$ is a non-trivial curve in W^c, rather than a single point. However, we can give a simple mild criterion on the orbit of a point x which rules out $\Gamma_\epsilon(x)$ being non-trivial, and we can argue that this criterion is satisfied for most points in our examples.

Lemma 1.3.2.1 *Let g be a partially hyperbolic diffeomorphism with a splitting $E^u \oplus E^c \oplus E^s$ such that E^c is 1-dimensional and integrable. Then there is $\epsilon_0 > 0$ such that $\Gamma_{\epsilon_0}(x) \subset W^c(x)$ for every x. Moreover, for every $\lambda > 0$ there is $\epsilon > 0$ such that*

$$\varlimsup_{n \to \infty} \frac{1}{n} \log \|Dg^{-n}|_{E^c(x)}\| > \lambda \quad \Rightarrow \quad \Gamma_\epsilon(x) = \{x\}. \tag{1.3.2.3}$$

Sketch of Proof Following the argument for expansivity in the uniformly hyperbolic setting, we choose ϵ_0 such that whenever $d(x, y) < \epsilon_0$, we can get from x to y by moving a distance d^s along a leaf of W^s, then a distance d^c along a leaf of W^c, then a distance d^u along a leaf of W^u. The argument given there shows that if $y \in \Gamma_{\epsilon_0}(x)$ then we must have $d^s(x, y) = d^u(x, y) = 0$, which implies that $y \in W^c(x)$. For (1.3.2.3), we observe that if the condition on Dg^{-n} is satisfied, then there are arbitrarily large n such that

$$\|Dg^{-n}|_{E^c(x)}\| > ce^{\lambda n}. \tag{1.3.2.4}$$

Choosing $\epsilon > 0$ sufficiently small that $|\log \|Dg|_{E^c(z)}\| - \log \|Dg|_{E^c(z')}\|| < \lambda/2$ whenever $d(z, z') < \epsilon$, we see that any $y \in \Gamma_\epsilon(x)$ satisfies

$$d(g^{-n}x, g^{-n}y) \geq ce^{\lambda n/2}d(x, y) \tag{1.3.2.5}$$

for all n satisfying (1.3.2.4). Since n can become arbitrarily large, this implies that $d(x, y) = 0$.

Remark 1.15 Replacing backwards time with forwards time, the analogous result for positive Lyapunov exponents is also true: $\overline{\lim} \frac{1}{n} \log \|Dg^n|_{E^c(x)}\| > \lambda$ implies that $\Gamma_\epsilon(x) = \{x\}$.

For the Mañé examples, we can use (1.3.2.2) to control $\|Dg^{-n}|_{E^c(x)}\|$ in terms of how much time the orbit of x spends outside $B(q, \rho)$; together with Lemma 1.3.2.1, this allows us to estimate the entropy of $NE(\epsilon)$. To formalize this, we write $\chi = \mathbf{1}_{\mathbb{T}^3 \setminus B(q,\rho)}$ and observe that by the definition of $\lambda_c(g)$ and $\lambda_s(g)$ in (1.3.2.1) and (1.3.2.2), we have

$$\|Dg^{-n}|_{E^c(x)}\| \geq \lambda_s(g)^{-s_n(x)}\lambda_c(g)^{-(n-s_n(x))} \quad \text{where } s_n(x) := \sum_{k=0}^{n-1} \chi(g^{-k}x).$$

It follows that

$$\varlimsup_{n \to \infty} \frac{1}{n} \log \|Dg^{-n}|_{E^c(x)}\| \geq -(\overline{r}(x) \log \lambda_s(g) + (1 - \overline{r}(x)) \log \lambda_c(g)) \tag{1.3.2.6}$$

where we write

$$\overline{r}(x) = \varlimsup_{n \to \infty} \frac{1}{n} s_n(x) = \varlimsup_{n \to \infty} \frac{1}{n} \sum_{k=0}^{n-1} \chi(g^{-k}x).$$

Fix $\lambda \in (0, -\log \lambda_s(g))$ and let $r > 0$ satisfy $-(r \log \lambda_s(g) + (1-r) \log \lambda_c(g)) > \lambda$. Then Lemma 1.3.2.1 and (1.3.2.6) show that for a sufficiently small $\epsilon > 0$, we have

$$NE(\epsilon) \subset \{x : \overline{r}(x) < r\}. \tag{1.3.2.7}$$

Since f_0 is Anosov, the uniform counting bounds in (1.2.4.11) give a constant Q such that $\Lambda(X, f_0, \epsilon, n) \leq Qe^{nh_{\text{top}}(X, f_0)}$ for all n. Using this together with (1.3.2.7) one can prove the following.

Lemma 1.3.2.2 ([14, §3.4]) *Writing $H(t) = -t \log t - (1 - t) \log(1 - t)$ for the usual bipartite entropy function, the Mañé examples satisfy*

$$h^{\perp}_{\exp}(g, \epsilon) < r(h_{\text{top}}(X, f_0) + \log Q) + H(2r).$$

Idea of Proof Given an ergodic measure μ that satisfies $\mu(\text{NE}(\epsilon))$ and thus satisfies $\overline{\lim} \frac{1}{n} S_n \chi(g^{-n}x) \leq r$ for μ-a.e. x, the Katok entropy formula [73] can be used to show that $h_{\mu}(f) \leq h(C)$, where

$$C := \{(x, n) \in \mathbb{T}^3 \times \mathbb{N} : S_n \chi(x) \leq rn\}. \tag{1.3.2.8}$$

To estimate $h(C)$, the idea is to partition an orbit segment $(x, n) \in C$ into pieces lying entirely inside or outside of $B(q, \rho)$. There can be at most rn pieces lying outside, so the number of transition times between inside and outside is at most $2rn$. The number of ways of choosing these transition times is thus at most

$$\binom{n}{2rn} = \frac{n!}{(2rn)!((1 - 2r)n)!} \approx e^{H(2r)n},$$

where the approximation can be made more precise using Stirling's formula or a rougher elementary integral estimate. This contributes the $H(2r)$ term to the estimate; the remaining terms are roughly due to the observation that given a pattern of transition times for which the segments lying outside $B(q, \rho)$ have lengths $k_1, \ldots, k_m$, the number of ϵ-separated orbit segments in C associated to this pattern is at most

$$\prod_{j=1}^{m} \Lambda(X, f_0, \epsilon, k_i) \leq \prod_{j=1}^{m} Q e^{k_i h_{\text{top}}(X, f_0)} \leq Q^m e^{rnh_{\text{top}}(X, f_0)} \leq (Q e^{h_{\text{top}}(X, f_0)})^{rn},$$

since no entropy is produced by the sojourns inside $B(q, \rho)$.

Since there is a semi-conjugacy from g to f_0, we have $h_{\text{top}}(X, g) \geq h_{\text{top}}(X, f_0)$. Thus we have $h^{\perp}_{\exp}(g) < h_{\text{top}}(g)$ whenever r satisfies

$$r(h_{\text{top}}(X, f_0) + \log Q) + H(2r) < h_{\text{top}}(X, f_0). \tag{1.3.2.9}$$

Recall that r must be chosen large enough such that $\lambda_s(g)^r \lambda_c(g)^{1-r} < 1$. Equivalently, for a given value of r, the perturbation must be chosen small enough for this to hold (that is, λ_c must be close enough to 1). Thus given f_0, we can find r small enough such that (1.3.2.9) holds, and then for any sufficiently small perturbation the above argument guarantees that $h^{\perp}_{\exp}(X, g) < h_{\text{top}}(X, g)$.

Remark 1.16 Since $\Gamma_\epsilon(x) \subset W^c(x)$, which is one-dimensional, it is not hard to show that $h_{\text{top}}(W^c(x)) = 0$, and thus $h_{\text{top}}(\Gamma_\epsilon(x)) = 0$ [15, 74]; in other words, f is *entropy expansive*. Entropy expansivity implies that $h_{\text{top}}(X, f, \epsilon) = h_{\text{top}}(X, f)$

[75], which for systems with (coarse) specification is sufficient for the construction of a Gibbs measure in Proposition 1.9. However, there does not seem to be any way to use entropy expansivity to carry out the arguments for ergodicity and uniqueness. The issue is that we need to use Bowen balls to construct adapted partitions which approximate Borel sets. When $\Gamma_\epsilon(x)$ is a point, the two-sided Bowen ball at x is a neighborhood of the point, which is key to the approximation argument. The analysis is significantly more difficult even when $\Gamma_\epsilon(x) \neq \{x\}$ has a simple explicit characterization, see Sect. 1.4.2.1 for more details in the flow case. If all we know about $\Gamma_\epsilon(x)$ is that $h(\Gamma_\epsilon(x)) = 0$ it is unclear how to proceed. On the other hand, for the Bonatti–Viana examples introduced in [76], entropy expansivity can fail [69] even while the condition $h_{\exp}^\perp < h_{\text{top}}$ is satisfied [14]. The Bonatti–Viana examples are 4-dimensional analogues of the Mañé examples that involve two separate perturbations and have a dominated splitting $T\mathbb{T}^4 = E^{cu} \oplus E^{cs}$ but are not partially hyperbolic. We were able to study their thermodynamic formalism in [14] despite these difficulties.

1.3.2.3   Specification for Mañé Examples

In order to apply Theorem 1.3.1.1 to the Mañé examples, one must investigate the specification property. Globally, specification at all scales certainly fails. Two approaches to deal with this are possible, and it is instructive to consider both—our choice is to work with a *coarse* specification property globally, or specification at all scales on a 'good collection of orbit segments'.

The key ingredient we are missing from the uniformly hyperbolic case is uniform contraction along W^{cs}, which is replacing W^s. We explain why we can obtain coarse specification globally. As explained in Remark 1.10, uniform contraction is not needed for the proof of specification; it suffices to know that

$$W_\delta^{cs}(x) \subset B_n(x, \delta) \text{ for all } x. \tag{1.3.2.10}$$

Since contraction in W^{cs} can fail for the Mañé example only in $B(q, \rho')$, one can easily show that (1.3.2.10) continues to hold as long as $\delta > 2\rho'$, and thus g has specification at these scales. Choosing ρ' to be small enough relative to ρ, Theorem 1.3.1.1 applies and establishes existence of a unique MME.

To see that the Mañé example does not have the specification property at all scales, we sketch a short argument which appears in much greater generality in [77]. Observe that for sufficiently small $\delta > 0$, the forward infinite Bowen ball $\Gamma_\delta^+(q)$ is the 1-dimensional local stable leaf $W_\delta^{ss}(q)$. Suppose that g has specification at scale δ with gap size τ, and let x be any point whose orbit never enters $B(q, \rho)$. Specification gives $y \in W_\delta^u(x)$ and $0 \le k \le \tau$ such that $f^k(y) \in W_\delta^{ss}(q)$.[13] In

---

[13]Use specification to get $y_n \in f^n(B_n(x, \delta)) \cap f^{-k_n}(B_n(q, \delta))$ for $0 \le k_n \le \tau$, choose k such that $k_n = k$ for infinitely many values of n, and let y be a limit point of the corresponding y_n.

other words, $f^{-\tau}(W_\delta^{ss}(q))$ intersects *every* local unstable leaf associated to an orbit that avoids $B(q, \rho)$. But this is impossible because the dimensions are wrong.[14]

Thus, if we want a global specification property, we must work at a fixed coarse scale, as described above. We explore the other option of returning to the ideas from Sect. 1.2.3 and recovering specification at all scales by restricting to a "good collection of orbit segments" in the next section.

1.3.3 The General Result for MMEs in Discrete-Time

Now we formulate a general result that combines the symbolic result using decompositions with Theorem 1.3.1.1 by allowing both expansivity and specification to fail, provided the obstructions have small entropy. This allows us to cover some new classes of examples, as we will see later, and is also important in dealing with nonzero potential functions.

Recall from Sect. 1.2.3 that a decomposition of the language $\mathcal{L}$ of a shift space consists of $C^p, G, C^s \subset \mathcal{L}$ such that every $w \in \mathcal{L}$ can be written as $w = u^p v u^s$ where $u^p \in C^p$, $v \in G$, and $u^s \in C^s$. As discussed in Sect. 1.2.4.1, for non-symbolic systems we replace $\mathcal{L}$ with the *space of orbit segments* $X \times \mathbb{N}$, where (x, n) corresponds to the orbit segment $x, f(x), f^2(x), \ldots, f^{n-1}(x)$.

Definition 1.3.3.1 A *decomposition* for $X \times \mathbb{N}$ consists of three collections $C^p, G, C^s \subset X \times \mathbb{N}_0$ for which there exist three functions $p, g, s : X \times \mathbb{N} \to \mathbb{N}_0$ such that for every $(x, n) \in X \times \mathbb{N}$, the values $p = p(x, n)$, $g = g(x, n)$, and $s = s(x, n)$ satisfy $p + g + s = n$, and

$$(x, p) \in C^p, \quad (f^p x, g) \in G, \quad (f^{p+s} x, s) \in C^s.$$

Given a decomposition, for each $M \in \mathbb{N}$ we write

$$G^M := \{(x, n) \in X \times \mathbb{N} : p(x, n) \le M \text{ and } s(x, n) \le M\}.$$

Theorem 1.3.3.1 (Non-uniform Bowen Hypotheses for Maps (MME Case)) *Let X be a compact metric space and $f : X \to X$ a continuous map. Suppose that $\epsilon > 40\delta > 0$ are such that $h_{\exp}^\perp(X, f, \epsilon) < h_{\text{top}}(X, f)$, and that the space of orbit segments $X \times \mathbb{N}$ admits a decomposition $C^p G C^s$ such that*

(I) *every collection G^M has specification at scale δ, and*
(II) $h(C^p \cup C^s, \delta) < h_{\text{top}}(X, f)$.

Then (X, f) has a unique measure of maximal entropy.

---

[14]Note that $f^{-\tau}(W_\delta^{ss}(q))$ intersects a local leaf of W^{cu} in at most finitely many points, and thus intersects at most finitely many of the corresponding local leaves of W^u; however, there are uncountably many of these corresponding to points that never enter $B(q, \rho)$.

The proof of Theorem 1.3.3.1 requires an extension of the counting arguments for decompositions (Sect. 1.2.3.1) to the general metric space setting, following similar ideas to those outlined in Sect. 1.2.4.4. Furthermore, the construction of a Gibbs measure and the proofs of ergodicity and uniqueness must be modified to reflect the fact that uniform lower bounds can only be obtained on $\mathcal{G}^M$. As in Sect. 1.2.3.1, we omit further discussion of these more technical aspects, referring to [2, 68] for complete details.

Remark 1.17 If $\mathcal{G}$ has specification at all scales, then a short continuity argument [2, Lemma 2.10] proves that every $\mathcal{G}^M$ does as well, which establishes (I).

1.3.4 Partially Hyperbolic Systems with One-Dimensional Center

Theorem 1.3.3.1 can be applied to a broad class of partially hyperbolic systems, which includes the Mañé examples. This result has not previously appeared elsewhere. We give an outline of the proof. Further details are analogous to the case of the Mañé examples, and we emphasize the key new points.

Theorem 1.3.4.1 *Let* $f : M \to M$ *be a partially hyperbolic diffeomorphism with* $TM - E^u \oplus E^c \oplus E^s$. *Assume that* $\dim E^c = 1$ *and that every leaf of the foliations* W^s *and* W^u *is dense in* M.

Let $\varphi^c(x) = \log \|Df|_{E^c(x)}\|$, *and given* $\mu \in \mathcal{M}_f^e(M)$, *let* $\lambda^c(\mu) = \int \varphi^c \, d\mu$ *be the* center Lyapunov exponent *of* μ. *Consider the quantities*

$$
\begin{aligned}
h^+ &:= \sup\{h_\mu(f) : \mu \in \mathcal{M}_f^e(M), \lambda^c(\mu) \geq 0\}, \\
h^- &:= \sup\{h_\mu(f) : \mu \in \mathcal{M}_f^e(M), \lambda^c(\mu) \leq 0\}.
\end{aligned}
\tag{1.3.4.1}
$$

Suppose that $h^+ \neq h^-$. *Then* f *has a unique MME.*

Remark 1.18 Since $h_{\text{top}}(X, f) = \max(h^+, h^-)$, the condition $h^+ \neq h^-$ is equivalent to the condition that either $h^+ < h_{\text{top}}(X, f)$ or $h^- < h_{\text{top}}(X, f)$. It would be interesting to investigate how typical this condition is. The only way for this condition to fail is if there is an ergodic MME with $\lambda^c = 0$, or if there are (at least) two ergodic MMEs for which λ^c takes both signs. See Sect. 1.3.5.4 for an interpretation of this condition in terms of topological pressure, and an extension of Theorem 1.3.4.1 to equilibrium states for nonzero potentials.

Remark 1.19 For 3-dimensional partially hyperbolic diffeomorphisms homotopic to Anosov, Ures [71] showed that there is a unique measure of maximal entropy. In this setting, Crisostomo and Tahzibi [78] gave some interesting criteria for uniqueness (and in some case finiteness) of equilibrium states. We note that our setting is a complementary regime to that of [79], which assumes compact center leaves, and in which non-uniqueness of the MME is typical.

First observe that arguments similar to those given for the Mañé example in Lemma 1.3.2.1 and Remark 1.15 show that $h_{\exp}^{\perp}(f) \leq \min(h^+, h^-)$, so the condition $h_{\exp}^{\perp}(f) < h_{\text{top}}(f)$ is satisfied whenever $h^+ \neq h^-$.

Remark 1.20 The upper bound on $h_{\exp}^{\perp}$ for the Mañé examples in Lemma 1.3.2.2 is actually an upper bound on h^+ in that setting, verifying that $h^+(g) < h_{\text{top}}(g)$ whenever the perturbation is small enough. Moreover, the leaves of W^u are all dense for these examples [80], so Theorem 1.3.4.1 applies to the Mañé examples.

The rest of the proof of Theorem 1.3.4.1 consists of finding a decomposition C^p, G, C^s for $X \times \mathbb{N}$ such that G has specification at all scales and $h(C^p \cup C^s) < h_{\text{top}}(X, f)$. We describe the general argument in the case when $h^+ < h_{\text{top}}(f)$, so intuitively, all of the large entropy parts of the system have negative central Lyapunov exponents.

1.3.4.1 A Small Collection of Obstructions

We take $C^s = \emptyset$. To describe C^p, we first observe that the condition $h^+ < h_{\text{top}}(f)$ implies that

$$\sup\{h_\mu(f) : \mu \in \mathcal{M}_f, \lambda^c(\mu) \geq 0\} < h_{\text{top}}(f),$$

where the difference is that now the supremum allows non-ergodic measures as well, and then a weak*-continuity argument gives $r > 0$ such that

$$\sup\{h_\mu(f) : \mu \in \mathcal{M}_f, \lambda^c(\mu) \geq -r\} < h_{\text{top}}(f). \tag{1.3.4.2}$$

We can relate the left-hand side of (1.3.4.2) to $h(C^p)$, where

$$C^p := \{(x, n) \in M \times \mathbb{N} : S_n\varphi^c(x) \geq -rn\}.$$

One relationship between these was mentioned when we bounded $h_{\exp}^{\perp}$ for the Mañé example (though the function being summed there was different). Here we want to go the other way and obtain an upper bound on $h(C^p)$. For this we observe that if we let $E_n \subset C_n^p$ be any (n, ϵ)-separated set, ν_n the equidistributed atomic measure on E_n, and $\mu_n = \frac{1}{n}\sum_{k=0}^{n-1} f_*^k \nu_n$, then half of the proof of the variational principle [50, Theorem 8.6] shows that any limit point of μ_n is f-invariant and has

$$h_\mu(f) \geq h(C^p, \epsilon).$$

Moreover, $\lambda^c(\mu) = \int \varphi^c \, d\mu(x) \geq -r$ by weak*-convergence and the definition of C^p. Together with (1.3.4.2), we conclude that $h(C^p) < h_{\text{top}}(f)$.

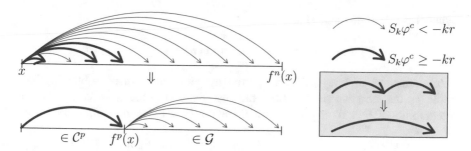

Fig. 1.7 A decomposition $C^p\mathcal{G}$ of the space of orbit segments

1.3.4.2   A Good Collection with Specification

We now describe a 'good' collection of orbit segments $\mathcal{G}$, and define a decomposition. To this end, take an arbitrary orbit segment $(x, n) \in M \times \mathbb{N}$, and remove the longest possible element of C^p from its beginning. That is, let $p = p(x, n)$ be maximal with the property that $(x, p) \in C^p$. Then we have

$$S_p\varphi^c(x) \geq -rp \text{ and } S_k\varphi^c(x) < -rk \text{ for all } p < k \leq n.$$

Subtracting the first from the second gives

$$S_{k-p}\varphi^c(f^p x) = S_k\varphi^c(x) - S_p\varphi^c(x) < -r(k - p),$$

which we can rewrite as

$$S_j\varphi^c(f^j x) < -rj \text{ for all } 0 \leq j \leq n - p.$$

In other words, as shown in Fig. 1.7, we have[15]

$$(f^p x, n - p) \in \mathcal{G} := \{(y, m) : S_j\varphi^c(y) < -rj \text{ for all } 0 \leq j \leq m\}.$$

Moreover, by choosing $\delta > 0$ sufficiently small that $|\varphi^c(y) - \varphi^c(z)| < r/2$ whenever $d(y, z) < \delta$, we see that if $(y, m) \in \mathcal{G}$ and $z \in B_m(y, \delta)$, then

$$\|Df^j|_{E^{cs}(z)}\| \leq e^{-rj/2} \text{ for all } 0 \leq j \leq m.$$

This is enough to prove the specification property for $\mathcal{G}$. If E^{cs} is integrable, then one can simply use the proof from the uniformly hyperbolic case verbatim,

---

[15]There is a clear analogy between what we are doing here and the notion of *hyperbolic time* introduced by Alves [81], and developed by Alves, Bonatti and Viana [82].

using (1.3.4.2) to guarantee that

$$W_\delta^{cs}(x) \subset B_n(x, \delta) \text{ whenever } (x, n) \in \mathcal{G}. \qquad (1.3.4.3)$$

Since questions of integrability in partial hyperbolicity can be subtle [83], we point out that one can still establish the specification property without assuming integrability of E^{cs}. To do this, fix $\theta > 0$ and consider the *center-stable cone*

$$K^{cs}(x) := \{v + w : v \in E^{cs}, w \in E^u, \|w\| < \theta\|v\|\} \subset T_x M;$$

then when establishing the "one-step specification" property in (1.2.4.7), one can take an *admissible* manifold $W \ni f^{n_2}(x_2)$ that has $T_y W \subset K^{cs}(x)$ at each $y \in W$, and replace $W_\delta^{cs}(x)$ with $f^{-n_2}(W) \cap B(x_2, \delta)$ in the argument. As long as $\theta > 0$ is sufficiently small, there will still be enough contraction along (x_2, n_2) for vectors in K^{cs} to guarantee that (1.3.4.3) holds.

1.3.5 Unique Equilibrium States

For the sake of simplicity, we have so far restricted our attention to measures of maximal entropy. However, the entire apparatus developed above works equally well for equilibrium states associated to "sufficiently regular" potential functions.

1.3.5.1 Topological Pressure

First we recall the notion of *topological pressure*. As with topological entropy in Sect. 1.2.4.1, we give a more general definition than is standard, defining pressure for collections of orbit segments $\mathcal{D} \subset X \times \mathbb{N}$; our definition reduces to the standard one when $\mathcal{D} = X \times \mathbb{N}$.

Definition 1.3.5.1 Given a continuous potential function $\varphi \colon X \to \mathbb{R}$ and a collection of orbit segments $\mathcal{D} \subset X \times \mathbb{N}$, for each $\epsilon > 0$ and $n \in \mathbb{N}$ we consider the *partition sum*

$$\Lambda(\mathcal{D}, \varphi, \epsilon, n) := \sup\left\{ \sum_{x \in E} e^{S_n\varphi(x)} : E \subset \mathcal{D}_n \text{ is } (n, \epsilon)\text{-separated} \right\},$$

where $S_n\varphi(x) = \sum_{k=0}^{n-1} \varphi(f^k x)$ is the nth Birkhoff sum. The *pressure* of φ on the collection $\mathcal{D}$ at scale $\epsilon > 0$ is

$$P(\mathcal{D}, \varphi, \epsilon) := \varlimsup_{n \to \infty} \frac{1}{n} \log \Lambda(\mathcal{D}, \varphi, \epsilon, n), \qquad (1.3.5.1)$$

and the pressure of φ on the collection $\mathcal{D}$ is

$$P(\mathcal{D}, \varphi) := \lim_{\epsilon \to 0} P(\mathcal{D}, \varphi, \epsilon). \tag{1.3.5.2}$$

As with entropy, in the case when $\mathcal{D} = Y \times \mathbb{N}$ we write $\Lambda(Y, \varphi, \epsilon, n)$, etc.

The variational principle for topological pressure states that

$$P(X, \varphi) = \sup_{\mu \in M_f(X)} \left(h_\mu(f) + \int \varphi \, d\mu \right). \tag{1.3.5.3}$$

A measure that achieves the supremum is called an *equilibrium state* for (X, f, φ).

As was the case with the MME, there is a standard construction from the proof of the variational principle that establishes existence of an equilibrium state in many cases: we have the following generalization of Proposition 1.4 and Corollary 1.2.4.1.

Proposition 1.12 (Building Approximate Equilibrium States) *With X, f, φ as above, fix $\epsilon > 0$, and for each $n \in \mathbb{N}$, let $E_n \subset X$ be an (n, ϵ)-separated set. Consider the Borel probability measures*

$$\nu_n := \frac{1}{\sum_{x \in E^n} e^{S_n \varphi(x)}} \sum_{x \subset E_n} \delta_x e^{S_n \varphi(x)}, \quad \mu_n := \frac{1}{n} \sum_{k=0}^{n-1} f_*^k \nu_n = \frac{1}{n} \sum_{k=0}^{n-1} \nu_n \circ f^{-k},$$

$$\tag{1.3.5.4}$$

Let μ_{n_j} be any subsequence that converges in the weak*-topology to a limiting measure μ. Then $\mu \in M_f(X)$ and

$$h_\mu(f) + \int \varphi \, d\mu \geq \overline{\lim_{j \to \infty}} \frac{1}{n_j} \log \sum_{x \in E_{n_j}} e^{S_{n_j} \varphi(x)}.$$

In particular, for every $\delta > 0$ there exists $\mu \in M_f(X)$ such that $h_\mu(f) + \int \varphi \, d\mu \geq P(X, f, \varphi, \delta)$.

Proof See [50, Theorem 9.10]. □

Corollary 1.3.5.1 *Let X, f be as above, and suppose that there is $\delta > 0$ such that $P(X, \varphi, \delta) = P(X, \varphi)$. Then there exists an equilibrium state for (X, f, φ). Indeed, given any sequence $\{E_n \subset X\}_{n=1}^{\infty}$ of maximal (n, δ)-separated sets, every weak*-limit point of the sequence μ_n from (1.3.5.4) is an equilibrium state.*

There is an analogue of Proposition 1.6 for pressure: if (X, f) is expansive at scale ϵ, then $P(X, \varphi, \epsilon) = P(X, \varphi)$, so Corollary 1.3.5.1 establishes existence of an equilibrium state, as well as a way to construct one. Then the goal becomes to prove uniqueness.

1.3.5.2 Regularity of the Potential Function: The Bowen Property

Even for uniformly hyperbolic systems, one should not expect every continuous potential function to have a unique equilibrium state. Indeed, for the full shift it is possible to show that given any finite set E of ergodic measures, there is a continuous potential function φ whose set of equilibrium states is precisely the convex hull of E; see [84, p. 117] and [85, p. 52].

For expansive systems (X, f) with specification, uniqueness of the equilibrium state can be guaranteed by the following regularity condition on the potential.

Definition 1.3.5.2 A continuous function $\varphi \colon X \to \mathbb{R}$ has the *Bowen property* at scale $\epsilon > 0$ if there is a constant $V > 0$ such that for every $(x, n) \in X \times \mathbb{N}$ and $y \in B_n(x, \epsilon)$, we have $|S_n\varphi(y) - S_n\varphi(x)| \leq V$.

The following generalization of Theorems 1.2.2.1 and 1.2.4.2 is the full statement of Bowen's original result from [5], with the slight modification that we make the scales explicit.

Theorem 1.3.5.1 *Let X be a compact metric space and $f \colon X \to X$ a continuous map. Suppose that there are $\epsilon > 40\delta > 0$ such that f is expansive or positively expansive at scale ϵ and has the specification property at scale δ. Then every continuous potential function $\varphi \colon X \to \mathbb{R}$ with the Bowen property at scale ϵ has a unique equilibrium state.*

The proof of Theorem 1.3.5.1 follows the argument outlined earlier for Theorems 1.2.2.1 and 1.2.4.2 in Sects. 1.2.2 and 1.2.4.4. The main difference is that now the computations involve Birkhoff sums. For example, if we consider the symbolic setting for a moment and recall the motivation from Sect. 1.2.2.2 for the Gibbs bound as the mechanism for uniqueness, we see that in addition to the use of the Shannon–McMillan–Breiman theorem in (1.2.2.2), it is natural to use the Birkhoff ergodic theorem and get

$$h_\mu(\sigma) + \int \varphi \, d\mu = \lim_{n \to \infty} \frac{1}{n}\big(-\log \mu[x_{[1,n]}] + S_n\varphi(x)\big).$$

For an equilibrium state, the left-hand side is $P(\varphi)$, and this can be rewritten as $P(\varphi) + \lim_{n \to \infty} \frac{1}{n}(\log \mu[x_{[1,n]}] - S_n\varphi(x)) = 0$, or equivalently,

$$\lim_{n \to \infty} \frac{1}{n} \log \left(\frac{\mu[x_{[1,n]}]}{e^{-nP(\varphi)+S_n\varphi(x)}}\right) = 0.$$

As with the Gibbs property for the MME, uniqueness of the equilibrium state can be guaranteed by requiring that the quantity inside the logarithm be bounded away from 0 and ∞.[16] Generalizing to arbitrary compact metric spaces by replacing cylinders

---

[16]Observe that this is impossible if φ does not satisfy the Bowen property.

with Bowen balls, we say that a measure μ has the *Gibbs property* for a potential φ at scale ϵ if there are constants $K > 0$ and $P \in \mathbb{R}$ such that for every $x \in X$ and $n \in \mathbb{N}$, we have

$$K^{-1}e^{-nP+S_n\varphi(x)} \leq \mu(B_n(x,\epsilon)) \leq Ke^{-nP+S_n\varphi(x)}. \tag{1.3.5.5}$$

If it is known that every equilibrium measure is almost expansive at scale ϵ (recall Definition 1.3.1.1)—in particular, if (X, f) is expansive at scale ϵ—and if μ is an ergodic Gibbs measure for φ, then the analogue of Proposition 1.1 holds: we have $P = P(\varphi) = h_\mu(f) + \int \varphi \, d\mu$, and μ is the unique equilibrium state for (X, f, φ). The proof is essentially the same, although now the computations involve Birkhoff sums.

Similarly, in the proof of the uniform counting bounds and the construction of an ergodic Gibbs measure using the procedure in Proposition 1.12, one encounters multiple steps where a Birkhoff sum $S_n\varphi(x)$ must be replaced with $S_n\varphi(y)$ for some y in the Bowen ball around x, and the Bowen property is required at these steps to guarantee "bounded distortion" in the estimates.

Recalling that topologically transitive locally maximal hyperbolic sets have expansivity and specification, it is natural to ask which potential functions have the Bowen property: how much does Theorem 1.3.5.1 extend Theorem 1.2.1.1?

Proposition 1.13 *If X is a locally maximal hyperbolic set for a diffeomorphism f, then every Hölder continuous function $\varphi: X \to \mathbb{R}$ has the Bowen property at scale ϵ, where ϵ is the scale of the local product structure.*

Proof Recalling the estimates (1.2.4.4) and (1.2.4.5) in the proof of Proposition 1.5, we see that for every $y \in B_n(x, \epsilon)$ and every $k \in \{0, 1, \ldots, n-1\}$, we have

$$d^u(f^k x, f^k y) \leq e^{-\lambda(n-k)}\epsilon \quad \text{and} \quad d^s(f^k x, f^k y) \leq e^{-\lambda k}\epsilon.$$

Writing C for the Hölder constant and γ for the Hölder exponent, we obtain

$$|\varphi(f^k x) - \varphi(f^k y)| \leq Cd(f^k x, f^k y)^\gamma$$
$$\leq C\big(2\max(d^u(f^k x, f^k y), d^s(f^k x, f^k y))\big)^\gamma$$
$$\leq C(2\epsilon)^\gamma \max(e^{-\lambda(n-k)\gamma}, e^{-\lambda k\gamma}),$$

and summing over $0 \leq k < n$ gives

$$|S_n\varphi(x) - S_n\varphi(y)| \leq \sum_{k=0}^{n-1} C(2\epsilon)^\gamma \max(e^{-\lambda(n-k)\gamma}, e^{-\lambda k\gamma})$$

$$\leq C(2\epsilon)^\gamma \sum_{k=0}^{n-1} e^{-\lambda\gamma(n-k)} + e^{-\lambda\gamma k} \leq 2C(2\epsilon)^\gamma \sum_{k=0}^{\infty} e^{-\lambda\gamma k} =: V.$$

This last quantity is finite and independent of x, y, n, which establishes the Bowen property for φ. $\square$

Remark 1.21 The theorem "Hölder potentials for uniformly hyperbolic systems have unique equilibrium states" is well-entrenched enough that it is worth stressing the following point: it is the dynamical Bowen property (bounded distortion), rather than the metric Hölder property, that is truly important here. In particular, if we consider a non-uniformly hyperbolic system that is conjugate to a uniformly hyperbolic one, such as the Manneville–Pomeau interval map or Katok map of the torus, then every potential with the Bowen property continues to have a unique equilibrium state, but there may be Hölder potentials with multiple equilibrium states. However, determining which potentials have the Bowen property may be a nontrivial task.

1.3.5.3   The Most General Discrete-Time Result

Recalling the weakened versions of expansivity and specification used in Theorem 1.3.3.1, it is natural to ask for a uniqueness result for equilibrium states that uses a weakened version of the Bowen property. Observe that the Bowen property can be formulated for a collection of orbit segments (rather than the entire system) by replacing $X \times \mathbb{N}$ in Definition 1.3.5.2 with $\mathcal{G} \subset X \times \mathbb{N}$.

Definition 1.3.5.3 A continuous function $\varphi \colon X \to \mathbb{R}$ has the *Bowen property* at scale $\epsilon > 0$ on a collection of orbit segments $\mathcal{G} \subset X \times \mathbb{N}$ if there is a constant $V > 0$ such that for every $(x, n) \in \mathcal{G}$ and $y \in B_n(x, \epsilon)$, we have $|S_n\varphi(y) - S_n\varphi(x)| \leq V$.

To formulate our most general discrete-time result on uniqueness of equilibrium states, we replace the entropy of obstructions to expansivity from Definition 1.3.1.2 with the *pressure of obstructions to expansivity at scale* ϵ:

$$P^{\perp}_{\exp}(\phi, \epsilon) := \sup \left\{ h_{\mu}(f) + \int \varphi \, d\mu : \mu \in \mathcal{M}^e_f(X) \text{ and } \mu(\mathrm{NE}(\epsilon)) > 0 \right\}.$$

Theorem 1.3.5.2 ([2, Theorem 5.6]) *Let X be a compact metric space, $f \colon X \to X$ a homeomorphism, and $\varphi \colon X \to \mathbb{R}$ a continuous potential function. Suppose that there are $\epsilon > 40\delta > 0$ such that $P^{\perp}_{\exp}(\varphi, \epsilon) < P(\varphi)$ and there exists a decomposition $(C^p, \mathcal{G}, C^s)$ for $X \times \mathbb{N}$ with the following properties:*

(I)   *every collection $\mathcal{G}^M$ has specification at scale δ,*
(II) *φ has the Bowen property on $\mathcal{G}$ at scale ϵ, and*
(III) *$P(C^p \cup C^s, \varphi, \delta) < P(\varphi)$.*

Then (X, f, φ) has a unique equilibrium state.

Remark 1.22 In applications to non-uniformly hyperbolic systems, it is very often the case that there is a natural collection of orbit segments $\mathcal{G}$ along which the dynamics is uniformly hyperbolic; this is the most common way of establishing specification for $\mathcal{G}$, as we saw in Sect. 1.3.2. In this case the proof of Proposition 1.13 shows that every Hölder potential φ has the Bowen property on $\mathcal{G}$. Then the question of uniqueness boils down to determining which Hölder potentials have the pressure gap properties (III) and $P_{\exp}^{\perp}(\varphi, \epsilon) < P(\varphi)$. It is often the case that one or both of these conditions fails for some Hölder potentials, as in the Manneville–Pomeau example.

1.3.5.4 Partial Hyperbolicity

For partially hyperbolic systems with one-dimensional center as in Sect. 1.3.4, Theorem 1.3.5.2 can be used to extend Theorem 1.3.4.1.

Theorem 1.3.5.3 *Let M, f, φ^c be as in Theorem 1.3.4.1. Given a Hölder continuous potential function $\varphi : M \to \mathbb{R}$, consider the quantities*

$$P^+ := \sup\left\{h_\mu(f) + \int \varphi \, d\mu : \mu \in \mathcal{M}_f^e(M), \lambda^c(\mu) \geq 0\right\},$$

$$P^- := \sup\left\{h_\mu(f) + \int \varphi \, d\mu : \mu \in \mathcal{M}_f^e(M), \lambda^c(\mu) \leq 0\right\}.$$

If $P^+ \neq P^-$, then (M, f, φ) has a unique equilibrium state.

Beyond the properties from Sect. 1.3.4, the only additional ingredient required for Theorem 1.3.5.3 is the fact that φ has the Bowen property on the collection of orbit segments $\mathcal{G}$ defined in (1.3.4.2), which follows from Remark 1.22 and the hyperbolicity estimate in (1.3.4.2); then uniqueness follows from Theorem 1.3.5.2.

It is worth noting that the condition $P^+ \neq P^-$ (and thus the condition $h^+ \neq h^-$) can be formulated in terms of the topological pressure function. The function $t \mapsto P(\varphi + t\varphi^c)$ is convex, being the supremum of the affine functions

$$P_\mu : t \mapsto h_\mu(f) + \int \varphi \, d\mu + t\lambda^c(\mu)$$

over all $\mu \in \mathcal{M}_f^e(M)$. Some of its possible shapes are shown in Fig. 1.8.

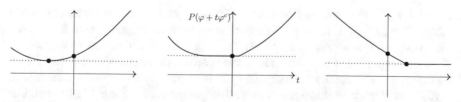

Fig. 1.8 Some possible graphs of $t \mapsto P(\varphi + t\varphi^c)$

Suppose there is $t > 0$ such that $P(\varphi + t\varphi^c) < P(\varphi)$, as in the third graph in Fig. 1.8. Then given any $\mu \in \mathcal{M}_f^e(M)$ with $\lambda^c(\mu) \geq 0$, we have

$$h_\mu(f) + \int \varphi \, d\mu = P_\mu(0) \leq P_\mu(t) \leq P(\varphi + t\varphi^c) < P(\varphi), \qquad (1.3.5.6)$$

and taking a supremum over all such μ gives $P^+ \leq P(\varphi + t\varphi^c) < P(\varphi)$, so that the condition of Theorem 1.3.5.3 is satisfied and (M, f, φ) has a unique equilibrium state, which has negative center Lyapunov exponent.

A similar argument holds if there is $t < 0$ such that $P(\varphi + t\varphi^c) < P(\varphi)$, as in the first graph in Fig. 1.8; (1.3.5.6) applies to all $\mu \in \mathcal{M}_f^e(M)$ with $\lambda^c(\mu) \leq 0$, so that $P^- < P(\varphi) = P^+$, and there is a unique equilibrium state, which has positive center Lyapunov exponent.

We see that the only way to have $P^+ = P^-$ is if the function $t \mapsto P(\varphi + t\varphi^c)$ has a global minimum at $t = 0$. Thus one could restate the last line of Theorem 1.3.5.3 as the conclusion that (M, f, φ) has a unique equilibrium state if there is $t \neq 0$ such that $P(\varphi + t\varphi^c) < P(\varphi)$. In particular, returning to Theorem 1.3.4.1, f has a unique MME if there is $t \neq 0$ such that $P(t\varphi^c) < P(0) = h_{\text{top}}(f)$.

1.4 Geodesic Flows

In this part, we focus on our geometric applications. In Sect. 1.4.1, we introduce some geometric background, and in Sect. 1.4.2 we describe the main results and some of the key ideas from the paper [3]. In Sect. 1.4.3, we discuss our approach to the Kolmogorov K-property. In Sect. 1.4.4, we give the main ideas of proof for the "pressure gap" for a wide class of potentials for geodesic flow on a rank 1 non-positive curvature manifold.

1.4.1 Geometric Preliminaries

1.4.1.1 Overview

Let $M = (M^n, g)$ be a closed connected C^∞ Riemannian manifold with dimension n, and $F = (f_t)_{t \in \mathbb{R}}$ denote the geodesic flow on the unit tangent bundle $X = T^1 M$. The geodesic flow is defined by picking a point and a direction (i.e. an element of $T^1 M$), and walking at unit speed along the geodesic determined by that data. More precisely, $f_t(v) = \dot{c}_v(t)$, where $c_v : \mathbb{R} \to M$ is the unique unit speed geodesic with $\dot{c}_v(0) = v$. Geodesic flows are of central importance in the theory of dynamical systems, and encode many important features of the geometry and topology of the underlying manifold M. For general background on geodesic flows, we refer to [86, 87].

If all sectional curvatures of M are negative at every point, then F is a transitive Anosov flow. In particular, the thermodynamic formalism is very well understood. To go beyond negative curvature, one generally needs the tools of non-uniform hyperbolicity. There are three further classes of manifolds that generally exhibit some kind of non-uniformly hyperbolic behaviour: nonpositive curvature; no focal points; and no conjugate points. The relationships are as follows:

negative curv. $\Rightarrow$ nonpositive curv. $\Rightarrow$ no focal points $\Rightarrow$ no conjugate points.

The reverse implications all fail in general.

The definition of nonpositive curvature is easy: all sectional curvatures are ≤ 0 at every point. No focal points and no conjugate points are defined in terms of Jacobi fields, which we will introduce shortly, but can be understood in terms of the growth of distance between geodesics which pass through the same point. If we work in the universal cover $\widetilde{M}$ and consider arbitrary geodesics c_1, c_2 with $c_1(0) = c_2(0)$, then non-positive curvature implies that $t \mapsto d(c_1(t), c_2(t))$ is convex, while no focal points is equivalent to the condition that $t \mapsto d(c_1(t), c_2(t))$ be nondecreasing for all such c_1, c_2, and no conjugate points is equivalent to the condition that this function never vanish for $t > 0$; in other words, there is at most one geodesic connecting any two points in $\widetilde{M}$. In Sect. 1.4.2.8, we will also briefly discuss geodesic flow on some classes of spaces beyond the Riemannian case: namely, CAT(-1) spaces (which generalize negative curvature) and CAT(0) spaces (which generalize non-positive curvature).

For intuition, negative curvature has the effect of spreading out geodesics which pass through the same point (think of a saddle), while positive curvature has the effect of bringing them back together after a finite amount of time (think of a sphere). As described in [88], one can imagine starting with a negatively curved surface and then "raising a bump of positive curvature"; at first the positive curvature effect is weak enough that the geodesic flow remains Anosov, but eventually the Anosov property is destroyed, and raising the bump far enough creates conjugate points.

In these notes, we focus on the case of equilibrium states for manifolds with nonpositive curvature using specification-based techniques as in [3]; this relies on a continuous-time version of Theorem 1.3.5.2, which we formulate in Sect. 1.4.2.1. This approach has been extended to manifolds without focal points by Chen et al. [18, 19].[17] We also state and sketch recent results by the first-named author, Knieper and War for the MME to surfaces with no conjugate points, and survey some relevant recent results for CAT(-1) and CAT(0) spaces.

In the remainder of this section we collect some geometric preliminaries. Some of the definitions are taken verbatim from [3] for notational consistency. For more details, we recommend recent works [3, 90], and more classical references [91–93].

---

[17] Another specification-based proof of uniqueness of the MME on surfaces without focal points was given by Gelfert and Ruggiero [89].

1.4.1.2 Surfaces

For purposes of exposition, we will often think about the surface case $n = 2$, although our approach applies in higher dimension too. By the Gauss–Bonnet theorem, the sphere has no metric of nonpositive curvature, and the only such metrics on the torus are flat everywhere; it can be easily verified that the corresponding geodesic flows have zero topological entropy and are not topologically transitive. Thus we are interested in studying surfaces of genus at least 2.

As a first example, we can think about a surface of genus 2 with an embedded flat cylinder, and negative curvature elsewhere. We could also consider the case where the flat cylinder collapses to a single closed geodesic on which the curvature vanishes, with strictly negative curvature elsewhere. In higher dimensions, much more complicated examples exist, such as the 3-dimensional *Gromov example* that we describe in Sect. 1.4.4.

Geodesic flow in non-positive curvature is a primary example of non-uniform hyperbolicity. The basic example of a surface containing a flat cylinder illustrates the primary difficulty: the co-existence of trajectories displaying hyperbolic behavior (geodesics in the negatively curved part of the surface) with trajectories displaying non-hyperbolic behavior (geodesics in the flat cylinder). More precisely, given a surface M of genus at least 2 with non-positive curvature, we let $K : M \to (-\infty, 0]$ be the Gaussian curvature, and $\pi : T^1 M \to M$ the natural projection of a tangent vector to its footpoint. Then we define the *singular set* to be

$$\text{Sing} := \{v \in T^1 M : K(\pi(f_t v)) = 0 \text{ for all } t \in \mathbb{R}\}. \tag{1.4.1.1}$$

That is, Sing is the set of v for which the corresponding geodesic γ_v experiences 0 curvature for all time. All other vectors are called *regular*:

$$\text{Reg} := T^1 M \setminus \text{Sing} = \{v \in T^1 M : K(\pi(f_t v)) < 0 \text{ for some } t \in \mathbb{R}\}. \tag{1.4.1.2}$$

Although the negative curvature encountered along regular geodesics guarantees some expansion/contraction, this may be arbitrarily weak because the geodesic can be arranged to experience 0 curvature for a long time (e.g., wrapping round an embedded flat cylinder) before hitting any negative curvature.

The set Sing is closed and flow-invariant, while the set Reg is open. The regular set is nonempty because M has genus at least 2, and in fact Reg is dense in $T^1 M$.

In higher dimensions one has a similar dichotomy between singular and regular vectors, which we will describe in the next section. This gives a partition of $T^1 M$ as Reg $\sqcup$ Sing, where Sing is closed and flow-invariant. As with surfaces, we will restrict our attention to the case when Reg $\neq \emptyset$; this *rank 1* assumption rules out examples such as direct products, and is the typical situation, as demonstrated by the higher rank rigidity theorem of Ballmann and Burns–Spatzier [94–96].

1.4.1.3 Invariant Foliations via Horospheres

Now let the dimension of M be any $n \geq 2$. We describe invariant stable and unstable foliations W^s and W^u of $X = T^1 M$ that are tangent to invariant subbundles E^s and E^u in $TX = TT^1 M$ along which we will eventually obtain the contraction and expansion estimates necessary to study uniqueness of equilibrium states.

We must be a little careful in defining these foliations: we cannot ask that $W^s(v)$ is the set of $w \in T^1 M$ so that $d(f_t v, f_t w) \to 0$ as $t \to \infty$ like we can in the uniformly hyperbolic setting. We must allow points that stay bounded distance apart (in the universal cover) for all forward time. However, this does not work as the definition of W^s because it does not distinguish the stable from the flow direction. To do things properly, there are two approaches.

- *Local approach:* Use stable and unstable orthogonal Jacobi fields to define E^s and E^u locally; see Sect. 1.4.1.4 below.
- *Global approach:* Define stable and unstable horospheres H^s and H^u in the universal cover $\widetilde{M}$ (this is typically done using Busemann functions) and use these to get W^s, W^u.

We outline this second approach here. Given $v \in T^1 M$, let $\tilde{v} \in T^1 \widetilde{M}$ be a lift of v, and construct $H^s(\tilde{v})$ as follows: for each $r > 0$ let

$$S^r(\tilde{v}, +) = \{x \in \widetilde{M} : d_{\widetilde{M}}(x, \pi(f_r \tilde{v})) - r\}$$

denote the set of points at distance r from $\pi(f_r \tilde{v}) = c_{\tilde{v}}(r)$, and let $H^s(\tilde{v})$ be the limit of $S^r(v, +)$ as $r \to \infty$. This defines a hypersurface that contains the point $\pi \tilde{v}$. Writing $W^s(\tilde{v})$ for the unit normal vector field to $H^s(\tilde{v})$ on the same side as $\tilde{v}$, the stable manifold $W^s(v)$ is the image of $W^s(\tilde{v})$ under the canonical projection $T^1 \widetilde{M} \to T^1 M$.

The unstable horosphere $H^u(\tilde{v})$ and the unstable manifold $W^u(v)$ are defined analogously, replacing $S^r(\tilde{v}, +)$ with

$$S^r(\tilde{v}, -) = \{x \in \widetilde{M} : d_{\widetilde{M}}(x, \pi(f_{-r} \tilde{v})) = r\}.$$

The horospheres are C^2 manifolds, so $W^s(v)$ and $W^u(v)$ are C^1 manifolds, and we can define the stable and unstable subspaces $E^s(v), E^u(v) \subset T_v T^1 M$ to be the tangent spaces of $W^s(v), W^u(v)$ respectively. The bundles E^s, E^u, which are both globally defined in this way, are respectively called the stable and unstable bundles. They are invariant and depend continuously on v; see [92, 97].

The following is equivalent to the standard definition of the regular set via Jacobi fields, which we will give in the next section.

Definition 1.4.1.1 A vector $v \in T^1 M$ is *regular* if $E^s(v) \cap E^u(v)$ is trivial (contains only the 0 vector in $T_v T^1 M$), and *singular* otherwise. Write Reg $\subset T^1 M$ for the set of regular vectors, and Sing $\subset T^1 M$ for the set of singular vectors.

On Reg, we obtain the expected splitting $T_v T^1 M = E^s(v) \oplus E^u(v) \oplus E^c(v)$, where $E^c(v)$ is the flow direction. This splitting degenerates on Sing.

Definition 1.4.1.2 The manifold M is *rank 1* if Reg $\neq \emptyset$.

Finally, we define a function which is of great importance in thermodynamic formalism. The *geometric potential* is the function that measures infinitesimal volume growth in the unstable distribution:

$$\varphi^u(v) = -\lim_{t \to 0} \frac{1}{t} \log \det(df_t|_{E^u(v)}) = -\frac{d}{dt}\Big|_{t=0} \log \det(df_t|_{E^u(v)}).$$

The potential φ^u is continuous and globally defined. When M has dimension 2, φ^u is Hölder along unstable leaves [97]. It is not known whether φ^u is Hölder along stable leaves. In higher dimensions, it is not known whether φ^u is Hölder continuous on either stable or unstable leaves. An advantage of our approach is that we sidestep the question of Hölder regularity for φ^u.

1.4.1.4 Jacobi Fields and Local Construction of Stables/Unstables

Now we give an alternate description of the stable and unstable subbundles and foliations, which can be shown to agree with the definitions in the previous section.

A *Jacobi field* along a geodesic γ is a vector field along γ obtained by taking a one-parameter family of geodesics that includes γ and differentiating in the parameter coordinate; equivalently, it is a vector field along γ satisfying

$$J''(t) + R(J(t), \dot\gamma(t))\dot\gamma(t) = 0, \tag{1.4.1.3}$$

where R is the Riemannian curvature tensor on M and $'$ represents covariant differentiation along γ.

We often want to remove the variations through geodesics in the flow direction from consideration. If $J(t)$ is a Jacobi field along a geodesic γ and both $J(t_0)$ and $J'(t_0)$ are orthogonal to $\dot\gamma(t_0)$ for some t_0, then $J(t)$ and $J'(t)$ are orthogonal to $\dot\gamma(t)$ for all t. Such a Jacobi field is an *orthogonal Jacobi field*.

A Jacobi field $J(t)$ along a geodesic γ is *parallel at t_0* if $J'(t_0) = 0$. A Jacobi field $J(t)$ is parallel if it is parallel for all $t \in \mathbb{R}$.

Definition 1.4.1.3 A geodesic γ is *singular* if it admits a nonzero parallel orthogonal Jacobi field, and *regular* otherwise.

If γ is singular in the sense of Definition 1.4.1.3, then every $\dot\gamma(t) \in T^1 M$ is singular in the sense of Definition 1.4.1.1, and similarly for regular.

We write $\mathcal{J}(\gamma)$ for the space of orthogonal Jacobi fields for γ; given $v \in T^1 M$ there is a natural isomorphism $\xi \mapsto J_\xi$ between $T_v T^1 M$ and $\mathcal{J}(\gamma_v)$, which has the property that

$$\|df_t(\xi)\|^2 = \|J_\xi(t)\|^2 + \|J'_\xi(t)\|^2. \tag{1.4.1.4}$$

An orthogonal Jacobi field J along a geodesic γ is *stable* if $\|J(t)\|$ is bounded for $t \geq 0$, and *unstable* if it is bounded for $t \leq 0$. The stable and the unstable Jacobi fields each form linear subspaces of $\mathcal{J}(\gamma)$, which we denote by $\mathcal{J}^s(\gamma)$ and $\mathcal{J}^u(\gamma)$, respectively. The corresponding stable and unstable subbundles of TT^1M are

$$E^u(v) = \{\xi \in T_v(T^1M) : J_\xi \in \mathcal{J}^u(\gamma_v)\},$$

$$E^s(v) = \{\xi \in T_v(T^1M) : J_\xi \in \mathcal{J}^s(\gamma_v)\}.$$

The bundle E^c is spanned by the vector field that generates the flow F. We also write $E^{cu} = E^c \oplus E^u$ and $E^{cs} = E^c \oplus E^s$. The subbundles have the following properties (see [92] for details):

- $\dim(E^u) = \dim(E^s) = n - 1$, and $\dim(E^c) = 1$;
- the subbundles are invariant under the geodesic flow;
- the subbundles depend continuously on v, see [92, 97];
- E^u and E^s are both orthogonal to E^c;
- E^u and E^s intersect non-trivially if and only if $v \in \text{Sing}$;
- E^σ is integrable to a foliation W^σ for each $\sigma \in \{u, s, cs, cu\}$.

It is proved in [98, Theorem 3.7] that the foliation W^s is minimal in the sense that $W^s(v)$ is dense in T^1M for every $v \in T^1M$. Analogously, the foliation W^u is also minimal.

1.4.2   Equilibrium States for Geodesic Flows

1.4.2.1   The General Uniqueness Result for Flows

We recall the general definitions of topological pressure, variational principle, and equilibrium states for flows, which are analogous to the discrete-time definitions from Sect. 1.3.5.1.

Given a compact metric space X and a continuous flow $F = (f_t)$ on X, we write $\mathcal{M}_F(X) = \bigcap_{t \in \mathbb{R}} \mathcal{M}_{f_t}(X)$ for the space of flow-invariant Borel probability measures on X, and $\mathcal{M}_F^e(X) \subset \mathcal{M}_F(X)$ for the set of ergodic measures.

For $\epsilon > 0$, $t > 0$, and $x \in X$, the *Bowen ball of radius ϵ and order t* is

$$B_t(x, \epsilon) = \{y \in X \mid d(f_s x, f_s y) < \epsilon \text{ for all } 0 \leq s \leq t\}.$$

A set $E \subset X$ is *(t, ϵ)-separated* if for all distinct $x, y \in E$ we have $y \notin \overline{B_t(x, \epsilon)}$.

Given a continuous potential function $\varphi \colon X \to \mathbb{R}$, we write $\Phi(x, t) = \int_0^t \varphi(f_s x)\, ds$ for the integral of φ along an orbit segment of length t. We interpret $\mathcal{D} \subset X \times [0, \infty)$ as a collection of finite-length orbit segments by identifying (x, t) with the orbit segment starting at x and lasting for time t. Writing $\mathcal{D}_t := \{x \in X : (x, t) \in \mathcal{D}\}$, the partition sums associated to $\mathcal{D}$ and φ are

$$\Lambda(\mathcal{D}, \varphi, \epsilon, t) = \sup\left\{ \sum_{x \in E} e^{\Phi(x, t)} : E \subset \mathcal{D}_t \text{ is } (t, \epsilon)\text{-separated}\right\}. \qquad (1.4.2.1)$$

The pressure of φ on the collection $\mathcal{D}$ is given by (1.3.5.1)–(1.3.5.2), replacing n with t:

$$P(\mathcal{D}, \varphi) = \lim_{\epsilon \to 0} P(\mathcal{D}, \varphi, \epsilon), \qquad P(\mathcal{D}, \varphi, \epsilon) = \varlimsup_{t \to \infty} \frac{1}{t} \log \Lambda(\mathcal{D}, \varphi, \epsilon, t).$$

We continue to write $P(Y, \varphi) = P(Y \times [0, \infty), \varphi)$ for $Y \subset X$, and often abbreviate $P(\varphi) = P(X, \varphi)$. The *variational principle for pressure* states that

$$P(\varphi) = \sup_{\mu \in \mathcal{M}_F(X)} \left(h_\mu(f_1) + \int \varphi\, d\mu\right).$$

A measure that achieves the supremum is an *equilibrium state* for (X, f, φ). When $\varphi = 0$, we recover the topological entropy $h(F)$, and an equilibrium state for $\varphi = 0$ is called a *measure of maximal entropy*.

Remark 1.23 As in the discrete-time case, if the entropy map $\mu \mapsto h_\mu$ is upper semi-continuous then equilibrium states exist for each continuous potential function. Geodesic flows in non-positive curvature are entropy-expansive due to the flat strip theorem [99]; this guarantees upper semi-continuity and thus existence.

In light of Remark 1.23, the real question is once again uniqueness. Our main tool will be a continuous-time analogue of Theorem 1.3.5.2, which gives non-uniform versions of specification, expansivity, and the Bowen property that are sufficient to give uniqueness.

The main novelty compared with the discrete-time case is the expansivity condition. For an expansive map, the set of points that stay close to x for all time is only the point x itself. For an expansive flow, this set is an orbit segment of x. Our set of non-expansive points for a flow is defined accordingly. For $x \in X$ and $\epsilon > 0$, we let the *bi-infinite Bowen ball* be

$$\Gamma_\epsilon(x) = \{y \in X : d(f_t x, f_t y) \le \epsilon \text{ for all } t \in \mathbb{R}\}.$$

The *set of non-expansive points at scale ϵ* is (compare this to Definition 1.3.1.1)

$$\mathrm{NE}(\epsilon, F) := \{x \in X \mid \Gamma_\epsilon(x) \not\subset f_{[-s, s]}(x) \text{ for any } s > 0\}, \qquad (1.4.2.2)$$

where $f_{[a,b]}(x) = \{f_t x : a \leq t \leq b\}$.[18] The *pressure of obstructions to expansivity* is

$$P_{\exp}^{\perp}(\varphi) := \lim_{\epsilon \to 0} P_{\exp}^{\perp}(\varphi, \epsilon),$$

where

$$P_{\exp}^{\perp}(\varphi, \epsilon) = \sup_{\mu \in \mathcal{M}_F^e(X)} \left\{ h_\mu(f_1) + \int \varphi \, d\mu : \mu(\text{NE}(\epsilon, \mathcal{F})) = 1 \right\}.$$

Remark 1.24 For rank 1 geodesic flow, a simple argument using the flat strip theorem guarantees that $\text{NE}(\epsilon, F) \subset \text{Sing}$, so we have $P_{\exp}^{\perp}(\varphi) \leq P(\text{Sing}, \varphi)$.

Our definitions of specification and the Bowen property are completely analogous to Definitions 1.2.4.4 and 1.3.5.3 from the discrete-time case. The specification property for flows was defined by Bowen in [101], and was used to prove uniqueness of equilibrium states by Franco [102].

Definition 1.4.2.1 A collection of orbit segments $G \subset X \times [0, \infty)$ *has the specification property at scale* $\delta > 0$ if there exists $\tau > 0$ such that for every $(x_1, t_1), \ldots, (x_k, t_k) \in G$, there exist $0 = T_1 < T_2 < \cdots < T_k$ and $y \in X$ such that $f^{T_i}(y) \in B_{t_i}(x_i, \delta)$ for all i, and moreover, writing $s_i = T_i + t_i$, we have $s_i \leq T_{i+1} \leq s_i + \tau$ for all i.

We say that G *has the specification property* if it has the specification property at scale δ for every $\delta > 0$.

Definition 1.4.2.2 A continuous function $\varphi : X \to \mathbb{R}$ has the *Bowen property at scale* $\epsilon > 0$ on a collection of orbit segments $G \subset X \times [0, \infty)$ if there is $V > 0$ such that for every $(x, t) \in G$ and $y \in B_t(x, \epsilon)$, we have $|\Phi(y, t) - \Phi(x, t)| \leq V$.

We say that φ has the *Bowen property on* G if there exists $\epsilon > 0$ such that φ has the Bowen property at scale ϵ on G.

An argument following the proof of Proposition 1.13 shows that for uniformly hyperbolic flows, any Hölder continuous function has the Bowen property. More generally, Remark 1.22 applies here as well: if the flow is uniformly hyperbolic along a collection of orbit segments $G \subset X \times [0, \infty)$, then every Hölder φ has the Bowen property on G.

---

[18] We note that the original formulation of expansivity for flows by Bowen and Walters [100] allows reparametrizations, which suggests that one might consider a potentially larger set in place of Γ_ϵ for expansive flows. The main motivation for allowing reparametrizations is to give a definition that is preserved under orbit equivalence. However, this is not relevant for our purposes. In our setup, the natural notion of expansivity would be to ask that there exists ϵ so that $\text{NE}(\epsilon, \mathcal{F}) = \emptyset$. This definition is sufficient for the uniqueness results, and strictly weaker than Bowen–Walters expansivity, although it is not an invariant under orbit equivalence. See the discussion of *kinematic expansivity* in [37].

As in Definition 1.3.3.1 for discrete time, a *decomposition for* $X \times [0, \infty)$ consists of three collections $\mathcal{P}, \mathcal{G}, \mathcal{S} \subset X \times [0, \infty)$ for which there exist three functions $p, g, s \colon X \times [0, \infty) \to [0, \infty)$ such that for every $(x, t) \in X \times [0, \infty)$, the values $p = p(x, t)$, $g = g(x, t)$, and $s = s(x, t)$ satisfy $t = p + g + s$, and

$$(x, p) \in \mathcal{P}, \quad (f_p(x), g) \in \mathcal{G}, \quad (f_{p+g}(x), s) \in \mathcal{S}.$$

The conditions we are interested in depend only on the collections $(\mathcal{P}, \mathcal{G}, \mathcal{S})$ rather than the functions p, g, s. However, we work with a fixed choice of (p, g, s) for the proof of the abstract theorem to apply.

One small difference from the discrete-time case is that we need to "fatten up" $\mathcal{P}$ and $\mathcal{S}$ slightly before imposing the smallness condition in the general uniqueness theorem. To this end, for a collection $\mathcal{D} \subset X \times [0, \infty)$, we define

$$[\mathcal{D}] := \{(x, k) \in X \times \mathbb{N} : (f_{-s}x, k + s + t) \in \mathcal{D} \text{ for some } s, t \in [0, 1]\}.$$

Theorem 1.4.2.1 (Non-uniform Bowen Hypotheses for Flows [2]) *Let* (X, F) *be a continuous flow on a compact metric space, and* $\varphi \colon X \to \mathbb{R}$ *be a continuous potential function. Suppose that* $P_{\exp}^{\perp}(\varphi) < P(\varphi)$ *and* $X \times [0, \infty)$ *admits a decomposition* $(\mathcal{P}, \mathcal{G}, \mathcal{S})$ *with the following properties:*

(I) $\mathcal{G}$ *has specification;*
(II) φ *has the Bowen property on* $\mathcal{G}$;
(III) $P([\mathcal{P}] \cup [\mathcal{S}], \varphi) < P(\varphi)$.

Then (X, F, φ) *has a unique equilibrium state* μ_{φ}.

Remark 1.25 The reason that in general we control the pressure of $[\mathcal{P}] \cup [\mathcal{S}]$ rather than the collection $\mathcal{P} \cup \mathcal{S}$ is a consequence of a technical step in the proof of the abstract result in [2] that required a passage from continuous to discrete time. This distinction does not matter for the λ-*decompositions* described in the next section, which cover all the applications we discuss here; see [103, Lemma 3.5].

1.4.2.2 Geodesic Flows in Non-positive Curvature

Now we return to the specific setting of geodesic flow in non-positive curvature. In Sect. 1.4.2.3 we explain why the outcome from the uniformly hyperbolic situation— a unique equilibrium state, whose support is all of $X = T^1 M$—cannot occur unless there is a *pressure gap* $P(\text{Sing}, \varphi) < P(\varphi)$. In Sect. 1.4.2.4 we formulate the main results on uniqueness given a pressure gap, ergodic properties of the unique equilibrium state, and how often the pressure gap occurs. In Sect. 1.4.2.5 we describe how the notion of periodic orbit equidistribution from Sect. 1.2.3.3 is adapted to this setting. The proof of the uniqueness result uses Theorem 1.4.2.1 and is outlined in Sect. 1.4.2.6. The proofs regarding ergodic properties, particularly the

Kolmogorov property, are described later in Sect. 1.4.3, and the pressure gap itself is discussed in Sect. 1.4.4.

1.4.2.3 Uniqueness Can Fail Without a Pressure Gap

For uniformly hyperbolic flows and Hölder continuous potentials, there is a unique equilibrium state, and this equilibrium state gives positive weight to every open set; it is *fully supported*. For geodesic flow in nonpositive curvature, this conclusion cannot hold unless there is a *pressure gap*, which we now describe.

Since the singular set Sing is closed and flow-invariant, we can apply the variational principle to the restriction of the flow to Sing, and obtain

$$P(\text{Sing}, \varphi) = \sup \left\{ h_\mu(f_1) + \int \varphi \, d\mu : \mu \in \mathcal{M}_F(\text{Sing}) \right\}.$$

As discussed in Remark 1.23, the geodesic flow is entropy-expansive and thus the entropy map $\mu \mapsto h_\mu(f_1)$ is upper semi-continuous. This guarantees that there exists $\nu \in \mathcal{M}_F(\text{Sing})$ with $h_\nu(f_1) + \int \varphi \, d\mu = P(\text{Sing}, \varphi)$.

If $P(\text{Sing}, \varphi) = P(\varphi)$, then ν is an equilibrium state for $(T^1 M, F, \varphi)$, and even if it happens that ν is the unique equilibrium state (which can be arranged, but is not generally expected), it is not fully supported. Thus in order to obtain the classical conclusion of unique equilibrium state and full support, we require a *pressure gap* $P(\text{Sing}, \varphi) < P(\varphi)$.

To see that the case $P(\text{Sing}, \varphi) = P(\varphi)$ can actually occur, we observe that there is a natural (f_t)-invariant volume measure μ_L on $X = T^1 M$ called the *Liouville measure*. Locally, μ_L is the product of the Riemannian volume on M and Haar measure on the unit sphere of dimension $n - 1$. Using the Ruelle–Margulis inequality, the Pesin entropy formula, and the fact that $-\int \varphi^u d\mu$ is the sum of the positive Lyapunov exponents for μ (where φ^u is the geometric potential), one can show that $P(\varphi^u) = 0$ and that μ_L is an equilibrium state for φ^u.

In negative curvature, φ^u is Hölder and μ_L is the unique equilibrium state. In non-positive curvature, however, μ_L often fails to be the unique equilibrium state.[19] For example, in the surface case, it is easily checked that $P(\text{Sing}, \varphi^u) = P(\varphi^u) = 0$, and any closed geodesic in Sing defines two equilibrium states for φ^u (one for each direction of travel around the geodesic).

Since a general uniqueness result for φ^u is impossible, we often turn our attention to the one-parameter family of potentials $q\varphi^u$, where $q \in \mathbb{R}$. Equilibrium states for these potentials are geometrically relevant, and a natural question is to identify the range of values for q so that uniqueness holds.

---

[19]We mention that $\mu_L(\text{Reg}) > 0$ and that $\mu_L|_{\text{Reg}}$ is known to be ergodic. Ergodicity of μ_L, which is a major open problem, is thus equivalent to the question of whether $\mu_L(\text{Sing}) = 0$.

1.4.2.4   Uniqueness Given a Pressure Gap

Our main result on uniqueness of equilibrium states for geodesic flow in non-positive curvature is the following.

Theorem 1.4.2.2 (Uniqueness of Equilibrium States for Rank 1 Geodesic Flow [3]) *Let* (f_t) *be the geodesic flow over a closed rank 1 manifold M and let* $\varphi \colon T^1M \to \mathbb{R}$ *be* $\varphi = q\varphi^u$ *or be Hölder continuous. If* φ *satisfies the* pressure gap

$$P(\mathrm{Sing}, \varphi) < P(\varphi), \qquad\qquad (1.4.2.3)$$

then φ *has a unique equilibrium state* μ. *This equilibrium state is hyperbolic, fully supported, and is the weak* limit of weighted regular closed geodesics in the sense of Sect. 1.4.2.5 below.*

Remark 1.26 Knieper used a Patterson–Sullivan type construction on the boundary at infinity to prove uniqueness of the MME (the case $\varphi = 0$) and deduce the entropy gap $h(\mathrm{Sing}) < h(T^1M)$ from this [99]. This construction has recently been extended to manifolds with no focal points by Fei et al. [105]. We work in the other direction: we need to first establish the gap (see Theorem 1.4.2.4 below), and then use this to prove uniqueness.

In Sect. 1.4.3 we discuss the following result on strengthened ergodic properties for the equilibrium states in Theorem 1.4.2.2, due to Ben Call and the second-named author.

Theorem 1.4.2.3 (K and Bernoulli Properties [103]) *Any unique equilibrium state provided by Theorem 1.4.2.2 has the K-property. The unique MME has the Bernoulli property.*

In dimension 2, the Margulis–Ruelle inequality gives $h(\mathrm{Sing}) = 0$, from which the pressure gap (1.4.2.3) follows when $\sup \varphi - \inf \varphi < h(X)$, via a soft argument based on the variational principle. In higher dimensions we may have $h(\mathrm{Sing}) > 0$ (see the Gromov example in Sect. 1.4.4), and the entropy gap $h(\mathrm{Sing}) < h(X)$ established by Knieper is nontrivial. In Sect. 1.4.4 we outline a direct proof of this gap that uses the specification property, and that generalizes to some nonzero potentials as follows.

Theorem 1.4.2.4 (Direct Proof of Entropy/Pressure Gap) *For geodesic flow on a closed rank 1 manifold M, every continuous potential* φ *that is locally constant on a neighbourhood of* Sing *satisfies the pressure gap condition* (1.4.2.3).

Remark 1.27 When Sing is a finite union of periodic orbits, which is the case for real analytic surfaces of non-positive curvature, Theorem 1.4.2.4 can be used to prove that the pressure gap holds for a C^0-open and dense set of potential functions.

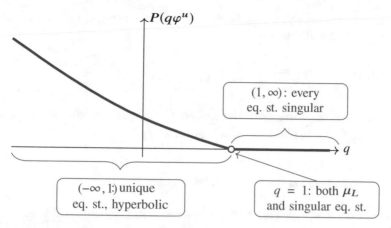

Fig. 1.9 Pressure for surfaces with non-positive curvature

For surfaces, the fact that $\varphi^u|_{\mathrm{Sing}} = 0$ and $h(\mathrm{Sing}) = 0$ implies that $P(\mathrm{Sing}, q\varphi^u) = 0$ for all $q \in \mathbb{R}$. It is an easy consequence of the Margulis–Ruelle inequality and Pesin's entropy formula that

$$P(q\varphi^u) > 0 \text{ for } q < 1,$$

and thus $q\varphi^u$ has a unique equilibrium state for all $q < 1$. We obtain the classic picture of the pressure function in non-uniform hyperbolicity, shown in Fig. 1.9. This is analogous to the familiar picture in the case of non-uniformly expanding interval maps with indifferent fixed points, e.g., the Manneville–Pomeau map [106–108].

1.4.2.5 Pressure and Periodic Orbits

We describe the sense in which the unique equilibrium state is the limit of periodic orbits, analogously to Sect. 1.2.3.3. For $a < b$, let $\mathrm{Per}_R(a, b]$ denote the set of closed regular geodesics with length in the interval $(a, b]$.[20] For each such geodesic γ, let $\Phi(\gamma)$ be the value given by integrating φ around γ; that is, $\Phi(\gamma) := \Phi(v, |\gamma|) = \int_0^{|\gamma|} \varphi(f_t v)\, dt$, where $v \in T^1 M$ is tangent to γ and $|\gamma|$ is the length of γ. Given $T, \delta > 0$, let

$$\Lambda^*_{\mathrm{Reg}}(\varphi, T, \delta) = \sum_{\gamma \in \mathrm{Per}_R(T-\delta, T]} e^{\Phi(\gamma)}.$$

---

[20]Here, we are following a notation convention of Katok: when we say a geodesic, we mean oriented geodesic, and we are considering γ as a periodic orbit living in $T^1 M$.

For a closed geodesic γ, let μ_γ be the normalized Lebesgue measure around the orbit. We consider the measures

$$\mu_{T,\delta}^{\text{Reg}} = \frac{1}{\Lambda_{\text{Reg}}^*(\varphi, T, \delta)} \sum_{\gamma \in \text{Per}_R(T-\delta, T]} e^{\Phi(\gamma)} \mu_\gamma.$$

We say that *regular closed geodesics weighted by φ equidistribute* to a measure μ if $\lim_{T\to\infty} \mu_{T,\delta}^{\text{Reg}} = \mu$ in the weak* topology for every $\delta > 0$.

1.4.2.6 Main Ideas of the Proof of Uniqueness

Theorem 1.4.2.2 is proved using the general result in Theorem 1.4.2.1. As observed in Remark 1.24, we have $P_{\text{exp}}^\perp(\varphi) \leq P(\text{Sing}, \varphi)$, so the condition $P_{\text{exp}}^\perp(\varphi) < P(\varphi)$ follows immediately from the pressure gap assumption (1.4.2.3), and it remains to find a decomposition of the space of orbit segments satisfying (I)–(III). We will do this using a function $\lambda \colon X \to [0, \infty)$ that measures 'hyperbolicity'. We want this function to be such that:

1. λ vanishes on Sing;
2. λ uniformly positive implies uniform hyperbolicity estimates.

There is a convenient geometrically-defined function which has the desired properties, whose definition in dimension 2 is simple: we let $\lambda(v)$ be the minimum of the curvature of the stable horosphere $H^s(v)$ and the unstable horosphere $H^u(v)$.[21]

If $v \in \text{Sing}$, then $\lambda(v) = 0$ due to the presence of a parallel orthogonal Jacobi field. The set $\{v \in \text{Reg} : \lambda(v) = 0\}$ may be non-empty, but it has zero measure for any invariant measure [3, Corollary 3.6].

If $\lambda(v) \geq \eta > 0$, then we have various uniform estimates at the point v, for example on the angle between $E^u(v)$ and $E^s(v)$, and on the growth of Jacobi fields at v. Thus, the function λ serves as a useful 'measure of hyperbolicity'. In particular, we get the following distance estimates: given $\eta > 0$ and $\delta = \delta(\eta) > 0$ sufficiently

---

[21]For manifolds M with $\text{Dim}(M) \geq 2$, we define $\lambda \colon T^1M \to [0, \infty)$ as follows. Let H^s, H^u be the stable and unstable horospheres for v. Let $\mathcal{U}_v^s \colon T_{\pi v}H^s \to T_{\pi v}H^s$ be the symmetric linear operator defined by $\mathcal{U}(v) = \nabla_v N$, where N is the field of unit vectors normal to H on the same side as v. This determines the second fundamental form of the stable horosphere H^s. We define $\mathcal{U}_v^u \colon T_{\pi v}H^u \to T_{\pi v}H^u$ analogously. Then $\mathcal{U}_v^u$ and $\mathcal{U}_v^s$ depend continuously on v, $\mathcal{U}^u$ is positive semidefinite, $\mathcal{U}^s$ is negative semidefinite, and $\mathcal{U}_{-v}^u = -\mathcal{U}_v^s$. For $v \in T^1M$, let $\lambda^u(v)$ be the minimum eigenvalue of $\mathcal{U}_v^u$ and let $\lambda^s(v) = \lambda^u(-v)$. Let $\lambda(v) = \min(\lambda^u(v), \lambda^s(v))$.

The functions λ^u, λ^s, and λ are continuous since the map $v \mapsto \mathcal{U}_v^{u,s}$ is continuous, and we have $\lambda^{u,s} \geq 0$. When M is a surface, the quantities $\lambda^{u,s}(v)$ are just the curvatures at πv of the stable and unstable horocycles, and we recover the definition of λ stated above.

small, $v \in T^1 M$, and $w, w' \in W_\delta^s(v)$, we have

$$d^s(f_t w, f_t w') \leq d^s(w, w') e^{-\int_0^t (\lambda(f_\tau v) - \eta/2)\, d\tau} \text{ for all } t \geq 0, \qquad (1.4.2.4)$$

where d^s is the distance on W^s. We get similar estimates for $w, w' \in W_\delta^u(v)$.

Now we use λ to define a decomposition. We give a general definition since the procedure here applies not just to geodesic flows, but to other examples including the partially hyperbolic systems in Sects. 1.3.2 and 1.3.4 (indeed, the decomposition in Sect. 1.3.4.2 is of this type); see [109].

Definition 1.4.2.3 Let X be a compact metric space and $F = (f_t)$ a continuous flow on X. Let $\lambda \colon X \to [0, \infty)$ be a bounded lower semicontinuous function[22] and fix $\eta > 0$. The λ-*decomposition* (with constant η) of $X \times [0, \infty)$ is given by defining

$$\mathcal{B}(\eta) = \left\{ (x, t) \mid \frac{1}{t} \int_0^t \lambda(f_s(x))\, ds < \eta \right\},$$

$$G(\eta) = \left\{ (x, t) \mid \frac{1}{\rho} \int_0^\rho \lambda(f_s(x))\, ds \geq \eta \right.$$

$$\left. \text{and } \frac{1}{\rho} \int_0^\rho \lambda(f_{-s} f_t(x))\, ds \geq \eta \text{ for all } \rho \in [0, t] \right\}$$

and then putting $\mathcal{P} = S = \mathcal{B}(\eta)$ and $G = G(\eta)$. We decompose an orbit segment (x, t) by taking the longest initial segment in $\mathcal{P}$ as the prefix, and the longest terminal segment in S as the suffix:[23] that is,

$$p(x, t) = \sup\{p \geq 0 : (x, p) \in \mathcal{P}\} \quad \text{and} \quad s(x, t) = \sup\{s \geq 0 : (f_{t-s} x, s) \in S\}.$$

The good core is what is left over; see Fig. 1.10.

For rank 1 geodesic flow, the decompositions associated to the horosphere curvature function λ have the following useful properties:

1. we can relate $P([\mathcal{P}] \cup [S], \varphi)$ to $P(\text{Sing}, \varphi)$;
2. the specification and Bowen properties hold for G and φ.

For the first of these, one can show that when $\eta > 0$ is small, $P(\mathcal{P} \cup S, \varphi)$ is close to the pressure of the set of orbit segments along which the integral of λ vanishes; this in turn can be shown to equal $P(\text{Sing}, \varphi)$. Thus the pressure gap

---

[22]This allows us to use indicator functions of open sets, which is helpful in some applications.

[23]We could also define the class of *one-sided* λ-*decompositions* by taking the longest initial segment in $\mathcal{B}(\eta)$, declaring what is left over to be good, and setting $S = \emptyset$, or conversely by putting $S = \mathcal{B}(\eta)$ and $\mathcal{P} = \emptyset$. This formalism is defined in [109]: the decompositions in Sect. 1.3.4.2 are examples of one-sided λ-decompositions.

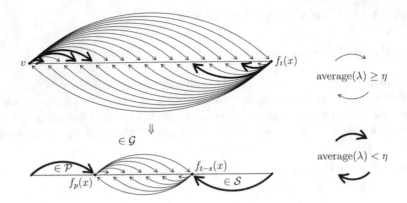

Fig. 1.10 A λ-decomposition

assumption (1.4.2.3) gives us $P([\mathcal{P}] \cup [\mathcal{S}], \varphi) < P(X, \varphi)$ for sufficiently small η, which is (III) in Theorem 1.4.2.1.

For the second of these, one can in fact prove the specification property for the larger collection

$$C(\eta) = \{(v, t) : \lambda(v) > \eta, \lambda(f_t v) > \eta\}; \tag{1.4.2.5}$$

this will be useful in Sect. 1.4.4. Observe that $\mathcal{G}(\eta) \subset C(\eta)$. The proof of the specification property is essentially the one from the uniformly hyperbolic case, as described in Sect. 1.2.4.3. See particularly Remark 1.10, and we refer to [3, §4] for the full proof. The key ingredient is uniformity of the local product structure at the end points of the orbit segments. This is provided by the condition that λ is uniformly positive at these points. Then we use uniform density of unstable leaves to transition between orbit segments. We additionally need some definite expansion along the unstable of each orbit segment, which follows from the uniformity of λ at the endpoints.

Remark 1.28 In fact, $C(\eta)$ satisfies a stronger version of specification than the one formulated in Definition 1.4.2.1: one can replace the conclusion that the shadowing can be accomplished

for *some* $0 = T_1 < T_2 < \cdots < T_k$ satisfying $T_{i+1} - T_i - t_i \in [0, \tau]$

with the stronger conclusion that it can be accomplished

for *every* $T_1 < T_2 < \cdots < T_k$ satisfying $T_{i+1} \geq T_i + \tau$.

That is, we are able to take all the transition times to be exactly τ, or any length at least τ that we choose. This stronger conclusion is important in both the K-property result in Sect. 1.4.3 and the entropy gap result discussed in Sect. 1.4.4.

Finally, for the Bowen property, the key is to use the distance estimate (1.4.2.4) to deduce that for every $(v, t) \in \mathcal{G}(\eta)$ and $w, w' \in W^s_\delta(v)$, we have

$$d^s(f_\tau w, f_\tau w') \le d^s(w, w')e^{-\tau\eta/2} \text{ for all } \tau \in [0, t],$$

with a similar estimate along the unstables (going backwards from the end of the orbit segment). Together with the local product structure, this allows the Bowen property on $\mathcal{G}$ for Hölder continuous potentials to be deduced from the same argument used in Proposition 1.13.

Remark 1.29 Since it is not known whether the geometric potential φ^u is Hölder continuous, an alternate proof is required to show that it satisfies the Bowen property on $\mathcal{G}$. This is one of the hardest parts of the analysis of [3], and relies on detailed estimates involving the Riccati equation.

Combining the ideas described above verifies the hypotheses of the abstract result in Theorem 1.4.2.1, so that the pressure gap (1.4.2.3) yields a unique equilibrium state.

1.4.2.7   Unique MMEs for Surfaces Without Conjugate Points

When M is merely assumed to have no conjugate points, life is substantially harder because many of the geometric tools used in the previous section are no long available, such as convexity of horospheres, monotonicity of the distance function, and continuity of the stable and unstable foliations of T^1M (cf. the "dinosaur" example of Ballmann et al. [110]).

Under the additional (strong) assumption that the flow is expansive, uniqueness of the MME was proved by Aurélien Bosché, a student of Knieper, in his Ph.D. thesis [111]. The following result says that at least in dimension 2, we can remove the assumption of expansivity.

Theorem 1.4.2.5 ([20]) *Let M be a closed manifold of dimension 2, with genus ≥ 2, equipped with a smooth Riemannian metric without conjugate points. Then the geodesic flow on T^1M has a unique measure of maximal entropy.*

Remark 1.30 A higher-dimensional version of Theorem 1.4.2.5 is available [20], but requires additional assumptions on M: existence of a 'background' metric with negative curvature; the divergence property; residually finite fundamental group; and a certain 'entropy gap' condition. All of these can be verified for every metric without conjugate points on a surface of genus 2.

Theorem 1.4.2.5 is proved using a coarse-scale expansivity and specification result. Issues of coarse scale did not arise in our non-positive curvature result, where we obtained the specification property at arbitrarily small scales. This removed a great deal of technicality from the analysis. We will not discuss the general coarse-scale analogue of Theorem 1.4.2.1, since we do not use it. Instead, we state the special case where $\varphi = 0$ and $G = X \times [0, \infty)$, which suffices for Theorem 1.4.2.5. This is the continuous-time analogue of Theorem 1.3.1.1.

Theorem 1.4.2.6 ([2]) *Let X be a compact metric space and $(f_t)\colon X \to X$ a continuous flow. Suppose that $\epsilon > 40\delta > 0$ are such that $h^{\perp}_{\exp}(X, (f_t), \epsilon) < h(X, (f_t))$, and that the system has the specification property at scale δ. Then $(X, (f_t))$ has a unique measure of maximal entropy.*

Note that Theorem 1.4.2.6 is stated using the hypothesis of specification for the entire system, without passing to a subcollection of orbit segments. The key tool in proving this fact for surfaces without conjugate points is the *Morse Lemma*, which states that if g, g_0 are two metrics on M such that g has no conjugate points and g_0 has negative curvature, then there is a constant $R > 0$ such that if c, α are geodesic segments w.r.t. g, g_0, respectively, in the universal cover $\widetilde{M}$ that agree at their endpoints, then they remain within a distance R for along their entire length.

Since M is a surface of genus ≥ 2, it admits a metric of negative curvature. Given an orbit segment $(v, t) \in T^1 M \times (0, \infty)$ for the g-geodesic flow, let p, q be the start and end points of some lift of the corresponding g-geodesic segment to the universal cover. Let $w \in T^1 M \times (0, \infty)$ lift to the unique unit tangent vector that begins a g_0-geodesic segment starting at p and ending at q, and let s be the g_0-length of this segment. Then $E\colon (v, t) \mapsto (w, s)$ defines a map from the space of g-orbit segments to the space of g_0-orbit segments with the property that (v, t) and $E(v, t)$ remain within R for their entire lengths.

Using this correspondence, one can take a finite sequence of g-orbit segments $(v_1, t_1), \ldots, (v_k, t_k)$, find g_0-orbit segments $E(v_i, t_i)$ that remain within R, and use the specification property for the (Anosov) g_0-geodesic flow to shadow these (w.r.t. g_0) by a single orbit segment (y, T). Then $E^{-1}(y, T)$ is a shadowing orbit (w.r.t. g) for the original segments (x_i, t_i), for which the transition times are uniformly bounded.

Writing down the details of the scales involved, one finds that the geodesic flow for g, has specification at scale[24] $\delta = 100A^3 R$, where $A \geq 1$ is such that $A^{-1} \leq \|v\|_g / \|v\|_{g_0} \leq A$ for all $v \in TM$. (Existence of A follows from compactness.)

To apply Theorem 1.4.2.6, it remains to prove that obstructions to expansivity at some scale $\epsilon > 40\delta$ have small entropy. The problem with this is that R itself, and especially $40\delta = 4000A^3 R$, is likely much larger than the diameter of M. So at this point, it looks like the previous paragraph is completely vacuous—*any* orbit segment of the appropriate length shadows the (v_i, t_i) segments to within δ.

The solution is to pass to a finite cover. By gluing together enough copies of a fundamental domain for M,[25] one can find a finite covering manifold N whose

---

[24]In fact one can improve this estimate, but the formula is more complicated [20].

[25]Formally, one needs to take a finite index subgroup of $\pi_1(M)$ that avoids all non-identity elements corresponding to a large ball in $\widetilde{M}$; this is possible because $\pi_1(M)$ is *residually finite*.

injectivity radius is $> 3\epsilon$. Observe that

- the geodesic flow on $T^1 M$ is a finite-to-1 factor of the geodesic flow on $T^1 N$, so there is an entropy-preserving bijection between their spaces of invariant measures, and in particular there is a unique MME for the geodesic flow over M if and only if there is a unique MME over N;
- the argument for specification that we gave above still works for the geodesic flow on N, with the same scale, because this scale comes from the Morse Lemma and is given at the level of the universal cover.

So it only remains to argue that $h_{\exp}^{\perp}(\epsilon) < h_{\text{top}}$ for the geodesic flow on N. This is done by observing that if $d(f_t v, f_t w) < \epsilon$ for all $t \in \mathbb{R}$ but w does not lie on the orbit of v, then lifting to geodesics on $\widetilde{M}$ and using the fact that we are below the injectivity radius of N allows us to conclude that the lifts of v, w are tangent to distinct geodesics between the same pair of points on the ideal boundary $\partial \widetilde{M}$. Thus if μ is any ergodic invariant measure that is *not* almost expansive at scale ϵ, then μ gives full weight to the set of vectors tangent to such "non-unique geodesics".

On the other hand, if $h_\mu > 0$, then μ is a hyperbolic measure by the Margulis–Ruelle inequality, and thus by Pesin theory, μ-a.e. v has transverse stable and unstable leaves. These leaves are the normal vector fields to the stable and unstable horospheres, and thus these horospheres meet at a single point, meaning that the geodesic through v is the *unique* geodesic between its endpoints on the ideal boundary. By the previous paragraph, this means that μ is almost expansive. It follows that $h_{\exp}^{\perp}(\epsilon) = 0 < h_{\text{top}}$, and so there is a unique MME by the coarse-scale result Theorem 1.4.2.6.

We remark that the proof technique sketched here does not extend to non-zero potentials, and a theory of equilibrium states for surfaces with no conjugate points beyond the MME case is currently not available.

1.4.2.8   Geodesic Flows on Metric Spaces

Another natural direction to extend the classical case of geodesic flow on a negative curvature manifold is to generalize beyond the Riemannian case. The geodesic flow on a compact locally $\text{CAT}(-1)$ metric space is one such generalization. Here, a geodesic is a curve that locally minimizes distance, and the flow acts on the space of bi-infinite geodesics parametrized with unit speed. In the Riemannian case this space is naturally identified with $T^1 M$. The $\text{CAT}(-1)$ property is a negative curvature condition which roughly says that a geodesic triangle is thinner than a comparison geodesic triangle in the model hyperbolic space with curvature -1. While one expects these flows to exhibit similar behavior to the classical case, branching phenomena and the lack of smooth structure are obstructions to some of the usual techniques.

More generally, one can study geodesic flow on a compact locally $\text{CAT}(0)$ metric space, in which geodesic triangles are thinner than Euclidean triangles. This is a generalization of geodesic flow in Riemannian non-positive curvature.

We survey some recent results in this direction. In the CAT(-1) case (allowing cusps), the MME has been well-studied using the boundary at infinity approach, see [112]. Constantine, Lafont and the second-named author studied the compact locally CAT(-1) case using the specification approach [21], and later using a symbolic dynamics approach [113], proving that every Hölder continuous potential has a unique equilibrium state, and obtaining many of the strong stochastic properties one expects from the classical case (e.g., Central Limit Theorem, Bernoullicity, Large Deviations). Broise-Alamichel, Paulin and Parkonnen [114] have extended the equilibrium state constructions and results of Paulin et al. [34] to the CAT(-1) case for a restricted class of potentials which includes the locally constant ones. (See §2.4 and §3.2 of [114] for a description of this class—in the compact case treated in [21], no such restrictions are required, as described in the introduction of [21].) The results of [114] give detailed information in the MME case for non-compact CAT(-1) spaces, and particularly for trees, which is the focus of their work.

The CAT(0) case has seen substantial recent advances in the MME case, notably by Ricks [115], who has proved uniqueness of the MME by extending Knieper's construction. A theory of equilibrium states for translation surfaces, which is an important class of CAT(0) examples, is currently being developed by Call, Constantine, Erchenko, Sawyer and Work [104]. A theory of equilibrium states for the general CAT(0) setting is currently open.

1.4.3   Kolmogorov Property for Equilibrium States

1.4.3.1   Moving Up the Mixing Hierarchy

We describe results of Ben Call and the second-named author on the Kolmogorov and Bernoulli properties [103].

A flow-invariant measure μ is said to have the *Kolmogorov property*, or *K-property*, if every time-t map has positive entropy with respect to any non-trivial partition ξ: that is, for every partition ξ that does not contain a set of full measure, and for every $t \neq 0$, we have $h_\mu(f_t, \xi) > 0$.[26]

Theorem 1.4.3.1 *Let $F = (f_t)$ be the geodesic flow over a closed rank 1 manifold M and let $\varphi \colon T^1 M \to \mathbb{R}$ be $\varphi = q\varphi^u$ or be Hölder continuous. If $P(\mathrm{Sing}, \varphi) < P(\varphi)$, then the unique equilibrium state μ_φ has the Kolmogorov property.*

In the case $\varphi = 0$, the mixing property for the unique MME was known due to work of Babillot [116]. Theorem 1.4.3.1 strengthens this. We recall the hierarchy of

---

[26]This can also be formulated in terms of the *Pinsker σ-algebra* for μ, which can be thought of as the biggest σ-algebra with entropy 0: the measure μ has the K-property if and only if the Pinsker σ-algebra for μ is trivial.

mixing properties (this is an "express train" version of the hierarchy):

Bernoulli $\Rightarrow$ K $\Rightarrow$ mixing of all orders $\Rightarrow$ mixing $\Rightarrow$ weak mixing $\Rightarrow$ ergodic.

When $\dim(M) = 2$, it was shown by Ledrappier et al. [117] that equilibrium states
are Bernoulli; their proof uses countable-state symbolic dynamics for 3-dimensional
flows. In higher dimensions, Theorem 1.4.3.1 gives the strongest known results.

The implications in the mixing hierarchy are not "if and only if"s in general.
However, in smooth settings with some hyperbolicity, a classic strategy for proving
the Bernoulli property is to move *up* the hierarchy, establishing K, and then proving
that K implies Bernoulli. This approach was notably carried out by Ornstein and
Weiss [118, 119], Pesin [120], and Chernov and Haskell [121]. In particular, a
major success of Pesin theory is his proof that the Liouville measure restricted to
the regular set is Bernoulli. We refer to the recent book of Ponce and Varão [122]
for more details on this process. Here we simply mention that this approach can be
carried out for the unique MME of rank 1 geodesic flow, and this is done in [103].

Theorem 1.4.3.2 (Bernoulli Property [103]) *Let* (f_t) *be the geodesic flow over a
closed rank 1 manifold* M. *The unique measure of maximal entropy is Bernoulli.*

1.4.3.2 Ledrappier's Approach

The main tool in the proof of Theorem 1.4.3.1 is a fantastic result of Ledrappier
[123], which deserves to be more widely known. Ledrappier's proof is about
one page long, and gives criteria for the K-property in terms of thermodynamic
formalism. The original result is for discrete-time systems. We state here a version
of it for flows; the proof is given in [103], and in more detail in [109].

Given a flow $F = (f_t)$ on a compact metric space X, the idea is to consider the
product flow $(X \times X, F \times F)$, i.e., the flow $(f_s \times f_s)_{s \in \mathbb{R}}$ given by

$$(f_s \times f_s)(x, y) = (f_s x, f_s y) \text{ for } s \in \mathbb{R}. \tag{1.4.3.1}$$

Theorem 1.4.3.3 (Criteria for K-Property) *Let* (X, F) *be a flow such that* f_t *is
asymptotically entropy expansive for all* $t \neq 0$, *and let* φ *be a continuous function
on* X. *Let* $(X \times X, F \times F)$ *be the product flow* (1.4.3.1), *and define* $\Phi: X \times X \to \mathbb{R}$
by $\Phi(x_1, x_2) = \varphi(x_1) + \varphi(x_2)$.

If Φ *has a unique equilibrium measure in* $\mathcal{M}_{F \times F}(X \times X)$, *then the unique
equilibrium state for* φ *in* $\mathcal{M}_F(X)$ *has the Kolmogorov property.*

The fact that (X, F, φ) has a unique equilibrium state when $(X \times X, F \times F, \Phi)$
does is a consequence of the following simple lemma.

Lemma 1.4.3.1 *Let* μ *be an equilibrium state for* (X, F, φ). *Then* $\mu \times \mu$ *is an
equilibrium state for* $(X \times X, F \times F, \Phi)$.

Proof Observe that

$$h_{\mu \times \mu}(f_1 \times f_1) = h_\mu(f_1) + h_\mu(f_1)$$

and

$$\int \Phi \, d(\mu \times \mu) = \int \varphi \, d\mu + \int \varphi \, d\mu.$$

Therefore, $h_{\mu \times \mu}(f_1 \times f_1) + \int \Phi \, d(\mu \times \mu) = 2P(X, F, \varphi) = P(X \times X, F \times F, \Phi)$. $\qquad\qquad\square$

From Lemma 1.4.3.1 we see that if μ, ν are distinct equilibrium states for $(X, \mathcal{F}, \varphi)$, then $\mu \times \mu$ and $\nu \times \nu$ are both equilibrium states for Φ. If Φ has a unique equilibrium state, then this means that $\mu \times \mu = \nu \times \nu$ and hence $\mu = \nu$; thus, we get uniqueness of the equilibrium state downstairs, and we see that if Φ has a unique equilibrium state, it must have the form $\mu \times \mu$ where μ is the unique equilibrium state for φ.

Now the main idea of Ledrappier's argument can be stated quite quickly: *By the argument above, if Φ has a unique equilibrium state, then so does φ. Write μ for this measure; then $\mu \times \mu$ is the unique equilibrium state for Φ. Now assume that μ is not K. Then μ has a non-trivial Pinsker σ-algebra. This can be used to define another equilibrium state for Φ. Contradiction.*

1.4.3.3 Decompositions for Products

Given Ledrappier's result, our strategy for proving the K property in Theorem 1.4.3.1 is now clear. We want to show that the product system of two copies of the geodesic flow has a unique equilibrium state for the class of potentials under consideration.

So let's find a decomposition for the product system.

Problem Lifting decompositions to products in general does not work well. One fact we do have in our favor is that if $\mathcal{G}$ has good properties, then so does $\mathcal{G} \times \mathcal{G}$. However, we need $\mathcal{G} \times \mathcal{G}$ to arise in a decomposition for $(X \times X, F \times F)$. In general this does not look at all promising: for example, the reader may try to do it for the S-gap shifts as studied in [1], and will quickly see the issue.

Idea Work with a nice class of decompositions that *does* behave well under products. We claim that the λ-decompositions from Definition 1.4.2.3 form such a class. To see this, suppose we have a λ-decomposition $(\mathcal{P}, \mathcal{G}, \mathcal{S})$ for a flow (X, F), and define $\tilde{\lambda} \colon X \times X \to [0, \infty)$ by

$$\tilde{\lambda}(x, y) = \lambda(x)\lambda(y). \tag{1.4.3.2}$$

This function inherits lower semicontinuity from λ, and we can consider the $\tilde{\lambda}$-decomposition $(\tilde{\mathcal{P}}, \tilde{\mathcal{G}}, \tilde{\mathcal{S}})$ for $(X \times X, F \times F)$.

Given $((x, y), t) \in \tilde{\mathcal{G}}$, it follows from (1.4.3.2) and boundedness of λ that we have $(x, t), (y, t) \in \mathcal{G}$ (with an appropriate choice of η), and thus $\tilde{\mathcal{G}} \subset \mathcal{G} \times \mathcal{G}$. This means that specification and the Bowen property for $\tilde{\mathcal{G}}$ can be deduced from the corresponding properties for $\mathcal{G}$.

But how big are $\tilde{\mathcal{P}}$ and $\tilde{\mathcal{S}}$? If $\lambda = 0$ on one of the coordinates, then anything is allowed on the other. Roughly, we can show that:

$$P(\tilde{\mathcal{P}} \cup \tilde{\mathcal{S}}, \Phi) \approx P(\varphi) + P(\mathcal{P} \cup \mathcal{S}, \varphi).$$

Recall that $P(\Phi) = 2P(\varphi)$. Thus, if we have $P(\mathcal{P} \cup \mathcal{S}, \varphi) < P(\varphi)$, then we expect to be able to obtain the estimate $P(\tilde{\mathcal{P}} \cup \tilde{\mathcal{S}}, \Phi) < P(\Phi)$. This is the strategy carried out in [103, 109].

1.4.3.4 Expansivity Issues

Specification and regularity are not the whole story; in fact, dealing with continuous time and related expansivity issues is the most difficult point in our analysis.

Recall from (1.4.2.2) that for flows we define

$$\mathrm{NE}(\epsilon, F) := \{x \in X \mid \Gamma_\epsilon(x) \not\subset f_{[-s,s]}(x) \text{ for any } s > 0\}.$$

For a product flow as in (1.4.3.1), the set $\Gamma_\epsilon(x, y)$ always contains $f_{[-s,s]}x \times f_{[-s,s]}y$. That is, we are considering a flow with a 2-dimensional center. The theory in Sect. 1.4.2.1 does not apply directly because $\mathrm{NE}(\epsilon, F \times F)$ as defined for a flow is the whole space! We have to build a new theory that uses information about

$$\mathrm{NE}^\times(\epsilon) := \{(x, y) \in X \times X \mid \Gamma_\epsilon(x, y) \not\subset f_{[-s,s]}(x) \times f_{[-s,s]}(y) \text{ for any } s > 0\}. \tag{1.4.3.3}$$

There are no new difficulties with counting estimates, but serious issues arise when we build adapted partitions. In the discrete time case, our adapted partition elements look like pixels and can be used to approximate sets. In the flow case, our adapted partition elements approach a small piece of orbit, so look like thin cigars. Collections of partition elements can thus be used to approximate flow-invariant sets. In the 'product of flows' case, the best we can do is approximate sets invariant under $f_s \times f_t$ for *all* $s, t \in \mathbb{R}$. This creates new technical obstacles that must be overcome in our uniqueness proof. In particular, to run our ergodicity proof, we need to be able to approximate sets which are invariant only under $f_s \times f_s$ for all $s \in \mathbb{R}$. This disconnect is a fundamental additional difficulty.

In [103], this difficulty is overcome by proving weak mixing for μ using a lower joint Gibbs estimate which gives a kind of partial mixing for sets that are flowed out by a small time interval. This can be used to prove weak mixing of μ by a spectral argument. This is equivalent to the desired ergodicity of $\mu \times \mu$.

1.4.4 Knieper's Entropy Gap

1.4.4.1 Entropy in the Singular Set

For the geodesic flow on a rank 1 non-positive curvature manifold, we have stated and discussed our main results on uniqueness of equilibrium states, and the K property for these equilibrium states. Our results hold under the hypothesis of the pressure gap $P(\text{Sing}, \varphi) < P(\varphi)$. Thus, being able to verify the pressure gap is of central importance for our results. In this section we outline the proof that the gap holds for $\varphi = 0$, when it reduces to the *entropy gap* $h(\text{Sing}) < h(X)$. The argument extends easily to potentials that are locally constant on a neighbourhood of Sing, as claimed in Theorem 1.4.2.4.

Our introduction of rank 1 manifolds in Sect. 1.4.1.2 focused on examples where Sing contains only periodic orbits and has 0 entropy, and indeed for any surface of nonpositive curvature, one can observe that every $\mu \in \mathcal{M}_F^e(\text{Sing})$ has $h_\mu(f_1) \leq \lambda^+(\mu) = \int -\varphi^u \, d\mu = 0$ by the Margulis–Ruelle inequality, where the last equality uses the fact that $\varphi^u|_{\text{Sing}} \equiv 0$ for surfaces. Then the variational principle give $h(\text{Sing}) = 0$, and since $h(X) > 0$ for all surfaces of genus at least 2, the entropy gap holds.

In higher dimensions, however, Sing can be more complicated[27] and it is not at all clear a priori that the entropy gap should always hold. The Gromov example described in [99, §6] demonstrates that starting in dimension 3, we may have $h(\text{Sing}) > 0$. To construct this example, let M_0 be a surface of constant negative curvature with one infinite cusp. Now cut off the cusp and flatten the end so that it is isometric to a flat cylinder with radius r. Take the product $M_1 = M_0 \times S$, where S is the circle of radius r. This defines a non-positive curvature 3-manifold with boundary, where the boundary is a flat torus $\partial M_1 = \partial M_0 \times S$. Now let $M_2 = S \times M_0$ so that $\partial M_2 = S \times \partial M_0$. Glue M_1 and M_2 along the boundaries (note that the order of the factors is reversed) to obtain a 3-manifold M.

One can show that the regular set in $T^1 M$ consists of all vectors in $T^1 M$ whose geodesic enters the non-flat part of both M_1 and M_2. The singular set is then the set of vectors whose geodesics stay entirely on one side (or in the flat cylinder). It is not hard to see that $h(\text{Sing}) > 0$. In fact, by defining M_0 using a cut arbitrarily high up the cusp, one can make $h(X) - h(\text{Sing})$ arbitrarily close to 0, and indeed it

---

[27] In dimension 2, it is in fact an open problem whether Sing can contain non-periodic orbits [124], but this does not affect the argument that $h(\text{Sing}) = 0$.

is not immediately obvious that this difference is non-zero. Why should there be an entropy gap at all?

Knieper's work in [99] proved that there is a unique MME for rank 1 geodesic flow, and that this measure is fully supported on T^1M. This in turn implies the entropy gap, as explained in Sect. 1.4.2.3.

Our argument in this section differs from Knieper's by being constructive, suitable for generalization, and (hopefully) shedding light on the mechanism that drives the 'entropy gap' phenomenon. In Sect. 1.4.4.2 we present the basic idea behind using the specification property to produce entropy in the symbolic setting, and then in Sect. 1.4.4.3 we discuss how this approach can be extended to geodesic flow in non-positive curvature. Full details of the argument are in [3].

1.4.4.2   Warm-Up: Shifts with Specification

The basic mechanism for using specification to produce entropy is simply to construct exponentially many orbit segments "by hand". This idea can be seen in its simplest form in the following result, which has been known since the 1970s, see [49].

Theorem 1.4.4.1 *Let (X, σ) be a shift space with the following strong specification property: there is $\tau \in \mathbb{N}$ such that for all $v, w \in \mathcal{L} = \mathcal{L}(X)$, there is $u \in \mathcal{L}_t$ such that $vuw \in \mathcal{L}$. If X has more than one point, then the strong specification property has positive entropy.*

Proof Fix $n \in \mathbb{N}$ such that there are $w^1, w^2 \in \mathcal{L}_n$ with $w^1 \neq w^2$. For each $k \geq 1$, define a map $\Phi \colon \{1, 2\}^k \to \mathcal{L}_{k(n+\tau)}$ by

$$\Phi(\underline{i}) = w^{i_1} v^1 w^{i_2} v^2 \cdots v^{k-1} w^{i_k} v^k,$$

where all the v^j have length τ and the expression on the right hand side is chosen to be in the language of X. The existence of such a word is guaranteed by the strong specification property.

Since $w^1 \neq w^2$, we can see that Φ is injective on $\{1, 2\}^k$, so $\#\mathcal{L}_{k(n+\tau)}(X) \geq 2^k$. Taking logs, dividing by $k(n + \tau)$, and sending $k \to \infty$ gives

$$h(X) \geq \lim_{k \to \infty} \frac{1}{k(n + \tau)} \log 2^k = \frac{1}{n + \tau} \log 2 > 0.$$

$\square$

We take this basic idea further, and sketch a proof of the following result about shifts with specification. The interest here is not so much in the statement, but rather in the fact that the proof contains the main entropy production idea that we will use for geodesic flow in the next section.

Theorem 1.4.4.2 *Consider a shift space* (X, σ) *with the strong specification property. Let* $Y \subset X$ *be a compact invariant proper subset. Then* $h(Y) < h(X)$.

Proof We use the specification property, words in $\mathcal{L}(Y)$ and a single word $w \notin \mathcal{L}(Y)$ to construct at least $e^{n(h(Y)+\epsilon)}$ words in $\mathcal{L}_n(X)$ for large n, giving the desired result.

Since $Y \neq X$, we can fix $w \notin \mathcal{L}(Y)$. Let t be the length of w, and τ the gap size in the strong specification property. We fix a "window size" $n > t + 2\tau$; given $N \in \mathbb{N}$, we divide the indices $\{1, 2, \ldots, nN\}$ into N "windows" of the form $\{kn + 1, kn + 2, \ldots, (k+1)n\}$ for $1 \leq k \leq N$. In particular, given $y \in \mathcal{L}_{nN}(Y)$, we consider the subwords of y that appear in each window, which have the form $u^k := y_{[kn+1,(k+1)n]}$ for $1 \leq k \leq n$.

Within each window, we can perform the following 'surgery' to replace u^k with a word that is in $\mathcal{L}_n(X)$ but not $\mathcal{L}(Y)$:

$$u^k \mapsto u^k_{[1,n-t-2\tau]} v^1 w v^2,$$

where the words v^1, v^2 of length τ are chosen as needed for the specification property.

In each of the N windows of length n, we can decide whether to do surgery or not. Given this choice, we use the specification property to create a new word of length nN; as long as we performed at least 1 surgery, this new word lies in $\mathcal{L}(X)$ but not in $\mathcal{L}(Y)$. In this way, from a single word $y_{[1,nN]}$, we can create $2^N - 1$ new words of length nN in $\mathcal{L}(X) \setminus \mathcal{L}(Y)$ by varying over all the possible choices of windows for doing this surgery procedure. Note that these words are all distinct because within each window, we can determine whether or not we did surgery by checking whether the word w appears.

This looks promising; however, it is too naive: we have to be careful as we vary over $y_{[1,nN]} \in \mathcal{L}(Y)$. In any window we selected for surgery, we are losing all the information on the last $t + 2\tau$ entries in the window. This means that up to $\#\mathcal{L}_{t+2\tau}$ distinct words could be mapped to the same word for *each* window we select for surgery. If we select too many windows, the gain in new words is far outweighed by the loss coming from this multiplicity estimate.

Fix *Carry out surgery on a small proportion of the windows, and argue that the number of new words created beats the loss of multiplicity.*

More precisely, fix $\alpha > 0$ small. Each surgery takes place at the boundary between two windows, so we consider the $N - 1$ internal boundary points of the N windows, i.e., the set

$$A = \{n, 2n, 3n, \ldots, (N-1)n\}.$$

Assuming for convenience that $\alpha N \in \mathbb{N}$, we declare $\alpha N - 1$ of the points in A to be "on",[28] and denote the set of "on" points by J. Let $\underline{J}_N^\alpha$ be the set of all such J, that is:

$$\underline{J}_N^\alpha = \{J \subset A : \#J = \alpha N - 1\}.$$

Note that since $\frac{N-k}{\alpha N - k} \geq \frac{1}{\alpha}$ for all $1 \leq k < \alpha N$, we have

$$\#\underline{J}_N^\alpha = \binom{N-1}{\alpha N - 1} = \prod_{k=1}^{\alpha N - 1} \frac{N-k}{\alpha N - k} \geq \left(\frac{1}{\alpha}\right)^{\alpha N - 1} = \alpha e^{(-\alpha \log \alpha)N}.$$

Fix $y = y_{[1,nN]} \in \mathcal{L}_{nN}(Y)$. Given $J \in \underline{J}_N^\alpha$, we carry out our surgery procedure on the windows whose boundaries are determined by J.[29] We obtain a new word $\Phi_J(y) \in \mathcal{L}_{nN}(X)$ which is definitely not in $\mathcal{L}(Y)$.

The set $\{\Phi_J(y) : J \in \underline{J}_N^\alpha\}$ is disjoint because we can recover J from $\Phi_J(y)$ by looking at which windows contain the "marker" w. Given J, the maximum number of words $y \in \mathcal{L}_{nN}(Y)$ that can have the same image $\Phi_J(y)$ is $C^{\alpha N - 1}$, where $C = \#\mathcal{L}_{t+2\tau}(Y)$ is independent of α and N. Thus if we carry out this procedure for each word in $\#\mathcal{L}_{nN}(Y)$ and each $J \in \underline{J}_N^\alpha$, we obtain

$$\#\left(\bigcup_{y_{[1,nN]} \in \mathcal{L}_{nN}(Y)} \bigcup_J \Phi_J(y)\right) \geq (C^{-1})^{\alpha N - 1} \binom{N-1}{\alpha N - 1} \#\mathcal{L}_{nN}(Y),$$

which gives

$$\#\mathcal{L}_{nN}(X) \geq \alpha e^{(-\alpha \log \alpha)N} e^{-\alpha N \log C} \#\mathcal{L}_{nN}(Y).$$

Taking logs, dividing by N, and sending $N \to \infty$, we see that

$$h(X) \geq h(Y) + \frac{\alpha}{n}(-\log \alpha - \log C).$$

If $\alpha > 0$ is chosen small enough, the quantity in brackets is positive, and thus $h(X) > h(Y)$. $\qquad \square$

---

[28] The idea is that we want to split a word $y_{[1,nN]}$ into αN subwords and perform surgeries near the points where it was split; these are the "on" points in A.

[29] Each such window determined by the set J has length some multiple of n. The surgery procedure is to remove the last $t + 2\tau$ symbols from each window and replace with a word of the form $v^1 w v^2$ where the words v^j are provided by the specification property to ensure that this procedure creates a word in $\mathcal{L}_{nN}(X)$.

1.4.4.3   Entropy Gap for Geodesic Flow

Now we return our attention to the geodesic flow on $X = T^1 M$ for a closed rank 1 non-positive curvature manifold M and outline the proof of the entropy gap $h(X) > h(\text{Sing})$.

We follow the same entropy production strategy described in the previous section. The singular set Sing $\subset X$ is a compact invariant proper subset. But how should we construct orbits? We do not expect that orbit segments contained in Sing will have the specification property. For example, orbit segments which are contained in the interior of a flat strip definitely do not have the specification property because of the flat geometry. If we stay ϵ-close inside the flat strip on the time interval $[0, t]$, the amount of additional time needed to escape the flat strip grows with t.

So we want to use a specification argument on orbit segments without specification, which does not immediately look promising. Let us recall what kind of orbits *do* have specification: it suffices to know that both the start and end of the orbit segment are 'uniformly' in the regular set.

More precisely, for any $\eta > 0$, we have the specification property on the collection

$$C(\eta) = \{(x, t) : x, f_t x \in \text{Reg}(\eta)\},$$

where $\text{Reg}(\eta) = \{x : \lambda(v) \geq \eta\}$. See Sect. 1.4.2.6 for the definition of λ and discussion of why the specification property holds on $C(\eta)$.

In order to make use of this fact, we require a reasonable way to approximate orbit segments in Sing by orbit segments in $C(\eta)$. This will be given by a map Π_t: Sing $\rightarrow$ Reg, which can be roughly summarized by the following slogan (which doesn't make sense as a rigorous statement):

Move the start of (v, t) along its stable into $\text{Reg}(\eta)$. *Move the end along an unstable into* $\text{Reg}(\eta)$.

We now explain the construction that makes this idea precise. In our approximation of (v, t), we ask that:

1. $\Pi_t(v), \Pi_t(f_t v) \in \text{Reg}(\eta)$.
2. there exists L so $f_s(\Pi_t v)$ and Sing are close for $s \in [L, t - L]$.

In the second property, one might hope to find L so $f_s(\Pi_t v)$ and $f_s v$ are close for $s \in [L, t - L]$; however, this is too much to ask for. We can see the issue if (v, t) is in the middle of a flat strip; the best we can hope for is that the orbit of $\Pi_t(v)$ approaches the *edge* of the flat strip; see Fig. 1.11, which also illustrates the following "regularizing" procedure.

We fix η_0 so $\text{Reg}(\eta_0)$ has nonempty interior. Then using density of stable and unstable leaves, together with a compactness argument, we show the following: There exists $R > 0$ such that for every $v \in T^1 M$ we have both $W^s_R(v) \cap \text{Reg}(\eta_0) \neq \emptyset$ and $W^u_R(v) \cap \text{Reg}(\eta_0) \neq \emptyset$.

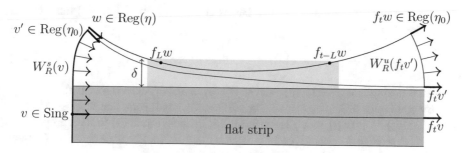

Fig. 1.11 The regularizing function $\Pi_t : v \mapsto w$

Using this fact, given $v \in \text{Sing}$, choose $v' \in W^s_R(v) \cap \text{Reg}(\eta_0)$. Then for $f_t(v')$, choose $f_t(w) \in W^u_R(f_t v') \cap \text{Reg}(\eta_0)$. Define $\Pi_t(v) := w$.

By continuity of λ, we have $\lambda(w) \geq \eta$ for an η slightly smaller than η_0. We can argue that the function $\lambda^u(f_t w)$ is small along all of the orbit segment except for an initial and terminal run of uniformly bounded length. This in turn implies that $d(f_t w, \text{Sing})$ is small, giving us condition (2). The reason $\lambda^u(f_t w)$ must be small away from the ends of the orbit segment is that otherwise small local stable and unstable manifolds centered here would get big too fast, contradicting that the endpoints of the orbit segment are in stable and unstable manifolds of size R. This is made precise by Proposition 3.13 of [3], which tells us that on a compact part of the regular set, for fixed ϵ and R, an ϵ-stable/unstable manifold grows in a uniform amount of time to cover a R-stable/unstable manifold.

In conclusion, we obtain the following properties:

Theorem 1.4.4.3 *For every $\delta > 0$ and $\eta \in (0, \eta_0)$, there exists $L > 0$ such that for every $v \in \text{Sing}$ and $t \geq 2L$, the image $w = \Pi_t(v)$ has the following properties:*

(1) $w, f_t(w) \in \text{Reg}(\eta)$;
(2) $d(f_s(w), \text{Sing}) < \delta$ *for all* $s \in [L, t - L]$;
(3) *for every* $s \in [L, t - L]$, $f_s(w)$ *and* v *lie in the same connected component of* $B(\text{Sing}, \delta) := \{w \in T^1 M : d(w, \text{Sing}) < \delta\}$.

This result is found in [3, Theorem 8.1], where the proof of (2) contains some typos: we take this opportunity to correct these typos by providing a complete proof here. (Most of this proof is word-for-word identical to the one in [3].)

Proof of Theorem 1.4.4.3 Let δ, η, η_0 be as in the statement of the theorem. For property (1), it is immediate from the definition of Π_t that $\lambda(f_t w) \geq \eta$. By uniform continuity of λ, we can take ϵ_0 sufficiently small such that if $v_2 \in W^u_{\epsilon_0}(v_1)$ and $\lambda(v_1) \geq \eta_0$, then $\lambda(v_2) \geq \eta$. By Burns et al. [3, Corollary 3.14], there exists $T_0 > 0$ such that if $t \geq T_0$ and $f_t(w) \in W^u_R(f_t v')$, then $w \in W^u_{\epsilon_0}(v')$. Thus, if $\lambda(v') \geq \eta_0$, then $\lambda(w) \geq \eta$. Thus, item (1) of the theorem holds for any $t \geq T_0$.

We turn our attention to item (2). Burns et al. [3, Proposition 3.4] tells us that there are η', $T_1 > 0$ such that

$$\text{if } \lambda^u(f_s v) \leq \eta' \text{ for all } |s| \leq T_1, \text{ then } d(v, \text{Sing}) < \delta. \qquad (1.4.4.1)$$

Given $v \in \text{Sing}$, we have $\Pi^s(v) = v' \in W_R^s(v)$, and $\lambda(f_s v) = 0$ for all s.

By continuity of λ^u, we can take ϵ_1 sufficiently small such that if $v_2 \in W_{\epsilon_1}^s(v_1)$, then $|\lambda^u(v_1) - \lambda^u(v_2)| < \eta'/2$. Applying [3, Proposition 3.13] to the compact set $\{v : \lambda^u(v) \geq \eta'/2\} \subset \text{Reg}$ gives $T_2 > 0$ such that if $\lambda^u(v_1) \geq \eta'/2$ and $\tau \geq T_2$, then $f_{-\tau} W_{\epsilon_1}^s(v_1) \supset W_R^s(f_{-\tau} v_1)$ and $f_\tau W_{\epsilon_1}^u(v_1) \supset W_R^u(f_\tau v_1)$.

Suppose for a contradiction that $\lambda^u(f_s v') \geq \eta'/2$ for some $s \geq T_2$. Applying the previous paragraph with $v_1 = f_s v'$ gives $f_s v \in f_s W_R^s(f_s v') \subset W_{\epsilon_1}^s(f_s v')$. By our choice of ϵ_1, this gives $\lambda^u(f_s v) > 0$, contradicting the fact that $v \in \text{Sing}$, and we conclude that $\lambda^u(f_s v') < \eta'/2$ for $s \geq T_2$.

Similarly, if there is $s \in [T_2, t - T_2]$ such that $\lambda^u(f_s w) \geq \eta'$, then the same argument with $v_1 = f_s w$ and $\tau = t - s$ gives $f_s v' \in f_{-(t-s)} W_R^u(f_t w) \subset W_{\epsilon_1}^u(f_s w)$, and our choice of ϵ_1 gives $\lambda^u(f_s v') \geq \lambda^u(f_s w) - \eta'/2 \geq \eta'/2$, a contradiction since $\lambda^u(f_s v') < \eta'/2$ for all $s \geq T_2$. Thus $\lambda^u(f_s w) < \eta'$ for all $s \in [T_2, t - T_2]$.

Applying (1.4.4.1) gives $d(f_s w, \text{Sing}) < \delta$ for all $s \in [T_2 + T_1, t - T_2 - T_1]$. Thus, taking $L = \max(T_0, T_1 + T_2)$, assertions (1) and (2) follow for $s \geq 2L$.

For item (3) of the theorem, we observe that v and w can be connected by a path $u(r)$ that follows first $W_R^s(v)$, then $f_{-t}(W_R^u(f_t v'))$ (see Figure 1.11), and that the arguments giving $d(f_s w, \text{Sing}) < \delta$ also give $d(f_s u(r), \text{Sing}) < \delta$ for every $s \in [L, t - L]$ and every r. We conclude that $f_s v$ and $f_s w$ lie in the same connected component of $B(\text{Sing}, \delta)$ for every such s.

The collection $\{(\Pi_t(v), t) : v \in \text{Sing}\}$ has the specification property. This is because an orbit segment $(\Pi_t(v), t)$ both starts and ends in $\text{Reg}(\eta)$. As discussed, the collection $C(\eta)$ of such orbit segments has the specification property.

We certainly do not expect the map Π_t to preserve separation of orbits. For example, in Fig. 1.11, we would expect a $v_2 \in \text{Sing}$ defining a geodesic parallel to γ_v (for example the arrow just above v in the picture) to be mapped to the same (or similar) point. However, using estimates in the universal cover, which we omit here, we can argue that Π_t has bounded multiplicity on a (t, ϵ) separated set, independent of t, in the following sense.

Proposition 1.14 *For every $\epsilon > 0$, there exists $C > 0$ such that if $E_t \subset \text{Sing}$ is a $(t, 2\epsilon)$-separated set for some $t > 0$, then for every $w \in T^1 M$, we have $\#\{v \in E_t \mid d_t(w, \Pi_t v) < \epsilon\} \leq C$.*

Now let us return to our entropy production argument. It is basically the argument we saw in Sect. 1.4.4.2, except that we need to apply the regularizing map Π_t before applying the specification property, as shown in Fig. 1.12.

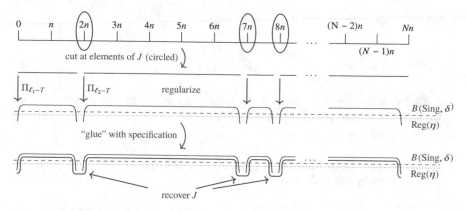

Fig. 1.12 Gluing singular orbits

As before, consider a time window $[0, nN]$. Given a subset J of $\alpha N - 1$ elements from the set $\{n, 2n, 3n, \ldots, (N-1)n\}$, we write $\ell_1, \ell_2, \ldots, \ell_{\alpha N}$ for the lengths of the intervals (in order) whose endpoints are determined by J.

For $(v_1, v_2, \ldots, v_{\alpha N}) \in \text{Sing}^{\alpha N}$, we apply the map $\Pi_{\ell_i - T}$ to each coordinate and glue the resulting orbit segments in $C(\eta)$ using specification (where T is the transition time in the specification property at a suitable scale).

Run this construction over $(g_{\ell_i - T}, \epsilon)$-separated sets for Sing in each coordinate, and for each choice of J, we construct exponentially more orbits than there are in Sing. The argument is analogous to our previous entropy production argument: for $\alpha > 0$ small, the growth from the $\binom{N-1}{\alpha N - 1}$ term beats the loss coming from multiplicity in the construction. In particular, we conclude that $h(X) > h(\text{Sing})$.

1.4.4.4 Other Applications of Pressure Production

The argument for entropy and pressure production described above is quite flexible, and can be used in many other contexts. For example, in [10] we used a variation on this argument to show that for a continuous potential φ with the Bowen property on the β-shift Σ_β,

$$\varlimsup_{n \to \infty} \frac{1}{n} \sum_{i=1}^{n-1} \varphi(\sigma^i w^\beta) < P(\Sigma_\beta, \varphi),$$

where w^β is the lexicographically maximal sequence in Σ_β; this in turn established a pressure gap condition leading to a uniqueness result, similar to the procedure described above for geodesic flow.

Another variation of the argument can be used to prove that a unique equilibrium state μ_φ coming from Bowen's original theorem (i.e., from the assumptions of expansivity, specification and the Bowen property) satisfies

$$P(\varphi) > \sup_{\mu \in M_f(X)} \int \varphi \, d\mu,$$

and thus that the entropy of μ_φ is positive.[30] Such a potential is often called *hyperbolic*. This idea was explored in [15, Theorem 6.1] and extended recently in the symbolic setting in [11].

Acknowledgments Vaughn Climenhaga is partially supported by NSF DMS-1554794. D.T. is partially supported by NSF DMS-1461163 and DMS-1954463.

References

1. V. Climenhaga, D.J. Thompson, Intrinsic ergodicity beyond specification: β-shifts, S-gap shifts, and their factors. Israel J. Math. **192**(2), 785–817 (2012)
2. V. Climenhaga, D.J. Thompson, Unique equilibrium states for flows and homeomorphisms with non-uniform structure. Adv. Math. **303**, 745–799 (2016)
3. K. Burns, V. Climenhaga, T. Fisher, D.J. Thompson, Unique equilibrium states for geodesic flows in nonpositive curvature. Geom. Funct. Anal. **28**(5), 1209–1259 (2018)
4. R. Bowen, Periodic points and measures for Axiom A diffeomorphisms. Trans. Am. Math. Soc. **154**, 377–397 (1971)
5. R. Bowen, Some systems with unique equilibrium states. Math. Syst. Theory **8**(3), 193–202 1974/1975
6. V. Climenhaga, Specification and towers in shift spaces. Commun. Math. Phys. **364**(2), 441–504 (2018)
7. B. Matson, E. Sattler, S-limited shifts. Real Anal. Exchange **43**(2), 393–415 (2018)
8. V. Climenhaga, R. Pavlov, One-sided almost specification and intrinsic ergodicity. Ergodic Theory Dynam. Syst. **39**(9), 2456–2480 2019
9. M. Shinoda, K. Yamamoto, Intrinsic ergodicity for factors of $(-\beta)$-shifts. Nonlinearity **33**(1), 598–609 (2020)
10. V. Climenhaga, D.J. Thompson, Equilibrium states beyond specification and the Bowen property. J. Lond. Math. Soc. **87**(2), 401–427 (2013)
11. V. Climenhaga, V. Cyr, Positive entropy equilibrium states. Israel J. Math. **232**(2), 899–920 (2019)
12. V. Climenhaga, D.J. Thompson, K. Yamamoto, Large deviations for systems with non-uniform structure. Trans. Am. Math. Soc. **369**(6), 4167–4192 (2017)
13. L. Carapezza, M. López, D. Robertson, Unique equilibrium states for some intermediate beta transformations. Stochastics Dyn. (to appear). https://doi.org/10.1142/S0219493721500350
14. V. Climenhaga, T. Fisher, D.J. Thompson, Unique equilibrium states for Bonatti-Viana diffeomorphisms. Nonlinearity **31**(6), 2532–2570 (2018)
15. V. Climenhaga, T. Fisher, D.J. Thompson, Equilibrium states for Mañé diffeomorphisms. Ergodic Theory Dynam. Syst. **39**(9), 2433–2455 (2019)
16. T. Wang, Unique equilibrium states, large deviations and Lyapunov spectra for the Katok map. Ergodic Theory Dynam. Syst. **41**(7), 2182–2219 (2021)

---

[30]https://vaughnclimenhaga.wordpress.com/2017/01/26/entropy-bounds-for-equilibrium-states/.

17. T. Fisher, K. Oliveira, Equilibrium states for certain partially hyperbolic attractors. Nonlinearity **33**, 3409–3423 (2020)
18. D. Chen, L.-Y. Kao, K. Park, Unique equilibrium states for geodesic flows over surfaces without focal points. Nonlinearity **33**, 1118–1155 (2020)
19. D. Chen, L.-Y. Kao, K. Park, Properties of equilibrium states for geodesic flows over manifolds without focal points. Adv. Math. **380**, 107564 (2021)
20. V. Climenhaga, G. Knieper, K. War, Uniqueness of the measure of maximal entropy for geodesic flows on certain manifolds without conjugate points. Adv. Math. **376**, 107452 (2021)
21. D. Constantine, J.-F. Lafont, D.J. Thompson, The weak specification property for geodesic flows on CAT(−1) spaces. Groups Geom. Dyn. **14**(1), 297–336 (2020)
22. P. Sun, Denseness of intermediate pressures for systems with the Climenhaga-Thompson structures. J. Math. Anal. Appl. **487**, 124027 (2020)
23. R. Pavlov, On intrinsic ergodicity and weakenings of the specification property. Adv. Math. **295**, 250–270 (2016)
24. R. Pavlov, On controlled specification and uniqueness of the equilibrium state in expansive systems. Nonlinearity **32**(7), 2441–2466 (2019)
25. L.-S. Young, Large deviations in dynamical systems. Trans. Am. Math. Soc. **318**(2), 525–543 (1990)
26. F. Takens, E. Verbitskiy, On the variational principle for the topological entropy of certain non-compact sets. Ergodic Theory Dynam. Syst. **23**(1), 317–348 (2003)
27. C.-E. Pfister, W.G. Sullivan, Large deviations estimates for dynamical systems without the specification property. Applications to the β-shifts. Nonlinearity **18**(1), 237–261 (2005)
28. C.-E. Pfister, W.G. Sullivan, On the topological entropy of saturated sets. Ergodic Theory Dynam. Syst. **27**(3), 929–956 (2007)
29. P. Varandas, Non-uniform specification and large deviations for weak Gibbs measures. J. Stat. Phys. **146**(2), 330–358 (2012)
30. A. Quas, T. Soo, Ergodic universality of some topological dynamical systems. Trans. Am. Math. Soc. **368**(6), 4137–4170 (2016)
31. T. Bomfim, P. Varandas, Multifractal analysis for weak Gibbs measures: from large deviations to irregular sets. Ergodic Theory Dynam. Syst. **37**(1), 79–102 (2017)
32. K. Yamamoto, On the weaker forms of the specification property and their applications. Proc. Am. Math. Soc. **137**(11), 3807–3814 (2009)
33. D. Kwietniak, M. Łącka, P. Oprocha, A panorama of specification-like properties and their consequences, in *Dynamics and Numbers*. Contemporary Mathematics, vol. 669. (American Mathematical Society, Providence, 2016), pp. 155–186
34. F. Paulin, M. Pollicott, B. Schapira, Equilibrium states in negative curvature. Astérisque **373**, (2015)
35. V. Climenhaga, Y. Pesin, Building thermodynamics for non-uniformly hyperbolic maps. Arnold Math. J. **3**(1), 37–82 (2017)
36. J. Buzzi, S. Crovisier, O. Sarig, Measures of maximal entropy for surface diffeomorphisms (2018, preprint). arXiv:1811.02240
37. T. Fisher, B. Hasselblatt, *Hyperbolic Flows*. Zurich Lectures in Advanced Mathematics, vol. 25 (European Mathematical Society Publishing House, Zürich, 2019), p. 737
38. V. Climenhaga, Y. Pesin, A. Zelerowicz, Equilibrium states in dynamical systems via geometric measure theory. Bull. Am. Math. Soc. **56**(4), 569–610 (2019)
39. V. Climenhaga, SRB and equilibrium measures via dimension theory. *A Vision for Dynamics – The Legacy of Anatole Katok* (Cambridge University Press, Cambridge, to appear). arXiv:2009.09260
40. R. Bowen, *Equilibrium States and the Ergodic Theory of Anosov Diffeomorphisms*. Lecture Notes in Mathematics, vol. 470 (Springer, Berlin, 2008)
41. W. Parry, M. Pollicott, Zeta functions and the periodic orbit structure of hyperbolic dynamics. Astérisque **187–188** (1990)
42. G. Keller, *Equilibrium States in Ergodic Theory*. London Mathematical Society Student Texts, vol. 42 (Cambridge University Press, Cambridge)

43. C. Beck, F. Schlögl, *Thermodynamics of Chaotic Systems*. Cambridge Nonlinear Science Series, vol. 4 (Cambridge University Press, Cambridge, 1993)
44. V. Baladi, *Positive Transfer Operators and Decay of Correlations*. Advanced Series in Nonlinear Dynamics, vol. 16 (World Scientific, River Edge, 2000)
45. V. Baladi, The magnet and the butterfly: thermodynamic formalism and the ergodic theory of chaotic dynamics, in *Development of Mathematics 1950–2000* (Birkhäuser, Basel, 2000), pp. 97–133
46. L.-S. Young, What are SRB measures, and which dynamical systems have them? J. Stat. Phys. **108**(5–6), 733–754 (2002)
47. C. Bonatti, L.J. Díaz, M. Viana, *Dynamics Beyond Uniform Hyperbolicity*. Encyclopaedia of Mathematical Sciences, vol. 102 (Springer, Berlin, 2005)
48. J.-R. Chazottes, Fluctuations of observables in dynamical systems: from limit theorems to concentration inequalities, in *Nonlinear Dynamics New Directions*. Nonlinear System Complexity, vol. 11 (Springer, Cham, 2015), pp. 47–85
49. M. Denker, C. Grillenberger, K. Sigmund, *Ergodic Theory on Compact Spaces*. Lecture Notes in Mathematics, vol. 527 (Springer, Berlin, 1976)
50. P. Walters, *An Introduction to Ergodic Theory*. Graduate Texts in Mathematics, vol. 79. (Springer, New York, 1982)
51. K. Petersen, *Ergodic Theory*. Cambridge Studies in Advanced Mathematics (Cambridge University Press, Cambridge, 1989)
52. M. Viana, K. Oliveira, *Foundations of Ergodic Theory*. Cambridge Studies in Advanced Mathematics, vol. 151 (Cambridge University Press, Cambridge, 2016)
53. M. Fekete, Über die Verteilung der Wurzeln bei gewissen algebraischen Gleichungen mit ganzzahligen Koeffizienten. Math. Z. **17**(1), 228–249 (1923)
54. T. Bomfim, M.J. Torres, P. Varandas, Topological features of flows with the reparametrized gluing orbit property. J. Differ. Equs. **262**(8), 4292–4313 (2017)
55. P. Sun, Zero-entropy dynamical systems with the gluing orbit property. Adv. Math. **372**, 107294 (2020)
56. W. Parry, Intrinsic Markov chains. Trans. Am. Math. Soc. **112**, 55–66 (1964)
57. R.L. Adler, B. Weiss, Entropy, a complete metric invariant for automorphisms of the torus. Proc. Nat. Acad. Sci. USA **57**, 1573–1576 (1967)
58. R.L. Adler, B. Weiss, Similarity of automorphisms of the torus. Memoirs of the American Mathematical Society, vol. 98 (American Mathematical Society, Providence, 1970)
59. B. Weiss, Subshifts of finite type and sofic systems. Monatsh. Math. **77**, 462–474 (1973)
60. R. Bowen, Maximizing entropy for a hyperbolic flow. Math. Syst. Theory **7**(4), 300–303 (1974)
61. A. Rényi, Representations for real numbers and their ergodic properties. Acta Math. Acad. Sci. Hungar **8**, 477–493 (1957)
62. W. Parry, On the β-expansions of real numbers. Acta Math. Acad. Sci. Hungar. **11**, 401–416 (1960)
63. F. Blanchard, β-expansions and symbolic dynamics. Theor. Comput. Sci. **65**(2), 131–141 (1989)
64. J. Schmeling, Symbolic dynamics for β-shifts and self-normal numbers. Ergodic Theory Dynam. Syst. **17**(3), 675–694 (1997)
65. F. Hofbauer, β-shifts have unique maximal measure. Monatsh. Math. **85**(3), 189–198 (1978)
66. P. Walters, Equilibrium states for β-transformations and related transformations. Math. Z. **159**(1), 65–88 (1978)
67. M. Boyle, Open problems in symbolic dynamics, in *Geometric and Probabilistic Structures in Dynamics*. Contemporary Mathematics, vol. 469 (American Mathematical Society, Providence, 2008), pp. 69–118
68. V. Climenhaga, D.J. Thompson, Intrinsic ergodicity via obstruction entropies. Ergodic Theory Dynam. Syst. **34**(6), 1816–1831 (2014)
69. J. Buzzi, T. Fisher, Entropic stability beyond partial hyperbolicity. J. Mod. Dyn. **7**(4), 527–552 (2013)

70. R. Mañé, Contributions to the stability conjecture. Topology **17**(4), 383–396 (1978)
71. R. Ures, Intrinsic ergodicity of partially hyperbolic diffeomorphisms with a hyperbolic linear part. Proc. Am. Math. Soc. **140**(6), 1973–1985 (2012)
72. J. Buzzi, T. Fisher, M. Sambarino, C. Vásquez, Maximal entropy measures for certain partially hyperbolic, derived from Anosov systems. Ergodic Theory Dynam. Syst. **32**(1), 63–79 (2012)
73. A. Katok, Lyapunov exponents, entropy and periodic orbits for diffeomorphisms. Inst. Hautes Études Sci. Publ. Math. **51**, 137–173 (1980)
74. W. Cowieson, L.-S. Young, SRB measures as zero-noise limits. Ergodic Theory Dynam. Syst. **25**(4), 1115–1138 (2005)
75. R. Bowen, Entropy-expansive maps. Trans. Am. Math. Soc. **164**, 323–331 (1972)
76. C. Bonatti, M. Viana, SRB measures for partially hyperbolic systems whose central direction is mostly contracting. Israel J. Math. **115**, 157–193 (2000)
77. N. Sumi, P. Varandas, K. Yamamoto, Partial hyperbolicity and specification. Proc. Am. Math. Soc. **144**(3), 1161–1170 (2016)
78. J. Crisostomo, A. Tahzibi, Equilibrium states for partially hyperbolic diffeomorphisms with hyperbolic linear part. Nonlinearity **32**(2), 584–602 (2019)
79. F. Rodriguez Hertz, M.A. Rodriguez Hertz, A. Tahzibi, R. Ures, Maximizing measures for partially hyperbolic systems with compact center leaves. Ergodic Theory Dynam. Syst. **32**(2), 825–839 (2012)
80. E. R. Pujals, M. Sambarino, A sufficient condition for robustly minimal foliations. Ergodic Theory Dynam. Syst. **26**(1), 281–289 (2006)
81. J.F. Alves, SRB measures for non-hyperbolic systems with multidimensional expansion. Ann. Sci. École Norm. Sup. **33**(1), 1–32 (2000)
82. J.F. Alves, C. Bonatti, M. Viana, SRB measures for partially hyperbolic systems whose central direction is mostly expanding. Invent. Math. **140**(2), 351–398 (2000)
83. F. Rodriguez Hertz, M.A. Rodriguez Hertz, R. Ures, A non-dynamically coherent example on $\mathbb{T}^3$. Ann. Inst. H. Poincaré Anal. Non Linéaire **33**(4), 1023–1032 (2016)
84. R.B. Israel, *Convexity in the Theory of Lattice Gases*. Princeton Series in Physics (Princeton University Press, Princeton, 1979)
85. D. Ruelle, Thermodynamic Formalism. Encyclopedia of Mathematics and its Applications, vol. 5. (Addison-Wesley, Reading, 1978)
86. J.M. Lee, *Introduction to Riemannian Manifolds*. Graduate Texts in Mathematics, vol. 176 (Springer, Cham, 2018)
87. K. Burns, M. Gidea, *Differential Geometry and Topology*. Studies in Advanced Mathematics (Chapman & Hall/CRC, Boca Raton, 2005)
88. R. Gulliver, On the variety of manifolds without conjugate points. Trans. Am. Math. Soc. **210**, 185–201 (1975)
89. K. Gelfert, R.O. Ruggiero, Geodesic flows modelled by expansive flows. Proc. Edinb. Math. Soc. **62**(1), 61–95 (2019)
90. K. Gelfert, B. Schapira, Pressures for geodesic flows of rank one manifolds. Nonlinearity **27**(7), 1575–1594 (2014)
91. W. Ballmann, Lectures on spaces of nonpositive curvature, DMV Seminar, vol. 25, With an appendix by Misha Brin, Birkhäuser Verlag, Basel (1995)
92. P. Eberlein, Geodesic flows in manifolds of nonpositive curvature, in *Smooth Ergodic Theory and Its Applications (Seattle, WA, 1999)*, Proceedings of Symposia in Pure Mathematics, vol. 69 (American Mathematical Society, Providence, 2001), pp. 525–571
93. P.B. Eberlein, *Geometry of Nonpositively Curved Manifolds*. Chicago Lectures in Mathematics (University of Chicago Press, Chicago, 1996)
94. W. Ballmann, Nonpositively curved manifolds of higher rank. Ann. Math. **122**(3), 597–609 (1985)
95. K. Burns, R. Spatzier, On topological Tits buildings and their classification. Inst. Hautes Études Sci. Publ. Math. **65**, 5–34 (1987)
96. K. Burns, R. Spatzier, Manifolds of nonpositive curvature and their buildings. Inst. Hautes Études Sci. Publ. Math. **65**, 35–59 (1987)

97. M. Gerber, A. Wilkinson, Hölder regularity of horocycle foliations. J. Differ. Geom. **52**(1), 41–72 (1999)
98. W. Ballmann, Axial isometries of manifolds of nonpositive curvature. Math. Ann. **259**(1), 131–144 (1982)
99. G. Knieper, The uniqueness of the measure of maximal entropy for geodesic flows on rank 1 manifolds. Ann. Math. **148**(1), 291–314 (1998)
100. R. Bowen, P. Walters, Expansive one-parameter flows. J. Differ. Equs. **12**, 180–193 (1972)
101. R. Bowen, Periodic orbits for hyperbolic flows. Am. J. Math. **94**, 1–30 (1972)
102. E. Franco, Flows with unique equilibrium states. Am. J. Math. **99**(3), 486–514 (1977)
103. B. Call, D.J. Thompson, Equilibrium states for products of flows and the mixing properties of rank 1 geodesic flows (2019, preprint). arXiv:1906.09315
104. B. Call, D. Constantine, A. Erchenko, N. Sawyer, G. Work, Unique equilibrium states for geodesic flows on flat surfaces with singularities (2021). arXiv: 2101.11806
105. F. Liu, F. Wang, W. Wu, On the Patterson-Sullivan measure for geodesic flows on rank 1 manifolds without focal points. Discrete Continuous Dynam. Syst. A **40**, 1517 (2020)
106. T. Prellberg, J. Slawny, Maps of intervals with indifferent fixed points: thermodynamic formalism and phase transitions. J. Stat. Phys. **66**(1–2), 503–514 (1992)
107. M. Urbański, Parabolic Cantor sets. Fund. Math. **151**(3), 241–277 (1996)
108. O.M. Sarig, Phase transitions for countable Markov shifts. Commun. Math. Phys. **217**(3), 555–577 (2001)
109. B. Call, The K-property for some unique equilibrium states in flows and homeomorphisms. J. Lond. Math. Soc. (to appear). arXiv:2007.00035
110. W. Ballmann, M. Brin, K. Burns, On surfaces with no conjugate points. J. Differ. Geom. **25**(2), 249–273 (1987)
111. A. Bosché, Expansive geodesic flows on compact manifolds without conjugate points. https://tel.archives-ouvertes.fr/tel-01691107
112. T. Roblin, Ergodicité et équidistribution en courbure négative. Mém. Soc. Math. Fr. **95**, 102 (2003)
113. D. Constantine, J.-F. Lafont, D.J. Thompson, Strong symbolic dynamics for geodesic flows on CAT(−1) spaces and other metric Anosov flows. J. Éc. Polytech. Math. **7**, 201–231 (2020)
114. A. Broise-Alamichel, J. Parkkonen, F. Paulin, *Equidistribution and Counting Under Equilibrium States in Negative Curvature and Trees*. Progress in Mathematics, vol. 329 (Birkhäuser/Springer, Cham, 2019)
115. R. Ricks, The unique measure of maximal entropy for a compact rank one locally CAT(0) space. Discrete Continuous Dynam. Syst. **41**(2), 507–523 (2021)
116. M. Babillot, On the mixing property for hyperbolic systems. Israel J. Math. **129**, 61–76 (2002)
117. F. Ledrappier, Y. Lima, O. Sarig, Ergodic properties of equilibrium measures for smooth three dimensional flows. Comment. Math. Helv. **91**(1), 65–106 (2016)
118. D.S. Ornstein, B. Weiss, Geodesic flows are Bernoullian. Israel J. Math. **14**, 184–198 (1973)
119. D. Ornstein, B. Weiss, On the Bernoulli nature of systems with some hyperbolic structure. Ergodic Theory Dynam. Syst. **18**(2), 441–456 (1998)
120. J.B. Pesin, Characteristic Ljapunov exponents, and smooth ergodic theory. Uspehi Mat. Nauk **32**(4), 55–112, 287 (1977)
121. N.I. Chernov, C. Haskell, Nonuniformly hyperbolic K-systems are Bernoulli. Ergodic Theory Dynam. Syst. **16**(1), 19–44 (1996)
122. G. Ponce, R. Varão, *An Introduction to the Kolmogorov-Bernoulli Equivalence*. Springer-Briefs in Mathematics (Springer, Cham, 2019)
123. F. Ledrappier, Mesures d'équilibre d'entropie complètement positive, in *Dynamical Systems, Vol. II—Warsaw*. Astérisque, vol. 50 (Soc. Math. France, 1977), pp. 251–272
124. K. Burns, V.S. Matveev, *Open Problems and Questions About Geodesics*. Ergodic Theory and Dynamical Systems (Cambridge University Press, Cambridge, 2019), pp. 1–44

Chapter 2
The Role of Continuity and Expansiveness on Leo and Periodic Specification Properties

Serge Troubetzkoy and Paulo Varandas

Abstract In this short note we prove that a continuous map of a compact manifold which is locally eventually onto and is expansive satisfies the periodic specification property. We also discuss the role of continuity as a key condition in the previous characterization. We include several examples to illustrate the relation between these concepts.

2.1 Introduction

There is a well known hierarchy of topological properties involving the topological indecomposability of a dynamical system, as transitivity, topological mixing, and the specification property, among many others. The relation between these and many others has been addressed by Akin, Auslander and Nagar [1]. The aim of this short note is to complement the above results, and to highlight the relation between the locally eventually onto (a dynamical property stronger than topological mixing) and the specification properties, and to make explicit the role of continuity on such characterization. The specification property was first introduced by Bowen [6], for a survey of specification-like properties we recommend the following article [13], while for a survey of mixing properties we recommend the article [1].

First let us recall some well known results. Blokh [5] showed that for a continuous map of the interval [0, 1] the periodic specification property is equivalent to topological mixing (see e.g., [7, 17]). So, while for continuous interval maps the picture is very well understood and most concepts of topological chaoticity coincide, this is no longer true for more general metric spaces or whenever continuity breaks down. Yet, the situation is well understood in the case of one-dimensional branched

S. Troubetzkoy (✉)
Aix Marseille Univ, CNRS, Centrale Marseille, I2M, Marseille, France
e-mail: serge.troubetzkoy@univ-amu.fr

P. Varandas
CMUP and Departamento de Matemática, Salvador, Brazil
e-mail: paulo.varandas@ufba.br

© The Author(s), under exclusive license to Springer Nature Switzerland AG 2021
M. Pollicott, S. Vaienti (eds.), *Thermodynamic Formalism*, Lecture Notes
in Mathematics 2290, https://doi.org/10.1007/978-3-030-74863-0_2

manifolds, where there is a characterization of transitive dynamics due to Blokh [4]. In brief terms, he established the following classification theorem: either a transitive map f of a graph has periodic points and it can be decomposed into n connected subgraphs with finite pairwise intersections which are cyclically permuted and f^n has the specification property, or f is aperiodic and it is just a cycle of n circles with f^n being an irrational rotation. We refer the reader to [4] for more details.

It is noticeable that while any locally eventually onto continuous map has dense periodic sets, it may not have periodic points (cf. [1, Theorem 2.30 and Example 2.31]). In particular, a locally eventually onto continuous map need not satisfy the periodic specification property. Two results complement this discussion. First, expansiveness play a key role to bridge between the specification and periodic specification properties: a topological dynamical system satisfying the specification property and whose natural extension is expansive satisfies the periodic specification property (see e.g., [13, Lemma 6]). Second, Yan et al. [20, Theorem 3.1] constructed an example of a topological mixing subshift, hence expansive, which does not even have the specification property.

The situation is well understood in the case of continuous, open and distance expanding maps on compact metric spaces. Indeed, since any such map satisfies the shadowing property and periodic points are dense in the non-wandering set, these admit a spectral decomposition theorem (see [16, Theorem 4.3.8]). Moreover, any such map is topologically mixing map if and only it is locally eventually onto. We refer the reader to [16, Sections 4.2 and 4.3] for more details. Similar, but slightly weaker results are known if we drop the openness assumption, instead assume shadowing [12].

In general, while the locally eventually onto property need not ensure the periodic specification property, the following result shows that expansiveness can act as a sufficient condition for it. We refer the reader to Sect. 2.2 for definitions.

Theorem 1.1 *Let X be a compact and connected topological manifold. If the topological dynamical system (X, f) is locally eventually onto and expansive then it has the periodic specification property.*

This result is no longer true if one replaces the condition of X being a compact topological manifold by the assumption of being an arbitrary compact metric space. We refer the reader to Example 2.4, where we present an expansive and locally eventually onto map for which the periodic specification fails.

Note that the specification property is a topological invariant, hence we can ask whether such a property holds for the continuous map f on (X, d) or on the metric space (X, d'), for a equivalent metric d'. In the case of compact and connected topological manifolds, Coven and Reddy [9] constructed adapted metrics, proving that every expansive dynamics is indeed expanding with respect to some equivalent metric. In particular, Theorem 1.1 is a direct consequence of the previous discussion together with the following:

Theorem 1.2 *Assume that the topological dynamical system (X, f) is expanding and locally eventually onto. Then (X, f) has the periodic specification property.*

The latter suggests that the failure of periodic specification for distance expanding maps is essentially related to the lack of periodic points (see Example 2.4), the non-compactness of the phase space (see Example 2.6), or that the dynamics is not mixing.

Given the previous result it is natural to ask whether any locally eventually onto continuous map satisfies the specification property.

Remark 2.1 It is worth mentioning that the situation is clear for continuous interval maps. Indeed, combining [8, Theorem B] and Blokh's theorem (cf. [17, Theorem 3.4]), it follows that the locally eventually onto property implies on the following conditions, which, for interval maps, are equivalent:

  (i) f^2 is transitive,
 (ii) f^n is transitive for every $n \geq 1$,
(iii) f is topologically mixing,
(iv) f satisfies the specification property.

While the converse holds in the case of piecewise monotone continuous interval maps (cf. [8, Lemma 4.1]), it fails for general continuous interval maps. In particular there are continuous interval maps satisfying the specification property for which the locally eventually onto property fails (see e.g., [2, Example 3]).

On the positive direction, we notice that the same strategy used in Blokh's theorem (cf. [17, Theorem 3.4]) can be used for conformal-like maps.

Theorem 1.3 *Every locally eventually onto, continuous and conformal-like map on a compact metric space satisfies the periodic specification property.*

Remark 2.2 In the definition of topological dynamical system, the assumption that the metric space is complete cannot be removed. Throughout $\mathbb{N}$ be the set of non-negative integers (hence containing 0). There exists a metric space $X \subset \{0, 1, 2\}^{\mathbb{N}}$ such that the shift map (X, σ) is locally eventually onto, it is clearly expansive, but fails even to present periodic points [1, Example 2.31].

Our second goal concerns describing the consequences of discontinuities on locally eventually onto maps. This is a problem dual to the one considered by Buzzi [7], the study of the specification property for piecewise monotone interval maps. In the case of piecewise monotone continuous interval maps f, the transitivity for f^2 ensures the following "almost" locally eventually onto property: for any open interval A and any closed interval $J \subset (0, 1)$ there exists $N \geq 1$ so that $f^N(A) \supset J$ (see [2, Theorem 6]). However, while the key step in this argument explores the density of periodic points, the classical argument that ensures the density of periodic points for expanding maps does not apply for transitive piecewise expanding interval maps given that dynamical balls may fail to grow to a large scale.

We shall focus on important classes of dynamical systems known as β-expansions and β-shifts (see e.g., [3]). These can be realized by geometric models in the interval; for each $\beta > 1$, the β-map is the C^∞-piecewise expanding interval map $T_\beta : [0, 1) \to [0, 1)$ given by

$$T_\beta(x) = \beta x - \lfloor \beta x \rfloor.$$

However, while the previous map is always expansive, and Markov for a countable set of parameters, T_β does not satisfy the specification property for Lebesgue almost every parameter $\beta > 1$ (cf. [7]). A characterization of the set of the values of β which lead to maps with specification can be found in [18]. The next result shows that continuity is essential in Theorem 1.2.

Theorem 1.4 *For Lebesgue almost every $\beta \in (1, +\infty)$ the map T_β:*

 (i) *is locally eventually onto;*
 (ii) *is expansive;*
 (iii) *does not satisfy the specification property [7].*

We complete this section with two final comments on the relation between the specification and the locally eventually onto properties for continuous maps in more general metric spaces. While any Anosov diffeomorphism satisfies the specification property (see e.g. [11]), every volume preserving Anosov diffeomorphism is clearly not locally eventually onto. Nevertheless, on the converse direction, locally eventually onto maps displaying non-uniform expansion often satisfy some measure-theoretical forms of specification (we refer the reader to [14, 19] for the precise formulations).

2.2   Definitions

Let (X, d) be a compact metric space, and $f : X \to X$ a continuous map. We refer to (X, f) as a *dynamical system*.

The map f is called *locally eventually onto* (*LEO*) if for every nonempty open set U there is an $n \in \mathbb{N} := \{0, 1, 2, \dots\}$ such that $f^n(U) = X$.

For integers $a \geq b \geq 0$ let $f^{[a,b]}(x) := \{f^j(x) : a \leq j \leq b\}$.

A family of orbit segments $\{f^{[a_j,b_j]}(x_j)\}_{j=1}^n$ is an *N-spaced specification* if $a_i - b_{i-1} \geq N$ for $2 \leq i \leq n$.

We say that a specification $\{f^{[a_j,b_j]}(x_j)\}_{j=1}^n$ is *ε-shadowed* by $y \in X$ if

$$d(f^k(y), f^k(x_i)) \leq \varepsilon \text{ for } a_i \leq k \leq b_i \text{ and } 1 \leq i \leq n.$$

We say that (X, f) has the *specification property* if for any $\varepsilon > 0$ there is a constant $N = N(\varepsilon)$ such that any N-spaced specification $\{f^{[a_j,b_j]}(x_j)\}_{j=1}^n$ is ε-shadowed by some $y \in X$. If additionally, y can be chosen in such a way that $f^{b_n-a_0+N}(y) = y$ then (X, f) has the *periodic specification property*.

The dynamical system (X, f) is *positively expansive* if there exists $\alpha > 0$, called *expansivity constant* of f, such that if $x, y \in X$ and $x \neq y$, then for some $n \geq 0$, $d(f^n n(x), f^n(y)) > \alpha$.

The dynamical system (X, f) is *expanding* if there are constants $\lambda > 1$ and $\delta_0 > 0$ such that, for all $x, y, z \in X$,

1. $d(f(x), f(y)) \geq \lambda d(x, y)$ whenever $d(x, y) < \delta_0$ and
2. $B(x, \delta_0) \cap f^{-1}(z)$ is a singleton whenever $d(f(x), z) < \delta_0$.

A dynamical system (X, f) satisfying condition (1) if called a *distance expanding map*. In any compact metrizable space, a continuous transformation is expanding if and only if it is open, i.e., maps open sets to open sets, and distance expanding (see [9, Lemma 1]). In [16] the authors describe the dynamical properties of such maps and obtaining, in particular, density of periodic points, the shadowing property and a spectral decomposition theorem (see [16, Section 4]).

The set $B_n(x, \varepsilon) := \{y \in X : d(f^i x, f^i y) < \varepsilon \text{ for } 0 \leq i < n\}$ is called a *Bowen ball*.

A dynamical system (X, f) is called *conformal-like* if the image of every ball is a ball. A conformal map is a map that preserves angles and orientation; in the special case of smooth dynamics, the Jacobian of a conformal map is a positive multiple of a rotation matrice. Hence linear conformal maps preserve balls and are thus conformal-like but not every linear conformal-like map is conformal; for example it could reverse orientation.

2.3 Proofs

2.3.1 Proof of Theorem 1.2

From the locally eventually onto property, for each $y \in X$, and $\varepsilon > 0$ there is an $N(y, \varepsilon) \geq 1$ such that $f^{N(y,\varepsilon)}(B(y, \varepsilon/3)) = X$. Morover, by compactness of X we can cover X by a finite collection of balls $\{B(y_i, \varepsilon/3)\}_i$. Let $N := \max_i\{N(y_i, \varepsilon)\}$. Then since any ball $B(y, \varepsilon)$ contains one of the $B(y_i, \varepsilon/3)$ we conclude that $f^N(B(y, \varepsilon)) = X$ for all $y \in X$.

Now since f is continuous and expanding, the image by f^m of a Bowen ball $B_m(x, \varepsilon)$ is $B(f^m(x), \varepsilon)$, for every $0 < \varepsilon < \delta_0$. Combining this with the previous paragraph yields $f^{m+N} B_m(x, \varepsilon) = X$ for each $x \in X$ and every $0 < \varepsilon < \delta_0$.

Fix $\varepsilon > 0$ and choose N as above. Consider an N-specification, i.e., a collection of orbit segments $\{f^{[a_j,b_j]}(x_j)\}_{j=1}^n$, with $a_i - b_{i-1} \geq N$ for $2 \leq i \leq n$. Setting $m_j := b_j - a_j$ and $N_j = m_j + N$ we have shown that

$$f^{N_j}(B_{m_j}(f^{a_j}(x_j), \varepsilon)) = X \supset B_{m_{j+1}}(f^{a_{j+1}}(x_{j+1}), \varepsilon),$$

and thus

$$B_{m_j}(f^{a_j}(x_j), \varepsilon) \cap f^{-N_j}(B_{m_{j+1}}(f^{a_{j+1}}(x_{j+1}), \varepsilon)) \neq \emptyset$$

hold for each $1 \leq j \leq n$. Iterating this, and noticing that f is expanding, yields that

$$\left\{ B_{m_1}(f^{a_1}(x_1), \varepsilon) \cap f^{-N_1-N_2-\cdots-N_i}(B_{m_i}(f^{a_i}(x_i), \varepsilon)) \right\}_{2 \leq i \leq n} \tag{2.3.1}$$

is a nested sequence of compact sets. Any point in the intersection of these sets ε-shadows the specification, and thus we have shown the specification property holds.

Finally we must show that the periodic specification property holds. Fix $\varepsilon > 0$ and consider an arbitrary N-specification $\{f^{[a_j, b_j]}(x_j)\}_{j=1}^{n}$ with N chosen as above. We extend this to a longer N specification by choosing $a_{n+1} = b_n + N$, $b_{n+1} = a_{n+1} + m_1$ and $x_{m+1} = f^{a_1-a_{n+1}}x_1$. Thus $m_{n+1} = m_1$ and $B_{m_{n+1}}(f^{a_{n+1}}x_{m+1}, \varepsilon) = B_{m_1}(f^{a_1}x_1, \varepsilon)$. Therefore the chain (2.3.1) of containments extends to

$$\left\{ B_{m_1}(f^{a_1}(x_1), \varepsilon) \cap f^{-N_1-N_2-\cdots-N_i}(B_{m_i}(f^{a_i}(x_i), \varepsilon)) \right\}_{2 \leq i \leq n+1}.$$

The closure of the intersection of the extended chain of containments must contain a point fixed by $f^{N_1+\cdots+N_m+1}$, hence the periodic specification property holds. $\quad\square$

2.3.2   Proof of Theorem 1.3

The strategy follows closely [7, Appendix A]. For that reason we just give a brief sketch of the proof. Let X be a compact metric space and $f : X \to X$ be a continuous, locally eventually onto conformal map. The key step is a uniform control on the images of Bowen balls. Indeed, while points in n-Bowen balls are within controlled distance to the original orbit during n iterates, it is the size of the image the Bowen ball by iteration of f^n which suggests how strong is the capability to obtain specification.

Claim For any $\varepsilon > 0$ there exists $\zeta(\varepsilon) > 0$ so that

$$\text{diam}(f^n(B_n(x, \varepsilon))) \geq \zeta(\varepsilon) \quad \text{for every } n \geq 1 \text{ and } x \in X.$$

Proof of the Claim Fix $x \in X$. By conformality, for each $n \geq 1$ the set $f^n(B_n(x, \varepsilon))$ is a ball around $f^n(x)$. Recall also that

$$B_{n+1}(x, \varepsilon) = \bigcap_{j=0}^{n} f^{-j}(B(f^j(x), \varepsilon)) = B_n(x, \varepsilon) \cap f^{-n}(B(f^n(x), \varepsilon)) \tag{2.3.2}$$

and clearly $B_n(x, \varepsilon) \cap f^{-n}(B(f^n(x), \varepsilon)) \subseteq B_n(x, \varepsilon)$.

In particular, by the conformality of f, for each $n \geq 1$ either: (i) the equality $B_{n+1}(x, \varepsilon) = B_n(x, \varepsilon)$ holds, or (ii) the set $B_n(x, \varepsilon) \cap f^{-n}(B(f^n(x), \varepsilon))$ is strictly contained in $B_n(x, \varepsilon)$. In the second case, there exists a point $y \in B_n(x, \varepsilon)$ so that $f^n(y) \notin B(f^n(x), \varepsilon)$. This shows that the ball $f^n(B_n(x, \varepsilon)) \supset B(f^n(x), \varepsilon)$, combining with (2.3.2) yields

$$f^n(B_{n+1}(x, \varepsilon)) = B(f^n(x), \varepsilon).$$

Altogether, this proves that for every $n \geq 1$ there exists $0 \leq j < n$ so that $f^n(B_n(x, \varepsilon)) = f^{n-j}(B(f^j(x), \varepsilon))$.

Thus, in order to prove the claim it is enough to show that the forward image of balls of a definite size do not degenerate: for any $\varepsilon > 0$ there exists $\zeta(\varepsilon) > 0$ such that $\operatorname{diam}(f^n(B(z, \varepsilon))) \geq \zeta(\varepsilon)$ for every $n \geq 1$ and every $z \in X$.

Indeed, since f is locally eventually onto, for any given $z \in X$ there exists $N(z, \varepsilon) > 0$ such that $f^{N(z,\varepsilon)}(B(z, \varepsilon)) = X$; hence there exists $\zeta_z(\varepsilon) > 0$ such that $\operatorname{diam}(f^n(B(z, \varepsilon))) \geq \zeta_z(\varepsilon)$ for every $n \geq 1$. The continuity of f and compactness of X ensures that $\min_{z \in X} \zeta_z(\varepsilon) > 0$, proving the claim. □

We now claim that f satisfies the periodic specification property. Indeed, given $\varepsilon > 0$ let $N = N(\varepsilon) \geq 1$ be such that $f^N(B(x, \zeta(\varepsilon))) = X$ for every $x \in X$. Such $N \geq 1$ does exists as f is locally eventually onto and X is compact. The proof of the periodic specification property now follows as in Theorem 1.2. □

2.3.3 Proof of Theorem 1.4

Since items (ii) and (iii) are known (see e.g., [7]) we need only prove that each T_β is locally eventually onto.

Fix $\beta > 1$ and take an arbitrary interval $J \subset [0, 1)$. We claim that there exists $N \geq 1$ so that $T_\beta^N(J) = [0, 1)$. We may assume without loss of generality that J is contained in some domain of smoothness for T_β. By the mean value theorem, $\operatorname{Leb}(T_\beta(J)) \geq \beta \operatorname{Leb}(J)$. If $T_\beta(J) \cap D_{T_\beta} = \emptyset$ then $\operatorname{Leb}(T_\beta^2(J)) \geq \beta^2 \operatorname{Leb}(J)$. Since the diameter is bounded, a recursive argument shows that $T_\beta^k(J) \cap D_{T_\beta} \neq \emptyset$ for some $k \geq 1$. In particular $T_\beta^k(J) \supset [0, a)$ for some $a \in (0, \frac{1}{\beta}]$. Since $T_\beta(0) = 0$, and T_β is monotone increasing in $[0, \frac{1}{\beta}]$ then there exists $N \geq 1$ so that $T_\beta^N(J) \supset [0, \frac{1}{\beta}]$. This assures that $T_\beta^{N+1}(J) = [0, 1)$. □

2.4 Examples

We finish with some examples. The first example is a simple examples of piecewise expanding continuous maps which need not be neither expansive nor transitive.

Example 2.1 Consider the continuous and piecewise expanding interval map $f_0 : [0, \frac{1}{2}] \to [0, \frac{1}{2}]$ given by

$$
f_0(x) = \begin{cases} 3x & \text{if } x \in [0, \frac{1}{6}] \\ -3x + 1 & \text{if } x \in (\frac{1}{6}, \frac{1}{3}] \\ 3x - 1 & \text{if } x \in (\frac{1}{3}, \frac{1}{2}]. \end{cases}
$$

Let $f : [0, 1] \to [0, 1]$ be obtained by replication of the dynamics f_0 in intervals of exponential decreasing growth accumulating 1, defined by the relation

$$
f(x) = 1 - 2^{-n} + 2^{-n} f_0(2^n(x - 1 + 2^{-n})), \qquad x \in (1 - 2^{-n}, 1 - 2^{-(n+1)}].
$$

and $f(0) = 0$, $f(1) = 1$. Clearly f is piecewise expanding, continuous, not expanding nor transitive.

The next example shows that transitivity is essential to avoid unattainable repelling points.

Example 2.2 Consider the continuous and C^1-piecewise expanding interval map $f : [0, 1] \to [0, 1]$ given by

$$
f(x) = \begin{cases} 3x & \text{if } x \in [0, \frac{1}{3}] \\ -2x + \frac{5}{3} & \text{if } x \in (\frac{1}{3}, \frac{2}{3}] \\ 2x - 1 & \text{if } x \in (\frac{2}{3}, 1]. \end{cases}
$$

The map is not transitive as $f([\frac{1}{3}, 1]) = [\frac{1}{3}, 1]$, in other words, $[\frac{1}{3}, 1]$ is an f-invariant domain. Thus f is not locally eventually onto. Nevertheless, the attractor $\Lambda := \bigcap_{n \geq 0} f^n((0, 1]]) = [\frac{1}{3}, 1]$ and $f \mid_\Lambda$ is locally eventually onto.

Finally we complete this note with an example showing that locally eventually onto is weaker than specification. We consider an example suggested by Lindenstrauss (cf. [1, Example 2.31]) of a locally eventualy onto map having no periodic points.

Example 2.3 Consider the subshift $Y_0 \subset \{0, 1, 2\}^{\mathbb{N}}$ consisting of the set of sequences that admit no consecutive 0's, let and let $\pi : Y_0 \to \{1, 2\}^{\mathbb{N}}$ be given by suppression of the 0's in the sequences belonging to Y_0. Endowing the shift spaces with the usual distances, π is a continuous map on a compact metric space, hence it is uniformly continuous.

Consider a minimal subshift $X \subset (\{1, 2\}^{\mathbb{N}}, \sigma)$ and let $Y = \pi^{-1}(X)$. Akin et al proved that (Y, σ) is locally eventually onto (cf. Example 2.31 in [1]). We claim that (Y, σ) does not satisfy the specification property. Recall that a factor of a map of a compact space with specification satisfies specification (cf. [10, Proposition 21.4]) This does not directly apply to our situation since we do not have compactness, however it is not hard to prove that the commuting diagram

$$
\begin{array}{ccc}
Y & \to_\sigma & Y \\
\downarrow\pi & & \downarrow\pi \\
X & \to_\sigma & X
\end{array}
$$

together with the uniform continuity of π ensures that if (Y, σ) satisfies the specification property then so does (X, σ). Second, (X, σ) does not satisfy the specification property. Indeed, if (X, T) has the specification property and its natural extension is expansive then (X, T) has the periodic specification property (see e.g., Lemma 6 in [13]). Altogether, this proves that (Y, σ) does not satisfy the specification property, as claimed.

The following example, suggested by F. Przytycki, describes an counter example to Theorem 1 if we do not assume that X is a topological manifold.

Example 2.4 Consider the circle $\mathbb{S}^1 = \mathbb{R}/\mathbb{Z}$ and the doubling map $f : \mathbb{S}^1 \to \mathbb{S}^1$ given by $f(x) = 2x \pmod 1$. This is an expanding map (with constants $\lambda = 2$ and $\delta_0 = \frac{1}{2}$), as $d(f(x), f(y)) = 2d(x, y)$ and $B(x, \frac{1}{2}) \cap f^{-1}(z)$ is a singleton whenever $d(f(x), z) < \frac{1}{2}$, for all $x, y, z \in \mathbb{S}^1$.

Consider an enumeration $(p_n)_{n \in \mathbb{N}}$ of the set of periodic points for f, in such a way that their sequence of periods is non-decreasing, and choose a sequence $(\eta_n)_{n \in \mathbb{N}}$ of positive real numbers converging quickly to zero in such a way that

$$
K = \mathbb{S}^1 \setminus \bigcup_{n \geq 0} \bigcup_{k \geq 0} f^{-k}(B(p_n, \eta_n))
$$

is a non-empty compact subset of $\mathbb{S}^1$, and thus $f(K) = K$.

We furthermore suppose that the sequence $(\eta_n)_{n \in \mathbb{N}}$ is chosen as follows. Since the periodic points of f are equidistributed in $\mathbb{S}^1$ (as f is semi-conjugated to the full shift on two symbols), for any $q_0 \in \mathbb{S}^1$ there exists $0 < \zeta_0 \ll 1$ so that $\bigcup_{k \geq 0} f^{-k}(B(q_0, \zeta_0)) \cap \mathrm{Per}(f) \neq \emptyset$. Set $q_0 = p_0 \in \mathrm{Per}(f)$. As all points in $\mathbb{S}^1$ have dense pre-orbits we conclude that

$$
K_0 = \mathbb{S}^1 \setminus \bigcup_{k \geq 0} f^{-k}(B(p_0, \zeta_0)) \supset K \neq \emptyset
$$

is a Cantor set. Let $n_1 = \inf\{\ell \geq 1 : p_\ell \in K_0\}$ and write $q_1 = p_{n_1}$. Choose $0 < \zeta_1 \ll \zeta_0$ such that $\bigcup_{k \geq 0} f^{-k}\left(B(q_0, \zeta_0) \cup B(q_1, \zeta_1)\right) \cap \mathrm{Per}(f) \neq \emptyset$. The previous condition can be assured by noting that any periodic point which intersects $B(q_0, \zeta_0) \cup B(q_1, \zeta_1)$ has combinatorics determined by either q_0 or q_1. Then

$$
K_1 = \mathbb{S}^1 \setminus \bigcup_{i=0}^{1} \bigcup_{k \geq 0} f^{-k}(B(q_i, \zeta_i)) \subset K_0
$$

is a Cantor set which does not contain any of the periodic points in the set $\{p_n : 0 \leq n \leq n_1\}$. Proceeding recursively, we obtain a strictly decreasing sequence $(\zeta_\ell)_\ell$ of positive real numbers, a strictly increasing sequence $(n_\ell)_\ell$ of positive integers and a nested sequence $(K_\ell)_\ell$ of Cantor sets such that

$$K_\ell = \mathbb{S}^1 \setminus \bigcup_{i=0}^{\ell} \bigcup_{k \geq 0} f^{-k}(B(q_i, \zeta_i))$$

contains some periodic point of f. Since the periodic points in $K_\ell \cap \mathrm{Per}(f)$ are dense K_ℓ we have that $f(K_\ell) = K_\ell$ for every $\ell \geq 1$. By construction, the set

$$K = \mathbb{S}^1 \setminus \bigcup_{i \geq 0} \bigcup_{k \geq 0} f^{-k}(B(q_i, \zeta_i)) \neq \emptyset$$

is a Cantor set having no periodic points and $f(K) = K$, as required.

Let us analyze the map $g := f \mid_K$. This is clearly an expansive, and distance expanding map. However, the following holds:

(a) g has no periodic points;
(b) g is not an open map;
(c) g is not an expanding map (i.e. condition (2) in Sect. 2.2 fails).

Property (a) is immediate from the construction. Property (b) follows because for every open, distance expanding map, periodic points are dense in the non-wandering set (see Corollaries 4.2.4 and 4.2.5 in [16]). Property (c) is a consequence of property (b), because of the equivalence between the notion of expanding with the notion of open, distance expanding on compact metric spaces (see [9, Lemma 1]). Moreover,

(d) g is locally eventually onto.

This property is not immediate for subshifts (see e.g., Example 2.5). In order to prove property (d) we will prove that each map $f \mid_{K_\ell}$ ($\ell \geq 1$) is locally eventually onto with uniform constants.

Fix any $\ell \geq 1$ and $\varepsilon > 0$. If $N = \lfloor \frac{-\log \varepsilon}{\log 2} \rfloor$ then f^N is a Markov map with 2^N full branches domains (injectivity domains), each of these with diameter larger than 2ε. In particular, $f^N(B(x, \varepsilon)) = \mathbb{S}^1$ for all $x \in \mathbb{S}^1$. We claim that

$$f^N\left(B(x, \varepsilon) \cap K_\ell\right) = K_\ell, \qquad \forall x \in \mathbb{S}^1. \tag{2.4.1}$$

The inclusion $\subseteq$ is immediate. For the converse inclusion $\supseteq$, without loss of generality we can choose $\varepsilon = 2^{-N}$ and thus $f^N \mid_{B(x,\varepsilon)}$ is a full branch for f^N Since K_ℓ has a dense set of the periodic points, every $z \in K_\ell$ is approximated by a sequence $(z_n)_n$ of periodic points in K_ℓ. In particular, each of the points in the set $f^{-N}(\{z\}) \cap B(x, \varepsilon)$ is an accumulation point of periodic points in K_ℓ. Since K_ℓ is compact, this assures that $f^{-N}(\{z\}) \cap B(x, \varepsilon) \in K_\ell$ and proves (2.4.1). Now, as the

Cantor sets are nested, one can use (2.4.1) to get

$$f^N\left(B(x,\varepsilon) \cap \bigcap_{\ell=1}^{n} K_\ell\right) = \bigcap_{\ell=1}^{n} K_\ell, \qquad \forall x \in \mathbb{S}^1, \ \forall n \geq 1.$$

This, together with the continuity of the map $P(\mathbb{S}^1) \ni A \mapsto f^N(A)$ (in the Hausdorff topology) implies that

$$f^N\left(B(x,\varepsilon) \cap K\right) = K, \qquad \forall x \in \mathbb{S}^1.$$

This proves that $g = f \mid_K$ is locally eventually onto.

Example 2.5 There are examples of strongly mixing subshifts which are not locally eventually onto. Indeed, Petersen [15] constructed a zero entropy, minimal and strongly mixing subshift $K \subset \{0, 1\}^{\mathbb{N}}$ We claim that the distance expanding map (K, σ) is not locally eventually onto.

Assume, by contradiction that (K, σ) is locally eventually onto. As K is compact, for any $\varepsilon > 0$ there exists $N = N(\varepsilon) \geq 1$ so that

$$\sigma^N(B_K(x, \varepsilon)) = K \quad \text{for every } x \in K. \tag{2.4.2}$$

This implies that for any $0 < \varepsilon < \frac{1}{2}\text{diam}(K)$ there exist points $x_1, x_2 \in K$ so that $d(\sigma^N(x_1), \sigma^N(x_2)) > \varepsilon$. Hence, if $s(n, \varepsilon)$ denote the maximal cardinality of (n, ε)-separated subsets of K, a recursive argument using (2.4.2) together with the observation that $\sigma^n(B(x, n, \varepsilon)) = B(\sigma^n(x), \varepsilon)$ for every $x \in K, n \geq 1$ and $\varepsilon > 0$ ensures that $s(kN, \varepsilon) \geq 2^k$ for every $k \geq 1$. Hence

$$h_{\text{top}}(\sigma \mid_K) = \lim_{\varepsilon \to 0} \limsup_{n \to \infty} \frac{1}{n} \log s(n, \varepsilon)$$

$$\geq \lim_{\varepsilon \to 0} \limsup_{k \to \infty} \frac{1}{kN} \log s(kN, \varepsilon) \geq \frac{1}{N} \log 2 > 0,$$

which leads to a contradiction. This proves that (K, σ) is not locally eventually onto.

The next simple example illustrates that compactness is an essential assumption in Theorem 1.2.

Example 2.6 The mixing and specification properties have been extensively studied in the case of symbolic dynamics (see e.g., [13, Section 8] and references therein). Here we give an example of a shift space, hence distance expanding, which is locally eventually onto, has dense periodic orbits but for which the specification property fails.

Consider the subshift $\Sigma_{\mathcal{G}} \subset \mathbb{N}^{\mathbb{N}}$ determined by the countable graph $\mathcal{G}$ with countable states $\mathbb{N}$ and whose allowed directed paths $v \to w$, $v, w \in \mathbb{N}$ are $0 \to w$ for every $w \in \mathbb{N}$, and the arrows $v \to w$ with $v \neq 0$ are admissible if and only if $w \in \{v - 1, v\}$. The cylinder sets are defined by $[v_0, v_1, \ldots, v_n] = \{(w_0, w_1, w_2, \ldots): w_i = v_i, \forall 0 \leq i \leq n\}$. The shift $\sigma : \Sigma_G \to \Sigma_G$ is locally eventually onto because

$$\sigma^{n+1}([v_0, v_1, \ldots, v_n]) \supset \sigma([1]) = \Sigma_G \quad \text{for every cylinder } [v_0, v_1, \ldots, v_n].$$

It is a simple exercise to show that the condition $\sigma^j([n]) \cap [1] = \emptyset$ for every $0 \leq j \leq n - 1$ is incompatible with the specification property.

In final example we prove an optimality of Theorem 1.2, in the sense that it fails if condition (2) in the definition of expanding map is removed.

Example 2.7 Let $\sigma : \{0, 1\}^{\mathbb{N}} \to \{0, 1\}^{\mathbb{N}}$ be the full shift, and let $g : [0] \to [0]$ be the first return map of σ to the cylinder $[0]$. More precisely, if $\tau : [0] \to \mathbb{N}$ is the first return time to $[0]$ given by

$$\tau(x_0, x_1, x_2, x_3, \ldots) = \inf\{k \geq 1: x_k = 0\}$$

then $g(\cdot) = \sigma^{\tau(\cdot)}(\cdot)$. Equivalently, if $x_0 = 0$ then

$$g(x_0, x_1, x_2, x_3, x_4, \ldots) = (x_k, x_{k+1}, x_{k+2}, \ldots)$$

where $k = \tau(x_0, x_1, x_2, x_3, \ldots)$. After identification of the set $\mathcal{S}$ of all finite words $(0, 1, 1, 1, \ldots, 1, 0)$ with the cylinder $[0, 1, 1, 1, \ldots, 1, 0] \subset \{0, 1\}^{\mathbb{N}}$, the map g acts as a full shift $\mathcal{S}^{\mathbb{N}}$. By the previous identification, we will consider $\mathcal{S}^{\mathbb{N}}$ as a subset of the cylinder $[0] \subset \{0, 1\}^{\mathbb{N}}$. This construction is often called the "Rome graph".

Let $\Sigma \subset \mathcal{S}^{\mathbb{N}}$ be the locally eventually onto subshift so that $g|_{\Sigma}: \Sigma \to \Sigma$ does not satisfy the specification property induced by Example 2.6. Indeed, just use the bijection

$$\mathbb{N} \to \mathcal{S} \quad \text{given by} \quad n \mapsto (0, \underbrace{1, 1, \ldots, 1}_{n}, 0) \tag{2.4.3}$$

to embed the subshift $\Sigma_G \subset \mathbb{N}^{\mathbb{N}}$ onto such subshift $\Sigma \subset \mathcal{S}^{\mathbb{N}}$. The possible unbounded amount of 1's in (2.4.3) makes the subshift $\Sigma \subset \mathcal{S}^{\mathbb{N}}$ not closed. Now, consider the σ-invariant and compact set $K \subset \{0, 1\}^{\mathbb{N}}$ obtained as the closure of the saturated set

$$\bigcup_{n \geq 1} \bigcup_{j=0}^{n-1} \sigma^j(\{w \in \Sigma: \tau(w) = n\}).$$

By construction and the fact that K is closed we get

$$K \cap [0] = \text{closure}(\mathcal{S}^{\mathbb{N}}) = \Sigma \cup \{01^{\infty}\}$$

and

$$K \cap [1] = \text{closure}\left(\bigcup_{n \geq 1} \bigcup_{j=1}^{n-1} \sigma^j \left(\{w \in \Sigma : \tau(w) = n\}\right)\right)$$

$$= \left(\bigcup_{n \geq 1} \bigcup_{j=1}^{n-1} \sigma^j \left(\{w \in \Sigma : \tau(w) = n\}\right)\right) \cup \{1^\infty\}$$

Note that the elements in $\{01^\infty, 1^\infty\}$ do not return to the cylinder $[0]$. Thus, using that $K \setminus \{01^\infty, 1^\infty\}$ is obtained by the union of the finite pieces of orbits of points in Σ until their first return time to Σ, the distance expanding map $(K, \sigma \mid_K)$ does not satisfy the specification property.

We claim that (K, σ) is locally eventually onto. Dealing with the induced topology, it is enough to prove that for any cylinder $[x_1, x_2, \ldots, x_n] \cap K \neq \emptyset$ there exists $N \geq 1$ so that $\sigma^N([x_1, x_2, \ldots, x_n]) = \{0, 1\}^{\mathbb{N}}$. This is a consequence of the fact that g is a Poincaré first return map of σ to the global cross-section $[0]$ and that g is locally eventually onto (recall Example 2.6). Then Theorem 1.2 implies that the long inverse branches condition (2) in the definition of expanding map fails.

Acknowledgments The authors are deeply grateful to A. Blokh and F. Przytycki for very useful comments on an early version of the paper. PV was partially supported by CMUP (UIDB/00144/2020), which is funded by FCT (Portugal) with national (MEC) and European structural funds through the programs FEDER, under the partnership agreement PT2020, and by Fundação para a Ciência e Tecnologia (FCT)—Portugal, through the grant CEECIND/03721/2017 of the Stimulus of Scientific Employment, Individual Support 2017 Call.

References

1. E. Akin, J. Auslander, A. Nagar, Variations on the concept of topological transitivity. Studia Math. **235**(3), 225–249 (2016)
2. M. Barge, J. Martin, Dense orbits on the interval. Michigan Math. J. **34**(1), 3–11 (1987)
3. F. Blanchard, β-expansions and symbolic dynamics. Theoret. Comput. Sci. **65**(2), 131–141 (1989)
4. A. Blokh, On transitive maps of one-dimensional branched manifolds. Differ. Equ. Probl. Math. Phys. Kiev **131**, 3–9 (1984)
5. A. Blokh, The spectral decomposition for one-dimensional maps. *Dynamics Reported* , vol. 4 (Springer, Berlin, 1995), pp. 1–59
6. R. Bowen, Periodic points and measures for Axiom A diffeomorphisms. Trans. Amer. Math. Soc. **154**, 377–397 (1971)
7. J. Buzzi, Specification of the interval. Trans. Amer. Math. Soc. **349**, 2737–2754 (1997)
8. E. Coven, I. Mulvey, Transitivity and the centre for maps of the circle. Ergod. Th. Dynam. Sys. **6**(1), 1–8 (1986)
9. E. Coven, W. Reddy, Positively expansive maps of compact manifolds, in *Global Theory of Dynamical Systems*, ed. by Z. Nitecki , C. Robinson. Lecture Notes in Mathematics, vol. 819 (Springer, Berlin, 1980)

10. M. Denker, C. Grillenberger, K. Sigmund, *Ergodic Theory on Compact Spaces*. Lecture Notes in Mathematics, vol. 527 (Springer, Berlin, 1976), iv+360 pp
11. A. Katok, B. Hasselblatt, *Introduction to the Modern Theory of Dynamical Systems* (Cambridge University Press, Cambridge, 1995)
12. M. Kulczycki, D. Kwietniak, P. Oprocha, On almost specification and average shadowing properties. Fund. Math. **224**(3), 241–278 (2014)
13. D. Kwietniak, M. Łacka, P. Oprocha, A panorama of specification-like properties and their consequences. Dyn. Numbers Contemp. Math. **669**, 155–186 (2016)
14. K. Oliveira, Every expanding measure has the nonuniform specification property. Proc. Amer. Math. Soc. **140**(4), 1309–1320 (2006)
15. K. Petersen, A topologically strongly mixing symbolic minimal set. Trans. Amer. Math. Soc. **148**(2), 603–612 (1970)
16. F. Przytycki, M. Urbański, *Conformal Fractals: Ergodic Theory Methods*. London Mathematical Society Lecture Note Series, vol. 371 (Cambridge University Press, Cambridge, 2010), x+354 pp
17. S. Ruette, *Chaos on the Interval*. University Lecture Series, vol. 67 (American Mathematical Society, Providence, 2017), xii+215 pp
18. J. Schmeling, Symbolic dynamics for β-shifts and self-normal numbers. Ergod. Th. Dyn. Sys. **17**(3), 675–694 (1997)
19. P. Varandas, Non-uniform specification and large deviations for weak Gibbs measures. J. Stat. Phys. **146**, 330–358 (2012)
20. Q. Yan, J. Yin, T. Wang, Some weak specification properties and strongly mixing. Chin. Ann. Math. Ser. B **38**(5), 1111–1118 (2017)

Part II
Low Dimensional Dynamics and Thermodynamics Formalism

Chapter 3
Thermodynamic Formalism and Geometric Applications for Transcendental Meromorphic and Entire Functions

Volker Mayer and Mariusz Urbański

Abstract In this survey we deal with transcendental meromorphic and entire functions. We thoroughly discuss in this context close relations between topological pressure and conformal measures for geometric potentials. For two more special classes of meromorphic functions, namely dynamically semi-regular and those in the class $\mathcal{D}$ that have negative spectrum, we then also discuss, by means of the appropriate transfer (or Perron-Frobenius) operator, the corresponding thermodynamic formalism for such potentials. It holds in its full classical version. At the end of the survey we discuss hyperbolic dimension, Bowen's Formula, and real analyticity of Hausdorff dimension.

3.1 Introduction

Originating from statistical physics, the dynamical theory of thermodynamic formalism was brought to mathematics, particularly to study expanding and hyperbolic dynamical systems, primarily by Bowen [20], Ruelle [85], Sinai [89], and Walters [108] in the 1970s. This theory provides an excellent framework for probabilistic description of the chaotic part of the dynamics and, in the context of smooth (particularly conformal) expanding/hyperbolic dynamical systems, gives a rich and detailed information about the geometry of expanding repellers, limit sets of Kleinian groups and iterated function systems, and Julia sets of holomorphic dynamical systems. More precisely, by establishing the existence and uniqueness of Gibbs and equilibrium states, and studying spectral and asymptotic properties

V. Mayer
Université de Lille, Département de Mathématiques, UMR 8524 du CNRS, Villeneuve d'Ascq Cedex, France
e-mail: volker.mayer@univ-lille.fr

M. Urbański (✉)
Department of Mathematics, University of North Texas, Denton, TX, USA
e-mail: urbanski@unt.edu; www.math.unt.edu/~urbanski

© The Author(s), under exclusive license to Springer Nature Switzerland AG 2021
M. Pollicott, S. Vaienti (eds.), *Thermodynamic Formalism*, Lecture Notes in Mathematics 2290, https://doi.org/10.1007/978-3-030-74863-0_3

of corresponding Perron-Frobenius operators, it permits to show that dynamical systems are "strongly" mixing (K-mixing, weak Bernoulli), have exponential decay of correlations, satisfy the Invariant Principle Almost Surely, in particular satisfy the Central Limit Theorem, and the Law of Iterated Logarithm. Furthermore, by studying the topological pressure function of geometric potentials, particularly its regularity properties (real analyticity, convexity), this theory gives a precise information about the fractal geometry of Julia and limit sets. Particularly, R. Bowen initially showed in [21] that the Hausdorff dimension of the limit set of a co-compact quasi–Fuchsian group is given by the unique zero of the appropriate pressure function. His result and its numerous versions commonly bear the name of Bowen's Formula ever since. Bowen's results easily carry through to the case of expanding (hyperbolic) rational functions providing a closed formula for the Hausdorff dimension of their Julia sets. D. Ruelle, positively answering a conjecture of D. Sullivan (see [96]–[101]), proved in [86] that this dimension depends in a real analytic way on the function.

For hyperbolic, and even much further beyond, rational functions, and more general distance expanding maps, the theory of thermodynamic formalism is now well developed and established, and its systematic account can be found in [78] (see also [49, 66, 109, 111]). The present text concerns transcendental entire and meromorphic functions. For these classes of functions many differences and new phenomena pop up that do not occur in the case of rational maps. The following two properties of transcendental functions show from the outset that the outlook of these classes is indeed totally different than the one of rational functions.

– Whereas the singularities of hyperbolic rational maps stay away form their Julia sets, for transcendental functions one always has to deal with the singularity at infinity.
– Transcendental functions have infinite degree.

One immediate consequence of the later fact is that for transcendental functions there is no measure of maximal entropy, which is one of the central objects in the theory of rational functions. Particularly for polynomials, where this measure coincides with harmonic measure viewing from infinity, and also for endomorphisms of higher dimensional projective spaces. Another consequence is that all Perron–Frobenius, or transfer, operators of a transcendental meromorphic functions are always defined by an infinite series. This is the reason that, even for such classical functions as exponential ones $f_\lambda(z) = \lambda e^z$, this operator taken in its most natural sense, is not even well-defined.

K. Barański first managed to overcome these difficulties and presented a thermodynamical formalism for the tangent family in [6]. Expanding the ideas from [6] led to [45], where Walters expanding maps and Barański maps were introduced and studied. One important feature of the maps treated in [6] and [45] was that all analytic inverse branches were well-defined at all points of Julia sets. This property dramatically fails for example for entire functions as $f_\lambda(z) = \lambda e^z$ (there are no well-defined inverse branches at infinity). To remedy this situation, the periodicity of f_λ was exploited to project the dynamics of these functions down to the cylinder and the appropriate thermodynamical formalism was developed in [104] and [105]. This approach has been adopted to other periodic transcendental functions; besides the papers cited above, see also [22–24, 47, 55, 106] and the survey [48].

The first general theory of thermodynamic formalism for transcendental mero-morphic and entire functions was laid down in the year 2008 in [56]. Mayer and Urbański [58] contains a complete treatment of this approach. It handled all the periodic functions cited above in a uniform way and went much farther beyond. The most important key point in these two papers was to replace the standard Euclidean metric by an appropriate Riemannian metric. Then the power series defining the Perron–Frobenius operators of geometric potentials becomes comparable to the Borel series and can be controlled by means of Nevanlinna's value distribution theory.

For a large class of transcendental entire functions whose set of singularities is bounded, quite an optimal approach to thermodynamic formalism was laid down and developed in [60].

It was observed in this paper that, for these entire functions, the transfer operator entirely depends on the geometry of the logarithmic tracts, in fact on the behavior of the boundary of the tracts near infinity. The best way to deal with the often fractal behavior of the tracts near infinity was by adapting the concept of integral means, a classical and powerful tool in the theory of conformal mappings.

This text provides an overview of the (geometric) thermodynamic formalism for transcendental meromorphic and entire functions with particular emphasis on geometric/fractal aspects such as Bowen's Formula expressing the hyperbolic dimension as a unique zero of a pressure function and the behavior of the latter when the transcendental functions vary in an analytic family.

There are some several important and interesting topics closely related to the subject matter of our exposition that will nevertheless not be treated at all or will be merely briefly mentioned in our survey. For example, this exposition only briefly indicates that thermodynamic formalism has been successfully developed for random transcendental dynamical systems; see [59, 63], comp. also [107] for non–hyperbolic random dynamics of transcendental functions. Non–hyperbolic functions will not be in the focus of our current exposition either but we would like to bring reader's attention to some relevant papers that include [57, 106] and [107]. Discussing all these topics at length and detail would increase the length of our survey substantially, making it too long, and would lead us too far beyond of what we intended to focus on in the current survey.

We would like to thank the referee for his valuable remarks which influenced the final version of the paper.

3.2 Notation

Frequently we have to replace Euclidean metric by some other Riemannian metric $d\sigma = \gamma \, |dz|$. A natural choice is the spherical metric in which case the density with respect to Euclidean metric is $\gamma(z) = 1/(1 + |z|^2)$. More generally, we consider metrics of the form

$$d\sigma(z) = d\sigma_\tau(z) = \frac{|dz|}{1 + |z|^\tau} \quad , \quad \tau \geq 0. \qquad (3.2.1)$$

They vary between euclidean and spherical metrics when $\tau \in [0, 2]$. If such a metric is used only away from the origin, then one can use the simpler form

$$d\tau(z) = |z|^{-\tau}|dz|. \qquad (3.2.2)$$

We denote by $D_\sigma(z, r)$ the open disk with center z and radius r with respect metric σ. If σ is the spherical metric then this disk is also denoted by $D_{sph}(z, r)$ and for the standard euclidean metric $\mathbb{D}(z, r)$. We also denote

$$\mathbb{D}_R = \mathbb{D}(0, R)$$

and

$$\mathbb{D}_R^* = \mathbb{C} \setminus \overline{\mathbb{D}}(z, R).$$

The symbol

$$A(r, R) := \mathbb{D}_R \setminus \overline{\mathbb{D}}_r$$

is used to denote the annulus centered at 0 with the inner radius r and the outer radius R.

The derivative of a function f with respect to a Riemannian metric $d\sigma = \gamma \, |dz|$ is given by

$$|f'(z)|_\sigma = \frac{d\sigma(f(z))}{d\sigma(z)} = |f'(z)|\frac{\gamma(f(z))}{\gamma(z)}. \qquad (3.2.3)$$

When the metric σ has the form (3.2.1) or (3.2.2) then $d\sigma$ only depends on τ and we will identify σ and τ and write $|f'(z)|_\tau$ instead of $|f'(z)|_\sigma$. Therefore,

$$|f'(z)|_\tau = \frac{|f'(z)|}{1 + |f(z)|^\tau}(1 + |z|^\tau) \quad \text{and} \quad |f'(z)|_\tau = \frac{|f'(z)|}{|f(z)|^\tau}|z|^\tau$$

in the case of the simpler form (3.2.2). When $\tau = 2$ then we also write $|f'(z)|_{sph}$.

Besides this, we use common notation such as $\mathbb{C}$ and $\hat{\mathbb{C}}$ for the Euclidean plane and the Riemann sphere respectively. Another common notation is

$$A \asymp B.$$

As usually, it means that the ratio A/B is bounded below and above by strictly positive and finite constants that do not depend on the parameters involved. The corresponding inequalities up to a multiplicative constant are denoted by

$$A \preceq B \quad \text{and} \quad A \succeq B.$$

Also,

$$\text{dist}(E, F)$$

denotes the Euclidean distance between the sets $E, F \subset \mathbb{C}$.

3.3 Transcendental Functions, Hyperbolicity and Expansion

We consider transcendental entire or meromorphic functions. Such a function $f : \mathbb{C} \to \hat{\mathbb{C}}$ can have two types of singularities: asymptotical and critical values. We refer to [17] for the classification of the different types of singularities, known as Iversen's classification, denote by $S(f)$ the closure of the set of critical values and finite asymptotic values of f.

Transcendental functions are very general and one is led, actually forced, to consider reasonable subclasses. The class $\mathcal{B}$ of bounded type functions consists of all meromorphic functions for which the set $S(f)$ is bounded. Bounded type entire functions have been introduced and studied in [37], $\mathcal{B}$ is also called the Eremenko–Lyubich class. It contains an important subclass, called Speiser class, which consists of all meromorphic functions for which the set $S(f)$ is finite.

3.3.1 Dynamical Preliminaries

For a general introduction of the dynamical aspects of meromorphic functions we refer to the survey article of Bergweiler [13] and the book [49]. We collect here some of its properties, primarily the ones we will need in the sequel. The Fatou set of a meromorphic function $f : \mathbb{C} \to \hat{\mathbb{C}}$ is denoted by $F(f)$. It is defined as usually to be the set of all points $z \in \mathbb{C}$ for which there exists a neighborhood U of z on which all the iterates f^k, $k \geq 1$, of the function f are defined and form a normal family. The complement of this set is the Julia set $\hat{\mathcal{J}}(f) = \hat{\mathbb{C}} \setminus F(f)$. We write

$$J(f) = \hat{\mathcal{J}}(f) \cap \mathbb{C}.$$

By Picard's theorem, there are at most two points $\xi \in \hat{\mathbb{C}}$ that have finite backward orbit $O^-(\xi) = \bigcup_{n \geq 0} f^{-n}(z_0)$. The set of these points is the exceptional set

$\mathcal{E}_f$. In contrast to the case of rational maps it may happen that $\mathcal{E}_f \subset \hat{\mathcal{J}}(f)$. Iversen's theorem [40, 68] asserts that every point $\xi \in \mathcal{E}_f$ is an asymptotic value. Consequently, $\mathcal{E}_f$ is contained in $S(f)$. the set of critical and finite asymptotic values of f. The post-critical set $\mathcal{P}(f)$ is defined to be the closure in the complex plane $\mathbb{C}$ of

$$\bigcup_{n \geq 0} f^n \big(S(f) \setminus f^{-n}(\infty) \big).$$

This set can contain the whole Julia set.

Definition 3.1 If $J(f) \setminus \mathcal{P}(f) \neq \emptyset$ then f is called tame.

The Julia set contains several dynamically important subsets. First, there is the *escaping set*

$$\mathcal{I}(f) = \{z \in \mathbb{C}; \ f^n(z) \text{ is defined for all } n \text{ and } \lim_{n \to \infty} f^n(z) = \infty\}.$$

This set is not always a subset of the Julia set, it may contain Baker domains. However, for entire functions of bounded type $\mathcal{I}(f) \subset J(f)$ ([37, Theorem 1]). More important for us is the following set.

Definition 3.2 The *radial (or conical)* Julia set $J_r(f)$ of f is the set of points $z \in J(f)$ such that there exist $\delta > 0$ and an unbounded sequence $(n_j)_{j=1}^{\infty}$ of positive integers such that the sequence $\big(|f^{n_j}(z)|\big)_{j=1}^{\infty}$ is bounded above and the map

$$f^{n_j} : U_j \longrightarrow \mathbb{D}(f^{n_j}(z), \delta)$$

is conformal, where U_j is the connected component of $f^{-n_j}(\mathbb{D}(f^{n_j}(z), \delta))$ containing z.

There are other definitions of radial sets in the literature. While the present definition is in the spirit of the one from [91], the radial points in [79] are defined by means of spherical disks. Namely, $z \in J_r^{sph}(f)$ if $z \in J(f)$ if there exist $\delta > 0$ and an unbounded sequence $(n_j)_{j=1}^{\infty}$ of positive integers such that

$$f^{n_j} : U_j \longrightarrow D_{sph}(f^{n_j}(z), \delta)$$

is conformal where U_j is the connected component of $f^{-n_j}(D_{sph}(f^{n_j}(z), \delta))$ containing z. Right from these definitions it is easy to see that $J_r(f) \subset J_r^{sph}(f)$. Also,

$$J_r(f) \subset J(f) \setminus \mathcal{I}(f). \tag{3.3.1}$$

The differences between all of these radial sets are dynamically insignificant in the sense that they all have the same Hausdorff dimension and this dimension coincides with the hyperbolic dimension, which we define right now.

Definition 3.3 The hyperbolic dimension of a meromorphic function $f : \mathbb{C} \to \hat{\mathbb{C}}$, denoted by $HD_{hyp}(f)$, is

$$HD_{hyp}(f) = \sup_K HD(K)$$

where the supremum is taken over all hyperbolic sets $K \subset \mathbb{C}$, i.e. over all compact sets $K \subset \mathbb{C}$ such that $f(K) \subset K$ and $f|_K$ is expanding.   □

Lemma 3.4 $HD_{hyp}(f) = HD(J_r(f)) = HD(J_r^{sph}(f))$.   □

Proof Let K be a hyperbolic set. Then, following [78, Section 5] especially Lemma 5.1.1, there exists $\eta > 0$ such that

$$f_{|\mathbb{D}(z,\eta)} \text{ injective and } \quad f(\mathbb{D}(z, \eta)) \supset \mathbb{D}(f(z), \eta) \quad \text{for all } z \in K.$$

This shows that $K \subset J_r(f)$ and thus $HD_{hyp}(f) \leq HD(J_r(f))$. Since $J_r(f) \subset J_r^{sph}(f)$ we also have $HD(J_r(f)) \leq HD(J_r^{sph}(f))$. The conclusion comes now from the result in [79] which says that $HD(J_r^{sph}(f)) = HD_{hyp}(f)$.   □

3.3.2 Hyperbolicity and Expansion

There are several notions of hyperbolic transcendental functions in the literature (see for example [110]). The following definition is used fairly frequently.

Definition 3.5 A meromorphic function $f : \mathbb{C} \to \hat{\mathbb{C}}$ is called hyperbolic if and only if

$$\mathcal{P}(f) \text{ is bounded and } \mathcal{P}(f) \cap J(f) = \emptyset. \tag{3.3.1}$$

Notice that then $f \in \mathcal{B}$, i.e. it is of bounded type. The following notion has been used by G. Stallard in [92]. Later it was considered in [56, 58] where the, somehow misleading, name topologically hyperbolic was used. Since it is based on the euclidean distance, let us call it *Euclidean hyperbolic* here.

Definition 3.6 A meromorphic function f is called E-*hyperbolic* if

$$\text{dist}(J(f), \mathcal{P}(f)) > 0.$$

Clearly, a hyperbolic function is E-hyperbolic but the later notion is much more general. For example, the function $f(z) = 2 - \log 2 + 2z - e^z$ is E-hyperbolic and has a Baker domain (see [14]). Other examples arise naturally in the context of Newton maps. This has been observed in [12] and $f(z) = z - \tan z$ is a different exemple of E-hyperbolic function that is not hyperbolic.

For E-hyperbolic functions every non-escaping point of the Julia set is a radial point. Together with (3.3.1) it follows that in this case we have equality between these type of points:

$$J_r(f) = J(f) \setminus I(f). \tag{3.3.2}$$

For rational functions, E-hyperbolicity is equivalent to the property of being expanding.

Definition 3.7 A meromorphic function $f : \mathbb{C} \to \hat{\mathbb{C}}$ is called *expanding* if and only if there are two constants $c > 0$ and $\gamma > 1$ such that

$$|(f^n)'(z)| \geq c\gamma^n \quad \text{for all } z \in J(f) \setminus f^{-n}(\infty) .$$

The function $f(z) = z - \tan z$ is E-hyperbolic and $|f'(z)| \to 1$ as $\Im f(z) \to \infty$. Since there are vertical lines in $J(f)$, it follows that this function is not expanding. Thus, contrary to the case of rational maps, E-hyperbolicity and expanding are not equivalent for transcendental functions. It is shown in [92] that every entire E-hyperbolic function f satisfies $\lim_{n\to\infty} |(f^n)'(z)| \to \infty$ for all $z \in \mathcal{J}_f$ and under some conditions the expanding property follows from E-hyperbolicity (see Proposition 4.4 in [58]).

Example 3.8 Let $0 < c < 1/e^3$. Then the Fatou function $f(z) = z - \log c + e^{-z}$ is not hyperbolic but it is E-hyperbolic and expanding.

In order to verify this statement, we recall the classical argument that f is semi-conjugate via $w = e^{-z}$ to the map $g(w) = cwe^{-w}$ (see for example [82]). By the choice of the constant $0 < c < 1/e^3$, the origin is an attracting fixed point of the map g and a simple estimation allows to check that

$$g(\overline{\mathbb{D}}_3) \subset \mathbb{D}_3. \tag{3.3.3}$$

Consequently, the half space $\{\Re z \geq -\log 3\}$ is contained in a Baker domain of f and the Julia set $J(f) \subset \{\Re z < -\log 3\}$. Now, a simple estimate shows that

$$|f'(z)| \geq 2 \quad \text{for all } z \text{ with } \Re z \leq -\log 3 .$$

Consequently f is expanding on its Julia set.

It remains to check that f is E-hyperbolic. The function f has no finite asymptotic value and its critical points are $c_k = 2\pi i k$, $k \in \mathbb{Z}$. It follows from (3.3.3) that there exists $\rho > -\log 3$ such that $\Re f^n(c_k) \geq \rho$ for every $n \geq 0$ and $k \in \mathbb{Z}$. This shows that f is indeed E-hyperbolic .

3.3.3 Disjoint Type Entire Functions

For entire functions there is a relevant strong form of hyperbolicity called *disjoint type*, the notion that first implicitly appeared in [7] and then was explicitly studied in several papers including [80, 84]. Disjoint type functions are of bounded type. So, let $f \in \mathcal{B}$ be an entire function and let $R > 0$ such that $S(f) \subset \mathbb{D}_R$. Up to normalization we can assume that $R = 1$. Then $f^{-1}(\mathbb{D}^*)$ consists of countably many mutually Jordan domains Ω_j with real analytic boundaries such that $f : \Omega_j \to \mathbb{D}^*$ is a covering map (see [37]). In terms of the [37]). In terms of the classification of singularities, this means that f has only *logarithmic singularities* over infinity. These connected components of $f^{-1}(\mathbb{D}^*)$ are called *tracts* and the restriction of f to any of these tracts Ω_j has the special form

$$f_{|\Omega_j} = \exp \circ \tau_j \quad \text{where} \quad \varphi_j = \tau_j^{-1} : \mathcal{H} = \left\{ z \in \mathbb{C} : \Re(z) > 0 \right\} \longrightarrow \Omega_j \quad (3.3.1)$$

is a conformal map. Later on we often assume that f has only finitely many tracts:

$$f^{-1}(\mathbb{D}^*) = \bigcup_{j=1}^{N} \Omega_j . \quad (3.3.2)$$

Notice that this is always the case if the function f has finite order. Indeed, if f has finite order then the Denjoy–Carleman–Ahlfors Theorem (see [67, p. 313]) states that f can have only finitely many direct singularities and so, in particular, only finitely many logarithmic singularities over infinity.

Definition 3.9 If $f \in \mathcal{B}$ is entire such that

$$S(f) \subset \mathbb{D} \quad \text{and} \quad \bigcup_j \overline{\Omega}_j \cap \overline{\mathbb{D}} = \emptyset \ , \quad \text{equivalently } f^{-1}(\overline{\mathbb{D}^*}) = \bigcup_j \overline{\Omega}_j \subset \mathbb{D}^*,$$

$$(3.3.3)$$

then f is called a *disjoint type* function.

This definition is not the original one but it is consistent with the disjoint type models in Bishop's paper [19]. The function f is then indeed of disjoint type in the sense of [7, 80, 84]. It is well known that for every $f \in \mathcal{B}$ the function λf, $\lambda \in \mathbb{C}^*$, is of disjoint type provided λ is small enough (see [8] and [80, p. 261]). Also, the Julia set of a disjoint type entire function is a subset of its tracts and therefore only the restriction of the function to these tracts is relevant for the study of dynamics of such a function near the Julia set.

Besides functions of class $\mathcal{S}$ and $\mathcal{B}$ we consider the following subclass of bounded type entire functions called class $\mathcal{D}$. In this definition,

$$Q_T = \left\{ 0 < \Re z < 4T \ ; \ -4T < \Im z < 4T \right\} \ , \quad T > 0. \quad (3.3.4)$$

Definition 3.10 An entire function $f : \mathbb{C} \to \mathbb{C}$ belongs to class $\mathcal{D}$ if it is of disjoint type, has only finitely many tracts (see (3.3.2)) and if, for every tract, the function φ of (3.3.1) satisfies

$$|\varphi(\xi)| \le M|\varphi(\xi')| \quad \text{for all} \quad \xi, \xi' \in Q_T \setminus Q_{T/8}, \tag{3.3.5}$$

for some constant $M \in (0, +\infty)$ and every $T \ge 1$.

3.4 Topological Pressure and Conformal Measures

This section is devoted to two crucial objects: the topological pressure and conformal measures. Compared to the case of rational functions, they both behave totally differently in the context of transcendental functions. For example, since transcendental functions have infinite degree, the topological pressure evaluated at zero is always infinite. Also, the existence of the pressure and, even more importantly, of conformal measures is not known in full generality for meromorphic functions.

3.4.1 Topological Pressure

A standard argument, based on mixing properties (see Lemma 5.8 in [58]), shows that for a E-hyperbolic meromorphic function $f : \mathbb{C} \to \mathbb{C}$ the following number, which might be finite or infinite,

$$P_\tau(t) := \limsup_{n \to \infty} \frac{1}{n} \log \sum_{f^n(z)=w} |(f^n)'(z)|_\tau^{-t} \tag{3.4.1}$$

does not depend on the point $w \in J(f)$. However, this number may depend on the metric τ and it clearly depends on the parameter $t > 0$.

Definition 3.11 Let $f : \mathbb{C} \to \mathbb{C}$ be E-hyperbolic meromorphic function. The topological pressure of f evaluated at $t > 0$ with respect to the metric τ as defined in (3.2.1) is the (possibly infinite) number $P_\tau(t)$ defined by formula (3.4.1). When $\tau = 2$, i.e. $d\sigma_\tau$ is the spherical metric, then we also write $P_\tau(t) = P_{sph}(t)$.

Given a meromorphic function $f : \mathbb{C} \to \mathbb{C}$, a number $\tau \ge 0$, and a parameter $t \ge 0$, we say that the topological pressure $P_\tau(t)$ exists if the number defined by formula (3.4.1) is independent of w for some "sufficiently large" set of points $w \in J(f)$.

The most general result on the existence of topological pressure going beyond E-hyperbolic functions is due to Barański et al. [10]. They work with spherical metric and call a meromorphic function f *exceptional* if and only if it has a (Picard) exceptional value a in the Julia set and f has a non-logarithmic singularity over a.

Theorem 3.12 ([10]) *Let $f : \mathbb{C} \to \mathbb{C}$ be either a meromorphic function in class $\mathcal{S}$ or a non–exceptional and tame function in class $\mathcal{B}$. Then the limit*

$$P_{sph}(t) := \lim_{n \to \infty} \frac{1}{n} \log \sum_{f^n(z) = w} |(f^n)'(z)|_{sph}^{-t}$$

exists (possibly equal to infinity) for all $t > 0$ and does not depend on w where w is a good pressure starting point $w \in \mathbb{C}$ whose precise meaning is given in [10, Section 4].

If f is tame then every $w \in J(f) \setminus \mathcal{P}(f)$ is such a good point. Also, if $f \in \mathcal{B}$ is E-hyperbolic, then each point $w \in J(f)$ is good.   □

It is also shown in [10] that the pressure function has the usual natural properties.

Proposition 3.13 ([10]) *Under the assumptions of Theorem 3.12, $P_{sph}(0) = +\infty$ and $P_{sph}(2) \leq 0$, and thus*

$$\Theta_{sph} := \inf\{t > 0 \ : \ P_{sph}(t) < \infty\} \in [0, 2] \tag{3.4.2}$$

In addition:

$$P_{sph}(t) = +\infty \quad \text{for all} \quad t < \Theta_{sph} \quad \text{and} \quad P_{sph}(t) < +\infty \quad \text{for all} \quad t > \Theta_{sph}.$$

The resulting function

$$(\Theta_{sph}, \infty) \ni t \longmapsto P_{sph}(t)$$

is non-increasing and convex, hence continuous.   □

Notice that this result does not provide any information about the behavior of the pressure function at the critical value $t = \Theta_{sph}$. For classical families, such as the exponential family, the pressure at Θ_{sph} is infinite. Curious examples of functions that behave differently at the critical value are provided in [61]. We will come back to such examples later in Theorem 3.37.

3.4.2 Conformal Measures and Transfer Operator

Conformal measures were first defined and introduced by Samuel Patterson in his seminal paper [70] (see also [71]) in the context of Fuchsian groups. Dennis Sullivan extended this concept to all Kleinian groups in [98, 99, 101]. He then, in the papers [96, 97, 100], defined conformal measures for all rational functions of the Riemann sphere $\hat{\mathbb{C}}$. He also proved their existence therein. Both Patterson and Sullivan came up with conformal measures in order to get an understanding of geometric measures, i.e. Hausdorff and packing ones. Although already Sullivan noticed that there are conformal measures for Kleinian groups that are not equal, nor even equivalent, to any Hausdorff and packing (generalized) measure, the main purpose to deal with them is still to understand Hausdorff and packing measures but goes beyond.

Conformal measures, in the sense of Sullivan have been studied in greater detail in [25], where, in particular, the structure of the set of their exponents was examined. We do this for our class of transcendental functions.

Since then conformal measures in the context of rational functions have been studied in numerous research works. We list here only very few of them appearing in the early stages of the development of their theory: [26, 30, 31]. Subsequently the concept of conformal measures, in the sense of Sullivan, has been extended to countable alphabet iterated functions systems in [52] and to conformal graph directed Markov systems in [53]. It was furthermore extended to transcendental meromorphic dynamics in [45, 104], and [56]. See also [58, 105], and [11]. Lastly, the concept of conformal measures found its place also in random dynamics; we cite only [59, 62], and [107].

Definition 3.14 Let $f : \mathbb{C} \to \hat{\mathbb{C}}$ be a meromorphic function. A Borel probability measure m_t on $J(f)$ is called $\lambda |f'|_\tau^t$-conformal if

$$m_t(f(E)) = \int_E \lambda |f'|_\tau^t dm_t.$$

for every Borel $E \subset J(f)$ such that the restriction $f_{|E}$ is injective. The scalar λ is called the conformal factor and, if $\lambda = 1$, then m_t is called a t-conformal measure.

If f has a $\lambda |f'|_\tau^t$-conformal measure m_t and if f is E-hyperbolic then, using Koebe's Distortion Theorem, we get for all $w \in J(f)$ that

$$1 \geq \sum_{z \in f^{-1}(w)} m_t(U_z) \asymp \lambda^{-1} m_t(\mathbb{D}(w, r)) \sum_{z \in f^{-1}(w)} |f'(z)|_\tau^{-t}, \tag{3.4.1}$$

where for every $z \in f^{-1}(w)$, U_z is the connected component of $f^{-1}(\mathbb{D}(w, r))$ containing z. Consequently, the series on the right hand side of (3.4.1) is well defined. This allows us to introduce the corresponding transfer, or Perron–Frobenius–Ruelle, operator. Its standard definition as an operator acting on the space $C_b(J(f))$ of continuous bounded functions on the Julia set $J(f)$ is the following.

Definition 3.15 Let $f : \mathbb{C} \to \hat{\mathbb{C}}$ be a E-hyperbolic meromorphic function. Fix $\tau > 0$ and $t > 0$. The transfer operator of f with (geometric) potential $\psi := -t \log |f'(z)|_\tau, t > 0$, is defined by

$$\mathcal{L}_t g(w) := \sum_{f(z)=w} e^{\psi(z)} g(z) = \sum_{f(z)=w} |f'(z)|_\tau^{-t} g(z), \quad w \in J(f), g \in C_b(J(f)).$$

$$\tag{3.4.2}$$

Note that (3.4.1) does not imply boundedness of the linear operator $\mathcal{L}_t$. This crucial issue will be discussed in the next section. Let us simply mention here that, iterating the inequality (3.4.1) (which is possible if we assume f to be E-hyperbolic) shows that we have the following relation between the pressure and the conformal factor λ:

$$P_\tau(t) \leq \log \lambda. \tag{3.4.3}$$

On the other hand, if the transfer operator, in fact its adjoint operator $\mathcal{L}_t^*$, is well defined, then m_t being a conformal measure equivalently means that m_t is an eigenmeasure of $\mathcal{L}_t^*$ with eigenvalue λ:

$$\mathcal{L}_t^* m_t = \lambda m_t.$$

As defined, the measure $m_t = m_{\tau,t}$ does depend on the metric τ. Given $\tau' \neq \tau$ and corresponding Riemannian metrics (see (3.2.2)), we then have

$$\frac{dm_{\tau',t}}{dm_{\tau,t}}(z) = \frac{|z|^{(\tau-\tau')t}}{\int_{J(f)} |\xi|^{(\tau-\tau')t} dm_{\tau,t}(\xi)} \tag{3.4.4}$$

provided the above integral is finite. For example, this allows one to get spherical conformal measures as soon as we have conformal measures $m_{\tau,t}$ for a τ-metric with $\tau \leq 2$; this will be the case later in the results Theorems 3.25 and 3.35. Indeed, the formula

$$\frac{dm_{sph,t}}{dm_{\tau,t}}(z) := \frac{|z|^{(\tau-\tau')t}}{\int_{J(f)} |\xi|^{(\tau-\tau')t} dm_{\tau,t}(\xi)} \tag{3.4.5}$$

defines a spherical conformal probability measure $m_{sph,t}$. But then it may happen that the corresponding density (Radon–Nikodym derivative) $d\mu_t/dm_{sph,t}$ in Theorem 3.25 or Theorem 3.35 is no longer a bounded function.

3.4.3 Existence of Conformal Measures

As we have already said, for rational functions, Denis Sullivan proved in [98] that every rational function admits a conformal measure with conformal factor $\lambda = 1$. For transcendental functions this is not so in full generality and this is again because of the singularity at infinity. In general, a conformal measure is obtained by a (weak) limit procedure and one has to make sure that the mass does not escape to infinity when passing to the limit.

There are two particular cases where natural t-conformal measures do exist. First of all, there are different types of meromorphic functions for which the (normalized) spherical Lebesgue measure is a 2-conformal measure. This is the case for functions f with $J(f) = \hat{\mathbb{C}}$ and for those having a Julia set of positive area such as the functions of the sine family. This is a result of Curtis McMullen [64]; we will come back to it and to its generalizations in greater detail in Sect. 3.4.5.

The other particular case is formed by meromorphic functions having as their Julia sets the real line $\mathbb{R}$ or a geometric circle, and thus having a natural 1-conformal measure. Functions of this type arise among inner functions studied by Aaronson [1] and Doering–Mané [35].

Coming now to the general case, the relation between topological pressure and the existence of conformal measures has been studied in [11]. The hypotheses of this paper are again those of Theorem 3.12 and thus it goes far beyond (E–) hyperbolic functions. Theorem C of that paper contains the following general statement for the existence of t-conformal measures.

Theorem 3.16 ([11]) *Let $f : \mathbb{C} \to \hat{\mathbb{C}}$ be either a meromorphic function in class $\mathcal{S}$ or a non-exceptional and tame function in class $\mathcal{B}$. If $P_{sph}(t) = 0$ for some $t > 0$, then f has a $|f'|_2^t$-conformal measure, i.e. a t-conformal measure, with respect to the spherical metric.*   □

For E-hyperbolic and expanding function there exists a general construction of conformal measures. It allows us to produce conformal measures, defined with respect to adapted τ-metrics, with various conformal factors λ. The proof of Proposition 8.7 in [60] along with Section 5.3 in [58] yield the following.

Theorem 3.17 *Let $f : \mathbb{C} \to \hat{\mathbb{C}}$ be E-hyperbolic and expanding. Assume that $t > 0$ and τ are such that*

$$\|\mathcal{L}_t \mathbb{1}\|_\infty < +\infty \quad and \quad \lim_{|w| \to \infty,\, w \in J(f)} \mathcal{L}_t \mathbb{1}(w) = 0.$$

Then there exist a $\lambda |f'|_\tau^t$-conformal measure with $\lambda = e^{P_\tau(t)}$.   □

The first hypothesis of this theorem tells us that we have a "good" well defined bounded linear transfer operator. The second hypothesis can be used to prove tightness of an appropriate sequence of purely atomic measure, which in turn allows us to produce, as its weak* limit, a desired conformal measure. Then Theorem 3.17 follows.

3.4.4 Conformal Measures on the Radial Set and Recurrence

For rational functions the behavior of conformal measures on the radial set is fairly well understood. For example, it has been studied in [32] and in [65, Section 5], and most of the arguments from these papers can be adapted to the transcendental case.

Theorem 3.18 *Let m_t be a $\lambda |f'|_\tau^t$-conformal measure of a meromorphic function $f : \mathbb{C} \to \hat{\mathbb{C}}$ such that $m_t(J_r(f)) > 0$. Then*

$$m_t(J_r(f)) = 1 \quad ,$$

m_t is ergodic, m_t almost every point has a dense orbit in $J(f)$ and m_t is a unique $\lambda |f'|_\tau^t$-conformal measure. More precisely, if m is a $\rho |f'|_\tau^t$-conformal measure then $\lambda = \rho$ and $m = m_t$. □

Proof The radial Julia set has been defined in Definition 3.2. For any $z \in J_r(f)$, let $\delta(z) > 0$ be the number δ and let $(n_j)_{j \geq 1}$ be the sequence associated to z, both according to Definition 3.2. Define then

$$J_r(f, \delta) := \left\{ z \in J_r(f) \; : \; \delta(z) \geq 2\delta \text{ and } \sup_{j \geq 1} \left\{ |f^{n_j}(z)| \right\} \leq 1/\delta \right\}.$$

Then

$$J_r(f) = \bigcup_{\delta > 0} J_r(f, \delta) \tag{3.4.1}$$

and, if $m_t(J_r(f)) > 0$, then $m_t(J_r(f, \delta)) > 0$ for some $\delta > 0$.

For all $z \in J_r(f)$, consider the blow up mappings

$$f^{n_j} : V_j(z) \longrightarrow \mathbb{D}\left(f^{n_j}(z), 2\delta \right), \quad j \geq 1,$$

where $V_j(z)$ is the connected component of $f^{-n_j}\left(\mathbb{D}\left(f^{n_j}(z), 2\delta \right) \right)$ containing z. Let

$$U_j(z) := V_j(z) \cap f^{-n_j}\left(\mathbb{D}\left(f^{n_j}(z), \delta \right) \right).$$

Then Koebe's Distortion Theorem applies for the map f^{n_j} on $U_j(z)$. In fact, what we need is a bounded distortion for the derivatives taken with respect to the Riemannian metric $d\tau$. This however is a straightforward consequence of Koebe's Theorem (see [58, Section 4.2]). Therefore

$$m_t(U_j(z)) \asymp \lambda^{-n_j} |(f^{n_j})'(z)|_\tau^{-t} m_t(\mathbb{D}(f^{n_j}(z), \delta)).$$

Now, since conformal measures are positive on all non-empty open sets relative to $J(f)$, we conclude that for every $\delta > 0$ there exists a constant $c > 0$ such that

$$m_t(\mathbb{D}(w, \delta)) \geq c$$

for every $w \in J(f) \cap B(0, 1/\delta)$. This shows that

$$m_t(U_j(z)) \asymp \lambda^{-n_j} |(f^{n_j})'(z)|_\tau^{-t} \tag{3.4.2}$$

for every $z \in J_r(f, \delta)$ and every $j \geq 1$ with comparability constants depending on δ only.

Having this estimate we now can proceed exactly as in [65, Theorem 5.1]. If v_t is any $\eta|f'|^t_\tau$-conformal measure then (3.4.2) also holds with v_t, η instead of m_t, ρ, and with other appropriated constants depending on δ only. Hence, for every $z \in J_r(f, \delta)$,

$$\frac{m_t(U_j(z))}{v_t(U_j(z))} \asymp 1 \quad \text{for every } j \geq 1.$$

Since in addition $\lim_{j\to\infty} \text{diam}(U_j(z)) = 0$ and all $U_j(z), z \in J_r(f, \delta), j \geq 1$, have shapes of not "too much" distorted balls, we conclude that the measures m_t and v_t are equivalent (mutually absolutely continuous) on $J_r(f, \delta)$. Invoking, (3.4.1), we deduce that these two measures are equivalent on $J_r(f)$. This is not the end of the proof yet but the interested reader is referred to the original proof in [65]. □

Recall that the Poincaré's Recurrence Theorem asserts that, given $T : X \to X$ measurable dynamical system preserving a finite measure, for every measurable set $F \subset X$ and almost every point $x \in F$, the point $T^n(x)$ is in F for infinitely many $n \geq 1$. A conformal measure m is called *recurrent* if the conclusion of the Poincaré recurrence theorem holds for it. In the case where the Perron–Frobenius–Ruelle theorem holds then, due to the existence of probability invariant measures, commonly called Gibbs states, equivalent to the conformal measure, the later is always recurrent. By Halmos' Theorem [39], recurrence is equivalent to *conservativity* which means that there does not exist a measurable wandering set of positive measure, i.e. a measurable set W with $m(W) > 0$ and such that

$$f^{-n}(W) \cap f^{-m}(W) = \emptyset$$

for all $n > m \geq 0$.

Theorem 3.19 *Assume that the transcendental function $f : C \to \hat{C}$ has m_t, a $\lambda|f'|^t_\tau$-conformal measure. Then*

– *m_t is recurrent and this holds if and only if $m_t(J_r(f)) = 1$ or*
– *m_t-almost every point is in $\mathcal{I}(f)$ or its orbit is attracted by $\mathcal{P}(f)$.*

□

Proof If $m_t(J(f)\backslash\mathcal{P}(f)) = 0$ then the second conclusion holds. So we may assume from now on that $m_t(J(f) \backslash \mathcal{P}(f)) > 0$. Notice that then f is tame and thus there exist $D = \mathbb{D}(w, r)$, a disk centered at some point $w \in J(f)$ and such that

$$\mathbb{D}(w, 2r) \cap \mathcal{P}(f) = \emptyset.$$

Assume that there exists $W \subset J(f) \setminus \mathcal{P}(f)$ a wandering set of positive measure. Since all omitted values are in $\mathcal{P}(f)$, there exists N such that

$$W' = f^N(D) \cap W$$

is a wandering set of positive measure. But then $W'' = f^{-N}(W') \cap D$ is a wandering set of positive measure contained in D. Conformality, bounded distortion, and the fact the W'' is wandering, give

$$1 \geq \sum_{n \geq 0} m_t(f^{-n}(W'')) \asymp m_t(W'') \sum_{n \geq 0} \mathcal{L}_t^n \mathbb{1}(z) \asymp \frac{m_t(W'')}{m_t(D)} \sum_{n \geq 0} m_t(f^{-n}(D)) .$$

The series in the middle is what is usually called the Poincaré series and we see that it is convergent for the exponent t. Now, a standard application of the Borel–Cantelli Lemma shows that a.e. z is in at most finitely many sets $f^{-n}(D)$ or, equivalently, only for finitely many n we have $f^n(z) \in D$. Since this true for every such disk D, it follows that

$$z \in I(f) \quad \text{or} \quad f^n(z) \to \mathcal{P}(f) \quad \text{for } m_t \text{ a.e. } z \in J(f). \tag{3.4.3}$$

This also shows that $m_t(J_r(f)) = 0$ in this case since, as we have seen in Theorem 3.18, if $m_t(J_r(f)) > 0$ then m_t a.e. orbit has a dense orbit in J_f which contradicts (3.4.3) since f is a tame function.

The other possibility is that $J(f) \setminus \mathcal{P}(f)$ does not contain a wandering set of positive measure. Then m_t is conservative hence recurrent on $J(f) \setminus \mathcal{P}(f)$. Let

$$V_\varepsilon(A) := \{z \in \mathbb{C} \ : \ dist(z, A) \leq \varepsilon\},$$

$$V_\varepsilon^c(A) := J(f) \setminus V_\varepsilon(A)$$

and consider the open set

$$U_\varepsilon = \mathbb{D}(0, 1/\varepsilon) \cap V_\varepsilon^c(\mathcal{P}(f)) \quad , \quad \varepsilon > 0.$$

If $\varepsilon > 0$ is small enough, $U_\varepsilon \cap J(f) \neq \emptyset$, and then $m_t(U_\varepsilon) > 0$. On the other hand, recurrence implies that

$$m_t(U_\varepsilon) = m_t(\{z \in U_\varepsilon \ : \ f^n(z) \in U_\varepsilon \text{ for infinitely many } n\text{'s}\}).$$

The set of points z such that, for some $\varepsilon > 0$, $z \in U_\varepsilon$ and $f^n(z) \in U_\varepsilon$ for infinitely many n's is a subset of $J_r(f)$. Therefore, $m_t(J_r(f)) \geq m_t(U_\varepsilon) > 0$ and then, by Theorem 3.18, $m_t(J_r(f)) = 1$. Notice also that then m_t is recurrent on the whole Julia set since $m_t(J(f) \setminus J_r(f)) = 0$. $\qquad \square$

Every rational function has a t-conformal measure of minimal exponent $t = \delta_p$; see [25]. Shishikura [88] gave the first examples of some polynomials p for which this exponent is maximal, i.e. $\delta_p = 2$. For them the corresponding conformal measures are not recurrent. Up to our best knowledge it is unknown whether there exist polynomials, even rational functions, p with $\delta_p < 2$ and with non-recurrent δ_p-conformal measures. However, there are such quadratic like examples; see Avila–Lyubich [4], and the first globally defined, i.e. on the whole complex plane, (transcendental meromorphic) functions having such behavior were produced in [61, Theorem 1.4]. Notice that these examples are even hyperbolic and their number Θ (see Theorem 3.37) is equal to the minimal exponent δ_p.

Theorem 3.20 ([61]) *There exist disjoint type entire functions $f : \mathbb{C} \to \hat{\mathbb{C}}$ of finite order, with $\Theta \in (1, 2)$, that do not have any recurrent Θ-conformal measure with conformal factor $\lambda = 1$.*   □

In fact [61, Theorem 1.4] states that these functions do not have Θ-conformal measures supported on the radial Julia set. But this is equivalent to non-recurrence by Theorem 3.19.

3.4.5   2-Conformal Measures

We finally discuss the special case of 2-conformal measures. As already mentioned above, for many transcendental, especially entire, functions the spherical Lebesgue measure m_{sph} of the Julia set is positive and thus it is a natural 2-conformal measure. In this case each of the following possibilities can occur:

- m_{sph} is recurrent and $m_{sph}(J_r(f)) = 1$.
- $m_{sph}(\mathcal{I}(f) \cap J(f)) = 1$.
- $m_{sph}(\mathcal{I}(f) \cap J(f)) = 0$ and $f^n(z) \to \mathcal{P}(f)$ for m_{sph}–a.e. $z \in \mathbb{C}$.

Let us first discuss the recurrent case for which the postcritically finite map $f(z) = 2\pi i e^z$ is a typical example having the property that $m_{sph}(J_r(f)) = 1$. For this function, the Julia set is the whole plane. From a classical zooming and Lebesgue density argument (see for example the proof of [37, Theorem 8]) follows that this always holds provided that the radial set is positively charged.

Proposition 3.21 *Let $f : \mathbb{C} \to \hat{\mathbb{C}}$ be a meromorphic function. If m_{sph} is a 2-conformal measure and if $m_{sph}(J_r(f)) > 0$, then $J(f) = \mathbb{C}$.*

For the third possibility, i.e. where the escaping set is not charged but where a.e. orbit is attracted by the post-critical set, we have some results due to Eremenko–Lyubich [37, Section 7].

Theorem 3.22 ([37]) *Let $f \in \mathcal{B}$ be an entire function of finite order having a finite logarithmic singular value. Then $m_{sph}(\mathcal{I}(f)) = 0$ and there exists $M > 0$ such that*

$$\liminf_{n \to \infty} |f^n(z)| < M \quad \text{for a.e. } z \in \mathbb{C}.$$

$\square$

Given this result combined with Theorem 3.19 we see that there are several possibilities. Assume that f satisfies the hypotheses of Theorem 3.22 and that the Julia set of f has positive area. Then, either the spherical Lebesgue measure is supported on the radial Julia set or a.e. orbit is attracted by the post-critical set.

As typical examples we can consider again the exponential family. As already mentioned, $f(z) = 2\pi i e^z$ is a recurrent example. Totally different is $f(z) = e^z$. Misiurewicz showed in [69] that $J(f) = \mathbb{C}$ and Lyubich proved in [50] that this function is not ergodic. Consequently $m_{sph}(J_r(f)) = 0$ and thus a.e. orbit is attracted by the orbit of 0, the only finite singular value.

Plenty of entire functions have the property $m_{sph}(\mathcal{I}(f) \cap J(f)) = 1$ (and $F(f) \neq \emptyset$). Initially, McMullen showed in [64] that the Julia set of every function from the sine–family $\alpha \sin(z) + \beta$, $\alpha \neq 0$, has positive area. This result has been generalized in many ways and to many types of entire functions; see [3, 15, 16, 90]. The authors of these papers did not really deal with Julia but with the escaping set and, as a matter of fact, they showed that

$$area(\mathcal{I}(f) \cap J(f)) > 0. \tag{3.4.1}$$

Since the escaping set is invariant, it suffices now to normalize properly the spherical Lebesgue measure restricted to $\mathcal{I}(f)$ in order to get the required 2-conformal measure that is entirely supported on the escaping set.

3.5 Perron–Frobenus–Ruelle Theorem, Spectral Gap and Applications

The whole thermodynamic formalism relies on the transfer operator and its properties. We recall that this operator has been introduced in Definition 3.15. In fact, this definition treats only the most relevant *geometric potentials*. More general potentials ψ were considered in [58]. They are obtained as a sum of a geometric potential plus an additional Hölder function. This class of potentials has its importance for the multifractal analysis (see the Chapters 8 and 9 of [58]) of conformal measures and their invariant versions. In the present text we restrict ourselves to geometric potentials, so to functions of the form

$$\psi := -t \log |f'| + b - b \circ f \tag{3.5.1}$$

for some appropriate function $b : J(f) \to \mathbb{R}$ (or $\mathbb{C}$). This coboundary is crucial since it allows us to deal with different Riemannian metrics on $\hat{\mathbb{C}}$. We start by investigating elementary examples to make this transparent.

We already have mentioned in the introduction that the "naive" transfer operator is not always well defined. Let us consider the simplest entire function $f(z) = \lambda e^z$ and a potential $\psi := -t \log |f'|$ without coboundary. Then, for all $w \neq 0$ and parameter t,

$$\mathcal{L}_t \mathbb{1}(w) = \sum_{z \in f^{-1}(w)} |f'(z)|^{-t} = \sum_{z \in f^{-1}(w)} |w|^{-t} = +\infty.$$

In other words, this operator is just not defined. This is the point where a coboundary b of (3.5.1) shows its significance.

We recall that the derivative of a function f with respect to a Riemannian metric $d\sigma = \gamma |dz|$ is given by Formula (3.2.3). The associated geometric potential is

$$\psi = -t \log |f'|_\sigma = -t \log |f'| + t \log \gamma - t \log \gamma \circ f.$$

Since the Euclidean metric plainly does not work, one can try the spherical metric $d\sigma = |dz|/(1 + |z|^2)$ which is another natural choice. Considering again $f(z) = \lambda e^z$, we get

$$\mathcal{L}_t \mathbb{1}(w) = \left(\frac{1 + |w|^2}{|w|} \right)^t \sum_{f(z)=w} (1 + |z|^2)^{-t}$$

which, this time, is finite provided that $t > 1/2$. In fact then, for large w, and with $x_0 = \log |w/\lambda|$,

$$\mathcal{L}_t \mathbb{1}(w) \asymp |w|^t \int_\mathbb{R} \frac{dy}{(1 + |x_0 + iy|^2)^t} = |w|^t (1 + x_0^2)^{1/2-t} \int_\mathbb{R} \frac{dy}{(1 + y^2)^t} < +\infty,$$

but

$$\lim_{w \to \infty} \mathcal{L}_t \mathbb{1}(w) = +\infty.$$

Thus, $\mathcal{L}_t$ is not a bounded operator.

It turns out that for the exponential family and in general for entire functions the logarithmic metric $d\sigma = 1/(1 + |z|)$ is best appropriate. This is a natural choice for several reasons. For example, this point of view is used in Nevanlinna's value distribution theory. Also, in the dynamics of entire functions from class $\mathcal{B}$, Eremenko–Lyubich [37] have introduced logarithmic coordinates, which now is a standard tool. Either working in these coordinates or considering derivatives with respect to the logarithmic metric are equivalent things.

3.5.1 Growth Conditions

The situation is different for meromorphic functions because of their behavior at poles. If $f : \mathbb{C} \to \hat{\mathbb{C}}$ is meromorphic and if b is a pole of multiplicity q, which is nothing else than a critical point of multiplicity $q \geq 1$ of f, then

$$|f'(z)| \asymp \frac{1}{|z - b|^{q+1}} = \frac{1}{|z - b|^{q(1+1/q)}} \asymp |f(z)|^{1+1/q} \quad \text{near} \quad b. \qquad (3.5.1)$$

Motivated by the exponential family λe^z, we introduced in [56] and [58] some classes of meromorphic functions for which there are relations between $|f'|$ and $|f|$. More precisely:

Definition 3.23 (Rapid Derivative Growth and Dynamical Semi-Regularity) A meromorphic function $f : \mathbb{C} \to \hat{\mathbb{C}}$ is said to have a rapid derivative growth if and only if there are $\underline{\alpha_2} > \max\{0, -\alpha_1\}$ and $\kappa > 0$ such that

$$|f'(z)| \geq \kappa^{-1}(1 + |z|)^{\alpha_1}(1 + |f(z)|^{\alpha_2}) \qquad (3.5.2)$$

for all finite $z \in J(f) \setminus f^{-1}(\infty)$. A E-hyperbolic and expanding meromorphic function of finite order ρ which satisfies the rapid derivative growth condition is called dynamically semi-regular.

Of course, $f(z) = \lambda e^z$ satisfies (3.5.2) with $\alpha_2 \equiv 1$ and $\alpha_1 = 0$. The reader can find many other families in Chapter 2 of [58] which are dynamically semi-regular.

For such functions there is a good choice of the coboundary b or, equivalently, of the Riemannian metric. We recall that we consider metrics of the form (3.2.1) and that we frequently use the simpler form of (3.2.2), namely $d\tau(z) = |z|^{-\tau}|dz|$. This is possible as soon as the Fatou set is not empty, which is the case for E-hyperbolic functions, since then we can assume without loss of generality that $0 \in F(f)$ and then ignore what happens near the origin.

If f has balanced growth then, setting $\hat{\tau} = \alpha_1 + \tau$,

$$|z|^{\hat{\tau}} \preceq |z|^{\hat{\tau}}|f(z)|^{\underline{\alpha_2}-\tau} \preceq |f'(z)|_\tau \preceq |z|^{\hat{\tau}}|f(z)|^{\overline{\alpha}_2-\tau}, \quad z \in J(f) \setminus f^{-1}(\infty), \qquad (3.5.3)$$

the right hand inequality being true under the weaker condition (3.5.2). Therefore, for a dynamically semi-regular function f, we get the estimate

$$\mathcal{L}_t \mathbb{1}(w) \preceq \frac{1}{|w|^{\underline{\alpha_2}-\tau}} \sum_{z \in f^{-1}(w)} |z|^{-\hat{\tau}t}, \quad w \in J(f).$$

This last sum, which also is called Borel sum, is very well known in Nevanlinna theory. The order ρ of f is precisely the critical exponent for this sum. Hence, if f has finite order ρ and if $\hat{\tau}t > \rho$, then it is a convergent series and in fact one has the crucial following property (see Proposition 3.6 in [58]).

Proposition 3.24 *If f is satisfies the rapid derivative growth condition, if $0 \in F(f)$, and if $\tau \in (0, \underline{\alpha}_2)$ then, for every $t > \rho/\hat{\tau}$, there exists M_t such that*

$$\mathcal{L}_t \mathbb{1}(w) \le M_t \quad \text{and} \quad \lim_{w \to \infty} \mathcal{L}_t \mathbb{1}(w) = 0 \quad , \quad w \in J(f), \tag{3.5.4}$$

Once having property (3.5.4), one can develop a full thermodynamic formalism provided that the function f is E-hyperbolic and expanding. The first issue is again about the existence of conformal measures. It is taken care of by Theorem 3.17. Therefore, for E-hyperbolic and expanding meromorphic functions satisfying the hypotheses of Proposition 3.24, we have good conformal measures for all $t > \Theta$.

We recall that dynamically semi-regular functions have been introduced in Definition 3.23. The following Perron–Frobenius–Ruelle Theorem is part of Theorem 1.1 in [57] and Theorem 5.15 of [58], which is true for a class of more general potentials.

Theorem 3.25 *If $f : \mathbb{C} \to \hat{\mathbb{C}}$ is a dynamically semi-regular meromorphic function then, for every $t > \frac{\rho}{\hat{\tau}}$, the following are true.*

(a) *The topological pressure $P(t) = \lim_{n \to \infty} \frac{1}{n} \mathcal{L}_t^n(\mathbb{1})(w)$ exists and is independent of $w \in J(f)$.*

(b) *There exists a unique $\lambda |f'|_{\hat{\tau}}^t$-conformal measure m_t and necessarily $\lambda = e^{P(t)}$.*

(c) *There exists a unique Gibbs state μ_t of the parameter t, where being Gibbs means that μ_t is a Borel probability f-invariant measure absolutely continuous with respect to m_t. Moreover, the measures m_t and μ_t are equivalent and are both ergodic and supported on the conical limit set of f.*

(d) *The Radon–Nikodym derivative $\psi_t = d\mu_t/dm_t : J(f) \to [0, +\infty)$ is a continuous nowhere vanishing bounded function satisfying $\lim_{z \to \infty} \psi_t(z) = 0$.*

□

Starting from this result, much more can be said but under the stronger growth condition (3.5.5). Namely, the Spectral Gap property along with its applications:

– The Spectral Gap [58, Theorem 6.5]

Theorem 3.26 *If f is a dynamically semi-regular function and if $t > \frac{\rho}{\hat{\tau}}$, then the following are true.*

(a) *The number 1 is a simple isolated eigenvalue of the operator $\hat{\mathcal{L}}_t := e^{-P(t)} \mathcal{L}_t : \mathbb{H}_\beta \to \mathbb{H}_\beta$, where $\beta \in (0, 1]$ is arbitrary and $\mathbb{H}_\beta$ is the Banach space of all complex–valued bounded Hölder continuous defined on $J(f)_f$, equipped with the corresponding Hölder norm. The rest of the spectrum of $\mathcal{L}_t$ is contained in a disk with radius strictly smaller than 1. In particular, the operator $\hat{\mathcal{L}}_t : \mathbb{H}_\beta \to \mathbb{H}_\beta$ is quasi–compact.*

(b) *More precisely: there exists a bounded linear operator $S : \mathbb{H}_\beta \to \mathbb{H}_\beta$ such that*

$$\hat{\mathcal{L}}_t = Q_1 + S,$$

where $Q_1 : \mathbb{H}_\beta \to \mathbb{C}\rho$ is a projector on the eigenspace $\mathbb{C}\rho$, given by the formula

$$Q_1(g) = \left(\int g \, dm_\phi \right) \rho_t,$$

$Q_1 \circ S = S \circ Q_1 = 0$ *and*

$$\|S^n\|_\beta \leq C\xi^n$$

for some constant $C > 0$, some constant $\xi \in (0, 1)$ and all $n \geq 1$.

$\square$

– [58, Corollary 6.6]

Corollary 3.27 *With the setting, notation, and hypothesis of Theorem 3.26 we have, for every integer $n \geq 1$, that $\hat{\mathcal{L}}^n = Q_1 + S^n$ and that $\hat{\mathcal{L}}^n(g)$ converges to $\left(\int g \, dm_\phi \right) \rho$ exponentially fast when $n \to \infty$. More precisely,*

$$\left\| \hat{\mathcal{L}}^n(g) - \left(\int g \, dm_\phi \right) \rho \right\|_\beta = \|S^n(g)\|_\beta \leq C\xi^n \|g\|_\beta \quad , \quad g \in H_\beta.$$

– Exponential Decay of Correlations [58, Theorem 6.16]

Theorem 3.28 *With the setting, notation, and hypothesis of Theorem 3.26 there exists a large class of functions ψ_1 such that for all $\psi_2 \in L^1(m_t)$ and all integers $n \geq 1$, we have that*

$$\left| \int (\psi_1 \circ f^n \cdot \psi_2) \, d\mu_t - \int \psi_1 \, d\mu_t \int \psi_2 \, d\mu_t \right| \leq O(\xi^n),$$

where $\xi \in (0, 1)$ comes from Theorem 3.26(b), while the big "O" constant depends on both ψ_1 and ψ_2.

$\square$

– Central Limit Theorem [58, Theorem 6.17]

Theorem 3.29 *With the setting, notation, and hypothesis of Theorem 3.26 there exists a large class of functions ψ such that the sequence of random variables*

$$\frac{\sum_{j=0}^{n-1} \psi \circ f^j - n \int \psi \, d\mu_t}{\sqrt{n}}$$

converges in distribution, with respect to the measure μ_t, to the Gauss (normal)
distribution $N(0, \sigma^2)$ with some $\sigma > 0$. More precisely, for every $t \in \mathbb{R}$,

$$\lim_{n \to \infty} \mu_t \left(\left\{ z \in J(f)_f : \frac{\sum_{j=0}^{n-1} \psi \circ f^j(z) - n \int \psi \, d\mu_t}{\sqrt{n}} \leq t \right\} \right)$$

$$= \frac{1}{\sigma \sqrt{2\pi}} \int_{-\infty}^{t} \exp\left(-\frac{u^2}{2\sigma^2}\right) du.$$

$\square$

– Variational Principle [58, Theorem 6.25]

Theorem 3.30 *With the setting, notation, and hypothesis of Theorem 3.26, we have*
that

$$P(t) = \sup \left\{ h_\mu(f) - t \int_{J(f)} \log |f'|_1 \, d\mu \right\},$$

where the supremum is taken over all Borel probability f-invariant ergodic mea-
sures μ with $\int_{J(f)} \log |f'|_1 \, d\mu > -\infty$. Furthermore, $\int_{J(f)} \log |f'|_1 \, d\mu_t > -\infty$
and μ_t is the only one among such measures satisfying the equality

$$P(t) = h_\mu(f) - t \int_{J(f)} \log |f'|_1 \, d\mu.$$

In the common terminology this means that the f-invariant measure μ_t is the only
equilibrium state of the potential $-t \log |f'|_1$. $\square$

In [58] appears also a stronger symmetric growth condition. It is the following
and it was used in order to get more geometric informations out of the thermodynamical formalism. The principal application of it was to obtain a Bowen's Formula
expressing the hyperbolic dimension as the zero of the topological pressure function.

Definition 3.31 (Balanced Growth and Dynamical Regularity) A meromorphic
function $f : \mathbb{C} \to \hat{\mathbb{C}}$ is balanced if and only if there are $\kappa > 0$, a bounded function
$\alpha_2 : J(f) \cap \mathbb{C} \to [\underline{\alpha_2}, \overline{\alpha}_2] \subset (0, \infty)$ and $\alpha_1 > -\underline{\alpha_2} = -\inf \alpha_2$ such that

$$\kappa^{-1}(1+|z|)^{\alpha_1}(1+|f(z)|^{\alpha_2(z)}) \leq |f'(z)| \leq \kappa(1+|z|)^{\alpha_1}(1+|f(z)|^{\alpha_2(z)}) \quad (3.5.5)$$

for all finite $z \in J(f) \setminus f^{-1}(\infty)$. A balanced E-hyperbolic and expanding
meromorphic function of finite order ρ is called dynamically regular.

In this stronger symmetric condition it is important that α_2 is a function
since (3.5.1) shows that at poles of a meromorphic function this exponent α_2
does depend on the multiplicity q. Typical meromorphic functions that satisfy the

balanced growth condition are all elliptic functions. Again, many other families
appear in Chapter 2 of [58].

3.5.2   Geometry of Tracts

For entire functions the thermodynamical formalism is known to hold in a much
larger setting than the functions that satisfy the growth conditions since we now have
a quite optimal approach of [60]. It shows that the geometry of the tracts determines
the behavior of the transfer operator. Let us briefly recall and explain this now.

As it was explained right after the Definition 3.9, in order to study the dynamics
of a disjoint type entire function f near the Julia set, only its restriction to the tracts
is relevant. Let us here consider the simplest case where $f \in \mathcal{B}$ has only one tract
Ω. Remember that $f_{|\Omega} = e^{\varphi^{-1}}$. A simple calculation gives

$$|f'|_1^{-1} = \frac{|\varphi'|}{|\varphi|} \circ \varphi^{-1}$$

in Ω. This gives that

$$\mathcal{L}_t \mathbb{1}(w) = \sum_{\xi \in \exp^{-1}(w)} \left| \frac{\varphi'(\xi)}{\varphi(\xi)} \right|^t \tag{3.5.1}$$

entirely does depend on the conformal representation φ of the tract and thus entirely
on the tract Ω itself. In fact, the operator $\mathcal{L}_t$ does depend on the geometry of Ω at
infinity. In order to study the behavior of this operator, one considers the rescaled
maps

$$\varphi_T := \frac{1}{|\varphi(T)|} \varphi \circ T : Q_1 \longrightarrow \frac{1}{|\varphi(T)|} \Omega_T$$

where Q_T, especially Q_1, has been defined in (3.3.4) and where for $T \geq 1$,

$$\Omega_T := \varphi(Q_T).$$

These maps behave especially well as soon as the tract has some nice geometric
properties.

3.5.2.1   Hölder Tracts

Loosely speaking, a Hölder domain is the image of the unit disk by a Hölder map.
But such domains are clearly bounded whereas logarithmic tracts are unbounded

domains. Following [54], we therefore consider natural exhaustions of the tract by Hölder domains and a scaling invariant notion of Hölder maps. A conformal map $h : Q_1 \to U$ is called (H, α)–Hölder if and only if

$$|h(z_1) - h(z_2)| \leq H|h'(1)||z_1 - z_2|^{\alpha} \quad \text{for all} \quad z_1, z_2 \in Q_1 . \tag{3.5.2}$$

Definition 3.32 The tract Ω is Hölder, if and only if(3.3.5) holds and the maps φ_T are uniformly Hölder, i.e. there exists (H, α) such that for every $T \geq 1$ the map φ_T satisfies (3.5.2).

Quasidisks and John domains serve as good examples of Hölder tracts.

3.5.2.2 Negative Spectrum

The boundary $\partial\Omega$ of a tract is an analytic curve. However, seen from infinity such a boundary may appear quite fractal. In order to quantify this property, we associate to a tract a version of integral means spectrum (see [51] and [72] for the classical case). In order to do so, let $h : Q_2 \to U$ be a conformal map onto a bounded domain U and define

$$\beta_h(r, t) := \frac{\log \int_I |h'(r + iy)|^t dy}{\log 1/r} \quad , \ r \in (0, 1) \ \text{and} \ t \in \mathbb{R}. \tag{3.5.3}$$

The integral is taken over $I = [-2, -1] \cup [1, 2]$ since this corresponds to the part of the boundary of U that is important for our purposes.

Applying this notion to the rescalings φ_T and then letting $T \to \infty$ leads to desired integral means of the tract Ω,

$$\beta_{\infty}(t) := \limsup_{T \to +\infty} \beta_{\varphi_T}(1/T, t), \tag{3.5.4}$$

and to the associated function

$$b_{\infty}(t) := \beta_{\infty}(t) - t + 1 \quad , \quad t \in \mathbb{R}. \tag{3.5.5}$$

It turns out that the function b_{∞} is convex, thus continuous, with $b_{\infty}(0) = 1$ and with $b_{\infty}(2) \leq 0$. Consequently, the function b_{∞} has at least one zero in $(0, 2]$ and we can introduce a number $\Theta \in (0, 2]$ by the formula

$$\Theta := \inf\{t > 0 \ : \ b_{\infty}(t) = 0\} = \inf\{t > 0 \ : \ b_{\infty}(t) \leq 0\}. \tag{3.5.6}$$

We only considered here the case of a single tract and the adaption for functions in $\mathcal{D}$ having finitely many tracts is straightforward (see [60]).

Definition 3.33 A function $f \in \mathcal{D}$ has negative spectrum if and only if, for every tract,

$$b_\infty(t) < 0 \quad \text{for all } t > \Theta.$$

A relation of the Hölder tracts property and the negative spectrum property is provided by the following.

Proposition 3.34 (Proposition 5.6 in [60]) *A function $f \in \mathcal{B}$ has negative spectrum if it has only finitely many tracts and all these tracts are Hölder.*

3.5.2.3   Back to the Thermodynamic Formalism and Its Applications

From now on we assume that f is a function of the class $\mathcal{D}$ and has negative spectrum. Let Θ be again the parameter introduced in (3.5.6).

Starting from the formula (3.5.1), one can express the transfer operator in terms of integral means (see [60, Proposition 4.3]) in the following way:

$$\mathcal{L}_t \mathbb{1}(w) \asymp (\log|w|)^{1-t} \left\{ \int_{-1}^{1} \left| \varphi'_{\log|w|}(1+iy) \right|^t dy + \sum_{n \geq 1} 2^{n\left(1-t+\beta_{\varphi_{2^n} \log|w|}(2^{-n}, t)\right)} \right\}$$

$$(3.5.7)$$

for every $t \geq 0$ and every $w \in \Omega$. The series appearing in this formula may diverge. Nevertheless, this formula very well describes the behavior of the transfer operator. It allows us to develop the thermodynamic formalism if the negative spectrum assumption holds. The first step is to verify again the conclusion of Proposition 3.24 which, we recall, is crucial for establishing the existence of conformal measures. Then one can adapt the arguments of [58] to get the following version of the Perron-Frobenius–Ruelle Theorem ([60, Theorem 1.2]).

Theorem 3.35 *Let $f \in \mathcal{D}$ be a function having negative spectrum and let $\Theta \in (0, 2]$ be the smallest zero of b_∞. Then, the following hold:*

- *For every $t > \Theta$, the whole thermodynamic formalism, along with its all usual consequences holds: the Perron-Frobenius-Ruelle Theorem, the Spectral Gap property along with its applications: Exponential Mixing, Exponential Decay of Correlations and Central Limit Theorem.*

- *For every $t < \Theta$, the series defining the transfer operator $\mathcal{L}_t$ diverges.*

$\square$

In many cases, by using a standard bounded distortion argument, this result and all its consequences can be extended beyond the class of disjoint type to larger subclasses of hyperbolic functions. For example, it does hold for all hyperbolic functions in class $\mathcal{S}$ having finitely many tracts and no necessarily being of disjoint type. This is for example the case for functions of finite order that satisfy (3.3.5). The later is a very general kind of quasi-symmetry condition.

Question

Is the assumption (3.3.5) necessary?

An important feature of the Hölder tract property is that it is a quasiconformal invariant notion. This has several important applications. Let us just mention one of them.

Theorem 3.36 (Theorem 1.3 in [60]) *Let M be an analytic family of entire functions in class S. Assume that there is a function $g \in M$ that has finitely many tracts over infinity and that all these tracts are Hölder. Then every function $f \in M$ has negative spectrum and the thermodynamic formalism holds for every hyperbolic map from M.* $\square$

Theorem 3.35 gives no information at the transition parameter $t = \Theta$. For all classical functions the transfer operator is divergent at Θ, and thus the pressure $P(\Theta) = +\infty$. This then implies that the pressure function has a zero $h > \Theta$. Functions with a completely different behavior have been found recently in [61].

Theorem 3.37 ([61]) *For every $1 < \Theta < 2$ there exists an entire function $f \in \mathcal{B}$ with the following properties:*

(a) The entire function f is of finite order and of disjoint type.
(b) The corresponding transfer operator has transition parameter Θ.
(c) The transfer operator is convergent at Θ and the property (3.5.4) holds.
(d) Consequently, the Perron–Frobenius–Ruelle Theorem 3.25 and its consequences hold at $t = \Theta$.
(e) The topological pressure at $t = \Theta$ is strictly negative.
(f) Consequently, the topological pressure of f has no zero.

$\square$

For the special case of $\Theta = 2$, the reader can find examples in [81].

Here are two more questions related to this section. First of all, we have seen in Proposition 3.34 that the Hölder tract property implies negative spectrum. For some special functions, Poincaré linearizer, both properties coincide ([60, Theorem 7.8]).

Question

Are all tracts of any entire function in class $\mathcal{B}$ with negative spectrum Hölder?

For Hölder tracts with corresponding Hölder exponent $\alpha \in (1/2, 1]$ it is known that $\Theta < 2$.

> **Question**
> What abut the general case? More precisely, if Ω is a Hölder tract with Hölder exponent $\alpha \in (0, 1]$, do we then have that $\Theta < 2$? If so, this would be an analogue of the Jones–Makarov Theorem [41] which states that the Hausdorff dimension of the boundary of an α-Hölder domain is less than two, furthermore, less than $2 - C\alpha$ where $C > 0$ is a universal constant.

3.6 Hyperbolic Dimension and Bowen's Formula

The Hausdorff dimension, and in fact all other fractal dimensions, of the Julia set of meromorphic functions have been studied a lot. The interested reader can consult the survey by Stallard [93]. Here we focus on the hyperbolic dimension.

3.6.1 Estimates for the Hyperbolic Dimension

We recall that the *hyperbolic dimension* $\mathrm{HD}_{\mathrm{hyp}}(f)$ of the function f is the supremum of the Hausdorff dimensions of all forward invariant compact sets on which the functions is expanding. Right from the definition,

$$\mathrm{HD}_{\mathrm{hyp}}(f) \leq \mathrm{HD}(J(f)).$$

It has recently been observed by Avila-Lyubich in [5] that there are polynomials for which there is strict inequality between these two dimension.

Theorem 3.38 ([5]) *There exists a Feigenbaum polynomial p for which*

$$HD_{hyp}(p) < HD(J(p)) = 2.$$

□

Although this result being rather exceptional for rational functions, it appears quite often for transcendental, especially entire, functions. Stallard [92] observed this implicitly and Urbański–Zdunik in [104].

Theorem 3.39 ([92, 104]) *There are (even hyperbolic) entire functions f of finite order and of class S for which*

$$HD_{hyp}(f) < HD(J(f)) = 2.$$

□

The equality $HD(J(f)) = 2$ goes back to McMullen's result [64]. In either case, of rational functions as well as of transcendental functions, we do not know any such example with the Hausdorff dimension of the Julia set equal to 2. Thus:

Question
Is there an entire or meromorphic function $f \in \mathcal{B}$ with a logarithmic tract over infinity and such that

$$HD_{hyp}(f) < HD(J(f)) < 2 \ ?$$

While the hyperbolic dimension of a meromorphic functions is often strictly smaller then the dimension of its Julia set. However, it can not be too small as long as the function has a logarithmic tract over infinity. In fact Barański, Karpińska and Zdunik [9] obtained the following very general result.

Theorem 3.40 ([9]) *The hyperbolic dimension of the Julia set of a meromorphic function with a logarithmic tract over infinity is greater than 1.*   □

For $f_\lambda(z) = \lambda e^z$, Karpinska [42] showed that the hyperbolic dimension goes to one as λ goes to zero. In this sense, the above estimate is sharp. However, if the logarithmic tracts have some regularity then one gets more information, see [54].

Theorem 3.41 ([54]) *If a meromorphic map f has a logarithmic tract over infinity and if this tract is Hölder, then*

$$HD_{hyp}(f) \geq \Theta \geq 1$$

where Θ is the number defined in (3.5.6).   □

In this result, one can not expect strict inequality except if $\Theta = 1$. Indeed, for every given $\Theta \in (1, 2)$ there is an entire function f with Hölder tract such that $HD_{hyp}(f) = \Theta$ (see [61]). On the other hand, the paper [54] provides a sufficient condition, expressed in terms of the boundary of the tract, which implies strict inequality.

The hyperbolic dimension can also be maximal. This has been shown by Rempe–Guillen [81]. He first constructs a local version, called now model, and then approximates it by entire functions. His approximation result is a very precise version of Arakelyan's approximation and is of its own interest.

Theorem 3.42 ([81]) *There exists a transcendental entire function f of disjoint type and finite order such that $HD_{hyp}(f) = 2$.*   □

3.6.2 Bowen's Formula

The pressure function $t \mapsto P_\tau(t)$ is convex, hence continuous and, when the map f is expanding, it is also strictly decreasing. Consequently, there exists a unique zero h of P_τ provided

$$P_\tau(t) \geq 0 \quad \text{for some } t.$$

It goes back to Bowen's paper [21] that this zero is of crucial importance when studying fractal dimensions of limit and Julia sets. Bowen showed that this number h is the Hausdorff dimension of the limit set for any co-compact quasifuchsian group. His result extends easily to the case of of Julia sets of hyperbolic rational functions. Since then his formula has been generalized in many various ways and it became transparent that for transcendental functions his formula detects the hyperbolic dimension rather than the Hausdorff dimension of the entire Julia sets.

The first result of this kind for transcendental functions is, up to our knowledge, was obtained in [104] and [105] while the most general Bowen's Formula for transcendental functions is due to Barański et al. [10]. Here again, we only formulate a version for E-hyperbolic functions while their result holds in much bigger generality.

Theorem 3.43 ([10]) *For every E-hyperbolic meromorphic function $f \in \mathcal{B}$ we have $P_{sph}(2) \leq 0$ and*

$$HD_{hyp}(f) = HD(J_r(f)) = \inf\left\{t > 0 \,;\; P_{sph}(t) \leq 0\right\}.$$

□

We recall that the authors showed the existence of the spherical pressure (Theorem 3.12) and that there exists Θ such that the pressure is finite for all $t > \Theta$ and infinite for all $t < \Theta$. If $P_{sph}(\Theta) \geq 0$, then the pressure has a smallest zero $h \geq \Theta$ and this number h turns out to be the hyperbolic dimension. Otherwise, so if $P_{sph}(\Theta) < 0$, then $HD_{hyp}(f) = \Theta$ and in fact such possibility does happen (Theorem 3.37).

Other versions of Bowen's formula, with pressure taken with respect to adapted Riemannian metrics, still of the form (3.2.1), are contained in [56, 58, 60] and also a version for random dynamics of transcendental functions in [59] and [107]. All these papers contain many other results related to Bowen's formula and formed an important step between [104, 105] and [9].

3.7 Real Analyticity of Fractal Dimensions

Bowen's Formula determines the hyperbolic dimension of a given "sufficiently hyperbolic" meromorphic function f. But

what happens to this dimension when the map f varies in an analytic family?

For rational functions, this has been explored in detail. In contrast to the case of entire functions, the radial and Julia sets of a hyperbolic rational function coincide and consequently also do the corresponding dimensions. Therefore, one is naturally interested in the behavior of the map

$$f \longmapsto \mathrm{HD}(J(f)).$$

In 1982, Ruelle [86] positively confirmed a conjecture of Sullivan and showed that the Hausdorff dimension of the Julia set of hyperbolic rational functions depends real–analytically on the map. The hyperbolicity hypothesis is essential here; see [88, Remark 1.4] and also [36].

The first result on analytic variation of the hyperbolic dimension of transcendental functions is due to Urbański and Zdunik [104] and concerns the exponential family λe^z. Since then this property has been obtained for many families of dynamically regular functions ([56, 91] and [58]; the last of these papers treating also real analyticity of appropriate multifractal spectra; for entire functions in class $\mathcal{D}$ see [60]). For the same kind of families, such analyticity is also true in the realm of random dynamics; see [63].

Instead of presenting a complete overview of all relevant, sometimes quite technical, results we now describe the general framework followed by two representative methods and results.

Similarly as the hyperbolicity hypothesis for rational functions, there are a number of conditions, in a sense necessary, needed to expect real analytic variation of the hyperbolic dimension in the transcendental case. They can be summarized as follows.

- $\mathcal{F}$ is an analytic family of meromorphic functions. The reader simply can assume that $\mathcal{F} = \{f_\lambda = \lambda f \ : \ \lambda \in \Lambda\}$ where f is a given meromorphic function and Λ an open subset of $\mathbb{C}^*$. Clearly there are more general settings. For example, in the case of entire functions in class $\mathcal{S}$ there is a natural notion of analytic family due to Eremenko–Lyubich [37]; they are in particular always finite dimensional.
- The functions of $\mathcal{F}$ are E-hyperbolic and expanding.
- The family $\mathcal{F}$ is structurally stable in the sense of holomorphic motions.

In most results the holomorphic motion is also assumed to have some uniform behavior which is for example implied by a condition called bounded deformation [58].

The last commonly used hypothesis is that the full thermodynamic formalism applies. Here appears a crucial fact which is specific to the transcendental case. The transfer operator of a transcendental function is usually not defined for small parameters $t > 0$. Let us follow the notation used in Theorem 3.35 and call again Θ the transition parameter. In fact, one must rather write Θ_f since this number can depend on a particular function f from a given family $\mathcal{F}$.

– The thermodynamic formalism holds for the functions in $\mathcal{F}$ with constant transition parameter $\Theta = \Theta_{f_\lambda}, \lambda \in \Lambda$.

In particular, Bowen's Formula applies to the functions we consider here and thus two cases appear: for $f \in \mathcal{F}$, either

$$\mathrm{HD}_{\mathrm{hyp}}(f) > \Theta \quad \text{or} \quad \mathrm{HD}_{\mathrm{hyp}}(f) = \Theta. \tag{3.7.1}$$

The first analyticity result we present here is due to Skorulski–Urbański obtained in [91].

Theorem 3.44 ([91]) *Suppose that* $\Lambda \subset \mathbb{C}$ *is an open set,* $\mathcal{F} = \{f_\lambda\}_{\lambda \in \Lambda}$ *is an analytic family of meromorphic functions and that, for some* $\lambda_0 \in \Lambda$, $f_{\lambda_0} : \mathbb{C} \to \hat{\mathbb{C}}$ *is a dynamically regular meromorphic function with* $\mathrm{HD}_{\mathrm{hyp}}\left(f_{\lambda_0}\right) > \Theta_{f_{\lambda_0}}$ *and which belongs to class S. Then the function*

$$\lambda \longmapsto \mathrm{HD}(J_r(f_\lambda))$$

is real–analytic in some open neighborhood of λ_0. □

Notice that here the main hypotheses are only imposed on the function f_{λ_0} and not on all functions in a neighborhood of it. The authors obtained this result by associating to the globally defined functions locally defined iterated functions systems (IFS). This is possible by employing so called nice sets whose existence in the transcendental case is due to Doobs [34] and which have been initially brought to complex dynamics by Rivera–Letelier in [83] and Przytycki and Rivera–Letelier in [76]. An open connected set $U \subset \mathbb{C}$ is called nice if and only if every connected component of $f^{-n}(U)$ is either contained in U or disjoint from U. If U is disjoint from the post–singular set, then one can consider all possible holomorphic inverse branches of iterates of f and the properties of the nice set imply that the inverse branches that land in U for the first time define a good countable alphabet conformal IFS in the sense of [52] and[53]. It turns out that the limit set of this IFS has the same dimension as the hyperbolic dimension of f [91, Theorem 3.4]. Thus it suffices to consider IFSs. The later have been extensively studied [53] providing many useful tools, and, especially, developing the full thermodynamic formalism, and introducing the concepts of regular, strongly regular, co-finitely regular and irregular conformal IFSs. One of the greatest challenges to apply Theorem 3.44 is to show that $\mathrm{HD}_{\mathrm{hyp}}\left(f_{\lambda_0}\right) > \Theta_{f_{\lambda_0}}$. In terms of the associated conformal IFSs this means that the IFS coming from f_{λ_0} is strongly regular.

The common underlying strategy for establishing real analytic variation of Hausdorff dimension of limit sets of conformal IFSs, see [56, 58, 60, 104] for ex., is to complexify the setting and to apply Kato–Rellich Perturbation Theorem. The later is possible thank's to the spectral gap property which means that $\exp(\mathrm{P}(t))$

is a leading isolated simple eigenvalue of the transfer operator and the rest of the spectrum of this operator is contained in a disk centered at 0 whose radius is strictly smaller than $\exp(P(t))$. An alternative powerful strategy is used in [63]. It is based on Birkhoff's approach [18] to the Perron–Frobenius Theorem via positive cones. This method has been successfully applied in various contexts. The paper [63] which deals with random dynamics, is based on ideas from Rugh's paper [87] who used complexified cones. This powerful method works well as soon as appropriate invariant cones are found and strict contraction of the transfer operator in the appropriate Hilbert metric has been shown. The following is a particular result in [63].

Theorem 3.45 ([63]) *Let* $f_\eta(z) = \eta e^z$ *and let* $a \in (\frac{1}{3e}, \frac{2}{3e})$ *and* $0 < r < r_{max}$, $r_{max} > 0$. *Suppose that* $\eta_1, \eta_2, ..$ *are i.i.d. random variables uniformly distributed in* $\mathbb{D}(a, r)$. *Let* $J_{\eta_1, \eta_2, ...}$ *denote the Julia set of the sequence of compositions*

$$f_{\eta_n} \circ f_{\eta_{n-1}} \circ \ldots \circ f_{\eta_2} \circ f_{\eta_1} : \mathbb{C} \longrightarrow \mathbb{C}, \quad n \geq 1,$$

and let

$$J_r(\eta_1, \eta_2, \ldots) = \left\{ z \in J_{\eta_1, \eta_2, ...} : \liminf_{n \to \infty} | f_{\eta_n} \circ \ldots \circ f_{\eta_1}(z)| < +\infty \right\}$$

be the radial Julia set of $\{f_{\eta_n} \circ \ldots \circ f_{\eta_1}\}_{n \geq 1}$. *Then, the Hausdorff dimension of* $J_r(\eta_1, \eta_2, \ldots)$ *is almost surely constant and depends real-analytically on the parameters* (a, r) *provided that* r_{max} *is sufficiently small.* $\square$

In contrast to the case of hyperbolic rational functions, analytic variation of the hyperbolic dimension can fail in the class of hyperbolic entire functions of bounded type. This has been recently proved in [61].

Theorem 3.46 ([61]) *There exists a holomorphic family* $\mathcal{F} = \{f_\lambda = \lambda f, \ \lambda \in \mathbb{C}^*\}$ *of finite order entire functions in class* $\mathcal{B}$ *such that the functions* f_λ, $\lambda \in (0, 1]$, *are all in the same hyperbolic component of the parameter space but the function*

$$\lambda \mapsto HypDim(f_\lambda)$$

is not analytic in $(0, 1]$. $\square$

In order to obtain this result, the authors exploited the dichotomy of (3.7.1). In fact, all positive analyticity results use, sometimes implicitly like in Theorem 3.45, the assumption $HD_{hyp}(f) > \Theta$. Using the formula (3.5.7) for the transfer operator, Mayer and Zdunik where able to construct in [61] entire functions for which $HD_{hyp}(f) = \Theta$, so obtaining the very special case of equality in (3.7.1). Moreover, among these functions there are some that have strictly negative pressure at Θ which is the key point not only for Theorem 3.46 but also for the absence of recurrent conformal measures in Theorem 3.20.

3.8 Beyond Hyperbolicity

For many kinds of non-hyperbolic holomorphic/conformal dynamical systems various forms of thermodynamical formalism have been also successfully developed and usually much earlier than for transcendental dynamics. This is the case for rational functions and generalized polynomial–like mappings having certain type of critical points in the Julia set so that the functions are no longer hyperbolic but sufficient expansion is maintained. Most notably this is so for parabolic rational functions, subexpanding rational functions, and most generally, for non-recurrent rational functions and topological Collet–Eckmann rational functions; see ex. [2, 4, 5, 25–32, 38, 73–77, 94, 95, 102, 103], and the references therein. Note that some of these papers such as [27, 73] and [75] for ex. deal with all rational functions, in particular with no restrictions on critical points at all.

But there is a substantial difference with the hyperbolic case. Except perhaps [27] and [73], the Perron–Frobenius (transfer) operator for the original system is then virtually of no use—no change of Riemaniann metric seems to work. The most relevant questions are then about the structure of conformal measures, most notably, their existence, uniqueness, and atomlessness, and about Borel probability invariant measures absolutely continuous with respect such conformal measures, their existence, uniqueness and stochastic properties. Also, application of such results to study the fractal structure of Julia sets.

Similarly as for non-expanding rational functions, also for non-hyperbolic non-expanding transcendental, entire and meromorphic, functions some forms of thermodynamic formalism have been developed. For the papers coping with critical points in the Julia sets, which is closest to rational functions, see for ex. [44, 46]. One class of trancendental meromorphic functions deserves here special attention. These are elliptic (doubly periodic) meromorphic functions. The first fully developed account of thermodynamic formalism for all elliptic functions and Hölder continuous potentials (satisfying some additional natural hypotheses) was presented in [55]. Up to our best knowledge all other contributions to thermodynamic formalism for elliptic functions deal with geometric potentials of the form $-t \log |f'|$. We would like to mention in this context the paper [47], and, especially, the book [49], which provides an extensive and fairly complete account of thermodynamic formalism for many special, but quite large, classes of elliptic functions with some sufficiently strong expanding features.

The main difficulty and main point of interest in the classes of meromorphic functions discussed in the last paragraph were caused by critical points lying in the Julia sets. Going beyond critical points, there are visible two directions of research. Both of them deal with transcendental entire functions where there are logarithmic singularities, in the form of asymptotic values, in the Julia sets.

One of them was initiated in [106] dealing with exponential functions λe^z, where 0, the asymptotic value, was assumed to escape to infinity sufficiently fast. The existence and uniqueness of conformal measures and the existence and uniqueness of Borel probability invariant measures absolutely continuous with respect to those conformal measures were proved therein. Its follow up was the paper [107] dealing with analogous classes of functions but iterated randomly. The full (random) thermodynamic formalism with respect to random conformal and invariant measures was laid down and developed therein.

The second direction of research initiated and developed in [57] aimed to analyze the contribution of non-recurrent logarithmic singularities. Indeed, the paper [57] by Mayer–Urbański considers the class of meromorphic functions with polynomial Schwarzian derivatives. For example the tangent family belongs to this class and in general such functions have no critical points and they have only finitely many logarithmic singularities. A surprising outcome of this paper was that the behavior of invariant measures absolutely continuous with respect to conformal measures did depend on the order of the function.

Theorem 3.47 *Let $f : \hat{\mathbb{C}} \to \hat{\mathbb{C}}$ be a meromorphic function f of polynomial Schwarzian derivative and assume that it is semi-hyperbolic in the following sense:*

- *All the asymptotic values are finite.*
- *The asymptotic values that belong to the Fatou set belong to attracting components.*
- *The asymptotic values that belong to the Julia set have bounded and non-recurrent forward orbits.*

Let $h := HD(J(f))$.

Then, a Patterson–Sullivan typ construction provides an atomless h-conformal measure and this measure is weakly metrically exact, hence ergodic and conservative. Moreover, there exists a σ-finite invariant measure μ absolutely continuous with respect to m and this measure

$$\mu \ \text{is finite} \quad \text{if and only if} \quad h > 3\frac{\rho}{\rho+1}$$

where $\rho = \rho(f)$ is the order of the function f. If μ is finite, then the dynamical systems (f, μ) it generates is metrically exact and, in consequence, its Rokhlin's natural extension is K-mixing. $\qquad\square$

Notice that $3\frac{\rho}{\rho+1} \geq 2$ if and only if the order $\rho \geq 2$. Consequently the measure μ is most often infinite. However, in the case of the tangent family, which is just one specific example among others, this invariant measure can be finite.

Acknowledgments Research of the second named author supported in part by the Simons Grant: 581668.

References

1. J. Aaronson, *An Introduction to Infinite Ergodic Theory*. Mathematical Surveys and Monographs, vol. 50 (American Mathematical Society, Providence, 1997)
2. J. Aaronson, M. Denker, M. urbański, Ergodic theory for Markov fibered systems and parabolic rational maps. Trans. Am. Math. Soc. **337**, 495–548 (1993)
3. M. Aspenberg, W. Bergweiler, Entire functions with Julia sets of positive measure. Math. Ann. **352**(1), 27–54 (2012)
4. A. Avila, M. Lyubich, Hausdorff dimension and conformal measures of Feigenbaum Julia sets. J. Am. Math. Soc. **21**, 305–363 (2008)
5. A. Avila, M. Lyubich, Lebesgue measure of Feigenbaum Julia sets. (2015, preprint). arXiv:1504.02986
6. K. Barański, Hausdorff dimension and measures on julia sets of some meromorphic maps. Fundam. Math. **147**(3), 239–260 (1995)
7. K. Barański, Trees and hairs for some hyperbolic entire maps of finite order. Math. Z. **257**(1), 33–59 (2007)
8. K. Barański, B. Karpińska, Coding trees and boundaries of attracting basins for some entire maps. Nonlinearity **20**, 391–415 (2007)
9. K. Barański, B. Karpińska, A. Zdunik, Hyperbolic dimension of Julia sets of meromorphic maps with logarithmic tracts. Int. Math. Res. Not. **4**, 615–624 (2009)
10. K. Barański, B. Karpińska, A. Zdunik, Bowen's formula for meromorphic functions. Ergodic Theory Dynam. Syst. **32**(4), 1165–1189 (2012)
11. K. Barański, B. Karpińska, A. Zdunik, Conformal measures for meromorphic maps. Ann. Acad. Sci. Fenn. Math. **43**(1), 247–266 (2018)
12. K. Barański, N. Fagella, X. Jarque, B. Karpińska, Fatou components and singularities of meromorphic functions. Proc. Royal Soc. Edinburgh **150**, 633–654 (2020)
13. W. Bergweiler, Iteration of meromorphic functions. Bull. Am. Math. Soc. **29**, 151–188 (1993)
14. W. Bergweiler, Invariant domains and singularities. Math. Proc. Cambridge Philos. Soc. **117**(3), 525–532 (1995)
15. W. Bergweiler, Lebesgue measure of Julia sets and escaping sets of certain entire functions. Fund. Math. **242**(3), 281–301 (2018)
16. W. Bergweiler, I. Chyzhykov, Lebesgue measure of escaping sets of entire functions of completely regular growth. J. Lond. Math. Soc. **94**(2), 639–661 (2016)
17. W. Bergweiler, A. Eremenko, Direct singularities and completely invariant domains of entire functions. Illinois J. Math. **52**(1), 243–259 (2008)
18. G. Birkhoff, Extensions of Jentzsch's theorem. Trans. Am. Math. Soc. **85**, 219–227 (1957)
19. C.J. Bishop, Models for the Eremenko-Lyubich class. J. Lond. Math. Soc. **92**(1), 202–221 (2015)
20. R. Bowen, *Equilibrium States and the Ergodic Theory of Anosov Diffeomorphisms*. Lecture Notes in Mathematics, vol. 470 (Springer, Berlin, 1975)
21. R. Bowen, Hausdorff dimension of quasicircles. Inst. Hautes Études Sci. Publ. Math. **50**, 11–25 (1979)
22. I. Coiculescu, B. Skorulski, Thermodynamic formalism of transcendental entire maps of finite singular type. Monatsh. Math. **152**(2), 105–123 (2007)
23. I. Coiculescu, B. Skorulski, Perturbations in the Speiser class. Rocky Mountain J. Math. **37**(3), 763–800 (2007)
24. A.M. Davie, M. Urbański, A. Zdunik, Maximizing measures of metrizable non-compact spaces. Proc. Edinb. Math. Soc. **50**(1), 123–151 (2007)
25. M. Denker, M. Urbański, On Sullivan's conformal measures for rational maps of the Riemann sphere. Nonlinearity **4**, 365–384 (1991)
26. M. Denker, M. Urbański, Hausdorff and conformal measures on Julia sets with a rationally indifferent periodic point. J. Lond. Math. Soc. **43**, 107–118 (1991)

27. M. Denker, M. Urbański, Ergodic theory of equilibrium states for rational maps. Nonlinearity **4**, 103–134 (1991)
28. M. Denker, M. Urbański, On absolutely continuous invariant measures for expansive rational maps with rationally indifferent periodic points. Forum Math. **3**, 561–579 (1991)
29. M. Denker, M. Urbański, On Hausdorff measures on Julia sets of subexpanding rational maps. Israel J. Math. **76**, 193–214 (1991)
30. M. Denker, M. Urbański, Geometric measures for parabolic rational maps. Ergodic Theory Dynam. Syst. **12**, 53–66 (1992)
31. M. Denker, M. Urbański, The capacity of parabolic Julia sets. Math. Zeitsch. **211**, 73–86 (1992)
32. M. Denker, D. Mauldin, Z. Nitecki, M. Urbański, Conformal measure for rational functions revisited. Fund. Math. **157**, 161–173 (1998)
33. A. DeZotti, L. Rempe-Gillen, Eventual hyperbolic dimension of entire functions and poincaré functions of polynomials (2020, preprint). arXiv:2001.06353
34. N. Dobbs, Nice sets and invariant densities in complex dynamics. Math. Proc. Cambridge Philos. Soc. **150**(1), 157–165 (2011)
35. C.I. Doering, R. Mané, *The Dynamics of Inner Functions*. Ensaios Matemáticos **3**, 5–79 (1989)
36. A. Douady, P. Sentenac, M. Zinsmeister, Implosion parabolique et dimension de Hausdorff. CR Acad. Sci. I **325**, 765–772 (1997)
37. A. Eremenko, M.Y. Lyubich, Dynamical properties of some classes of entire functions. Ann. l'institut Fourier **42**(4), 989–1020 (1992)
38. P. Grzegorczyk, F. Przytycki, W. Szlenk, On iterations of Misiurewicz's rational maps on the Riemann sphere. Ann. Inst. Henri Poincaré, **53**, 431–434 (1990)
39. P.R. Halmos, Invariant measures. Ann. Math. **48**, 735–754 (1947)
40. F. Iversen, Recherche sur les fonctions inverses des fonctions méromorphes. Thèse de Helsingfors (1914)
41. P.W. Jones, N.G. Makarov, Density properties of harmonic measure. Ann. Math. Second Ser. **142**(3), 427–455 (1995)
42. B. Karpińska, Area and Hausdorff dimension of the set of accessible points of the Julia sets of λe^z and $\lambda \sin z$. Fund. Math. **159**(3), 269–287 (1999)
43. J. Kotus, G. Świątek, No finite invariant density for Misiurewicz exponential maps. Comptes Rendus de l'Académie de Sciences - Série Mathématique **346**, 559–562 (2008)
44. J. Kotus, G. Świątek, Invariant measures for meromorphic Misiurewicz maps. Math. Proc. Cambridge Philos. Soc. **145**, 685–697 (2008)
45. J. Kotus, M. Urbański, Conformal, geometric and invariant measures for transcendental expanding functions. Math. Ann. **324**(3), 619–656 (2002)
46. J. Kotus, M. Urbański, Existence of invariant measures for transcendental subexpanding functions. Math. Zeitschrift **243**, 25–36 (2003)
47. J. Kotus, M. Urbański, Geometry and ergodic theory of non–recurrent elliptic functions. J. Anal. Math. **93**, 35–102 (2004)
48. J. Kotus, M. Urbański, Fractal measures and ergodic theory of transcendental meromorphic functions, in *Transcendental Dynamics and Complex Analysis*. London Mathematical Society Lecture Note Series, vol. 348 (Cambridge University Press, Cambridge, 2008), pp. 251–316
49. J. Kotus, M. Urbański, Ergodic Theory, Geometric Measure Theory, Conformal Measures and the Dynamics of Elliptic Functions (2020, preprint). arxiv
50. M.Y. Lyubich, Mesurable dynamics of the exponential. Sibirskii Mataticheskii Zhurnal **28**(5), 128–133 (1987)
51. N.G. Makarov, Fine structure of harmonic measure. Algebra i Analiz **10**(2), 1–62 (1998)
52. R.D. Mauldin, M. Urbański, Dimensions and measures in infinite iterated function systems. Proc. Lond. Math. Soc. **73**(3), 105–154 (1996)
53. R.D. Mauldin, M. Urbański, *Graph Directed Markov Systems*. Cambridge Tracts in Mathematics, vol. 148 (Cambridge University Press, Cambridge, 2003)

54. V. Mayer, A lower bound of the hyperbolic dimension for meromorphic functions having a logarithmic hölder tract. Conformal Geom. Dynam. **22**, 62–77 (2018)
55. V. Mayer, M. Urbański, Gibbs and equilibrium measures for elliptic functions. Math. Z. **250**(3), 657–683 (2005)
56. V. Mayer, M. Urbański, Geometric thermodynamic formalism and real analyticity for meromorphic functions of finite order. Ergodic Theory Dynam. Syst. **28**(3), 915–946 (2008)
57. V. Mayer, M. Urbański, Ergodic properties of semi-hyperbolic functions with polynomial Schwarzian derivative. Proc. Edinb. Math. Soc. **53**(2), 471–502 (2010)
58. V. Mayer, M. Urbański, Thermodynamical formalism and multifractal analysis for meromorphic functions of finite order. Mem. Am. Math. Soc. **203**(954), vi+107 (2010)
59. V. Mayer, M. Urbański, Random dynamics of transcendental functions. J. d'Analyse Math. **134**(1), 201–235 (2018)
60. V. Mayer, M. Urbański, Thermodynamic formalism and integral means spectrum of logarithmic tracts for transcendental entire functions. Trans. Amer. Math. Soc. **373**, 7669–7711 (2020).
61. V. Mayer, A. Zdunik, The failure of Ruelle's property for entire functions. Adv. Math. **384** (to appear, 2021)
62. V. Mayer, B. Skorulski, M. Urbanski, *Distance Expanding Random Mappings, Thermodynamical Formalism, Gibbs Measures and Fractal Geometry*. Lecture Notes in Mathematics, vol. 2036 (Springer, Heidelberg, 2011)
63. V. Mayer, M. Urbański, A. Zdunik, Real analyticity for random dynamics of transcendental functions. Ergodic Theory Dynam. Syst. **40**, 490–520 (2018)
64. C. McMullen, Area and Hausdorff dimension of Julia sets of entire functions. Trans. Am. Math. Soc. **300**(1), 329–342 (1987)
65. C.T. McMullen, Hausdorff dimension and conformal dynamics ii: geometrically finite rational maps. Comment. Math. Helv. **75**, 535–593 (2000)
66. S. Munday, M. Roy, M. Urbański, Non–Invertible Dynamical Systems (in preparation)
67. R. Nevanlinna, *Eindeutige Analytische Funktionen* (Springer, Berlin, 1974). Zweite Auflage, Reprint, Die Grundlehren der mathematischen Wissenschaften, Band 46
68. R. Nevanlinna, *Analytic Functions*. Translated from the second German edition by Phillip Emig. Die Grundlehren der mathematischen Wissenschaften, Band, vol. 162 (Springer, New York, 1970)
69. On iterates of e^z. Ergodic Theory Dynam. Syst. **1**, 103–106 (1981)
70. S.J. Patterson, The limit set of a Fuchsian group. Acta Math. **136**, 241–273 (1976)
71. S.J. Patterson, Lectures on measures on limit sets of Kleinian groups, in *Analytical and Geometric Aspects of Hyperbolic Space*. London Mathematical Society, Lecture Notes, vol. 111 (Cambridge University Press, Cambridge, 1987)
72. C. Pommerenke, *Boundary Behaviour of Conformal Maps*. Grundlehren der Mathematischen Wissenschaften [Fundamental Principles of Mathematical Sciences], vol. 299 (Springer, Berlin, 1992)
73. F. Przytycki, On the Perron–Frobenius-Ruelle operator for rational maps on the Riemann sphere and for Hölder continuous functions. Bol. Soc. Bras. Mat. **20**, 95–125 (1990)
74. F. Przytycki, Iterations of holomorphic Collet-Eckmann maps: conformal and invariant measures. Trans. Am. Math. Soc. **350**(2), 717–742 (1998)
75. F. Przytycki, Geometric pressure in real and complex 1-dimensional dynamics via trees of preimages and via spanning sets. Monatshefte fur Mathematik **185**(1), 133–158 (2018)
76. F. Przytycki, J. Rivera-Letelier, Statistical properties of topological Collet-Eckmann maps. Ann. Sci. École Norm. Sup. **40**(1), 135–178 (2007)
77. F. Przytycki, J. Rivera-Letelier, Nice inducing schemes and the thermodynamics of rational maps. Commun. Math. Phys. **301**(3), 661–707 (2011)
78. F. Przytycki, M. Urbański, *Conformal Fractals: Ergodic Theory Methods*. London Mathematical Society Lecture Note Series, vol. 371 (Cambridge University Press, Cambridge, 2010)
79. L. Rempe, Hyperbolic dimension and radial julia sets of transcendental functions. Proc. Am. Math. Soc. **137**(4), 1411–1420 (2009)

80. L. Rempe, Rigidity of escaping dynamics for transcendental entire functions. Acta Math. **203**(2), 235–267 (2009)
81. L. Rempe-Gillen, Hyperbolic entire functions with full hyperbolic dimension and approximation by eremenko-lyubich functions. Proc. Lond. Math. Soc. **108**(5), 1193–1225 (2014)
82. P.J. Rippon, Baker domains, in *Transcendental Dynamics and Complex Analysis*. London Mathematical Society Lecture Note Series, vol. 348 (Cambridge University Press, Cambridge, 2008), pp. 371–395
83. J. Rivera–Letelier, A connecting lemma for rational maps satisfying a no-growth condition. Ergod. Theory Dynam. Syst. **27**, 595–636 (2007)
84. G. Rottenfusser, J. Rückert, L. Rempe, D. Schleicher, Dynamic rays of bounded-type entire functions. Ann. Math. **173**(1), 77–125 (2011)
85. D. Ruelle, *Thermodynamic Formalism*. Encyclopedia of Mathematics and its Applications, vol. 5 (Addison-Wesley, Reading, 1978). The mathematical structures of classical equilibrium statistical mechanics, With a foreword by Giovanni Gallavotti and Gian-Carlo Rota
86. D. Ruelle, Repellers for real analytic maps. Ergodic Theory Dynam. Syst. **2**(1), 99–107 (1982)
87. H.H. Rugh, On the dimension of conformal repellors. randomness and parameter dependency. Ann. Math. **168**(3), 695–748 (2008)
88. M. Shishikura, The Hausdorff dimension of the boundary of the Mandelbrot set and Julia sets. Ann. Math. **147**, 225–267 (1998)
89. Y.G. Sinai, Gibbs measures in eergodic theory. Russian Math. Surv. **27**, 21–70 (1972)
90. D.J. Sixsmith, Julia and escaping set spiders' webs of positive area. Int. Math. Res. Not. **2015**(19), 9751–9774 (2015)
91. B. Skorulski, M. Urbański, Finer fractal geometry for analytic families of conformal dynamical systems. Dyn. Syst. **29**(3), 369–398 (2014)
92. G.M. Stallard, Entire functions with Julia sets of zero measure. Math. Proc. Cambridge Philos. Soc. **108**(3), 551–557 (1990)
93. G.M. Stallard, Dimensions of Julia sets of transcendental meromorphic functions, in *Transcendental Dynamics and Complex Analysis*. London Mathematical Society Lecture Note Series, vol. 348 (Cambridge University Press, Cambridge, 2008), pp. 425–446
94. B. Stratmann, M. Urbański, Real analyticity of topological pressure for parabolically semihyperbolic generalized polynomial-like maps. Indag. Math. **14**, 119–134 (2003)
95. B. Stratmann, M. Urbański, Multifractal analysis for parabolically semihyperbolic generalized polynomial-like maps. IP New Stud. Adv. Math. **5**, 393–447 (2004)
96. D. Sullivan, The density at infinity of a discrete group. Inst. Hautes Etudes Sci. Pub. Math. **50**, 171–202 (1979)
97. D. Sullivan, Disjoint spheres, approximation by imaginary quadratic numbers and the logarithmic law for geodesics. Acta Math. **149**, 215–237 (1982)
98. D. Sullivan, Seminar on conformal and hyperbolic geometry. Notes by M. Baker and J. Seade, Preprint IHES (1982), 92 pp
99. D. Sullivan, Conformal dynamical systems, in *Geometric Dynamics*. Lecture Notes in Mathematics, vol. 1007 (Springer, Berlin, 1983), pp. 725–752
100. D. Sullivan, Entropy, Hausdorff measures old and new, and the limit set of a geometrically finite Kleinian groups. Acta. Math. **153**, 259–277 (1984)
101. D. Sullivan, Quasiconformal homeomorphisms in dynamics, topology, and geometry, in *Proceedings of International Congress of Mathematics, Berkeley* (American Mathematical Society, Providence, 1986)
102. M. Urbański, Rational functions with no recurrent critical points. Ergod. Theory Dynam. Syst. **14**, 391–414 (1994)
103. M. Urbański, Geometry and ergodic theory of conformal non–recurrent dynamics. Ergod. Theory Dynam. Syst. **17**, 1449–1476 (1997)
104. M. Urbański, A. Zdunik, The finer geometry and dynamics of the hyperbolic exponential family. Michigan Math. J. **51**(2), 227–250 (2003)
105. M. Urbański, A. Zdunik, Real analyticity of Hausdorff dimension of finer Julia sets of exponential family. Ergodic Theory Dynam. Syst. **24**(1), 279–315 (2004)

106. M. Urbański, A. Zdunik, Geometry and ergodic theory of non–hyperbolic exponential maps. Trans. Am. Math. Soc. **359**(8), 3973–3997 (2007)
107. M. Urbański, A. Zdunik, Random non–hyperbolic exponential maps. (2018, preprint). To appear Journal d'Analyse Mathématique
108. P. Walters, A variational principle for the pressure of continuous transformations. Am. J. Math. **97**, 937–971 (1975)
109. P. Walters, *An Introduction to Ergodic Theory*. Graduate Texts in Mathematics, vol. 79 (Springer, New York, 1982)
110. J.-H. Zheng, Dynamics of hyperbolic meromorphic functions. Discrete Contin. Dyn. Syst. **35**(5), 2273–2298 (2015)
111. M. Zinsmeister, *Thermodynamic Formalism and Holomorphic Dynamical Systems*. SMF/AMS Texts and Monographs, vol. 2 (American Mathematical Society, Providence; Société Mathématique de France, Paris, 2000). Translated from the 1996 French original by C. Greg Anderson

Part III
Probability Theory Ergodicity and Thermodynamic Formalism

Chapter 4
Recurrent Sets for Ergodic Sums of an Integer Valued Function

Jean-Pierre Conze

Abstract For an ergodic measure preserving dynamical system $(X, \mathcal{B}, \mu, T)$ and an integrable function f with values in $\mathbb{Z}^d$, let $(S_n f(x) := \sum_{k=0}^{n-1} f(T^k x), n \geq 1)$ be the process of ergodic sums of f. Given a finite or infinite subset $\mathcal{L}$ of $\mathbb{Z}^d$, a question is whether $\mathcal{L}$ is recurrent for the process in the sense that $S_n f(x) \in \mathcal{L}$ infinitely often for a.e. x. We will survey various examples, for non centered or centered functions f in dimension $d = 1$ or ≥ 1. For example, for $d = 1$, one can estimate the number of visits before time n to the set of squares in $\mathbb{Z}$ when $\mu(f) \neq 0$ (consequence of J. Bourgain's result (1989)). But if $\mathcal{L}$ has unbounded gaps and if $0 \notin \mathcal{L}$, over rotations there are simple integrable centered functions f generating "non regular" cocycles such that $(S_n f)$ does not intersect $\mathcal{L}$. For a transient random walk in $\mathbb{Z}^d$, $d \geq 3$, we give examples of infinite recurrent sets and infinite transient sets.

4.1 Introduction

In what follows $(X, \mathcal{B}, \mu)$ is a probability space without atoms and T is an *ergodic* measure preserving transformation acting on X. Let f be a measurable function on X with values in $\mathbb{R}^d$, $d \geq 1$. Its ergodic sums under the iteration of T are

$$S(T, f, n, x) = S_n f(x) := \sum_{k=0}^{n-1} f(T^k x), n \geq 1.$$

The sequence $(S(T, f, n, x), n \geq 1)$ is a "cocycle" denoted by (T, f). We denote by $S(T, f, x)$, or simply $S(f, x)$, the set $\{S(T, f, n, x), n \geq 1\}$. For $d = 1$, we call the cocycle (T, f) *positive* if $f(x) > 0$, for a.e. $x \in X$.

J.-P. Conze (✉)
University of Rennes, CNRS, IRMAR - UMR 6625, Rennes, France
e-mail: conze@univ-rennes1.fr

Unless explicitly stated, we assume that f takes its values in $\mathbb{Z}^d$ and that, for any integer $a > 1$, f does not take its values in $a\,\mathbb{Z}^d$, for μ-a.e. x.

Our aim is to discuss the following question: if $\mathcal{L}$ is a subset of $\mathbb{Z}^d$, do the ergodic sums $S_n f(x)$ visit $\mathcal{L}$ infinitely often for a.e. x and is there a quantitative estimate?

As it is known, for the process $(S_n f)_{n \geq 1}$ generated by an ergodic transformation, there is a dichotomy: it is either transient ($S_n f(x) \to \infty$, for a.e. x) or recurrent (the $\mathbb{Z}^d$-valued process satisfies $S_n f(x) = \underline{0}$, infinitely often, for a.e. x). In the previous question, the set $\mathcal{L}$ should be an infinite set in the transient case, while in the recurrent case, the discussion is more about the "recurrence set" of the cocycle.

Definition 4.1 For a cocycle (T, f) with values in $\mathbb{R}^d$, we call $\underline{r} \in \mathbb{R}^d$ a recurrent value, if $\mu(x : \|S(T, f, n, x) - \underline{r}\| < \varepsilon \text{ i.o.}) = 1$, for all $\varepsilon > 0$. The recurrence set is the set $\mathcal{R}(T, f)$ of recurrent values.

So, for $f : X \to \mathbb{Z}^d$, $\mathcal{R}(T, f)$ is the set of values which are visited infinitely often.

In the recurrent case, we will see that the recurrence set is related to the "regularity" of the cocycle (Sect. 4.3). A cocycle (T, f) is transient if and only if $\mathcal{R}(T, f)$ is empty. The question of recurrence into infinite sets in the sense of the next definition is adapted to the transient case, even it can be asked for a recurrent process.

Definition 4.2 A subset $\mathcal{L}$ of $\mathbb{Z}^d$ is called *recurrent for a cocycle* (T, f) if, for a.e. x, $S(T, f, n, x) \in \mathcal{L}$ for infinitely many n.

Sets of Recurrence in Ergodic Theory

The question of recurrence for a subset $\mathcal{L}$ of $\mathbb{Z}^d$ appeared for random walks in the sixties [30]. Some years later, it became an important topic in ergodic theory.

Definition 4.3 (Furstenberg [16]) A set of positive integers $\mathcal{L} = \{\ell_1 < \ell_2 < \ldots < \ell_k < \ldots\}$ is called a "set of recurrence" (or a recurrent sequence, or a Poincaré set) if, for all dynamical systems $(Y, \mathcal{A}, \nu, S)$ and all subsets $A \in \mathcal{A}$ of positive measure, there are infinitely many $\ell \in \mathcal{L}$ such that $\nu(A \cap S^{-\ell}A) > 0$.

An equivalent property is that, for every system $(Y, \mathcal{A}, \nu, S)$ and every subset $A \in \mathcal{A}$ with $\nu(A) > 0$, the intersection $\mathcal{L} \cap \{\ell : S^\ell y \in A\}$ is infinite for a.e. $y \in A$.

In Definition 4.2, we are interested in the question of recurrence with respect to a given cocycle (T, f). The answer for a given set depends on the spectral or stochastic properties of T. For $d = 1$, Definition 4.3 expresses a "universal" property of recurrence for a set $\mathcal{L} \subset \mathbb{N}$, which is equivalent to recurrence in the sense of Definition 4.2 for all positive integrable cocycles (T, f) (see below).

Let us mention also another related point of view: ergodic theorems along a subsequence and "universally representative" sampling schemes, a topic developed in the 1970s and later. The pointwise ergodic theorem along recurrence times proved by Bourgain [6] shows that the sequence of recurrence times in a set for a dynamical system is almost surely a "universally good sequence" of summation and this can be used as in [25].

Our aim is merely to survey various examples. We are going to take for $\mathcal{L}$ sequences of positive density or polynomial growth, sequences of ergodic sums for another dynamical system (Y, ν, S), polynomial sequences, the sequence of prime numbers and also arbitrary strictly increasing sequences.

The content of the paper is the following. Section 4.2 concerns the case $d = 1$, $\int f \, d\mu \neq 0$, in relation with special maps and return times into a set. In Sect. 4.3, we take $d \geq 1$, f integrable and centered. The relation between recurrence set, set of essential values and "regularity" of the cocycle is discussed. Sets which are not recurrent for some non regular cocycles, are constructed. In Sect. 4.4, for transient cocycles defined by random walks in $\mathbb{Z}^d$, $d \geq 3$, we give examples of infinite recurrent sets and infinite transient sets.

Acknowledgments: I would like to thank V. Bergelson, Y. Coudène, Y. Guivarc'h and E. Lesigne for fruitful discussions, as well as the referee for the very helpful remarks.

This paper is dedicated to the memory of Michael Boshernitzan, with whom some of the topics presented here have been discussed some years ago.

4.2 Non Centered Case for $d = 1$

In this section, $\mathcal{L} = \{\ell_1 < \ell_2 < \ldots < \ell_n < \ldots\}$ is a strictly increasing sequence of positive integers. We begin by preliminaries on special maps.

4.2.1 Preliminaries

4.2.1.1 Special Map T_f

With the notation of the introduction, let $f : X \to \mathbb{Z}$ be integrable and ≥ 1 (the cocycle is said to be *positive*). The *(discrete time) special map* T_f is defined on

$$\tilde{X} := \{(x, k), \ x \in X, k = 0, \ldots, f(x) - 1\}$$

by $T_f(x, k) := (x, k+1)$, if $0 \leq k < f(x) - 1$, $:= (Tx, 0)$, if $k = f(x) - 1$.

The probability measure $\tilde{\mu}$ is defined on $\tilde{X}$ by $\tilde{\mu}(A \times \{k\}) = \mu(f)^{-1} \mu(A)$, for $k \geq 0$ and $A \subset \{x : k \leq f(x) - 1\}$. The space X can be identified with the subset $B_0 = \{(x, 0), x \in X\}$ of $\tilde{X}$ with normalized measure. The set B_0 is the *basis* and $f - 1$ the *ceiling* function of the special map T_f.

For $x \in B_0$, $f(x)$ is the first return time of $(x, 0)$ in B_0 for T_f. The n-th return time is $S(T, f, n, x)$. Therefore, it holds:

$$T_f^n(x, 0) \in B_0 \iff n \in \mathcal{S}(T, f, x). \tag{4.2.1}$$

This shows that formally, for $d = 1$, the study of the values of the ergodic sums for a positive cocycle reduces to that of sets of recurrence for the associated special map.

Induced Cocycle For a dynamical system $(X, \mathcal{B}, \mu, T)$ and a set of positive measure $B \in \mathcal{B}$, let T_B be the induced map on B, $R^B(x)$ the first return time of x in B and $R_n^B(x) := \sum_{k=0}^{n-1} R^B(T_B^k x)$ the n-th return time of x in B.

We have $\{R_n^B(x), n \geq 1\} = \mathcal{S}(T_B, R^B, x)$ and, since T is assumed to be ergodic,

$$\lim_n \frac{1}{n} R_n^B(x) = \mu(B)^{-1}, \quad \text{for a.e. } x \in B. \tag{4.2.2}$$

If f is a measurable function on X, the *induced cocycle* on B is defined on B by

$$f^{T,B}(x) = f^B(x) := f(x) + f(Tx) + \ldots + f(T^{R^B(x)-1} x).$$

When f is integrable, then f^B is integrable. We have the inclusion $\mathcal{S}(T_B, f^B, x) \subset \mathcal{S}(T, f, x)$ for $x \in B$. The special maps T_f and $(T_B)_{f^B}$ coincide.

4.2.1.2 Aperiodicity

Definition 4.4 Let (T, f) be any 1-dimensional cocycle (with $f : X \to \mathbb{Z}$). Let us consider the coboundary multiplicative equation in $t \in \mathbb{R}$ and h measurable with modulus 1:

$$h(Tx) = e^{2\pi i t \, f(x)} h(x), \quad \text{for a.e. } x. \tag{4.2.3}$$

We say that (T, f) is *aperiodic* if (4.2.3) has no measurable solution for $t \in \mathbb{R} \setminus \mathbb{Z}$, and *r-aperiodic* if (4.2.3) has no measurable solution for $t \in \mathbb{Q} \setminus \mathbb{Z}$.

See also Definition 4.6 for $d \geq 1$. Remark that the terminology "aperiodic" is also associated to the equation (in h measurable and ρ constant) $Th = \rho \, e^{2\pi i t \, f} h$, (see [22] for non uniformly expanding Markov maps).

Let B be a set of positive measure in X. For any f a cocycle (T, f) is aperiodic if and only if (T_B, f^B) is aperiodic. Given t, there is a solution of (4.2.3) for (T, f), if and only if there is a solution of (4.2.3) for (T_B, f^B).

Eigenvalues of T_f The map T_f is ergodic if and only if T is ergodic. Equation (4.2.3) has a solution h if and only if $e^{2\pi i t}$ is an eigenvalue of T_f.

It follows that T_f is weakly mixing, if and only if (T, f) is aperiodic. It is totally ergodic, i.e., T_f^k is ergodic for every $k \geq 1$, if and only if (T, f) is r-aperiodic.

Examples Cocycles (T, f) with T a Bernoulli scheme and f a function of the first coordinate yield simple examples of aperiodic cocycles. Another examples are step functions over a 1-dimensional irrational rotation $x \rightarrow x + \alpha$ mod 1 with discontinuities satisfying some Diophantine condition with respect to α (cf. [18]).

Recall that the *Kronecker factor* of an ergodic dynamical system is its maximal factor with discrete spectrum. The *rational Kronecker factor* is generated by the eigenfunctions with eigenvalues roots of unity.

4.2.1.3   From $\mu(f) > 0$ to $f \geq 1$

For the non centered case $\mu(f) > 0$, we show how to carry back to the case $f \geq 1$. First let us recall the proof of the known fact that, on a dynamical system, a function with positive integral is the sum of a non negative function and a coboundary.

Lemma 4.1 *If $\int f \, d\mu > 0$, there are h measurable, g integrable non negative with $\mu(g \geq 1) > 0$, such that: $f = g + Th - h$. If f is integer valued, so are g and h.*

Proof Let $m_n(x) := \min_{1 \leq k \leq n} S_k f(x)$, $n \geq 1$. We have $m_{n+1}(x) = \min(f(x), f(x) + m_n(Tx)) = f(x) - m_n^-(Tx)$, if $m_n(Tx) \leq 0, = f(x) = f(x) - m_n^-(Tx)$, if $m_n(Tx) > 0$, which implies $m_{n+1}(x) = f(x) - m_n^-(Tx)$.

The limit $m_\infty(x) := \lim_n m_n(x)$ is a.e. finite, because $S_n f(x) \rightarrow +\infty$ by the ergodic theorem. It follows: $f(x) = m_\infty^-(Tx) - m_\infty^-(x) + m_\infty^+(x)$, for a.e. x.

If $m_\infty^+(x) = 0$ for a.e. x, then f is a coboundary. By considering the return times on a set on which m_∞^- is bounded, we get a contradiction with $S_n f(x) \rightarrow +\infty$.

By construction, $m_n(x) \leq f(x)$, which implies $m_\infty \leq f(x) \leq |f(x)|$, hence $m_\infty^-(x) \leq |f(x)|$ and m_∞^+, as f, is integrable. Moreover m_n (and therefore m_∞) has values in $\mathbb{Z}$ as f. Putting $h := m_\infty^-$, $g := m_\infty^+$, we get the result. $\square$

We say that a class C of cocycles is *closed by induction* if, for any (T, f) in C, the induced cocycle $(T_B, f^{T,B})$ is in C for every B of positive measure. Aperiodic cocycles or r-aperiodic cocycles are examples of such classes.

Proposition 4.1 *Let C be a class of cocycles closed by induction. If $\mathcal{L}$ is recurrent for every positive cocycle (T, f) in C, then $\mathcal{L}$ is recurrent for (T, f) in C such that $\mu(f) > 0$.*

Proof

(1) First we show that, if $\mu(f) > 0$, there is a partition of X in measurable sets of positive measure on which the induced cocycle is ≥ 1.

By the previous lemma, we can write $f = g + Th - h$, where h, g have values in $\mathbb{Z}^+$, the measure of the set $B := (g \geq 1)$ is positive and g is integrable. Let $D_j := \{h = j\}$ for $j \geq 0$. Since $S_n f = S_n g + T^n h - h$, it holds: $S_n f(x) = S_n g(x)$, if $x, T^n x \in D_j$. Let $C_{r,j} = T^{-r} B \cap D_j$. By ergodicity of T, we have $D_j = \bigcup_{r \geq 0} C_{r,j}$.

On $C_{r,j}$, the ergodic sums generated by $f^{C_{r,j}}(x)$ and $g^{C_{r,j}}(x)$ under the action of the induced map $T_{C_{r,j}}$ coincide, since $S_n f(x) = S_n g(x)$ if x and $T^n x \in C_{r,j}$.

Now, by Rohlin lemma, up to a set of zero measure, each set $C_{r,j}$ can be cut in a countable family of subsets $C_{r,j,\ell}$ of positive measure such that $R^{C_{r,j,\ell}}(x) > r$. For $x \in C_{r,j}$ we have: $T^r x \in B = \{g \geq 1\}$ and the induced cocycle on $C_{r,j,\ell}$ satisfies:

$$g^{C_{r,j,\ell}}(x) = g(x) + g(Tx) + \ldots + g(T^{R_{C_{r,j,\ell}}(x)-1}x) \geq g(T^r x) \geq 1, \text{ for } x \in C_{r,j,\ell}.$$

(2) Since $X = \cup_{r,j,\ell} C_{r,j,\ell}$, it suffices to prove the result for the restriction of f to the sets $C_{r,j,\ell}$ (called simply C). For $x \in C$, the ergodic sums of g^C and f^C for the induced map T_C coincide and belong to the sets of ergodic sums of g and f for T.

Let (T, f) in a class C closed by induction. If $\mathcal{L}$ is recurrent for positive cocycles in C, then $\mathcal{L}$ is recurrent for the induced cocycle (T_C, f^C), hence for (T, f). $\qquad\square$

4.2.2 Sequence of Positive Density

Let us assume that $\mathcal{L} = \{\ell_1 < \ell_2 < \ldots < \ell_n < \ldots\}$ has a positive density. Let C be a finite constant such that $\ell_n \leq Cn, \forall n \geq 1$.

Lemma 4.2

(1) If a dynamical system $(Y, \mathcal{A}, \nu, S)$ is weakly mixing, it holds

$$\frac{1}{n} \sum_{k=1}^{n} S^{\ell_k} \varphi \xrightarrow[n \to \infty]{L^2(\nu)} \int \varphi \, d\nu, \forall \varphi \in L^2(\nu). \tag{4.2.1}$$

(2) If the dynamical system has the Lebesgue spectrum property, then convergence a.e. holds in (4.2.1) for any $\varphi \in L^1(\nu)$.

Proof

(1) Let φ be in $L^2(\nu)$. Without loss of generality, we can assume $\int \varphi \, d\nu = 0$ and $\int |\varphi|^2 \, d\nu = 1$. There is a set J in $\mathbb{N}$ of zero density such that $\lim_{k \to \infty, k \notin J} \langle S^k \varphi, \varphi \rangle = 0$. Therefore, for $\varepsilon > 0$, there is $N_\varepsilon, L_\varepsilon$ such that, for $N \geq N_\varepsilon$, $|J \cap [1, N]| \leq \varepsilon N$ and $|\langle S^j \varphi, \varphi \rangle| \leq \varepsilon$, if $j \notin J$ and $j \geq L_\varepsilon$. Hence, we have

$$\| \sum_{k=1}^{n} S^{\ell_k} \varphi \|_2^2 \leq n + 2 \sum_{1 \leq k' < k \leq n} |\langle S^{\ell_k - \ell_{k'}} \varphi, \varphi \rangle| = n + (A) + (B) + (C) \text{ with}$$

$$(A) = 2 \sum_{\substack{L_\varepsilon < \ell_k - \ell_{k'} \in J^c \\ 1 \le k' < k \le n}} |\langle S^{\ell_k - \ell_{k'}} \varphi, \varphi \rangle| \le 2\varepsilon n^2,$$

$$(B) = 2 \sum_{\substack{L_\varepsilon \ge \ell_k - \ell_{k'} \in J^c \\ 1 \le k' < k \le n}} |\langle S^{\ell_k - \ell_{k'}} \varphi, \varphi \rangle| \le 2 L_\varepsilon^2,$$

$$(C) = 2 \sum_{\substack{\ell_k - \ell_{k'} \in J \\ 1 \le k' < k \le n}} |\langle S^{\ell_k - \ell_{k'}} \varphi, \varphi \rangle|$$

$$\le 2 \sum_{k' \le n} \mathrm{Card}\{k \le n : \ell_k \in J + \ell_{k'}\}$$

$$\le 2 \sum_{k' \le n} |(J + \ell_{k'}) \cap [1, \ell_n]| \le 2 \sum_{k' \le n} |J \cap [-\ell_{k'}, \ell_n - \ell_{k'}]|$$

$$\le 2n|J \cap [1, \ell_n]|$$

$$\le 2C\varepsilon n^2, \text{ for } \ell_n \ge N_\varepsilon \text{ and since } \ell_n \le Cn.$$

This implies $\dfrac{1}{n^2} \| \sum_{k=1}^{n} S^{\ell_k} \varphi \|_2^2 \le (n^{-1} + 2L_\varepsilon^2 n^{-2}) + 2(1 + C)\varepsilon$, for $n \ge N_\varepsilon$.

(2) Recall that a proof is the following: for the ergodic sums along a sequence of positive density, a maximal inequality holds; hence the set of φ such that pointwise convergence holds in (4.2.1) is closed in $L^1(\nu)$. As this set contains linear combinations of functions in $L^2(\nu)$ with orthogonal images by S^n, the result follows.

□

Theorem 4.1 *Let $\mathcal{L}$ be a sequence of positive density and $\mathcal{L}_n := \mathcal{L} \cap [1, n]$.*

(1) If (T, f) is an aperiodic positive cocycle, then

$$\frac{\mathrm{Card}(\mathcal{L}_n \cap S(T, f, \cdot))}{\mathrm{Card}(\mathcal{L}_n)} \xrightarrow[n \to \infty]{L^2(\mu)} \mu(f)^{-1}. \tag{4.2.2}$$

(2) $\mathcal{L}$ is recurrent for every aperiodic cocycle (T, f) such that $\mu(f) > 0$.
(3) If T_f has the Lebesgue spectrum property, then convergence a.e. holds in (4.2.2).

Proof

(1) With $S = T_f$, Lemma 4.2 implies, if T_f is weakly mixing:

$$\frac{1}{n} \sum_{k=1}^{n} 1_{B_0}(T_f^{\ell_k}(x, 0)) \xrightarrow[n \to \infty]{L^2} \tilde{\mu}(B_0) = \mu(f)^{-1}.$$

As $\sum_{k=1}^{n} 1_B(T_f^{\ell_k}(x,0)) = \sum_{k=1}^{n} 1_{\ell_k \in \mathcal{S}(T,f,x)} = \text{Card}([1,\ell_n] \cap \mathcal{L} \cap \mathcal{S}(T,f,x))$, the convergence (4.2.2) holds if (T,f) is aperiodic and $f \geq 1$. This implies:

$$\limsup_n \frac{\text{Card}(\mathcal{L}_n \cap \mathcal{S}(T,f,x))}{\text{Card}(\mathcal{L}_n)} > 0, \text{ for a.e. } x.$$

(2) By Proposition 4.1, this last property still holds if the cocycle is aperiodic and $\mu(f) > 0$.

(3) The last statement follows from 2) in Lemma 4.2.

$\square$

A Counter-Example for (T,f) Non Aperiodic The following construction gives a simple example of a sequence with positive density disjoint from the set of values of a positive non aperiodic cocycle.

Let $\alpha \in]0,1[$ be an irrational number and T_α the irrational rotation $x \to x + \alpha \mod 1$ on $[0,1[$. Let a be an integer ≥ 2, and define f by $f(x) = a$, for $x \in [0, 1-\alpha[, = a+1$, for $x \in]1-\alpha, 1[$. The special map over T_α with ceiling function f is the rotation $T_\gamma : x \to x + \gamma \mod 1$ on the unit interval, with $\gamma = \frac{1}{a+\alpha}$.

To show it, let us start with the rotation T_γ and consider the induced map on $]1 - \gamma, 1[$. Since $1 - \gamma < a\gamma < 1$, the interval $]2 - (1+a)\gamma, 1[$ is mapped mod 1 on $]1 - \gamma, a\gamma[$ after a iterations, and the interval $]1 - \gamma, 2 - (1+a)\gamma[$ is mapped on $]a\gamma, 1[$ after $a + 1$ iterations.

Therefore the induced map is the permutation on $]1 - \gamma, 1[$ of the two sub-intervals $]1 - \gamma, a\gamma[$ and $]a\gamma, 1[$. After normalisation by γ^{-1} and translation by $1 - \gamma^{-1}$, we obtain the isometric exchange of the sub-intervals $]0, 1 - \alpha[$ and $]1 - \alpha, 1[$ of $]0, 1[$, which is the rotation by α.

Observe that $1 - 2\gamma > 0$. Let $J :=]0, 1 - 2\gamma[$ and $\gamma_1 := 1 - \gamma$. There is an integer $p \geq 1$ such that $J, T_\gamma^{-1} J, \ldots, T_\gamma^{-p+1} J$ covers the unit interval. It follows that in each interval of integers $\{kp, \ldots, kp + p - 1\}, k \geq 1$, there is ℓ_k such that $T_\gamma^{\ell_k} \gamma_1 \in J$: the increasing sequence $\mathcal{L} = \{\ell_k \geq 1 : T_\gamma^{\ell_k} \gamma_1 \in J\}$ has positive density. It holds $T_\gamma^{\ell_k}(]\gamma_1, 1[) \subset [0, \gamma_1[$, which implies $\mathcal{L} \cap \mathcal{S}(T_\alpha, f, x) = \emptyset$.

Polynomial Growth

The property of recurrence with respect to an aperiodic cocycle is satisfied by classes of sequences with polynomial growth. For conciseness reason, we will only present an example.

Let us assume that $\mathcal{L} = \{\ell_1 < \ell_2 < \ldots < \ell_n < \ldots\}$ is such that the following property holds:

Property 4.1 For all $h \geq 1$, the sequences $(\ell_{n+h} - \ell_n)_{n \geq 1}$ are strictly increasing (for n large enough) and with positive density, i.e., for a finite constant C_h, $\ell_{n+h} - \ell_n < C_h n, \forall n \geq 1$.

An example is $\ell_n = n^2 + p_n$, with $p_n = O(n)$.

Lemma 4.3 *Let $\mathcal{L} = \{\ell_1 < \ell_2 < \ldots < \ell_n < \ldots\}$ be an sequence of integers which satisfies Property 4.1. If $(Y, \mathcal{A}, v, S)$ is weakly mixing, it holds:*

$$\frac{1}{n} \sum_{k=1}^{n} S^{\ell_k} \varphi \xrightarrow[n \to \infty]{L^2(v)} \int \varphi \, dv, \forall \varphi \in L^2(v). \tag{4.2.3}$$

Proof By Van der Corput inequality, for $N \geq 1$, all integers $H \in [1, N]$ and all $t \in \mathbb{R}$, it holds:

$$\left|\frac{1}{N} \sum_{k=1}^{N} e^{it\ell_k}\right|^2 \leq \frac{2}{H} + \frac{4}{H} \sum_{h=1}^{H-1} \left|\frac{1}{N} \sum_{k=1}^{N-h} e^{it(\ell_{k+h} - \ell_k)}\right|.$$

Let φ be in $L^2(v)$ with integral 0. Denoting by η_φ the spectral measure of φ, it follows:

$$\left\|\frac{1}{N} \sum_{k=1}^{N} S^{\ell_k} \varphi\right\|_2^2 = \int_0^1 \left|\frac{1}{N} \sum_{k=1}^{N} e^{2\pi i \ell_k t}\right|^2 d\eta_\varphi(t)$$

$$\leq \frac{2}{H} + \frac{4}{H} \sum_{h=1}^{H-1} \int_0^1 \left|\frac{1}{N} \sum_{k=1}^{N-h} e^{2\pi i t(\ell_{k+h} - \ell_k)}\right| d\eta_\varphi(t)$$

$$\leq \frac{2}{H} + \frac{4}{H} \sum_{h=1}^{H-1} \left(\int_0^1 \left|\frac{1}{N} \sum_{k=1}^{N-h} e^{2\pi i t(\ell_{k+h} - \ell_k)}\right|^2 d\eta_\varphi(t)\right)^{\frac{1}{2}}$$

$$= \frac{2}{H} + \frac{4}{H} \sum_{h=1}^{H-1} \left\|\frac{1}{N} \sum_{k=1}^{N} S^{\ell_{k+h} - \ell_k} \varphi\right\|_2.$$

For $\varepsilon > 0$, let $H := 1 + [2\varepsilon^{-1}]$. If $h \leq H$, by Lemma 4.2 there is $N(h)$ such that $\|\frac{1}{N} \sum_{k=1}^{N-h} S^{\ell_{k+h} - \ell_k} \varphi\|_2 \leq \varepsilon$, if $N \geq N_h$. Therefore $\|\frac{1}{N} \sum_{k=1}^{N} S^{\ell_k} \varphi\|_2^2 \leq \varepsilon + 4\varepsilon$, for $N \geq \max_{h \leq H} N_h$. $\qquad\square$

Corollary 4.1 *If $\mathcal{L}$ satisfies Property 4.1, it is recurrent for every aperiodic cocycle (T, f) such that $\mu(f) > 0$.*

4.2.3 Arithmetic Sequences

Furstenberg proved in [16] that arithmetic polynomial sequences are recurrent sequences. In 1988, J. Bourgain gave a pointwise result for these sequences, i.e., an ergodic theorem along polynomial sequences. Let $(Y, \mathcal{A}, \nu, S)$ be a dynamical system.

Theorem 4.2 ([5, 6]) *If P is a polynomial with integer coefficients, then, for $\varphi \in L^r(\nu)$, $r > 1$, $\frac{1}{n} \sum_{k=1}^{n} \varphi(S^{P(k)} y)$ converges a.e. for $n \to \infty$.*

For polynomial sequences, as well as in Sect. 4.2.5, in order to apply the theorem to our problem of values of ergodic sums, we need the positivity of the limit on A, when $\varphi = 1_A$.

Theorem 4.3 *If P is a polynomial without constant term, then, for all $A \in \mathcal{A}$,*

$$\lim_n \frac{1}{n} \sum_{k=1}^{n} 1_A(S^{P(k)} y) > 0, \text{ for a.e. } y \in A. \qquad (4.2.1)$$

If S is totally ergodic, in particular weakly mixing, then the limit is $\nu(A)$.

Proof In the totally ergodic case, the limit is $\nu(A)$ by Weyl's equirepartition theorem. If S is not totally ergodic, we take into account the rational spectrum of S and the proof is as in Theorem 3.5 in [16]. □

Corollary 4.2 *Let P be a polynomial of degree r with integer coefficients and without constant term. Let $\Phi_P(f, N, x)$ be the number of terms less than N in $S(T, f, x)$ of the form $P(k)$.*

(a) *If (T, f) is a positive cocyle, for a.e. x there exists a constant $c(x) > 0$ (depending on P) such that $\Phi_P(N, f, x) \sim c(x) N^{\frac{1}{r}}$. If (T, f) is r-aperiodic (cf. 4.2.1.1), then $c(x) = \mu(f)^{-\frac{1}{r}}$.*

(b) *For a cocyle such that $\mu(f) > 0$, $\Phi_P(N, f, x) \geq d(x) N^{\frac{1}{r}}$, with $d(x) > 0$, for a.e. x.*

Proof For (a), we apply the previous results to $S = T_f$, the special map, and to the set $A = B_0$, the basis of the special map. The result follows then from Theorem 4.2.3.

For (b), as in Proposition 4.1, we decompose X in a countable family of subsets C such that the induced cocycle f^C on C is ≥ 1 and we apply a) to f^C and the induced map T_C: there is a constant $c(x) > 0$ such that, for a.e. $x \in C$, the number of terms in $S(T_C, f^C, x)$ of the form $P(k)$ less than N is $\sim c(x) N^{\frac{1}{r}}$ when $N \to \infty$.

For $N \geq 1$, there is $\tau(N, x)$ such that $R^C_{\tau(N,x)}(x) \leq N < R^C_{\tau(N+1,x)}(x)$. Therefore the number of terms in $S(T, f, x)$ of the form $P(k)$ less than $\tau(N, x)$ is asymptotically $\geq c(x) \tau(N, x)^{\frac{1}{r}}$ when $N \to \infty$.

By (4.2.2), we have: $\lim_N \frac{1}{N} \tau(N, x) = \mu(C)$. We conclude that the number of terms in $S(T, f, x)$ of the form $P(k)$ less than N is $\geq d(x) N^{\frac{1}{r}}$, with $d(x) > 0$. □

An analogous result holds when $\mathcal{L} = \mathcal{P}$, the set of prime numbers. We use the following result (J. Bourgain [4] for $p > \frac{1+\sqrt{3}}{2}$, M. Wierdl [31] for $p > 1$): if $p > 1$ and $\varphi \in L^p(Y, \nu)$, $\dfrac{1}{\mathrm{Card}\{k \in \mathcal{P}, k < N\}} \displaystyle\sum_{k \in \mathcal{P},\, k < N} \varphi(T^k y)$ converges a.e. as $N \to \infty$.

If S is totally ergodic, by Vinogradov's equirepartition theorem, the limit is $\nu(\varphi)$. It follows that, if (T, f) is r-aperiodic and $\mu(f) > 0$, for a.e. x there exists $d(x) > 0$ such that

$$\mathrm{Card}\{k \in \mathcal{P} \textstyle\bigcap S(f, T, x), k < N\} \geq d(x)\, \mathrm{Card}\{k \in \mathcal{P}, k < N\}, \quad \text{for a.e. } x.$$

4.2.4 Arbitrary Sequences and Mixing Special Flows

Proposition 4.2 *Any strictly increasing sequence of integers $\mathcal{L} = \{\ell_1 < \ell_2 < \ldots < \ell_n < \ldots\}$ is recurrent for a positive cocycle (T, f) such that T_f is strongly mixing.*

Proof Let $(Y, \mathcal{A}, \nu, S)$ be a strongly mixing dynamical system. By a theorem of Blum-Hanson ([3]), it holds: $\lim_N \| \frac{1}{N} \sum_{i=0}^{N-1} 1_A \circ T^{\ell_i} - \nu(A) \|_2 = 0$, for all $A \in \mathcal{A}$.

If T_f is strongly mixing, this implies that, for the basis B_0 of the special map T_f, along a strictly increasing subsequence (N_k):

$$\lim_k \frac{1}{N_k} \sum_{i=0}^{N_k-1} 1_{B_0}(T_f^{\ell_i}(x, 0)) = \tilde{\mu}(B) = \mu(f)^{-1}, \quad \text{for } \mu\text{-a.e. } x.$$

As $1_{B_0}(T_f^{\ell_i}(x, 0)) \neq 0 \iff \ell_i \in S(T, f, x)$, this implies, for a.e. $x \in X$:

$$\mathrm{Card}\big(\{\ell_1, \ell_2, \ldots, \ell_{N_k}\} \textstyle\bigcap S(f, T, x)\big) \sim \mu(f)^{-1} N_k.$$

□

Remark The set of differences of the elements of a strictly increasing sequence is recurrent for any cocycle (T, f) such that $\mu(f) > 0$. On the contrary, as shown in [16], if $\mathcal{L}$ is a lacunary increasing sequence of integers (i.e, $\inf_k \frac{\ell_{k+1}}{\ell_k} > 1$), it is not a set of recurrence and there is a cocycle (T, f) for which $\mathcal{L}$ is not recurrent.

Examples of Mixing Special Maps

The previous proposition leads to the question of finding families of dynamical systems (X, T) and functions f such that T_f is strongly mixing.

An obvious example is given by $(X, \mathcal{A}, \mu, T)$ a strongly mixing dynamical system: for any set $B \in \mathcal{A}$ with $\mu(B) > 0$, for all infinite sequences of positive integers $\mathcal{L}$, for a.e. $x \in B$, the sequence $R_n^B(x)$ of return times in B visits infinitely often $\mathcal{L}$. More interesting, we would like to find classes of cocycles (T, f) such that the associated special map T_f is strongly mixing or (stronger property) a K-system.

Guivarc'h and Hardy [21] gave examples of special flows which are mixing. Let us take a subshift of finite type with a Gibbs measure (Σ, T, μ). By the corollary in [21, page 96], if f is Hölderian positive on Σ and under a condition of aperiodicity, then the special map T_f is mixing. The proof is related to a renewal theorem for a stationary process satisfying a spectral gap property. Since we are considering here functions with values in $\mathbb{Z}$, we restrict to "locally constant functions" over a subshift, i.e., functions depending on a finite number of coordinates. It follows from [21] that any infinite subset of $\mathbb{N}$ is a recurrence set for the cocycle (T, f) if T is a subshift of finite type with a Gibbs measure and $f > 0$ a locally constant function, under an aperiodicity condition on (T, f).

In this direction, the question of the K-property for a special flow has been studied by B. M. Gurevich [19, 20] and by Blanchard [2]. The following result gives a sufficient condition for special flows T_f to be K and so strongly mixing.

Theorem 4.4 ([2]) *Let T_f be a special flow (with discrete time) over a dynamical system $(X, \mathcal{B}, \mu, T)$, with an integer valued ceiling function $f \geq 1$.*

Suppose that f is $\mathcal{B}^- \cap \mathcal{B}^+$ measurable, where $\mathcal{B}^-$ and $\mathcal{B}^+$ are two σ-algebras, respectively T-increasing and T-decreasing and such that $\mathcal{B}^- \cap T\mathcal{B}^+$ is trivial.

Then, if the values of f are not contained in $a\mathbb{Z}$ for any $a > 1$, the dynamical system $(\tilde{X}, \tilde{\mu}, T_f)$ is a K-system.

4.2.5 Intersection of Cocycles

We consider now the question of the intersection of the sets of values of two cocycles. This is based on J. Bourgain's results on return times.

4.2.5.1 Return Times Theorem

Theorem 4.5 (Bourgain [8]) *Let $(X, \mathcal{B}, \mu, T)$ be a dynamical system and let g be in $L^\infty(X, \mu)$ orthogonal to the eigenfunctions of T. There exists a set X_0 of full μ-measure in X such that, for $x \in X_0$, for every dynamical system $(Y, \mathcal{A}, \nu, S)$, for all $h \in L^\infty(\nu)$, $\lim_N \frac{1}{N} \sum_{n=1}^N g(T^n x) h(S^n y) = 0$, for ν-almost every y.*

We need a slight extension of this result to any g in $L^\infty(X, \mu)$ and an information on the positivity of the limit. According to Theorem 4.5, it suffices to consider the Kronecker factor $\mathcal{K}_T$ of T and to show the positivity of the limit when $g \geq 0$ is replaced by its projection on $\mathcal{K}_T$ (which is > 0 a.e. on $(g > 0)$).

Hence, we suppose in the following lemma that $(X, \mathcal{B}, \mu, T)$ is a rotation on a compact abelian group, with μ its Haar measure. We denote by $\hat{X}$ the dual group of characters $\{\chi_j, j \in J\}$ and by $\lambda_j \colon T\chi_j = \lambda_j \chi_j$, the corresponding eigenvalue. By using the ergodic decomposition of ν, we can assume $(Y, \mathcal{A}, \nu, S)$ to be ergodic. We show the positivity of the limit for x in a set of full measure in $(g > 0)$ (independent from the choice of the system S) and for almost all y in $(h > 0)$.

Lemma 4.4 *Let g be a function in $L^2(X, \mu)$ of the form $g = \sum_{j \in J} c_j \chi_j$.*

(1) There is a set X_1 of full measure in X such that, for every $x \in X_1$, for any dynamical system (Y, ν, S) and any h in $L^2(Y, \nu)$,

$$\lim_N \frac{1}{N} \sum_{k=0}^{N-1} g(T^k x)\, h(S^k y) = \sum_{j \in J} c_j \chi_j(x)\, (\Pi_j h)(y), \quad \text{for a.e. } y \in Y,$$

(4.2.1)

where $(\Pi_j h)(y)$ is the projection of h on the eigenspace of eigenvalue $\bar{\lambda}_j$ in $L^2(\nu)$.

(2) If g, h are non negative, there is a set X_0 of full measure in $(g > 0)$ such that the limit in (4.2.1) is > 0 for all $x \in X_0$ and a.e. $y \in (h > 0)$.

Proof

(1) Let $(J_p)_{p \geq 1}$ be an increasing family of finite sets of characters whose union is the set of all characters on X. Let $g_p = \sum_{j \in J_p} c_j \chi_j$. We have:

$$\frac{1}{N} \sum_{k=0}^{N-1} g_p(T^k x)\, h(S^k y) = \sum_{j \in J_p} c_j \chi_j(x)\, [\frac{1}{N} \sum_{k=0}^{N-1} \lambda_j^k\, h(S^k y)].$$

By the ergodic theorem, there is a set D_j of measure 1 of points $y \in Y$ for which the mean between brackets converges to the projection $(\Pi_j h)(y)$ of h on the eigenspace of eigenvalue $\bar{\lambda}_j$ in $L^2(\nu)$. The limit is 0 when h is orthogonal to the subspace generated in $L^2(\nu)$ by the eigenfunctions of eigenvalues λ_j. The set $D = \cap_j D_j$ has also full measure in Y.

Let $(g - g_p)^*$ be the maximal function $\sup_N \frac{1}{N} \sum_{k=0}^{N-1} |g(T^k x) - g_p(T^k x)|$. Suppose first h bounded. The ergodic maximal lemma implies

$$\left| \frac{1}{N} \sum_{k=0}^{N-1} g(T^k x)\, h(S^k y) - \frac{1}{N} \sum_{k=0}^{N-1} g_p(T^k x)\, h(S^k y) \right|$$

$$\leq \|h\|_\infty \frac{1}{N} \sum_{k=0}^{N-1} |g(T^k x) - g_p(T^k x)|$$

$$\leq \|h\|_\infty\, (g - g_p)^*(x), \text{ with } \|(g - g_p)^*\|_2 \leq 2\|g - g_p\|_2.$$

Let (p_r) be a sequence such that $\sum_r \|g - g_{p_r}\|_2 < +\infty$. Then $(g - g_{p_r})^*(x) \to 0$ for x in a set E of full μ-measure.

Moreover we have $(\Pi_j h)(y) = d_i \zeta_j(y)$, where d_j is a constant and ζ_j is an eigenfunction for S of modulus 1 with eigenvalue $\bar{\lambda}_j$, and the series $\sum_j |d_j|^2$ converges. Therefore $\sum_j |c_j||d_j| < +\infty$ and $\sum_{j \in J_p} c_j \chi_j(x) (\Pi_j h)(y) \to \sum_{j \in J} c_j d_j \chi_j(x) \zeta_j(y)$. Putting

$$\Delta_N(x, y) := |\frac{1}{N} \sum_{k=0}^{N-1} g(T^k x) h(S^k y) - \sum_{j \in J} c_j d_j \chi_j(x) \zeta_j(y)|,$$

we have for $r \geq 1$: $\limsup_N \Delta_N(x, y) \leq (A_r) + (B_r) + (C_r)$, where

$$(A_r) := \limsup_N |\frac{1}{N} \sum_{k=0}^{N-1} g(T^k x) h(S^k y)$$

$$- \frac{1}{N} \sum_{k=0}^{N-1} g_{p_r}(T^k x) h(S^k y)| \leq \|h\|_\infty (g - g_{p_r})^*(x),$$

$$(B_r) := \limsup_N |\frac{1}{N} \sum_{k=0}^{N-1} g_{p_r}(T^k x) h(S^k y)$$

$$- \sum_{j \in J_{p_r}} c_j \chi_j(x) (\Pi_j h)(y)| = 0, \text{ for } y \in D,$$

and $(C_r) := \sum_{j \in J \setminus J_{p_r}} |c_j||d_j|$. It follows

$$\limsup_N \Delta_N(x, y) \leq (A_r) + 0 + (C_r) \to 0,$$

when $r \to +\infty$, if x is in the set of full μ-measure E and y in D.

If h is not bounded, but in $L^2(\nu)$, we use again a maximal inequality:

$$\sup_N \frac{1}{N} \sum_{k=0}^{N-1} |g(T^k x)| |h(S^k y)|$$

$$\leq (\sup_N \frac{1}{N} \sum_{k=0}^{N-1} |g(T^k x)|^2)^{\frac{1}{2}} (\sup_N \frac{1}{N} \sum_{k=0}^{N-1} |h(S^k y)^2)^{\frac{1}{2}}$$

$$\leq [(g^2)^*(x)]^{\frac{1}{2}} [(h^2)^*(y)]^{\frac{1}{2}},$$

where $(g^2)^*$, $(h^2)^*$ are the maximal functions of g^2 for T and h^2 for S, which satisfy the (weak) ergodic maximal inequality valid for L^1 functions. Then we observe that, for $x \in E \cap ((g^2)^* < +\infty)$, the set of h such that $\Delta_N(x, y) \to 0$ for a.e y contains L^∞ and is closed in $L^2(\nu)$ by the maximal inequality.

(2) By what precedes, the limit is unchanged if h is replaced by its projection on the Kronecker factor of $(Y, \mathcal{A}, \nu, S)$ and even by its projection $\tilde{h}$ on the factor generated by the eigenfunctions with eigenvalues which appear in the representation of g. Up to an isomorphism this factor can be viewed as a factor of $(X, \mathcal{B}, \mu, T)$ and $\tilde{h}$ as a function on X. Now we have the Fourier series representations on X: $g(x) = \sum_{j \in J} c_j \chi_j(x)$ and $\tilde{h}(y) = \sum_{j \in J} d_j \overline{\chi}_j(y)$ and

$$\lim \frac{1}{N} \sum_{k=0}^{N-1} g(T^k x) \tilde{h}(S^k y) = \sum_j c_j d_j \chi_j(x - y) = \int_X g(t + x) \tilde{h}(t + y) \, dt.$$

On the compact abelian group X, there is a distance invariant by translation. Denote by $\beta(x, \delta)$ the ball centered at x with radius $\delta > 0$. Almost every point x in $(g > 0)$, resp. y in $(\tilde{h} > 0)$, is a Lebesgue density point in the respective sets. It means that, for such x (resp. y), for every $\varepsilon > 0$ there is $\delta(\varepsilon) > 0$ such that the intersection $(g > 0) \cap \beta(x, \delta(\varepsilon))$ (resp. $(\tilde{h} > 0) \cap \beta(y, \delta(\varepsilon)))$ has a measure which is a $1 - \varepsilon$ proportion of the measure of the corresponding ball. Therefore $\mu(\beta(0, \delta(\varepsilon)) \cap [(g > 0) - x] \cap [(h > 0) - y]) \geq (1 - 2\varepsilon) \mu(\beta(0, \delta(\varepsilon)))$.

For $\delta_1 = \delta(\frac{1}{4})$, the measure of $\beta(0, \delta_1) \cap [(g > 0) - x] \cap [(\tilde{h} > 0) - y]$ is positive. If t belongs to the previous set of positive measure, then $g(t + x) \tilde{h}(t + y) > 0$. It follows:

$$\int g(t + x) \tilde{h}(t + y) \, dt \geq \int_{\beta(0, \delta_1)} g(t + x) \tilde{h}(t + y) \, dt > 0.$$

If X_0 is the set of Lebesgue density points of $X_1 \cap (g > 0)$, this shows the positivity of the limit for $x \in X_0$ and for a.e. y in $(\tilde{h} > 0)$, hence in $(h > 0)$.   □

Corollary 4.3 *Let $(X, \mathcal{B}, \mu, T)$ be an ergodic dynamical system and let $B \in \mathcal{B}$. Then there is a set of full measure in X for which for every ergodic dynamical system $(Y, \mathcal{A}, \nu, S)$ and all $A \in \mathcal{A}$, the following limit exists for ν-a.e. $y \in A$:*

$$c(x, y) = \lim_N \frac{1}{N} \sum_{n=1}^{N} 1_B(T^n x) 1_A(S^n y).$$

There is B_0 with full measure in B (not depending on $(Y, \mathcal{A}, \nu, S)$) such that $c(x, y) > 0, \forall x \in B_0$ and for ν-a.e. $y \in A$. If $(X, \mathcal{B}, \mu, T)$ is weakly mixing, then $c(x, y) = \mu(B) \mathbb{E}(1_A | \mathcal{J}_S(y))$.

Corollary 4.4 *Let $f : X \to \mathbb{N}$ be an integrable function such that $f \geq 1$. There is a set of full measure X_0 in X such that, for every dynamical system $(Y, \mathcal{A}, \nu, S)$, for*

$x \in X_0$ and $h \geq 1$, $h : Y \rightarrow \mathbb{N}$ integrable, the following limit exists:

$$c_h(x, y) := \lim_N \frac{1}{N} Card\{k \leq N : S_k f(x) \in S(S, h, y)\}$$

and is > 0 for a.e. $y \in (h > 0)$. If (T, f) is aperiodic and (Y, ν, S) ergodic, then $c_h(x, y) = \mu(f)^{-1} \nu(h)^{-1}$.

If we assume only $\mu(f) > 0$ and $\nu(g) > 0$, then $Card(S(T, f, x) \bigcap S$ $(S, h, y)) = \infty$, for μ-a.e x and ν-a.e. y.

4.2.6 Cocycles for T and T^{-1}

If T and S are commuting measure preserving maps on the same space, a question is to compare $S(T, f, x)$ and $S(S, f, x)$ for the same x. In this direction, we take $S = T^{-1}$ and use a pointwise result of Bourgain [7] for the means $\frac{1}{n} \sum_{k=0}^{n-1} f(T^k x) g(T^{mk} x)$, $m \neq 0, 1$ (a special case of the means $\frac{1}{n} \sum_{k=0}^{n-1} f_1(T^k x) f_2(T^{2k} x) \ldots f_m(T^{mkx})$ considered by H. Furstenberg in the study of "multiple recurrence").

By Bourgain [7], for a dynamical system (X, μ, T) and for $f, g \in L^\infty$, the means $\frac{1}{n} \sum_{k=0}^{n-1} f(T^k x) g(T^{-k} x)$ converge to 0, for a.e. x, when g is orthogonal to the eigenfunctions. Therefore, denoting by $\widetilde{f}$ and $\widetilde{g}$ the projections of f and g on the Kronecker factor, it holds: $\frac{1}{n} \sum_{k=0}^{n-1} f(T^k x) g(T^{-k} x) - \frac{1}{n} \sum_{k=0}^{n-1} \widetilde{f}(T^k x) \widetilde{g}(T^{-k} x) \rightarrow 0$.

If g is an eigenfunction, the means converge by Birkhoff's theorem and this can be extended to the closed linear span of the eigenfunctions by using: for f, g in L^2, $\frac{1}{n} \sum_{k=0}^{n-1} |f(T^k x)| |g(T^{-k} x)| \leq [(f^2)^*(x)]^{\frac{1}{2}} [(g^2)^*(x)]^{\frac{1}{2}}$, where $(f^2)^*, (g^2)^*$ are the maximal functions of f^2, g^2 for T and T^{-1}. If (e_j) is an orthonormal system of eigenfunctions for T, it follows: $\dfrac{1}{n} \sum_{k=0}^{n-1} \widetilde{f}(T^k x) \widetilde{g}(T^{-k} x) \rightarrow \sum_j \langle f, e_j \rangle \langle g, e_j \rangle$.

This shows that, for a.e. x, $\frac{1}{n} \sum_{k=0}^{n-1} f(T^k x) g(T^{-k} x)$ has a limit $\gamma(f, g, x)$ which does not change if f, g are replaced by their projection on the Kronecker factor of T. When T is weakly mixing, then $\gamma(f, g, x) = \mu(f) \mu(g)$. It remains to show the positivity of the limit.

Theorem 4.6

(1) For a.e. $x \in A$, $\lim_N \frac{1}{N} \sum_{n=0}^{N-1} 1_A(T^n x) 1_A(T^{-n} x) > 0$.

(2) If T is totally ergodic, for a.e. $x \in A$, $\lim_N \frac{1}{N} \sum_{n=0}^{N-1} 1_A(T^n x) 1_{A^c}(T^{-n} x) > 0$.

Proof With the previous notation, let $B_D := \{x : \gamma(1_A, 1_D, x) = 0\}$, for $D \subset X$. By invariance of μ, we get:

$$0 = \lim_N \int 1_{B_D}(x) \frac{1}{N} \sum_{n=0}^{N-1} 1_A(T^n x) \, 1_D(T^{-n} x) \, d\mu(x)$$

$$= \lim_N \frac{1}{N} \int \sum_{n=0}^{N-1} 1_{B_D}(T^n x) \, 1_A(T^{2n} x) \, 1_D(x) \, d\mu(x). \qquad (4.2.1)$$

(1) For $D = A$, to show the positivity, for a.e. $x \in A$, it suffices to show that $\mu(B_A \cap A) = 0$. By (4.2.1), this follows from Theorem 3.5 in [15].

(2) Let us consider the Kronecker factor of the dynamical system (X, μ, T). With additive notation, it can be represented as a translation $t \to t + \theta$ by an element θ on a compact abelian group G endowed with its Haar measure λ.

The map $x \to 2x = x + x = R(x)$ is a continuous endomorphisms of G, since $d(R(x), R(y)) \le 2d(x, y)$, if d is a G-invariant distance. By totally ergodicity, the trivial character is the only character χ which equals 1 on $2G$ (i.e., $\chi(g) = \pm 1$, $\forall g$). By duality, this implies that R is surjective. Hence it leaves the Haar measure λ on G invariant: indeed, the measure $f \to \lambda_2(f) := \int_G f(Rx) \, d\lambda(x)$ is invariant by translation by $2y$, $\forall y \in G$, hence by any translation because R is surjective, which implies that λ_2 is the Haar measure λ.

Now, taking $D = A^c$ in (4.2.1), let $B := B_{A^c} = \{x : \lim_n \frac{1}{n} \sum_{k=0}^{n-1} 1_A(T^k x) \, 1_{A^c}(T^{-k} x) = 0\}$. We want to show that $B \cap A$ is negligible. Suppose that $\mu(B) > 0$. Putting $\tilde{B} := \{(P 1_B)(t) > 0\}$, we have $\lambda(\tilde{B}) > 0$ (where P is the conditional expectation on the Kronecker factor) and the limit (namely 0) is the same as for the projection by P (cf. Furstenberg [15, Lemma 3.4]). This implies

$$0 = \lim_N \frac{1}{N} \int \sum_{n=0}^{N-1} (P 1_B)(t + k\theta) \, (P 1_A)(t + 2k\theta) \, (P 1_{A^c})(t) \, d\lambda(t)$$

$$= \int (P 1_B)(t + u) \, (P 1_A)(t + 2u) \, (P 1_{A^c})(t) \, du \, d\lambda(t)$$

$$= \int (P 1_B)(t) \, (P 1_A)(t + u) \, (P 1_{A^c})(t - u) \, du \, d\lambda(t).$$

Let us show that $2\tilde{B}$ generates an open subgroup of G. We use the invariance of λ by R. The set $2\tilde{B}$ is measurable. Since $\tilde{B} \subset R^{-1}(2\tilde{B})$ and $\lambda(R^{-1}(2\tilde{B})) = \lambda(2\tilde{B})$), it follows $\lambda(2\tilde{B}) \ge \lambda(\tilde{B}) > 0$. Now $1_{(2\tilde{B})} * 1_{(-2\tilde{B})}$ is a continuous function not identically 0, hence > 0 on a non empty open set contained in $(2\tilde{B}) + (-2\tilde{B}) = (2\tilde{B}) - (2\tilde{B})$. This shows the claim.

Let t be in $\tilde{B}$. We have: $\int (P1_A)(t+u)\,(P1_{A^c})(t-u)\,du = 0$, hence:

$$\int (P1_A)(u)\,(P1_A)(2t-u)\,du = \int (P1_A)(u)\,du = \mu(A).$$

Since $\int (P1_A)^2(u)\,du \leq \int (P1_A)(u)\,du = \mu(A)$ and $\int (P1_A)^2(2t-u)\,du \leq \int (P1_A)(2t-u)\,du = \mu(A)$, we have equality in the Cauchy-Schwarz Inequality, which implies: $P1_A(u) = P1_A(2t-u)$, λ-a.e. Therefore,

$$(P1_A)(u) = \sum_{\chi \in \hat{G}} \overline{c_\chi}\chi(-u) = \sum_{\chi \in \hat{G}} c_\chi \chi(u)$$

$$= \sum_{\chi \in \hat{G}} c_\chi \chi(2t-u) = \sum_{\chi \in \hat{G}} c_\chi \chi(2t)\chi(-u).$$

Let χ be such that $c_\chi \neq 0$. It holds $\chi(2t) = \overline{c_\chi}/c_\chi$; hence $\chi \equiv 1$ on the open set $(2\tilde{B}) - (2\tilde{B})$. So the character χ is equal to 1 on the open subgroup of G generated by $(2\tilde{B}) - (2\tilde{B})$. But the quotient is finite and this implies that χ is a root of unity. By the assumption that T is totally ergodic, it follows that $\chi \equiv 1$ on G. This implies $P1_A = \mu(A)$, hence $0 = \int (P1_B)(u)\,du = \mu(B)$. $\square$

Remark Theorem 4.6 is related to the following result of Cuny and Derriennic ([12]): *Let f be a measurable function on an ergodic invertible dynamical system (X, μ, T). The set of points x for which the strict inequality $f(T^{-n}x) < f(T^n x)$ holds for all n large enough is negligible.*

Theorem 4.6 applied to the special map T_f implies:

Corollary 4.5 *If (T, f) is positive r-aperiodic, it holds for a.e. x:*

$$\lim_N \frac{1}{N} Card([1, N] \cap S(T, f, x) \cap S(T^{-1}, f, x)) > 0, \qquad (4.2.2)$$

$$\lim_N \frac{1}{N} Card([1, N] \cap S(T, f, x) \cap S(T^{-1}, f, x)^c) > 0. \qquad (4.2.3)$$

Proof The set of ergodic sums $(S(T^{-1}, f, n)(x), n \geq 1)$ coincides with the set of return times of $(x, 0)$ in the basis B_0 for the inverse T_f^{-1} of the special map T_f. Therefore the result follows from Theorem 4.6 above by the equivalences

$$(T_f^n(x, 0) \in B_0 \text{ and } T_f^{-n}(x, 0) \in B_0) \Leftrightarrow n \in S(T, f, x) \cap S(T^{-1}, f, x),$$

$$(T_f^n(x, 0) \in B_0 \text{ and } T_f^{-n}(x, 0) \in B_0^c) \Leftrightarrow n \in S(T, f, x) \cap S(T^{-1}, f, x)^c.$$

If T_f is weakly mixing $((T, f)$ aperiodic), the limit is the constant $\mu(B_0)^2 = \mu(f)^{-2}$ in (4.2.2) and $\mu(f)\,(1 - \mu(f)^{-1}) > 0$ in (4.2.3). $\square$

4.3   Centered Case, $d \geq 1$

We consider now the case $d \geq 1$, f integrable and centered (centering is obviously a necessary condition for recurrence). We start by recalling briefly some definitions and facts about essential values and regularity of a cocycle (cf. [29]).

4.3.1   Values of a Regular Cocycle

Recurrent and Essential Values
   Let $(X, \mathcal{B}, \mu, T)$ be a dynamical system and (T, f) a cocycle, where f has values in $G = \mathbb{Z}^d$ or $\mathbb{R}^d$.

Essential Value   An element $v \in \overline{G} = G \cup \infty$ is called an *essential value* for (T, f), if for each open neighborhood $U \ni v$ in $\overline{G}$, for each $B \in \mathcal{B}$ of positive measure, there exists $n \geq 1$ such that $\mu(B \cap T^{-n} B \cap [f_n \in U]) > 0$. We denote the set of essential values by $\overline{\mathcal{E}}(T, f)$ and by $\mathcal{E}(T, f) := \overline{\mathcal{E}}(T, f) \cap G$ the set of finite essential values.

For a recurrent cocycle (T, f) (i.e. such that the origin belongs to $\mathcal{R}(T, f)$), the set $\mathcal{E}(T, f)$ is a subgroup of $\mathbb{Z}^d$. If f and g differs by a coboundary ($f = g \mid Th - h$, for h measurable), then $\overline{\mathcal{E}}(T, f) = \overline{\mathcal{E}}(T, g)$.

The recurrence set $\mathcal{R}(T, f)$ contains the set of finite essential values. The converse is false as shown by the example 1 below.

In the discrete case, the set $\mathcal{E}(T, f)$ can be defined equivalently as $\mathcal{E}(T, f) = \bigcap_B \mathcal{R}(T_B, f^{T,B})$, where the intersection is taken over the family of measurable sets of positive measure. In other words, $\underline{k} \in \mathbb{Z}^d$ is a finite essential value of the cocycle (T, f) if the ergodic sums for all induced cocycles (cf. 4.2.1.1) visit $\underline{k}$ infinitely often.

A cocycle (T, f) is ergodic if the *skew product* from $X \times G$ to itself defined by $T_f : (x, z) \to (Tx, z + f(x))$ is ergodic. Recall also the notion of "regular cocycle". One of the equivalent definitions is:

Definition 4.5   A cocycle (T, f) is *regular* if there is a closed subgroup H of G and a measurable function $u : X \to G$ such that $\psi := f - u \circ T + u$ takes μ-a.e. its values in H and $T_\psi : (x, h) \to (Tx, h + \psi(x))$ is ergodic for the product measure $\mu \otimes \lambda_H$ on $X \times H$.

Observe that for a regular cocycle the group H above is $\mathcal{E}(T, f)$. Clearly an ergodic cocycle (T, f) is regular.

In general, a question is to find whether a given cocycle is regular (and not a coboundary) or not. This depends both on T and f. In the hyperbolic case, one can expect that (T, f) is regular for a "smooth" f, whereas, for T a rotation, construction of simple functions generating non regular cocyles can be done. We will illustrate this point below and in the two following subsections. Right now, let us mention two examples:

Example 4.1 A simple example of ergodic cocycle (T, f) is given by T an irrational rotation and $f = 1_{[0,\frac{1}{2}[} - 1_{[\frac{1}{2},1[}$.

Example 4.2 Let us recall the construction of a cocycle with increment ± 1, oscillating between $\pm\infty$, hence with set of recurrence $\mathbb{Z}$, but such that $\mathcal{E}(T, f) = \{0\}$ (cf. [10]).

For $\beta, r \in \mathbb{R}$, let T_r be the rotation $x \to x + r \mod 1$ and $\varphi_{\beta,r} := 1_{[0,\beta]} - 1_{[0,\beta]} \circ T_r$. Let $\alpha \in]0, 1[$ with unbounded partial quotients. Using results in [18], one can show that there are $\beta \notin \alpha\mathbb{Z} + \mathbb{Z}$, $s \notin \mathbb{Q}$, γ of modulus 1 and ψ measurable of modulus 1 such that $e^{2\pi i s 1_{[0,\beta]}} = \gamma \, T\psi/\psi$.

Hence $e^{2\pi i \varphi_{\beta,r}}$ is a multiplicative coboundary for every r. Equivalently, $\varphi_{\beta,r}$ has values in $s^{-1}\mathbb{Z}$, up to an additive coboundary, which implies that $\mathcal{E}(T_\alpha, \varphi_{\beta,r}) \subset \mathbb{Z} \cap s^{-1}\mathbb{Z} = \{0\}$. Moreover, one easily shows that there are values of r such that $\varphi_{\beta,r}$ is not an additive coboundary.

For such a value of r, the function $\varphi_{\beta,r}$ satisfies: $\overline{\mathcal{E}}(T_\alpha, \varphi_{\beta,r}) = \{0, \infty\}$, hence is non regular.

Let us make some remarks about the question of recurrence for a given set $\mathcal{L}$ in $\mathbb{Z}^d$. If f is a coboundary, i.e., $f = Tu - u$, for a measurable u, $S(T, f, x)$ coincides with $R(u) - u(x)$, where $R(u)$ is the "range" of u. Nothing can be said a priory for this range. For a regular cocycle (T, f) which is not a coboundary, if $\mathcal{L}$ contains a non zero element in each non trivial subgroup of $\mathbb{Z}^d$ (a non zero multiple of every integer if $d = 1$), then it follows from Definition 4.5 that $\mathcal{L}$ intersects the set $S(T, f, x)$.

Examples of non regular 1-dimensional cocycles with unbounded gaps were constructed by Lemańczyk [26]. (A real 1-dimensional cocycle (T, f) has unbounded gaps, if there exists a sequence of open intervals I_n such that $|I_n| \to \infty$ and $S_k f(x) \notin I_n$ for all $x \in X, k \in \mathbb{Z}, n \geq 1$.) We are going to construct in Sect. 4.3.3 examples in dimension $d \geq 1$ over rotations on $\mathbb{T}^r$.

4.3.2 Hyperbolic Models

2-d Random Walks and 2-d Hyperbolic Cocycles

A random walk on $\mathbb{Z}^2$, reduced, centered and with a second order moment, yields an example of recurrent regular cocycle. The same type of result should hold in models like the planar Markov random walks studied in [23].

Analogously, regularity in the sense of Definition 4.5 for a 2-dimensional centered cocycle over a dynamical system of hyperbolic type (like an Anosov map) seems to hold. It would be interesting to have a proof in some generality.

Below, using a CLT (which implies recurrence) and a Hopf argument we give a sketch of proof for a simple model of cocycle (T, f) where T is a hyperbolic automorphism of a 2-dimensional torus X, μ is the Lebesgue measure on X and f belongs to a space $\mathcal{F}$ of functions from X to $G = \mathbb{Z}^2$ defined as follows.

For f with range of values $R(f) \subset G$ and for $a \in R(f)$, let $\partial B_f(a)$ be the boundaries of the sets $B_f(a) := \{x : f(x) = a\}$. The space $\mathcal{F}$ consists of the G-valued centered functions f which satisfy (for a constant C): $\sum_{a \in R(f)} \mu(\{x : d(x, \partial B_f(a)\} < \delta) \leq C\delta$, for all $\delta > 0$.

We will need a notion of aperiodicity extending in dimension $d \geq 1$ the property given in Definition 4.4 for $d = 1$:

Definition 4.6

(a) Let (T, f) be a cocycle, where f has values in $\mathbb{Z}^d$ or $\mathbb{R}^d$. It is *aperiodic* if, whenever $f - Th + h \in H$, with H is a closed subgroup of $\mathbb{R}^d$ and h a measurable function, then $H = \mathbb{R}^d$.

(b) When f has values in $\mathbb{Z}^d$, the cocycle (T, f) is *r-aperiodic* if, whenever $f - Th + h \in H$, with H a subgroup of $\mathbb{Z}^d$, then $H = \mathbb{Z}^d$.

Let $f_n(x) = \sum_{j=0}^{n-1} f(T^j x)$ denote simply the ergodic sum of f. For a function φ on $X \times G$, we denote the ergodic sums of φ for the action of the skew-product T_f by $S_n\varphi(x, u) := \sum_{j=0}^{n-1} \varphi(T^j x, u + f_j(x))$.

Lemma 4.5 *Suppose that there is a rate of contraction $\lambda < 1$ along stable leaves for the action of T. If f is in $\mathcal{F}$, for all x in a set X_0 of full measure in X, for y in the stable leaf of x, there is $N(x, y)$ such that $f(T^n x) = f(T^n y)$, for $n \geq N(x, y)$,*

Proof Take $\delta \in]\lambda, 1[$. Since $\sum_{a \in R(f)} \mu(\{x : d(x, \partial B_f(a)\} < \delta) \leq C\delta$ (here X is the 2-dimensional torus), it holds by invariance of the measure:

$$\mu\left(\bigcap_N \bigcup_{n \geq N} \bigcup_{a \in R(\varphi)} \{x : d(T^n x, \partial B_f(a)) < \delta^n\}\right) \leq \lim_N \frac{C\delta^N}{1 - \delta} = 0.$$

Therefore, for a set X_0 of full measure in X, for all $x \in X_0$ there is $N(x)$ such that for $n \geq N(x)$ $d(T^n x, \partial B_f(a)) \geq \delta^n$, for all $a \in R(f)$. Let x be in X_0 and y in the stable leaf of x. If n is large enough, $d(T^n x, T^n y) < \lambda^n$, so that, if $T^n x$ belongs to $B_f(a)$ and $T^n y \notin B_f(a)$, then $T^n x$ must be at a distance $< \lambda^n < \delta^n$ to $\partial B_f(a)$, contrary to what precedes. This implies the result. $\qquad\square$

Theorem 4.7 *Under the aperiodicity condition 4.6 (b), the cocycle (T, f), where T is a hyperbolic automorphism of a 2-dimensional torus X and $f : X \to \mathbb{Z}^2$ belongs to $\mathcal{F}$, is ergodic.*

Proof

(1) For the functional space $\mathcal{F}$ the Local Limit Theorem is a question, but the simpler question of the Central Limit Theorem can be solved for instance by

using the method of foliation and martingale as in [11]. The CLT insures the recurrence of the process $(S_n f)$ for f with values in $\mathbb{Z}^2$.

Now we will use a method like Hopf's argument for the proof of ergodicity of the geodesic flow. Let y be in the stable leaf of x. By Lemma 4.5, we can write $f_j(y) - f_j(x) = F^+(x, y) + \varepsilon_j^+(x, y)$, with F^+ not depending on j and ε_j^+ such that there is $N(x, y)$ a.e. finite for which $\varepsilon_j^+(x, y) = 0$, if $j \geq N(x, y)$. It follows:

$$T_f^j(y, v) = (T^j y, v + f_j(y)) = (T^j y, v + f_j(x) + F^+(x, y) + \varepsilon_j^+(x, y))$$

$$= (T^j y, v + f_j(x) + F^+(x, y)), \text{ for } j \geq N(x, y).$$

Taking $v = u - F^+(x, y)$, we get for $j \geq N(x, y)$:

$$d(T_f^j(y, u - F^+(x, y)), T_f^j(x, u))$$

$$= d((T^j y, u + f_j(x)), (T^j x, u + f_j(x)) \to 0.$$

Therefore $(y, u - F^+(x, y)$ belongs to the stable leaf of (x, u) for the action of T_f on $X \times G$ and if φ is Lipschitzian with respect to the first coordinate, then $\sum_n |\varphi(T_f^n(y, u - F^+(x, y))) - \varphi(T_f^n(x, u))| < +\infty$.

The same observation holds for the unstable leaf with T_f^{-1} and a function $F^-(x, y)$.

(2) Let h on G such that $h(g) > 0$ everywhere on G and $\sum_{g \in G} h(g) = 1$. If φ has compact support on $X \times G$, there is $c > 0$ such that $|\varphi(x, g)| \leq ch(g)$, $\forall (x, g) \in X \times G$. The recurrence of the skew product implies: $\lim S_n h(x, u) = +\infty$.

Denote by $\mathcal{J}$ the σ-algebra of T_f-invariant sets and $\tilde{\mu}$ the probability $h\mu \times dz$ on $X \times G$, where dz is the counting measure on G.

We consider an integrable function $\varphi(x, u)$ on $X \times \mathbb{Z}^2$ and denote by L_φ the limit in the ergodic theorem (with $S_n h$ as denominator), $L_\varphi(x, u) = \lim_n \dfrac{S_n \varphi(x, u)}{S_n h(x, u)} = \mathbb{E}_{\tilde{\mu}}(\dfrac{\varphi}{h}|\mathcal{J})$.

The formula shows that the limit is the same for the action by T_f^{-1} on $X \times G$. To prove ergodicity, it suffices to check that $L_\varphi(x, u)$ is a.e. constant, for every φ with compact support on $X \times G$. We can assume that φ is Lipschitzian with respect to x.

Let x, y be on the same stable leaf. We have, with $A_n = S_n \varphi(x, u)$, $B_n = S_n h(x, u)$, $C_n = S_n \varphi(y, u + F^+(x, y))$, $D_n = S_n h(y, u + F^+(x, y))$,

$$|\frac{S_n \varphi(x, u)}{S_n h(x, u)} - \frac{S_n \varphi(y, u + F^+(x, y))}{S_n h(y, u + F^+(x, y))}|$$

$$= |\frac{A_n}{B_n} - \frac{C_n}{D_n}| \leq |\frac{C_n}{D_n}||\frac{D_n - B_n}{B_n}| + |\frac{A_n - C_n}{B_n}|.$$

As $\left|\dfrac{D_n - B_n}{B_n}\right|, \left|\dfrac{A_n - C_n}{B_n}\right|$ tend to 0 and $\left|\dfrac{C_n}{D_n}\right|$ is bounded, it follows $L_\varphi(x, u) = L_\varphi(y, u + F^+(x, y))$ and similarly $L_\varphi(y, u) = L_\varphi(z, u + F^-(y, z))$, for x, y (resp. y, z) on the same stable (resp. unstable) leaf.

Starting from a point x_0, one can reach any point z by traveling along stable and unstable leaves (from x_0 to y, then from y to z). Thus we have: $L_\varphi(x_0, u) = L_\varphi(z, u + F^+(x_0, y) + F^-(y, z))$. We fix x_0 and get $L_\varphi(x_0, u + M(z)) = L_\varphi(z, u)$, with $M(z) := -[F^+(x_0, y) + F^-(y, z)]$.

We have shown that the T_φ-invariant function L_φ has the form: $L_\varphi(z, u) = r(u + M(z))$, for some function r. Let H_φ be the subgroup of periods of L_φ (with respect to the second coordinate). The function f satisfies: $r(u + f(z) + M(Tz)) = r(u + M(z))$. Therefore $f(z) + M(Tz) - M(z) \in H_\varphi$.

If f is aperiodic (cf. Definition 4.6), then $H_\varphi = G$, so that L_φ depends only on the first coordinate. As it is invariant, it is a constant by ergodicity of T. This shows that the cocycle is ergodic if f is aperiodic.

$\square$

4.3.3   A Cocycle Disjoint from a Sequence with Unbounded Gaps (UGB)

For $d \geq 1$, let $\mathcal{L}$ be a non empty subset of $\mathbb{Z}^d$ with $0 \notin \mathcal{L}$ and "unbounded gaps" (UBG), i.e., such that for every $R > 0$ there is a ball of radius R disjoint from $\mathcal{L} \cup -\mathcal{L}$ (for example the set of squares for $d = 1$).

The construction of a cocycle such that its ergodic sums never take values in $\mathcal{L}$ can be done following the method of Rohlin's towers as in [26] for a general aperiodic dynamical system.

Below we will construct an explicit example over a rotation. For $r \geq 1$, let $\alpha = (\alpha_1, \ldots, \alpha_r) \in \mathbb{R}^r$ and let $T = T_\alpha : x \to x + \alpha$ mod 1 be the corresponding rotation on the torus $\mathbb{T}^r$. We suppose T_α ergodic on $\mathbb{T}^r$ endowed with the Lebesgue measure denoted by μ (equivalently $\sum_{i=1}^r k_i \alpha_i \in \mathbb{Z} \Rightarrow k_i = 0, \forall i$). We are going to construct $\varphi : \mathbb{T}^r \to \mathbb{Z}^d$, integrable and centered, not a coboundary, for which the cocycle generated over T_α is recurrent and such that $S(\varphi, T_\alpha, x)$, for a.e. x is disjoint from $\mathcal{L} \subset \mathbb{Z}^d$.

The construction will yields examples of non regular cocycles. There are sets $\mathcal{L}$ (dimension 1) with the property (UBG) which contain multiples of any integer. So, this latter condition is sufficient to intersect the ergodic sums when (T, f) is a regular cocycle and not a coboundary, but is not sufficient in general.

Notation $|v|$ denotes the sup norm of an element $v \in \mathbb{R}^d$; for $x \in \mathbb{R}^d$, $\delta(x) := \inf_{z \in \mathbb{Z}^d} |x - z|$ is the distance from x to $\mathbb{Z}^d$. We denote by C a generic constant.

We use r-dimensional Diophantine approximations (cf. the presentation of [9]). Recall that, for the norm $|.|$, a positive integer q is called a best (simultaneous

Diophantine) approximation denominator for α, if

$$\delta(q\alpha) < \delta(k\alpha), \ \forall k \in \{1, \ldots, q-1\}. \tag{4.3.1}$$

Let $(q_n)_{n\geq 0} = (q_n(\alpha))_{n\geq 0}$ denote the sequence of the best approximation denominators for α in increasing order. Then it holds for all $n \in \mathbb{N}$:

$$\delta(q_n\alpha) \leq q_{n+1}^{-\frac{1}{d}}. \tag{4.3.2}$$

An information on the growth of q_n is given by the inequality:

$$q_{n+2^{d+1}} \geq 2q_{n+1} + q_n, \ \forall n \geq 1. \tag{4.3.3}$$

Let A be a subset of $\mathbb{T}^r$ (for instance a ball of radius $\frac{1}{2}$) for which there are $C, c > 0$ such that

$$\|1_A - 1_A(.+t)\|_1 \leq C|t|, \ \text{for } t \in \mathbb{T}^r \text{ and } |t| \text{ small}, \tag{4.3.4}$$

$$\|1_A - 1_A(.+t)\|_1 \geq c, \ \text{if } \tfrac{1}{2} \leq \delta(t) \leq \tfrac{3}{4}. \tag{4.3.5}$$

Let $\mathcal{L}$ be UBG in $\mathbb{Z}^d$. We choose (a_n) and (r_n) resp. in $\mathbb{Z}^d$ and $\mathbb{N}$ as follows:

- (a_n) is such that $a_1 = 0$ and, for a constant $\lambda > 2c^{-1}$ (with c the constant in (4.3.5)),

$$|a_n| > \lambda \sum_{k=1}^{n-1} |a_k|, \ d(a_n, \mathcal{L} \cup -\mathcal{L}) > \sum_{k=1}^{n-1} |a_k|, \ \text{for } n \geq 2; \tag{4.3.6}$$

- $(r_n) = (q_{k_n})$ is a sub-sequence of the sequence of best approximations for α with k_n chosen such that

$$\delta(r_{n+1}\alpha)^{-1} \geq \max(n^2 |a_{n+1}| \delta(r_n\alpha)^{-1}, |a_{n+2}|^{d+1}), \ \forall n \geq 1. \tag{4.3.7}$$

The construction of the sequence (r_n) is possible according to (4.3.2).

Let $\theta_n(x) := a_n \, 1_A(r_n x \bmod 1)$ and let $\varphi_{\mathcal{L}}$ (denoted also φ) be defined by

$$\varphi_{\mathcal{L}}(x) := \sum_{k=1}^{\infty} \big(\theta_k(x) - \theta_k(x+\alpha)\big). \tag{4.3.8}$$

Denoting simply by T the rotation T_α and by Tv the composition $v \circ T_\alpha$, we write $\varphi = \sum_{p\geq1} u_p = V_{n-1} + u_n + R_n$, with

$$u_p = \theta_p - T\theta_p, \quad V_{n-1} = \sum_{p=1}^{n-1} u_p, \quad R_n = \sum_{p\geq n+1} u_p.$$

By the first condition in (4.3.6) on (a_n), we have

$$\|S_j V_{n-1}\|_\infty = \|\sum_{p=1}^{n-1}(\theta_p - T^j\theta_p)\|_\infty \leq \sum_{p=1}^{n-1} |a_p| \leq \lambda^{-1}|a_n|, \qquad (4.3.9)$$

The invariance of the Lebesgue measure under the transformations $x \to r_k x \bmod 1$, for $r_k \in \mathbb{N}^*$, and (4.3.4) imply: $\|\theta_n - \theta_n(. + \alpha)\|_1 \leq C |a_n| \, \delta(r_n\alpha)$. Therefore Inequality (4.3.7) implies: convergence of the series, $\varphi \in L^1$ and, with $C_1 = C\sum_{p\geq1} p^{-2}$,

$$\|R_n\|_1 \leq \sum_{p\geq n+1} \|\theta_p - \theta_p(.+\alpha)\|_1 \leq C \sum_{p\geq n+1} |a_p|\, \delta(r_p\alpha) \leq C_1\, \delta(r_n\alpha).$$

$$(4.3.10)$$

Moreover, the measure of the support D_n of $\theta_n - \theta_n(. + \alpha)$ satisfies: $\mu(D_n) \leq C\delta(r_n\alpha)$. It follows that, for a.e. x there is $N(x)$ such that $\varphi(x) = \sum_1^{N(x)}(\theta_k(x) - \theta_k(x + \alpha))$.

Proposition 4.3

(a) *Let $\mathcal{L}$ be a set in $\mathbb{Z}^d$ with unbounded gaps and $0 \notin \mathcal{L}$. The function $\varphi = \varphi_{\mathcal{L}}$ defined by (4.3.8) is integrable centered, the cocycle (T_α, φ) is recurrent and φ is not a coboundary. The set $S(T_\alpha, \varphi, x)$ is disjoint from $\mathcal{L}$ for a.e. x.*

(b) *If $\mathcal{L}$ intersects any non trivial sub-lattice, then $\overline{\mathcal{E}}(T_\alpha, \varphi) = \{0, \infty\}$ and (T_α, φ) is a non regular cocycle.*

Proof

(1) *(Recurrence of $(S_n\varphi)$)* For $d = 1$, the centering of the integrable function φ suffices to ensure the recurrence. For $d > 1$, a sufficient condition for recurrence is: $\|S_n\varphi\|_1 = o(n^{\frac{1}{d}})$. Let us show that it is satisfied.

By (4.3.9) and (4.3.10), for all $n, t \geq 1$, it holds $\|S_n\varphi\|_1 \leq \|S_n V_t\|_1 + \|S_n R_t\|_1 \leq \lambda^{-1}|a_t|+n\,\delta(r_t\alpha)$. To check the above condition, we will find a non decreasing sequence (t_n) such that $|a_{t_n}| \leq \varepsilon_n n^{\frac{1}{d}}$, $n\,\delta(r_{t_n}\alpha) \leq 1$ and $\varepsilon_n \to 0$.

Let us take t_n such that $|a_{t_n}| \leq (\ln n)^{-1} n^{\frac{1}{d}} < |a_{t_n+1}|$. By (4.3.7), we have $\delta(r_{t_n}\alpha)^{-1} \geq |a_{t_n+1}|^{d+1} \geq (\ln n)^{-(d+1)} n^{1+\frac{1}{d}} \geq n$, for n big enough. This shows that (T_α, φ) is a recurrent d-dimensional cocycle.

(2) Suppose that φ is a coboundary: $\varphi = \psi - \psi \circ T_\alpha$, where ψ is a measurable function. With c the positive constant in (4.3.5), for M large enough, the measure of the set $B = \{x : |\psi(x)| \leq \frac{1}{2}M\}$ is $> 1 - \frac{1}{4}c$.

Let j be an integer. We have $|S_j\varphi| \leq M$ on the set $B_j := B \cap T_\alpha^{-j} B$ which has a measure $> 1 - \frac{1}{2}c$. From (4.3.9), it follows $|S_j\varphi - S_j V_{n-1}| \leq M + \lambda^{-1} |a_n|$ on B_j. On the other side, we have

$$\|S_j u_n\|_1 = \|S_j(\theta_n - T\theta_n)\|_1$$

$$= |a_n| \int_{\mathbb{T}^r} |1_A(r_n x \bmod 1) - 1_A(r_n x + r_n j\alpha \bmod 1)| \, d\mu$$

$$= |a_n| \int_{\mathbb{T}^r} |1_A(x \bmod 1) - 1_A(x + r_n j\alpha \bmod 1)| \, d\mu.$$

There is $j = j_n \leq \delta(r_n\alpha)^{-1}$ such that $\delta(j_n r_n \alpha) \in [\frac{1}{2}, \frac{3}{4}]$. According to (4.3.5), we get: $\|S_{j_n}(\theta_n - T\theta_n)\|_1 \geq c|a_n|$, hence:

$$\int_{B_{j_n}} |S_{j_n} u_n| d\mu \geq \int_{\mathbb{T}^r} |S_{j_n} u_n| d\mu - \mu(B_{j_n}^c) |a_n| \geq \frac{1}{2}c|a_n|.$$

For the remainder R_n, (4.3.10) implies: $\|S_{j_n} R_n\|_1 \leq j_n \|R_n\|_1 \leq C_1 j_n \delta(r_n\alpha) \leq C_1$.

Finally, we get a contradiction since the previous inequalities imply, with $\lambda^{-1} > \frac{1}{2}c$, for all $n \geq 1$,

$$M + \lambda^{-1} |a_n| \geq \left(\int_{B_{j_n}} |S_{j_n}\varphi - S_{j_n} V_{n-1}| d\mu \right)$$

$$= \left(\int_{B_{j_n}} |S_{j_n} u_n + S_{j_n} R_n| d\mu \right) \geq \frac{1}{2}c |a_n| - \|S_{j_n} R_n\|_1 \geq \frac{1}{2}c |a_n| - C_1.$$

(3) We claim that $\forall n \geq 1, j \geq 1, \forall x \in X, \sum_1^n (\theta_k(x) - \theta_k(x + j\alpha)) \notin \mathcal{L}$.
The proof is by induction on n. For $n = 1$, $\theta_1(x) - \theta_1(x + j\alpha) = 0 \notin \mathcal{L}$.
Assume that $\sum_{k=1}^{n-1} (\theta_k(x) - \theta_k(x + j\alpha))$ does not belong to $\mathcal{L}$. Suppose now that $\sum_{k=1}^n (\theta_k(x) - \theta_k(x + j\alpha)) \in \mathcal{L}$. Then one of the following cases occurs:

(a) $\theta_n(x) - \theta_n(x + j\alpha) = 0$, hence $\sum_{k=1}^{n-1} (\theta_k(x) - \theta_k(x + j\alpha)) + 0 \in \mathcal{L}$, which is excluded by induction hypothesis;
(b) $\sum_1^{n-1} (\theta_k(x) - \theta_k(x + j\alpha)) \pm a_n \in \mathcal{L}$, hence $d(a_n, \mathcal{L} \cup -\mathcal{L}) \leq \sum_{k=1}^{n-1} |a_k|$. This is contrary to (4.3.6) in the construction of (a_n). It shows the claim.

(4) We have $S_j \varphi(x) = \sum_{k=1}^{M_j(x)} (\theta_k(x) - \theta_k(x + j\alpha))$, with $M_j(x) = \sup_{\ell \leq j} N(T_\alpha^\ell x)$. From 3), the sums $\sum_{k=1}^{n} (\theta_k(x) - \theta_k(x + j\alpha))$ never take values in $\mathcal{L}$. It follows, that $S_j \varphi(x)$, as well, never takes values in $\mathcal{L}$, for a.e. x.

(5) To prove (b), observe that, if $\mathcal{E}(T_\alpha, \varphi) \neq \{0\}$, then the set of finite essential values is a non trivial lattice which intersects $\mathcal{L}$ according to the assumption on $\mathcal{L}$ in b). This is impossible, since the ergodic sums do not intersect $\mathcal{L}$. As φ is not a coboundary, it remains the case $\overline{\mathcal{E}}(T_\alpha, \varphi) = \{0, \infty\}$ and the cocycle is non regular.

□

4.4   Recurrent Sets for Random Walks

For $d = 1$ and non centered cocycles, Sect. 4.2 was based on results about recurrent sets for a transformation. For $d > 1$, a method to show transience of a set in $\mathbb{Z}^d$ can be based on limit theorems in distribution like the local limit theorem if such a result is available. This is the case for the cocycle generated by a random walk or for some cocycles (X, T) when T has strong stochastic properties and f belongs to a suitable functional space.

A Sufficient Condition   Let $\mathcal{L}$ be a subset of $\mathbb{Z}^d$. If $\sum_n \sum_{a \in \mathcal{L}} \mu\{x : S_n f(x) = a\} < \infty$, then for a.e. x, the number of visits of $S_n f(x)$ to $\mathcal{L}$ is finite. This is the easy direction of Borel–Cantelli lemma and can be used when an estimate of $\mu\{x : S_n f(x) = a\}$ is known. For dynamical systems with hyperbolicity, there are results for some integer valued functions [17], but we will restrict to the example of random walks.

Random Walks

Let (Z_n) be a random walk starting from 0: $Z_n = X_0 + \ldots + X_{n-1}$, where $(X_n, n \geq 0)$ is a sequence of iid random variables with values in $G = \mathbb{Z}^d, d \geq 1$, and distribution p. If T is the shift acting on the product space $\Omega = G^{\mathbb{Z}}$ endowed with the product measure $p^{\mathbb{Z}}$, the random walk (Z_n) defines a cocycle (T, f) with $f : \omega \to X_0(\omega)$ and the problem of recurrence of a set $\mathcal{L} \in \mathbb{Z}^d$, studied in the sixties, fits in the framework discussed here.

The question (cf [30]) was to find whether for the r.w. a given set $\mathcal{L} \subset \mathbb{Z}^d$ is recurrent ($Z_n(\omega) \in \mathcal{L}$ infinitely often for a.e. ω) or transient (finite number of visits to $\mathcal{L}$ for a.e. ω). The problem of recurrence to 0 for the random walk is a special case, but can be extended to an infinite set $\mathcal{L}$.

For $d = 1, 2$ recurrence (to 0) holds for a simple centered random walk and the model is relevant of the previous section. We will consider the case when (Z_n) is transient, $d \geq 3$, strictly aperiodic (in the sense of random walks) with a finite second moment.

(1) *Examples of transient sets for random walks* $(d \geq 3)$

Recall that if (Z_n) is a strictly aperiodic centered random walk with finite second moment, then, for a constant C:

$$\mathbb{P}(Z_n = \underline{k}) \leq Cn^{-\frac{d}{2}}, \ \forall \underline{k} \in \mathbb{Z}^d. \tag{4.4.11}$$

Let $B(0, R)$ denote the ball of center 0 and radius R. By the law of iterated logarithm, there is a constant $c > 0$ such that, for a.e. ω, the inequality $\|Z_n(\omega)\| > c (n \ \mathrm{Log} \ \mathrm{Log} \ n)^{\frac{1}{2}}$ is satisfied only for finitely many values of n. Hence, for a.e. ω, there is $N(\omega)$ such that $\|Z_n(\omega)\| > c (n \ \mathrm{LogLog} \ n)^{\frac{1}{2}}$, for $n \geq N(\omega)$. By (4.4.11) this implies the sufficient condition of transience for a set $\mathcal{L} \subset \mathbb{Z}^d$:

$$\sum_{n \geq 1} n^{-\frac{d}{2}} \ \mathrm{Card}(\mathcal{L} \cap B(0, n^{\frac{1}{2}} \ (\mathrm{Log} \ \mathrm{Log} n)^{\frac{1}{2}})) < +\infty.$$

If $\mathcal{L}$ satisfies: $\mathrm{Card}(\mathcal{L} \cap B(0, R)) \leq CR^\alpha$ for some constants C, α, the series converges if $-\frac{d}{2} + \frac{\alpha}{2} < -1$. The set $\mathcal{L}$ is transient if $\alpha < d - 2$.

Example 4.3 Let us consider the subset

$$\mathcal{L}^{d,p} := \{([k_1^p], [k_2^p], \ldots, [k_d]^p), k_1, \ldots, k_d \in \mathbb{N}\} \in \mathbb{Z}^d, \ d \geq 3. \tag{4.4.12}$$

For $\mathcal{L}^{d,p}$ we have $\alpha = \frac{d}{p}$ and transience holds if $p > \frac{d}{d-2}$.

(2) *Examples of recurrent sets for random walks*

Consider a simple (centered) random walk in $\mathbb{Z}^d$, $d \geq 3$. It is transient and a criterion of recurrence for subsets of $\mathbb{Z}^d$ (called *Wiener's test for recurrence*) has been given in terms of *capacity* [24, 27, 30]:

For $n \geq 1$, let $\mathcal{L}_n$ denote the set $\mathcal{L} \cap \{\underline{k} \in \mathbb{Z}^d : 2^n \leq \|\underline{k}\| < 2^{n+1}\}$ and let $\mathrm{Cap} \ (\mathcal{L}_n)$ denote its capacity. Then the set $\mathcal{L}$ is recurrent if and only if

$$\sum_{n=1}^{\infty} 2^{-(d-2)n} \ \mathrm{Cap} \ (\mathcal{L}_n) = +\infty. \tag{4.4.13}$$

Therefore a method to prove the recurrence of a set $\mathcal{L}$ is to obtain a lower bound for the capacity $\mathrm{Cap} \ (\mathcal{L}_n)$.

Let $A \subset \mathbb{Z}^d$ be a finite set. Denote by $G(., .)$ the Green function of the random walk and by E_A the escape probability function of A. It holds $\sum_{y \in A} G(x, y) E_A(y) = 1$, for $x \in A$, and $\mathrm{Cap} \ (A) = \sum_{y \in A} E_A(y)$ (cf. [30] or [27]).

Moreover, for a simple random walk in $\mathbb{Z}^d$, for a constant factor c, we have $G(x, y) \leq c\|x - y\|^{-(d-2)}$, if $x \neq y$. It follows:

$$\text{Card}(A) = \sum_{x \in A}[\sum_{y \in A} G(x, y)E_A(y)] \leq c \sum_{y \in A}[\sum_{x \in A, x \neq y} \|x - y\|^{-(d-2)}] E_A(y)$$

$$\leq c[\sup_{y \in A} \sum_{x \in A, x \neq y} \|x - y\|^{-(d-2)}] \sum_{y \in A} E_A(y) = c\gamma(A) \text{ Cap } (A)$$

$$\text{with } \gamma(A) := \sup_{y \in A} \sum_{x \in A, x \neq y} \|x - y\|^{-(d-2)}.$$

Example 4.4 Let us consider again $\mathcal{L} = \mathcal{L}^{d,p}$.

To show recurrence using (4.4.13), a method is to estimate from below $\gamma(\mathcal{L}_n)$, where $\mathcal{L}_n = \mathcal{L}_n^{d,p} = \mathcal{L}^{d,p} \bigcap \{\underline{k} : 2^n \leq \|\underline{k}\| < 2^{n+1}\}$.

We start with some remarks:

(1) In the estimation of $\gamma(A)$ for a finite set A, changing the norm on $\mathbb{Z}^d$ modifies the bounds only by a constant factor. We use the sup norm: $\|x - y\| = \max(|x_i - y_i|, i = 1, \ldots, d)$ and, putting $\lambda_n = 2^{n/p}$, we express $\gamma(\mathcal{L}_n)$, as

$$\sup_{1 \leq r_i \leq \lambda_n, i=1,\ldots,d} {\sum_{1 \leq k_i \leq \lambda_n, i=1,\ldots,d}}' \frac{1}{[\max(|[k_i^p] - [r_i^p]|, i = 1, \ldots, d)]^{d-2}}.$$

Above and below, $\sum'$ means that $(0, \ldots, 0)$ is excluded in the denominator of the sum.

(2) We will use the inequality:

$$|(u + t)^q - u^q| \geq |t|^q, \forall u \geq 0, t \geq -u, q \geq 1. \tag{4.4.14}$$

(3) Observe also that if we perturb each r and k by a small perturbation: $r \to r + \alpha_r$, $k \to k + \beta_r$, the sum above is modified only up to a bounded factor. Therefore, after replacing the coordinates $[k_i^p]$ of the elements in $\mathcal{L}_n$ by k_i^p, we have to bound

$$\gamma_0(\mathcal{L}_n) := \sup_{1 \leq r_i \leq \lambda_n, i=1,\ldots,d} {\sum_{1 \leq k_i \leq \lambda_n, i=1,\ldots,d}}' \frac{1}{[\max(|k_i^p - r_i^p|, i = 1, \ldots, d)]^{d-2}}. \tag{4.4.15}$$

Lemma 4.6 *We have the bound*

$$\gamma_0(\mathcal{L}_n^{d,p}) \leq C \frac{p}{p - \frac{d}{d-2}}, \text{ if } p > \frac{d}{d - 2}, \leq Cdn, \text{ if } p = \frac{d}{d - 2}. \tag{4.4.16}$$

Proof

(a) Using (4.4.14) and putting $q = p(d-2)$, we get that the sum in (4.4.15), for $1 \le r_i \le \lambda_n, i = 1, \ldots, d$, is bounded by

$$\sideset{}{'}\sum_{-r_i+1 \le t_i \le \lambda_n - r_i,\, i=1,\ldots,d} \frac{1}{[\max(|(r_i + t_i)^p - r_i^p|,\, i = 1, \ldots, d)]^{d-2}}$$

$$\le \sideset{}{'}\sum_{-r_i+1 \le t_i \le \lambda_n - r_i,\, i=1,\ldots,d} \frac{1}{[\max(|t_i|^p,\, i = 1, \ldots, d)]^{d-2}}$$

$$\le \sideset{}{'}\sum_{0 \le t_i \le \lambda_n,\, i=1,\ldots,d} \frac{2^d}{\max(|t_i|^q,\, i = 1, \ldots, d)}$$

$$\le \sum_{j=1,\ldots,d} \sum_{1 \le t_i \le \lambda_n,\, i=j,\ldots,d} \frac{2^d}{\max(|t_i|^q,\, i = j, \ldots, d)} .$$

(b) A bound for the sum is given by bounding an integral:

Let $q \ge d$ and $L \ge 1$. Put $J(q,d) := \int_1^L \cdots \int_1^L \frac{1}{\sup(t_1^q,\ldots,t_d^q)} \, dt_1 \ldots dt_d$.

Using the inequality $J(q,d) \le (1) + (2)$ with

$$(1) = \int_1^L \cdots \int_1^L \left[\int_1^{\sup(t_2,\ldots,t_d)} \frac{1}{\sup(t_2^q,\ldots,t_d^q)} \, dt_1 \right] dt_2 \ldots dt_d$$

$$\le \int_1^L \cdots \int_1^L \frac{1}{\sup(t_2^{q-1},\ldots,t_d^{q-1})} \, dt_2 \ldots dt_d,$$

$$(2) = \int_1^L \cdots \int_1^L \left[\int_{\sup(t_2,\ldots,t_d)}^L \frac{1}{t_1^q} \, dt_1 \right] dt_2 \ldots dt_d$$

$$\le (q-1)^{-1} \int_1^L \cdots \int_1^L \frac{1}{\sup(t_2^{q-1},\ldots,t_d^{q-1})} \, dt_2 \ldots dt_d,$$

we get by iteration: $J(q,d) \le \frac{q}{q-d}$, if $q > d$, $\le Cq \ln L$, if $q = d$.

It follows: $\gamma_0(\mathcal{L}_n^{d,p}) \le J(q,d) \le C \frac{p}{p - \frac{d}{d-2}}$, if $p > \frac{d}{d-2}$, $\le Cdn$, if $p = \frac{d}{d-2}$.

$\square$

Proposition 4.4 For $d \ge 3$, $\mathcal{L}^{d,p}$ is recurrent if $1 \le p \le \frac{d}{d-2}$, transient if $p > \frac{d}{d-2}$.

Proof With $p_0 = \frac{d}{d-2}$, Lemma 4.6 implies   Cap $(\mathcal{L}_n^{d,p_0}) \geq \frac{C}{n}$ Card$(\mathcal{L}_n^{d,p_0})$. As Card$(\mathcal{L}_n^{d,p_0})$ is of order $(2^{n/p_0})^d = 2^{n(d-2)}$, we have $\sum_{n=1}^{\infty} 2^{-n(d-2)}$ Cap $(\mathcal{L}_n) = +\infty$, Condition (4.4.13) is satisfied and $\mathcal{L}^{d,p}$ is recurrent for $p = p_0 = \frac{d}{d-2}$.

For $p > p_0$, transience of the set $\mathcal{L}^{d,p}$ has been shown previously.

If $p < p_0$, let $\mathcal{L}'$ be the subset of $\mathcal{L}^{d,p}$ defined by

$$\{(\ell(k_i)^p, i = 1, \ldots, d), (k_1, \ldots, k_d) \in \mathbb{N}^d\}, \text{ where } \ell(k_i)^p \leq k_i^{p_0} < (\ell(k_i) + 1)^p,$$

i.e., $\ell(k_i) = [k_i^{p_0/p}]$. As $\ell(k_i)^p = k_i^{p_0}(1 + o(1))$, we can apply Remark 3 before Lemma 4.6 and find that $\mathcal{L}'$ is recurrent as $\mathcal{L}^{d,p_0}$. Therefore the set $\mathcal{L}^{d,p}$, which contains $\mathcal{L}'$, is recurrent for $p < p_0$.   □

Example 4.5 Let $\overline{\mathcal{L}} = (\ell_1 < \ell_2 < \ldots)$ be a strictly increasing sequence of integers and let $\mathcal{L}$ be the set in $\mathbb{Z}^3$ defined by $\{\ell = (0, 0, \ell), \ell \in \overline{\mathcal{L}}\}$.

The problem of recurrence for $\mathcal{L}$ can be interpreted as follows. The r.w. in $\mathbb{Z}^3$ is transient, but its restriction to the two first coordinates is recurrent in $\mathbb{Z}^2$. Therefore, recurrence for the set $\mathcal{L}$ with respect to the r.w. is equivalent to recurrence for the subset $\overline{\mathcal{L}}$ of $\mathbb{Z}$ and the induced (non integrable) cocycle (induction on $0 \times 0 \times \mathbb{Z} \subset \mathbb{Z}^3$).

When $\mathcal{L}$ is the set Q in $\mathbb{Z}^3$ of the points $(0, 0, \ell)$ with ℓ prime, McKean [28] has shown that with probability 1 the standard 3-dimensional random walk visits Q infinitely often. Erdós [13] has shown that the number of points in Q with $\ell \leq n$ visited by the random walk is a.s. $\sim c \ln \ln n$.

Some Problems

Finally, let us mention some problems related to the topics presented in the paper.

(1) Construct (new) families of special flows which are K-flows.
(2) Show in some generality the regularity of recurrent 2-dimensional cocycles over dynamical systems of hyperbolic type.
(3) Prove a local limit theorem for f in a space of functions with discrete values like the space $\mathcal{F}$ introduced in 4.3.2 when T has hyperbolic type.
(4) Extend the results about recurrent sets valid for random walks in dimension $d \geq 3$ to more general classes of cocycles.
(5) For the billiards in the plane with $\mathbb{Z}^2$-periodic rectangular obstacles, the position of the ball yields a cocycle with values in $\mathbb{Z}^2$. It has been shown that, generically, the cocycle is recurrent [1] and non regular [14]. A question is to find the set of recurrent obstacles.

References

1. A. Avila, P. Hubert, Recurrence for the wind-tree model. Ann. de l'Inst. H. Poincaré AN **37**, 1–11 (2020)
2. F. Blanchard, K-flots et théorème de renouvellement. Z. Wahrscheinlichkeitstheorie und Verw. Gebiete **36**(4), 345–358 (1976)
3. J.R. Blum, D.L. Hanson, On the mean ergodic theorem for subsequences. Bull. Am. Math. Soc. **66**, 308–311 (1960)
4. J. Bourgain, *An Approach to Pointwise Ergodic Theorems, GAFA-Seminar*. Lecture Notes in Mathematics (Springer, Berlin, 1987)
5. J. Bourgain, On the pointwise ergodic theorem on L^p for arithmetic sets. Israel J. Math. **61**(1), 73–84 (1988)
6. J. Bourgain, Pointwise ergodic theorems for arithmetic sets. IHES Publ. Math. **69**, 5–45 (1989)
7. J. Bourgain, Double recurrence and almost sure convergence. J. R. Angew. Math. **404**, 140–161 (1990)
8. J. Bourgain, H. Furstenberg, Y. Katznelson, D. Ornstein, Appendix on return-time sequences. Publ. Math. IHES **69**, 42–45 (1989)
9. N. Chevallier, Best simultaneous Diophantine approximations and multidimensional continued fraction expansions. Mosc. J. Comb. Number Theory **3**(1), 3–56 (2013)
10. J.-P. Conze, Recurrence, ergodicity and invariant measures for cocycles over a rotation. Contemp. Math. **485**, 45–70 (2009)
11. J.-P. Conze, S. Le Borgne, Méthode de martingale et flot géodésique sur une surface de courbure constante négative. Ergodic Theory Dynam. Syst. **21**(2), 421–441 (2001)
12. C. Cuny, Y. Derriennic, On the convergence of bilateral ergodic averages. Colloq. Math. **164**(2), 327–339 (2021)
13. P. Erdos, A problem about prime numbers and the random walk, II. Illinios J. Math. **5**, 352–353 (1961)
14. K. Frączek, C. Ulcigrai, Non-ergodic $\mathbb{Z}$-periodic billiards and infinite translation surfaces. Invent. Math. **197**(2), 241–298 (2014)
15. H. Furstenberg, Ergodic behavior of diagonal measures and a theorem of Szemerédi on arithmetic progressions. J. d'analyse Math. **31**, 204–256 (1977)
16. H. Furstenberg, Poincaré recurrence and number theory. Bull. Am. Math. Soc. **5**(3), 211–234 (1981)
17. S. Gouëzel, Berry-Esseen theorem and local limit theorem for non uniformly expanding maps. Ann. de l'Inst. H. Poincaré PR **41**, 997–1024 (2005)
18. M. Guénais, F. Parreau, Valeurs propres de transformations liées aux rotations irrationnelles et aux fonctions en escalier (Eigenvalues of transformations arising from irrational rotations and step functions) (2006). arXiv:math/0605250
19. B.M. Gurevich, Certain conditions for the existence of K-decompositions for special flows. (Russian) Trudy Moskov. Mat. Obšč. **17**, 89–116 1967
20. B.M. Gurevich, A certain condition for the existence of a K-decomposition for a special flow. (Russian) Uspehi Mat. Nauk **24**(5), 233–234 1969
21. Y. Guivarc'h et J. Hardy, Théorèmes limites pour une classe de chaînes de Markov et applications aux difféomorphismes d'Anosov. Annales de l'Inst. H. Poincaré PR **24**(1), 73–98 (1988)
22. S. Gouëzel, Regularity of coboundaries for non uniformly expanding Markov maps. Proc. Am. Math. Soc. **134**(2), 391–401 (2005)
23. L. Hervé, F. Pène, On the recurrence set of planar Markov random walks. J. Theor. Probab. **26**, 169–197 (2013). https://doi.org/10.1007/s10959-012-0414-7
24. K. Itô, H.P. McKean, Potentials and the random walk. Illin. J. Math. **4**, 119–132 (1960)
25. M. Lacey, K. Petersen, M. Wierdl, D. Rudolph, Random ergodic theorems with universally representative sequences. Ann. Inst. H. Poincaré PR **30**(3), 353–395 (1994)

26. M. Lemańczyk, Analytic nonregular cocycles over irrational rotations. Comment. Math. Univ. Carolin. **36**(4), 727–735 (1995)
27. G.F. Lawler, V. Limic, *Random Walk: A Modern Introduction*. Cambridge Studies in Advanced Mathematics, vol. 123 (2010)
28. H.P. McKean Jr., A problem about prime numbers and the random walk, I. Illinios. J. Math. **5**, 351 (1961)
29. K. Schmidt, *Lectures on Cocycles of Ergodic Transformations Groups*. Lectures Notes in Mathematics (Mc Millan, London, 1977)
30. F. Spitzer, *Principles of Random Walk*. The University Series in Higher Mathematics (D. Van Nostrand, Princeton, 1964)
31. M. Wierdl, Pointwise ergodic theorem along the prime numbers. Isreal J. Math. **64**(3), (1988)

Chapter 5
Almost Sure Invariance Principle for Random Distance Expanding Maps with a Nonuniform Decay of Correlations

Davor Dragičević and Yeor Hafouta

Abstract We prove a quenched almost sure invariance principle for certain classes of random distance expanding dynamical systems which do not necessarily exhibit uniform decay of correlations.

5.1 Introduction

The aim of this note is to establish an almost sure invariance principle (ASIP) for certain classes of random dynamical systems. More precisely, similarly to the setting introduced in [16], the dynamics is formed by compositions

$$f_\omega^n := f_{\sigma^{n-1}\omega} \circ \ldots \circ f_{\sigma\omega} \circ f_\omega, \ \omega \in \Omega$$

of locally distance expanding maps f_ω satisfying certain topological assumptions which are driven by an invertible, measure preserving transformation σ on some probability space $(\Omega, \mathcal{F}, \mathbb{P})$. Then, under suitable assumptions and for Hölder continuous observables $\psi_\omega : X \to \mathbb{R}$, $\omega \in \Omega$ we establish a quenched ASIP. Namely, we prove that for $\mathbb{P}$-a.e. $\omega \in \Omega$, the random Birkhoff sums $\sum_{j=0}^{n-1} \psi_{\sigma^j\omega} \circ f_\omega^j$ can be approximated in the strong sense by a sum of Gaussian independent random variables $\sum_{j=0}^{n-1} Z_j$ with the error being negligible compared to $n^{\frac{1}{2}}$. In comparison with the previous results dealing with the ASIP for random or sequential dynamical systems, the main novelty of our work is that we do not require that our dynamics exhibits uniform (with respect to ω) decay of correlations.

D. Dragičević (✉)
Department of Mathematics, University of Rijeka, Rijeka, Croatia
e-mail: ddragicevic@math.uniri.hr

Y. Hafouta
Department of Mathematics, The Ohio State University, Columbus, OH, USA
e-mail: yeor.hafouta@mail.huji.ac.il

© The Author(s), under exclusive license to Springer Nature Switzerland AG 2021
M. Pollicott, S. Vaienti (eds.), *Thermodynamic Formalism*, Lecture Notes
in Mathematics 2290, https://doi.org/10.1007/978-3-030-74863-0_5

In a more general setting and under suitable assumptions, Kifer proved in [13] a central limit theorem (CLT) and a law of iterated logarithm (LIL). As Kifer remarks, his arguments (see [13, Remark 4.1]) also yield an ASIP when there is an underlying random family of σ-algebras which are sufficiently fast well mixing in an appropriate (random) sense (i.e. in the setup of [13, Theorem 2.1]). In the context of random dynamics, Kifer's results can be applied to random expanding maps which admit a (random) symbolic representation. One of the main ingredients in [13] is a certain inducing argument, an approach that we also follow in the present paper. The main idea is that an ASIP for the original system will follow from an ASIP for a suitably constructed induced system.

For some classical work devoted to ASIP, we refer to [3, 20]. In addition, we stress that there are quite a few works whose aim is to establish ASIP for deterministic dynamical systems. In this direction, we refer to the works of Field, Melbourne and Török [8], Melbourne and Nicol [17, 18], and more recently to Korepanov [14, 15]. In [9], Gouëzel developed a new spectral technique for establishing ASIP, which was applied to certain classes of deterministic dynamical systems with the property that the corresponding transfer operator exhibits a spectral gap.

Gouëzel's method was also used in [1] to obtain the annealed ASIP for certain classes of piecewise expanding random dynamical systems. In [6] the authors proved for the first time (we recall that Kifer in [13] only briefly commented that his methods also yield an ASIP) a quenched ASIP for piecewise expanding random dynamical systems, by invoking a recent ASIP for (reverse) martingales due to Cuny and Merlevede [5] (which was also applied in many other deterministic and sequential setups; see for example [12]). While the type of maps f_ω considered in [6] is more general than the ones considered in the present paper, in contrast to [6] in the present paper we do not assume a uniform decay of correlations. Moreover, the methods used in this paper can be extended to vector-valued observables ψ_ω (see Remark 5.1). On the other hand, it is unclear if the techniques in [6] can be extended to the vector-valued case since the results in [5] deal exclusively with scalar-valued observables. Finally, we mention our previous work [7], where we have obtained a quenched ASIP for certain classes of hyperbolic random dynamical systems. In addition, we have improved the main result from [6]. However, the classes of dynamics we have considered again exhibit uniform decay of correlations.

Our techniques for establishing ASIP (besides the already mentioned inducing arguments), rely on a certain adaptation of the method of Gouëzel [9] which is of independent interest. Indeed, we first need to modify Gouëzel's arguments and show that they yield an ASIP for non-stationary sequences of random variables, which are not necessarily bounded in some L^p space.

We stress that our error term in ASIP is of order $n^{1/4+O(1/p)}$, where p comes from certain L^p-regularity conditions we impose for the induced system. This is rather close to the $n^{1/4}$ rate for deterministic uniformly expanding systems [9], when $p \to \infty$ (although this rate was significantly improved by Korepanov [15]).

5.2   Random Distance Expanding Maps

Let $(\Omega, \mathcal{F}, \mathbb{P})$ be a complete probability space. Furthermore, let $\sigma : \Omega \to \Omega$ be an invertible $\mathbb{P}$-preserving transformation such that $(\Omega, \mathcal{F}, \mathbb{P}, \sigma)$ is ergodic. Moreover, let (X, ρ) be a compact metric space normalized in size so that $\operatorname{diam} X \leq 1$ together with the Borel σ-algebra $\mathcal{B}$, and let $\mathcal{E} \subset \Omega \times X$ be a measurable set (with respect to the product σ-algebra $\mathcal{F} \times \mathcal{B}$) such that the fibers

$$\mathcal{E}_\omega = \{x \in X : (\omega, x) \in \mathcal{E}\}, \quad \omega \in \Omega$$

are compact. Hence (see [4, Chapter III]), it follows that the map $\omega \to \mathcal{E}_\omega$ is measurable with respect to the Borel σ-algebra induced by the Hausdorff topology on the space $\mathcal{K}(X)$ of compact subspaces of X. Moreover, the map $\omega \mapsto \rho(x, \mathcal{E}_\omega)$ is measurable for each $x \in X$. Finally, the projection map $\pi_\Omega(\omega, x) = \omega$ is measurable and it maps any $\mathcal{F} \times \mathcal{B}$-measurable set to an $\mathcal{F}$-measurable set (see [4, Theorem III.23]).

Let $f_\omega \colon \mathcal{E}_\omega \to \mathcal{E}_{\sigma\omega}$, $\omega \in \Omega$ be a family of surjective maps such that the map $(\omega, x) \to f_\omega(x)$ is measurable with respect to the σ-algebra $\mathcal{P}$ which is the restriction of $\mathcal{F} \times \mathcal{B}$ on $\mathcal{E}$. Consider the skew product transformation $F : \mathcal{E} \to \mathcal{E}$ given by

$$F(\omega, x) = (\sigma\omega, f_\omega(x)). \tag{5.1}$$

For $\omega \in \Omega$ and $n \in \mathbb{N}$, set

$$f_\omega^n := f_{\sigma^{n-1}\omega} \circ \ldots \circ f_\omega \colon \mathcal{E}_\omega \to \mathcal{E}_{\sigma^n\omega}.$$

Let us now introduce several additional assumptions for the family f_ω, $\omega \in \Omega$. More precisely, we require that:

- (*topological exactness*) there exist a constant $\xi > 0$ and a random variable $\omega \mapsto n_\omega \in \mathbb{N}$ such that for $\mathbb{P}$-a.e. $\omega \in \Omega$ and any $x \in \mathcal{E}_\omega$ we have that

$$f_\omega^{n_\omega}(B_\omega(x, \xi)) = \mathcal{E}_{\sigma^{n_\omega}\omega}, \tag{5.2}$$

where $B_\omega(x, r)$ denotes an open ball in $\mathcal{E}_\omega$ centered at x with radius r;
- (*pairing property*) there exist random variables $\omega \mapsto \gamma_\omega > 1$ and $\omega \mapsto D_\omega \in \mathbb{N}$ such that for $\mathbb{P}$-a.e. $\omega \in \Omega$ and for any $x, x' \in \mathcal{E}_{\sigma\omega}$ with $\rho(x, x') < \xi$ (ξ comes from the previous assumption), we have that

$$f_\omega^{-1}(\{x\}) = \{y_1, \ldots, y_k\}, \quad f_\omega^{-1}(\{x'\}) = \{y_1', \ldots, y_k'\}, \tag{5.3}$$

$$k = k_{\omega, x} = |f_\omega^{-1}(\{x\})| \leq D_\omega$$

and

$$\rho(y_i, y_i') \le (\gamma_\omega)^{-1}\rho(x, x'), \quad \text{for } 1 \le i \le k. \tag{5.4}$$

The above assumptions were considered in [10], and they hold true in the setup of distance expanding maps considered in [16]. We note that all the results stated in [16] hold true under these assumptions (see [16, Chapter 7]) and not only under the assumptions from [16, Section 2]. For $\omega \in \Omega$ and $n \in \mathbb{N}$, set

$$\gamma_{\omega,n} := \prod_{i=0}^{n-1} \gamma_{\sigma^i \omega} \quad \text{and} \quad D_{\omega,n} := \prod_{i=0}^{n-1} D_{\sigma^i \omega}. \tag{5.5}$$

By induction, it follows from the pairing property that for $\mathbb{P}$-a.e. $\omega \in \Omega$ and for any $x, x' \in \mathcal{E}_{\sigma^n \omega}$ with $\rho(x, x') < \xi$, we have that

$$(f_\omega^n)^{-1}(\{x\}) = \{y_1, \ldots, y_k\} \quad \text{and} \quad (f_\omega^n)^{-1}(\{x'\}) = \{y_1', \ldots, y_k'\}, \tag{5.6}$$

where

$$k = k_{\omega,x,n} = |(f_\omega^n)^{-1}(\{x\})| \le D_{\omega,n},$$

and

$$\rho\big(f_\omega^j y_i, f_\omega^j y_i'\big) \le (\gamma_{\sigma^j \omega, n-j})^{-1}\rho(x, x'), \text{ for } 1 \le i \le k \text{ and } 0 \le j < n. \tag{5.7}$$

Let $g : \mathcal{E} \to \mathbb{C}$ be a measurable function. For any $\omega \in \Omega$, consider the function $g_\omega := g(\omega, \cdot) : \mathcal{E}_\omega \to \mathbb{C}$. For any $0 < \alpha \le 1$, set

$$v_{\alpha,\xi}(g_\omega) := \inf\{R > 0 : |g_\omega(x) - g_\omega(x')| \le R\rho^\alpha(x, x') \text{ if } \rho(x, x') < \xi\},$$

and let

$$\|g_\omega\|_{\alpha,\xi} = \|g_\omega\|_\infty + v_{\alpha,\xi}(g_\omega),$$

where $\| \cdot \|_\infty$ denotes the supremum norm and $\rho^\alpha(x, x') := \big(\rho(x, x')\big)^\alpha$. We emphasize that these norms are $\mathcal{F}$-measurable (see [10, p. 199]).

Let $\mathcal{H}_\omega^{\alpha,\xi} = (\mathcal{H}_\omega^{\alpha,\xi}, \| \cdot \|_{\alpha,\xi})$ denote the space of all $h : \mathcal{E}_\omega \to \mathbb{C}$ such that $\|h\|_{\alpha,\xi} < \infty$. Moreover, let $\mathcal{H}_{\omega,\mathbb{R}}^{\alpha,\xi}$ be the space of all real-valued functions in $\mathcal{H}_\omega^{\alpha,\xi}$.

Take a random variable $H : \Omega \to [1, \infty)$ such that

$$\int_\Omega \ln H_\omega \, d\mathbb{P}(\omega) < \infty,$$

where $H_\omega := H(\omega)$. Moreover, let $\mathcal{H}^{\alpha,\xi}(H)$ be the set of all measurable functions $g : \mathcal{E} \to \mathbb{C}$ satisfying $v_{\alpha,\xi}(g_\omega) \le H_\omega$ for $\omega \in \Omega$. Furthermore, for $\omega \in \Omega$ set

$$\mathcal{H}_\omega^{\alpha,\xi}(H) := \{g : \mathcal{E}_\omega \to \mathbb{C} : g \text{ measurable and } v_{\alpha,\xi}(g) \le H_\omega\}$$

and

$$Q_\omega(H) = \sum_{j=1}^{\infty} H_{\sigma^{-j}\omega}(\gamma_{\sigma^{-j}\omega,j})^{-\alpha}. \tag{5.8}$$

Since $\omega \mapsto \ln H_\omega$ is integrable, we have (see [16, Chapter 2]) that $Q_\omega(H) < \infty$ for $\mathbb{P}$-a.e. $\omega \in \Omega$. The following simple distortion property is a direct consequence of (5.7).

Lemma 5.1 *Take $\omega \in \Omega$, $n \in \mathbb{N}$ and $\varphi = (\varphi_0, \ldots, \varphi_{n-1})$, where $\varphi_i \in \mathcal{H}_{\sigma^i\omega}^{\alpha,\xi}(H)$ for $0 \le i \le n-1$. Set*

$$S_n^\omega \varphi := \sum_{j=0}^{n-1} \varphi_j \circ f_\omega^j.$$

Furthermore, take $x, x' \in \mathcal{E}_{\sigma^n\omega}$ such that $\rho(x, x') < \xi$ and let y_i, y_i', $1 \le i \le k$ be as in (5.6). Then, for any $1 \le i \le k$ we have that

$$|S_n^\omega \varphi(y_i) - S_n^\omega \varphi(y_i')| \le \rho^\alpha(x, x') Q_{\sigma^n\omega}(H).$$

5.2.1 Transfer Operators

Let us take an observable $\psi : \mathcal{E} \to \mathbb{R}$ such that $\psi \in \mathcal{H}^{\alpha,\xi}(H)$. We consider the associated random Birkhoff sums

$$S_n^\omega \psi = \sum_{i=0}^{n-1} \psi_{\sigma^i\omega} \circ f_\omega^i, \quad \text{for } n \in \mathbb{N} \text{ and } \omega \in \Omega.$$

Furthermore, suppose that $\phi : \mathcal{E} \to \mathbb{R}$ also belongs to $\mathcal{H}^{\alpha,\xi}(H)$. For $\omega \in \Omega$, $z \in \mathbb{C}$ and $g : \mathcal{E}_\omega \to \mathbb{C}$, we define

$$\mathcal{L}_\omega^z g(x) = \sum_{y \in f_\omega^{-1}(\{x\})} e^{\phi_\omega(y) + z\psi_\omega(y)} g(y). \tag{5.1}$$

It follows from [10, Theorem 5.4.1.] that $\mathcal{L}_\omega^z : \mathcal{H}_\omega^{\alpha,\xi} \to \mathcal{H}_{\sigma\omega}^{\alpha,\xi}$ is a well-defined and bounded linear operator for each $\omega \in \Omega$ and $z \in \mathbb{C}$. Moreover, the map $z \mapsto \mathcal{L}_\omega^z$ is analytic for each $\omega \in \Omega$.

Let us denote $\mathcal{L}_\omega^0$ simply by $\mathcal{L}_\omega$. It follows from [16, Theorem 3.1.] that for $\mathbb{P}$-a.e. $\omega \in \Omega$, there exists a triplet $(\lambda_\omega, h_\omega, \nu_\omega)$ consisting of a positive number $\lambda_\omega > 0$, a strictly positive function $h_\omega \in \mathcal{H}_\omega^{\alpha,\xi}$ and a probability measure ν_ω on $\mathcal{E}_\omega$ so that

$$\mathcal{L}_\omega h_\omega = \lambda_\omega h_{\sigma\omega}, \quad (\mathcal{L}_\omega)^* \nu_{\sigma\omega} = \lambda_\omega \nu_\omega, \quad \nu_\omega(h_\omega) = 1,$$

and that maps $\omega \mapsto \lambda_\omega$, $\omega \mapsto h_\omega$ and $\omega \mapsto \nu_\omega$ are measurable. We can assume without any loss of generality that $\lambda_\omega = 1$ for $\mathbb{P}$-a.e. $\omega \in \Omega$ (since otherwise we can replace $\mathcal{L}_\omega$ with $\mathcal{L}_\omega/\lambda_\omega$). For $\mathbb{P}$-a.e. $\omega \in \Omega$, let μ_ω be a measure on $\mathcal{E}_\omega$ given by $d\mu_\omega := h_\omega d\nu_\omega$. We recall (see [16, Lemma 3.9]) that these measures satisfy the so-called *equivariant property*, i.e. we have that

$$f_\omega^* \mu_\omega = \mu_{\sigma\omega}, \quad \text{for } \mathbb{P}\text{-a.e. } \omega \in \Omega. \tag{5.2}$$

Moreover, these measures give rise to a measure μ on $\Omega \times \mathcal{E}$ with the property that for any $A \in \mathcal{F} \times \mathcal{B}$,

$$\mu(A) = \int_\Omega \mu_\omega(A_\omega) d\mathbb{P}(\omega),$$

where $A_\omega = \{x \in \mathcal{E}_\omega; (\omega, x) \in A\}$. Then, μ is invariant for the skew-product transformation F given by (5.1). Moreover, μ is ergodic.

For $\bar{t} = (t_0, \ldots, t_{n-1}) \in \mathbb{R}^n$, set

$$\mathcal{L}_\omega^{\bar{t},n} := \mathcal{L}_{\sigma^{n-1}\omega}^{it_{n-1}} \circ \ldots \circ \mathcal{L}_{\sigma\omega}^{it_1} \circ \mathcal{L}_\omega^{it_0}.$$

Moreover, let $\mathcal{L}_\omega^n := \mathcal{L}_\omega^{\bar{0},n}$, where $\bar{0} = (0, \ldots, 0) \in \mathbb{R}^n$. Note that

$$\|\mathcal{L}_\omega^n \mathbf{1}\|_\infty \leq (\deg f_\omega^n) \cdot e^{\|S_n^\omega \phi\|_\infty} \leq D_{\omega,n} e^{\|S_n^\omega \phi\|_\infty} < \infty,$$

where $\mathbf{1}$ is the function taking constant value 1 and

$$\deg f_\omega^n := \sup_{x \in \mathcal{E}_{\sigma^n\omega}} |(f_\omega^n)^{-1}(\{x\})|.$$

Lemma 5.2 *For any $\mathbb{P}$-a.e. $\omega \in \Omega$ we have that for any $n \in \mathbb{N}$, $T > 0$, $\bar{t} = (t_0, \ldots, t_{n-1}) \in [-T, T]^n$ and $g \in \mathcal{H}_\omega^{\alpha,\xi}$,*

$$v_{\alpha,\xi}(\mathcal{L}_\omega^{\bar{t},n} g) \leq \|\mathcal{L}_\omega^n \mathbf{1}\|_\infty \big(v_{\alpha,\xi}(g)(\gamma_{\omega,n})^{-\alpha} + 2Q_{\sigma^n\omega}(H)(1+T)\|g\|_\infty \big).$$

Consequently,

$$\|\mathcal{L}_{\omega}^{\bar{t},n} g\|_{\alpha,\xi} \leq \|\mathcal{L}_{\omega}^{n} \mathbf{1}\|_{\infty} \left(v_{\alpha,\xi}(g)(\gamma_{\omega,n})^{-\alpha} + (1 + 2Q_{\sigma^n\omega}(H))(1+T)\|g\|_{\infty} \right).$$
(5.3)

Proof The proof is similar to the proof of [10, Lemma 5.6.1.], but for reader's convenience all the details are given. The idea is to apply Lemma 5.1 for $\varphi = (\varphi_0, \dots, \varphi_{n-1})$ given by

$$\varphi_j := \phi_{\sigma^j\omega} + it_j \psi_{\sigma^j\omega}, \quad \text{for } 0 \leq j \leq n-1.$$

Set $A_n^{\omega} = \sum_{j=0}^{n-1} t_j \psi_{\sigma^j\omega} \circ f_{\omega}^j$. Firstly, by the definition of $\mathcal{L}_{\omega}^n$ we have

$$\|\mathcal{L}_{\omega}^{\bar{t},n} g\|_{\infty} \leq \|g\|_{\infty} \|\mathcal{L}_{\omega}^n \mathbf{1}\|_{\infty}.$$
(5.4)

In order to complete the proof of the lemma we need to approximate $v_{\alpha,\xi}(\mathcal{L}_{\omega}^{\bar{t},n} g)$. Let $x, x' \in \mathcal{E}_{\sigma^n\omega}$ be such that $\rho(x, x') < \xi$ and let $y_1, \dots, y_k$ and $y_1', \dots, y_k'$ be the points in $\mathcal{E}_{\omega}$ satisfying (5.3) and (5.4). We can write

$$\left| \mathcal{L}_{\omega}^{\bar{t},n} g(x) - \mathcal{L}_{\omega}^{\bar{t},n} g(x') \right|$$

$$= \left| \sum_{q=1}^{k} \left(e^{S_n^{\omega}\phi(y_q) + iA_n^{\omega}(y_q)} g(y_q) - e^{S_n^{\omega}\phi(y_q') + iA_n^{\omega}(y_q')} g(y_q') \right) \right|$$

$$\leq \sum_{q=1}^{k} e^{S_n^{\omega}\phi(y_q)} \left| e^{iA_n^{\omega}(y_q)} g(y_q) - e^{iA_n^{\omega}(y_q')} g(y_q') \right|$$

$$+ \sum_{q=1}^{k} \left| e^{iA_n^{\omega}(y_q')} g(y_q') \right| \cdot \left| e^{S_n^{\omega}\phi(y_q)} - e^{S_n^{\omega}\phi(y_q')} \right| =: I_1 + I_2.$$

In order to estimate I_1, observe that for any $1 \leq q \leq k$,

$$\left| e^{iA_n^{\omega}(y_q)} g(y_q) - e^{iA_n^{\omega}(y_q')} g(y_q') \right|$$

$$\leq |g(y_q)| \cdot \left| e^{iA_n^{\omega}(y_q)} - e^{iA_n^{\omega}(y_q')} \right| + |g(y_q) - g(y_q')| =: J_1 + J_2.$$

By the mean value theorem and then by Lemma 5.1,

$$J_1 \leq 2T\|g\|_{\infty} Q_{\sigma^n\omega}(H)\rho^{\alpha}(x, x'),$$

while by (5.7),

$$J_2 \leq v_{\alpha,\xi}(g)\rho^{\alpha}(y_q, y_q') \leq v_{\alpha,\xi}(g)(\gamma_{\omega,n})^{-\alpha}\rho^{\alpha}(x, x').$$

It follows that

$$I_1 \leq \mathcal{L}_\omega^n \mathbf{1}(x)\big(2T\|g\|_\infty Q_{\sigma^n\omega}(H) + v_{\alpha,\xi}(g)(\gamma_{\omega,n})^{-\alpha}\big)\rho^\alpha(x, x').$$

Next, we estimate I_2. By the mean value theorem and Lemma 5.1,

$$|e^{S_n^\omega \phi(y_q)} - e^{S_n^\omega \phi(y_q')}| \leq Q_{\sigma^n\omega}(H) \cdot \max\{e^{S_n^\omega \phi(y_q)}, e^{S_n^\omega \phi(y_q')}\}\rho^\alpha(x, x')$$

and therefore

$$I_2 \leq \|g\|_\infty(\mathcal{L}_\omega^n \mathbf{1}(x) + \mathcal{L}_\omega^n \mathbf{1}(x'))Q_{\sigma^n\omega}(H)\rho^\alpha(x, x')$$

$$\leq 2\|g\|_\infty \|\mathcal{L}_\omega^n \mathbf{1}\|_\infty Q_{\sigma^n\omega}(H)\rho^\alpha(x, x'),$$

yielding the first statement of the lemma and (5.3) follows from (5.4), together with the first statement. □

By Lemma 5.2, together with the observation that $(\gamma_{\omega,n})^{-\alpha} \leq 1$, we conclude that there exists a random variable $C\colon \Omega \to [1, \infty)$ such that for $\mathbb{P}$-a.e. $\omega \in \Omega$, $n \in \mathbb{N}$ and for any $\bar{t} = (t_0, t_1, \ldots, t_{n-1}) \in [-1, 1]^n$, we have that

$$\|\mathcal{L}_\omega^{\bar{t},n}\|_{\alpha,\xi} \leq C(\sigma^n\omega)\|\mathcal{L}_\omega^n \mathbf{1}\|_\infty, \tag{5.5}$$

where $\|\mathcal{L}_\omega^{\bar{t},n}\|_{\alpha,\xi}$ denotes the operator norm of $\mathcal{L}_\omega^{\bar{t},n}$ when considered as a linear operator from $\mathcal{H}_\omega^{\alpha,\xi}$ to $\mathcal{H}_{\sigma^n\omega}^{\alpha,\xi}$. Note that we can just take $C(\omega) = 4(1 + Q_\omega)$. For $\mathbb{P}$-a.e. $\omega \in \Omega$, we define $\hat{\mathcal{L}}_\omega\colon \mathcal{H}_\omega^{\alpha,\xi} \to \mathcal{H}_{\sigma\omega}^{\alpha,\xi}$ by

$$\hat{\mathcal{L}}_\omega g = \mathcal{L}_\omega(gh_\omega)/h_{\sigma\omega}, \quad g \in \mathcal{H}_\omega^{\alpha,\xi}.$$

Moreover, for $n \in \mathbb{N}$, set

$$\hat{\mathcal{L}}_\omega^n := \hat{\mathcal{L}}_{\sigma^{n-1}\omega} \circ \ldots \circ \hat{\mathcal{L}}_{\sigma\omega} \circ \hat{\mathcal{L}}_\omega.$$

Clearly,

$$\hat{\mathcal{L}}_\omega^n g = \mathcal{L}_\omega^n(gh_\omega)/h_{\sigma^n\omega}, \quad \text{for } g \in \mathcal{H}_\omega^{\alpha,\xi} \text{ and } n \in \mathbb{N}.$$

We need the following result which is a direct consequence of [16, Lemma 3.18.].

Lemma 5.3 *There exist $\lambda > 0$ and a random variable $K\colon \Omega \to (0, \infty)$ such that*

$$\|\hat{\mathcal{L}}_\omega^n g\|_\infty \leq \max(1, 1/Q_\omega)K(\sigma^n\omega)e^{-\lambda n}\|g\|_{\alpha,\xi},$$

for $\mathbb{P}$-a.e. $\omega \in \Omega$, $n \in \mathbb{N}$ and $g \in \mathcal{H}_\omega^{\alpha,\xi}$ such that $\int_{\mathcal{E}_\omega} g \, d\mu_\omega = 0$.

Applying Lemma 5.3 with the function $g = 1/h_\omega - 1$, and taking into account that $\mathcal{L}^n_\omega h_\omega = h_{\sigma^n \omega}$ (since $\lambda_\omega = 1$), it follows from (5.5) that for $\mathbb{P}$-a.e. $\omega \in \Omega, n \in \mathbb{N}$ and for any $\bar{t} = (t_0, t_1, \dots, t_{n-1}) \in [-1, 1]^n$,

$$\|\mathcal{L}^{\bar{t},n}_\omega\|_{\alpha,\xi} \leq (1 + U(\omega))K(\sigma^n \omega)C'(\sigma^n \omega) \tag{5.6}$$

where $C'(\omega) = C(\omega)\|h_\omega\|_\infty$ and $U(\omega) = \max(1, 1/Q_\omega) \cdot (1 + \|1/h_\omega\|_{\alpha,\xi})$.

5.3 A Refined Version of Gouëzel's Theorem

In this section we present a more general version of Gouëzel's almost sure invariance principle for non-stationary processes [9, Theorem 1.3.]. This result will than be used in the next section to obtain the almost sure invariance principle for random distance expanding maps.

Let $(A_1, A_2, \dots)$ be an $\mathbb{R}$-valued process on some probability space $(\Omega, \mathcal{F}, \mathbb{P})$. We first recall the condition that we denote (following [9]) by (H): there exist $\varepsilon_0 > 0$ and $C, c > 0$ such that for any $n, m > 0, b_1 < b_2 < \dots < b_{n+m+k}, k > 0$ and $t_1, \dots, t_{n+m} \in \mathbb{R}$ with $|t_j| \leq \varepsilon_0$, we have that

$$\left| \mathbb{E}\left(e^{i \sum_{j=1}^n t_j (\sum_{\ell=b_j}^{b_{j+1}-1} A_\ell) + i \sum_{j=n+1}^{n+m} t_j (\sum_{\ell=b_j+k}^{b_{j+1}+k-1} A_\ell)}\right) \right.$$

$$\left. - \mathbb{E}\left(e^{i \sum_{j=1}^n t_j (\sum_{\ell=b_j}^{b_{j+1}-1} A_\ell)}\right) \cdot \mathbb{E}\left(e^{i \sum_{j=n+1}^{n+m} t_j (\sum_{\ell=b_j+k}^{b_{j+1}+k-1} A_\ell)}\right) \right|$$

$$\leq C(1 + \max |b_{j+1} - b_j|)^{C(n+m)} e^{-ck}.$$

Theorem 5.1 *Suppose that $(A_1, A_2, \dots)$ is an $\mathbb{R}$-valued centered process on the probability space $(\Omega, \mathcal{F}, \mathbb{P})$ that satisfies (H). Furthermore, assume that:*

- *there exist $u > 0$ and $L \in \mathbb{N}$ such that for any $n, m \in \mathbb{N}, m \geq L$ we have that*

$$Var\left(\sum_{j=n+1}^{n+m} A_j \right) \geq um; \tag{5.1}$$

- *there exist constants $p \geq 6$ and $a, C > 0$ such that for any $n \in \mathbb{N}$ we have*

$$\|A_n\|_{L^p} \leq an^{\frac{1}{p}}. \tag{5.2}$$

In addition, for any $n, m \in \mathbb{N}$ the finite sequence $(A_i/(n+m)^{1/p})_{n+1 \leq i \leq n+m}$ also satisfies condition (H) with the same constants ε_0, C and c.

Then for any $\delta > 0$, there exists a coupling between (A_j) and a sequence (B_j) of independent centered normal random variables such that

$$\left| \sum_{j=1}^{n} (A_j - B_j) \right| = o(n^{a_p + \delta}) \quad a.s., \tag{5.3}$$

where

$$a_p = \frac{p}{4(p-1)} + \frac{1}{p}.$$

Moreover, there exists a constant $C > 0$ such that for any $n \in \mathbb{N}$,

$$\left\| \sum_{j=1}^{n} A_j \right\|_{L^2} - C n^{a_p + \delta} \le \left\| \sum_{j=1}^{n} B_j \right\|_{L^2} \le \left\| \sum_{j=1}^{n} A_j \right\|_{L^2} + C n^{a_p + \delta}. \tag{5.4}$$

Finally, there exists a coupling between (A_j) and a standard Brownian motion $(W_t)_{t \ge 0}$ such that

$$\left| \sum_{j=1}^{n} A_j - W_{\sigma_n^2} \right| = o(n^{\frac{1}{2}a_p + \frac{1}{4} + \delta}) \quad a.s.,$$

where

$$\sigma_n = \left\| \sum_{j=1}^{n} A_j \right\|_{L^2}.$$

Remark 5.1 The above result (together with its proof) is similar to [9, Theorem 1.3]. However, we stress that [9, Theorem 1.3] requires that the process $(A_1, A_2, \ldots)$ is bounded in L^p, while the above Theorem 5.1 works under the assumption that (5.2) holds. Consequently, the estimate for the error term in (5.3) is different from that in [9, Theorem 1.3].

Note also that our condition (5.1) replaces condition (1.3) in [9, Theorem 1.3]. This, of course, makes it impossible to get a precise formula for the variance of the approximating Gaussian random variables $\sum_{j=1}^{n} B_j$, as in [9]. However, in our context we have the estimate (5.4). Observe that (5.4) together with (5.1) ensures that

$$\lim_{n \to \infty} \frac{\left\| \sum_{j=1}^{n} B_j \right\|_{L^2}}{\left\| \sum_{j=1}^{n} A_j \right\|_{L^2}} = 1.$$

Therefore, Theorem 5.1 yields a corresponding almost sure version of the CLT for the sequence $\frac{1}{a_n} \sum_{j=1}^{n} A_j$, where $a_n = \| \sum_{j=1}^{n} A_j \|_{L^2}$. As we have mentioned, a precise formula for the variance of the approximating Gaussian random variables in the context of [9, Theorem 1.3] was obtained in [9, Lemma 5.7]. Hence, in our modification of the proof of [9, Theorem 1.3] we will not need an appropriate version of [9, Lemma 5.7] (and instead we will prove (5.4) directly).

We also note that our modification of the arguments in [9] also yields a certain convergence rate for $p \in (4, 6)$, but in order to keep our exposition as simple as possible we have formulated the results only under the assumption that $p \geq 6$.

Finally, we remark that like in [9] we can consider processes taking values in $\mathbb{R}^d$ and that Theorem 5.1 holds in this case also. We prefer to work with processes in $\mathbb{R}$ to keep our exposition as simple as possible.

Proof of Theorem 5.1 We follow step by step the proof of [9, Theorem 1.3] by making necessary adjustments. Firstly, applying [9, Proposition 4.1] with the finite sequence $(A_i / (n+m)^{1/p})_{n+1 \leq i \leq n+m}$, we get that for each $\eta > 0$ there exists $C > 0$ such that

$$\left\| \sum_{j=n+1}^{n+m} A_j \right\|_{L^{p-\eta}} \leq Cm^{\frac{1}{2}} (n+m)^{1/p}, \quad \text{for } m, n \geq 0. \tag{5.5}$$

We note that although [9, Proposition 4.1] was formulated for an infinite sequence, the proof for a finite sequence proceeds by using the same arguments. We consider the so-called big and small blocks as introduced in [9, p.1659]. Fix $\beta \in (0, 1)$ and $\varepsilon \in (0, 1 - \beta)$. Furthermore, let $f = f(n) = \lfloor \beta n \rfloor$. Then, Gouëzel decomposes $[2^n, 2^{n+1})$ into a union of $F = 2^f$ intervals $(I_{n,j})_{0 \leq j < F}$ of the same length, and F gaps $(J_{n,j})_{0 \leq j < F}$ between them. In other words, we have

$$[2^n, 2^{n+1}) = J_{n,0} \cup I_{n,0} \cup J_{n,1} \cup I_{n,1} \cup \ldots \cup J_{n,F-1} \cup I_{N,F-1}.$$

Let us outline the construction of this decomposition. For $1 \leq j < F$, we write j in the form $j = \sum_{k=0}^{f-1} \alpha_k(j) 2^k$ with $\alpha_k \in \{0, 1\}$. We then take the smallest r with the property that $\alpha_r(j) \neq 0$ and take $2^{\lfloor \varepsilon n \rfloor} 2^r$ to be the length of $J_{n,j}$. In addition, the length of $J_{n,0}$ is $2^{\lfloor \varepsilon n \rfloor} 2^f$. Finally, the length of each interval $I_{n,j}$ is $2^{n-f} - (f+2) 2^{\lfloor \varepsilon n \rfloor - 1}$.

In addition, we recall some notations from [9] which we will also use. We define a partial order on $\{(n, j) : n \in \mathbb{N}, 0 \leq j < F(n)\}$ by writing $(n, j) \prec (n', j')$ if the interval $I_{n,j}$ is to the left of $I_{n',j'}$. Observe that a sequence $((n_k, j_k))_k$ tends to infinity if and only if $n_k \to \infty$. Moreover, let

$$X_{n,j} := \sum_{\ell \in I_{n,j}} A_\ell$$

and

$$\mathcal{I} := \bigcup_{n,j} I_{n,j} \quad \text{and} \quad \mathcal{J} := \bigcup_{n,j} J_{n,j}.$$

The rest of the proof will be divided (following again [9]) into six steps.

First Step We first prove the following version of [9, Proposition 5.1].

Proposition 5.1 *There exists a coupling between* $(X_{n,j})$ *and* $(Y_{n,j})$ *such that, almost surely, when* (n, j) *tends to infinity,*

$$\left| \sum_{(n',j') \prec (n,j)} X_{n',j'} - Y_{n',j'} \right| = o(2^{(\beta+\varepsilon)n/2}).$$

Here, $(Y_{n,j})$ *is a family of independent random variables such that* $Y_{n,j}$ *and* $X_{n,j}$ *are equally distributed.* □

Before we outline the proof of Proposition 5.1, we will first introduce some preparatory material. Let $\tilde{X}_{n,j} = X_{n,j} + V_{n,j}$, where the $V_{n,j}$'s are independent copies of the random variable V constructed in [9, Proposition 3.8], which are independent of everything else (enlarging our probability space if necessary). Write $X_n = (X_{n,j})_{0 \leq j < F(n)}$ and $\tilde{X}_n = (\tilde{X}_{n,j})_{0 \leq j < F(n)}$. Then, we have the following version of [9, Lemma 5.2].

Lemma 5.4 *Let* $\tilde{Q}_n$ *be a random variable distributed like* $\tilde{X}_n$, *but independent of* $(\tilde{X}_1, \ldots, \tilde{X}_{n-1})$. *We have*

$$\pi\big((\tilde{X}_1, \ldots, \tilde{X}_{n-1}, \tilde{X}_n), (\tilde{X}_1, \ldots, \tilde{X}_{n-1}, \tilde{Q}_n)\big) \leq C4^{-n}, \tag{5.6}$$

where $\pi(\cdot, \cdot)$ *is the Prokhorov metric (see [9, Definition 3.3]) and* $C > 0$ *is some constant not depending on* n. □

Proof of Lemma 5.4 The proof is carried out by repeating the proof of [9, Lemma 5.2] with one slight modification. For reader's convenience we provide a complete proof.

The random process $(X_1, \ldots, X_n)$ takes its values in $\mathbb{R}^D$, where $D = \sum_{m=1}^{n} F(m) \leq C2^{\beta n}$. Moreover, each component in $\mathbb{R}$ of this process is one of the $X_{n,j}$, hence it is a sum of at most 2^n consecutive variables A_ℓ. On the other hand, the interval $J_{n,0}$ is a gap between $(X_j)_{j<n}$ and X_n, and its length k is $C^{\pm 1}2^{\varepsilon n + \beta n}$. Let ϕ and γ denote the respective characteristic functions of $(X_1, \ldots, X_{n-1}, X_n)$ and $(X_1, \ldots, X_{n-1}, Q_n)$, where Q_n is distributed like X_n and is independent of $(X_1, \ldots, X_{n-1})$. The assumption (H) ensures that for Fourier parameters $t_{m,j}$ all bounded by ε_0, we have

$$|\phi - \gamma| \leq C(1 + 2^n)^{CD} e^{-ck} \leq Ce^{-c'2^{\beta n + \varepsilon n}},$$

if n is large enough. Let $\tilde{\phi}$ and $\tilde{\gamma}$ be the characteristic functions of, respectively, $(\tilde{X}_1, \ldots, \tilde{X}_n)$ and $(\tilde{X}_1, \ldots, \tilde{X}_{n-1}, \tilde{Q}_n)$: they are obtained by multiplying ϕ and γ by the characteristic function of V is each variable. Since this function is supported in $\{|t| \leq \varepsilon_0\}$, we obtain, in particular, that

$$|\tilde{\phi} - \tilde{\gamma}| \leq C e^{-c2^{\beta n + \varepsilon n}}.$$

We then use [9, Lemma 3.5.] with $N = D$ and $T' = e^{2^{\varepsilon n/2}}$ to obtain that

$$\pi((\tilde{X}_1, \ldots, \tilde{X}_n), (\tilde{X}_1, \ldots, \tilde{X}_{n-1}, \tilde{Q}_n))$$
$$\leq \sum_{m \leq n} \sum_{j < F(m)} \mathbb{P}(|\tilde{X}_{m,j}| \geq e^{2^{\varepsilon n/2}}) + e^{CD2^{\varepsilon n/2}} e^{-c2^{\beta n + \varepsilon n}}.$$

So far our arguments were identical to those in the proof of [9, Lemma 5.2]. In the rest of the proof we will introduce the above mentioned modification of the arguments from [9]. Using the Markov inequality, we obtain that

$$\mathbb{P}(|\tilde{X}_{m,j}| \geq e^{2^{\varepsilon n/2}}) \leq e^{-2^{\varepsilon n/2}} \mathbb{E}|\tilde{X}_{m,j}|.$$

However, since $\|A_l\|_{L^p} \leq a l^{1/p}$ for every $l \in \mathbb{N}$ (and for some constant $a > 0$), we have that $\mathbb{E}|\tilde{X}_{m,j}| \leq C 2^{n + \frac{n}{p}}$. Summing the resulting upper bounds for $\mathbb{P}(|\tilde{X}_{m,j}| \geq e^{2^{\varepsilon n/2}})$, we obtain the desired result. □

The following result follows from Lemma 5.4 exactly in the same way as [9, Corollary 5.3] follows from [9, Lemma 5.2].

Corollary 5.1 *Let $\tilde{R}_n = (\tilde{R}_{n,j})_{j < F(n)}$ be distributed like $\tilde{X}_n$ and such that the $\tilde{R}_n$ are independent of each other. Then there exist $C > 0$ and a coupling between $(\tilde{X}_1, \tilde{X}_2, \ldots)$ and $(\tilde{R}_1, \tilde{R}_2, \ldots)$ such that for all (n, j),*

$$\mathbb{P}(|\tilde{X}_{n,j} - \tilde{R}_{n,j}| \geq C 4^{-n}) \leq C 4^{-n}.$$

We also need the following version of [9, Lemma 5.4].

Lemma 5.5 *For any $n \in \mathbb{N}$, we have*

$$\pi\left(((\tilde{R}_{n,j})_{0 \leq j < F(n)}, (\tilde{Y}_{n,j})_{0 \leq j < F(n)}\right) \leq C 4^{-n}$$

where $\tilde{Y}_{n,j} = Y_{n,j} + V_{n,j}$. □

Proof of Lemma 5.5 We follow the proof of [9, Lemma 5.4]. We define $\tilde{Y}^i_{n,j}$ for $0 \leq i \leq f$ as follows: for $0 \leq k < 2^{f-i}$, the random vector $\tilde{\mathcal{Y}}^i_{n,k} := (\tilde{Y}^i_{n,j})_{k2^i \leq j < (k+1)2^i}$ is distributed as $(\tilde{X}_{n,j})_{k2^i \leq j < (k+1)2^i}$, and $\tilde{\mathcal{Y}}^i_{n,k}$ is independent of $\tilde{\mathcal{Y}}^i_{n,k'}$ when $k \neq k'$.

Set $\tilde{Y}^i = (\tilde{Y}^i_{n,j})_{0 \le j < F}$, for $0 \le i \le f$. By Gouëzel [9, (5.7)], we have that

$$\pi(\tilde{Y}^i, \tilde{Y}^{i-1}) \le \sum_{k=0}^{2^{f-i}-1} \pi(\tilde{\mathcal{Y}}^i_{n,k}, (\tilde{\mathcal{Y}}^{i-1}_{n,2k}, \tilde{\mathcal{Y}}^{i-1}_{n,2k+1})), \tag{5.7}$$

for $1 \le i \le f$. As in the proof of [9, Lemma 5.4], as a consequence of the condition (H), the difference between the characteristic functions of $\tilde{\mathcal{Y}}^i_{n,k}$ and $(\tilde{\mathcal{Y}}^{i-1}_{n,2k}, \tilde{\mathcal{Y}}^{i-1}_{n,2k+1})$ is at most $Ce^{-c'2^{\varepsilon n+i}}$ for n large enough. Hence, by applying [9, Lemma 3.5] with $N = 2^i$ and $T' = e^{2^{\varepsilon n/2}}$ we obtain that

$$\pi(\tilde{\mathcal{Y}}^i_{n,k}, (\tilde{\mathcal{Y}}^{i-1}_{n,2k}, \tilde{\mathcal{Y}}^{i-1}_{n,2k+1}))$$

$$\le \sum_{j=k2^i}^{(k+1)2^i-1} \mathbb{P}(|\tilde{X}_{n,j}| \ge e^{2^{\varepsilon n/2}}) + Ce^{2^{\varepsilon n/2+i}} e^{-c'2^{\varepsilon n+i}}.$$

By estimating $\mathbb{P}(|\tilde{X}_{n,j}| \ge e^{2^{\varepsilon n/2}})$ as in the proof of Lemma 5.4, we conclude that

$$\pi(\tilde{\mathcal{Y}}^i_{n,k}, (\tilde{\mathcal{Y}}^{i-1}_{n,2k}, \tilde{\mathcal{Y}}^{i-1}_{n,2k+1})) \le Ce^{-2^{\delta n}}, \tag{5.8}$$

for some $\delta > 0$. The conclusion of the lemma now follows from (5.7) and (5.8) by summing over i and noting that the process $(\tilde{Y}^f_{n,j})_{0 \le j < F}$ coincides with $(\tilde{R}_{n,j})_{0 \le j < F}$ and that $(\tilde{Y}^0_{n,j})_{0 \le j < F}$ coincides with $(\tilde{Y}_{n,j})_{0 \le j < F}$.   □

Finally, relying on Corollary 5.1 and Lemma 5.5, the proof of Proposition 5.1 is completed exactly as in [9].   □

Second Step We now establish the version of [9, Lemma 5.6]. We first recall the following result (see [22, Corollary 3] or [9, Proposition 5.5]).

Proposition 5.2 *Let* $Y_0, \ldots, Y_{b-1}$ *be independent centered* $\mathbb{R}^d$*-valued random vectors. Let* $q \ge 2$ *and set* $M = \left(\sum_{j=0}^{b-1} \mathbb{E}|Y_j|^q\right)^{1/q}$. *Assume that there exists a sequence* $0 = m_0 < m_1 < \ldots < m_s = b$ *such that with* $\zeta_k = Y_{m_k} + \ldots + Y_{m_{k+1}-1}$ *and* $B_k = Cov(\zeta_k)$, *for any* $v \in \mathbb{R}^d$ *and* $0 \le k < s$ *we have that*

$$100M^2|v|^2 \le B_k v \cdot v \le 100CM^2|v|^2, \tag{5.9}$$

where $C \ge 1$ *is some constant. Then, there exists a coupling between* $(Y_0, \ldots, Y_{b-1})$ *and a sequence of independent Gaussian random vectors* $(S_0, \ldots, S_{b-1})$ *such that* $Cov(S_j) = Cov(Y_j)$ *for each* $j \in \mathbb{N}$ *and*

$$\mathbb{P}\left(\max_{0 \le i \le b-1} \left|\sum_{j=0}^i Y_j - S_j\right| \ge Mz\right) \le C'z^{-q} + \exp(-C'z), \tag{5.10}$$

for all $z \geq C' \log s$. Here, C' is a positive constant which depends only of C, d and q.

Lemma 5.6 *Suppose that $p > 2 + 2/\beta$. Then for any $n \in \mathbb{N}$, there exists a coupling between $(Y_{n,0}, \ldots, Y_{n,F(n)-1})$ and $(S_{n,0}, \ldots, S_{n,F(n)-1})$, where the $S_{n,j}$'s are independent centered Gaussian random variables with $Var(S_{n,j}) = Var(Y_{n,j})$, such that*

$$\sum_n \mathbb{P} \left(\max_{1 \leq i \leq F(n)} \left| \sum_{j=0}^{i-1} Y_{n,j} - S_{n,j} \right| \geq 2^{((1-\beta)/2 + (\beta+1)/p + \varepsilon/2)n} \right) < \infty. \tag{5.11}$$

Proof of Lemma 5.6 Take $q \in (2, p)$. By (5.5), we have that

$$\|Y_{n,j}\|_{L^q} \leq C2^{(1-\beta)n/2 + n/p}, \tag{5.12}$$

where we have used that the right end point of each $I_{n,j}$ does not exceed 2^{n+1} and that $X_{n,j}$ and $Y_{n,j}$ are equally distributed. It follows from (5.12) that

$$M := \left(\sum_{j=0}^{F-1} \|Y_{n,j}\|_{L^q}^q \right)^{\frac{1}{q}}$$

satisfies

$$M \leq C2^{n/p + \beta n/q + (1-\beta)n/2}.$$

Therefore, if q is sufficiently close to p then M^2 is much smaller than 2^n, where we have used that $p > 2 + 2/\beta$. On the other hand, by (5.1) we have

$$Var(Y_{n,j}) = Var(X_{n,j}) \geq u2^{(1-\beta)n} \tag{5.13}$$

for some constant $u > 0$ which does not depend on n and j. Here we have taken into account that the length of each $I_{n,j}$ is of magnitude $2^{(1-\beta)n}$. By (5.13) we have

$$Var\left(\sum_{j=0}^{F-1} Y_{n,j} \right) = \sum_{j=0}^{F-1} Var(Y_{n,j}) \geq c2^n, \tag{5.14}$$

where $c > 0$ is some constant.

Next, set $v_j = v_{n,j} = Var(Y_{n,j})$. Then $v_j \leq \|Y_{n,j}\|_{L^q}^2 \leq M^2$. Let u_1 be the largest index such that

$$v_0 + \ldots + v_{u_1-1} \geq 100M^2.$$

Such index exists since $\sum_{j=0}^{F-1} v_j$ is much larger than M^2 (see (5.14)). Notice now that

$$v_0 + \ldots + v_{u_1-1} \leq v_0 + \ldots + v_{u_1-2} + M^2 \leq 101 M^2.$$

This gives us the first block $\{Y_{n,0}, \ldots, Y_{n,u_1-1}\}$ of consecutive $Y_{n,j}$'s from the proof of [9, Lemma 5.6] such that (5.9) holds. We can continue by forming $k+1$ consecutive blocks, namely

$$\{Y_{n,0}, \ldots, Y_{n,u_1-1}\}, \ldots, \{Y_{n,u_k}, \ldots, Y_{n,u_{k+1}-1}\},$$

where k is the first step in the construction such that

$$v_{u_{k+1}} + \ldots + v_F < 100 M^2.$$

Then, we add $Y_{n,u_{k+1}}, \ldots, Y_{n,F}$ to the last block $\{Y_{u_k}, \ldots, Y_{n,u_{k+1}-1}\}$ we have constructed. This means that we can always assume that the sum of the variances of the random variables $Y_j = Y_{n,j}$ along successive blocks is not less than $100 M^2$ and that it doesn't exceed $201 M^2$. The statement of the lemma now follows by applying Proposition 5.2 with $z = 2^{\varepsilon n/2}$, taking into account that the number of blocks is trivially bounded by $F = F(n)$.

Third Step It follows from the previous two steps of the proof that, when $p > 2 + 2/\beta$ there exists a coupling between $(A_n)_{n \in I}$ and a sequence $(B_n)_{n \in I}$ of independent centered normal random variables so that when (n, j) tends to infinity, we have

$$\left| \sum_{\ell < i_{n,j}, \ell \in I} (A_\ell - B_\ell) \right| = o(2^{(\beta+\varepsilon)n/2} + 2^{((1-\beta)/2+(\beta+1)/p+\varepsilon)n}),$$

where $i_{n,j}$ denotes the smallest element of $I_{n,j}$. We note that we have also used the so-called Berkes–Philipp lemma (see [3, Lemma A.1] or [9, Lemma 3.1]).

Fourth Step We now establish the version of [9, Lemma 5.8]. However, before we do that we need the following result, which is a consequence of [19, Theorem 1] (see also [21, Corollary B1]).

Lemma 5.7 *Let $Y_1, \ldots, Y_d$ be a finite sequence of random variables. Let $v > 2$ be finite and assume that there exist constants $C_1, C_2 > 0$ such that $\|Y_i\|_{L^v} \leq C_1$ for every $i \in \{1, \ldots, d\}$. Moreover, assume that for any $a, n \in \mathbb{N}$ satisfying $a + n \leq d$, we have that*

$$\|S_{a,n}\|_{L^v} \leq C_2^2 n^{\frac{1}{2}},$$

where

$$S_{a,n} = \sum_{i=a+1}^{a+n} Y_i.$$

Then, there exists a constant $K > 0$ (depending only on C_1, C_2 and v) such that for any a and n,

$$\|M_{a,n}\|_{L^v} \leq K n^{\frac{1}{2}}, \tag{5.15}$$

where

$$M_{a,n} = \max\{|S_{a,1}|, \ldots, |S_{a,n}|\}.$$

The following is the already announced version of [9, Lemma 5.8].

Lemma 5.8 *We have that as $(n, j) \to \infty$,*

$$\max_{m < |I_{n,j}|} \left| \sum_{\ell=i_{n,j}}^{i_{n,j}+m} A_\ell \right| = o(2^{((1-\beta)/2+\beta/p+1/p+\varepsilon)n}) \quad a.s. \tag{5.16}$$

Proof of Lemma 5.8 Let $q \in (2, p)$. Consider the finite sequence

$$Y_k = A_k/(i_{n,j} + |I_{n,j}|)^{1/p}, \quad k \in I_{n,j}.$$

Then, by (5.2) there exists a constant $C_1 > 0$ which does not depend on n and j so that $\|Y_k\|_{L^q} \leq C_1$, for any $k \in I_{n,j}$. Moreover, by (5.5), there exists a constant $C_2 > 0$ which does not depend on n and j so that for any relevant a and b,

$$\left\| \sum_{k=a+1}^{a+b} Y_k \right\|_{L^q} \leq C_2 b^{\frac{1}{2}}.$$

Using the same notation as in statement of Lemma 5.7, we observe that it follows from (5.15) that

$$\|M_{n,b}\|_{L^q} \leq K b^{\frac{1}{2}},$$

for some constant $K > 0$ (which depends only C_1, C_2 and q).

In particular, by setting $v = (1 - \beta)/2 + \beta/p + \varepsilon/2$, we have that

$$\mathbb{P}(M_{i_{n,j},|I_{n,j}|} \geq 2^{vn}) \leq \| M_{i_{n,j},|I_{n,j}|} \|_{L^q}^q / 2^{vnq} \leq K |I_{n,j}|^{q/2} / 2^{vnq}.$$

Moreover, observe that

$$\sum_{n,j} |I_{n,j}|^{q/2}/2^{vnq} \leq \sum_n 2^{\beta n} 2^{(1-\beta)nq/2-vnq}.$$

Notice that the above sum is finite if q is sufficiently close to p. Applying the Borel-Cantelli lemma yields that, as $(n, j) \to \infty$,

$$\max_{m<|I_{n,j}|} \left| \sum_{\ell=i_{n,j}}^{i_{n,j}+m} Y_\ell \right| = o(2^{((1-\beta)/2+\beta/p+\varepsilon)n}),$$

which implies that (5.16) holds (since the right end point of $I_{n,j}$ does not exceed 2^{n+1}).

Fifth Step By combining the last two steps, we derive that when k tends to infinity,

$$\left| \sum_{\ell<k,\, \ell\in I} (A_\ell - B_\ell) \right| = o(k^{(\beta+\varepsilon)/2} + k^{(1-\beta)/2+(\beta+1)/p+\varepsilon})$$

assuming that $p > 2 + 2/\beta$.

Sixth Step Fix some n and consider the finite sequence $Y_i = A_i/n^{1/p}$ where $i \in \{1, \ldots, n\}$. It follows from our assumptions that $(Y_i)_i$ satisfies property (H) (with constants that do not depend on n). Applying [9, Lemma 5.9] with the finite sequence (Y_i) (instead of A_i there), we see that for any $\alpha > 0$, there exists $C = C_\alpha$ (which does not depend on n) such that for any interval $J \subset [1, n]$ we have

$$n^{-2/p}\mathbb{E}\left| \sum_{\ell\in J\cap\mathcal{J}} A_i \right|^2 = \mathbb{E}\left| \sum_{\ell\in J\cap\mathcal{J}} Y_i \right|^2 \leq C|J\cap\mathcal{J}|^{1+\alpha}. \tag{5.17}$$

We recall the following version of the Gal-Koksma law of large numbers, which is a direct consequence of [19, Theorem 3] together with some routine estimates (as those given in the proof of [19, Theorem 6]). We also note that the lemma can be proved by an easy adaptation of the arguments in the proof of [20, Theorem A1].

Lemma 5.9 *Let $Y_1, Y_2, \ldots$ be a sequence of random variables such that with some constants $\sigma \geq 1$, $C > 0$, $p > 1$ and for any $m, n \in \mathbb{N}$ we have that*

$$\left\| \sum_{j=m+1}^{m+n} Y_j \right\|_{L^2}^2 \leq C\big((n+m)^\sigma - m^\sigma\big) \cdot (n+m)^{\frac{2}{p}}.$$

Then, for any $\delta > 0$ we have that $\mathbb{P}$-a.s. as $n \to \infty$,

$$\sum_{j=1}^{n} Y_j = o(n^{\sigma/2+1/p} \ln^{3/2+\delta} n).$$

Relying on (5.17) and Lemma 5.9, one can now repeat the arguments appearing after the statement of [9, Lemma 5.9] with the finite sequence $\left(A_i/k^p\right)_{1 \leq i \leq k}$ (instead of $(A_i)_i$), and conclude that

$$\sum_{\ell < k,\, \ell \in \mathcal{J}} A_\ell/k^{\frac{1}{p}} = o(k^{\beta/2+\varepsilon}).$$

Finalizing the Proof Combining the estimates from the previous steps we get a coupling of (A_ℓ) with independent centered normal random variables (B_ℓ) such that

$$\left|\sum_{\ell < k}(A_k - B_k)\right| = o(k^{\beta/2+\varepsilon+\frac{1}{p}} + k^{(1-\beta)/2+(\beta+1)/p+\varepsilon}), \quad \text{a.s.}$$

Taking $\beta = p/(2p - 2)$, we obtain (5.3). Observe that for this choice of β we have $p > 2+2/\beta$ since $p \geq 6$. When $4 < p < 6$ we can make a different choice of β and obtain a slightly less attractive rate. To complete the proof of Theorem 5.1, it remains to estimate the variance of the approximating Gaussian $G_n = \sum_{j=1}^{n} B_j$. Firstly, by applying [7, Proposition 9] with the finite sequence $\left(A_i/2^{(n+1)/p}\right)_{1 \leq i \leq 2^{n+1}}$ replacing $(A_i)_i$, we obtain that

$$\left\| \sum_{(n',j')\prec(n,j)} X_{n',j'} - Y_{n',j'} \right\|_{L^2} \leq C2^{\beta n/2 + n/p},$$

where $(Y_{n',j'})$ are given by Proposition 5.1. Since $Y_{n',j'}$ and $S_{n',j'}$ have the same variances, we conclude that

$$\left| \left\| \sum_{(n',j')\prec(n,j)} X_{n',j'} \right\|_{L^2} - \left\| \sum_{(n',j')\prec(n,j)} S_{n',j'} \right\|_{L^2} \right| \leq C2^{\beta n/2 + n/p}. \tag{5.18}$$

Take $n \in \mathbb{N}$, and let N_n be such that $2^{N_n} \leq n < 2^{N_n+1}$. Furthermore, let j_n be the largest index such that the left end point of I_{N_n, j_n} is smaller than n. In the case when

$n \in I_{N_n, j_n}$ we have

$$\sum_{i=1}^{n} A_i - \sum_{(n',j') \prec (N_n, j_n)} X_{n',j'} = \sum_{(n',j') \prec (N_n, j_n)} \sum_{i \in J_{n',j'}} A_i + \sum_{i \in J_{N_n, j_n}} A_i$$

$$+ \sum_{i=i_{N_n, j_n}}^{n} A_i$$

$$= \sum_{i \leq n, i \in J} A_i + \sum_{i=i_{N_n, j_n}}^{n} A_i$$

$$=: I_1 + I_2.$$

Recall next that by Gouëzel [9, (5.1)] the cardinality of $\mathcal{J} \cap [1, 2^{N_n+1}]$ does not exceed $C 2^{\varepsilon(N_n+1)} 2^{\beta N_n} (\varepsilon N_n + 2)$, which for our specific choice of N_n is at most $C n^{\beta+3\varepsilon/2}$ (where C denotes a generic constant independent of n). Using (5.17) with a sufficiently small α we derive that

$$\|I_1\|_{L^2} \leq C n^{1/p + \beta/2 + \varepsilon}.$$

On the other hand, applying (5.5) we obtain that

$$\|I_2\|_{L^2} \leq C |I_{N_n, j_n}|^{\frac{1}{2}} 2^{N_n/p}$$
$$\leq C 2^{N_n(1-\beta)/2 + N_n/p} \leq C n^{(1-\beta)/2 + 1/p} \leq C n^{\beta/2 + 1/p}$$

where we have used that for our specific choice of β we have $(1 - \beta)/2 = \beta/2 - \beta/p < \beta/2$. We conclude that there exists a constant $C' > 0$ so that for any $n \geq 1$,

$$\left\| \sum_{j=1}^{n} A_j - \sum_{(n',j') \prec (N_n, j_n)} X_{n',j'} \right\|_{L^2} \leq C' n^{\beta/2 + \varepsilon + 1/p}.$$

The proof of (5.4) in the case when $n \in I_{N_n, j_n}$ is completed now using (5.18). The case when $n \notin I_{N_n, j_n}$ is treated similarly. We first write

$$\sum_{i=1}^{n} A_i - \sum_{(n',j') \prec (N_n, j_n)} X_{n',j'} = \sum_{j \in \mathcal{J}, j \leq n} A_i + X_{N_n, j_n} := I_1 + I_2.$$

Then the L^2-norms of I_1 and I_2 are bounded exactly as in the case when $n \in I_{N_n, j_n}$, and the proof of (5.4) is complete. Finally, the last conclusion in the statement of the theorem follows directly from (5.3), (5.4) together with [11, Theorem 3.2A],

[3, Lemma A.1] (seel also [9, Lemma 3.1]) and the so-called Strassen–Dudley theorem [2, Theorem 6.9] (see also [9, Theorem 3.4]).

5.4 Main Result

The goal of this section is to establish the quenched almost sure invariance principle for random distance expanding maps satisfying suitable conditions. This is done by applying Theorem 5.1.

Without any loss of generality, we can suppose that our observable $\psi : \mathcal{E} \to \mathbb{R}$ is fiberwise centered, i.e. that $\int_{\mathcal{E}_\omega} \psi_\omega \, d\mu_\omega = 0$ for $\mathbb{P}$-a.e. $\omega \in \Omega$. Indeed, otherwise we can simply replace ψ with $\tilde{\psi}$ given by

$$\tilde{\psi}_\omega = \psi_\omega - \int_{\mathcal{E}_\omega} \psi_\omega \, d\mu_\omega, \quad \omega \in \Omega.$$

In what follows, $\mathbb{E}_\omega(\varphi)$ will denote the expectation of a measurable $\varphi : \mathcal{E}_\omega \to \mathbb{R}$ with respect to μ_ω. The proof of the following result can be obtained by repeating the arguments from [6, Lemma 12.] and [6, Proposition 3.] (see also [13, Theorem 2.3.])

Proposition 5.3 *We have the following:*

1. there exists $\Sigma^2 \geq 0$ such that

$$\lim_{n \to \infty} \frac{1}{n} \mathbb{E}_\omega \left(\sum_{k=0}^{n-1} \psi_{\sigma^k \omega} \circ f_\omega^k \right)^2 = \Sigma^2, \quad \text{for } \mathbb{P}\text{-a.e. } \omega \in \Omega; \tag{5.1}$$

2. $\Sigma^2 = 0$ if and only if there exists $\varphi \in L^2_\mu(\mathcal{E})$ such that

$$\psi = \varphi - \varphi \circ F.$$

From now on we shall assume that $\Sigma^2 > 0$. For any integer $L \geq 1$ consider the set

$$A_L = \left\{ \omega \in \Omega : \frac{1}{n} \mathbb{E}_\omega \left(\sum_{k=0}^{n-1} \psi_{\sigma^k \omega} \circ f_\omega^k \right)^2 \geq \frac{1}{2} \Sigma^2, \quad \forall n \geq L \right\}.$$

Then $A_L \subset A_{L'}$ if $L \leq L'$ and the union of the A_L's has probability 1. Due to measurability of Q_ω, $C(\omega)$, $K(\omega)$, and $\omega \mapsto h_\omega$, for any $C_0 > 0$ and $L \in \mathbb{N}$ the set

$$E := \{ \omega \in \Omega : \max\{C(\omega), K(\omega), \|h_\omega\|_\infty, \|1/h_\omega\|_{\alpha,\xi}, 1/Q_\omega\} \leq C_0 \} \cap A_L \tag{5.2}$$

is measurable, and when C_0 and L are sufficiently large we have that $\mathbb{P}(E) > 0$. Fix some large enough C_0 and L, and for $\omega \in \Omega$, let

$$m_1(\omega) := \inf\{n \in \mathbb{N} : \sigma^n \omega \in E\}.$$

For $k > 1$ we inductively define

$$m_k(\omega) := \inf\{n > m_{k-1}(\omega) : \sigma^n \omega \in E\}.$$

Due to ergodicity of $\mathbb{P}$, we have that $m_k(\omega)$ is well-defined for $\mathbb{P}$-a.e. $\omega \in \Omega$ and every $k \in \mathbb{N}$. Let us consider the associated induced system $(E, \mathcal{F}_E, \mathbb{P}_E, \iota)$, where $\mathcal{F}_E = \{A \cap E : A \in \mathcal{F}\}$, $\mathbb{P}_E(A) = \frac{\mathbb{P}(A)}{\mathbb{P}(E)}$, $A \in \mathcal{F}_E$ and $\iota(\omega) = \sigma^{m_1(\omega)}\omega$ for $\omega \in E$. We recall that $\mathbb{P}_E$ is invariant for ι and in fact ergodic.

It follows from Birkhoff's ergodic theorem that

$$\lim_{n \to \infty} \frac{k_n(\omega)}{n} = \mathbb{P}(E) \quad \text{for } \mathbb{P}\text{-a.e. } \omega \in \Omega, \tag{5.3}$$

where

$$k_n(\omega) := \max\{k \in \mathbb{N} : m_k(\omega) \leq n\}.$$

Moreover, Kac's lemma implies that

$$\lim_{n \to \infty} \frac{m_n(\omega)}{n} = \frac{1}{\mathbb{P}(E)}, \quad \text{for } \mathbb{P}\text{-a.e. } \omega \in \Omega.$$

By combining the last two equalities, we conclude that

$$\lim_{n \to \infty} \frac{m_{k_n(\omega)}(\omega)}{n} = 1, \quad \text{for } \mathbb{P}\text{-a.e. } \omega \in \Omega.$$

For $\mathbb{P}$ a.e. $\omega \in \Omega$, set

$$\Psi_\omega := \sum_{j=0}^{m_1(\omega)-1} \psi_{\sigma^j \omega} \circ f_\omega^j.$$

We assume that there exists $p \geq 6$, so that

the map $\omega \mapsto A(\omega) := \|\Psi_\omega\|_\infty$ belongs to $L^p(\Omega, \mathcal{F}, \mathbb{P})$. $\tag{5.4}$

Finally, let $L_\omega := \mathcal{L}_\omega^{m_1(\omega)}$ and $F_\omega := f_\omega^{m_1(\omega)}$, for $\omega \in \Omega$.

We are now in a position to state the main result of our paper (recall our assumption that $\Sigma^2 > 0$).

Theorem 5.2 *For $\mathbb{P}$-a.e. $\omega \in \Omega$ and arbitrary $\delta > 0$, there exists a coupling between $(\psi_{\sigma^i\omega} \circ f_\omega^i)_i$, considered as a sequence of random variables on $(\mathcal{E}_\omega, \mu_\omega)$, and a sequence $(Z_k)_k$ of independent centered (i.e. of zero mean) Gaussian random variables such that*

$$\left| \sum_{i=1}^n \psi_{\sigma^i\omega} \circ f_\omega^i - \sum_{i=1}^n Z_i \right| = o(n^{a_p+\delta}), \quad a.s., \tag{5.5}$$

where

$$a_p = \frac{p}{4(p-1)} + \frac{1}{p}.$$

Moreover, there exists $C = C(\omega) > 0$ so that for any $n \geq 1$,

$$\left\| \sum_{i=1}^n \psi_{\sigma^i\omega} \circ f_\omega^i \right\|_{L^2} - Cn^{a_p+\delta} \leq \left\| \sum_{i=1}^n Z_i \right\|_{L^2} \leq \left\| \sum_{i=1}^n \psi_{\sigma^i\omega} \circ f_\omega^i \right\|_{L^2} + Cn^{a_p+\delta}.$$
$$\tag{5.6}$$

Finally, there exists a coupling between $(\psi_{\sigma^i\omega} \circ f_\omega^i)_i$ and a standard Brownian motion $(W_t)_{t\geq 0}$ such that

$$\left| \sum_{i=1}^n \psi_{\sigma^i\omega} \circ f_\omega^i - W_{\sigma_{\omega,n}^2} \right| = o(n^{\frac{1}{2}a_p+\frac{1}{4}+\delta}) \quad a.s.,$$

where

$$\sigma_{\omega,n} = \left\| \sum_{i=1}^n \psi_{\sigma^i\omega} \circ f_\omega^i \right\|_{L^2}.$$

Remark 5.2 Observe that $a_p \to \frac{1}{4}$ as $p \to \infty$. We note that our proof also yields convergence rate when $4 < p < 6$, which has a slightly less attractive form in terms of p. In addition, we emphasize that $\left\| \sum_{i=1}^n Z_i \right\|_{L^2}$ depends on ω but that it is asymptotically deterministic. More precisely, it follows from (5.1) and (5.6) that

$$\lim_{n\to\infty} \frac{\left\| \sum_{i=1}^n Z_i \right\|_{L^2}^2}{n\Sigma^2} = 1.$$

Proof of Theorem 5.2 Our strategy proceeds as follows. Firstly, we will apply Theorem 5.1 to establish the invariance principle for the induced system. Secondly, we extend the invariance principle to our original system. Throughout the proof, $C > 0$ will denote a generic constant independent on ω and other parameters involved in the estimates.

For $\omega \in E$ (recall that E is given by (5.2)), set $A_n = \Psi_{\iota^n \omega} \circ F_\omega^n$, $n \in \mathbb{N}$. Obviously, A_n depends also on ω but in order to make the notation as simple as possible, we do not make this dependence explicit.

Observe that it follows from (5.4) and Birkhoff's ergodic theorem that there exists a random variable $R \colon E \to (0, \infty)$ such that:

$$\|A_n\|_{L^p} \le R(\omega) n^{1/p} \quad \text{for } \mathbb{P}\text{-a.e. } \omega \in E \text{ and } n \in \mathbb{N}. \tag{5.7}$$

It follows easily from (5.2) and (5.1) that for any $k \in \mathbb{N}$, $n \ge L$ and $\omega \in E$,

$$\frac{1}{n} Var\left(\sum_{j=0}^{n-1} A_{j+k}\right) \ge \frac{1}{2}\Sigma^2, \tag{5.8}$$

where we have used that $m_n(\iota^k(\omega)) \ge n$. We conclude from (5.7) and (5.8) that the processes $(A_n)_{n \in \mathbb{N}}$ satisfies (5.2) and (5.1), respectively.

Hence, in order to apply Theorem 5.1, we need to show that $(A_n)_{n \in \mathbb{N}}$ satisfies property (H) and, in addition, that for any $n < m$ the finite sequence $(A_i/(n+m)^{1/p})_{n+1 \le i \le n+m}$ also satisfies (H) (with uniform constants). In fact, we will prove the following: the process $(a_n A_n)_{n \in \mathbb{N}}$ satisfies (H) for any sequence $(a_n)_{n \in \mathbb{N}} \subset (0, 1]$ (and with uniform constants). Let us begin by introducing some auxiliary notations. For $\mathbb{P}$ a.e. $\omega \in \Omega$ and $z \in \mathbb{C}$, let

$$\hat{\mathcal{L}}_\omega^z g := \hat{\mathcal{L}}_\omega(g e^{z \psi_\omega}) = \mathcal{L}_\omega(g e^{z \psi_\omega} h_\omega)/h_{\sigma \omega}, \quad \text{for } g \in \mathcal{H}_\omega^{\alpha, \xi}.$$

Furthermore, for $z \in \mathbb{C}$ and $n \in \mathbb{N}$, set

$$\hat{\mathcal{L}}_\omega^{z,n} := \hat{\mathcal{L}}_{\sigma^{n-1}\omega}^z \circ \ldots \circ \hat{\mathcal{L}}_\omega^z.$$

It is easy to verify that

$$\hat{\mathcal{L}}_\omega^{z,n} g = \mathcal{L}_\omega^n (g e^{z S_n^\omega \psi} h_\omega)/h_{\sigma^n \omega} = \mathcal{L}_\omega^{z,n}(g h_\omega)/h_{\sigma^n \omega}.$$

Finally, for $\omega \in \Omega$, $n \in \mathbb{N}$ and $\bar{t} = (t_0, t_1, \ldots, t_{n-1}) \in \mathbb{R}^n$, let

$$L_\omega^{\bar{t},n} = \hat{\mathcal{L}}_{\iota^{n-1}\omega}^{it_{n-1}, m_n(\omega) - m_{n-1}(\omega)} \circ \ldots \circ \hat{\mathcal{L}}_{\iota \omega}^{it_1, m_2(\omega) - m_1(\omega)} \circ \hat{\mathcal{L}}_\omega^{it_0, m_1(\omega)}.$$

Observe that

$$L_\omega^{\bar{t},n} g = (\mathcal{L}_{\iota^{n-1}\omega}^{it_{n-1}, m_n(\omega) - m_{n-1}(\omega)} \circ \ldots \circ \mathcal{L}_{\iota \omega}^{it_1, m_2(\omega) - m_1(\omega)} \circ \mathcal{L}_\omega^{it_0, m_1(\omega)})(g h_\omega)/h_{\iota^n \omega},$$

for any $g \in \mathcal{H}_\omega^{\alpha,\xi}$. It follows from (5.6), (5.2) and the above formula that for $n \in \mathbb{N}$ and $\bar{t} \in [-1, 1]^n$, we have that

$$\|L_\omega^{\bar{t},n}\|_{\alpha,\xi} \leq C. \tag{5.9}$$

For $\omega \in \Omega$ and $g \in \mathcal{H}_\omega^{\alpha,\xi}$, set

$$\Pi_\omega g := \left(\int_{\mathcal{E}_\omega} g \, d\mu_\omega \right) \mathbf{1}$$

where $\mathbf{1}$ denotes the function which takes the constant value 1, regardless of the space on which it is defined. Since $L_\omega^{\bar{0},k} = \hat{\mathcal{L}}_\omega^{m_k(\omega)}$ and $m_k(\omega) \geq k$, it follows from Lemma 5.3 and (5.2) that

$$\|(L_\omega^{\bar{0},k} - \Pi_\omega)g\|_\infty \leq Ce^{-\lambda k}\|g\|_{\alpha,\xi}, \tag{5.10}$$

for $\omega \in E$, $g \in \mathcal{H}_\omega^{\alpha,\xi}$ and $k \in \mathbb{N}$.

Take now $n, m, k \in \mathbb{N}$, $b_1 < b_2 < \ldots < b_{n+m+k}$ and $t_1, \ldots, t_{n+m} \in \mathbb{R}$ with $|t_j| \leq 1$. We have that

$$\mathbb{E}_{\mu_\omega}\left(e^{i \sum_{j=1}^n t_j (\sum_{\ell=b_j}^{b_j+1-1} B_\ell) + i \sum_{j=n+1}^{n+m} t_j (\sum_{\ell=b_j+k}^{b_{j+1}+k-1} B_\ell)} \right)$$

$$= \mathbb{E}_{\mu_{\iota^{b_n+m+1+k}\omega}} \left(L_{\iota^{b_n+1+k}\omega}^{\bar{t},b_{n+m+1}-b_{n+1}} L_{\iota^{b_n+1}\omega}^{\bar{0},k} L_{\iota^{b_1}\omega}^{\bar{s},b_{n+1}-b_1} \mathbf{1} \right),$$

where $B_n = a_n A_n$,

$$\bar{s} = (a_{b_1}t_1, \ldots, a_{b_2-1}t_1, a_{b_2}t_2, \ldots, a_{b_3-1}t_2, \ldots, a_{b_n}t_n, \ldots, a_{b_{n+1}-1}t_n),$$

and

$$\bar{t} = (a_{b_{n+1}+k}t_{n+1}, \ldots, a_{b_{n+2}+k-1}t_{n+1}, \ldots, a_{b_{n+m}+k}t_{n+m}, \ldots, a_{b_{n+m+1}+k-1}t_{n+m}).$$

Consequently,

$$\mathbb{E}_{\mu_\omega}\left(e^{i \sum_{j=1}^n t_j (\sum_{\ell=b_j}^{b_{j+1}-1} B_\ell) + i \sum_{j=n+1}^{n+m} t_j (\sum_{\ell=b_j+k}^{b_{j+1}+k-1} B_\ell)} \right)$$

$$= \mathbb{E}_{\mu_{\iota^{b_n+m+1+k}\omega}} \left(L_{\iota^{b_n+1+k}\omega}^{\bar{t},b_{n+m+1}-b_{n+1}} \left(L_{\iota^{b_n+1}\omega}^{\bar{0},k} - \Pi_{\iota^{b_n+1}\omega} \right) L_{\iota^{b_1}\omega}^{\bar{s},b_{n+1}-b_1} \mathbf{1} \right)$$

$$+ \mathbb{E}_{\mu_{\iota^{b_n+m+1+k}\omega}} \left(L_{\iota^{b_n+1+k}\omega}^{\bar{t},b_{n+m+1}-b_{n+1}} \Pi_{\iota^{b_n+1}\omega} L_{\iota^{b_1}\omega}^{\bar{s},b_{n+1}-b_1} \mathbf{1} \right)$$

$$=: I_1 + I_2.$$

We claim next that

$$|I_1| \le C e^{-\lambda k}. \tag{5.11}$$

Indeed, set

$$A := L_{\iota^{b_{n+1}+k}\omega}^{\bar{\iota},b_{n+m+1}-b_{n+1}}, \quad B := L_{\iota^{b_{n+1}}\omega}^{\bar{0},k} - \Pi_{\iota^{b_{n+1}}\omega} \quad \text{and} \quad g := L_{\iota^{b_1}\omega}^{\bar{s},b_{n+1}-b_1}\mathbf{1}.$$

Then,

$$\|A\|_\infty := \sup_{f:\|f\|_\infty=1} \|Af\|_\infty \le \|L_{\iota^{b_{n+1}+k}\omega}^{\bar{0},b_{n+m+1}-b_{n+1}}\mathbf{1}\|_\infty = \|\mathbf{1}\|_\infty = 1,$$

and therefore

$$|I_1| \le \|A(Bg)\|_\infty \le \|A\|_\infty \cdot \|Bg\|_\infty \le \|Bg\|_\infty.$$

Applying (5.9) we have

$$\|g\|_{\alpha,\xi} \le C,$$

and thus it follows from (5.10) that

$$|I_1| \le \|Bg\|_\infty \le C e^{-\lambda k}.$$

We conclude that (5.11) holds.

On the other hand,

$$I_2 = \mathbb{E}_\omega\big(e^{i\sum_{j=1}^n t_j(\sum_{\ell=b_j}^{b_{j+1}-1} B_\ell)}\big) \cdot \mathbb{E}_\omega\big(e^{i\sum_{j=n+1}^{n+m} t_j(\sum_{\ell=b_j+k}^{b_{j+1}+k-1} B_\ell)}\big).$$

We conclude that the process $(B_n)_{n\in\mathbb{N}}$ satisfies property (H) with constants that do not depend on the sequence (a_l). Thus, Theorem 5.1 yields the almost sure invariance principle for the process $(\Psi_{\iota^n\omega} \circ F_\omega^n)_{n\in\mathbb{N}}$.

It remains to observe that the conclusion of Theorem 5.2 now follows from the Berkes–Philipp lemma (see [3, Lemma A.1] or [9, Lemma 3.1]) and the following lemma which together with (5.3), ensures that (5.6) holds true.

Lemma 5.10 *There exists a random variable* $U : \Omega \to (0, \infty)$ *such that*

$$\Bigg\| \sum_{j=0}^{n-1} \psi_{\sigma^j\omega} \circ f_\omega^j - \sum_{j=0}^{k_n(\omega)-1} \Psi_{\iota^j\omega} \circ F_\omega^j \Bigg\|_\infty \le U(\omega)n^{1/p},$$

for $\mathbb{P}$-a.e. $\omega \in \Omega$ *and* $n \in \mathbb{N}$. $\qquad\qquad\qquad\qquad\qquad\qquad\qquad\qquad \square$

Proof of the Lemma If $n = m_{k_n(\omega)}(\omega)$ then there is nothing to prove, and so we assume that $m_{k_n(\omega)}(\omega) < n$. Observe that

$$\sum_{j=0}^{n-1} \psi_{\sigma^j \omega} \circ f_\omega^j - \sum_{j=0}^{k_n(\omega)-1} \Psi_{\iota^j \omega} \circ F_\omega^j = \sum_{j=m_{k_n(\omega)}(\omega)}^{n-1} \psi_{\sigma^j \omega} \circ f_\omega^j$$

$$= \sum_{j=m_{k_n(\omega)}(\omega)}^{m_{k_n(\omega)+1}(\omega)-1} \psi_{\sigma^j \omega} \circ f_\omega^j - \sum_{j=n}^{m_{k_n(\omega)+1}(\omega)-1} \psi_{\sigma^j \omega} \circ f_\omega^j$$

$$= \Psi_{\sigma^{k_n(\omega)}\omega} \circ f_\omega^{k_n(\omega)} - \Psi_{\sigma^n \omega} \circ f_\omega^n$$

and thus

$$\left\| \sum_{j=0}^{n-1} \psi_{\sigma^j \omega} \circ f_\omega^j - \sum_{j=0}^{k_n(\omega)-1} \Psi_{\iota^j \omega} \circ F_\omega^j \right\|_\infty \leq \|\Psi_{\sigma^{k_n(\omega)}\omega}\|_\infty + \|\Psi_{\sigma^n \omega}\|_\infty,$$

where we have used that $\sigma^j \omega \notin E$ when $m_{k_n(\omega)}(\omega) < j < m_{k_n(\omega)+1}(\omega)$. Hence, the conclusion of the lemma follows directly from Birkhoff's ergodic theorem, (5.3) and (5.4). □

Acknowledgments We would like to thank the anonymous referee for his/hers constructive and illuminating comments that helped us to improve our paper. D. D. was supported in part by Croatian Science Foundation under the project IP-2019-04-1239 and by the University of Rijeka under the projects uniri-prirod-18-9 and uniri-prprirod-19-16.

References

1. R. Aimino, M. Nicol, S. Vaienti, Annealed and quenched limit theorems for random expanding dynamical systems. Probab. Theory Related Fields **162**, 233–274 (2015)
2. P. Billingsley, *Convergence of Probability Measures*, 2nd edn. (Wiley, New York, 1999)
3. J. Berkes, W. Philipp, Approximation theorems for independent and weakly dependent random vectors. Ann. Probab. **7**, 29–54 (1979)
4. C. Castaing, M. Valadier, *Convex Analysis and Measurable Multifunctions*. Lecture Notes in Mathematics, vol. 580 (Springer, New York, 1977)
5. C. Cuny, F. Merlevède, Strong invariance principles with rate for "reverse" martingales and applications. J. Theor. Probab. **28**, 137–183 (2015)
6. D. Dragičević, G. Froyland, C. Gonzalez-Tokman, S. Vaienti, Almost sure invariance principle for random piecewise expanding maps. Nonlinearity **31**, 2252–2280 (2018)
7. D. Dragičević, Y. Hafouta, A vector-valued almost sure invariance principle for random hyperbolic and piecewise-expanding maps. https://arxiv.org/abs/1912.12332
8. M. Field, I. Melbourne, A. Török, Decay of correlations, central limit theorems and approximation by Brownian motion for compact Lie group extensions. Ergodic Theory Dynam. Syst. **23**, 87–110 (2003)

9. S. Gouëzel, Almost sure invariance principle for dynamical systems by spectral methods. Ann. Probab. **38**, 1639–1671 (2010)
10. Y. Hafouta, Y. Kifer, *Nonconventional Limit Theorems and Random Dynamcis* (World Scientific, New-Jersey, 2018)
11. D.L. Hanson, R.P. Russo, Some results on increments of the wiener process with applications to lag sums of I.I.D. Random variables. Ann. Probab. **11**, 609–623 (1983)
12. N. Haydn, M. Nicol, A. Török, S. Vaienti, Almost sure invariance principle for sequential and non-stationary dynamical systems. Trans. Am. Math. Soc. **369**, 5293–5316 (2017)
13. Y. Kifer, Limit theorems for random transformations and processes in random environments. Trans. Am. Math. Soc. **350**, 1481–1518 (1998)
14. A. Korepanov, Equidistribution for nonuniformly expanding dynamical systems. Commun. Math. Phys. **359**, 1123–1138 (2018)
15. A. Korepanov, Rates in almost sure invariance principle for dynamical systems with some hyperbolicity. Commun. Math. Phys. **363**, 173–190 (2018)
16. V. Mayer, B. Skorulski, M. Urbański, *Distance Expanding Random Mappings, Thermodynamical Formalism, Gibbs Measures and Fractal Geometry*. Lecture Notes in Mathematics, vol. 2036 (Springer, Berlin, 2011)
17. I. Melbourne, M. Nicol, Almost sure invariance principle for nonuniformly hyperbolic systems. Commun. Math. Phys. **260**, 131–146 (2005)
18. I. Melbourne, M. Nicol, A vector-valued almost sure invariance principle for hyperbolic dynamical systems. Ann. Probab. **37**, 478–505 (2009)
19. F. Mórciz, Moment inequalities and the strong laws of large numbers. Z. Wahrsch. Verw. Gebiete **35**, 299–314 (1976)
20. W. Philip, W.F. Stout, Almost sure invariance principle for sums of weakly dependent random variables. Memoirs of the American Mathematical Society, vol. 161 (American Mathematical Society, Providence, 1975)
21. R.J. Serfling, Moment inequalities for the maximum cumulative sum. Ann. Math. Stat. **41**, 1227–1234 (1970)
22. A.Y. Zaitsev, Estimates for the rate of strong approximation in the multidimensional invariance principle. J. Math. Sci. **145**, 4856–4865 (2007)

Chapter 6
Limit Theorem for Reflected Random Walks

Hoang-Long Ngo and Marc Peigné

Abstract Let $\xi_n, n \in \mathbb{N}$ be a sequence of i.i.d. random variables with values in $\mathbb{Z}$. The associated random walk on $\mathbb{Z}$ is $S(n) = \xi_1 + \cdots + \xi_{n+1}$ and the corresponding "reflected walk" on $\mathbb{N}_0$ is the Markov chain $X = (X(n))_{n \geq 0}$ given by $X(0) = x \in \mathbb{N}_0$ and $X(n+1) = |X(n) + \xi_{n+1}|$ for $n \geq 0$. It is well know that the reflected walk $(X(n))_{n \geq 0}$ is null-recurrent when the ξ_n are square integrable and centered. In this paper, we prove that the process $(X(n))_{n \geq 0}$, properly rescaled, converges in distribution towards the reflected Brownian motion on $\mathbb{R}^+$, when $\mathbb{E}[\xi_n^2] < +\infty$, $\mathbb{E}[(\xi_n^-)^3] < +\infty$ and the ξ_n are aperiodic and centered.

6.1 Introduction and Notations

Let $(\xi_n)_{n \geq 1}$ be a sequence of $\mathbb{Z}$-valued, independent and identically distributed random variables, with common law μ defined on a probability space $(\Omega, \mathcal{F}, \mathbb{P})$. We denote $S = (S(n))_{n \geq 0}$ the classical random walks with steps ξ_k defined by $S(0) = 0$ and $S(n) = \xi_1 + \ldots + \xi_n$ for any $n \geq 1$.

Throughout this paper, we denote $\mathbb{N}_0$ the set of non-negative integers and we consider the *reflected random walk* $(X(n))_{n \geq 0}$ on $\mathbb{N}_0$ defined by

$$X(n+1) = |X(n) + \xi_{n+1}|, \quad \text{for } n \geq 0,$$

where $X(0)$ is a $\mathbb{N}_0$-valued random variables. When $X(0) = x\,\mathbb{P}$-a.s., with $x \in \mathbb{N}_0$, the process $(X(n))_{n \geq 0}$ is also denoted by $(X^x(n))_{n \geq 0}$. It evolves as the random walk $x + S(n)$ as long as it stays non negative. When $x + S(n)$ enters the set of negative

H.-L. Ngo
Hanoi National University of Education, Cau Giay, Vietnam
e-mail: ngolong@hnue.edu.vn

M. Peigné (✉)
Institut Denis Poisson UMR 7013, Université de Tours, Université d'Orléans, CNRS, Orléans, France
e-mail: peigne@univ-tours.fr

© The Author(s), under exclusive license to Springer Nature Switzerland AG 2021
M. Pollicott, S. Vaienti (eds.), *Thermodynamic Formalism*, Lecture Notes in Mathematics 2290, https://doi.org/10.1007/978-3-030-74863-0_6

integers, the sign of its value is changed; the same construction thus applies starting from $|x + S(n)|, \ldots$ and so on.

The process $(X^x(n))_{n \geq 0}$ is a Markov chain on $\mathbb{N}_0$ starting from x. Several papers describing its stochastic behavior have been published; we refer to [17] where the recurrence of the reflected random walk is studied under some conditions which are nearly to be optimal. The reader may find also several references therein.

Firstly, $(X^x(n))_{n \geq 0}$ has some similarities with the classical random walk on $\mathbb{R}$; for instance, a strong law of large numbers holds, namely

$$\lim_{n \to +\infty} \frac{X^x(n)}{n} = 0 \quad \mathbb{P}\text{-a.s.}$$

when $\mathbb{E}[|\xi_n|] < +\infty$ and $\mathbb{E}[\xi_n] = 0$ (see Lemma 6.3.1 in section 3). Nevertheless, in contrast to what holds for the classical random walk on $\mathbb{R}$, this does not yield to the recurrence of $(X^x(n))_{n \geq 0}$. In [17], it is proved that the process $(X^x(n))_{n \geq 0}$ is null-recurrent when $\mathbb{E}[|\xi_n|^{3/2}] < +\infty$ and $\mathbb{E}[\xi_n] = 0$ and that $(X^x(n))_{n \geq 0}$ may be transient when $\mathbb{E}[|\xi_n|^{3/2}] = +\infty$, even if $\mathbb{E}[|\xi_n|^{3/2-\epsilon}] < +\infty$ for any $\epsilon > 0$. The reader can find in [12] a necessary and sufficient condition for the recurrence of $(X^x(n))_{n \geq 0}$ (see Theorem 4.6) but this condition cannot be reduced to the existence of some moments.

Once the strong law of large number holds, it is natural to study the oscillations of the process around its expectation. Let us state our result.

Theorem 6.1.1 *Let $(\xi_n)_{n \geq 1}$ be a sequence of $\mathbb{Z}$-valued i.i.d. random variables such that*

A1. $\mathbb{E}[\xi_n^2] = \sigma^2 < +\infty$ *and* $\mathbb{E}[(\xi_n^-)^3] < +\infty$;[1]
A2. $\mathbb{E}[\xi_n] = 0$;
A3. *The distribution of the ξ_n is strongly aperiodic, i.e. the support of the distribution of ξ_n is not included in the coset of a proper subgroup of $\mathbb{Z}$.*

Let $(X(t))_{t \geq 0}$ be the continuous time process constructed from the sequence $(X(n))_{n \geq 0}$ by linear interpolation between the values at integer points. Then, as $n \to +\infty$, the sequence of stochastic processes $(X_n(t))_{n \geq 1}$, defined by

$$X_n(t) := \frac{1}{\sigma \sqrt{n}} X(nt), \quad n \geq 1, 0 \leq t \leq 1,$$

weakly converges in the space of continuous functions on $[0, 1]$ to the absolute value $(|B(t)|)_{t \geq 0}$ of the Brownian motion on $\mathbb{R}$.

Let us insist on the fact that $X^x(n)$ coincides with $x + S(n)$ as long as it stays non-negative, but after it may differ drastically. The sequence of successive reflection times of $(X^x(n))_{n \geq 0}$ introduces some strong inhomogeneity on time and makes it

---

[1] $\xi_n^- = \max(0, -\xi_n)$ denotes the negative part of ξ_n.

necessary to adopt a totally different approach to prove an invariance principle as stated above.

A model which is quite similar to $(X^n(x))_{n \geq 0}$ is the queuing process $(W^x(n))_{n \geq 0}$, also called the *Lindley process*, corresponding to the waiting times in a single server queue. We think to $(W^x(n))_{n \geq 0}$ as an absorbing random walk on $\mathbb{N}_0$; as $W^x(n)$, it evolves as the random walk $x + S(n)$ as long as it stays non-negative and, when it attempts to cross 0 and become negative, the new value is reset to 0 before continuing. We refer to [15] for precise descriptions and variations on this process and follow the same strategy to obtain the invariance principle.

The excursions of $(W^x(n))_{n \geq 0}$ and $(X^x(n))_{n \geq 0}$ between two consecutively times of absorption-reflection coincide with some parts of the trajectory of $(S(n))_{n \geq 0}$, up to a translation; thus, their study is related to the fluctuations of $(S(n))_{n \geq 0}$. Hence, as in [15], we introduce the sequence of strictly descending ladder epochs $(\ell_l)_{l \geq 0}$ of the random walk $(S(n))_{n \geq 0}$ defined inductively by $\ell_0 = 0$ and, for any $l \geq 1$,

$$\ell_{l+1} := \min\{n > \ell_l \mid S(n) < S(\ell_l)\}.$$

When $\mathbb{E}[|\xi_n|] < +\infty$ and $\mathbb{E}[\xi_n] = 0$, the random variables $\ell_1, \ell_2 - \ell_1, \ell_3 - \ell_2, \dots$ are $\mathbb{P}$-a.s. finite and i.i.d. and the same property holds for the random variables $S(\ell_1), S(\ell_2) - S(\ell_1), S(\ell_3) - S(\ell_2), \dots$. In other words, the processes $(\ell_l)_{l \geq 0}$ and $(S(\ell_l))_{l \geq 0}$ are random walks on $\mathbb{N}_0$ and $\mathbb{Z}$ with respective distribution $\mathcal{L}(\ell_1)$ and $\mathcal{L}(S(\ell_1))$.

Let us briefly point out the main difference between $(W^x(n))_{n \geq 0}$ and $(X(n))_{n \geq 0}$. At an absorption time, the value of the process $W^x(n)$ is reset to 0 before continuing as a classical random walk for a while: there is a total loss of memory of the past after each absorption. Rather, at a reflection time, the process $X^x(n)$ equals the absolute value of $x + S(n)$. This value is the "new" starting point of the process, for a while, and has a great influence on the next reflection time; in other words, the process always captures some memory of the past at any time of reflection. This phenomenon has to be taken into account and requires a precise study of the sub-process $(X(r_k))_{k \geq 0}$ of $(X(n))_{n \geq 0}$ corresponding to these successive times $(r_k)_{k \geq 0}$ of reflection; our strategy consists in studying the spectrum of the transition probabilities matrix $\mathcal{R}$ of $(X(r_k))_{k \geq 0}$, acting on some Banach space $\mathcal{B} = \mathcal{B}_\alpha$ of functions from $\mathbb{N}_0$ to $\mathbb{C}$ with growth less than x^α at infinity, for some $\alpha > 0$ to be fixed. In particular, in order to apply recent results on renewal sequences [9], we need precise estimates on the tail of distribution of the reflection times; this is the main reason of the restrictive assumption $\mathbb{E}[(\xi_n^-)^3] < +\infty$ instead of moment of order 2, as we could expect. More precisely, throughout the paper, we need the following properties to be satisfied:

(i) The operator $\mathcal{R}$ acts on $\mathcal{B}_\alpha$.
   This holds when $\mathbb{E}[|S(\ell_1)|^{1+\alpha}] < +\infty$ and yields to the condition $\mathbb{E}[(\xi_n^-)^{2+\alpha}] < +\infty$ (see Proposition 6.1).
(ii) The function $\mathbb{N}_0 \to \mathbb{N}_0, x \mapsto x$, belongs to $\mathcal{B}_\alpha$; this imposes the condition $\alpha \geq 1$ (see Proposition 6.2).

Eventually, we fix $\alpha = 1$ from Sect. 6.1.1 on.

Notations *Throughout the text, we use the following notations. Let $u = (u_n)_{n\geq 0}$ and $v = (v_n)_{n\geq 0}$ be two sequences of positive reals; we write*

- $u \overset{c}{\preceq} v$ *(or simply $u \preceq v$) when $u_n \leq c v_n$ for some constant $c > 0$ and n large enough;*
- $u_n \sim v_n$ *when $\lim_{n\to +\infty} \frac{u_n}{v_n} = 1$.*
- $u_n \approx v_n$ *when $\lim_{n\to +\infty}(u_n - v_n) = 0$.*

6.2 Fluctuations of Random Walks and Auxiliary Estimates

6.2.1 On the Fluctuation of Random Walks

Let h be the Green function of the random walk $(S(\ell_l))_{l\geq 6}$, called sometimes the "descending renewal function" of S, defined by

$$h(x) = \begin{cases} \sum_{l=0}^{+\infty} \mathbb{P}[S(\ell_l) \geq -x] & \text{if } x \geq 0, \\ 0 & \text{otherwise.} \end{cases}$$

The function h is harmonic for the random walk $(S(n))_{n\geq 0}$ killed when it reaches the negative half line $(-\infty; 0]$; namely, for any $x \geq 0$,

$$\mathbb{E}[h(x + \xi_1); x + \xi_1 > 0] = h(x).$$

This holds for any oscillating random walk, possible without finite second moment.

Similarly, we denote $\tilde{h}$ the ascending renewal function of the random walk $(S(n))_{n\geq 0}$ (i.e. the descending renewal function of $(-S(n))_{n\geq 0}$).

Both functions h and $\tilde{h}$ are increasing, $h(0) = \tilde{h}(0) = 1$ and $h(x) = O(x), \tilde{h}(x) = O(x)$ as $x \to +\infty$ (see [1], p. 648 and [2]).

We have also to take into account the fact that the random walk S does not always start from the origin; hence, for any $x \geq 0$, we set $\tau^S(x) := \inf\{n \geq 1 : x + S(n) < 0\}$; it holds

$$[\tau^S(x) > n] = [L_n \geq -x],$$

where $L_n = \min(S(1), \ldots, S(n))$. The following result is a combination of Theorem 2 and Proposition 11 in [7] and Theorem A in [13] (see also Theorems II.6 and II.7 in [14]).

Lemma 6.2.1 *For any $x \geq 0$,*

1.

$$\mathbb{P}[\tau^S(x) > n] \sim c_1 \frac{h(x)}{\sqrt{n}} \qquad as \quad n \to +\infty,$$

where $c_1 = \dfrac{\mathbb{E}[-S_{\ell_1}]}{\sigma\sqrt{2\pi}}$. Moreover, there exists a constant $C_1 > 0$ such that for any $x \geq 0$ and $n \geq 1$,

$$\mathbb{P}[\tau^S(x) > n] \leq C_1 \frac{h(x)}{\sqrt{n}}.$$

2. For any $x, y \geq 0$,

$$\mathbb{P}[\tau^S(x) > n, x + S(n) = y] \sim \frac{1}{\sigma\sqrt{2\pi}} \frac{h(x)\tilde{h}(y)}{n^{3/2}} \qquad as \quad n \to +\infty,$$

and there exists a constant $C_2 > 0$ such that, for any $x, y \geq 0$ and $n \geq 1$,

$$\mathbb{P}[\tau^S(x) > n, x + S(n) = y] \leq C_2 \frac{h(x)\tilde{h}(y)}{n^{3/2}}.$$

These assertions yield a precise estimate of the probability $\mathbb{P}[\tau^S(x) = n]$ itself, and not only the tail of the distribution of τ^S. As a direct consequence, the sequence of descending ladder epochs $(\ell_l)_{l \geq 1}$ of the random walk $(S(n))_{n \geq 0}$ satisfies some renewal theorem [7]. Let us state these two consequences which enlighten the next section where similar statements concerning the successive epochs of reflections of the reflected random are proved.

Corollary 6.2.1 *For any $x \geq 0$,*

$$\mathbb{P}[\tau^S(x) = n] \sim \frac{c_1}{2} h(x) \frac{1}{n^{3/2}} \qquad as \quad n \to +\infty,$$

and there exists a constant $C_3 > 0$ such that, for any $x \geq 0$ and $n \geq 1$,

$$\mathbb{P}[\tau^S(x) = n] \leq C_3 \frac{h(x)}{n^{3/2}}.$$

Furthermore,

$$\sum_{l=0}^{+\infty} \mathbb{P}[\ell_l = n] \sim \frac{1}{c_1 \pi} \frac{1}{\sqrt{n}} \qquad as \quad n \to +\infty.$$

6.2.2 Conditional Limit Theorems

The following statement corresponds to Lemma 2.3 in [1]; the symbol "$\Rightarrow$" means "weak convergence".

Lemma 6.2.2 *Assume* $\mathbb{E}(\xi_i^2) < +\infty$ *and* $\mathbb{E}(\xi_i) = 0$. *Then, for any* $x \geq 0$,

$$\mathcal{L}\left(\left(\frac{S([nt])}{\sigma\sqrt{n}}\right)_{0 \leq t \leq 1} \middle| \min\{S(1), \ldots, S(n)\} \geq -x\right) \Rightarrow \mathcal{L}(L^+) \quad as \ n \to +\infty,$$

where L^+ *is the Brownian meander.*

In particular, for any bounded and Lipschitz continuous function $\phi : \mathbb{R} \to \mathbb{R}$,

$$\lim_{n \to +\infty} \mathbb{E}\left[\phi\left(\frac{x + S(n)}{\sigma\sqrt{n}}\right) \middle| \tau^S(x) > n\right] = \int_0^{+\infty} \phi(z)z e^{-z^2/2} dz.$$

This Lemma is useful in the sequel to control the fluctuations of the excursions of the process $(X(n))_{n \geq 0}$ between two successive times of reflection. In order to control also the higher dimensional distributions of these excursions, we need some invariance principle for random walk bridges conditioned to stay positive. The following result corresponds in our setting to Corollary 2.5 in [5].

Lemma 6.2.3 *For any bounded, Lipschitz continuous function* $\phi : \mathbb{R} \to \mathbb{R}$, *any* $x, y \geq 0$, *and any* $t > s > 0$,

$$\lim_{n \to +\infty} \mathbb{E}\left[\phi\left(\frac{x + S([ns])}{\sigma\sqrt{n}}\right) \middle| \tau^S(x) > [nt], x + S([nt]) = y\right]$$

$$= \int_0^{+\infty} 2\phi(u\sqrt{s}) \exp\left(-\frac{u^2}{2\frac{s}{t}\frac{t-s}{t}}\right) \frac{u^2}{\sqrt{2\pi \frac{s^3}{t^3} \frac{(t-s)^3}{t^3}}} du.$$

6.3 On the Sub-process of Reflections

We present briefly some results from [8] and [17]. The reflected times r_n, $n \geq 0$, of the random walk $(X(n))_{n \geq 0}$ are defined by: for any $x \geq 0$,

$$r_0 = r_0(x) = 0 \quad \text{and} \quad r_{n+1} = \inf\{m > r_n \mid X(r_n) + \xi_{r_n+1} + \cdots + \xi_m < 0\}.$$

Notice that these random variables are $\mathbb{N}_0 \cup \{+\infty\}$-valued stopping times with respect to the filtration $(\mathcal{G}_n)_{n \geq 0}$.

When $\mathbb{E}[|\xi_n|] < +\infty$ and $\mathbb{E}[\xi_n] = 0$, the random walk $(S(n))_{n \geq 0}$ is oscillating, hence the r_n, $n \geq 0$, are all finite $\mathbb{P}$-a.s. and $S(n)/n$ converges $\mathbb{P}$-a.s. towards 0. The strong law of large numbers is still true for the reflected random walk $(X^x(n))_{n \geq 0}$ on $\mathbb{N}_0$ but does not derive directly.

Lemma 6.3.1 *If* $\mathbb{E}[|\xi_n|] < +\infty$ *and* $\mathbb{E}[\xi_n] = 0$, *then, for any* $x \in \mathbb{N}_0$,

$$\lim_{n \to +\infty} \frac{X^x(n)}{n} = 0 \quad \mathbb{P}\text{-}a.s.$$

Proof For any $n \geq 1$, there exists a (random) integer $k_n \geq 1$ such that $r_{k_n} \leq n < r_{k_n+1}$. It holds

$$X^x(n) = X^x(r_{k_n}) + \left(\xi_{r_{k_n}+1} + \cdots + \xi_n\right) = X^x(r_{k_n}) + S(n) - S(r_{k_n}),$$

so that

$$0 \leq \frac{X^x(n)}{n} = \frac{X^x(r_{k_n})}{n} + \frac{S(n)}{n} - \frac{S(r_{k_n})}{n} \leq \frac{\max\{|\xi_1|, \ldots, |\xi_n|\}}{n} + \frac{S(n)}{n} - \frac{S(r_{k_n})}{n}.$$

The first term on the right hand side converges $\mathbb{P}$-a.s. towards 0 since $\mathbb{E}[|\xi_n|] < +\infty$.

By the strong law of large number, the second term tends $\mathbb{P}$-a.s. to 0.

At last, the same property holds for the last term, since $\left|\dfrac{S(r_{k_n})}{n}\right| = \left|\dfrac{S(r_{k_n})}{r_{k_n}}\right| \times$

$\dfrac{r_{k_n}}{n} \leq \left|\dfrac{S(r_{k_n})}{r_{k_n}}\right|$. $\qquad\qquad\qquad\qquad\qquad\qquad\qquad\qquad\qquad\qquad\qquad\qquad \square$

It follows from Lemma 2.3 in [16] that the sub-process of reflections $(X(r_k))_{k \geq 0}$ is a Markov chain on $\mathbb{N}_0$ with transition probability $\mathcal{R}$ given by: for all $x, y \in \mathbb{N}_0$,

$$\mathcal{R}(x, y) = \begin{cases} 0 & \text{if } y = 0 \\ \sum_{w=0}^{x} U^*(-w)\mu^*(w - x - y) & \text{if } y \geq 1, \end{cases} \tag{6.3.1}$$

where μ^* is the distribution of $S(\ell_1)$ and $U^* = \displaystyle\sum_{n=0}^{+\infty}(\mu^*)^{*n}$ denotes its potential.

Set $C := \sup\{y \geq 1 : \mu(-y) > 0\}$. The support of μ^* equals $\mathbb{Z}^- = \mathbb{Z} \cap (-\infty, 0)$ when $C = +\infty$, otherwise it is $\{-C, \ldots, -1\}$; furthermore, $U^*(-w) > 0$ for any $w \geq 0$. Then, $\mathcal{R}(x, y) > 0$ if and only if $y \in \mathbb{S}_r$, where $\mathbb{S}_r = \mathbb{N}_0 \backslash \{0\}$ when $C = +\infty$ and $\mathbb{S}_r = \{1, \ldots, C\}$ otherwise. Consequently, the set $\mathbb{S}_r$ is the unique irreducible and ergodic class of the Markov chain $(X(r_k))_{k \geq 0}$ and this chain is aperiodic on $\mathbb{S}_r$.

The measure ν on $\mathbb{N}_0$ defined by

$$\nu(x) = \sum_{y=1}^{+\infty} \left(\frac{1}{2}\mu^*(-x) + \mu^*((-x - y, -x)) + \frac{1}{2}\mu^*(-x - y)\right)\mu^*(-y),$$

is, up to a multiplicative constant, the unique stationary measure for $(X(r_k))_{k\geq 0}$; its support equals $\mathbb{S}_r$ (see Theorem 3.6 [16]).

Notice that this measure ν is finite when $\mathbb{E}[\xi_n] = 0$ and $\mathbb{E}[|S(\ell_1)|^{1/2}] < +\infty$ (and in particular when $\mathbb{E}[\xi_n] = 0$ and $\mathbb{E}[|\xi_n|^{3/2}] < +\infty$ [17]). **In this case, we normalize ν it in such a way it is a probability measure.**

6.3.1 On the Spectrum of the Transition Probabilities Matrix $\mathcal{R}$

Let us recall some spectral properties of the matrix $\mathcal{R} = (\mathcal{R}(x, y))_{x,y\in\mathbb{N}_0}$. By Property 2.3 in [8], the matrix $\mathcal{R}$ is quasi-compact on the space $L^\infty(\mathbb{N}_0)$ of bounded functions on $\mathbb{N}_0$, with 1 as the unique (and simple) dominant eigenvalue; in particular, the rest of the spectrum of $\mathcal{R}$ is included in a disc with radius < 1.

It is of interest in the next section to let $\mathcal{R}$ act on a bigger space than $L^\infty(\mathbb{N}_0)$. For instance, following [8], we may fix $K > 1$ and consider the Banach space

$$L_K(\mathbb{N}_0) := \{\phi : \mathbb{N}_0 \to \mathbb{C} : \|\phi\|_K := \sup_{x\geq 0} |\phi(x)|/K^x < +\infty\}$$

endowed with the norm $\|\cdot\|_K$. By Property 2.3 in [8], if $\sum_{x\geq 0} K^x \mu(x) < +\infty$ then $\mathcal{R}$ acts as a compact operator on $L_K(\mathbb{N}_0)$.

In this article, we only assume that μ has a finite moment of order 2 and its negative part has moment of order 3. Consequently, we consider a smaller Banach space $\mathcal{B}_\alpha$ adapted to these hypotheses and defined by: for $\alpha > 0$ fixed,

$$\mathcal{B}_\alpha := \left\{\phi : \mathbb{N}_0 \to \mathbb{C} : |\phi|_\alpha := \sup_{x\geq 0} \frac{|\phi(x)|}{1 + x^\alpha} < +\infty\right\}.$$

Endowed with the norm $|\cdot|_\alpha$, the space $\mathcal{B}_\alpha$ is a Banach space on $\mathbb{C}$.

Proposition 6.1 *Fix $\alpha > 0$ and assume $\mathbb{E}[\xi_n^2] + \mathbb{E}[(\xi_n^-)^{2+\alpha}] < +\infty$ and $\mathbb{E}[\xi_n] = 0$. Then, the operator $\mathcal{R}$ acts on $\mathcal{B}_\alpha$ and $\mathcal{R}(\mathcal{B}_\alpha) \subset L^\infty(\mathbb{N}_0)$. Furthermore,*

1. *$\mathcal{R}$ is compact on $\mathcal{B}_\alpha$ with spectral radius 1;*
2. *1 is the unique eigenvalue of $\mathcal{R}$ with modulus 1, it is simple with corresponding eigenspace $\mathbb{C}\mathbf{1}$;*
3. *the rest of the spectrum of $\mathcal{R}$ on $\mathcal{B}_\alpha$ is included in a disc with radius < 1.*

Let Π be the projection from $\mathcal{B}_\alpha$ onto the eigenspace $\mathbb{C}\mathbf{1}$ corresponding to this spectral decomposition, i.e. such that $\Pi\mathcal{R} = \mathcal{R}\Pi = \Pi$. In other words, there exists a bounded operator Q on $\mathcal{B}_\alpha$ with spectral radius < 1 such that $\mathcal{R}$ may be decomposed as follows:

$$\mathcal{R} = \Pi + Q, \quad \Pi Q = Q\Pi = 0 \quad \text{with} \quad \Pi(\cdot) = \nu(\cdot)\mathbf{1}.$$

In the next section, we require that $\mathcal{B}_\alpha$ does contain the descending and ascending renewal functions h and $\tilde{h}$ of the random walk S. This imposes in particular that α is greater or equal to 1.

Proof

(1) By (6.3.1), for any $\phi \in \mathcal{B}_\alpha$ and $x \geq 0$,

$$\mathcal{R}\phi(x) = \sum_{y \geq 1} \sum_{w=0}^{x} U^*(-w)\mu^*(w - x - y)\phi(y)$$

with $U^*(-w) = \sum_{n=0}^{+\infty} \mathbb{P}[S(l_n) = -w] = \mathbb{P}\Big[\cup_{n \geq 0} [S(l_n) = -w] \Big] \leq 1.$

Therefore,

$$|\mathcal{R}\phi(x)| \leq \sum_{y \geq 1} \sum_{w=0}^{x} \mu^*(w - x - y)|\phi(y)|$$

$$\leq \sum_{y \geq 1} \mu^*((-\infty, -y))|\phi(y)|$$

$$\leq \left(\sum_{y \geq 1} (1 + y^\alpha)\mu^*((-\infty, -y)) \right) |\phi|_\alpha.$$

By Theorem 1 in [6], the condition $\mathbb{E}[(\xi_n^-)^{2+\alpha}] < +\infty$ implies $\mathbb{E}\big[|S(\ell_1)|^{1+\alpha}\big] < +\infty$; hence,

$$\sum_{y \geq 1} (1 + y^\alpha)\mu^*((-\infty, -y)) \leq \mathbb{E}[|S(\ell_1)|] + \mathbb{E}\big[|S(\ell_1)|^{1+\alpha}\big] < +\infty.$$

Consequently,

$$|\mathcal{R}\phi|_\alpha \leq |\mathcal{R}\phi|_\infty \leq \left(\mathbb{E}[|S(\ell_1)|] + \mathbb{E}\big[|S(\ell_1)|^{1+\alpha}\big] \right)|\phi|_\alpha \qquad (6.3.2)$$

which proves that $\mathcal{R}$ acts on $\mathcal{B}_\alpha$ when $\mathbb{E}[(\xi_n^-)^{2+\alpha}] < +\infty$. More precisely, the operator $\mathcal{R}$ is bounded from $\mathcal{B}_\alpha$ into $L^\infty(\mathbb{N}_0)$ and since the canonical injection $L^\infty(\mathbb{N}_0) \hookrightarrow \mathcal{B}_\alpha$ is compact, the operator $\mathcal{R}$ is compact on $\mathcal{B}_\alpha$.

Let us now check that $\mathcal{R}$ has spectral radius $\rho_\alpha = 1$ on $\mathcal{B}_\alpha$. On the one hand, the equality $\mathcal{R}\mathbf{1} = \mathbf{1}$, with $\mathbf{1} \in \mathcal{B}_\alpha$, yields $\rho_\alpha \geq 1$. On the other hands, $\mathcal{R}$ is a power bounded operator on $\mathcal{B}_\alpha$, which readily implies $\rho_\alpha \leq 1$; indeed, for any

$n \geq 1$,

$$|\mathcal{R}^n \phi(x)| \leq \sum_{z=0}^{+\infty} \mathcal{R}^{n-1}(x, z)|\mathcal{R}\phi(z)| \leq |\mathcal{R}\phi|_\infty \sum_{z=0}^{+\infty} \mathcal{R}^{n-1}(x, z) = |\mathcal{R}\phi|_\infty,$$

which yields, combining with (6.3.2),

$$|\mathcal{R}^n \phi|_\alpha \leq |\mathcal{R}^n \phi|_\infty \leq \left(\mathbb{E}[|S(\ell_1)|] + \mathbb{E}\left[|S(\ell_1)|^{1+\alpha}\right]\right)|\phi|_\alpha.$$

Consequently, denoting $\|\mathcal{R}^n\|_\alpha$ the norm of $\mathcal{R}^n$ on $\mathcal{B}_\alpha$, it holds

$$\sup_{n \geq 0} \|\mathcal{R}^n\|_\alpha \leq \left(\mathbb{E}[|S(\ell_1)|] + \mathbb{E}\left[|S(\ell_1)|^{1+\alpha}\right]\right) < +\infty.$$

This achieves the proof of assertion 1.

(2) Let us control the peripherical spectrum of $\mathcal{R}$ in $\mathcal{B}_\alpha$. Let $\theta \in \mathbb{R}$ and $\phi \in \mathcal{B}_\alpha$ such that $\mathcal{R}\phi = e^{i\theta}\phi$.

By (6.3.2), the function $\mathcal{R}\phi$ is bounded, so is ϕ. Furthermore, the operator $\mathcal{R}$ being positive, it holds $|\phi| \leq \mathcal{R}|\phi|$. Consequently, the function $|\phi|_\infty - |\phi|$ is super-harmonic and non-negative, hence constant since the Markov chain $(X(r_n))_{n\geq0}$ is irreducible and recurrent on this set.

Without loss of generality, we may assume $|\phi| = 1$ on $\mathbb{S}_r$, i.e. $\phi(x) = e^{i\varphi(x)}$ for any $x \in \mathbb{S}_r$, with $\varphi : \mathbb{S}_r \to \mathbb{R}$. Equality $\mathcal{R}\phi = e^{i\theta}\phi$ may be rewritten as: for any $x \in \mathbb{S}_r$,

$$\sum_{y \in \mathbb{S}_r} e^{i(\varphi(y)-\varphi(x))}\mathcal{R}(x, y) = e^{i\theta}.$$

Recall that $\mathcal{R}(x, y) > 0$ for any $x, y \in \mathbb{S}_r$; thus, by convexity, $e^{i(\varphi(y)-\varphi(x))} = e^{i\theta}$ for any $x, y \in \mathbb{S}_r$. Thus, $e^{i\theta} = 1$ and the function ϕ is harmonic on $\mathbb{S}_r$, hence constant. Eventually, the function ϕ is constant on $\mathbb{N}_0$: this is the consequence of equality $\mathcal{R}\phi(x) = e^{i\theta}\phi(x) = \phi(x)$, valid for any $x \in \mathbb{N}_0$, combined with the facts that $\mathcal{R}(x, y) > 0$ if and only if $y \in \mathbb{S}_r$ and that ϕ is constant on $\mathbb{S}_r$.

(3) Assertion 3 is a consequence of assertion 2 and the compactness of $\mathcal{R}$ on $\mathcal{B}_\alpha$. □

6.3.2   A Renewal Limit Theorem for the Times of Reflections

In this section, we prove the analogous of Corollary 6.2.1 for the process $(r_n)_{n\geq0}$. Let us introduce some notations and conventions.

From now on, we focus on the process $(X(n))_{n\geq0}$ and denote

$$((\mathbb{N}^0)^{\otimes\mathbb{N}}, (\mathcal{P}(\mathbb{N}^0))^{\otimes\mathbb{N}}, (X(n))_{n\geq0}, (\mathbb{P}_x)_{x\in\mathbb{N}_0}, \theta)$$

the canonical space associated to this process, that is the space of trajectories of the Markov chain $(X(n))_{n\geq 0}$. In particular, $\mathbb{P}_x, x \in \mathbb{N}_0$, denotes the *conditional probability* with respect to the event $[X(0) = x]$ and $\mathbb{E}_x$ the corresponding *conditional expectation*. The operator θ is the classical *shift transformation* defined by: for any $(x_k)_{k\geq 0} \in (\mathbb{N}^0)^{\otimes \mathbb{N}}$,

$$\theta((x_k)_{k\geq 0}) = ((x_{k+1})_{k\geq 0}.$$

For $n \geq 1$ and $x, y \geq 0$, set

$$R_n(x, y) := \mathbb{P}_x[r_1 = n, X(n) = y],$$

and

$$\Sigma_n(x, y) := \sum_{k=1}^{+\infty} \mathbb{P}_x[r_k = n, X(n) = y].$$

We are interested in the behavior as $n \to +\infty$ of these quantities. It has been already studied in [15] (see Lemma 7) for the Lindley process. For the reflected random walk, the argument is more complicated since the position at time r_k may vary, so that the excursions of the random walk $(X(n))_{n\geq 0}$ between two successive reflection times are not independent. This explain why we focus here on the reflection process and it is of interest to express quantities $R_n(x, y)$ and $\Sigma_n(x, y)$ in terms of operators and product of operators related to this sub-process.

We consider the linear operators $R_n : L^\infty(\mathbb{N}_0) \to L^\infty(\mathbb{N}_0), n \geq 0$, defined by: for any $\phi \in L^\infty(\mathbb{N}_0)$ and $x \geq 0$,

$$R_n\phi(x) = \sum_{y\geq 1} R_n(x, y)\phi(y) = \mathbb{E}_x[r_1 = n; \phi(X(n))].$$

In particular, $R_n(x, y) = R_n \mathbf{1}_{\{y\}}(x)$. The quantity $\Sigma_n(x, y)$ is also expressed in terms of the R_k as follows:

$$\Sigma_n(x, y) = \sum_{k=1}^{+\infty} \mathbb{P}_x[r_k = n, X(n) = y]$$

$$= \sum_{k=1}^{+\infty} \sum_{j_1+\cdots+j_k=n} \mathbb{P}_x[r_1 = j_1, r_2 - r_1 = j_2, \ldots, r_k - r_{k-1} = j_k, X(n) = y]$$

$$= \sum_{k=1}^{+\infty} \sum_{j_1+\cdots+j_k=n} R_{j_1} \ldots R_{j_k} \mathbf{1}_{\{y\}}(x) \tag{6.3.3}$$

Firstly, let us check that the R_n act on $\mathcal{B}_\alpha$.

Lemma 6.3.2 *There exists a positive constant C_4 such that, for any $n \geq 1$ and $\alpha > 0$,*

$$|R_n|_\alpha \leq C_4 \frac{\mathbb{E}\left[(\xi_n^-)^{2+\alpha}\right]}{n^{3/2}}.$$

Proof For any $\phi \in \mathcal{B}_\alpha$ and $x \geq 0$,

$$|R_n\phi(x)| \leq \sum_{y \geq 1} |\phi(y)| \mathbb{P}_x[r_1 = n, X(n) = y]$$

$$= \sum_{y \geq 1} \sum_{z \geq 0} |\phi(y)| \mathbb{P}[\tau^S(x) \geq n - 1, x + S(n-1) = z, z + \xi_n = -y]$$

$$= \sum_{y \geq 1} \sum_{z \geq 0} |\phi(y)| \mathbb{P}[\tau^S(x) \geq n - 1, x + S(n-1) = z] \mathbb{P}[\xi_n = -y - z].$$

Hence, by Lemma 6.2.1,

$$\frac{|R_n\phi(x)|}{1 + x^\alpha} \preceq \frac{1}{n^{3/2}} \sum_{y \geq 1} \sum_{z \geq 0} |\phi(y)| \frac{h(x)}{1 + x^\alpha} \tilde{h}(z) \mathbb{P}[\xi_1 = -y - z].$$

Since $h(x) = O(x)$ and $\tilde{h}(z) = O(z)$,

$$\frac{|R_n\phi(x)|}{1 + x^\alpha} \preceq \frac{|\phi|_\alpha}{n^{3/2}} \sum_{y \geq 1} \sum_{z \geq 0} (1 + y^\alpha) \tilde{h}(z) \mathbb{P}[\xi_1 = -y - z]$$

$$\preceq \frac{|\phi|_\alpha}{n^{3/2}} \sum_{y \geq 1} \sum_{z \geq 0} (1 + y^\alpha) z \mathbb{P}[\xi_1 = -y - z]$$

$$= \frac{|\phi|_\alpha}{n^{3/2}} \sum_{t \geq 1} \sum_{y=1}^{t} (1 + y^\alpha)(t - y) \mathbb{P}[\xi_1 = -t]$$

$$\preceq \frac{|\phi|_\alpha}{n^{3/2}} \sum_{t \geq 1} t^{2+\alpha} \mathbb{P}[\xi_1 = -t],$$

which achieves the proof. □

Hence, $\sum_{n \geq 1} |R_n|_\alpha < +\infty$; in particular, the sequence $(\sum_{n=1}^{N} R_n)_{N \geq 1}$ converges in $\mathcal{B}_\alpha$. Note that its limit equals $\mathcal{R}$ in $\mathcal{B}_\alpha$; indeed,

$$\sum_{n \geq 1} R_n\phi(x) = \sum_{n \geq 1} \mathbb{E}_x[\phi(X(n)), r_1 = n] = \mathbb{E}_x[\phi(X(r_1))] = \mathcal{R}\phi(x).$$

We can write $\mathcal{R} = \sum_{n \geq 1} R_n$ and, for any $z \in \overline{\mathbb{D}} := \{z \in \mathbb{C} : |z| \leq 1\}$, we set

$$\mathcal{R}(z) = \sum_{n \geq 1} z^n R_n.$$

Proposition 6.2 *Fix $\alpha > 0$ and assume $\mathbb{E}[\xi_n^2] + \mathbb{E}[(\xi_n^-)^{2+\alpha}] < +\infty$ and $\mathbb{E}[\xi_n] = 0$. The sequence $(R_n)_{n \geq 0}$ is an **aperiodic renewal sequence of operators**, i.e. it satisfies the following properties (see [9]):*

(R1). *The operator $\mathcal{R} = \mathcal{R}(1)$ has a simple eigenvalue at 1 and the rest of its spectrum is contained in a disk of radius < 1.*

(R2). *For any $n \geq 1$, set $\mathbf{r}_n := \nu R_n \mathbf{1} = \sum_{x \geq 1} \nu(x) \mathbb{P}_x(r_1 = n)$; hence,*

$$\Pi R_n \Pi = \mathbf{r}_n \Pi,$$

where Π denotes the eigenprojection of $\mathcal{R}$ for the eigenvalue 1.

(R3). *There exists a constant $\mathbf{C} > 0$ such that $\quad |R_n|_\alpha \leq \frac{\mathbf{C}}{n^{3/2}}$.*

(R4). *$\sum_{j>n} \mathbf{r}_j \sim \frac{\mathbf{c}}{\sqrt{n}}$ with $\mathbf{c} = c_1 \nu(h)$, where c_1 is the positive constant given by Lemma 6.2.1 and h is the descending renewal function of the random walk S.*

(R5). *The spectral radius of $\mathcal{R}(z)$ is strictly less than 1 for $z \in \overline{\mathbb{D}} \setminus \{1\}$.*

Proof (R1) is a direct consequence of Proposition 6.1.

(R2) Recall that $\Pi \phi = \nu(\phi)\mathbf{1}$ for any $\phi \in \mathcal{B}_\alpha$. Hence, setting $g_n(x) := \mathbb{P}_x(r_1 = n)$, it holds $R_n \Pi \phi = \nu(\phi) g_n$, thus

$$\Pi R_n \Pi \phi = \nu(\phi) \Pi(g_n) = \sum_{x \geq 1} \nu(x) \mathbb{P}_x(r_1 = n) \nu(\phi) \mathbf{1},$$

which is the expected result.

(R3) follows from Lemma 6.3.2.

(R4) Thanks to Lemma 6.2.1,

$$\sum_{j \geq n} \mathbf{r}_j = \sum_{x \geq 1} \sum_{j \geq n} \nu(x) \mathbb{P}_x[r_1 = j] = \sum_{x \geq 1} \nu(x) \mathbb{P}_x[r_1 \geq n] \sim c_1 \frac{\nu(h)}{\sqrt{n}} \quad \text{as} \quad n \to +\infty.$$

Notice that $0 < \nu(h) < +\infty$ since $\mathbb{E}[|S(\ell_1)|] < +\infty$; indeed, $1 \leq h(x) = O(x)$ and

$$\sum_{x \geq 1} x\nu(x) \leq \sum_{x \geq 1} \sum_{y \geq 1} \sum_{w=x}^{x+y} \mu^*(-w)\mu^*(-y)x = \sum_{y \geq 1} \sum_{w \geq 1} \mu^*(-w)\mu^*(-y) \sum_{x=(w-y)\vee 0}^{w} x$$

$$\leq \sum_{y \geq 1} \sum_{w \geq 1} yw\mu^*(-w)\mu^*(-y)$$

$$= \left(\sum_{y \geq 1} y\mu^*(-y)\right)^2 = (\mathbb{E}[|S(\ell_1)|])^2 < +\infty.$$

(R5) The argument is the same as the one used to control the peripherical spectrum of $\mathcal{R}$ in Proposition 6.1. For any $z \in \overline{\mathbb{D}} \setminus \{1\}$, the operators $\mathcal{R}(z)$ are compact on $\mathcal{B}_\alpha$, with spectral radius $\rho_z \leq 1$.

If $\rho_z = 1$, there exist $\theta \in \mathbb{R}$ and $\phi \in \mathcal{B}_\alpha$ such that $\mathcal{R}(z)\phi = e^{i\theta}\phi$. Hence $|\phi| = |\mathcal{R}(z)\phi| \leq \mathcal{R}|\phi|$ and since $\mathcal{R}(\mathcal{B}_\alpha) \subset L^\infty(\mathbb{N}_0)$, the function $|\phi|$ is bounded on $\mathbb{N}_0$, thus constant on $\mathbb{S}_r$.

Without loss of generality, we may assume $|\phi| = 1$ on $\mathbb{S}_r$, i.e. $\phi(x) = e^{i\varphi(x)}$ for any $x \in \mathbb{S}_r$, with $\varphi : \mathbb{S}_r \to \mathbb{R}$. Equality $\mathcal{R}(z)\phi = e^{i\theta}\phi$ may be rewritten as: for any $x \in \mathbb{S}_r$,

$$\sum_{n \geq 1} \sum_{y \in \mathbb{S}_r} z^n e^{i\varphi(y)} \mathbb{P}_x(r_1 = n; X(n) = y) = e^{i\theta} e^{i\varphi(x)}.$$

By convexity, since $\sum_{n \geq 1} \sum_{y \in \mathbb{S}_r} \mathbb{P}_x(r_1 = n; X(n) = y) = 1$, we obtain: for all $n \geq 1$ and $x, y \in \mathbb{S}_r$,

$$z^n e^{i\varphi(y)} = e^{i\theta} e^{i\varphi(x)}.$$

Setting $x = y$, it yields $z^n = e^{i\theta}$, so that z^n does not depend on n. Finally $z = 1$. Thus, $\rho_z < 1$ when $z \in \overline{\mathbb{D}} \setminus \{1\}$. $\square$

By (R5), for $|z| < 1$, the operator $T(z) := (I - \mathcal{R}(z))^{-1}$ is well defined in $\mathcal{B}_\alpha$; a direct formal computation yields $T(z) = \sum_{n=0}^{+\infty} T_n z^n$, where the T_n are bounded operators on $\mathcal{B}_\alpha$ defined by:

$$T_0 = I \quad \text{and} \quad T_n = \sum_{k=1}^{+\infty} \sum_{j_1 + \cdots + j_k = n} R_{j_1} \cdots R_{j_k} \quad \text{for} \quad n \geq 1.$$

The so-called *renewal equation* $T(z) := (I - R(z))^{-1}$ is of fundamental importance to understand the asymptotics of the T_n, several functional analytic tools can be brought into play. Such sequences of operators $(R_n)_{n \geq 0}$ and $(T_n)_{n \geq 0}$ have been the object of many studies, related to renewal theory in a non-commutative setting. We refer to the paper [9], which fits perfectly here. The following statement is analogous of the last assertion of Corollary 6.2.1 for the reflected random walk.

Corollary 6.3.1 *The sequence $(\sqrt{n}T_n)_{n \geq 1}$ converges in $\mathcal{B}_\alpha$ towards the operator* $\frac{1}{\pi c_1 \nu(h)} \Pi$.

Proof Apply Theorem 1.4 in [9] with $\beta = 1/2$ and $\ell(n) = \mathbf{c} = c_1 \nu(h)$. $\square$

As a direct consequence, by equality (6.3.3), it holds

$$\lim_{n \to +\infty} \sqrt{n} \Sigma_n(x, y) = \frac{\nu(y)}{\pi c_1 \nu(h)}.$$

In the next section, we have to consider and study some modifications of the $\Sigma_n(x, y)$ which we introduce now. For any $x \geq 0$ and $0 < s < t < 1$,

$$\widehat{\Sigma}_n(x, t, s) := n \sum_{l \geq 0} \mathbb{P}_x[r_l = [ns], r_{l+1} > [nt]],$$

and

$$\widetilde{\Sigma}_n(x, t, s) := n^2 \sum_{l=0}^{+\infty} \mathbb{P}_x [r_l = [ns], r_{l+1} = [nt]].$$

These quantities appear in a natural way to control the finite distribution of the process $(X_n(t))_{n \geq 0}$.

6.4 Proof of Theorem 6.1.1

From now on, we fix $\alpha = 1$; this implies that $h \in \mathcal{B}_\alpha$, which is necessary from now on (see Lemmas 6.4.2 and 6.4.4).

6.4.1 One-Dimensional Distribution

We fix a bounded and Lispchitz continuous function $\phi : \mathbb{R} \to \mathbb{R}$.

Lemma 6.4.1 *For any $t \in [0, 1]$ and $x \geq 0$, it holds*

$$\lim_{n \to +\infty} \mathbb{E}_x [\phi(X_n(t))] = \int_0^{+\infty} \phi(u) \frac{2e^{-u^2/2t}}{\sqrt{2\pi t}} du = \mathbb{E}[\phi(|B_t|)],$$

where B is a standard Brownian motion.

Proof We fix $t \in (0, 1)$ and decompose the expectation $\mathbb{E}\left[\phi\left(\dfrac{X([nt])}{\sigma\sqrt{n}}\right)\right]$ as follows:

$$\mathbb{E}_x\left[\phi\left(\frac{X([nt])}{\sigma\sqrt{n}}\right)\right]$$

$$\approx \sum_{k=0}^{[nt]-1} \sum_{l \geq 0} \mathbb{E}_x\left[\phi\left(\frac{X([nt])}{\sigma\sqrt{n}}\right);\right.$$

$$\left. r_l = k, X(k) + \xi_{k+1} \geq 0, \dots, X(k) + \xi_{k+1} + \cdots + \xi_{[nt]} \geq 0\right]$$

$$= \sum_{k=0}^{[nt]-1} \sum_{y \geq 0} \Sigma_k(x, y) \mathbb{E}\left[\phi\left(\frac{y + \xi_{k+1} + \ldots + \xi_{[nt]}}{\sigma\sqrt{n}}\right);\right.$$

$$\left. y + \xi_{k+1} \geq 0, \ldots, y + \xi_{k+1} + \cdots + \xi_{[nt]} \geq 0\right]$$

$$= \sum_{k=0}^{[nt]-1} \sum_{y \geq 0} \Sigma_k(x, y) \mathbb{E}\left[\phi\left(\frac{y + S([nt] - k)}{\sigma\sqrt{n}}\right) \mid \tau^S(y) > [nt] - k\right]$$

$$\times \mathbb{P}\left[\tau^S(y) > [nt] - k\right].$$

For each $k = 2, \ldots, [nt] - 4$ and any $s \in [\frac{k}{n}, \frac{k+1}{n})$,

$$f_n(s) = n \sum_{y \geq 0} \Sigma_{[ns]}(x, y) \mathbb{E}\left[\phi\left(\frac{y + S([nt] - [ns])}{\sigma\sqrt{n}}\right) \mid \tau^S(y) > [nt] - [ns]\right]$$

$$\times \mathbb{P}\left[\tau^S(y) > [nt] - [ns]\right],$$

and $f_n(s) = 0$ on $[0, \frac{2}{n})$ and $[\frac{[nt]-1}{n}, t)$. Hence,

$$\mathbb{E}_x\left[\phi\left(\frac{X([nt])}{\sigma\sqrt{n}}\right)\right] = \int_0^t f_n(s)ds + O\left(\frac{1}{\sqrt{n}}\right).$$

Now, let us set : for $n \geq 1$ and any $y \in \mathbb{N}_0$,

$$a_n(y) = \Sigma_{[ns]}(x, y)\mathbb{P}\left[\tau^S(y) > [nt] - [ns]\right],$$

$$b_n(y) = \mathbb{E}\left[\phi\left(\frac{y + S([nt] - [ns])}{\sigma\sqrt{n}}\right) \mid \tau^S(y) > [nt] - [ns]\right].$$

For any $n \geq 1$, it holds

$$\sum_{y \geq 0} a_n(y) = n \sum_{l \geq 0} \mathbb{P}_x[r_l = [ns], r_{l+1} > [nt]] =: \widehat{\Sigma}_n(x, t, s),$$

and $|b_n(y)| \leq |\phi|_\infty$. The two following lemmas allow us to control the behavior as $n \to +\infty$ of the integral $\int_0^t f_n(s)ds$; the proof of Lemma 6.4.2 is postponed to the last section, the one of Lemma 6.4.3 is straightforward.

Lemma 6.4.2 *For each $0 < s < t < 1$,*

$$\lim_{n \to +\infty} \widehat{\Sigma}_n(x, t, s) = \frac{1}{\pi\sqrt{s(t - s)}}.$$

Moreover, there exists a positive constant C_5 such that

$$\widehat{\Sigma}_n(x, t, s) \leq C_5 \frac{1 + x}{\sqrt{s(t - s)}} \quad \text{for all } 0 < s < t < 1 \text{ and } x \in \mathbb{N}.$$

Lemma 6.4.3 *Let $(a_n(y))_{y \in \mathbb{N}_0^k}$, $(b_n(y))_{y \in \mathbb{N}_0^k}$ be arrays of real numbers for some integer $k \geq 1$. Suppose that*

- $a_n(y) \geq 0$;
- $\lim\limits_{n \to +\infty} \sum\limits_{y \in \mathbb{N}_0^k} a_n(y) = A$;
- $\lim\limits_{n \to +\infty} b_n(y) = B$ *for all* $y \in \mathbb{N}_0^k$;
- $\sup\limits_{n \geq 1, y \in \mathbb{N}_0^k} |b_n(y)| < +\infty$. $\qquad\qquad\qquad\qquad\qquad$ $\square$

Then

$$\lim_{n \to +\infty} \sum_{y \geq 0} a_n(y) b_n(y) = AB.$$

Lemmas 6.2.2, 6.4.2 and 6.4.3 combined altogether yield: for any $s \in (0, t)$,

$$\lim_{n \to +\infty} f_n(s) = \frac{1}{\pi} \frac{1}{\sqrt{s(t - s)}} \int_0^{+\infty} \phi(z\sqrt{t - s}) z e^{-z^2/2} dz.$$

Moreover,

$$\sup_n |f_n(s)| \leq C_5 \frac{1 + x}{\sqrt{s(t - s)}} |\phi|_\infty =: \hat{f}(s).$$

Since $\hat{f} \in L^1[0, t]$, the Lebesgue dominated convergence theorem yields

$$\lim_{n \to +\infty} \mathbb{E}\left[\phi\left(\frac{X([nt])}{\sigma\sqrt{n}} \right) \right] = \lim_{n \to +\infty} \int_0^t f_n(s) ds$$

$$= \frac{1}{\pi} \int_0^t \frac{1}{\sqrt{s(t - s)}} \left(\int_0^{+\infty} \phi(z\sqrt{t - s}) z e^{-z^2/2} dz \right) ds$$

$$= \int_0^{+\infty} \phi(u) \frac{2 e^{-u^2/2t}}{\sqrt{2\pi t}} du,$$

where the last equation follows from the identity ([11], p. 17)

$$\int_0^{+\infty} \frac{1}{\sqrt{t}} \exp\left(-\alpha t - \frac{\beta}{t} \right) dt = \sqrt{\frac{\pi}{\alpha}} e^{-2\sqrt{\alpha\beta}} \quad (\alpha, \beta > 0) \qquad (6.4.1)$$

and some change of variable computation. We achieve the proof of Lemma 6.4.1 by noting that, since ϕ is Lipschitz continuous (with Lipschitz coefficient $[\phi]$),

$$
\left| \mathbb{E}_x \left[\phi \left(\frac{X([nt])}{\sigma \sqrt{n}} \right) \right] - \mathbb{E}_x \left[\phi \left(X_n(t) \right) \right] \right| \leq [\phi] \mathbb{E}_x \left[\left| \frac{X([nt])}{\sigma \sqrt{n}} - X_n(t) \right| \right]
$$

$$
\leq \frac{1}{\sigma \sqrt{n}} [\phi] \mathbb{E} \left[|\xi_{[nt]+1}| \right] \to 0 \text{ as } n \to +\infty.
$$

(6.4.2)

$\square$

6.4.2 Two-Dimensional Distributions

The convergence of the finite-dimensional distributions of $(X_n(t))_{n \geq 1}$ is more delicate. We detail the argument for two-dimensional ones, the general case may be treated in a similar way.

Let us fix $0 < s < t, n \geq 1$ and denote

$$
\kappa = \kappa(n, s) = \min\{k > [ns] : X(k-1) + \xi_k < 0\}.
$$

We decompose $\mathbb{E}_x \left[\phi_1 \left(\frac{X([ns])}{\sigma \sqrt{n}} \right) \phi_2 \left(\frac{X([nt])}{\sigma \sqrt{n}} \right) \right]$ as

$$
\underbrace{\sum_{k=[ns]+1}^{[nt]} \mathbb{E}_x \left[\phi_1 \left(\frac{X([ns])}{\sigma \sqrt{n}} \right) \phi_2 \left(\frac{X([nt])}{\sigma \sqrt{n}} \right) \mathbf{1}_{\{\kappa=k\}} \right]}_{A_1(n)}
$$

$$
quad + \underbrace{\mathbb{E}_x \left[\phi_1 \left(\frac{X([ns])}{\sigma \sqrt{n}} \right) \phi_2 \left(\frac{X([nt])}{\sigma \sqrt{n}} \right) \mathbf{1}_{\{\kappa>[nt]\}} \right]}_{A_2(n)}.
$$

The term $A_1(n)$ deals with the trajectories of the process X which reflect between $[ns] + 1$ and $[nt]$ while $A_2(n)$ concerns the others trajectories.

6.4.2.1 Estimate of $A_1(n)$

As in the previous section, we decompose $A_1(n)$ as

$$A_1(n) = \sum_{k_1=0}^{[ns]-1} \sum_{k_2=[ns]}^{[nt]} \sum_{l=0}^{+\infty} \sum_{y\geq 1} \sum_{z\geq 1} \sum_{w\geq 0}$$

$$\mathbb{E}_x\left[\phi_1\left(\frac{X([ns])}{\sigma\sqrt{n}}\right)\phi_2\left(\frac{X([nt])}{\sigma\sqrt{n}}\right)\right];$$

$$r_l = k_1, X(k_1) = z, z + \xi_{k_1+1} \geq 0, \ldots, z + \xi_{k_1+1} + \cdots + \xi_{k_2-2} \geq 0,$$

$$z + \xi_{k_1+1} + \cdots + \xi_{k_2-1} = w, w + \xi_{k_2} = -y\Big]$$

$$= \sum_{k_1=0}^{[ns]-1} \sum_{k_2=[ns]}^{[nt]} \sum_{l=0}^{+\infty} \sum_{y\geq 1} \sum_{z\geq 1} \sum_{w\geq 0} \mathbb{E}_x\left[\phi_1\left(\frac{z + \xi_{k_1+1} + \cdots + \xi_{[ns]}}{\sigma\sqrt{n}}\right)\right.$$

$$\left.\times \phi_2\left(\frac{y + \xi_{k_2+1} + \cdots + \xi_{[nt]}}{\sigma\sqrt{n}}\right)\right];$$

$$r_l = k_1, X(k_1) = z, z + \xi_{k_1+1} \geq 0, \ldots, z + \xi_{k_1+1} + \cdots + \xi_{k_2-2} \geq 0,$$

$$z + \xi_{k_1+1} + \cdots + \xi_{k_2-1} = w, w + \xi_{k_2} = -y\Big]$$

$$= \sum_{k_1=0}^{[ns]-1} \sum_{k_2=[ns]}^{[nt]} \sum_{l=0}^{+\infty} \sum_{y\geq 1} \sum_{z\geq 1} \sum_{w\geq 0}$$

$$\mathbb{E}_x\left[\phi_2\left(\frac{y + \xi_{k_2+1} + \cdots + \xi_{[nt]}}{\sigma\sqrt{n}}\right)\right] \mathbb{P}_x[r_l = k_1, X(k_1) = z]$$

$$\times \mathbb{E}_x\left[\phi_1\left(\frac{z + \xi_{k_1+1} + \cdots + \xi_{[ns]}}{\sigma\sqrt{n}}\right)\right],$$

$$z + \xi_{k_1+1} \geq 0, \ldots, z + \xi_{k_1+1} + \cdots + \xi_{k_2-2} \geq 0,$$

$$z + \xi_{k_1+1} + \cdots + \xi_{k_2-1} = w, w + \xi_{k_2} = -y\Big].$$

Using the fact that the ξ_k are i. i. d., we obtain

$$A_1(n) = \sum_{k_1=0}^{[ns]-1} \sum_{z\geq 1} \Sigma_{k_1}(x, z) \sum_{k_2=[ns]}^{[nt]} \sum_{y\geq 1} \sum_{w\geq 0} \mathbb{E}_y\left[\phi_2\left(\frac{X([nt] - k_2)}{\sigma\sqrt{n}}\right)\right]$$

$$\times \mathbb{E}\left[\phi_1\left(\frac{z + S([ns] - k_1)}{\sigma\sqrt{n}}\right) \mid \tau^S(z) > k_2 - k_1 - 1,\right.$$

$$z + S(k_2 - k_1 - 1) = w\Big]$$

$$\times \mathbb{P}[\tau^S(z) > k_2 - k_1 - 1, z + S(k_2 - k_1 - 1) = w]\mathbb{P}[\xi_1 = -w - y].$$

For any $2 \le k_1 < [ns] - 6$ and $[ns] \le k_2 \le [nt]$ and any $s_1 \in [\frac{k_1}{n}, \frac{k_1+1}{n})$ and $s_2 \in [\frac{k_2}{n}, \frac{k_2+1}{n})$, we write

$$f_n(s_1, s_2) = n^2 \sum_{z \ge 1} \Sigma_{[ns_1]}(x, z) \sum_{y \ge 1} \sum_{w \ge 0} \mathbb{E}_y\left[\phi_2\left(\frac{X([nt] - [ns_2])}{\sigma\sqrt{n}}\right)\right]$$

$$\times \mathbb{E}\left[\phi_1\left(\frac{z + S([ns] - [ns_1])}{\sigma\sqrt{n}}\right) \mid \tau^S(z) > [ns_2] - [ns_1] - 1,\right.$$

$$z + S([ns_2] - [ns_1] - 1) = w\Big]$$

$$\times \mathbb{P}\left[\tau^S(z) > [ns_2] - [ns_1] - 1, z + S([ns_2] - [ns_1] - 1) = w\right]$$

$$\times \mathbb{P}[\xi_1 = -w - y],$$

and $f_n(s_1, s_2) = 0$ for the others values of k_1, such that $0 \le k_1 \le [ns]$. Hence,

$$A_1(n) = \int_0^s ds_1 \int_s^t ds_2\, f_n(s_1, s_2) + O\left(\frac{1}{\sqrt{n}}\right).$$

It follows from Lemma 6.2.3 that, for each $z, w \ge 0$,

$$\lim_{n \to +\infty} \mathbb{E}\left[\phi_1\left(\frac{z + S([ns] - [ns_1])}{\sigma\sqrt{n}}\right) \mid \tau^S(z) > [ns_2] - [ns_1] - 1,\right.$$

$$z + S([ns_2] - [ns_1] - 1) = w\Big]$$

$$= \int_0^{+\infty} 2\phi_1(u\sqrt{s_2 - s_1}) \exp\left(-\frac{u^2}{2\frac{s-s_1}{s_2-s_1}\frac{s_2-s}{s_2-s_1}}\right) \frac{u^2}{\sqrt{2\pi \frac{(s-s_1)^3}{(s_2-s_1)^3} \frac{(s_2-s)^3}{(s_2-s_1)^3}}} du$$

$$= \frac{2}{\sqrt{2\pi}} \int_0^{+\infty} \phi_1(v) \exp\left(-\frac{v^2}{2\frac{(s-s_1)(s_2-s)}{s_2-s_1}}\right) \frac{v^2}{\sqrt{\frac{(s-s_1)^3(s_2-s)^3}{(s_2-s_1)^3}}} dv.$$

By Lemma 6.4.1,

$$\lim_{n \to +\infty} \mathbb{E}_y \left[\phi_2 \left(\frac{X([nt] - [ns_2])}{\sigma \sqrt{n}} \right) \right] = \int_0^{+\infty} \phi_2(u) \frac{2 e^{-u^2/2(t - s_2)}}{\sqrt{2\pi (t - s_2)}} du.$$

We set

$$a_n(x, y, z, w) = n^2 \Sigma_{[ns_1]}(x, z)$$
$$\times \mathbb{P}\left[\tau^S(z) > [ns_2] - [ns_1] - 1, z + S([ns_2] - [ns_1] - 1) = w \right]$$
$$\times \mathbb{P}[\xi_1 = -w - y],$$

$$b_n(y, z, w) = \mathbb{E}_y \left[\phi_2 \left(\frac{X([nt] - [ns_2])}{\sigma \sqrt{n}} \right) \right]$$
$$\times \mathbb{E}\left[\phi_1 \left(\frac{z + S([ns] - [ns_1])}{\sigma \sqrt{n}} \right) | \tau^S(z) > [ns_2] - [ns_1] - 1, \right.$$
$$\left. z + S([ns_2] - [ns_1] - 1) = w \right]$$

Note that $\sum_{z \geq 1} \sum_{y \geq 1} \sum_{w \geq 0} a_n(x, y, z, w) = \widetilde{\Sigma}_n(x, s_2, s_1)$. The behavior as $n \to +\infty$ of the quantity $\widetilde{\Sigma}_n(x, s_2, s_1)$ is given by the following Lemma, whose proof is postponed to the last section.

Lemma 6.4.4 *For all $0 < s < t < 1$, it holds*

$$\lim_{n \to +\infty} \widetilde{\Sigma}_n(x, t, s) = \frac{1}{2\pi \sqrt{s(t - s)^3}}.$$

Moreover, there exists a positive constant C_6 such that, for all $0 < s < t < 1$ and $n \geq 0$,

$$\widetilde{\Sigma}_n(x, t, s) \leq C_6 \frac{1 + x}{\pi \sqrt{s(t - s)^3}}.$$

By Lemmas 6.4.4 and 6.4.3, we get $\lim_{n \to +\infty} f_n(s_1, s_2) = f(s_1, s_2)$ where

$$f(s_1, s_2) = \frac{1}{\pi^2 \sqrt{s_1}} \int_0^{+\infty} \phi_1(v) \exp\left(-\frac{v^2}{2 \frac{(s_2 - s)(s - s_1)}{s_2 - s_1}} \right) \frac{v^2}{\sqrt{(s - s_1)^3 (s_2 - s)^3}} dv$$

$$\times \int_0^{+\infty} \phi_2(u) \frac{e^{-u^2/2(t - s_2)}}{\sqrt{t - s_2}} du.$$

Moreover, following the argument in the proof of Lemma 6.4.1, we can show that the sequence $(|f_n|)_{n\geq 1}$ is uniformly bounded by a function which is integrable with respect to Lebesgue measure on $[0, s] \times [s, t]$. Hence, using again the Lebesgue dominated convergence theorem, we get

$$\lim_{n\to+\infty} A_1(n) = \int_0^s ds_1 \int_s^t ds_2 f(s_1, s_2)$$

$$= \frac{1}{\pi^2} \int_0^s \frac{ds_1}{\sqrt{s_1}} \int_s^t ds_2 \int_0^{+\infty} \phi_1(v) \exp\left(-\frac{v^2}{2\frac{(s_2-s)(s-s_1)}{s_2-s_1}}\right) \frac{v^2}{\sqrt{(s-s_1)^3(s_2-s)^3}}$$

$$\times \int_0^{+\infty} \phi_2(u) \frac{e^{-u^2/2(t-s_2)}}{\sqrt{t-s_2}} du\, dv,$$

which yields, using again (6.4.1),

$$\lim_{n\to+\infty} A_1(n) = \frac{2}{\pi\sqrt{s(t-s)}} \int_0^{+\infty} \int_0^{+\infty} \phi_1(v)\phi_2(u) e^{-v^2/2s} e^{-\frac{(u+v)^2}{2(t-s)}} du\, dv.$$

$$(6.4.3)$$

6.4.2.2 Estimate of $A_2(n)$

We decompose $A_2(n)$ as

$$\sum_{y=0}^{+\infty} \sum_{k\leq[ns]} \sum_{l\geq 0} \mathbb{E}_x\left[\phi_1\left(\frac{X([ns])}{\sigma\sqrt{n}}\right)\phi_2\left(\frac{X([nt])}{\sigma\sqrt{n}}\right);\right.$$

$$\left. r_l = k, X(k) = y, y + \xi_{k+1} \geq 0, \ldots, y + \xi_{k+1} + \cdots + \xi_{[nt]} \geq 0\right].$$

$$= \sum_{y=0}^{+\infty} \sum_{k\leq[ns]} \mathbb{E}_x\left[\phi_1\left(\frac{y + \xi_{k+1} + \cdots + \xi_{[ns]}}{\sigma\sqrt{n}}\right)\phi_2\left(\frac{y + \xi_{k+1} + \cdots + \xi_{[nt]}}{\sigma\sqrt{n}}\right);\right.$$

$$\left. y + \xi_{k+1} \geq 0, \ldots, y + \xi_{k+1} + \cdots + \xi_{[nt]} \geq 0\right]$$

$$\times \sum_{l\geq 0} \mathbb{P}_x[r_l = k, X(k) = y].$$

Since (ξ_k) is a i.i.d. sequence,

$$A_2(n) = \sum_{y=0}^{+\infty} \sum_{k \leq [ns]} \Sigma_k(x, y) \mathbb{E}\left[\phi_1\left(\frac{y + S([ns] - k)}{\sigma\sqrt{n}}\right)\phi_2\left(\frac{y + S([nt] - k)}{\sigma\sqrt{n}}\right);\right.$$
$$\left. \tau^S(y) > [nt] - k\right].$$

For $u \in (0, s]$, we denote

$$g_n(u) = n \sum_{y=0}^{+\infty} \Sigma_{[nu]}(x, y) \mathbb{E}\left[\phi_1\left(\frac{y + S([ns] - [nu])}{\sigma\sqrt{n}}\right)\phi_2\left(\frac{y + S([nt] - [nu])}{\sigma\sqrt{n}}\right);\right.$$
$$\left. \tau^S(y) > [nt] - [nu]\right].$$

Now, let us compute the pointwise limit on $(0, s]$ of the sequence $(g_n)_{n\geq 1}$. We write $g_n(u)$ as

$$g_n(u) = n \sum_{y=0}^{+\infty} \Sigma_{[nu]}(x, y)$$
$$\times \mathbb{E}\left[\phi_1\left(\frac{y + S([ns] - [nu])}{\sigma\sqrt{n}}\right)\phi_2\left(\frac{y + S([nt] - [nu])}{\sigma\sqrt{n}}\right)\right.$$
$$\times \left|\tau^S(y) > [nt] - [nu]\right]$$
$$\times \mathbb{P}_y\left[\tau^S(y) > [nt] - [nu]\right].$$

We set

$$a_n(x, y) = n\Sigma_{[nu]}(x, y)\mathbb{P}_y\left[\tau^S(y) > [nt] - [nu]\right],$$

and

$$b_n(y) = \mathbb{E}\left[\phi_1\left(\frac{y + S([ns] - [nu])}{\sigma\sqrt{n}}\right)\phi_2\left(\frac{y + S([nt] - [nu])}{\sigma\sqrt{n}}\right)\left|\tau^S(y) > [nt] - [nu]\right].$$

Note that $\sum_{y=0}^{+\infty} a_n(y) = \widehat{\Sigma}_{[nu]}(x, t, u)$. Since ϕ_1, ϕ_2 are bounded and continuous on $\mathbb{R}$, it follows from Theorem 3.2 in [4] and Theorems 2.23 and 3.4 in [10] that

$$
\lim_{n \to +\infty} b_n(y) = \lim_{n \to +\infty} \mathbb{E}\left[\phi_1\left(\frac{y + S([ns] - [nu])}{\sigma\sqrt{[nt] - [nu]}} \frac{\sqrt{[nt] - [nu]}}{\sqrt{n}}\right)\right.
$$
$$
\times \phi_2\left(\frac{y + S([nt] - [nu])}{\sigma\sqrt{[nt] - [nu]}} \frac{\sqrt{[nt] - [nu]}}{\sqrt{n}}\right)\Big| \tau^S(y) > [nt] - [nu]\right]
$$
$$
= \int_0^{+\infty} \int_0^{+\infty} \phi_1(y\sqrt{t - u})\phi_2(z\sqrt{t - u})\left(\frac{t - u}{s - u}\right)^{3/2} y e^{-\frac{t-u}{2(s-u)}y^2}
$$
$$
\times \frac{e^{-\frac{1}{2}\frac{t-u}{t-s}(z-y)^2} - e^{-\frac{1}{2}\frac{t-u}{t-s}(z+y)^2}}{\sqrt{2\pi\left(1 - \frac{s-u}{t-u}\right)}} dy\, dz
$$
$$
= \frac{1}{\sqrt{2\pi(t - s)}} \int_0^{+\infty} \int_0^{+\infty} \phi_1(y')\phi_2(z')\frac{\sqrt{t - u}}{(s - u)^{3/2}} y' e^{-\frac{y'^2}{2(s-u)}}
$$
$$
\times \left(e^{-\frac{(z'-y')^2}{2(t-s)}} - e^{-\frac{(z'+y')^2}{2(t-s)}}\right) dy'\, dz'.
$$

Again, we can use the argument in the proof of Lemma 6.4.1 to show that the sequence (g_n) converges point wise to g with

$$
g(u) = \frac{1}{\pi^{3/2}\sqrt{2(t - s)}} \frac{1}{\sqrt{u(s - u)^3}}
$$
$$
\times \int_0^{+\infty} \int_0^{+\infty} \phi_1(y')\phi_2(z')y' e^{-\frac{y'^2}{2(s-u)}}\left(e^{-\frac{(z'-y')^2}{2(t-s)}} - e^{-\frac{(z'+y')^2}{2(t-s)}}\right) dy'\, dz',
$$

and (g_n) is also dominated by a function which is integrable on $[0, s]$ with respect to the Lebesgue measure. Lebesgue's dominated convergence theorem yields

$$
\lim_{n \to +\infty} A_2(n) = \lim_{n \to +\infty} \frac{1}{n} \sum_{k \le [ns]} g_n(k/n) = \int_0^s g(u)\, du
$$
$$
= \frac{1}{\pi^{3/2}\sqrt{2(t - s)}} \int_0^s du \int_0^{+\infty} dy' \int_0^{+\infty} dz'
$$
$$
\times \phi_1(y')\phi_2(z')\frac{e^{-\frac{y'^2}{2(s-u)}}}{\sqrt{u(s - u)^3}} \frac{y'}{\sqrt{2\pi(t - s)}}\left(e^{-\frac{(z'-y')^2}{2(t-s)}} - e^{-\frac{(z'+y')^2}{2(t-s)}}\right)
$$
$$
= \frac{1}{\pi^{3/2}s\sqrt{2(t - s)}} \int_0^{+\infty} dy' \int_0^{+\infty} dz'\phi_1(y')\phi_2(z')
$$

$$\times \left(e^{-\frac{(z'-y')^2}{2(t-s)}} - e^{-\frac{(z'+y')^2}{2(t-s)}} \right) \left(\int_0^1 \frac{y'}{\sqrt{v(1-v)^3}} e^{-\frac{y'^2}{2s(1-v)}} dv \right)$$

$$= \frac{1}{\pi \sqrt{s(t-s)}} \int_0^{+\infty} dy' \int_0^{+\infty} dz' \phi_1(y') \phi_2(z') e^{-y'^2/2s} \left(e^{-\frac{(z'-y')^2}{2(t-s)}} - e^{-\frac{(z'+y')^2}{2(t-s)}} \right).$$

$$(6.4.4)$$

6.4.2.3   Conclusion

Combining (6.4.3) and (6.4.4), we may write

$$\lim_{n \to +\infty} \mathbb{E}\left[\phi_1 \left(\frac{X([ns])}{\sigma \sqrt{n}} \right) \phi_2 \left(\frac{X([nt])}{\sigma \sqrt{n}} \right) \right]$$

$$= \frac{1}{\pi \sqrt{s(t-s)}} \int_0^{+\infty} dy' \int_0^{+\infty} dz' \phi_1(y') \phi_2(z') e^{-y'^2/2s}$$

$$\times \left(e^{-\frac{(z'-y')^2}{2(t-s)}} + e^{-\frac{(z'+y')^2}{2(t-s)}} \right)$$

$$= \mathbb{E}[\phi_1(|B_s|)\phi_2(|B_t|)].$$

Using a similar estimate as the one in (6.4.2), we get

$$\lim_{n \to +\infty} \mathbb{E}[\phi_1(X_n(s))\phi_2(X_n(t))] = \mathbb{E}[\phi_1(|B_s|)\phi_2(|B_t|)],$$

which concludes the convergence of (X_n) in two-dimensional marginal distribution to a reflected Brownian motion.

6.4.3   Finite Dimensional Distributions

The convergence of d-dimensional marginal distributions of $(X_n(t))_{n \geq 1}$ for any $d \geq 2$ may be done by induction on d. Let us fix $n \geq 1, d \geq 3$, then reals $0 < s_1 < \cdots < s_d$ and $\phi_1, \ldots, \phi_d$ bounded and Lipschitz continuous real valued functions defined on $\mathbb{R}$.

Let κ denote the first reflection time after $[ns_1]$, i.e., $\kappa = \kappa(n, s_1) = \min\{k >$
$[ns_1] : X(k-1) + \xi_k < 0\}$. We decompose $\mathbb{E}_x\left[\prod_{i=1}^{d}\phi_i\left(\dfrac{X([ns_i])}{\sigma\sqrt{n}}\right)\right]$ as

$$\sum_{j=1}^{d-1}\sum_{k=[ns_j]+1}^{[ns_{j+1}]}\mathbb{E}_x\left[\prod_{i=1}^{d}\phi_i\left(\frac{X([ns_i])}{\sigma\sqrt{n}}\right); \kappa = k\right]$$

$$+ \mathbb{E}_x\left[\prod_{i=1}^{d}\phi_i\left(\frac{X([ns_i])}{\sigma\sqrt{n}}\right); \kappa > [ns_d]\right].$$

Then we can deal with the terms

$$\mathbb{E}_x\left[\prod_{i=1}^{d}\phi_i\left(\frac{X([ns_i])}{\sigma\sqrt{n}}\right); \kappa = k\right] \quad \text{and} \quad \mathbb{E}_x\left[\prod_{i=1}^{d}\phi_i\left(\frac{X([ns_i])}{\sigma\sqrt{n}}\right); \kappa > [ns_d]\right]$$

in the same ways as we do for A_1 and A_2, respectively.

More precisely, for each $1 \le j \le d-1$ and $k \in \{[ns_j]+1, \ldots, [ns_{j+1}]\}$, we write

$$\mathbb{E}_x\left[\prod_{i=1}^{d}\phi_i\left(\frac{X([ns_i])}{\sigma\sqrt{n}}\right); \kappa = k\right]$$

$$= \sum_{k_1=0}^{[ns_1]-1}\sum_{l\ge0}\sum_{y\ge1}\sum_{z\ge1}\sum_{w\ge0}\mathbb{E}_x\left[\prod_{i=1}^{d}\phi_i\left(\frac{X([ns_i])}{\sigma\sqrt{n}}\right); r_l = k_1, X(k_1) = z,\right.$$

$$z + \xi_{k_1+1} \ge 0, \ldots, z + \xi_{k_1+1} + \cdots + \xi_{k-2} \ge 0,$$

$$\left. z + \xi_{k_1+1} + \cdots + \xi_{k-1} = w, w + \xi_k = -y\right]$$

$$= \sum_{k_1=0}^{[ns_1]-1}\sum_{l\ge0}\sum_{y\ge1}\sum_{z\ge1}\sum_{w\ge0}\mathbb{E}_x\left[\prod_{i_1=1}^{j}\phi_{i_1}\left(\frac{z + \xi_{k_1+1} + \cdots + \xi_{[ns_j]}}{\sigma\sqrt{n}}\right)\right.$$

$$\times \prod_{i_2=j+1}^{d}\phi_{i_2}\left(\frac{y + \xi_{k+1} + \cdots + \xi_{[ns_j]}}{\sigma\sqrt{n}}\right); r_l = k_1, X(k_1) = z, z + \xi_{k_1+1} \ge 0, \ldots,$$

$$\left. z + \xi_{k_1+1} + \cdots + \xi_{k-2} \ge 0, z + \xi_{k_1+1} + \cdots + \xi_{k-1} = w, w + \xi_k = -y\right]$$

$$= \sum_{k_1=0}^{[ns_1]-1}\sum_{z\ge1}\Sigma_{k_1}(x, z)\sum_{y\ge1}\sum_{w\ge0}\mathbb{E}_y\left[\prod_{i_2=j+1}^{d}\phi_{i_2}\left(\frac{X([ns_j]-k_2)}{\sigma\sqrt{n}}\right)\right]$$

$$\times \mathbb{E}\left[\prod_{i_1=1}^{j} \phi_{i_1}\left(\frac{z + S([ns_j] - k_1)}{\sigma\sqrt{n}}\right)\right.$$

$$\times \left|\tau^S(z) > k - k_1 - 1, z + S(k - k_1 - 1) = w\right]$$

$$\times \mathbb{P}[\tau^S(z) > k - k_1 - 1, z + S(k - k_1 - 1) = w]\mathbb{P}[\xi_1 = -w - y].$$

Now we can use the induction hypothesis and Corollary 2.5 in [5] to deal with the first and the second expectations.

6.4.4   Tightness

Recall that the modulus of continuity of a function $f : [0, 1] \to \mathbb{R}$ is defined by

$$w_f(\delta) = \sup_{t,s\in[0,1],|t-s|<\delta} |f(t) - f(s)|.$$

It is clear that $w_X(\delta) \leq w_S(\delta)$. Using Theorem 7.3 in [3], the tightness of X follows directly from the one of the classical random walk $(S(n))_{n\geq 0}$. We achieve the proof of Theorem 6.1.1, applying Theorem 7.1 in [3].

6.5   Auxiliary Proofs

Proof of Lemma 6.4.2 By setting $h_n(y) = \sqrt{n}\mathbb{P}_y[r_1 > n]$, the Markov property yields

$$\widehat{\Sigma}_n(x, t, s) = n \sum_{l\geq 0} \mathbb{E}_x\left[\mathbb{P}_{X(r_l)}[r_1 \circ \theta^{r_l} > [nt] - [ns]]; r_l = [ns]\right]$$

$$= \frac{\sqrt{n}}{\sqrt{[nt] - [ns]}} \sqrt{n} \sum_{l\geq 0} \mathbb{E}_x\left[h_{[nt]-[ns]}(X(r_l)); r_l = [ns]\right]$$

$$= \frac{1 + o(n)}{\sqrt{s(t - s)}} \sqrt{[ns]}T_{[ns]}(h_{[nt]-[ns]})(x).$$

Let us prove that $\sqrt{[ns]}T_{[ns]}(h_{[nt]-[ns]})(x) \to \frac{1}{\pi}$ as $n \to +\infty$. Indeed,

$$\left|\sqrt{[ns]}T_{[ns]}(h_{[nt]-[ns]})(x) - \frac{1}{\pi}\right| \leq B_1(n) + B_2(n),$$

with

$$B_1(n) = \left| \sqrt{[ns]} T_{[ns]}(h_{[nt]-[ns]})(x) - \frac{1}{\pi v(h)} v(h_{[nt]-[ns]}) \right|, \quad \text{and}$$

$$B_2(n) = \frac{1}{\pi v(h)} \left| v(h_{[nt]-[ns]}) - v(h) \right|.$$

By Lemma 6.2.1, it holds $0 \le h_n(y) \le C_1 h(y)$, with $h(y) = O(y)$, so that the sequence $(h_n)_{n\ge 1}$ is bounded in $\mathcal{B}_\alpha$. Thus, Corollary 6.3.1 yields

$$B_1(n) \le (1+x) \left| \sqrt{[ns]} T_{[ns]} - \frac{1}{\pi v(h)} \Pi \right|_\alpha |h_{[nt]-[ns]}|_\alpha \longrightarrow 0 \quad \text{as} \quad n \to +\infty.$$

Similarly, by Lemma 6.2.1 and the dominated convergence theorem,

$$\lim_{n\to+\infty} \left| v(h_{[nt]-[ns]}) - v(h) \right| = 0,$$

so that $B_2(n) \longrightarrow 0$ as $n \to +\infty$.

$\square$

Proof of Lemma 6.4.4 By setting $\tilde{h}_n(y) = n^{3/2} \mathbb{P}_y[r_1 = n]$, the Markov property yields

$$\widetilde{\Sigma}_n(x,s,t) = n^2 \sum_{l\ge 0} \mathbb{E}_x \left[\mathbb{P}_{X(r_l)}[r_1 \circ \theta^{r_l} = [nt] - [ns]]; r_l = [ns] \right]$$

$$= \frac{n^{3/2}}{([nt]-[ns])^{3/2}} \sqrt{n} \sum_{l\ge 0} \mathbb{E}_x \left[\tilde{h}_{[nt]-[ns]}(X(r_l)); r_l = [ns] \right]$$

$$= \frac{1+o(n)}{\sqrt{s}(t-s)^{3/2}} \sqrt{[ns]} T_{[ns]}(\tilde{h}_{[nt]-[ns]})(x).$$

By Corollary 6.2.1, it holds $0 \le \tilde{h}_n(y) \le C_3 h(y)$, with $h(y) = O(y)$, so that the sequence $(\tilde{h}_n)_{n\ge 1}$ is bounded in $\mathcal{B}_\alpha$. We conclude as above to prove Lemma 6.4.2.

$\square$

Acknowledgments H.-L. Ngo thanks the University of Tours for generous hospitality in the Institue Denis Poisson (IDP) and financial support in May 2019. This article is a result of the research team with the title "Quantitative Research Methods in Economics and Finance", Foreign Trade University, Ha Noi, Vietnam.

M. Peigné thanks the Vietnam Institute for Advanced Studies in Mathematics (VIASM) and the Vietnam Academy of Sciences And Technology (VAST) in Ha Noi for their kind and friendly hospitality and accommodation in June 2018.

Thanks are also due to M. Pollicott who proposed to publish this article in the Chaire Jean Morlet Series.

Both authors thank the referee for many helpful comments that improved the text and some proofs. Duy Tran Vo also pointed them several misprints.

References

1. V.I. Afanasyev, J. Geiger, G. Kersting, V.A. Vatutin, Criticality for branching processes in random environment. Anna. Probab. **33**(2), 645–673 (2005)
2. V.I. Afanasyev, C. Böinghoff, G. Kersting, V.A. Vatutin, Limit theorems for weakly subcritical branching processes in random environment. J. Theor. Probab. **25**(3), 703–732 (2012)
3. P. Billingsley, *Convergence of Probability Measures* (Willey, New York, 1968)
4. E. Bolthausen, On a functional central limit theorem for random walks conditioned to stay positive. Ann. Probab. **4**(3), 480–485 (1976)
5. F. Caravenna, L. Chaumont, An invariance principle for random walk bridges conditioned to stay positive. Electron. J. Probab. **18**(60), 1–32 (2013)
6. Y.S. Chow, T.L. Lai, Moments of ladder variables for driftless random walks. Z. Wahrsch. Verw. Gebiete **48**, 253–257 (1979)
7. R.A. Doney, Local behaviour of first passage probabilities. Probab. Theory Related Fields **152**(3–4), 559–588 (2012)
8. R. Essifi, M. Peigné, Return probabilities for the reflected random walk on $\mathbb{N}_0$. J. Theor. Probab. **28**(1), 231–258 (2015)
9. S. Gouëzel, Correlation asymptotics from large deviations in dynamical systems with infinite measure. Colloquium Math. **125**, 193–212 (2011)
10. D.L. Iglehart, Functional central limit theorems for random walks conditioned to stay positive. Ann. Probab. **2**(4), 608–619 (1974)
11. K. Itô, H.P. McKean Jr., *Diffusion Processes and Their Sample Paths* (Springer Science Business Media, Berlin, 2012)
12. J.H.B. Kemperman, The oscillating random walk. Stoch. Process Their Appl. **2**, 1–29 (1974)
13. M.V. Kozlov, On the asymptotic behavior of the probability of non-extinction for critical branching processes in a random environment. Theory Probab. Appl. **21**(4), 791–804 (1976)
14. E. Le Page, M. Peigné, A local limit theorem on the semi-direct product of $\mathbb{R}^{*+}$ and $\mathbb{R}^d$. Annal. de l'I.H.P. Probab. et Stat., **33**(2), 223–252 (1997)
15. H-L. Ngo, M. Peigné, Limit theorem for perturbed random walks. Theory Stoch. Process. **24**(2), 61–78 (2019)
16. M. Peigné, W. Woess, On recurrence of a reflected random walk on the half line (2006). http://arxiv.org/abs/ math/0612306
17. M. Peigné, W. Woess, Stochastic dynamical systems with weak contractivity I. Strong and local contractivity. Colloq. Math. **125**, 1–54 (2011)

Chapter 7
The Strong Borel–Cantelli Property in Conventional and Nonconventional Setups

Yuri Kifer

Abstract We study the strong Borel–Cantelli property both for events and for shifts on sequence spaces considering both a conventional and a nonconventional setups. Namely, under certain conditions on events $\Gamma_1, \Gamma_2, \ldots$ we show that with probability one

$$\left(\sum_{n=1}^{N} \prod_{i=1}^{\ell} P(\Gamma_{q_i(n)})\right)^{-1} \sum_{n=1}^{N} \prod_{i=1}^{\ell} \mathbb{I}_{\Gamma_{q_i(n)}} \to 1 \text{ as } N \to \infty$$

where $q_i(n)$, $i = 1, \ldots, \ell$ are integer valued functions satisfying certain assumptions and $\mathbb{I}_\Gamma$ denotes the indicator of Γ. When $\ell = 1$ (called the conventional setup) this convergence can be established under ϕ-mixing conditions while when $\ell > 1$ (called a nonconventional setup) the stronger ψ-mixing condition is required. These results are extended to shifts T of sequence spaces where $\Gamma_{q_i(n)}$ is replaced by $T^{-q_i(n)} C_n^{(i)}$ where $C_n^{(i)}$, $i = 1, \ldots, \ell$, $n \geq 1$ is a sequence of cylinder sets. As an application we study the asymptotical behavior of maximums of certain logarithmic distance functions and of (multiple) hitting times of shrinking cylinders.

7.1 Introduction

The classical second Borel–Cantelli lemma states that if $\Gamma_1, \Gamma_2, \ldots$ is a sequence of independent events such that

$$\sum_{n=1}^{\infty} P(\Gamma_n) = \infty \tag{7.1.1}$$

Y. Kifer (✉)
Institute of Mathematics, The Hebrew University, Jerusalem, Israel
e-mail: kifer@math.huji.ac.il

© The Author(s), under exclusive license to Springer Nature Switzerland AG 2021
M. Pollicott, S. Vaienti (eds.), *Thermodynamic Formalism*, Lecture Notes
in Mathematics 2290, https://doi.org/10.1007/978-3-030-74863-0_7

then with probability one infinitely many of events Γ_i occur, i.e.

$$\sum_{n=1}^{\infty} \mathbb{I}_{\Gamma_n} = \infty \quad \text{almost surely (a.s.)} \tag{7.1.2}$$

where $\mathbb{I}_\Gamma$ is the indicator of a set (event) Γ.

There is a long list of papers, starting probably with [13], providing conditions which replace the independency by a weaker assumption and which still yield (7.1.2) (see, for instance, [4] and references there). On the other hand, it was shown in Theorem 3 of [16] that under ϕ-mixing with a summable coefficient ϕ the condition (7.1.1) yields the stronger version of the second Borel–Cantelli lemma in the form

$$\frac{S_N}{\mathcal{E}_N} \to 1 \quad \text{almost surely (a.s.) as } N \to \infty \tag{7.1.3}$$

where $S_N = \sum_{n=1}^{N} \mathbb{I}_{\Gamma_n}$ and $\mathcal{E}_N = \sum_{n=1}^{N} P(\Gamma_n)$.

The same paper [16] started another line of research, known now under the name dynamical Borel–Cantelli lemmas, where (7.1.3) is proved for $S_N = \sum_{n=1}^{N} \mathbb{I}_{\Gamma_n} \circ T^n$ where T is a measure preserving transformation on a probability space (Ω, P) and Γ_n, $n \geq 1$ is a sequence of measurable sets. For such S_N's the convergence (7.1.3) was proved, in particular, for the Gauss map $Tx = \frac{1}{x} \pmod 1$, $x \in (0, 1]$ preserving the Gauss measure $P(\Gamma) = \frac{1}{\ln 2} \int_\gamma \frac{dx}{1+x}$. This line of research became quite popular in the last two decades. In particular, [3] proves (7.1.3) in the dynamical setup considering T being the so called subshift of finite type on a sequence space where Γ_n, $n \geq 1$ is a sequence of cylinders while another series of papers dealt with uniformly and non-uniformly hyperbolic dynamical systems as a transformation T and with geometric balls as Γ_n's (see, for instance, [7, 10] and references there).

In this paper we consider, in particular, "nonconventional" extensions of some of the above results aiming to prove that under certain conditions (7.1.3) holds true with $S_N = \sum_{n=1}^{N} (\prod_{i=1}^{\ell} \mathbb{I}_{\Gamma_{q_i(n)}})$ and $\mathcal{E}_N = \sum_{n=1}^{N} \prod_{i=1}^{\ell} P(\Gamma_{q_i(n)})$ where $q_i(n)$, $i = 1, \ldots, \ell$ functions taking on positive integer values on positive integers and satisfying certain assumptions valid, in particular, for certain polynomials with integer coefficients. When $\ell = 1$ (conventional setup) the ϕ-mixing with a summable coefficient ϕ suffices for our result, while for $\ell > 1$ we have to impose stronger ψ-mixing conditions.

In the dynamical systems setup we consider $S_N = \sum_{n=1}^{N} (\prod_{i=1}^{\ell} \mathbb{I}_{C_n^{(i)}} \circ T^{q_i(n)})$ and $\mathcal{E}_N = \sum_{n=1}^{N} \prod_{i=1}^{\ell} P(C_n^{(i)})$ where T is the left shift on a sequence space $\mathcal{A}^{\mathbb{Z}}$ with a finite or countable alphabet while $C_n^{(i)}$, $i = 1, \ldots, \ell$, $n \geq 1$ is a sequence of cylinder sets. As an application we study the asymptotic behaviors of expressions $M_N = \max_{1 \leq n \leq N}(\min_{1 \leq i \leq \ell} \Phi_{\tilde{\omega}^{(i)}} \circ T^{q_i(n)})$ where $\Phi_{\tilde{\omega}}(\omega) = -\ln(d(\omega, \tilde{\omega}))$, $\omega, \tilde{\omega} \in \mathcal{A}^{\mathbb{N}}$ and $d(\cdot, \cdot)$ is the natural distance on the sequence space.

Our results extend some of the previous work in the following aspects. First, the strong Borel–Cantelli property in the nonconventional setup $\ell > 1$ was not studied before at all. Secondly, even in the conventional setup $\ell = 1$ considering rather general functions $q(n) = q_1(n)$ (not necessarily strictly increasing) in place of just $q(n) = n$ seems to be new, as well. Thirdly, we extend for shifts some of the results from [3] considering sequence spaces with countable alphabets and ϕ-mixing invariant measures rather than just subshifts of finite type with Gibbs measures which are exponentially fast ψ-mixing (see [1]). This allows to apply our results, for instance, to Gibbs-Markov maps and to Markov chains with a countable state space satisfying the Doeblin condition since both examples are exponentially fast ϕ-mixing, see [14] and [2], respectively.

In the next section we will formulate precisely our setups and assumptions and state our main results. In Sect. 7.3 we will prove the strong Borel–Cantelli property for events under the ϕ-mixing condition in the conventional setup $\ell = 1$ and under ψ-mixing condition in the nonconventional setup $\ell > 1$. In Sects. 7.4 and 7.5 we extend the strong Borel–Cantelli property to shifts under the ϕ-mixing when $\ell = 1$ and under ψ-mixing when $\ell > 1$, respectively. In Sect. 7.6 we exhibit applications to the asymptotic behaviors of maximums along shifts of logarithmic distance functions while in the last Sect. 7.7 we apply the strong Borel–Cantelli property to derive the asymptotics of multiple hitting times of shrinking cylinder sets.

7.2 Preliminaries and Main Results

We start with a probability space $(\Omega, \mathcal{F}, P)$ and a two parameter family of σ-algebras $\mathcal{F}_{mn}$ indexed by pairs of integers $-\infty \leq m \leq n \leq \infty$ and such that $\mathcal{F}_{mn} \subset \mathcal{F}_{m'n'} \subset \mathcal{F}$ if $m' \leq m \leq n \leq n'$. Recall that the ϕ and ψ dependence coefficient between two σ-algebras $\mathcal{G}$ and $\mathcal{H}$ can be written in the form (see [2]),

$$\phi(\mathcal{G}, \mathcal{H}) = \sup_{\Gamma \in \mathcal{G},\, \Delta \in \mathcal{H}} \{|\tfrac{P(\Gamma \cap \Delta)}{P(\Gamma)} - P(\Delta)|,\; P(\Gamma) \neq 0\} \qquad (7.2.1)$$

$$= \tfrac{1}{2} \sup\{\|E(g|\mathcal{G}) - Eg\|_{L^\infty} :\; g \text{ is } \mathcal{H}\text{-measurable and } \|g\|_{L^\infty} \leq 1\}$$

and

$$\psi(\mathcal{G}, \mathcal{H}) = \sup_{\Gamma \in \mathcal{G},\, \Delta \in \mathcal{H}} \{|\tfrac{P(\Gamma \cap \Delta)}{P(\Gamma)P(\Delta)} - 1|,\; P(\Gamma)P(\Delta) \neq 0\} \qquad (7.2.2)$$

$$= \tfrac{1}{2} \sup\{\|E(g|\mathcal{G}) - Eg\|_{L^\infty} :\; g \text{ is } \mathcal{H}\text{-measurable and } E|g| \leq 1\},$$

respectively. The ϕ-dependence (mixing) and the ψ-dependence (mixing) in the family $\mathcal{F}_{mn}$ is measured by the coefficients

$$\phi(k) = \sup_m \phi(\mathcal{F}_{-\infty,m}, \mathcal{F}_{m+k,\infty}) \text{ and } \psi(k) = \sup_m \psi(\mathcal{F}_{-\infty,m}, \mathcal{F}_{m+k,\infty}),$$

$$(7.2.3)$$

respectively, where $k = 0, 1, 2, \ldots$. The probability measure P is called ϕ-mixing or ψ-mixing with respect to the family of σ-algebras $\mathcal{F}_{mn}$ if $\phi(n) \to 0$ or $\psi(1) < \infty$ and $\psi(n) \to 0$ as $n \to \infty$, respectively.

Our setup includes also functions $q_1(n)$, $q_2(n)$, $\ldots$, $q_\ell(n)$ with $\ell \geq 1$ taking on nonnegative integer values on integers $n \geq 0$ and satisfying

Assumption 7.2.1 There exists a constant $K > 0$ such that

(i) for any $i \neq j$, $1 \leq i, j \leq \ell$ and every integer k the number of integers $n \geq 0$ satisfying at least one of the equations

$$q_i(n) - q_j(n) = k \quad \text{and} \quad q_i(n) = k \tag{7.2.4}$$

does not exceed K (when $\ell = 1$ only the second equation in (7.2.4) should be taken into account);

(ii) the cardinality of the set $\mathcal{N}$ of all pairs $n > m \geq 0$ satisfying

$$\max_{1 \leq i \leq \ell} q_i(n) \leq \max_{1 \leq i \leq \ell} q_i(m) \tag{7.2.5}$$

does not exceed K.

Observe that Assumption 7.2.1 is satisfied if q_i, $i = 1, \ldots, \ell$ are essentially distinct nonconstant polynomials (i.e. $|q_i(n) - q_j(n)| \to \infty$ as $n \to \infty$ for any $i \neq j$) with integer coefficients taking on nonnegative values on nonnegative integers. Indeed, $q_i(n) - q_j(n)$ and $q_i(n)$ are nonconstant polynomials, and so the number of n's solving one of equations in (7.2.4) is bounded by the degree of the corresponding polynomial. In order to show that (7.2.5) can hold true in the polynomial case only for finitely many pairs $m < n$ observe that there exists $n_0 \geq 1$ such that all polynomials $q_1(n)$, $q_2(n)$, $\ldots$, $q_\ell(n)$ are strictly increasing on $[n_0, \infty)$. Hence, if $n > m \geq n_0$ then (7.2.5) cannot hold true. If $0 \leq m < n_0$ and $n \geq n_0$ then there exists $n_1 \geq n_0$ such that for all $n \geq n_1$ (7.2.5) cannot hold true, as well. The remaining case $0 \leq m < n_0$ and $0 \leq n < n_1$ concerns less than $n_0 n_1$ pairs $m < n$.

Next, we will state our result concerning sequences of events. Let $\Gamma_1, \Gamma_2, \ldots \in \mathcal{F}$ be a sequence of events and each σ-algebra $\mathcal{F}_{mn}$, $1 \leq m \leq n < \infty$ be generated by the events $\Gamma_m, \Gamma_{m+1}, \ldots, \Gamma_n$. Set also $\mathcal{F}_{mn} = \mathcal{F}_{1n}$ for $-\infty \leq m \leq 0$ and $n \geq 1$, $\mathcal{F}_{mn} = \{\emptyset, \Omega\}$ for $m, n \leq 0$ and $\mathcal{F}_{m,\infty} = \sigma\{\Gamma_m, \Gamma_{m+1}, \ldots\}$. Set

$$S_N = \sum_{n=1}^{N} (\prod_{i=1}^{\ell} \mathbb{I}_{\Gamma_{q_i(n)}}) \quad \text{and} \quad \mathcal{E}_N = \sum_{n=1}^{N} (\prod_{i=1}^{\ell} P(\Gamma_{q_i(n)})). \tag{7.2.6}$$

Theorem 7.2.2 *Let ϕ and ψ be dependence coefficients defined by (7.2.3) for the above σ-algebras $\mathcal{F}_{mn}$. Assume that $\phi(n)$, $n \geq 0$ is summable in the case $\ell = 1$ and $\psi(n)$, $n \geq 0$ is summable in the case $\ell > 1$. Suppose that the functions $q_1(n), \ldots, q_\ell(n)$ satisfy Assumption 7.2.1(i) and*

$$\mathcal{E}_N \to \infty \quad as \quad N \to \infty. \tag{7.2.7}$$

Then, with probability one,

$$\lim_{N \to \infty} \frac{S_N}{\mathcal{E}_N} = 1 \quad as \quad N \to \infty. \tag{7.2.8}$$

Next, we will present our results concerning shifts. Here $\Omega = \mathcal{A}^{\mathbb{Z}}$ is the space of sequences $\omega = (\ldots, \omega_{-1}, \omega_0, \omega_1, \ldots)$ with terms ω_i from a finite or countable alphabet $\mathcal{A}$ which is not a singleton with the index i running along integers (or along natural numbers $\mathbb{N}$ which can also be considered requiring very minor modifications). We assume that the basic σ-algebra $\mathcal{F}$ is generated by all cylinder sets while the σ-algebras $\mathcal{F}_{mn}$, $n \geq m$ are generated by the cylinder sets of the form $\{\omega = (\omega_i)_{-\infty < i < \infty} : \omega_i = a_i \text{ for } m \leq i \leq n\}$ for some $a_m, a_{m+1}, \ldots, a_n \in \mathcal{A}$. The setup includes also the left shift $T : \Omega \to \Omega$ acting by $(T\omega)_i = \omega_{i+1}$ and a T-invariant probability measure P on $(\Omega, \mathcal{F})$, i.e. $P(T^{-1}\Gamma) = P(\Gamma)$ for any measurable $\Gamma \subset \Omega$. In this setup ϕ and ψ-dependence coefficients defined by (7.2.3) will be considered with respect to the family of σ-algebras $\mathcal{F}_{mn}$, $m \leq n$ defined above. Without loss of generality we assume that the probability of each 1-cylinder $[a] = \{\omega = (\omega_i)_{i \in \mathbb{Z}} : \omega_0 = a\}$ is positive, i.e. $P([a]) > 0$ for any $a \in \mathcal{A}$, and since $\mathcal{A}$ is not a singleton we have also that $\sup_{a \in \mathcal{A}} P([a]) < 1$.

Each cylinder C is defined on an interval of integers $\Lambda = [l, r]$, $l \leq r$, i.e. $C = \{\omega = (\omega_i)_{-\infty < i < \infty} : \omega_i = a_i, i = l, l+1, \ldots, r\}$ for some $a_l, \ldots, a_r \in \mathcal{A}$. Given a constant $D > 0$ call an interval of integers $\Lambda_1 = [l_1, r_1]$ to be right D-nested in the interval of integers $\Lambda_2 = [l_2, r_2]$ if $[l_1, r_1] \subset (-\infty, r_2 + D)$, i.e. $r_1 < r_2 + D$. Such an interval Λ_1 will be called D-nested in Λ_2 if $[l_1, r_1] \subset (l_2 - D, r_2 + D)$. The latter notion was used also in [3].

Let $C_n^{(j)}$, $j = 1, \ldots, \ell$, $n = 1, 2, \ldots$ be a sequence of cylinder sets defined on intervals of integers Λ_n, $n = 1, 2, \ldots$ so that $C_n^{(j)}$, $j = 1, \ldots, \ell$ are defined on Λ_n for each $n \geq 1$. Set

$$S_N = \sum_{n=1}^{N} \left(\prod_{i=1}^{\ell} \mathbb{I}_{C_n^{(i)}} \circ T^{q_i(n)} \right) \quad \text{and} \quad \mathcal{E}_N = \sum_{n=1}^{N} \prod_{i=1}^{\ell} P(C_n^{(i)}). \tag{7.2.9}$$

Theorem 7.2.3 *Suppose that the functions $q_1(n), \ldots, q_\ell(n)$ satisfy Assumption 7.2.1 and*

$$\mathcal{E}_N \to \infty \quad as \quad N \to \infty. \tag{7.2.10}$$

Let $C_n^{(j)}$, $j = 1, \ldots, \ell$, $n \geq 1$ be a sequence of cylinder sets defined on intervals $\Lambda_n \subset \mathbb{Z}$ as described above and $D > 0$ be a constant.

(i) If $\ell = 1$ assume that the ϕ-dependence coefficient is summable and that for all $m < n$ the interval Λ_m is right D-nested in Λ_n. Then, with probability one,

$$\lim_{N \to \infty} \frac{S_N}{\mathcal{E}_N} = 1 \quad as \quad N \to \infty. \tag{7.2.11}$$

(ii) *If* $\ell > 1$ *assume that the* ψ-*dependence coefficient is summable and that for all* $m < n$ *the interval* Λ_m *is D-nested in* Λ_n. *Then with probability one (7.2.11) holds true, as well.*

As in most papers on the strong Borel–Cantelli property both Theorems 7.2.2 and 7.2.3 rely on the following basic result.

Theorem 7.2.4 *Let* $\Gamma_1, \Gamma_2, \ldots$ *be a sequence of events such that for any* $N \geq M \geq 1$,

$$\sum_{m,n=M}^{N} (P(\Gamma_m \cap \Gamma_n) - P(\Gamma_m)P(\Gamma_n)) \leq c \sum_{n=M}^{N} P(\Gamma_n) \qquad (7.2.12)$$

where a constant $c > 0$ *does not depend on* M *and* N. *Then for each* $\varepsilon > 0$ *almost surely*

$$S_N = \mathcal{E}_N + O(\mathcal{E}_N^{1/2} \log^{\frac{3}{2}+\varepsilon} \mathcal{E}_N) \qquad (7.2.13)$$

where

$$S_N = \sum_{n=1}^{N} \mathbb{I}_{\Gamma_n} \quad and \quad \mathcal{E}_N = \sum_{n=1}^{N} P(\Gamma_n).$$

In particular, if

$$\mathcal{E}_N \to \infty \quad as \quad N \to \infty$$

then with probability one

$$\lim_{N\to\infty} \frac{S_N}{\mathcal{E}_N} = 1 \quad as \quad N \to \infty.$$

This result (as well as the part of Theorem 7.2.2 for $\ell = 1$ and $q_1(n) = n$) appears already in Theorem 3 from [16] and in a slightly more general (analytic) form it is proved as Lemma 10 in §7 of Ch.1 from [18]. Both sources refer to [17] as the origin of this result.

We observe that Theorem 7.2.3 extends Theorem 2.1 from [3] in several directions. First, for $\ell = 1$ we prove the result for arbitrary ϕ-mixing probability measures with a summable coefficient ϕ on a shift space with a countable alphabet and not just for subshifts of finite type with Gibbs measures. Secondly, the case $\ell > 1$ and rather general functions $q_i(n)$ in place of just $\ell = 1$ and $q_1(n) = n$ were not considered before both in the setups of Theorems 7.2.2 and 7.2.3.

A direct application of Theorem 7.2.3 yields corresponding strong Borel–Cantelli property for dynamical systems which have symbolic representations by means of finite or countable partitions, for instance, hyperbolic dynamical systems (see, for instance, [1]) where sequences of cylinders in Theorem 7.2.3 should be replaced by corresponding sequences of elements of joins of iterates of the partition. By a slight modification (just by considering cylinder sets defined on intervals of nonnegative integers only) Theorem 7.2.3 remains valid for one-sided shifts and then it can be applied to noninvertible dynamical systems having a symbolic representation via their finite or countable partitions such as expanding transformations, the Gauss map of the interval and more general transformations generated by f-expansions (see [8]).

In Sect. 7.6 we apply Theorem 7.2.3 to some limiting problems obtaining a symbolic version of results from [9] which dealt with dynamical systems on $\mathbb{R}^d$ or manifolds and not with shifts. Namely, in the setup of Theorem 7.2.3 introduce the distance between $\omega = (\omega_i)_{i \in \mathbb{Z}}$ and $\tilde{\omega} = (\tilde{\omega}_i)_{i \in \mathbb{Z}}$ from Ω by

$$d(\omega, \tilde{\omega}) = \exp(-\gamma \min\{i \geq 0 : \omega_i \neq \tilde{\omega}_i \text{ or } \omega_{-i} \neq \tilde{\omega}_{-i}\}), \ \gamma > 0. \qquad (7.2.14)$$

Set

$$\Phi_{\tilde{\omega}}(\omega) = -\ln(d(\omega, \tilde{\omega})) \text{ for } \omega, \tilde{\omega} \in \Omega \text{ and} \qquad (7.2.15)$$

$$M_{N,\tilde{\omega}}(\omega) = M_{N,\tilde{\omega}^{(1)},\ldots,\tilde{\omega}^{(\ell)}} = \max_{1 \leq n \leq N} \min_{1 \leq i \leq \ell} (\Phi_{\tilde{\omega}^{(i)}} \circ T^{q_i(n)}(\omega))$$

for some fixed ℓ-tuple $\tilde{\omega} = (\tilde{\omega}^{(1)}, \ldots, \tilde{\omega}^{(\ell)})$, $\tilde{\omega}^{(i)} \in \Omega$, $i = 1, \ldots, \ell$.

Theorem 7.2.5 *Assume that the entropy of the partition into 1-cylinders is finite, i.e.*

$$-\sum_{a \in \mathscr{A}} P([a]) \ln P([a]) < \infty. \qquad (7.2.16)$$

Then, under the conditions of Theorem 7.2.3 for almost all $\tilde{\omega}^{(1)}, \ldots, \tilde{\omega}^{(\ell)} \in \Omega$ with probability one,

$$\frac{M_{N,\tilde{\omega}^{(1)},\ldots,\tilde{\omega}^{(\ell)}}}{\ln N} \to \frac{\gamma}{2\ell h} \text{ as } N \to \infty \qquad (7.2.17)$$

where h is the Kolmogorov–Sinai entropy of the shift T on the probability space $(\Omega, \mathcal{F}, P)$ and, as in Theorem 7.2.3, if $\ell = 1$ we assume only ϕ-mixing with a summable coefficient ϕ and if $\ell > 1$ we assume ψ-mixing with a summable coefficient ψ (and in both cases $h > 0$ by Lemma 3.1 in [12] and Lemma 3.1 in [11]).

In Sect. 7.7 we demonstrate another application of Theorem 7.2.3 deriving the asymptotical behavior of multiple hitting times of shrinking cylinders. Namely, set

$$\tau_{C_n(\tilde{\omega})} = \min\{k \geq 1 : \prod_{i=1}^{\ell} \mathbb{I}_{C_n(\tilde{\omega})} \circ T^{q_i(k)}(\omega) = 1\}$$

where $\omega, \tilde{\omega} \in \Omega$ and $C_n(\omega) = \{\omega = (\omega_i)_{i \in \mathbb{Z}} \in \Omega : \omega_i = \tilde{\omega}_i \text{ provided } |i| \leq n\}$.

Theorem 7.2.6 *Assume that (7.2.16) holds true. Then under the conditions of Theorem 7.2.3 for $P \times P$-almost all pairs $(\omega, \tilde{\omega}) \in \Omega \times \Omega$,*

$$\lim_{n \to \infty} \frac{1}{n} \ln \tau_{C_n(\tilde{\omega})}(\omega) = 2\ell h. \tag{7.2.18}$$

We observe that (7.2.18) was proved in [11] under the ψ-mixing assumption assuming additionally stronger conditions than here while the ϕ-mixing case was not treated there at all. The proof of Theorem 7.2.6 here is different from [11] as it relies on the Borel–Cantelli lemma and the strong Borel–Cantelli property which is an adaptation to our symbolic (and nonconventional) setup of proofs from [5] and [6]. We note that both Theorems 7.2.5 and 7.2.6 remain valid (with essentially the same proof) for one sided shifts just by deleting 2 in (7.2.17) and (7.2.18).

7.3   Proof of Theorem 7.2.2

7.3.1   The Case $\ell = 1$

Let $N \geq M$ and fix an m between M and N. By Assumption 7.2.1 for each k there exists at most K of integers n such that $q(n) - q(m) = k$ where $q(n) = q_1(n)$. If $q(n) - q(m) = k \geq 1$ then by the definition of the ϕ-dependence coefficient

$$|P(\Gamma_{q(m)} \cap \Gamma_{q(n)}) - P(\Gamma_{q(m)})P(\Gamma_{q(n)})| \leq \phi(k)P(\Gamma_{q(m)}). \tag{7.3.1}$$

Hence,

$$\sum_{N \geq n \geq M,\, q(n) > q(m)} |P(\Gamma_{q(m)} \cap \Gamma_{q(n)}) - P(\Gamma_{q(m)})P(\Gamma_{q(n)})| \leq K P(\Gamma_{q(m)}) \sum_{k=1}^{\infty} \phi(k). \tag{7.3.2}$$

Since the coefficient ϕ is summable and that similar inequalities hold true when $q(m) > q(n)$ we conclude that the condition (7.2.12) of Theorem 7.2.4 is satified with $\Gamma_{q(n)}$ in place of Γ_n, $n = 1, 2, \ldots$ there, and so (7.2.8) follows in the case $\ell = 1$ assuming (7.2.7).

7.3.2   The Case $\ell > 1$

We start with the following counting arguments concerning the functions q_i, $i = 1, \ldots, \ell$ satisfying Assumption 7.2.1. Introduce

$$q(n) = \min_{1 \leq i \neq j \leq \ell} |q_i(n) - q_j(n)|.$$

By Assumption 7.2.1(i) for each pair $i \neq j$ and any k there exists at most K nonnegative integers n such that $q_i(n) - q_j(n) = k$, and so

$$\#\{n > 0 : q(n) = k\} < K\ell^2 \tag{7.3.3}$$

where $\#$ stands for "the number of ...". We will need also the following semi-metric between integers $k, l > 0$,

$$\delta(k, l) = \min_{1 \leq i, j \leq \ell} |q_i(k) - q_j(l)|.$$

It follows from Assumption 7.2.1(i) that for any integers $m > 0$ and $k \geq 0$,

$$\#\{n > 0 : \delta(m, n) = k\} < 2K^2\ell^2. \tag{7.3.4}$$

Indeed, the number of m's such that $q_j(m) = q_i(n) - k$ for a fixed i, j, n and k does not exceed K by Assumption 7.2.1(i) and (7.3.4) follows since $1 \leq i, j \leq \ell$.

In order to prove Theorem 7.2.2 for $\ell > 1$ we will estimate first

$$|E(X_m X_n) - EX_m EX_n| = |P(\cap_{i=1}^{\ell}(\Gamma_{q_i(m)} \cap \Gamma_{q_i(n)}))$$
$$- P(\cap_{i=1}^{\ell} \Gamma_{q_i(m)}) P(\cap_{i=1}^{\ell} \Gamma_{q_i(n)})| \tag{7.3.5}$$

where $m, n > 0$ and $X_k = \prod_{i=1}^{\ell} \mathbb{I}_{\Gamma_{q_i(k)}}$. If $\delta(m, n) = k \geq 1$ then by Lemma 3.3 in [11] and the definition of the ψ-dependence coefficient

$$|E(X_m X_n) - EX_m EX_n| \leq 2^{2\ell+2}\psi(k)(2 - (1 + \psi(k))^{\ell}) - 2EX_m EX_n \tag{7.3.6}$$

where we assume, in fact, that k is large enough so that $\psi(k) < 2^{1/\ell} - 1$. Thus, let $k_0 = \min\{k : \psi(k) < 2^{1/\ell} - 1\}$. Then by (7.3.4) and (7.3.6),

$$\sum_{N \geq n \geq M} |E(X_m X_n) - EX_m EX_n| \leq cEX_m \tag{7.3.7}$$

where

$$c = 2K^2\ell^2 \Big(1 + 2^{2\ell+2}(2 - (1 + \psi(k_0))^{\ell})^{-2} \sum_{k=k_0}^{\infty} \psi(k)\Big)$$

where we took into account that

$$|EX_m^2 - (EX_m)^2| \leq EX_m.$$

Summing in (7.3.7) in m between M an N we obtain the condition (7.2.12) of Theorem 7.2.4 with $\cap_{i=1}^{\ell} \Gamma_{q_i(n)}$ in place of Γ_n there. Hence if

$$\sum_{n=1}^{\infty} P(\cap_{i=1}^{\ell} \Gamma_{q_i(n)}) = \infty \qquad (7.3.8)$$

then Theorem 7.2.4 yields that with probability one

$$\frac{S_N}{\tilde{\mathcal{E}}_N} \to 1 \quad \text{as} \quad N \to \infty \qquad (7.3.9)$$

where $\tilde{\mathcal{E}}_N = \sum_{n=1}^{N} P(\cap_{i=1}^{\ell} \Gamma_{q_i(n)})$.

Since we assume (7.2.10) and not (7.3.8), it remains to show that under our conditions,

$$\frac{\tilde{\mathcal{E}}_N}{\mathcal{E}_N} \to 1 \quad \text{as} \quad N \to \infty. \qquad (7.3.10)$$

By Lemma 3.2 from [11] we obtain when $q(n) = k \geq 1$ that

$$\left| P(\cap_{i=1}^{\ell} \Gamma_{q_i(n)}) - \prod_{i=1}^{\ell} P(\Gamma_{q_i(n)}) \right| \leq ((1 + \psi(k))^{\ell} - 1) \prod_{i=1}^{\ell} P(\Gamma_{q_i(n)}). \qquad (7.3.11)$$

For $q(n) = 0$ we estimate the left hand side of (7.3.11) just by 1. Hence, by (7.3.3),

$$|\tilde{\mathcal{E}}_N - \mathcal{E}_N| \leq K\ell^2 + \sum_{n=1, q(n) \geq 1}^{N} \left(((1 + \psi(q(n)))^{\ell} - 1) \prod_{i=1}^{\ell} P(\Gamma_{q_i(n)}) \right) \qquad (7.3.12)$$

$$\leq K\ell^2 + \sum_{n=1, q(n) \geq 1}^{N} ((1 + \psi(q(n)))^{\ell} - 1)$$

$$\leq K\ell^2 + K\ell^2 \sum_{n=1}^{\infty} ((1 + \psi(q(n)))^{\ell} - 1) \leq C < \infty$$

for some constant $C > 0$, since the coefficient ψ is summable. Dividing (7.3.12) by $\mathcal{E}_N$ and taking into account (7.2.10) we obtain (7.3.10) and complete the proof of Theorem 7.2.2.

7.4   Proof of Theorem 7.2.3(i)

Here $\ell = 1$, and so we set $C_n = C_n^{(1)}$ and $q(n) = q_1(n)$. Consider cylinder sets C_m and C_n, $1 \le m < n$ defined on intervals of integers $\Lambda_m = [l_m, r_m]$ and $\Lambda_n = [l_n, r_n]$ with Λ_m right D-nested in Λ_n implying that $r_m < r_n + D$. Let $k = q(n) - q(m)$. By Assumption 7.2.1(i) for each m and k this equality can hold true only for at most K of n's and by Assumption 7.2.1(ii) for no more than K of n's we may have $q(n) \le q(m)$. Next, we can write

$$r_n + q(n) > r_m + q(m) + k - D. \tag{7.4.1}$$

Assume first that

$$l_n + q(n) \le r_m + q(m) \quad \text{and} \quad r_n + q(n) > r_m + q(m). \tag{7.4.2}$$

Let $C_n = [a_{l_n}, a_{l_n+1}, \ldots, a_{r_n}]$ and $\hat{C}_{m,n} = [a_{t_{m,n}}, a_{t_{m,n}+1}, \ldots, a_{r_n}]$ where we assume that $r_n > l_n$,

$$t_{m,n} = s_{m,n} + [\tfrac{1}{2}(r_n - s_{m,n} + 1)] \quad \text{and}$$

$$a_{m,n} = l_n + (r_m + q(m) - l_n - q(n)) + 1 = r_m + q(m) - q(n) + 1.$$

It follows that

$$r_n - t_{m,n} + 1 \ge [\tfrac{1}{2}(k - D)] \quad \text{and} \quad t_{m,n} + q(n) - r_m + q(m) \ge [\tfrac{1}{2}(k - D)] - 1. \tag{7.4.3}$$

Assuming that $k \ge D+4$ we obtain by the definition of the ϕ-dependence coefficient that

$$P(T^{-q(m)}C_m \cap T^{-q(n)}C_n) \le P(T^{-q(m)}C_m \cap T^{-q(n)}\hat{C}_{m,n}) \tag{7.4.4}$$

$$\le P(C_m)P(\hat{C}_{m,n}) + \phi([\tfrac{1}{2}(k - D)] - 1)P(C_m).$$

To make the estimate (7.4.4) suitable for our purposes we recall that according to Lemma 3.1 in [12] there exists $\alpha > 0$ such that any cylinder set C defined on an interval of integers $\Lambda = [l, r]$ satisfies

$$P(C) \le e^{-\alpha(r-l)}, \tag{7.4.5}$$

and so

$$P(\hat{C}_{m,n}) \le \exp(-\alpha([\tfrac{1}{2}(k - D)] - 1)). \tag{7.4.6}$$

In addition to (7.4.4) we can write also

$$P(C_m)P(C_n) \le e^{-\alpha(r_n - l_n)} P(C_m) \le e^{-\alpha(k-D)} P(C_m) \tag{7.4.7}$$

where we used that by (7.4.1),

$$r_n - l_n \ge r_n - s_{m,n} + 1 = r_n + q(n) - r_m - q(m) > k - D.$$

Observe that by Assumption 7.2.1 there exists at most $K(D+1)$ of n's for which $q(n) - q(m) = k \le D$, and so by (7.4.1) the second inequality in (7.4.2) may fail only for at most $K(D+1)$ of n's. For such n's we use the trivial estimate

$$|P(T^{-q(m)}C_m \cap T^{-q(n)}C_n) - P(C_m)P(C_n)| \le P(C_m). \tag{7.4.8}$$

Now if

$$l_n + q(n) > r_m + q(m) \tag{7.4.9}$$

then by the definition of the ϕ-dependence coefficient we can write by (7.4.1) that

$$|P(T^{-q(m)}C_m \cap T^{-q(n)}C_n) - P(C_m)P(C_n)| \tag{7.4.10}$$
$$\le \phi(l_n + q(n) - r_m - q(m))P(C_m) \le \phi(k - D - (r_n - l_n))P(C_m)$$

but this may not suffice for our purposes when $r_n - l_n$ is large. In this case we proceed as in (7.4.4), (7.4.6) and (7.4.7) where we take $\hat{C}_n = [a_{t_n}, a_{t_n+1}, \ldots, a_{r_n}]$ with $t_n = l_n + [\frac{1}{2}(r_n - l_n)] + 1$. Then

$$t_n + q(n) - r_m - q(m) > [\frac{1}{2}(r_n - l_n)] + 1 \quad \text{and} \quad r_n - t_n \ge [\frac{1}{2}(r_n - l_n)] - 1,$$

and so

$$P(T^{-q(m)}C_m \cap T^{-q(n)}C_n) \le P(T^{-q(m)}C_m \cap T^{-q(n)}\hat{C}_n) \tag{7.4.11}$$
$$\le P(C_m)P(\hat{C}_n) + \phi([\frac{1}{2}(r_n - l_n)] + 1)P(C_m)$$
$$\le \left(e^{-\alpha([\frac{1}{2}(r_n - l_n)] - 1)} + \phi([\frac{1}{2}(r_n - l_n)])\right)P(C_m).$$

Thus, when (7.4.9) holds true we use (7.4.10) if $r_n - l_n \le \frac{k-D}{2}$ and (7.4.11) when $r_n - l_n > \frac{k-D}{2}$. In both cases we will obtain the estimate

$$|P(T^{-q(m)}C_m \cap T^{-q(n)}C_n) - P(C_m)P(C_n)| \tag{7.4.12}$$
$$\le \left(e^{-\alpha([\frac{1}{4}(k-D)] - 1)} + \phi([\frac{1}{4}(k-D)])\right)P(C_m).$$

Finally, taking into account that $q(n) - q(m) = k \leq D$ can occur only for at most $K(D+1)$ of n's and for each k the equality $q(n) - q(m) = k$ may hold true for at most K of n's we conclude from (7.4.4), (7.4.6)–(7.4.8), (7.4.12) and from the summability of the coefficient ϕ that for any $m = M, M+1, \ldots, N$,

$$\sum_{n=M}^{N} |P(T^{-q(m)}C_m \cap T^{-q(n)}C_n) - P(C_m)P(C_n)| \leq cP(C_m) \qquad (7.4.13)$$

for some constant $c > 0$ independent of M and N. Summing in m between M and N we conclude that the condition (7.2.12) of Theorem 7.2.4 is satisfied with $\Gamma_n = T^{-q(n)}C_n$, and so assuming (7.2.10) we obtain (7.2.11) completing the proof of Theorem 7.2.3(i).

7.5 Proof of Theorem 7.2.3(ii)

Observe that if $\delta(n, m) = k$, $n > m \geq 0$ and the pair n, m does not belong to the exceptional set $\mathcal{N}$ having cardinality at most K then by Assumption 7.2.1(ii) for some $i_0, j_0 \leq \ell$,

$$q_{j_0} = \max_{1 \leq j \leq \ell} q_j(n) \geq q_{i_0}(m) + k = \max_{1 \leq i \leq \ell} q_i(m) + k. \qquad (7.5.1)$$

Let C_m and C_n be cylinder sets defined on $\Lambda_m = [l_m, r_m]$ and $\Lambda_n = [l_n, r_n]$, respectively. Since C_m is D-nested in C_n, $r_m \leq r_n + D$, and so by (7.5.1),

$$r_m + q_{i_0}(m) \leq r_n + q_{j_0}(n) - k + D. \qquad (7.5.2)$$

Assume first that

$$l_n + q_{j_0}(n) \leq r_m + q_{i_0}(m) \quad and \quad r_n + q_{j_0}(n) > r_m + q_{i_0}(m). \qquad (7.5.3)$$

Let $C_n = [a_{l_n}, a_{l_n+1}, \ldots, a_{r_n}]$ and $\hat{C}_{m,n} = [a_{s_{m,n}}, a_{s_{m,n}+1}, \ldots, a_{r_n}]$ where

$$s_{m,n} = l_n + (r_m + q_{i_0}(m) - l_n - q_{j_0}) + 1 = r_m + q_{i_0}(m) - q_{j_0}(n) + 1, \qquad (7.5.4)$$

and so $\hat{C}_{m,n}$ is defined on the interval $[s_{m,n}, r_n]$ of the length

$$r_n - s_{m,n} + 1 = r_n + q_{j_0}(n) - r_m - q_{i_0}(m) \geq k - D \qquad (7.5.5)$$

where the last inequality follows from (7.5.2). Hence, by the definition of the ψ-dependence coefficient

$$P\big(\cap_{i=1}^{\ell}(T^{-q_i(m)}C_m^{(i)}\cap T^{-q_i(n)}C_n^{(i)})\big) \tag{7.5.6}$$

$$\leq P\big(\cap_{i=1}^{\ell}(T^{-q_i(m)}C_m^{(i)}\cap T^{-q_{j_0}(n)}\hat{C}_{m,n}^{(j_0)})\big)$$

$$\leq (1+\psi(1))P(\cap_{i=1}^{\ell}T^{-q_i(m)}C_m^{(i)})P(T^{-q_{j_0}(n)}\hat{C}_{m,n}^{(j_0)})$$

$$\leq (1+\psi(1))e^{-\alpha(k-D)}P(\cap_{i=1}^{\ell}T^{-q_i(m)}C_m^{(i)})$$

where $\hat{C}_{m,n}^{(j_0)}$ is constructed as above with $C_n = C_n^{(j_0)}$.

We can write also that

$$P(\cap_{i=1}^{\ell}T^{-q_i(m)}C_m^{(i)})P(\cap_{i=1}^{\ell}T^{-q_i(n)}C_n^{(i)}) \leq P(C_n^{(1)})P(\cap_{i=1}^{\ell}T^{-q_i(m)}C_m^{(i)}) \tag{7.5.7}$$

$$\leq e^{-\alpha(r_n-l_n)}P(\cap_{i=1}^{\ell}T^{-q_i(m)}C_m^{(i)}).$$

Since $r_n - l_n \geq r_n - s_{m,n} + 1 \geq k - D$, it follows that under the condition (7.5.3),

$$|P\big(\cap_{i=1}^{\ell}(T^{-q_i(m)}C_m^{(i)}\cap T^{-q_i(n)}C_n^{(i)})\big) \tag{7.5.8}$$

$$- P(\cap_{i=1}^{\ell}T^{-q_i(m)}C_m^{(i)})P(\cap_{i=1}^{\ell}T^{-q_i(n)}C_n^{(i)})|$$

$$\leq (1+\psi(1))e^{-\alpha(k-D)}P(\cap_{i=1}^{\ell}T^{-q_i(m)}C_m^{(i)}).$$

On the other hand, if

$$l_n + q_{j_0}(n) > r_m + q_{i_0}(m), \tag{7.5.9}$$

then by the definition of the ψ-dependence coefficient we obtain similarly to the above that

$$|P\big(\cap_{i=1}^{\ell}(T^{-q_i(m)}C_m^{(i)}\cap T^{-q_i(n)}C_n^{(i)})\big) \tag{7.5.10}$$

$$- P(\cap_{i=1}^{\ell}T^{-q_i(m)}C_m^{(i)})P(\cap_{i=1}^{\ell}T^{-q_i(n)}C_n^{(i)})|$$

$$\leq (1+\psi(l_n + q_{j_0}(n) - r_m - q_{i_0}(m)))P(\cap_{i=1}^{\ell}T^{-q_i(m)}C_m^{(i)})P(C_n^{(1)})$$

$$\leq (1+\psi(1))e^{-\alpha(r_n-l_n)}P(\cap_{i=1}^{\ell}T^{-q_i(m)}C_m^{(i)}).$$

Let a number $d_0 \geq 1$ be such that

$$\psi(d_0) < 2^{1/\ell} - 1 \quad \text{and} \quad k - (r_n - l_n + 2D) > d_0. \tag{7.5.11}$$

Since $r_n - l_n \geq r_m - l_m - 2D$ by D-nesting, it follows by (7.4.5) and Lemma 3.3 from [11] that

$$|P\big(\cap_{i=1}^{\ell} (T^{-q_i(m)} C_m^{(i)} \cap T^{-q_i(n)} C_n^{(i)})\big) \tag{7.5.12}$$

$$- P(\cap_{i=1}^{\ell} T^{-q_i(m)} C_m^{(i)}) P(\cap_{i=1}^{\ell} T^{-q_i(n)} C_n^{(i)})|$$

$$\leq 2^{2\ell+2} \psi(k - \max(r_n - l_n, r_m - l_m))$$

$$\times (2 - (1 + \psi(k - \max(r_n - l_n, r_m - l_m)))^{\ell})^{-2} P(\cap_{i=1}^{\ell} T^{-q_i(m)} C_m^{(i)})$$

$$\times P(\cap_{i=1}^{\ell} T^{-q_i(n)} C_n^{(i)}) \leq 2^{2\ell+2} \psi(k - (r_n - l_n + 2D))(2 - (1 + \psi(d_0))^{\ell})^{-2}$$

$$\times e^{-\alpha(r_n - l_n)} P(\cap_{i=1}^{\ell} T^{-q_i(m)} C_m^{(i)}).$$

Since the cardinality of $\mathcal{N}$ does not exceed K we have

$$\sum_{(n,m)\in\mathcal{N}} |P\big(\cap_{i=1}^{\ell} (T^{-q_i(m)} C_m^{(i)} \cap T^{-q_i(n)} C_n^{(i)})\big) \tag{7.5.13}$$

$$- P(\cap_{i=1}^{\ell} T^{-q_i(m)} C_m^{(i)}) P(\cap_{i=1}^{\ell} T^{-q_i(n)} C_n^{(i)})|$$

$$\leq K P(\cap_{i=1}^{\ell} T^{-q_i(m)} C_m^{(i)}).$$

Next, we estimate now the remaining sum

$$\sum_{n>m,(n,m)\notin\mathcal{N}} |P\big(\cap_{i=1}^{\ell} (T^{-q_i(m)} C_m^{(i)} \cap T^{-q_i(n)} C_n^{(i)})\big) \tag{7.5.14}$$

$$- P(\cap_{i=1}^{\ell} T^{-q_i(m)} C_m^{(i)}) P(\cap_{i=1}^{\ell} T^{-q_i(n)} C_n^{(i)})|.$$

For the part of the sum in n's satisfying (7.5.3) we apply the inequality (7.5.8) which yields the contribution to the total sum estimated using (7.3.4) by

$$2K\ell^2(1 + \psi(1)) P(\cap_{i=1}^{\ell} T^{-q_i(m)} C_m^{(i)}) \sum_{k=0}^{\infty} e^{-\alpha(k-D)} \tag{7.5.15}$$

$$= 2K\ell^2 e^{\alpha D}(1 + \psi(1))(1 - e^{-\alpha})^{-1} P(\cap_{i=1}^{\ell} T^{-q_i(m)} C_m^{(i)}).$$

For the parts of the sum (7.5.14) which correspond to n's satisfying (7.5.9) but not (7.5.11) we obtain that

$$e^{-\alpha(r_n - l_n)} \leq e^{-\alpha k} e^{-\alpha(2D - d_0)}, \tag{7.5.16}$$

and so taking into account (7.3.4) the summation in (7.5.14) over n's satisfying (7.5.9) can be estimated by

$$2K\ell^2(1 + \psi(1)) e^{-\alpha(2D - d_0)} P(\cap_{i=1}^{\ell} T^{-q_i(m)} C_m^{(i)}) \sum_{k=0}^{\infty} e^{-\alpha k} \tag{7.5.17}$$

$$= 2K\ell^2(1 + \psi(1)) e^{-\alpha(2D - d_0)}(1 - e^{-\alpha}) P(\cap_{i=1}^{\ell} T^{-q_i(m)} C_m^{(i)}).$$

It remains to estimate the part of the sum (7.5.14) which corresponds to n's satisfying (7.5.11) where we use (7.5.12). We observe that

$$\psi(k - (r_n - l_n + 2D))e^{-\alpha(r_n - l_n)} \tag{7.5.18}$$

$$= e^{2\alpha D}\psi(k - (r_n - l_n + 2D))e^{-\alpha(r_n - l_n + 2D)}$$

$$\leq e^{2\alpha D} \max(\psi([k/2]), \psi(1)e^{-\alpha[k/2]}) \leq e^{2\alpha D}(\psi([k/2]) + \psi(1)e^{-\alpha[k/2]})$$

since either $r_n - l_n + 2D \geq k/2$ or $k - (r_n - l_n + 2D) \geq k/2$. Both summands in the right hand side of (7.5.18) are summable in k (the first one by the assumption) which gives an estimate for the part of the sum (7.5.14) corresponding to n's satisfying (7.5.11) in the form

$$cP(\cap_{i=1}^{\ell} T^{-q_i(m)} C_m^{(i)}) \tag{7.5.19}$$

where $c > 0$ does not depend on m. By estimates (7.5.8), (7.5.12), (7.5.13), (7.5.15) and (7.5.17)–(7.5.19) above we conclude that the whole sum consisting of the part appearing in (7.5.13) plus the part displayed by (7.5.14) can be estimated by the expression (7.5.19) with another constant $c > 0$ independent of m. It follows that there exists $\tilde{c} > 0$ such that for all $N > M \geq 1$,

$$\sum_{n,m=M}^{N} |P(\cap_{i=1}^{\ell} (T^{-q_i(m)} C_m^{(i)} \cap T^{-q_i(n)} C_n^{(i)})) \tag{7.5.20}$$

$$- P(\cap_{i=1}^{\ell} T^{-q_i(m)} C_m^{(i)}) P(\cap_{i=1}^{\ell} T^{-q_i(n)} C_n^{(i)})|$$

$$\leq \tilde{c} \sum_{m=M}^{N} P(\cap_{i=1}^{\ell} T^{-q_i(m)} C_m^{(i)}).$$

If, in addition,

$$\sum_{m=1}^{\infty} P(\cap_{i=1}^{\ell} T^{-q_i(m)} C_m^{(i)}) = \infty \tag{7.5.21}$$

then by Theorem 7.2.4 we obtain that with probability one

$$\frac{\sum_{n=1}^{N}(\prod_{i=1}^{\ell} \mathbb{I}_{C_n^{(i)}} \circ T^{q_i(n)})}{\sum_{n=1}^{N} P(\cap_{i=1}^{\ell} T^{-q_i(n)} C_n^{(i)})} \to 1 \text{ as } N \to \infty. \tag{7.5.22}$$

It remains to show that under the condition (7.2.10) with probability one,

$$\frac{\sum_{n=1}^{N} P(\cap_{i=1}^{\ell} T^{-q_i(n)} C_n^{(i)})}{\sum_{n=1}^{N} \prod_{i=1}^{\ell} P(C_n^{(i)})} \to 1 \quad \text{as} \quad N \to \infty. \tag{7.5.23}$$

Observe again that

$$P(\cap_{i=1}^{\ell} T^{-q_i(n)} C_n^{(i)}) \leq P(C_n^{(1)}) \leq e^{-\alpha(r_n - l_n)}. \tag{7.5.24}$$

Next, we split the sum in the left hand side of (7.5.21) into two sums

$$S_1 = \sum_{n:\,(r_n - l_n) \leq \frac{2}{\alpha} \ln n} P(\cap_{i=1}^{\ell} T^{-q_i(n)} C_n^{(i)})$$

$$\text{and }\ S_2 = \sum_{n:\,(r_n - l_n) > \frac{2}{\alpha} \ln n} P(\cap_{i=1}^{\ell} T^{-q_i(n)} C_n^{(i)}).$$

By (7.5.24),

$$S_2 \leq \sum_{n=1}^{\infty} n^{-2} < \infty \text{ and also} \qquad \sum_{n:\,(r_n - l_n) > \frac{2}{\alpha} \ln n} \prod_{i=1}^{\ell} P(C_n^{(i)}) < \infty.$$

Hence, it suffices to show that under the condition (7.2.10) with probability one,

$$\frac{\sum_{n \leq N:\,(r_n - l_n) \leq \frac{2}{\alpha} \ln n} P(\cap_{i=1}^{\ell} T^{-q_i(n)} C_n^{(i)})}{\sum_{n \leq N:\,n:\,(r_n - l_n) \leq \frac{2}{\alpha} \ln n} \prod_{i=1}^{\ell} P(C_n^{(i)})} \to 1 \quad \text{as}\quad N \to \infty. \tag{7.5.25}$$

Set $q(n) = \min_{i \neq j} |q_i(n) - q_j(n)|$. Observe that by Assumption 7.2.1(i) for each k,

$$\#\{n :\, q(n) = k\} \leq K\ell^2. \tag{7.5.26}$$

Consider first n's satisfying

$$q(n) \leq r_n - l_n. \tag{7.5.27}$$

In this case by (7.5.24),

$$P(\cap_{i=1}^{\ell} T^{-q_i(n)} C_n^{(i)}) \leq e^{-\alpha q(n)} \tag{7.5.28}$$

and relying on (7.5.26) we conclude that

$$\sum_{n:\,q(n) \leq r_n - l_n} P(\cap_{i=1}^{\ell} T^{-q_i(n)} C_n^{(i)}) \leq K\ell^2 \sum_{k=0}^{\infty} e^{-\alpha k} = K\ell^2 (1 - e^{-\alpha})^{-1}$$

and the same estimate holds true for $\sum_{n:\,q(n) \leq r_n - l_n} \prod_{i=1}^{\ell} P(C_n^{(i)})$. Hence, the sum over such n's does not influence the asymptotical behavior in (7.5.23) and (7.5.25) since the denominators there tend to ∞.

It remains to consider the sums over n's satisfying

$$q(n) > r_n - l_n. \tag{7.5.29}$$

In this case we can apply Lemma 3.2 from [11] to obtain that

$$\left| P(\cap_{i=1}^{\ell} T^{-q_i(n)} C_n^{(i)}) - \prod_{i=1}^{\ell} P(C_n^{(i)}) \right| \tag{7.5.30}$$

$$\leq \left((1 + \psi(q(n) - (r_n - l_n)))^{\ell} - 1 \right) \prod_{i=1}^{\ell} P(C_n^{(i)})$$

$$\leq \left((1 + \psi(q(n) - (r_n - l_n)))^{\ell} - 1 \right) e^{-\ell\alpha(r_n-l_n)}.$$

Now observe that either $r_n - l_n$ or $q(n) - (r_n - l_n)$ is greater or equal to $\frac{1}{2}q(n)$. Denote by $\mathcal{N}_1$ the set of n's for which $r_n - l_n \geq \frac{1}{2}q(n)$ and by $\mathcal{N}_2$ the set of n's for which $q(n) - (r_n - l_n) \geq \frac{1}{2}q(n)$. Taking into account (7.5.26) and (7.5.29) we obtain that

$$\sum_{n \in \mathcal{N}_1} \left((1 + \psi(q(n) - (r_n - l_n)))^{\ell} - 1 \right) e^{-\ell\alpha(r_n-l_n)} \tag{7.5.31}$$

$$\leq ((1 + \psi(1))^{\ell} - 1) \sum_{n \in \mathcal{N}_1} e^{-\frac{1}{2}\ell\alpha q(n)}$$

$$\leq K\ell^2 ((1 + \psi(1))^{\ell} - 1) \sum_{k=0}^{\infty} e^{-\frac{1}{2}\ell\alpha k}$$

$$= K\ell^2 ((1 + \psi(1))^{\ell} - 1)(1 - e^{-\frac{1}{2}\ell\alpha})^{-1} < \infty.$$

Next, taking into account that $\psi(k)$ is summable we see that

$$\sum_{n \in \mathcal{N}_2} \left((1 + \psi(q(n) - (r_n - l_n)))^{\ell} - 1 \right) e^{-\ell\alpha(r_n-l_n)} \tag{7.5.32}$$

$$\leq \sum_{n \in \mathcal{N}_2} \left((1 + \psi(\max(1, [\tfrac{1}{2}q(n)])))^{\ell} - 1 \right)$$

$$\leq 2K\ell^2 \sum_{k=1}^{\infty} ((1 + \psi(k))^{\ell} - 1) = 2K\ell^2 \sum_{k=1}^{\infty} \sum_{m=1}^{\ell} \binom{\ell}{m} (\psi(k))^m < \infty.$$

Hence,

$$\left| \sum_{n=1}^{\infty} \left(P(\cap_{i=1}^{\ell} T^{-q_i(n)} C_n^{(i)}) - \prod_{i=1}^{\ell} P(C_n^{(i)}) \right) \right| < \infty \tag{7.5.33}$$

and since $\sum_{n=1}^{\infty} \prod_{i=1}^{\ell} P(C_n^{(i)}) = \infty$, we obtain (7.5.25), and so (7.5.23), as well, completing the proof of Theorem 7.2.3(ii).

7.6   Asymptotics of Maximums of Logarithmic Distance Functions

In this section we will prove Theorem 7.2.5. Let $\tilde{\omega}^{(j)} = (\tilde{\omega}_i^{(j)})_{i \in \mathbb{Z}} \in \Omega$ and $C_n(\tilde{\omega}^{(j)})$, $j = 1, \ldots, \ell$, $n = 1, 2, \ldots$ be a sequence of cylinder sets such that

$$C_n(\tilde{\omega}^{(j)}) = \{\omega = (\omega_i)_{i \in \mathbb{Z}} \in \Omega : \omega_i = \tilde{\omega}_i^{(j)} \text{ provided } |i| \le r_n\}$$

where $r_n \uparrow \infty$ as $n \uparrow \infty$ is a sequence of integers. Observe that by the Shannon–McMillan–Breiman theorem (see, for instance, [15]) for almost all $\tilde{\omega} \in \Omega$,

$$\lim_{n \to \infty} \frac{1}{2r_n} \ln P(C_n(\tilde{\omega})) = -h \tag{7.6.1}$$

where h is the Kolmogorov–Sinai entropy of the shift T with respect to P since the latter measure is ergodic whether we assume ϕ or ψ-mixing.

Now suppose that

$$\sum_{n=1}^{\infty} \prod_{i=1}^{\ell} P(C_n(\tilde{\omega}^{(i)})) < \infty. \tag{7.6.2}$$

It follows from (7.5.32) that (7.6.2) implies also

$$\sum_{n=1}^{\infty} P(\cap_{i=1}^{\ell} T^{-q_i(n)} C_n(\tilde{\omega}^{(i)})) < \infty \tag{7.6.3}$$

which is, of course, a tautology if $\ell = 1$. It follows from the first Borel–Cantelli lemma that for almost all $\omega \in \Omega$ only finitely many events $\{T^{q_i(n)} \omega \in C_n(\tilde{\omega}^{(i)}), i = 1, \ldots, \ell\}$ can occur. But if the latter event does not hold true then

$$T^{q_j(n)} \notin C_n(\tilde{\omega}^{(j)}) \text{ for some } 1 \le j \le \ell,$$

and so

$$d(T^{q_j(n)}\omega, \tilde{\omega}^{(j)}) > e^{-\gamma r_n} \text{ i.e. } \Phi_{\tilde{\omega}^{(j)}}(T^{q_j(n)}\omega) < \gamma r_n \quad (7.6.4)$$

where the distance $d(\cdot, \cdot)$ and the function Φ were defined in (7.2.14) and (7.2.15). It follows that in this case there exists $N_{\tilde{\omega}}$, $\tilde{\omega} = (\tilde{\omega}^{(1)}, \ldots, \tilde{\omega}^{(\ell)})$ finite with probability one and such that for all $N > N_{\tilde{\omega}}(\omega)$,

$$M_{N,\tilde{\omega}}(\omega) < \gamma r_N,$$

where $M_{N,\tilde{\omega}}(\omega)$ was defined in (7.2.15). Hence,

$$\limsup_{N\to\infty} \frac{M_{N,\tilde{\omega}}}{\ln N} \le \gamma \limsup_{N\to\infty} \frac{r_N}{\ln N} \text{ a.s.} \quad (7.6.5)$$

Next, assume that

$$\mathcal{E}_{N,\tilde{\omega}} = \sum_{n=1}^{N} \prod_{i=1}^{\ell} P(C_n(\tilde{\omega}^{(i)})) \to \infty \text{ as } N \to \infty \quad (7.6.6)$$

which by (7.5.32) implies also that

$$\sum_{n=1}^{\infty} P(\cap_{i=1}^{\ell} T^{-q_i(n)} C_n(\tilde{\omega}^{(i)})) = \infty. \quad (7.6.7)$$

Set

$$L_{n,\tilde{\omega}}(\omega) = \max\{m \le n : T^{q_i(m)}\omega \in C_m(\tilde{\omega}^{(i)}) \text{ for } i = 1, \ldots, \ell\}.$$

It follows from Theorem 7.2.3 that under (7.6.6) for almost all $\omega \in \Omega$,

$$L_{n,\tilde{\omega}}(\omega) \to \infty \text{ as } n \to \infty.$$

Observe also that

$$S_N(\omega) = \sum_{n=1}^{N} (\prod_{i=1}^{\ell} \mathbb{I}_{C_n(\tilde{\omega}^{(i)})} \circ T^{q_i(n)}(\omega)) = S_{L_{n,\tilde{\omega}}(\omega)}. \quad (7.6.8)$$

By (7.4.13), (7.5.19) and (7.5.32) we can use (7.2.13) which yields that for almost all $\omega \in \Omega$,

$$0 \le \mathcal{E}_{N,\tilde{\omega}} - \mathcal{E}_{L_{n,\tilde{\omega}},\tilde{\omega}} \le O(\mathcal{E}_{N,\tilde{\omega}}^{1/2} \ln^{\frac{3}{2}+\varepsilon} \mathcal{E}_{N,\tilde{\omega}}), \quad (7.6.9)$$

and so for almost all ω,

$$\lim_{N\to\infty} \frac{\mathcal{E}_{L_{n,\tilde{\omega}}(\omega),\tilde{\omega}}}{\mathcal{E}_{N,\tilde{\omega}}} = 1. \tag{7.6.10}$$

Next, observe that if $m = L_{n,\tilde{\omega}}(\omega)$ then for each $i = 1, \ldots, \ell$,

$$d(T^{q_i(m)}\omega, \tilde{\omega}^{(i)}) \leq e^{-\gamma r_m} \text{ i.e. } \Phi_{\tilde{\omega}^{(i)}}(T^{q_i(m)}\omega) \geq \gamma r_m,$$

and so $M_{m,\tilde{\omega}}(\omega) \geq \gamma r_m$. It follows that

$$M_{N,\tilde{\omega}}(\omega) \geq M_{L_{N,\tilde{\omega}}(\omega)}(\omega) \geq \gamma r_{L_{N,\tilde{\omega}}(\omega)}, \tag{7.6.11}$$

and so

$$\liminf_{N\to\infty} \frac{M_{N,\tilde{\omega}}(\omega)}{\ln N} \geq \gamma (\liminf_{N\to\infty} \frac{r_N}{\ln N}) \liminf_{N\to\infty} \frac{\ln L_{N,\tilde{\omega}}(\omega)}{\ln N}. \tag{7.6.12}$$

Next, in order to complete the proof of Theorem 7.2.5, we will choose sequences r_n, $n = 1, 2, \ldots$ for appropriate upper and lower bounds. For the upper bound we will take $r_n = [\frac{1+\delta}{2\ell h} \ln n]$ for some $\delta > 0$. Then by (7.6.1) for almost all $\tilde{\omega}^{(1)}, \ldots, \tilde{\omega}^{(\ell)} \in \Omega$,

$$\ln \prod_{i=1}^{\ell} P(C_n(\tilde{\omega}^{(i)})) \sim -(1+\delta)\ln n \quad \text{as} \quad n \to \infty,$$

and so the series (7.6.2) converges as needed. Substituting such r_N's to (7.6.5) and letting $\delta \to 0$ we obtain

$$\limsup_{N\to\infty} \frac{M_{N,\tilde{\omega}}}{\ln N} \leq \frac{\gamma}{2\ell h} \quad \text{a.s.} \tag{7.6.13}$$

Now we deal with the lower bound choosing $r_n = [\frac{1-\delta}{2\ell h} \ln n]$. Then by (7.6.1) for almost all $\tilde{\omega}^{(1)}, \ldots, \tilde{\omega}^{(\ell)} \in \Omega$ as $n \to \infty$,

$$\ln \prod_{i=1}^{\ell} P(C_n(\tilde{\omega}^{(i)})) \sim -(1-\delta)\ln n, \tag{7.6.14}$$

and so the series (7.6.6) diverges as needed. For such r_N's we have that

$$\liminf_{N\to\infty} \frac{r_N}{\ln N} = \frac{1-\delta}{2\ell h} \tag{7.6.15}$$

and letting $\delta \to 0$ the proof of Theorem 7.2.5 will be completed by (7.6.12), (7.6.13) and (7.6.15) once we show that for almost all $\omega \in \Omega$,

$$\liminf_{N \to \infty} \frac{\ln L_{N,\tilde{\omega}}(\omega)}{\ln N} = 1. \tag{7.6.16}$$

By (7.6.14) there exists a random variable $n(\tilde{\omega}) < \infty$ a.s. such that if $n \geq n(\tilde{\omega})$ then

$$n^{-(1-\frac{3}{4}\delta)} \leq \prod_{i=1}^{\ell} P(C_n(\tilde{\omega}^{(i)})) \leq n^{-(1-\frac{4}{3}\delta)}. \tag{7.6.17}$$

If $L_{N,\tilde{\omega}}(\omega) \geq n(\omega)$ then we obtain from (7.6.9) and (7.6.17) that

$$\frac{4}{3\delta}(N^{\frac{3}{4}\delta} - (L_{N,\tilde{\omega}}(\omega) + 1)^{\frac{3}{4}\delta}) \tag{7.6.18}$$

$$\leq \sum_{n=L_{N,\tilde{\omega}}(\omega)+1}^{N} n^{-(1-\frac{3}{4}\delta)}$$

$$\leq O\Big((n(\omega) + \sum_{n=n(\omega)}^{N} n^{-(1-\frac{4}{3}\delta)})^{1/2} \ln^{\frac{3}{2}+\varepsilon}(n(\omega) + \sum_{n=n(\omega)}^{N} n^{-(1-\frac{4}{3}\delta)})\Big)$$

$$\leq O\Big(n(\omega) + \frac{3}{4\delta}N^{\frac{4}{3}\delta})^{1/2} \ln^{\frac{3}{2}+\varepsilon}(n(\omega) + \frac{3}{4\delta}N^{\frac{4}{3}\delta})\Big).$$

Dividing these inequalities by $N^{\frac{3}{4}\delta}$, letting $N \to \infty$ and taking into account that $N \geq L_{N,\tilde{\omega}}(\omega)$ by the definition, we see that

$$\frac{L_{N,\tilde{\omega}}(\omega)}{N} \to 1, \quad \text{and so } \ln N - \ln L_{N,\tilde{\omega}}(\omega) \to 0 \text{ a.s. as } N \to \infty$$

implying (7.6.16) and completing the proof of Theorem 7.2.5.   □

7.7   Asymptotics of Hitting Times

In this section we will prove Theorem 7.2.6 deriving first that for $P \times P$-almost all pairs $(\omega, \tilde{\omega})$,

$$\liminf_{n \to \infty} \frac{1}{n} \ln \tau_{C_n(\tilde{\omega})} \geq 2\ell h. \tag{7.7.1}$$

Let $\lambda k \leq n \leq \lambda(k+1)$ for some $\lambda > 0$. Then

$$\frac{\ln \tau_{C_{\lambda k}(\tilde{\omega})}}{\lambda(k+1)} \leq \frac{\ln \tau_{C_n(\tilde{\omega})}}{n} \leq \frac{\ln \tau_{C_{\lambda(k+1)}(\tilde{\omega})}}{\lambda k},$$

and so

$$\liminf_{n \to \infty} \frac{\ln \tau_{C_n(\tilde{\omega})}}{n} = \liminf_{k \to \infty} \frac{\ln \tau_{C_{\lambda k}(\tilde{\omega})}}{\lambda k} \qquad (7.7.2)$$

where we alert the reader that the definition of the cylinder $C_n(\tilde{\omega})$ here agrees with the corresponding definition in Sect. 7.6 provided $r_n = n$ there.

Next, assume that $\lambda > (2\ell h)^{-1}$ and set

$$I_k(\tilde{\omega}) = \cup_{j=1}^{e^k} \cap_{i=1}^{\ell} T^{-q_i(j)} C_{\lambda k}(\tilde{\omega}).$$

Then

$$P(I_k(\tilde{\omega})) \leq \sum_{j=1}^{e^k} P(\cap_{i=1}^{\ell} T^{-q_i(j)} C_{\lambda k}(\tilde{\omega})) \qquad (7.7.3)$$

and we are going to show that for P-almost all $\tilde{\omega}$,

$$\sum_{k=1}^{\infty} \sum_{j=1}^{e^k} P(\cap_{i=1}^{\ell} T^{-q_i(j)} C_{\lambda k}(\tilde{\omega})) < \infty. \qquad (7.7.4)$$

Indeed, applying the Shannon-McMillan-Breiman theorem we obtain that for P-almost all $\tilde{\omega}$ and each $\varepsilon > 0$ there exists $k(\varepsilon, \tilde{\omega})$ such that if $k \geq k(\varepsilon, \tilde{\omega})$ then

$$P(C_{\lambda k}(\tilde{\omega})) \leq \exp(-k(2\lambda h - \varepsilon)). \qquad (7.7.5)$$

When $\ell = 1$ we employ (7.7.5) for $k \geq k(\varepsilon, \tilde{\omega})$ and (7.4.5) for $k < k(\varepsilon, \tilde{\omega})$ which yields the estimate of the left hand side of (7.7.4) by

$$\sum_{1 \leq k \leq k(\varepsilon, \tilde{\omega})} e^k e^{-\alpha(2\lambda k - 1)} + \sum_{k=1}^{\infty} e^{-k(2\lambda h - \varepsilon - 1)}. \qquad (7.7.6)$$

The first sum in (7.7.6) contains finitely many terms, and so it is bounded, while the second sum in (7.7.6) is also bounded since $2\lambda h - \varepsilon > 1$ by the choice of λ provided $\varepsilon > 0$ is small enough.

Next, we will deal with the case $\ell > 1$. First, recall the notation $q(n) = \min_{i \neq j} |q_i(n) - q_j(n)|$ and observe that by (7.5.6),

$$\#\{n : q(n) \leq 2\lambda k + 2\} \leq 2(\lambda k + 1)K\ell^2. \tag{7.7.7}$$

Now we split the sum in the left hand side of (7.7.4) into two sums

$$S_1 = \sum_{k=1}^{\infty} \sum_{j:j \leq e^k, \, q(j) \leq 2\lambda k + 2} P(\cap_{i=1}^{\ell} T^{-q_i(j)} C_{\lambda k}(\tilde{\omega})) \tag{7.7.8}$$

$$\leq 2K\ell^2 \sum_{k=1}^{\infty}(\lambda k + 1)P(C_{\lambda k}(\tilde{\omega})) \leq 2K\ell^2 \sum_{k=1}^{\infty}(\lambda k + 1)e^{-2\alpha(\lambda k - 1)} < \infty,$$

where we use (7.4.5), and

$$S_2 = \sum_{k=1}^{\infty} \sum_{j:j \leq e^k, \, q(j) > 2\lambda k + 2} P(\cap_{i=1}^{\ell} T^{-q_i(j)} C_{\lambda k}(\tilde{\omega})) \tag{7.7.9}$$

$$\leq k(\varepsilon, \tilde{\omega})e^{k(\varepsilon, \tilde{\omega})} + \sum_{k=k(\varepsilon, \tilde{\omega})}^{\infty} \sum_{j:j \leq e^k, \, q(j) > 2\lambda k + 2} P(\cap_{i=1}^{\ell} T^{-q_i(j)} C_{\lambda k}(\tilde{\omega})).$$

If $q(j) > 2\lambda k + 2$ and $k \geq k(\varepsilon, \tilde{\omega})$ then employing Lemma 3.2 from [11] and (7.7.5) above we obtain

$$P(\cap_{i=1}^{\ell} T^{-q_i(j)} C_{\lambda k}(\tilde{\omega})) \leq (1 + \psi(1))^{\ell}(P(C_{\lambda k}(\tilde{\omega})))^{\ell}$$

$$\leq (1 + \psi(1))^{\ell} \exp(-k(2\lambda h\ell - \varepsilon\ell)) \tag{7.7.10}$$

where ψ is the dependence coefficient from (7.2.3). For $\varepsilon > 0$ small enough $2\lambda h\ell - \varepsilon\ell > 1$ by the choice of λ, and so by (7.7.9) and (7.7.10),

$$S_2 \leq k(\varepsilon, \tilde{\omega})e^{k(\varepsilon, \tilde{\omega})} + \sum_{k=1}^{\infty} \exp(-k(2\lambda h\ell - \varepsilon\ell - 1)) < \infty$$

which together with (7.7.8) yields (7.7.4).

Hence, by the (first) Borel–Cantelli lemma there exists $K(\omega) = K(\omega, \tilde{\omega}) < \infty$ a.s. such that for all $k \geq K(\omega)$ there are no events

$$T^{q_i(j)}\omega \in C_{\alpha k}(\tilde{\omega}) \text{ for all } i = 1, \ldots, \ell \text{ and some } 1 \leq j \leq e^k.$$

It follows that for P-almost all ω and $k \geq K(\omega)$.

$$\tau_{C_{\alpha k}(\tilde{\omega})}(\omega) > e^k.$$

This together with (7.7.2) yields that for $P \times P$-almost all pairs $(\omega, \tilde{\omega})$,

$$\liminf_{n \to \infty} \frac{\ln \tau_{C_n(\tilde{\omega})}(\omega)}{n} \geq \lambda^{-1}. \qquad (7.7.11)$$

Since λ can be chosen arbitrarily close to $(2\ell h)^{-1}$ we obtain (7.7.1).

Next, we will prove that for $P \times P$-almost all pairs $(\omega, \tilde{\omega})$,

$$\limsup_{n \to \infty} \frac{\ln \tau_{C_n(\tilde{\omega})}(\omega)}{n} \leq 2\ell h. \qquad (7.7.12)$$

Choose $\varepsilon' > \varepsilon > 0$ small and $\beta > 0$ close to $(2\ell h)^{-1}$ so that

$$\beta(2\ell h + \varepsilon) < 1 \text{ and } \beta(2\ell h + \varepsilon') - \frac{1 - \beta(2\ell h - \varepsilon)}{1 - \beta(2\ell h + \varepsilon)} > 0 \qquad (7.7.13)$$

which implies, in particular, that $\beta(2\ell h + \varepsilon') > 1$.

Set

$$\Gamma = \{(\omega, \tilde{\omega}) \in \Omega : \limsup_{n \to \infty} \frac{\ln \tau_{C_n(\tilde{\omega})}(\omega)}{n} > 2\ell h + \varepsilon'\}.$$

If $(\omega, \tilde{\omega}) \in \Gamma$ then for infinitely many n's,

$$\tau_{C_{\beta \ln n}(\tilde{\omega})}(\omega) > n^{\beta(2\ell h + \varepsilon')}. \qquad (7.7.14)$$

For n's satisfying (7.7.14),

$$\omega \notin \bigcup_{1 \leq j \leq n^{\beta(2\ell h + \varepsilon')}} \bigcap_{1 \leq i \leq \ell} T^{-q_i(j)} C_{\beta \ln n}(\tilde{\omega}) \supset \bigcup_{n \leq j \leq n^{\beta(2\ell h + \varepsilon')}} \bigcap_{1 \leq i \leq \ell} T^{-q_i(j)} C_{\beta \ln j}(\tilde{\omega})$$

which implies that there exists a sequence $n_k \to \infty$ as $k \to \infty$ such that

$$\sum_{1 \leq j \leq n_k} \prod_{1 \leq i \leq \ell} \mathbb{I}_{C_{\beta \ln j}(\tilde{\omega})} \circ T^{q_i(j)}(\omega) = \sum_{1 \leq j \leq n_k^{\beta(2\ell h + \varepsilon')}} \prod_{1 \leq i \leq \ell} \mathbb{I}_{C_{\beta \ln j}(\tilde{\omega})} \circ T^{q_i(j)}(\omega)$$

$$(7.7.15)$$

for each k.

By the Shannon–McMillan–Breiman theorem there are $\tilde{\Omega} \subset \Omega$ with $P(\tilde{\Omega}) = 1$ and a random variable J finite on $\tilde{\Omega}$ such that for any $j \geq J(\tilde{\omega})$,

$$j^{-\beta(2h + \frac{\varepsilon}{\ell})} = e^{-(2h + \frac{\varepsilon}{\ell})\beta \ln j} \leq P(C_{\beta \ln j}(\tilde{\omega})) < e^{-(2h - \frac{\varepsilon}{\ell})\beta \ln j} = j^{-\beta(2h - \frac{\varepsilon}{\ell})}.$$

Hence, there are random variables k_1 and k_2 such that for all n large enough,

$$k_1(\tilde{\omega}) n^{1-\beta(2\ell h+\varepsilon)} \leq \sum_{j=1}^{n} \left(P(C_{\beta \ln j}(\tilde{\omega}))\right)^{\ell} \leq k_2(\tilde{\omega}) n^{1-\beta(2\ell h-\varepsilon)}. \tag{7.7.16}$$

It follows that for $(\omega, \tilde{\omega}) \in \Gamma$, $\tilde{\omega} \in \tilde{\Omega}$ and all k large enough

$$\frac{\sum_{1 \leq j \leq n_k} \left(P(C_{\beta \ln j}(\tilde{\omega}))\right)^{\ell}}{\sum_{1 \leq j \leq n_k^{\beta(2\ell h+\varepsilon')}} \left(P(C_{\beta \ln j}(\tilde{\omega}))\right)^{\ell}} \tag{7.7.17}$$

$$\leq \frac{k_2(\tilde{\omega})}{k_1(\tilde{\omega})} n_k^{(1-\beta(2\ell h-\varepsilon)-\beta(2\ell h+\varepsilon')(1-\beta(2\ell h+\varepsilon))} \to 0 \text{ as } k \to \infty$$

since by the choice of ε, ε' and β,

$$(1 - \beta(2\ell h - \varepsilon) - \beta(2\ell h + \varepsilon')(1 - \beta(2\ell h + \varepsilon))$$

$$= (1 - \beta(2\ell h + \varepsilon)(\frac{1 - \beta(2\ell h - \varepsilon)}{1 - \beta(2\ell h + \varepsilon)} - \beta(2\ell h + \varepsilon') < 0.$$

By (7.7.15) we obtain from (7.7.17) that

$$\frac{\sum_{1 \leq j \leq n_k} \prod_{1 \leq i \leq \ell} \mathbb{I}_{C_{\beta \ln j}(\tilde{\omega})} \circ T^{q_i(j)}(\omega)}{\sum_{1 \leq j \leq n_k} \left(P(C_{\beta \ln j}(\tilde{\omega}))\right)^{\ell}} \tag{7.7.18}$$

$$\times \frac{\sum_{1 \leq j \leq n_k^{\beta(2\ell h+\varepsilon')}} \left(P(C_{\beta \ln j}(\tilde{\omega}))\right)^{\ell}}{\sum_{1 \leq j \leq n_k^{\beta(2\ell h+\varepsilon')}} \prod_{1 \leq i \leq \ell} \mathbb{I}_{C_{\beta \ln j}(\tilde{\omega})} \circ T^{q_i(j)}(\omega)} \to \infty \text{ as } k \to \infty.$$

By (7.7.16) for all $\tilde{\omega} \in \tilde{\Omega}$,

$$\sum_{1 \leq j \leq n} \left(P(C_{\beta \ln j}(\tilde{\omega}))\right)^{\ell} \to \infty \text{ as } n \to \infty,$$

and so by Theorem 7.2.3 for P-almost all ω,

$$\frac{\sum_{1 \leq j \leq n} \prod_{1 \leq i \leq \ell} \mathbb{I}_{C_{\beta \ln j}(\tilde{\omega})} \circ T^{q_i(j)}(\omega)}{\sum_{1 \leq j \leq n} \left(P(C_{\beta \ln j}(\tilde{\omega}))\right)^{\ell}} \to 1 \text{ as } n \to \infty.$$

Thus, (7.7.18) can hold true only for a set of pairs $(\omega, \tilde{\omega})$ having $P \times P$-measure zero, and so $P \times P(\Gamma) = 0$. Since ε and ε' can be chosen arbitrarily close to zero, (7.7.12) follows for $P \times P$-almost all $(\omega, \tilde{\omega})$, which together with (7.7.1) completes the proof of Theorem 7.2.6.

References

1. R. Bowen, *Equilibrium States and the Ergodic Theory of Anosov Diffeomorphisms*, Lecture Notes in Mathematics, vol. 470 (Springer, Berlin, 1975)
2. R.C. Bradley, *Introduction to Strong Mixing Conditions* (Kendrick Press, Heber City, 2007)
3. N. Chernov, D. Kleinbock. Dynamical Borel–Cantelli lemmas for Gibbs measures. Israel J. Math. **122**, 1–27 (2001)
4. J. Dedecker, F. Merlevéde, E. Rio, Criteria for Borel–Cantelli lemmas with applications to Markov chains and dynamical systems, Preprint, HAL-02088063. arXiv: 1904.01850
5. S. Galatolo, Dimension via waiting time and recurrence. Math. Res. Lett. **12**, 377–386 (2005)
6. S. Galatolo, Dimension and hitting time in rapidly mixing systems. Math. Res. Lett. **14**, 797–805 (2007)
7. N. Haydn, M. Nicol, T. Persson, S. Vaienti, A note on Borel–Cantelli lemmas for non-uniformly hyperbolic dynamical systems. Ergod. Theory Dyn. Syst. **33**, 475–498 (2013)
8. L. Heinrich, Mixing properties and central limit theorem for a class of non-identical piecewise monotonic C^2-transformations. Math. Nachricht. **181**, 185–214 (1996)
9. M.P. Holland, M. Nicol, A. Török, Almost sure convergence of maxima for chaotic dynamical systems. Stoch. Proc. Appl. **126**, 3145–3170 (2016)
10. C. Gupta, A Borel–Cantelli lemma for nonuniformly expanding systems. Nonlinearity **23**, 1991–2008 (2010)
11. Yu. Kifer, A. Rapaport, Poisson and compound Poisson approximations in conventional and nonconventional setups. Probab. Th. Relat. Fields **160**, 797–831 (2014)
12. Yu. Kifer, F. Yang, Geometric law for numbers of returns until a hazard under ϕ-mixing. Israel J. Math. **12**, (2019, to appear). arXiv: 1812.09927
13. P. Lévy, *Théorie de l'addition des variables aléatoirs. XVII* (Gauthier-Villars, Paris, 1937)
14. I. Melbourne, M. Nicol, Almost sure invariance principle for nonuniformly hyperbolic systems. Commun. Math. Phys. **260**, 131–146 (2005)
15. K. Petersen, *Ergodic Theory* (Cambridge University Press, Cambridge, 1983)
16. W. Philipp, Some metrical theorems in number theory. Pacific J. Math. **20**, 109–127 (1967)
17. W.M. Schmidt, Metrical theorems on fractional parts of sequences. Trans. Amer. Math. Soc. **110**, 493–518 (1964)
18. V.G. Spindzuk, *Metric Theory of Diophantine Approximations* (V.H.Winston, Washington, 1979)

Chapter 8
Application of the Convergence of the Spatio-Temporal Processes for Visits to Small Sets

Françoise Pène and Benoît Saussol

Abstract The goal of this article is to point out the importance of spatio-temporal processes in different questions of quantitative recurrence. We focus on applications to the study of the number of visits to a small set before the first visit to another set (question arising from a previous work by Kifer and Rapaport), the study of high records, the study of line processes, the study of the time spent by a flow in a small set. We illustrate these applications by results on billiards or geodesic flows. This paper contains in particular new result of convergence in distribution of the spatio temporal processes associated to visits by the Sinai billiard flow to a small neighbourhood of arbitrary points in the billiard domain.

8.1 Introduction

Let $(\Omega, \mathcal{F}, \mu, T)$ or $(\Omega, \mathcal{F}, \mu, Y = (Y_t)_{t \geq 0})$ be a probability preserving dynamical system in discrete or continuous times. Let $(A_\varepsilon)_{\varepsilon > 0}$ be a family of measurable subsets of Ω with $\mu(A_\varepsilon) \to 0+$ as $\varepsilon \to 0$. Given a family $(h_\varepsilon)_{\varepsilon > 0}$ of positive real numbers and a family $(H_\varepsilon)_{\varepsilon > 0}$ of measurable normalization functions $H_\varepsilon : A_\varepsilon \to V$ where V is a locally compact metric space endowed with its Borel σ-algebra $\mathcal{V}$, we study the family of spatio-temporal point processes $(\mathcal{N}_\varepsilon)_{\varepsilon > 0}$ on $[0, +\infty) \times V$ given by

$$\mathcal{N}_\varepsilon(x) := \mathcal{N}(T, A_\varepsilon, h_\varepsilon, H_\varepsilon) := \sum_{n \geq 1 \,:\, T^n(x) \in A_\varepsilon} \delta_{(nh_\varepsilon, H_\varepsilon(T^n(x)))} \quad \text{for a map } T$$

$$(8.1.1)$$

F. Pène (✉) · B. Saussol
Univ Brest, Université de Brest, LMBA, Laboratoire de Mathématiques de Bretagne Atlantique, CNRS UMR 6205, Brest, France
e-mail: francoise.pene@univ-brest.fr; benoit.saussol@univ-brest.fr

© The Author(s), under exclusive license to Springer Nature Switzerland AG 2021
M. Pollicott, S. Vaienti (eds.), *Thermodynamic Formalism*, Lecture Notes
in Mathematics 2290, https://doi.org/10.1007/978-3-030-74863-0_8

or

$$\mathcal{N}_\varepsilon(x) := \mathcal{N}(Y, A_\varepsilon, h_\varepsilon, H_\varepsilon) = \sum_{t>0 \,:\, Y_t \text{ enters } A_\varepsilon} \delta_{(th_\varepsilon, H_\varepsilon(Y_t(x)))} \quad \text{for a flow } Y.$$

$$(8.1.2)$$

We are interested in results of convergence in distribution of $(\mathcal{N}_\varepsilon)_{\varepsilon>0}$ to a point process $\mathcal{P}$ as $\varepsilon \to 0$ with a particular focus on applications of results of such kind. Various results of convergence of such processes to Poisson point processes have been proved in [14, 21] for billiard maps and flows.

Let us point out the fact that these spatio-temporal processes contain a lot of information: they do not only contain information on the visit time but they also contain informations on the spatial position at these visit times. For these reasons, one may extract further information from results of convergence of these processes. Among the applications that have already been studied, let us mention:

- Study of the visits to a small neighborhood of an hyperbolic periodic point of a transformation (see [21, Section 5], with application to Anosov maps).
  Such visits occurs by clusters (once a point visits such a neighbourhood, it stays close to the periodic point during an unbounded time before living this area). The idea we used to study these clusters was to consider a process $\mathcal{N}_\varepsilon$ corresponding to the last (or first) position of the clusters.
- Convergence of a normalized Birkhoff sum processes

$$\left(\left(n^{-\frac{1}{\alpha}} \sum_{k=0}^{\lfloor nt \rfloor - 1} f \circ T^k \right)_{t \geq 0} \right)_{n \geq 1}$$

to an α-stable process. In [25] Tyran-Kamińska provided criteria ensuring such a result. One of the conditions is the convergence of

$$\mathcal{N}_{1/n} = \mathcal{N}(T, \{|f| > \gamma n^{\frac{1}{\alpha}}\}, 1/n, n^{-\frac{1}{\alpha}} f(\cdot))$$

(for every $\gamma > 0$) to some Poisson point process. The general results of [21] combined with the criteria of [25] have been used in [14] to prove convergence to a Lévy process for the Birkhoff sum process of Hölder observable of billiards in dispersing domains with cusps.

We won't detail again the above applications. Our goal here is to emphasize on further ones.

After recalling in Sect. 8.2 below the general results of convergence of spatio-temporal point processes to Poisson point processes established in [21], we present in the remaining sections four other important applications of such convergence results:

- The number of visits to (or of the time spent in) a small set before the first visit to a second small set (motivated by Kifer and Rapaport [16]), with application to the Sinai billiard flow with finite horizon,
- The evolution of the number of records larger than some threshold, with an application to billiards with corners and cusps of order larger than 2,
- The Line process of random geodesics (motivated by Athreya, Lalley, Sapir and Wroten [2]),
- The time spent by a flow in a small set, with application to the Sinai billiard flow with finite horizon.

Appendix contains a new theorem of convergence of point processes for the Sinai billiard flow and for neighborhoods of arbitrary positions in the billiard domain, which is used in the examples that illustrate the applications above. Finally we also present an application to the closest approach by the billiard flow.

8.2   Convergence Results for Transformations and Special Flows

We set $E := [0, +\infty) \times V$ and we endow it with its Borel σ-algebra $\mathcal{E} = \mathcal{B}([0, +\infty)) \otimes \mathcal{V}$. We also consider the family of measures $(m_\varepsilon)_{\varepsilon>0}$ on $(V, \mathcal{V})$ defined by

$$m_\varepsilon := \mu(H_\varepsilon^{-1}(\cdot)|A_\varepsilon) \tag{8.2.1}$$

and a family $\mathcal{W}$ closed under finite unions and intersections of relatively compact open subsets of V, that generates the σ-algebra $\mathcal{V}$. Let λ be the Lebesgue measure on $[0, \infty)$.

We will approximate the point process defined by (8.1.1) or (8.1.2) by a Poisson point process on E. Given a σ-finite measure η on $(E, \mathcal{E})$, recall that a process $\mathcal{N}$ is a Poisson point process on E of intensity η if

(i) $\mathcal{N}$ is a point process (i.e. $\mathcal{N} = \sum_i \delta_{x_i}$ with x_i being E-valued random variables),
(ii) For every pairwise disjoint Borel sets $B_1, \ldots, B_n \subset E$, the random variables $\mathcal{N}(B_1), \ldots, \mathcal{N}(B_n)$ are independent Poisson random variables with respective parameters $\eta(B_1), \ldots, \eta(B_n)$.

Let $M_p(E)$ be the space of all point measures defined on E, endowed with the topology of vague convergence; it is metrizable as a complete separable metric space. A family of point processes $(\mathcal{N}_\varepsilon)_{\varepsilon>0}$ converges in distribution to $\mathcal{N}$ if for any bounded continuous function $f : M_p(E) \to \mathbb{R}$ the following convergence holds true

$$\mathbb{E}(f(\mathcal{N}_\varepsilon)) \to \mathbb{E}(f(\mathcal{N})), \quad \text{as } \varepsilon \to 0. \qquad (8.2.2)$$

For a collection $\mathcal{A}$ of measurable subsets of Ω, we define the following quantity:

$$\Delta(\mathcal{A}) := \sup_{A \in \mathcal{A}, B \in \sigma(\cup_{n=1}^\infty T^{-n}\mathcal{A})} |\mu(A \cap B) - \mu(A)\mu(B)|. \qquad (8.2.3)$$

We set λ for the Lebesgue measure on $[0, \infty)$.

Theorem 8.2.1 (Convergence Result for Transformations [21, Theorem 2.1])
We assume that

(i) for any finite subset $\mathcal{W}_0$ of $\mathcal{W}$ we have $\Delta(H_\varepsilon^{-1}\mathcal{W}_0) = o(\mu(A_\varepsilon))$,
(ii) there exists a measure m on $(V, \mathcal{V})$ such that for every $F \in \mathcal{W}$, $m(\partial F) = 0$
and $\lim_{\varepsilon \to 0} \mu(H_\varepsilon^{-1}(F)|A_\varepsilon)$ converges to $m(F)$.

Then the family of point processes $(\mathcal{N}_\varepsilon)_{\varepsilon>0}$ converges strongly[1] in distribution, as $\varepsilon \to 0$, to a Poisson point process $\mathcal{P}$ of intensity $\lambda \times m$.

In particular, for every relatively compact open $B \subset E$ such that $(\lambda \times m)(\partial B) = 0$, $(\mathcal{N}_\varepsilon(B))_{\varepsilon>0}$ converges in distribution, as $\varepsilon \to 0$, to a Poisson random variable with the parameter $(\lambda \times m)(B)$.

Let us explain roughly the strategy used in [21] to apply Theorem 8.2.1. First the measure m appears as the limit of $(\mu(H_\varepsilon^{-1}(\cdot)|A_\varepsilon))_{\varepsilon>0}$. Second, we construct $\mathcal{W}$ as the union of finer and finer finite partitions of $\mathcal{V}$ with boundary neglectable with respect to m. Finally we obtain (i) as a consequence of some decorrelation result combined with the neglectability of fast returns.

Theorem 8.2.2 (Convergence Result for Special Flows [21, Theorem 2.3])
Assume $(\Omega, \mu, Y = (Y_t)_t)$ can be represented as a special flow over a probability preserving dynamical system (M, ν, F) with a roof function $\tau : M \to (0, +\infty)$ with $M \subset \Omega$ and set $\Pi : \Omega \to M$ for the projection such that $\Pi(Y_s(x)) = x$ for all $x \in M$ and all $s \in [0, \tau(x))$.

Assume moreover that Y enters A_ε at most once between two consecutive visits to M and that there exists a family of measurable normalization functions $G_\varepsilon : M \to V$ such that the family of point processes $(\mathcal{N}(F, \Pi(A_\varepsilon), h_\varepsilon, G_\varepsilon))_{\varepsilon>0}$ converges in distribution, as $\varepsilon \to 0$ and with respect to some probability measure $\tilde\nu \ll \nu$, to a Poisson point process of intensity $\lambda \times m$, where m is some measure on $(V, \mathcal{V})$. Then the family of point processes $(\mathcal{N}(Y, A_\varepsilon, h_\varepsilon/\mathbb{E}_\nu[\tau], G_\varepsilon \circ \Pi))_{\varepsilon>0}$ converges in distribution, as $\varepsilon \to 0$ (with respect to any probability measure absolutely continuous with respect to μ), to a Poisson process $\mathcal{P}$ of intensity $\lambda \times m$.

---

[1]I.e. with respect to any probability measure absolutely continuous w.r.t. μ.

8.3 Number of Visits to a Small Set Before the First Visit to a Second Small Set

Suppose B_ε^0 and B_ε^1 are two disjoint sets. We define the spatio-temporal process $\mathcal{N}_\varepsilon$ with $A_\varepsilon = B_\varepsilon^0 \cup B_\varepsilon^1$, $H_\varepsilon(x) = \ell$ if $x \in B_\varepsilon^\ell$, $\ell = 0, 1$, that is on $[0, +\infty) \times \{0, 1\}$

$$\mathcal{N}_\varepsilon(x) = \sum_{n=1}^\infty \sum_{\ell=0}^1 \delta_{(n\mu(A_\varepsilon),\ell)} 1_{B_\varepsilon^\ell}(T^n x) \qquad (8.3.1)$$

in the case of a transformation T or

$$\mathcal{N}_\varepsilon(x) = \sum_{t>0} \sum_{\ell=0}^1 \delta_{(th_\varepsilon,\ell)} 1_{Y_t \text{ enters } B_\varepsilon^\ell} \qquad (8.3.2)$$

in the case of a flow Y. In [16] Kifer and Rapaport studied the distribution of a (multiple) event $T^n x \in B_\varepsilon^1$ until a (multiple) hazard $T^n(x) \in B_\varepsilon^0$. We stick here to single event and hazard and define, in the case of a transformation T,

$$\mathcal{M}_\varepsilon(x) := \sum_{n=1}^{\tau_{B_\varepsilon^0}(x)} 1_{B_\varepsilon^1}(T^n x), \qquad (8.3.3)$$

where we set $\tau_B(x) := \inf\{n \geq 1 : T^n(x) \in B\}$ or, in the case of a flow Y:

$$\mathcal{M}_\varepsilon(x) := \sum_{t \in (0, \tau_{B_\varepsilon^0}(x))} 1_{Y_t \text{ enters } B_{\varepsilon^1}}, \qquad (8.3.4)$$

where we set $\tau_B(x) := \inf\{t > 0 : Y_t(x) \in B\}$. The process $\mathcal{M}_\varepsilon$ counts the number of entrances of the flow in the 1-set before its first visit to the 0-set.

In the case of a flow, it is also natural to consider the following process $\mathcal{M}_\varepsilon'$ measuring the time spent by the flow in the 1-set before its first visit to the 0-set:

$$\mathcal{M}_\varepsilon'(x) := \int_0^{\tau_{B_\varepsilon^0}(x)} 1_{B_\varepsilon^1} \circ Y_s(x) \, ds. \qquad (8.3.5)$$

In view of the study of this last process, we will consider the following process measuring the time spent by the flow in each set:

$$\left(\mathcal{L}_\varepsilon := \sum_{j=0}^1 \sum_{t : Y_t \text{ enters } B_\varepsilon^j} \delta_{th_\varepsilon, j, a_\varepsilon D_{B_\varepsilon^j} \circ Y_t} \right)_{\varepsilon > 0}$$

with $D_A := \tau_{\Omega \setminus A}$.

Theorem 8.3.1 *Let $p \in (0, 1)$ and $\mathbb{P}$ be a probability measure on Ω. Assume, in the case of a flow, that $\lim_{\varepsilon \to 0} \mathbb{P}(B_\varepsilon^0 \cup B_\varepsilon^1) = 0$.*

If the spatio-temporal process $\mathcal{N}_\varepsilon$ defined as in (8.3.1) or (8.3.2) converges, with respect to $\mathbb{P}$, to a PPP of intensity $\lambda \times \mathcal{B}(p)$ where $\mathcal{B}(p)$ denotes the Bernoulli measure with parameter p (for a transformation we expect $p = \lim_{\varepsilon \to 0} \mu(B_\varepsilon^1)/\mu(A_\varepsilon)$), then the process $(\mathcal{M}_\varepsilon)_{\varepsilon > 0}$ has asymptotically geometric distribution, more precisely it converges in distribution to $\mathcal{M}$ with $\mathbb{P}(\mathcal{M} = k) = p^k(1 - p)$ for any $k \geq 0$; in particular the asymptotic value for the commitor *function is*

$$\lim_{\varepsilon \to 0} \mathbb{P}(\tau_{B_\varepsilon^0} < \tau_{B_\varepsilon^1}) = \lim_{\varepsilon \to 0} \mathbb{P}(\mathcal{M}_\varepsilon = 0) = 1 - p.$$

In the case of a flow, if $(a_\varepsilon \tau_{\Omega \setminus B_\varepsilon^1})_{\varepsilon > 0}$ converges in probability $\mathbb{P}$ to 0 and if $(\mathcal{L}_\varepsilon)_{\varepsilon > 0}$ supported on $[0, +\infty) \times \{0, 1\} \times \bar{\mathbb{R}}_+$ converges in distribution with respect to $\mathbb{P}$ to a PPP $\mathcal{L}_0$ with intensity $\lambda \times \sum_{j=0}^{1} p_j(\delta_j \times m'_j)$ where the m'_j are probability measures, then $(a_\varepsilon \mathcal{M}'_\varepsilon)_{\varepsilon > 0}$ converges to $\sum_{i=1}^{\mathcal{M}} X_i$ where $(X_i)_i$ is a sequence of i.i.d. random variables with distribution m'_1 and independent of $\mathcal{M}$ where $\mathcal{M}$ is as above.

Proof We first observe that the mapping

$$J : \xi \in M_p([0, +\infty) \times \{0, 1\}) \mapsto \xi([0, \tau^0] \times \{1\})$$

is continuous where $\tau^0 = \sup\{t \geq 0 : \xi([0, t] \times \{0\}) = 0\}$ is continuous at a.e. realization ξ of $\chi := PPP(\lambda \times \mathcal{B}(p))$. Indeed, $\xi(\cdot \times \{0\})$ and $\xi(\cdot \times \{1\})$ are the realization of two homogeneous independent Poisson process hence τ^0 is a.s. not an atom of $\xi(\cdot \times \{1\})$. Observe that, in the case of a transformation, $\mathcal{M}_\varepsilon = J(\mathcal{N}_\varepsilon)$ and in the case of a flow $\mathbb{P}(\mathcal{M}_\varepsilon \neq J(\mathcal{N}_\varepsilon)) = \mathbb{P}(Y_0 \in B_\varepsilon^0 \cup B_\varepsilon^1) \to 0$. Therefore, by the continuous mapping theorem, $(\mathcal{M}_\varepsilon)_{\varepsilon > 0}$ converges in distribution to $G := J(\chi)$ as ε goes to 0.

We now compute the law of G. The first hazard τ^0 has an exponential distribution with parameter $1 - p$, while $\chi^1(\cdot) := \chi(\cdot \times \{1\})$ is a Poisson point process with intensity $p\lambda$, and the two are independent. Therefore, for any $k \in \mathbb{N}$

$$\mathbb{P}(G = k) = \mathbb{P}(\chi^1([0, \tau^0]) = k)$$

$$= \int_0^\infty e^{-pt} \frac{(pt)^k}{k!}(1 - p)e^{-(1-p)t} \, dt = (1 - p)p^k.$$

This ends the proof of the first part of the Theorem. Let us now prove the last one. We use the fact that the mapping $J : \xi \in M_p([0, +\infty) \times \{0, 1\} \times \bar{\mathbb{R}}_+) \mapsto \int_{[0, \tau^0] \times \{1\} \times [0, K_0]} z \, d\xi(t, j, z)$ is continuous at a.e. realization ξ of χ and conclude the proof as above by the continuous mapping theorem and the Slutsky theorem since $a_\varepsilon \mathcal{M}'_\varepsilon = \mathbf{1}_{\{Y_0 \notin B_\varepsilon^0\}} \left(J(\mathcal{L}_\varepsilon) + a_\varepsilon \tau_{\Omega \setminus B_\varepsilon^1} \right)$. $\square$

Example 8.1 Consider the billiard flow $(Y_t)_t$ associated to a Sinai billiard with finite horizon in a domain $Q \subset \mathbb{T}^2$ (see Appendix for details). Let $\mathbb{P}$ be any probability measure on $\Omega := Q \times S^1$ absolutely continuous with respect to Lebesgue. We fix two distinct point positions $q_0, q_1 \in Q$ and two positive real numbers $r_0, r_1 > 0$. Set $B^i_\varepsilon := B(q_i, r_i\varepsilon) \times S^1$ and $d_i = 2 - \mathbf{1}_{q_i \in \partial Q}$.

Then $(\mathcal{M}_\varepsilon)_{\varepsilon>0}$ converges in distribution with respect to $\mathbb{P}$ to $\mathcal{M}$ with $\mathbb{P}(\mathcal{M} = k) = p^k(1 - p)$ for any $k \geq 0$ and with $p = \frac{d_1 r_1}{d_0 r_0 + d_1 r_1}$.

Moreover $(\varepsilon^{-1}\mathcal{M}'_\varepsilon)_{\varepsilon>0}$ converges in distribution with respect to $\mathbb{P}$ to $2r_1 \sum_{i=1}^{\mathcal{M}} Y_i$ where $(Y_i)_{i\geq 1}$ is a sequence of i.i.d. random variables with density $y \mapsto \frac{y}{\sqrt{1-y^2}}\mathbf{1}_{[0,1]}(y)$ independent of $\mathcal{M}$, with $\mathcal{M}$ as above.

Proof Recall that the billiard flow Y preserves the normalized Lebesgue measure μ on $Q \times S^1$. In view of applying Theorem 8.3.1, observe first that $\lim_{\varepsilon \to 0} \mathbb{P}(B^0_\varepsilon \cup B^1_\varepsilon) = 0$ and $\mathbb{E}[\varepsilon^{-1}\tau_{\Omega \setminus B^1_\varepsilon}] \leq 2r_1\mathbb{P}(B^1_\varepsilon)$, thus $(\varepsilon\tau_{\Omega \setminus B^1_\varepsilon})_{\varepsilon>0}$ converges in probability $\mathbb{P}$ to 0.

As a direct consequence of Theorem 8.6.2, the family of spatio-temporal processes $(\mathcal{N}_\varepsilon)_{\varepsilon>0}$ given by (8.3.2), with $h_\varepsilon = \frac{(d_0 r_0 + d_1 r_1)\varepsilon}{Area(Q)}$, converges in distribution to a PPP of intensity $\lambda \times \mathcal{B}(\frac{d_1 r_1}{d_0 r_0 + d_1 r_1})$ and so the first conclusion of Theorem 8.3.1 holds true with $p = \frac{d_1 r_1}{d_0 r_0 + d_1 r_1}$. This ends the proof of the convergence of $(\mathcal{M}_\varepsilon)_{\varepsilon>0}$. Due to Theorem 8.6.1, $(\mathcal{L}_\varepsilon)_{\varepsilon>0}$ (with $a_\varepsilon = \varepsilon$ and h_ε as before) converges in distribution to a PPP with intensity $\lambda \times \sum_{j=0}^{1} p_j(\delta_j \times m'_j)$ where $p_j := \frac{d_j r_j}{d_0 r_0 + d_1 r_1}$ and where m'_j has density $y \mapsto \frac{y}{2r_j\sqrt{4r_j^2 - y^2}}\mathbf{1}_{[0,2r_j]}(y)$. Thus the last conclusion of Theorem 8.3.1 holds also true with these notations. We conclude by taking $Y_i = X_i/(2r_1)$.

8.4 Number of High Records

We define the high records point process by

$$\mathcal{R}_f(u, \ell) = \sum_{k=1}^{\infty} \delta_{ku}\mathbf{1}_{\{f\circ T^k > \max(\ell, f, \dots, f\circ T^{k-1})\}} .$$

The successive times of records of an observable along an orbit are obviously tractable from the time and values of the observations along this orbit. The following proposition states that this is still the case for the corresponding asymptotic distributions. This has already been noticed in [13], in particular in the context of Extremal events. Our result is similar to the proof of [13, Theorem 3.1] from [13, Theorem 5.1].

Proposition 8.1 *Let $(\Omega, \mathcal{F}, \mu, T)$ be a probability preserving dynamical system and $f : \Omega \to [0, +\infty)$ be a measurable function. Assume the family $\big(\mathcal{N}_\varepsilon = \mathcal{N}(T, \{f > \varepsilon^{-1}\}, h_\varepsilon, 1/(\varepsilon f))\big)_{\varepsilon>0}$ of point processes on $[0, +\infty) \times [0, 1]$ converges in distribution with respect to $\mathcal{P}$ to a Poisson point process of intensity $\lambda \times$*

m with m a probability measure on $[0, 1]$ *without any atom. Then* $\left(\mathcal{R}_f(h_\varepsilon, \varepsilon^{-1})\right)_{\varepsilon>0}$
converges in distribution, as $\varepsilon \to 0$ *to a Point process* $\mathcal{R} = \sum_{\ell=1}^{\infty} Z_\ell \delta_{T_\ell}$ *where*
$T_\ell = \sum_{i=1}^{\ell} X_i$, *the* X_i *are independent standard exponential random variable*
and the Z_ℓ *are independent random variable having Bernoulli distribution with*
respective parameters ℓ^{-1}, *and the two sequences are independent.*

Proof Define the mapping

$$F : \xi = \sum_i \delta_{(t_i, v_i)} \in M_p([0, \infty) \times [0, 1]) \mapsto \sum_{i \in I(\xi)} \delta_{t_i},$$

where $I(\xi)$ are the records of ξ, defined by those i such that for any j one has
$t_j < t_i \implies v_j > v_i$. The map F is continuous at each ξ such that the t_i's, and
the v_i's, are distincts. This is the case for a.e. realization ξ of a Poisson process
of intensity $\lambda \times m$. Therefore by the continuous mapping theorem $\mathcal{R}_f(h_\varepsilon, \varepsilon^{-1}) =
F(\mathcal{N}_\varepsilon)$ converges to $\chi = F(PPP(\lambda \times m))$.

We are left to compute the distribution. Observe that $PPP(\lambda \times m)$ is distributed
as $\sum_{\ell=1}^{\infty} \delta_{(T_\ell, W_\ell)}$ with (T_ℓ) as in the statement and the W_ℓ are i.i.d. with distribution
m, the two sequences being independent. Let $Z_\ell = 1_{\{W_\ell \text{ is a record}\}}$. By Resnick [23,
Proposition 4.3] the Z_ℓ are independent, have probability $1/\ell$, and when $Z_\ell = 1$ we
keep the point T_ℓ.   □

In particular, for every $t > 0$ the number of records exceeding the value ε^{-1} before
the time th_ε^{-1} corresponds to $\mathcal{R}_f(h_\varepsilon, \varepsilon^{-1})([0, t])$ and the conclusion of Proposition 8.1 implies that it converges to $\sum_{\ell=1}^{N_t} Z_\ell$ where Z_ℓ are as in Proposition 8.1 and
where $(N_s)_{s \geq 0}$ is a standard Poisson Process independent of $(Z_\ell)_{\ell \geq 1}$.

Example 8.2 Consider a dispersive billiard with corners and cusps of maximal order
$\beta_* > 2$ as in [14]. Consider the induced system (Ω, μ, T) corresponding to the
successive reflection times outside a neighbourhood $\mathcal{U}$ of cusps and write $R(x)$ for
the number of reflections in $\mathcal{U}$ starting from x. Set $\alpha = \frac{\beta_*}{\beta_* - 1} \in (1, 2)$.

Setting $A_\varepsilon := \{R \circ T^{-1} > \varepsilon^{-1}\}$, it has been proved in [14, Lemma 4.5] that there
exists an explicit $c_0 > 0$ such that $\mu(A_\varepsilon) \sim c_0 \varepsilon^\alpha$ as $\varepsilon \to 0$.

The assumptions of Proposition 8.1 hold true with $f = R \circ T^{-1}$ and $h_\varepsilon = \mu(A_\varepsilon) \sim c_0 \varepsilon^\alpha$. So the same assumptions hold true with $h_\varepsilon = c_0 \varepsilon^\alpha$.

Furthermore the number R_n of records of R higher than $n^{1/\alpha}$ before the n-th
reflection outside cusps converges to $\sum_{\ell=1}^{N} Z_\ell$ where Z_ℓ are as in Proposition 8.1
and where N is a Poisson random variable of parameter c_0 and independent of
$(Z_\ell)_{\ell \geq 1}$.

Proof It follows from the proof of [14, Lemma 4.8] that[2] the family of point
processes $(\mathcal{N}(T, A_\varepsilon, \mu(A_\varepsilon), \varepsilon R \circ T^{-1}))_{\varepsilon>0}$ on $[0, +\infty) \times [1, +\infty]$ converges in

---

[2]Jung et al. [14, Lemma 4.8] states that this convergence is true in the set of point processes on
$[0, +\infty) \times [1, +\infty)$, but its proof can be adapted in a straightforward way to obtain our purpose
by considering not only intervals of the form (c, c') but also intervals of the form $(c, +\infty]$.

distribution to a PPP with intensity of density $(t, y) \mapsto \alpha y^{-\alpha-1} \mathbf{1}_{y>0}$ with respect to the Lebesgue measure.

Therefore the assumptions of Proposition 8.1 hold true with $f = R \circ T^{-1}$ and $h_\varepsilon = \mu(A_\varepsilon) \sim c_0 \varepsilon^\alpha$. So the same assumptions hold true with $h_\varepsilon = c_0 \varepsilon^\alpha$. This ends the proof of the first part.

For the second one, we apply Proposition 8.1 with $\varepsilon = n^{-\frac{1}{\alpha}}$. $\square$

8.5 Line Process of Random Geodesics

We study the line process generated by a geodesic as in [2] and recover their main result. Let N be a compact Riemannian surface of negative curvature. The geodesic flow $(Y_t)_t$ on the unit tangent bundle $\Omega = T^1 N$ preserves the Liouville measure μ. Let $\pi_N : T^1 N \to N$ be the canonical projection $(q, v) \mapsto q$. We denote by $D(q, \varepsilon)$ the ball in N of radius ε. We now state the main theorem, postponing the details and precise definitions thereafter.

Theorem 8.5.1 *Fix $q_0 \in N$. For any $a > 0$, the intersection of the neighborhood $D(q_0, \varepsilon)$ with the geodesic segment $\pi_N(\{Y_t(x), 0 \le t \le a\varepsilon^{-1}\})$, where x is taken at random on (Ω, μ), converges in distribution, after normalization, as $\varepsilon \to 0$, to a Homogeneous Poisson line process in the unit disk of intensity $a/Area(N)$.*

A Poisson line process in the unit disk D of the plane, of intensity $\kappa \in (0, \infty)$, is a probabilistic process which draw lines in the disk. Each line L is parametrized by $(r, \theta) \in [-1, 1] \times [0, \pi]$ where

$$L = \{(x, y) \in D : r = x \cos\theta + y \sin\theta\},$$

and the parameters (r, θ) are produced by a Poisson point process of intensity $\frac{\kappa}{\pi} dr d\theta$ on $[-1, 1] \times [0, \pi]$. Equivalently, changing the parametrization to (s, φ) where $s \in \partial D =: S$ is one point of intersection of the line with the unit circle and φ is the angle between the line L (directed into the disk) and the normal at s pointing inside the disk (see Fig. 8.1), gives a Poisson point process of intensity $\frac{\kappa \cos\varphi}{2\pi} ds d\varphi$ (the jacobian is $\cos\varphi$ and each line has two representations in this parametrization).

Fig. 8.1 Parametrization of the line L by (r, θ) or (s, φ)

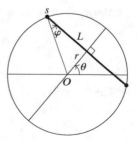

Fig. 8.2 A geodesic arc γ
entering the ball $D(q_0, \varepsilon)$

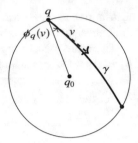

The intensity κ in the theorem is equal to $a/Area(N)$, therefore the intensity
in this parametrization will be $\frac{a}{2\pi Area(N)} \cos\varphi ds d\varphi = \frac{a}{Vol(T^1N)} \cos\varphi ds d\varphi$. The
convergence of a point process in this parametrization implies it in the original one
(by continuity of the change of parameter; see [23, Proposition 3.18]).

The exponential map $\exp_{q_0}$ is a local diffeomorphism on a neighborhood $U \subset$
$T_{q_0}N$ of 0. Thus its inverse is well defined on $D(q_0, \varepsilon)$ for ε small enough so that
$B(0, \varepsilon) \subset U$. We identify $T_{q_0}N$ with $\mathbb{R}^2$. Set $V = S \times [-\frac{\pi}{2}, \frac{\pi}{2}]$. For $q \in D(q_0, \varepsilon)$
we let $s_\varepsilon(q) = \varepsilon^{-1} \exp_{q_0}^{-1}(q)$ and for $q \in \partial D(q_0, \varepsilon)$ and $v \in T_q N$ we denote by
$\phi_q(v)$ the angle between the normal at q pointing inside the disk and v (see Fig. 8.2).

The intersection $\mathcal{I}_\varepsilon^a(x) := \pi_N(Y_{[0,a\varepsilon^{-1}]}(x)) \cap D(q_0, \varepsilon)$ consists of finitely many
geodesic arcs $\gamma_i := \pi_N(Y_{[t_i, t_i + \ell_i]}(x))$, where ℓ_i is the length of the arc; we drop
the dependence on x and ε for simplicity. The arcs γ_i are fully crossing the ball,
except possibly for the two extremities (at $t = 0$ or $t = a\varepsilon^{-1}$) which could give an
incomplete arc. The later happens with a vanishing probability as $\varepsilon \to 0$, therefore
we will ignore this eventuality. The arc γ_i enters the ball at the position q_i with
direction v_i where $(q_i, v_i) := Y_{t_i}(x)$.

When $\varepsilon \to 0$, the geodesic arcs γ_i which compose the intersection $\mathcal{I}_\varepsilon^a$
become more and more straight. This justifies the definition of the convergence in
distribution of $\mathcal{I}_\varepsilon^a$ as the convergence in distribution of the point process

$$\sum_i \delta_{(s_\varepsilon(q_i), \phi_q(v_i))}. \tag{8.5.1}$$

Loosely speaking, we identify the images $s_\varepsilon(\gamma_i)$ with the chord of the unit disk D
originated in $s_\varepsilon(q_i)$ and direction v_i.

We now proceed with the proof of the theorem. Let $\mathcal{A}_\varepsilon \subset T^1N$ be the set of
points (q, v) such that $q \in \partial D(q_0, \varepsilon)$ and v is pointing inside the ball. We define on
$\mathcal{A}_\varepsilon$

$$\mathcal{H}_\varepsilon(q, v) = (s_\varepsilon(q), \phi_q(v)) \in V. \tag{8.5.2}$$

The theorem is a byproduct of the following result for the geodesic flow.

Proposition 8.2 *The process of entrances in the ball for the position for the geodesic flow* $N(Y, \mathcal{A}_\varepsilon, 2\varepsilon/Area(N), \mathcal{H}_\varepsilon)$ *on* $[0, +\infty[\times V$ *converges to a Poisson point process with intensity* $\frac{1}{4\pi} \cos \varphi \, dt \, ds \, d\varphi$.

Proof of Theorem 8.5.1 The counting process

$$\mathcal{L}_\varepsilon^a(\cdot) := N(Y, \mathcal{A}_\varepsilon, 2\varepsilon/Area(N), \mathcal{H}_\varepsilon)([0, 2a/Area(N)] \times \cdot) \qquad (8.5.3)$$

produces a point (s, φ) each time that the geodesic flow Y_t enters in $D(q_0, \varepsilon)$ for some t such that $2\varepsilon t/Area(N) \le 2a/Area(N)$, that is $t \le a\varepsilon^{-1}$. By Proposition 8.2 and the continuous mapping theorem the point process $\mathcal{L}_\varepsilon^a$ converges to a Poisson point process of intensity $\frac{2a}{Area(N)} \frac{1}{4\pi} \cos \varphi \, ds \, d\varphi$. By the above discussion, in particular (8.5.1), this completes the proof of the theorem. □

We emphasize that this proof only uses the convergence stated in Proposition 8.2, therefore it applies for more general 'geodesic-like flows', for instance the argument applies immediately to billiards systems, using Theorem 8.6.2 in place of Proposition 8.2.

Proof of Proposition 8.2 The first step is to construct a Markov section for the geodesic flow, subordinated to a finite family of disks $D_i \subset T^1 N$. Fix some $\delta > 0$ sufficiently small. By Bowen [3] there exists a Markov section $(X_i)_i$ of size δ, in particular $\operatorname{diam} X_i < \delta$ and $T^1 N = \cup_i Y_{[-\delta,0]}(X_i)$. One can choose the disks $D_i \supset X_i$ in such a way that

$$D_i \subset \{(q, v) : q \in Q_i, |\angle(n_q, v)| > \frac{\pi}{2} - \delta\}$$

where Q_i are C^2 curve in N and n_q is the normal vector to Q_i at q (with $q \mapsto n_q$ continuous). Without loss of generality we assume that $q_0 \notin \cup_i Q_i$.

The flow (Y_t) is represented by a special flow over the Poincaré section $M := \cup_i X_i$, with a C^2 roof function τ. Let Π be the projection onto M along the flow in backward time. The flow $(T^1 N, (Y_t), \mu)$ projects down to a system (M, F, ν), conjugated to a subshift of finite type with a Gibbs measure of a Hölder potential. In order to apply Theorem 8.2.2 we need to check that the set $A_\varepsilon := \Pi \mathcal{A}_\varepsilon$ and $H_\varepsilon(x) := \mathcal{H}_\varepsilon(Y_s(x))$ where $s > 0$ is the minimal time such that $Y_s(x) \in \mathcal{A}_\varepsilon$ fulfill the hypotheses of Theorem 8.2.1. For that we will apply [21, Proposition 3.2]. The Poincaré map F has a hyperbolic structure with an exponential rate, thus it satisfies the setting of [21, Proposition 3.2] with any polynomial rate α, in particular $\alpha = 4$ works. Here the boundary is meant in the induced topology on M. It suffices to prove that for some $p_\varepsilon = o(\nu(A_\varepsilon))$ one has (i) $\nu(\tau_{A_\varepsilon} \le p_\varepsilon) = o(1)$ and (ii) $\nu((\partial A_\varepsilon)^{[p_\varepsilon^{-\alpha}]}) = o(\nu(A_\varepsilon))$, the two other assumptions being trivially satisfied in our situation.

Measure of A_ε: The Liouville measure μ is the product of the normalized surface on N times the Haar measure on $T^1 N$. Its projection ν to the Poincaré section satisfies $d\nu = c_\nu \cos \varphi \, dr \, d\varphi$ for some normalizing constant

$c_\nu = \left(\sum_i \int_{X_i} \cos\varphi dr d\varphi \right)^{-1}$, where r is the curvilinear abscissa on Q_i and φ the angle between the velocity and the normal to Q_i. Moreover we have $d\mu = (\int_M \tau d\nu)^{-1} d\nu \times dt|_{M_\tau}$ where $M_\tau = \{(x,t) : x \in M, 0 \leq t < \tau(x)\}$.

The geodesic flow preserves the measure $\cos\varphi dr d\varphi$ from $A_\varepsilon \subset M$ to $\mathcal{A}_\varepsilon$, therefore

$$\nu(A_\varepsilon) = c_\nu \int_{A_\varepsilon} \cos\varphi dr d\varphi = c_\nu \int_{\mathcal{A}_\varepsilon} \cos\varphi dr d\varphi = c_\nu \int_{\partial D(q_0,\varepsilon)} dr \int_{-\pi/2}^{\pi/2} \cos\varphi d\varphi$$

$$\sim c_\nu 4\pi\varepsilon.$$

Short returns: For any $q \in D(q_0, \varepsilon)$, let $R_\varepsilon(q)$ be the set of $v \in T_q^1 N$ such that the geodesic segment $\gamma_{[0,\varepsilon^{-1/2}]}(q,v)$ enters again $D(q_0, \varepsilon)$ after leaving $D(q_0, 2\varepsilon)$. The result of [2, Lemma 5.3] ensures the existence of $K > 0$ such that for any $q \in D(q_0, \varepsilon)$

$$Leb(R_\varepsilon(q)) \leq K\varepsilon^{1-1/2} = K\sqrt{\varepsilon}.$$

Therefore, setting $\hat{\mathcal{A}}_\varepsilon = \{(q,v) \in \mathcal{A}_\varepsilon : v \in R_\varepsilon(q)\}$ we get that the two dimensional Lebesgue measure of $\hat{\mathcal{A}}_\varepsilon$ is $O(\varepsilon^{3/2})$. A fortiori since the projection Π preserves the measure $\cos\varphi dr d\varphi$ we get

$$\nu(\Pi\hat{\mathcal{A}}_\varepsilon) = c_\nu \int_{\Pi\hat{\mathcal{A}}_\varepsilon} \cos\varphi dr d\varphi = c_\nu \int_{\hat{\mathcal{A}}_\varepsilon} \cos\varphi dr d\varphi = O(\varepsilon^{3/2}).$$

Let $p_\varepsilon = \lfloor (\max \tau)^{-1} \varepsilon^{-1/2} \rfloor$ and notice that $A_\varepsilon \cap \{\tau_{A_\varepsilon} \leq p_\varepsilon\} \subset \Pi\hat{\mathcal{A}}_\varepsilon$. By the previous estimates we get

$$\nu(A_\varepsilon \cap \{\tau_{A_\varepsilon} \leq p_\varepsilon\}) = O(\varepsilon^{3/2}).$$

Hence

$$\nu(\tau_{A_\varepsilon} \leq p_\varepsilon | A_\varepsilon) = o(1).$$

This is the assumption (i).

We now prove (ii). The boundary of A_ε in the induced topology of M is included in the set of $\Pi(q,v)$ where v is tangent to the boundary of $\partial D(q_0, \varepsilon)$. This defines for each i such that $X_i \cap A_\varepsilon$ is nonempty and contains at most two C^2 curves in D_i of finite length (by transversality), therefore its ε^2-neighborhood has a measure $O(\varepsilon^2)$.

Finally, the measure $dm_\varepsilon = (H_\varepsilon)_* \nu(\cdot|A_\varepsilon)$ is equal to the measure $dm := \frac{1}{4\pi} \cos\varphi ds d\varphi$, since the measure $\cos\varphi dr d\varphi$ is preserved by the inverse of the projection Π from A_ε to $\mathcal{A}_\varepsilon$ and H_ε has constant jacobian ε in these coordinates. By Theorem 8.2.1 the point process $N(F, A_\varepsilon, \nu(A_\varepsilon), H_\varepsilon)$ converges to a Poisson point process of intensity $\lambda \times m$. Applying Theorem 8.2.2 with $h_\varepsilon = c_\nu 4\pi\varepsilon$ and $h'_\varepsilon = h_\varepsilon/E_\nu(\tau)$ we get that $\mathcal{N}(Y, \mathcal{A}_\varepsilon, h'_\varepsilon, \mathcal{H}_\varepsilon)$ converges to a Poisson point process of intensity $\lambda \times m$. In addition,

$$\int_M \tau d\nu = c_\nu \int_M \tau \cos\varphi dr d\varphi = c_\nu \int_{M_\tau} \cos\varphi dt dr d\varphi = c_\nu Vol(T^1 N).$$

Thus, since $Vol(T^1 N) = 2\pi Area(N)$ we get that $h'_\varepsilon = \frac{2\varepsilon}{Area(N)}$, proving the proposition.   □

8.6 Time Spent by a Flow in a Small Set

Given a flow $Y = (Y_t)_t$ defined on Ω and a set $A \subset \Omega$, a very natural question is to study the time spent by the flow in the set A, that is the local time $L_T(A)$ given by following quantity :

$$L_T(A) := \lambda \left(\{t \in [0, T] : Y_t \in A\}\right).$$

This quantity measures the time spent by the flow Y in the set A between time 0 and time T (the symbol L refers to the local time). We also write $D_A := \inf\{t > 0 : Y_t \notin A\}$ for the duration of the present visit to the set A.

Proposition 8.3 *Let $J \geq 1$ and $Y = (Y_t)_{t \geq 0}$ be a flow defined on $(\Omega, \mathcal{F}, \mathbb{P})$. Assume that $(\mathcal{N}_\varepsilon = N(Y, A_\varepsilon, h_\varepsilon, \mathcal{H}_\varepsilon))_{\varepsilon > 0}$ converges in distribution (with respect to $\mathbb{P}$) to a PPP $\mathcal{N}_0$ of intensity $\lambda \times m$ with $H_\varepsilon(A_\varepsilon) \subset V = \{1, \ldots, J\} \times W$ where $m = \sum_{j=1}^J (p_j \delta_j \times m_j)$, with $\sum_{j=1}^J p_j = 1$ and where m_j are probability measure on some separable metric space W. Suppose in addition that, for some a_ε and each x entering in A_ε, $a_\varepsilon D_{A_\varepsilon}(x) = \mathcal{D}_\varepsilon(H_\varepsilon(x))$ with $\lim_{\varepsilon \to 0} \mathcal{D}_\varepsilon(j, w) =: \mathcal{D}_j(w)$ uniformly in $w \in W$, where $\mathcal{D}_j : W \to \mathbb{R}_+$ is continuous.
Then*

$$\left(\mathcal{L}_\varepsilon = \sum_{t : Y_t(x) \text{ enters } A_\varepsilon} \delta_{t h_\varepsilon, H_\varepsilon^{(1)}(Y_t(x)), a_\varepsilon D_{A_\varepsilon} \circ Y_t(x)}\right)_{\varepsilon > 0}$$

converges in distribution with respect to $\mathbb{P}$ to a PPP $\mathcal{L}_0$ on $[0, +\infty) \times \{1, \ldots, J\} \times \bar{\mathbb{R}}_+$ with intensity $\lambda \times \sum_{j=1}^J (p_j \delta_j \times (\mathcal{D}_j)_*(m_j))$.

If moreover $a_\varepsilon D_{A_\varepsilon} \overset{\mathbb{P}}{\to} 0$, setting $L_T^{(i)}(A_\varepsilon) := L_T(A_\varepsilon \cap H_\varepsilon^{-1}(\{i\} \times W))$, then, for every $T > 0$, $((a_\varepsilon L_{\lfloor t/h_\varepsilon \rfloor}^{(j)}(A_\varepsilon))_{t \in [0,T], j=1,\ldots,J})_{\varepsilon > 0}$ converges in distribution to

$$\left(\sum_{k=1}^{N_t^{(j)}} X_k^{(j)}\right)_{t \in [0,T], j=1,\ldots,J}$$

as $\varepsilon \to 0$, where $(N_t^{(j)})_{t > 0}$ are independent Poisson process with parameter p_j and where $(X_k^{(j)})_{k \geq 1}$ are independent sequences of *independent identically distributed random variables with distribution* $(\mathfrak{D}_j)_*(m_j)$ *independent of* $(N_t^{(j)})_{t>0}$

Proof Observe that, for every $\epsilon \geq 0$, $\mathcal{L}_\varepsilon = (\psi_\varepsilon)_*(\mathcal{N}_\varepsilon)$ with $\psi_\varepsilon : (t, j, w) \mapsto (t, j, \mathfrak{D}_\varepsilon(j, w))$ if $\varepsilon > 0$ and with $\psi_0 : (t, j, w) \mapsto (t, j, \mathfrak{D}_j(w))$. Using [23, Proposition 3.13] we prove the first statement.

Assume now that $a_\varepsilon D_{A_\varepsilon} \xrightarrow{\mathbb{P}} 0$. Then

$$\left(a_\varepsilon L_{t/h_\varepsilon}^{(j)} = a_\varepsilon \min(t/h_\varepsilon, D_{A_\varepsilon \cap H_\varepsilon^{-1}(\{j\}\times W)}) + \int_{[0,t]\times\{j\}\times\bar{\mathbb{R}}_+} z\, d\mathcal{L}_\varepsilon(s, i, z) \right)_{t\in[0,T], j=1,...,J}$$

which converges to $\left(a_\varepsilon L_{t/h_\varepsilon}^{(j)} = \int_{[0,t]\times\{j\}\times\bar{\mathbb{R}}_+} z\, d\mathcal{L}_0(s, i, z) \right)_{t\in[0,T], j=1,...,J}$. $\square$

We apply the previous result to the dispersive billiard flow in a Sinai billiard with finite horizon.

Theorem 8.6.1 (Time Spent by the Billiard Flow in a Shrinking Ball for the Position) *Consider the billiard flow associated to a Sinai billiard with finite horizon in a domain $Q \subset \mathbb{T}^2$ (see Appendix for details). Recall that this flow preserves the normalized Lebesgue measure on $Q \times S^1$. Let J be a positive integer. Let $q_1, \ldots, q_J \in Q$ be a J pairwise distinct fixed position in the billiard domain and $r_1, \ldots, r_J$ be J positive real numbers. We set $d_j = 2$ if $q_j \notin \partial Q$ and $d_j = 1$ if $q_j \in \partial Q$ and $d := \sum_{j=1}^J d_j r_j$ and also*

$$\left(\mathcal{L}_\varepsilon = \sum_{j=1}^J \sum_{t\,:\,Y_t(x)\ \text{enters}\ B(q_j, r_j\varepsilon)\times S^1} \delta_{\frac{d\varepsilon t}{Area(Q)}, j, \varepsilon^{-1} D_{B(q_j, r_j\varepsilon)\times S^1} \circ Y_t(x)} \right)_{\varepsilon>0}$$

and

$$L_{t/\varepsilon}^{(j)} := \int_0^{\frac{t}{\varepsilon}} \mathbf{1}_{\{Y_s(\cdot)\in B(q_j, r_j\varepsilon)\times S^1\}}\, ds.$$

Then, $(\mathcal{L}_\varepsilon)_{\varepsilon>0}$ converges strongly in distribution to a PPP $\mathcal{L}_0$ with intensity $\lambda \times \sum_{j=1}^J \frac{d_j r_j}{d}(\delta_j \times m_j')$ where m_j' is the distribution of $r_j X$ with X a random variable of density $y \mapsto \frac{y}{4}. \arccos'(\frac{y}{2})\mathbf{1}_{[0,2]}(y) = \frac{y}{2\sqrt{4-y^2}}\mathbf{1}_{[0,2]}(y)$.

Moreover, for every $T > 0$, $((\varepsilon^{-1} L_{t/\varepsilon}^{(j)})_{t\in[0,T], j=1,...,J})_{\varepsilon>0}$ converges strongly in distribution to $\left(r_j \sum_{k=1}^{N_t^{(j)}} X_k^{(j)} \right)_{t\in[0,T], j=1,...,J}$ as $\varepsilon \to 0$, where $(N_t^{(j)})_{t>0}$ are independent Poisson process with parameter $\frac{d_j Area(Q)}{d^2\pi}$, where $(X_k^{(j)})_{k\geq 1}$ is a sequence of independent identically distributed random variables with density $x \mapsto \frac{x}{2\sqrt{4-x^2}}\mathbf{1}_{[0,2]}(x)$ independent of $(N_t)_{t>0}$.

Proof of Theorem 8.6.1 Due to Theorem 8.6.2, we know that the family of processes

$$\sum_{j=1}^{J} \sum_{t \,:\, (Y_s(y))_s \text{ enters } B(q_j,\varepsilon)\times S^1 \text{ at time } t} \delta\left(\frac{\det}{Area(Q)}, \frac{\Pi_Q(Y_t(y))-q_j}{\varepsilon}, \Pi_V(Y_t(y))\right)$$

converges in distribution (when y is distributed with respect to any probability measure absolutely continuous with respect to the Lebesgue measure on $\mathcal{M}$) as $\varepsilon \to 0$ to a Poisson Point Process with intensity $\lambda \times \tilde{m}_0$ where $\tilde{m}_0$ is the probability measure on $\{1,\dots,J\} \times S^1 \times S^1$ with density $(j,p,\mathbf{u}) \mapsto \sum_{j=1}^{J} \frac{r_j d_j}{d} \frac{1}{2d_j\pi} \langle(-p),\mathbf{u}\rangle^+ \mathbf{1}_{\{\langle p, \mathbf{n}_{q_j}\rangle\geq 0\}}$ with $d := \sum_{j=1}^{J} d_j r_j$.

We will apply Proposition 8.3 with $A_\varepsilon := \bigcup_{j=1}^{J} B(q_j,\varepsilon r_j) \times S^1$ and $H_\varepsilon(q,\mathbf{v}) = \left(j, \frac{\overrightarrow{q_j q}}{r_j\varepsilon}, \mathbf{v}\right)$ if $q \in \partial B(q_j, r_j\varepsilon)$.

Let $x = (q, \mathbf{v})$ entering in $B(q_j,\varepsilon) \times S^1$. If the billiard flow crosses $B(q_J, \varepsilon) \times S^1$ before any collision with ∂Q, then

$$\varepsilon^{-1} D_{B(q_j,r_j\varepsilon)\times S^1}(q,\mathbf{v}) = 2\varepsilon^{-1}(\overrightarrow{qq_j},\mathbf{v}) = D_0(H_\varepsilon(x)),$$

with $D_0(j,p,\mathbf{u}) = 2r_j(\overrightarrow{-p,\mathbf{u}})$. This is always the case if $q_j \notin \partial Q$. But, if $q_j \in \partial Q$, it can also happen that the billiard flow collides with ∂Q at a point $q' \in B(q_j,\varepsilon)$ before exiting $B(q_j,\varepsilon) \times S^1$. Then the point q' is at distance in $O(\varepsilon^2)$ of the tangent line to ∂Q at q_j, and the tangent line of ∂Q at q' makes an angle in $O(\varepsilon)$ with the tangent line of ∂Q at q_j. In this case

$$\varepsilon^{-1} D_{B(q_j,r_j\varepsilon)\times S^1}(q,\mathbf{v}) = 2\varepsilon^{-1}(\overrightarrow{qq_j},\mathbf{v}) + O(\varepsilon) = D_0(H_\varepsilon(x)) + O(\varepsilon),$$

uniformly in $x = (q,\mathbf{v})$ and ε. In any case, we set $a_\varepsilon = \varepsilon^{-1}$ and $\mathfrak{D}_\varepsilon = D_0 + O(\varepsilon)$.

Applying now Proposition 8.3, we infer that $(\mathcal{L}_\varepsilon)_{\varepsilon>0}$ converges strongly in distribution to a PPP $\mathcal{L}_0$ with intensity $\lambda \times \sum_{j=1}^{J} \frac{d_j r_j}{\sum_{j'=1}^{J} d_{j'}r_{j'}} (\delta_j \times (D_j)_*(m_j))$, with $D_j(p,\mathbf{u}) = 2r_j(\overrightarrow{-p,\mathbf{u}})$ and m_j the probability measure on $S^1 \times S^1$ with density

$$(p,\mathbf{u}) \mapsto \frac{d_j}{2\pi}\langle(-p),\mathbf{u}\rangle^+ \mathbf{1}_{\{\langle p, \mathbf{n}_{q_j}\rangle\geq 0\}}.$$

It remains to identify the distribution $(D_j)_*(m_j)$. By the transfer formula, we obtain

$$\int_0^\infty h(D_j(p,\mathbf{u}))\, dm_j(p,\mathbf{u}) = \frac{1}{2d_j\pi} \int_{S^1\times S^1} h(r_j\langle-2p,\mathbf{u}\rangle)\langle(-p),\mathbf{u}\rangle^+ \mathbf{1}_{\{\langle p, \mathbf{n}_{q_j}\rangle\geq 0\}}\, dp\, d\mathbf{u}$$

$$= \frac{1}{2}\int_{-\frac{\pi}{2}}^{\frac{\pi}{2}} h(2r_j\cos\varphi)\cos\varphi\, d\varphi$$

$$= \int_0^{\frac{\pi}{2}} h(2r_j \cos \varphi) \cos \varphi \, d\varphi$$

$$= \int_0^2 h(r_j y)(\arccos(\cdot/2))'(y) \frac{y}{2} \, dy.$$

Thus we have proved that the probability distribution $(D_j)_* \tilde{m}_j$ is the distribution of $r_j X$ with X a random variable of density $y \mapsto \frac{y}{4} \cdot \arccos'(\frac{y}{2}) \mathbf{1}_{[0,2]}(y) = \frac{y}{2\sqrt{4-y^2}} \mathbf{1}_{[0,2]}(y)$.

We can apply the last point of Proposition 8.3 since $\varepsilon^{-1} D_{A_\varepsilon} \le 2 \max_j r_j \mathbf{1}_{A_\varepsilon} \overset{\mathbb{P}}{\to} 0$ for any probability measure $\mathbb{P}$ absolutely continuous with respect to the Lebesgue measure on $Q \times S^1$. $\qquad\qquad\qquad\qquad\qquad\qquad\qquad\qquad\qquad\qquad\qquad\qquad\qquad\square$

Appendix: Visits by the Sinai Flow to a Finite Union of Balls in the Billiard Domain

In this appendix we are interested in spatio temporal processes for the Sinai billiard flow with finite horizon.

Let us start by recalling the model and introducing notations. We consider a finite family $\{O_i, \ i = 1, \ldots, I\}$ of convex open sets of the two-dimensional torus $\mathbb{T}^2 = \mathbb{R}^2/\mathbb{Z}^2$. We consider the billiard domain $Q = \mathbb{T}^2 \setminus \bigcup_{i=1}^I O_i$ and call the O_i obstacles. We assume that these obstacles have C^3-smooth boundary with non null curvature and that their closures are pairwise disjoint. We consider a point particle moving in Q in the following way: the point particle goes straight at unit speed in Q and obeys the classical Descartes reflexion law when it collides with an obstacle. We then define the billiard flow $(Y_t)_{t \in \mathbb{R}}$ as follows. $Y_t(q, \mathbf{v}) = (q_t, v_t)$ is the couple position-velocity of the point particle at time t if the particle has position q and velocity $\mathbf{v}$ at time 0. To avoid any confusion, we consider the billiard flow being defined on the quotient $(Q \times S^1)/\mathcal{R}$, with $\mathcal{R}$ is the equivalence relation corresponding to the identification of pre-collisional and post-collisional vectors at a reflection time:

$$(q, \mathbf{v}) \mathcal{R} (q', \mathbf{v}') \iff (q, \mathbf{v}) = (q', \mathbf{v}') \quad \text{or} \quad \mathbf{v}' = \mathbf{v} - 2\langle \mathbf{n}_q, \mathbf{v} \rangle \mathbf{n}_q,$$

where $\mathbf{n}_q$ is the unit normal vector to ∂Q at q directed inward Q if $q \in \partial Q$, with convention $\mathbf{n}_q = 0$ if $q \notin \partial Q$. This flow preserves the normalized Lebesgue measure μ on $Q \times S^1$.

We assume moreover that every billiard trajectory meets ∂Q (finite horizon assumption).

Let us write $\Pi_Q : Q \times S^1 \to Q$ and $\Pi_V : Q \times S^1 \to S^1$ for the canonical projections given respectively by $\Pi_Q(q, \mathbf{v}) = q$ and $\Pi_V(q, \mathbf{v}) = \mathbf{v}$.

Theorem 8.6.2 (Visits of the Billiard Flow to a Finite Union of Shrinking Balls in the Billiard Domain) *Let $q_1, \ldots, q_J \in Q$ be pairwise distinct positions in the billiard domain and $r_1, \ldots, r_j$ be positive real numbers. We set $d_j = 2$ if $q_j \notin \partial Q$ and $d_j = 1$ if $q_j \in \partial Q$ and $d = \sum_{j=1}^{J} d_j r_j$.*
Then, the family of processes

$$
\sum_{j=1}^{J} \sum_{t\,:\,(Y_s(y))_s \text{ enters } B(q_j, \varepsilon r_j) \times S^1 \text{ at time } t} \delta_{\left(\frac{d\varepsilon t}{Area(Q)}, j, \frac{\Pi_Q(Y_t(y)) - q_j}{r_j \varepsilon}, \Pi_V(Y_t(y)) \right)}
$$

converges in distribution (when y is distributed with respect to any probability measure absolutely continuous with respect to the Lebesgue measure on $\mathcal{M}$) as $\varepsilon \to 0$ to a Poisson Point Process with intensity $\lambda \times \tilde{m}_0$ where $\tilde{m}_0$ is the probability measure on $V := \{1, \ldots, J\} \times S^1 \times S^1$ with the density $(j, p, \mathbf{u}) \mapsto \frac{r_j}{2d\pi} \langle (-p), \mathbf{u} \rangle^+ \mathbf{1}_{\{\langle p, \mathbf{n}_{q_j} \rangle \geq 0\}}$.

Observe that if $q_j \in \partial Q$, the set of $p \in S^1$ satisfying $\langle p, \mathbf{n}_{q_j} \rangle \geq 0$ is a semicircle, whereas it is the full circle S^1 when q_j is in the interior of Q.

This result has already been proved in [21, Theorem 4.4] for $J = 1$ and Lebesgue-almost every position q_1. The extension to a finite number of points is relatively easy. The most difficult part is to treat all the possible positions in the billiard domain.

Along the paper we provided various applications of this theorem to different questions. We present here a result on the closest approaches to a given point in the billiard table by the orbit of the billiard flow.

Example H.3 Consider the billiard flow associated with the Sinai billiard having finite horizon in a domain $Q \subset \mathbb{T}^2$. Consider a fixed position $q_0 \in Q$. Set $d = 2 - \mathbf{1}_{q_0 \in \partial Q}$. During each visit of the flow to $B(q_0, \varepsilon)$, the closest distance to q_0 is given by $L_0(q, \mathbf{v}) := \varepsilon |\sin \angle (\overrightarrow{qq_0}, \mathbf{v})|$ where $(q, \mathbf{v})$ is the entry point.

Then the family of closest approach point process

$$
\left(C_\varepsilon := \mathcal{N}(Y, B(q_0, \varepsilon) \times S^1, d\varepsilon / Area(Q), \varepsilon^{-1} L_0) \right)_{\varepsilon > 0}
$$

on $[0, +\infty) \times [0, 1]$ converges in distribution (with respect to any probability measure absolutely continuous with respect to the Lebesgue measure on $Q \times S^1$) to a PPP with intensity 1.

Proof Due to Theorem 8.6.2, the family of spatio-temporal processes

$$
(\mathcal{N}_\varepsilon := \mathcal{N}(Y, B(q_0, \varepsilon) \times S^1, d\varepsilon / Area(Q), H_\varepsilon)_{\varepsilon > 0}
$$

with $H_\varepsilon(q, \mathbf{v}) = (\varepsilon^{-1} \overrightarrow{q_0 q}, \mathbf{v})$ converges in distribution (with respect to any probability measure absolutely continuous with respect to Lebesgue on $Q \times S^1$) to a PPP of intensity $\lambda \times \tilde{m}_0$ where $\tilde{m}_0$ is the probability measure on $S^1 \times S^1$ with the density $(p, \mathbf{u}) \mapsto \frac{1}{2d\pi} \langle (-p), \mathbf{u} \rangle^+ \mathbf{1}_{\{\langle p, \mathbf{n}_{q_0} \rangle \geq 0\}}$ (where $\mathbf{n}_{q_0}$ is the unit normal vector to ∂Q at q_0 directed inward Q if $q_0 \in \partial Q$, $\mathbf{n}_{q_0} = 0$ otherwise).

Observe that

$$C_\varepsilon = \tilde{G}(\mathcal{N}_\varepsilon),$$

with $\tilde{G}(t, p, \mathbf{u}) = (t, G(p, \mathbf{u}))$ where $G(p, \mathbf{u}) = (t, |\sin \angle(-p, \mathbf{u})|)$. Thus $(C_\varepsilon)_{\varepsilon > 0}$ converges strongly in distribution to the PPP with intensity $\lambda \times G_*(\tilde{m}_0)$ and it remains to identify $\tilde{m}_1 = G_*(\tilde{m}_0)$. Due to the transfer formula, we obtain

$$\int_0^\infty h(G(p, \mathbf{u})) \, d\tilde{m}_0(p, \mathbf{u}) = \frac{1}{2d\pi} \int_{S^1 \times S^1} h(|\sin \angle(-2p, \mathbf{u})|)$$
$$\times (\cos \angle(-p, \mathbf{u}))^+ \mathbf{1}_{\{\langle p, \mathbf{n}_{q_j} \rangle \geq 0\}} \, dp \, d\mathbf{u}$$
$$= \frac{1}{2} \int_{-\frac{\pi}{2}}^{\frac{\pi}{2}} h(|\sin \varphi|) \cos \varphi \, d\varphi$$
$$= \int_0^{\frac{\pi}{2}} h(\sin \varphi) \cos \varphi \, d\varphi = \int_0^1 h(y) \, dy.$$

Proof of Theorem 8.6.2 Due to [28, Theorem 1], it is enough to prove the result for the convergence in distribution with respect to μ. Assume $\varepsilon > \min_{j \neq j'} \frac{q_j q_{j'}}{4}$. We use the representation of the billiard flow as a special flow over the discrete time billiard system (M, ν, F) corresponding to collision times and with τ the length of the free flight before the next collision.

Set $\tilde{A}_\varepsilon = \bigcup_{j=1}^J \tilde{A}_\varepsilon^{(j)}$, where $\tilde{A}_\varepsilon^{(j)}$ is the set of the configurations entering in $A_\varepsilon^{(j)} := (Q \cap B(q_j, \varepsilon)) \times S^1$, i.e. $\tilde{A}_\varepsilon^{(j)}$ is the set of $(q, \mathbf{v}) \in (Q \cap \partial B(q_j, \varepsilon]) \times S^1$ s.t. $\langle \mathbf{q}\mathbf{q_0}, \mathbf{v} \rangle > 0$. Set also $A_\varepsilon := \bigcup_{j=1}^J A_\varepsilon^{(j)}$.

Set $h'_\varepsilon := d\varepsilon / Area(Q)$ and $H_\varepsilon(q', \mathbf{v}) = (j, \frac{\overrightarrow{q_j q'}}{r_j \varepsilon}, \mathbf{v})$ if $q' \in \partial B(q_j, r_j \varepsilon)$. Here M is the set of reflected unit vectors based on ∂Q, ν is the probability measure with the density proportional to $(q, \mathbf{v}) \mapsto \langle \mathbf{n}(q), \mathbf{v} \rangle$, where $\mathbf{n}(q)$ is the unit vector normal to ∂Q at q directed towards Q and $F : M \to M$ is the transformation mapping a configuration at a collision time to the configuration corresponding to the next collision time.

The normalizing function G_ε is given by $G_\varepsilon(x) = H_\varepsilon(Y_{\tau_{\tilde{A}_\varepsilon}^{(Y)}(x)}(x))$ with $\tau_{\tilde{A}_\varepsilon}^{(Y)}(y) := \inf\{t > 0 : Y_t(y) \in \tilde{A}_\varepsilon\}$.

As in the setting of Theorem 8.2.2, we write Π for the projection on M, that is $\Pi(q', \mathbf{v}) = (q, \mathbf{v})$ is the post-collision vector at the previous collision time. We take here $h_\varepsilon := \nu(\Pi(\tilde{A}_\varepsilon))$.

As for [21, Theorem 4.4], we will apply [21, Proposition 3.2] after checking its assumptions. We define $\tilde{A}_\varepsilon^{(j)} := \{(q, \mathbf{v}) \in \partial B(q_j, \varepsilon) \times S^1 : \langle \overrightarrow{qq_j}, \mathbf{v} \rangle \geq 0\}$.

(i) *Measure of the set.* We have to adapt slightly the first item of the proof of [21, Theorem 4.4] which deals with the asymptotic behaviour of $\nu(B_\varepsilon)$ with $B_\varepsilon :=$ $\Pi(\widetilde{A}_\varepsilon)$. Observe that $B_\varepsilon = \bigcup_{j=1}^J B_\varepsilon^{(j)}$ with $B_\varepsilon^{(j)} := \Pi(\widetilde{A}_\varepsilon^{(j)})$, i.e. $B_\varepsilon^{(j)}$ is the set of configurations $(q, \mathbf{v}) \in M$ such that the billiard trajectory $(Y_t(q))_{t \geq 0}$ will enter $B(q_j, \varepsilon r_j)$ before touching ∂Q. As seen in [19, Lemma 5.1],

$$\text{if } q_j \in Q \setminus \partial Q, \quad \nu(B_\varepsilon^{(j)}) = \frac{|Q \cap \partial B(q_j, r_j \varepsilon)|}{|\partial Q|} = \frac{2\pi r_j \varepsilon}{|\partial Q|}.$$

With exactly the same proof, we obtain that

$$\text{if } q_j \in \partial Q, \quad \nu(B_\varepsilon^{(j)}) = \frac{|Q \cap \partial B(q_j, r_j \varepsilon)|}{|\partial Q|} \sim \frac{\pi r_j \varepsilon}{|\partial Q|}.$$

Moreover, for every distinct j, j', $B_\varepsilon^{(j)} \cap B_\varepsilon^{(j')}$ is contained in $\Pi(B(x_{j,j'}, K_{j,j'}\varepsilon) \cup B(x_{j',j}, K_{j,j'}\varepsilon))$ where $x_{j,j'} = \left(q_j, \overrightarrow{q_j q_j' q_j q_j'}\right)$ and $K_{j,j'} = \max\left(1, \frac{3}{q_j q_{j'}}\right)$. So, due to [19, Lemma 5.1], $\nu(B_\varepsilon^{(j)} \cap B_\varepsilon^{(j')}) = O(\varepsilon^2) = o(\varepsilon)$. Hence we conclude that

$$\nu(B_\varepsilon) \sim \sum_{j=1}^J \nu(B_\varepsilon^{(j)}) \sim \frac{d\pi\varepsilon}{|\partial Q|},$$

as $\varepsilon \to 0$.

(ii) Observe that

$$\mathcal{N}(Y, A_\varepsilon, h'_\varepsilon, H_\varepsilon) = \sum_{j=1}^J \mathcal{N}(Y, A_\varepsilon^{(j)}, h'_\varepsilon, H_\varepsilon) \geq \mathcal{N}(Y, A'_\varepsilon, h'_\varepsilon, H_\varepsilon),$$

where $A'_\varepsilon = \bigcup_{j=1}^J \Pi^{-1}(\Pi(\widetilde{A}_\varepsilon^{(j)})) \setminus \bigcup_{j' \neq j} \Pi^{-1}(\Pi(\widetilde{A}_\varepsilon^{(j')}))$ and that, for all $T > 0$,

$$\mathbb{E}_\mu \left[\left(\mathcal{N}(Y, A_\varepsilon, h'_\varepsilon, H_\varepsilon) - \mathcal{N}(Y, A'_\varepsilon, h'_\varepsilon, G_\varepsilon) \right) ([0, T] \times V) \right]$$
$$\leq \frac{T \max \tau}{2 h_\varepsilon (\min \tau)^2} \sum_{j,j' : j \neq j'} \nu\left(\widetilde{A}_\varepsilon^{(j)} \cap \widetilde{A}_\varepsilon^{(j')}\right) = o(1),$$

where we used the representation of Y as a special flow over (M, ν, F) due to the fact, proved in the previous item, that for any distinct labels j, j', $\nu\left(\widetilde{A}_\varepsilon^{(j)} \cap \widetilde{A}_\varepsilon^{(j')}\right) = o(\varepsilon)$. Thus it is enough to prove the convergence in distribution of $\mathcal{N}(Y, A'_\varepsilon, h'_\varepsilon, H_\varepsilon)$ with respect to μ.

(iii) The same argument ensures that, with respect to ν, the convergence in distribution of $\mathcal{N}(F, B_\varepsilon, h_\varepsilon, G_\varepsilon)$ to $\mathcal{P}$ is equivalent to the convergence in distribution of $\mathcal{N}(F, B'_\varepsilon, h_\varepsilon, G_\varepsilon)$, with $B'_\varepsilon := \Pi(A'_\varepsilon)$.

(iv) Note that $\nu((\partial B_\varepsilon)^{[\varepsilon^\delta]}) = o(\nu(B_\varepsilon))$, for every $\delta > 1$.

(v) Due to Lemma 8.6.1, for every $\sigma > 1$, $\nu(\tau_{B_\varepsilon} \leq \varepsilon^{-\sigma}|B_\varepsilon) = o(1)$, where τ_B is here the first time $k \geq 1$ at which $F^k(\cdot) \in B$.

(vi) Now let us prove that $(\nu(G_\varepsilon^{-1}(\cdot)|B_\varepsilon))_{\varepsilon>0}$ converges to $\tilde{m}_0$ as $\varepsilon \to 0$.
Let us consider the measure $\tilde{\mu}$ on $\{1, \ldots, J\} \times S^1 \times S^1$ with the density $(j, p, \mathbf{u}) \mapsto r_j \langle (-p), \mathbf{u} \rangle^+$.
Observe first that $\tilde{m}_0 = \tilde{\mu}(\cdot|A)$ with $A := \bigcup_{j=1}^J A^{(j)}$ and

$$A^{(j)} := \left\{ (p, \mathbf{u}) \in S^1 \times S^1 : \langle (-p), \mathbf{u} \rangle \geq 0, \ \langle p, \mathbf{n}_{q_j} \rangle \geq 0 \right\}$$

and second that $\nu(G_\varepsilon^{-1}(\cdot)|B_\varepsilon) = \tilde{\mu}(\cdot|G_\varepsilon(B_\varepsilon))$. But

$$\tilde{\mu}(A \setminus G_\varepsilon(B_\varepsilon)) \leq \sum_{j=1}^J \tilde{\mu}\left(H_\varepsilon \left(Y_{\tau_{\tilde{A}_\varepsilon^{(j)}}^{(Y)}(\cdot)} \left(\bigcup_{j' \neq j} (B_\varepsilon^{(j)} \cap B_\varepsilon^{(j')}) \right) \right) \right)$$

$$\leq \sum_{j=1}^J 2 \max \tau |\partial Q| r_j \varepsilon \nu \left(\bigcup_{j \neq j'} (B_\varepsilon^{(j)} \cap B_\varepsilon^{(j')}) \right) = o(\nu(B_\varepsilon))$$

and $G_\varepsilon(B_\varepsilon) \setminus A$ corresponds to points $(p, \mathbf{u}) \in S^1 \times S^1$ with $q_j \in \partial Q$ with $0 < \langle p, \mathbf{u} \rangle \leq O(\varepsilon)$, thus

$$\tilde{\mu}(G_\varepsilon(B_\varepsilon) \setminus A) = O(\varepsilon).$$

This ends the proof of the convergence in distribution of the family of measures $(\nu(G_\varepsilon^{-1}(\cdot)|B_\varepsilon))_{\varepsilon>0}$ to $\tilde{m}_0$ as $\varepsilon \to 0$.

(vii) For the construction of $\mathcal{W}$ we use [21, Proposition 3.4]. $\qquad\square$

Thus, due to [21, Proposition 3.2], we conclude the convergence of distribution with respect to ν of $(\mathcal{N}(F, B_\varepsilon, h_\varepsilon, G_\varepsilon))_{\varepsilon>0}$ and so, due to (ii), of $(\mathcal{N}(F, B'_\varepsilon, h_\varepsilon, G_\varepsilon))_{\varepsilon>0}$ to a PPP $\mathcal{P}$ with intensity $\lambda \times \tilde{m}_0$. Applying now Theorem 8.2.2, we deduce the strong convergence in distribution of $(\mathcal{N}(F, A'_\varepsilon, h_\varepsilon/\mathbb{E}_\nu[\tau], H_\varepsilon))_{\varepsilon>0}$ to $\mathcal{P}$ and so, due to (iii), the convergence in distribution with respect to μ of $(\mathcal{N}(F, A_\varepsilon, h_\varepsilon/\mathbb{E}_\nu[\tau], H_\varepsilon))_{\varepsilon>0}$ to $\mathcal{P}$. Now we conclude by Zweimüller [28, Theorem 1] and by noticing that

$$\frac{h_\varepsilon}{\mathbb{E}_\nu[\tau]} = \frac{d\pi\varepsilon}{|\partial Q|\mathbb{E}_\nu[\tau]} = \frac{d\varepsilon}{Area(Q)} = h'_\varepsilon.$$

Lemma 8.6.1

$$\forall \sigma \in (0, 1), \quad \nu(\tau_{B_\varepsilon} \leq \varepsilon^{-\sigma} | B_\varepsilon) = o(1) \tag{8.6.1}$$

Proof This point corresponds to the second item of the proof of [21, Theorem 4.4], which for Lebesgue-almost every point came from [19, Lemma 6.4]. To prove (8.6.1), we write

$$\nu(\tau_{B_\varepsilon} \leq \varepsilon^{-\sigma} | B_\varepsilon) \leq \sum_{k=1}^{\lfloor \varepsilon^{-\sigma} \rfloor} \nu(F^{-n}(B_\varepsilon) | B_\varepsilon). \tag{8.6.2}$$

Thus our goal is to bound $\nu(F^{-n}(B_\varepsilon) | B_\varepsilon)$.

Step 1: Useful Notations
We parametrize M by $\bigcup_{i=1}^{I} \{i\} \times (\mathbb{R} / |\partial O_i| \mathbb{Z}) \times [-\frac{\pi}{2}; \frac{\pi}{2}]$. A reflected vector $(q, \mathbf{v}) \in M$ is represented by (i, r, φ) if $q \in \partial \Gamma_i$ as curvilinear absciss $r \partial O_i$ and if φ is the angular measure in $[-\pi/2, \pi/2]$ of $(\mathbf{n}(q), \mathbf{v})$ where $\mathbf{n}(q)$ is the normal vector to ∂Q at q.
For any C^1-curve γ in M, we write $\ell(\gamma)$ for the euclidean length in the (r, φ) coordinates of γ. If moreover γ is given in coordinates by $\varphi = \phi(r)$, then we also write $p(\gamma) := \int_\gamma \cos(\phi(r)) \, dr$. We define the time until the next reflection in the future by

$$\tau(q, \mathbf{v}) := \min\{s > 0 \, : \, q + s\mathbf{v} \in \partial Q\}.$$

It will be useful to define $\mathcal{S}_0 := \{\varphi = \pm\pi/2\}$. Recall that, for every $k \geq 1$, F^k defines a C^1-diffeomorphism from $M \backslash \mathcal{S}_{-k}$ to $M \backslash \mathcal{S}_k$ with $\mathcal{S}_{-k} := \bigcup_{m=0}^{k} F^{-m}(\mathcal{S}_0)$ and $\mathcal{S}_k := \bigcup_{m=0}^{k} F^m(\mathcal{S}_0)$.

Step 2: Geometric Study of B_ε and of $F(B_\varepsilon)$
Moreover the boundary of each connected component of B_ε (resp. $F(B_\varepsilon)$) is made with a bounded number of C^1 curves of the following forms:

- curves of $\mathcal{S}_0$, corresponding, in (r, φ)-coordinates, to $\{\varphi = \pm\frac{\pi}{2}\}$.
- C^1 curves of $F^{-1}(\mathcal{S}_0)$ (resp. $F(\mathcal{S}_0)$), which have the form $\varphi = \phi(r)$ with ϕ a C^1 decreasing (resp. increasing) function satisfying $\min \kappa \leq |\phi'(r)| \leq \max \kappa + \frac{1}{\min \tau}$, where $\kappa(q)$ is the curvature of ∂Q at $q \in \partial Q$ and where τ is the free flight length before the next collision time.
- if $q_0 \notin \partial Q$: C^1 curves, corresponding to the set of points $x = (q, \mathbf{v}) \in M$ (resp. $F(x)$) such that $[\Pi_Q(x), \Pi_Q(F(x))]$ is tangent to $\partial B(q_0, \varepsilon)$. These curves have the form $\varphi = \phi_\varepsilon(r)$ with ϕ_ε a decreasing (resp. increasing) function satisfying $\min \kappa \leq |\phi'_\varepsilon(r)| \leq \max \kappa + \frac{1}{d(q_0, \partial Q) - \varepsilon} \leq \max \kappa + \frac{2}{\tau_0}$), with $\tau_0 := d(q_0, \partial Q)$ as soon as $\varepsilon < \frac{\tau_0}{2}$.
- if $q_0 \in \partial Q$: C^1 curves, corresponding to the set of points $x = (q, \mathbf{v}) \in M$ (resp. $F(x)$) such that $[\Pi_Q(x), \Pi_Q(F(x))]$ is tangent to $\partial B(q_0, \varepsilon)$ or such that

$\Pi_Q(F(x))$ is an extremity of $B(q_0, \varepsilon) \cap Q$ and $[\Pi_Q(x), \Pi_Q(F(x))]$ contains no other point of $B(q_0, \varepsilon)$. These curves have the form $\varphi = \phi_\varepsilon(r)$ with ϕ_ε a decreasing (resp. increasing) function satisfying $\min \kappa \le |\phi'_\varepsilon(r)|$.

The points $x = (q, \mathbf{v}) \in M$, with $d(q, q_0) \ll 1$ almost immediately entering (resp. exiting) $B(q_0, \varepsilon) \times S^1$ are contained in a union R_ε of two rectangles of width $O(\varepsilon^{1/2})$ for the position (around q_0) and of width $O(\varepsilon)$ for the velocity direction (around the tangent vectors to ∂Q at q_0).

In $B_\varepsilon \setminus R_\varepsilon$ (resp. $F(B_\varepsilon) \setminus (R_\varepsilon \cup \Pi_Q^{-1}(B(q_0, \varepsilon)))$) we also have $|\phi'_\varepsilon(r)| \le \max \kappa + \frac{2}{\tau_0}$ with $\tau_0 := \min \tau$ as soon as $\varepsilon < \frac{\tau_0}{2}$.

We say that a **curve γ of M satisfies assumption (C)** if it is given by $\varphi = \phi(r)$ with ϕ being C^1-smooth, increasing and such that $\min \kappa \le \phi' \le \max \kappa + \frac{2}{\tau_0}$. We recall the following facts.

- There exist $C_0, C_1 > 0$ and $\lambda_1 > 1$ such that, for every γ satisfying Assumption (C) and every integer m such that $\gamma \cap S_{-m} = \emptyset$, $F^m \gamma$ is a C^1-smooth curve satisfying assumption (C) and $C_1 p(F^m \gamma) \ge \lambda_1^m p(\gamma)$ and $\ell(\gamma) \le C_0 \sqrt{p(F\gamma)}$.
- There exist $C_2 > 0$ and $\lambda_2 > \lambda_1^{1/2}$ such that, for every integer m, the number of connected components of $M \setminus S_{-m}$ is less than $C_2 \lambda_2^m$. Moreover S_{-m} is made of curves $\varphi = \phi(r)$ with ϕC^1-smooth and strictly decreasing.
- If $\gamma \subset M \setminus S_{-1}$ is given by $\varphi = \phi(r)$ or $r = \mathfrak{r}(\varphi)$ with ϕ or $\mathfrak{r}$ increasing and C^1 smooth, then $F\gamma$ is C^1, is given by $\varphi = \phi_1(r)$ with $\min \kappa \le \phi'_1 \le \max \kappa + \frac{1}{\min \tau}$. Moreover $\int_{F\gamma} d\varphi \ge \int_\gamma d\varphi$.

We observe that there exist $K'_0 > 0$ and $\varepsilon_0 > 0$ such that, for every $\varepsilon \in (0, \varepsilon_0)$, $F(B_\varepsilon) \setminus R_\varepsilon$ is made of a bounded number of connected components $V_\varepsilon^{(i)}$ each of which is a strip of width at most $K'_0 \varepsilon$ of the following form in (r, φ)-coordinates:

- $\{(r, \varphi) : r \in J, \phi_1^{(i)}(r) \le \varphi \le \phi_2^{(i)}(r)\}$ (with J an interval) and is delimited by two continuous piecewise C^1 curves γ_j given by $\varphi = \phi_j(r)$ satisfying assumption (C) and $\|\phi_1^{(i)} - \phi_2^{(i)}\|_\infty \le K'_0 \varepsilon$.
- or possibly, if $q_0 \in \partial Q$, $\{(r, \varphi) : r_{1,\varepsilon}^{(i)} \le r \le r_{2,\varepsilon}^{(i)}\}$ with $|r_{1,\varepsilon}^{(i)} - r_{2,\varepsilon}^{(i)}| \le K'_0 \varepsilon$.

In particular, with the previous notations, any connected component $V_\varepsilon^{(i)}$ of $F(B_\varepsilon) \setminus R_\varepsilon$ has the form $\bigcup_{u \in [0,1]} \widetilde{\gamma}_u^{(i)}$, where $\widetilde{\gamma}_u^{(i)}$ corresponds to the graph $\{\psi^{(i)}(u, r) = (r, u\phi_1^{(i)}(r) + (1-u)\phi_2^{(i)}(r)) : r \in J_i\}$ (or possibly $\{\psi^{(i)}(u, \varphi) = (ur_{1,\varepsilon}^{(i)} + (1-u)r_{2,\varepsilon}^{(i)}, \varphi), \varphi \in J_i\}$ if $q_0 \in \partial Q$). Thus

$$\forall E \in \mathcal{B}(M), \quad \nu(E \cap F(B_\varepsilon \setminus R_\varepsilon)) \le \frac{Leb(E \cap F(B_\varepsilon \setminus R_\varepsilon))}{2|Q|}$$

$$\le \sum_i \frac{1}{2|\partial Q|} \int_{J_i \times [0,1]} \mathbf{1}_{\psi^{(i)}(u,s) \in E} \left| \frac{\partial}{\partial u} \psi^{(i)}(u, s) \right| \, ds \, du$$

$$\le \frac{K'_0 \varepsilon}{2|\partial Q|} \sup_{[0,1]} \ell(E \cap \widetilde{\gamma}_u). \tag{8.6.3}$$

Step 3: Scarcity of Very Quick Returns
Let us prove the existence of $K_1 > 0$ such that,

$$\forall s \geq 1, \ \forall \varepsilon < \frac{\tau_0}{2}, \quad \nu(F^{-s-1}(B_\varepsilon)|B_\varepsilon) \leq K_1 (\lambda_2/\lambda_1^{\frac{1}{2}})^s \varepsilon^{\frac{1}{2}}. \tag{8.6.4}$$

Let $u \in (0, \varepsilon)$. We define γ to be a connected component of $\widetilde{\gamma}_u \cap F(B_\varepsilon) \cap F^{-s}(B_\varepsilon)$. The curve γ satisfies Assumption (C) or is vertical. In any case, any connected component of $F(\gamma)$ satisfies Assumption (C) and $\ell(\gamma) \leq C_0 \sqrt{p(F(\gamma))}$ (indeed, if γ is vertical, then $\ell(\gamma) \leq \frac{1}{\min \tau} p(F(\gamma))$). It follows

$$\ell(\gamma) \leq C_0 \sqrt{p(F(\gamma))} \leq C_0 \sqrt{C_1 \lambda_1^{1-s} p(F^s \gamma)} \leq C_0' \sqrt{C_1 \lambda_1^{1-s} K_0' \varepsilon}$$

using first the fact that $F(\gamma)$ is an increasing curve contained in $M \setminus S_{-s}$ and secondly, the fact that $F^s \gamma$ is an increasing curve satisfying Condition (C) and contained in B_ε. Since $F(\widetilde{\gamma}_u) \setminus S_s$ contains at most $C_2 \lambda_2^s$ connected components, using (8.6.3), we obtain

$$\nu(F^{-s-1}(B_\varepsilon) \cap B_\varepsilon \setminus R_\varepsilon) = \nu(F^{-s}(B_\varepsilon) \cap F(B_\varepsilon \setminus R_\varepsilon)) \leq \frac{K_0' \varepsilon}{2|\partial Q|} \sup_{[0,1]} C_2 \lambda_2^s C_0' \sqrt{C_1} \lambda_1^{\frac{s}{2}} \varepsilon^{\frac{1}{2}}.$$

We conclude by using the fact that $\nu(B_\varepsilon) = \frac{d\pi\varepsilon}{|\partial Q|}$ and that $\nu(R_\varepsilon) = O(\varepsilon^{\frac{3}{2}})$.

Step 4: Scarcity of Intermediate Quick Returns
We prove now that for any $a > 0$, there exists $s_a > 0$ such that

$$\sum_{n=-a \log \varepsilon}^{\varepsilon^{-s_a}} \nu(B_\varepsilon \cap F^{-n} B_\varepsilon) = o(\nu(B_\varepsilon)). \tag{8.6.5}$$

Since $\nu(B_\varepsilon) \approx \varepsilon$ and $\nu(R_\varepsilon) = O(\varepsilon^{\frac{3}{2}})$, up to adding the condition $s_a < 1/2$, it remains to prove (8.6.5) with $\nu(B_\varepsilon \cap F^{-n} B_\varepsilon)$ replaced by $\nu((B_\varepsilon \setminus R_\varepsilon) \cap F^{-n}(B_\varepsilon))$.

If $q_0 \in \partial Q$ and if $\widetilde{\gamma}_u$ is vertical, we replace it in the argument below by the connected components of $F(\widetilde{\gamma}_u)$ and will conclude by noticing that; for any measurable set A, $\ell(\widetilde{\gamma}_u \cap F^{-1}(A)) \leq C_0'' \ell(F(\widetilde{\gamma}_u \cap A))$.

We denote the kth homogeneity strip[1] by $\mathbb{H}_k$ for $k \neq 0$ and set $\mathbb{H}_0 = \cup_{|k|<k_0} \mathbb{H}_k$ for some fixed k_0. Set $s := \min(-a \log \theta, 1)/3$. Let $k_\varepsilon = \varepsilon^{-s}$ and $H^\varepsilon = \cup_{|k| \leq k_\varepsilon} \mathbb{H}_k$. For any $u \in [0, 1]$, we set $\widetilde{\gamma}_{k,u} = \widetilde{\gamma}_u \cap \mathbb{H}_k$. Each $\widetilde{\gamma}_{k,u}$ is a weakly homogeneous unstable curve.

We cut each curve $\widetilde{\gamma}_{k,u}$ into small pieces $\widetilde{\gamma}_{k,u,i}$ such that each $F^j \widetilde{\gamma}_{k,u,i}$, $j = 0, \ldots, n$ is contained in a homogeneity strip and a connected component of $M \setminus S_1$.

---

[1] See [6] for notations and definitions.

For $x \in \widetilde{\gamma}_{k,u,i}$ we denote by $r_n(x)$ the distance (in $F^n \widetilde{\gamma}_u$) of $F^n(x)$ to the boundary of $F^n \widetilde{\gamma}_{k,u,i}$.

Recall that the growth lemma [6, Theorem 5.52] ensures the existence of $\theta \in (0, 1)$, $c > 0$ such that, for any weakly homogeneous unstable curve γ one has

$$\ell(\gamma \cap \{r_n < \delta\}) \le c\theta^n \delta + c\delta\ell(\gamma). \tag{8.6.6}$$

Therefore,

$$\ell(\widetilde{\gamma}_u \cap F^{-n}(B_\varepsilon) \setminus \mathbb{H}_\varepsilon)$$

$$\le \sum_{|k| \le k_\varepsilon} \ell(\cap\{r_n \ge \varepsilon^{1-s}\} \cap F^{-n}(B_\varepsilon)) + \ell(\widetilde{\gamma}_{u,k} \cap \{r_n < \varepsilon^{1-s}\}).$$

The first term inside the above sum is bounded by the sum $\sum_i \ell(\widetilde{\gamma}_{u,k,i} \cap F^{-n}(B_\varepsilon))$ over those i's such that $F^n(\widetilde{\gamma}_{u,k,i})$ is of size larger than ε^{1-s}. In particular $\ell(\widetilde{\gamma}_{u,k,i}) \ge \varepsilon^{1-s}$. On the other hand, by transversality

$$\ell(F^n(\widetilde{\gamma}_{u,k,i}) \cap B_\varepsilon) \le c\varepsilon.$$

By distortion (See Lemma 5.27 in [6]) we obtain

$$\ell(\widetilde{\gamma}_{u,k,i} \cap F^{-n}(B_\varepsilon)) \le c\varepsilon^s \ell(\widetilde{\gamma}_{u,k,i}).$$

Summing up over these i gives the first term inside the sum is bounded by

$$\ell(\widetilde{\gamma}_{u,k} \cap \{r_n \ge \varepsilon^{1-s}\} \cap F^{-n}(B_\varepsilon)) \le c\varepsilon^s \ell(\widetilde{\gamma}_{u,k,i}).$$

Thus

$$\ell(\widetilde{\gamma}_{u,k} \cap \{r_n < \varepsilon^{1-s}\}) \le c\theta^n \varepsilon^{1-s} + c\varepsilon^{1-s} \ell(\widetilde{\gamma}_{u,k}).$$

A final summation over k gives

$$\ell(\widetilde{\gamma}_u \cap F^{-n}(B_\varepsilon) \setminus \mathbb{H}_\varepsilon) \le c(\varepsilon^s + \varepsilon^{1-s})\ell(\widetilde{\gamma}_u) + ck_\varepsilon \theta^n \varepsilon^{1-s}.$$

This combined with (8.6.3) leads to

$$\nu(F(B_\varepsilon \setminus R_\varepsilon) \cap F^{-n}(B_\varepsilon)) \le \nu(F(B_\varepsilon \setminus R_\varepsilon) \cap \mathbb{H}_\varepsilon) + O(\varepsilon^{1+s}) = O(\varepsilon^s \nu(B_\varepsilon)).$$

where we use the fact that $B_\varepsilon \setminus \mathbb{H}_\varepsilon$ is contained in a uniformly bounded union of rectangles of horizontal width $O(\varepsilon)$ and contained in the $k_\varepsilon^{-2} = \varepsilon^{2s}$-neighbourhood of $\mathcal{S}_0$. We take $s_a < \min(s, \frac{1}{2})$.

Step 5: End of the Proof of (8.6.1)

Choose $a = 1/(4\log(\lambda_2/\lambda_1^{1/2}))$. Observe that, due to (8.6.4), we have

$$\sum_{s=1}^{-a\log\varepsilon} \mu(F^{-s}A_\varepsilon|A_\varepsilon) \le \frac{K_1}{\lambda_2/\lambda_1^{\frac{1}{2}} - 1}(\lambda_2/\lambda_1^{\frac{1}{2}})^{-a\log\varepsilon}\varepsilon^{1/2} \le \frac{K_1}{\lambda_2/\lambda_1^{\frac{1}{2}} - 1}\varepsilon^{1/4}.$$

This combined with (8.6.5) leads to

$$\sum_{n=1}^{\varepsilon^{-sa}} \nu(F^{-n}B_\varepsilon|B_\varepsilon) = o(1). \tag{8.6.7}$$

Let $\sigma > 1$. In view of (8.6.2), it remains to control $\nu(F^{-n}B_\varepsilon|B_\varepsilon)$ for the intermediate integers n such that $\varepsilon^{-sa} \le n \le \varepsilon^{-\sigma}$. We approximate the set B_ε by the union $\widetilde{B}_\varepsilon$ of connected components of $M \setminus (S_{-k(\varepsilon)} \cup S_{k(\varepsilon)})$ that intersects B_ε, with $k(\varepsilon) = \lfloor |\log\varepsilon|^2\rfloor$. There exists $\widetilde{C} > 0$ and $\widetilde{\theta} \in (0,1)$ such that, for any positive integer k, the diameter of each connected component of $M \setminus (S_{-k} \cup S_k)$ is less than $\widetilde{C}\widetilde{\theta}^k$.

Thus $B_\varepsilon \subset \widetilde{B}_\varepsilon$ and $\nu(\widetilde{B}_\varepsilon \setminus B_\varepsilon) \le \nu\left((\partial B_\varepsilon)^{[\widetilde{C}\widetilde{\theta}^{k(\varepsilon)}]}\right) = O(\varepsilon\widetilde{\theta}^{k(\varepsilon)})$. But, due to [20, Lemma 4.1], we also have

$$\forall m > 1, \quad \forall n \ge 2k(\varepsilon), \quad \nu\left(\widetilde{B}_\varepsilon \cap F^{-n}\widetilde{B}_\varepsilon\right) = \nu(\widetilde{B}_\varepsilon)^2 + O(n^{-m}\nu(\widetilde{B}_\varepsilon)).$$

Since $k(\varepsilon) = o(\varepsilon^{-sa})$ and thus

$$\forall m > 1, \quad \sum_{n=\varepsilon^{-sa}}^{\varepsilon^{-\sigma}} \nu(F^{-n}B_\varepsilon|B_\varepsilon) \le O\left(\varepsilon^{1-\sigma} + \varepsilon^{sa(m-1)-\sigma} + \widetilde{\theta}^{k(\varepsilon)}\right) = o(1),$$

as $\varepsilon \to 0$, since $\sigma < 1$, $\widetilde{\theta} \in (0,1)$, $k(\varepsilon) \to +\infty$ and by taking $m > 1 + \frac{\sigma}{sa}$. This combined with (8.6.7) and (8.6.2) ends the proof of (8.6.1). $\qquad\square$

References

1. J.F. Alves, D. Azevedo, Statistical properties of diffeomorphisms with weak invariant manifolds. Discrete Contin. Dyn. Syst. **36**, 1–41 (2016)
2. J.-S. Athreya, S.-P. Lalley, J. Sapir, M. Wroten, Local geometry of random geodesics on negatively curved surfaces. Ann. Henri Lebesgue **4**, 187–226 (2021)
3. R. Bowen, Symbolic dynamics for hyperbolic flows. Amer. J. Math. **95**, 429–459 (1973)
4. M. Carney, M. Nicol, H.K. Zhang, Compound Poisson law for hitting times to periodic orbits in two-dimensional hyperbolic systems. J. Stat. Phys. **169**(4), 804–823 (2017)
5. J.-R. Chazottes, P. Collet, Poisson approximation for the number of visits to balls in nonuniformly hyperbolic dynamical systems. Erg. Th. Dyn. Syst. **33**(1), 49–80 (2013)

6. N. Chernov, R. Markarian, Chaotic billiards, in *Mathematical Surveys and Monographs*, vol. 127 (American Mathematical Society, Providence, 2006), xii+316 pp
7. J. De Simoi, I.P. Toth, An expansion estimate for dispersing planar billiards with corner points. Ann. H. Poincaré **15**, 1223–1243 (2014)
8. W. Doeblin, Remarques sur la théorie métrique des fractions continues (French). Compositio Math. **7**, 353–371 (1940)
9. A.C.M. Freitas, J.M. Freitas, M. Magalhães, Complete convergence and records for dynamically generated stochastic processes. Trans. Am. Math. Soc. **373**, 435–478 (2020)
10. J. Freitas, A. Freitas, M. Todd, The compound Poisson limit ruling periodic extreme behaviour of non-uniformly hyperbolic dynamics. Commun. Math. Phys. **321**, 483–527 (2013)
11. N. Haydn, S. Vaienti, The compound Poisson distribution and return times in dynamical systems. Probabil. Theory Related Field **144**(3/4), 517–542 (2009)
12. N. Haydn, K. Wasilewska, Limiting distribution and error terms for the number of visits to balls in non-uniformly hyperbolic dynamical systems. Discrete Contin. Dyn. Syst. **36**, 2585–2611 (2016)
13. M. Holland, M. Todd, Weak convergence to extremal processes and record events for non-uniformly hyperbolic dynamical systems. Ergodic Theory Dyn. Syst. **39**(4), 980–1001 (2019)
14. P. Jung, F. Pène, H.-K. Zhang, Convergence to α-stable Lévy motion for chaotic billiards with cusps at flat points. Nonlinearity **33**(2), 807–839 (2019)
15. A. Katok, B. Hasselblatt, *Introduction to the Modern Theory of Dynamical Systems* (Cambridge University Press, Cambridge, 1995)
16. Y. Kifer, A. Rapaport, Geometric law for multiple returns untils a hazard. Nonlinearity **32**, 1525 (2019)
17. V. Lucarini, D. Faranda, A.C. Moreira Freitas, J.M. Freitas, T. Kuna, M. Holland, M. Nicol, M. Todd, S. Vaienti, *Extremes and Recurrence in Dynamical Systems*. Pure and Applied Mathematics (Hoboken) (Wiley, Hoboken, 2016). xi+295 pp
18. I. Melbourne, R. Zweimüller, Weak convergence to stable Lévy processes for nonuniformly hyperbolic dynamical systems. Annales de l'Institut Henri Poincaré, Probabilités et Statistiques **51**(2), 545–556 (2015)
19. F. Pène, B. Saussol, Back to balls in billiards. Comm. Math. Phys. **293**(3), 837–866 (2010)
20. F. Pène, B. Saussol, Poisson law for some non-uniformly hyperbolic dynamical systems with polynomial rate of mixing. Ergodic Theory Dyn. Syst. **36**(8), 2602–2626 (2016)
21. F. Pène, B. Saussol, Spatio-temporal Poisson processes for visits to small sets. Israel J. Math. **240**, 625–665 (2020)
22. B. Pitskel, Poisson limit theorem for Markov chains. Egodic Theory Dyn. Syst. **11**(3), 501–513 (1001)
23. S. Resnick, *Extreme Values, Regular Variation, and Point Processes*. Springer Series in Operation Research and Financial Engineering (Springer, Berlin, 2008)
24. B. Saussol, S. Troubetzkoy, S. Vaiénti, Recurrence, dimensions and Lyapunov exponents. J. Statist. Phys. **106**, 623–634 (2002)
25. M. Tyran-Kamińska, Weak convergence to Lévy stable processes in dynamical systems. Stochast. Dyn. **10**(02), 263–289 (2010)
26. L.-S. Young, Statistical properties of dynamical systems with some hyperbolicity. Ann. Math. **147**, 585–650 (1998)
27. L.-S. Young, Recurrence times and rates of mixing. Israel J. Math. **110**, 153–188 (1999)
28. R. Zweimüller, Mixing limit theorems for ergodic transformations. J. Theoret. Probabil. **20**, 1059–1071 (2007)

Part IV
Geometry and Thermodynamics Formation

Chapter 9
Rate of Mixing for Equilibrium States in Negative Curvature and Trees

Anne Broise-Alamichel, Jouni Parkkonen, and Frédéric Paulin

Abstract In this survey based on the recent book by the three authors, we recall the Patterson-Sullivan construction of equilibrium states for the geodesic flow on negatively curved orbifolds or tree quotients, and discuss their mixing properties, emphasizing the rate of mixing for (not necessarily compact) tree quotients via coding by countable (not necessarily finite) topological shifts. We give a new construction of numerous nonuniform tree lattices such that the (discrete time) geodesic flow on the tree quotient is exponentially mixing with respect to the maximal entropy measure: we construct examples whose tree quotients have an arbitrary space of ends or an arbitrary (at most exponential) growth type.

9.1   A Patterson-Sullivan Construction of Equilibrium States

We refer to [22, Chap. 3, 6, 7] and [3, Chap. 2, 3, 4] for details and complements on this section.

Let X be (see [3] for a more general framework)

- either a complete, simply connected Riemannian manifold $\widetilde{M}$ with dimension m at least 2 and pinched sectional curvature at most -1,
- or (the geometric realisation of) a simplicial tree $\mathbb{X}$ whose vertex degrees are uniformly bounded and at least 3. In this case, we respectively denote by $E\mathbb{X}$ and $V\mathbb{X}$ the sets of vertices and edges of $\mathbb{X}$. For every edge e, we denote by $o(e), t(e), \overline{e}$ its original vertex, terminal vertex and opposite edge.

Let us fix an indifferent basepoint x_* in $\widetilde{M}$ or in $V\mathbb{X}$.

A. Broise-Alamichel · F. Paulin (✉)
Laboratoire de mathématique d'Orsay, UMR 8628 CNRS, Université Paris-Saclay, Orsay Cedex, France
e-mail: anne.broise@universite-paris-saclay.fr; frederic.paulin@universite-paris-saclay.fr

J. Parkkonen
Department of Mathematics and Statistics, University of Jyväskylä, Finland
e-mail: jouni.t.parkkonen@jyu.fi

Recall (see for instance [2]) that a geodesic ray or line in X is an isometric embedding from $[0,+\infty[$ or $\mathbb{R}$ respectively into X, that two geodesic rays are *asymptotic* if they stay at bounded distance one from the other, and that the *boundary at infinity* of X is the space $\partial_\infty X$ of asymptotic classes of geodesic rays in X endowed with the quotient topology of the compact-open topology. When $X = \widetilde{M}$, up to a translation factor, two asymptotic geodesic rays converge exponentially fast one to the other, and $\partial_\infty \widetilde{M}$ is homeomorphic to the sphere $\mathbb{S}_{m-1}$ of dimension $m-1$. When X is a tree, up to a translation factor, two asymptotic geodesic rays coincide after a certain time, and $\partial_\infty \widetilde{M}$ is homeomorphic to a Cantor set.

For every x in X, the Gromov-Bourdon *visual distance* d_x on $\partial_\infty X$ seen from x (inducing the topology of $\partial_\infty X$) is defined by

$$d_x(\xi, \eta) = \lim_{t \to +\infty} e^{\frac{1}{2}(d(\xi_t, \eta_t) - d(x, \xi_t) - d(x, \eta_t))},$$

where $\xi, \eta \in \partial_\infty X$ and $t \mapsto \xi_t, \eta_t$ are any geodesic rays converging to ξ, η respectively. The visual distances seen from two points of X are Lipschitz equivalent.

Let Γ be a discrete group of isometries of X which is *nonelementary*, that is, does not preserve a subset of cardinality at most 2 in $X \cup \partial_\infty X$. When $X = \widetilde{M}$, this is equivalent to Γ not being virtually nilpotent. When X is a tree, we furthermore assume that X has no nonempty proper invariant subtree (this is not an important restriction, as one may always replace X by its unique minimal nonempty invariant subtree), and that Γ does not map an edge to its opposite one.

The *limit set* $\Lambda\Gamma$ of Γ is the smallest nonempty closed invariant subset of $\partial_\infty X$, which is the complement of the orbit Γx_* in its closure $\overline{\Gamma x_*}$, in the compactification $X \cup \partial_\infty X$ of X by its boundary at infinity.

Examples

(1) Let $\widetilde{M}$ be a symmetric space with negative curvature, e.g. the real hyperbolic plane $\mathbb{H}^2_\mathbb{R}$, and let Γ be an arithmetic lattice in $\mathrm{Isom}(\widetilde{M})$, e.g. $\Gamma = \mathrm{PSL}_2(\mathbb{Z})$ acting by homographies on the upper halfplane model of $\mathbb{H}^2_\mathbb{R}$ with constant curvature -1 (see for instance [10], and [17] for a huge amount of examples).

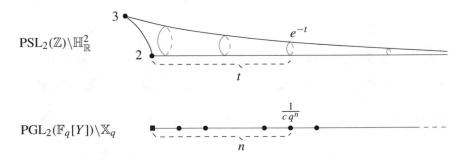

(2) For every prime power q, let $\mathbb{X}$ be the regular tree of degree $q + 1$, and let $\Gamma = \mathrm{PGL}_2(\mathbb{F}_q[Y])$, acting on $\mathbb{X}$ seen as the Bruhat-Tits tree $\mathbb{X}_q$ of PGL_2 over the local field $\mathbb{F}_q((Y^{-1}))$ (see for example [25], and [1] for a huge amount of examples).

Note that the pictures of the quotients $\Gamma \backslash X$ are very similar in the above two special examples, in particular

- the lengths of the closed horocycle quotients in $\mathrm{PSL}_2(\mathbb{Z}) \backslash \mathbb{H}^2_{\mathbb{R}}$ go exponentially to 0 (they are equal to e^{-t} where t is the distance of the horocycle quotient to the orbifold point of order 2),
- the orders of the vertex stabilisers along a geodesic ray in $\mathbb{X}_q$ lifting the quotient ray $\mathrm{PGL}_2(\mathbb{F}_q[Y]) \backslash \mathbb{X}_q$ increase exponentially (they are equal to $c\, q^n$ where c is a constant and n is the distance of the vertex to the origin of the ray), see for instance [3, §15.2].

Remark 9.1 Note that we allow torsion in Γ, as this is in particular important in the tree case; we allow $\Gamma \backslash X$ to be noncompact; and we allow Γ not to be a lattice. These allowances give in the tree case the possibility to have almost any (metrisable, compact, totally disconnect) space of ends and almost any type of asymptotic growth of the quotient $\Gamma \backslash X$ (linear, polynomial, exponential, etc.), see [1] and Section 9.3.3.

Recall that Γ is a lattice in X if either the Riemannian volume $\mathrm{Vol}(\Gamma \backslash \widetilde{M})$ of the quotient orbifold $\Gamma \backslash \widetilde{M}$ is finite, or if the *graph of groups volume*

$$\mathrm{Vol}(\Gamma \backslash\!\backslash \mathbb{X}) = \sum_{[x] \in \Gamma \backslash V\mathbb{X}} \frac{1}{\mathrm{Card}(\Gamma_x)}$$

(where Γ_x is the stabiliser of x in Γ) of the quotient graph of groups $\Gamma \backslash\!\backslash \mathbb{X}$ is finite. Note the analogy, in the two special examples above, between the computation of (most of) the volume of $\mathrm{PSL}_2(\mathbb{Z}) \backslash \mathbb{H}^2_{\mathbb{R}}$ as a converging integral of the lengths of the closed horocycle quotients and of the volume of $\mathrm{PGL}_2(\mathbb{F}_q[Y]) \backslash\!\backslash \mathbb{X}_q$ (which does converge by a geometric mean argument).

The Phase Space   Let $\mathscr{G}X$ be the space of geodesic lines $\ell : \mathbb{R} \to X$ in X, such that, when X is a tree, $\ell(0)$ is a vertex, endowed with the $\mathrm{Isom}(X)$-invariant distance (inducing its topology) defined by

$$d(\ell, \ell') = \int_{-\infty}^{+\infty} d(\ell(t), \ell'(t))\, e^{-2|t|}\, dt\;,$$

and with the $\mathrm{Isom}(X)$-equivariant *geodesic flow*, which is the one-parameter group of homeomorphisms

$$\mathfrak{g}^t : \ell \mapsto \{s \mapsto \ell(s + t)\}$$

for all $\ell \in \mathscr{G}X$, with continuous time parameter $t \in \mathbb{R}$ if $X = \widetilde{M}$ and discrete time parameter $t \in \mathbb{Z}$ if X is a tree. We again call *geodesic flow* and denote by $(\mathfrak{g}^t)_t$ the quotient flow on the *phase space* $\Gamma \backslash \mathscr{G}X$.

Note that the map from the unit tangent bundle $T^1\widetilde{M}$ endowed with Sasaki's metric to $\mathscr{G}\widetilde{M}$, which associates to a unit tangent vector v the unique geodesic line whose tangent vector at time $t = 0$ is v, is an $\mathrm{Isom}(\widetilde{M})$-equivariant bi-Hölder-continuous[1] homeomorphism, by which we identify the two spaces from now on.

Potentials on the Phase Space We now introduce the supplementary data (with physical origin) that we will consider on our phase space. Assume first that $X = \widetilde{M}$. Let $\widetilde{F} : T^1\widetilde{M} \to \mathbb{R}$ be a *potential*, that is, a Γ-invariant, bounded[2] Hölder-continuous real map on $T^1\widetilde{M}$. Two potentials $\widetilde{F}, \widetilde{F}^* : T^1\widetilde{M} \to \mathbb{R}$ are *cohomologous* (see for instance [16]) if there exists a Hölder-continuous, bounded, differentiable along flow lines, Γ-invariant function $\widetilde{G} : T^1\widetilde{M} \to \mathbb{R}$, such that, for every $v \in T^1\widetilde{M}$,

$$\widetilde{F}^*(v) - \widetilde{F}(v) = \frac{d}{dt}_{|t=0} \widetilde{G}(\mathfrak{g}^t v) \,.$$

For every $x, y \in \widetilde{M}$, let us define (with the obvious convention of being 0 if $x = y$) the integral of $\widetilde{F}$ between x and y, called the *amplitude* of $\widetilde{F}$ between x and y, to be

$$\int_x^y \widetilde{F} = \int_0^{d(x,y)} \widetilde{F}(\mathfrak{g}^t v)\, dt$$

and v is the tangent vector to the geodesic segment from x to y.

Now assume that X is a tree. Let $\widetilde{c} : E\mathbb{X} \to \mathbb{R}$ be a (logarithmic) *system of conductances* (see for instance [29]), that is, a Γ-invariant, bounded real map on $E\mathbb{X}$. Two systems of conductances $\widetilde{c}, \widetilde{c}^* : E\mathbb{X} \to \mathbb{R}$ are *cohomologous* if there exists a Γ-invariant function $\widetilde{f} : V\mathbb{X} \to \mathbb{R}$, such that for every $e \in E\mathbb{X}$

$$\widetilde{c}^*(e) - \widetilde{c}(e) = f(t(e)) - f(o(e)) \,.$$

For every $\ell \in \mathscr{G}\mathbb{X}$, we denote by $e_0^+(\ell) = \ell([0, 1]) \in E\mathbb{X}$ the first edge followed by ℓ, and we define $\widetilde{F} : \mathscr{G}\mathbb{X} \to \mathbb{R}$ as the map $\ell \mapsto \widetilde{c}(e_0^+(\ell))$. For every $x, y \in V\mathbb{X}$, we now define the *amplitude* of $\widetilde{F}$ between x and y, to be

$$\int_x^y \widetilde{F} = \sum_{i=1}^k \widetilde{c}(e_i)\, dt$$

---

[1] In order to deal with noncompactness issues, a map f between two metric spaces is *Hölder-continuous* if there exist $c, c' > 0$ and $\alpha \in \,]0, 1]$ such that for every x, y in the source space, if $d(x, y) \leq c$, then $d(f(x), f(y)) \leq c' d(x, y)^\alpha$.

[2] See [3, §3.2] for a weakening of this assumption.

if $(e_1, e_2, \ldots, e_k)$ is the geodesic edge path in $\mathbb{X}$ between x and y.

In both cases, we will denote by $F : \Gamma\backslash\mathscr{G}X \to \mathbb{R}$ the function on the phase space induced by $\widetilde{F}$ by taking the quotient modulo Γ, that we call the *potential* on $\Gamma\backslash\mathscr{G}X$. Note that we make no assumption of reversibility on F.

Cohomological Invariants Let us now introduce three cohomological invariants of the potentials on the phase space.

The *pressure* of F is the physical complexity associated with the potential F defined by

$$P_F = \sup_{\mu \ (\mathfrak{g}^t)_t\text{-invariant proba on } \Gamma\backslash\mathscr{G}X} \left(h_\mu + \int_{\Gamma\backslash\mathscr{G}X} F \, d\mu \right)$$

where h_μ is the metric entropy[3] of μ for the time 1 map $\mathfrak{g}^1$ of the geodesic flow.

The *critical exponent* of F is the weighted (by the exponential amplitudes) orbital growth rate of the group Γ, defined by

$$\delta_F = \lim_{n\to+\infty} \frac{1}{n} \ln \left(\sum_{\gamma\in\Gamma, \ n-1 < d(\imath_*, \gamma\imath_*) \leq n} \exp\left(\int_{x_*}^{\gamma x_*} \widetilde{F} \right) \right).$$

Note that the critical exponent δ_0 of the zero potential is the usual critical exponent of the group Γ (see for instance [21]). We have $\delta_F \in \,]-\infty, +\infty[$ since

$$\delta_0 + \inf \widetilde{F} \leq \delta_F \leq \delta_0 + \sup \widetilde{F} \, .$$

Note that $\delta_{F\circ\iota} = \delta_F$ where $\iota : \mathscr{G}X \to \mathscr{G}X$ is the involutive *time reversal map* defined by $\ell \mapsto \{t \mapsto \ell(-t)\}$.

The *period* for the potential F of a periodic orbit $\mathscr{O}$ of the geodesic flow $(\mathfrak{g}^t)_t$ on $\Gamma\backslash\mathscr{G}X$ is $\int_{\mathscr{O}} F = \int_{\ell(0)}^{\ell(t_{\mathscr{O}})} \widetilde{F}$ where $\ell \in \mathscr{G}X$ maps to $\mathscr{O}$ and

$$t_{\mathscr{O}} = \inf\{t > 0 \ : \ \Gamma\mathfrak{g}^t\ell = \Gamma\ell\}$$

---

[3]The metric entropy h_μ is the upper bound, for all measurable countable partitions ξ of $\Gamma\backslash\mathscr{G}X$, of

$$\lim_{k\to+\infty} \frac{1}{k} H_\mu(\xi \vee \cdots \vee g^{-k}\xi)$$

where $H_\mu(\xi) = -\sum_{E\in\xi} \mu(E) \ln \mu(E)$ is Shannon's entropy of the countable partition ξ, see for instance [11], and the join $\xi \vee \xi'$ of two partitions ξ and ξ' is the partition by the nonempty intersections of an element of ξ and an element of ξ'.

is the *length* of the periodic orbit $\mathcal{O}$. The *Gurevich pressure* of F is the growth rate of the exponentials of periods for F of the periodic orbits, defined by

$$\mathscr{P}_F^{\mathrm{Gur}} = \lim_{n \to +\infty} \frac{1}{n} \ln \sum_{\mathcal{O} \,:\, t_{\mathcal{O}} \leq n,\ \mathcal{O} \cap W \neq \emptyset} \exp\left(\int_{\mathcal{O}} F\right),$$

where the sum is taken over the periodic orbits $\mathcal{O}$ of $(\mathfrak{g}^t)_t$ on $\Gamma \backslash \mathscr{G} X$ with length at most n and meeting W, where W is any relatively compact open subset of $\Gamma \backslash \mathscr{G} X$ meeting the nonwandering set of the geodesic flow (recall that we made no assumption of compactness on the phase space).

Note that the above three limits exist, and are independent of the choices of x_* and W, and depend only on the cohomology class of the potential F.

The following result proved in [22, Theo. 4.1 and 6.1] extends the case of the zero potential due to Otal and Peigné [20].

Theorem 9.2 (Paulin-Pollicott-Schapira) *If $X = \widetilde{M}$ has pinched sectional curvatures with uniformly bounded derivatives,[4] then*

$$P_F = \delta_F = \mathscr{P}_F^{\mathrm{Gur}}. \qquad \qquad \Box$$

Note that the dynamics of the geodesic flow $(\mathfrak{g}^t)_t$ on the phase space $\Gamma \backslash \mathscr{G} X$ is very chaotic. In particular, there are lots of $(\mathfrak{g}^t)_t$-invariant measures on $\Gamma \backslash \mathscr{G} X$. We give two basic examples, and we will then construct, using potentials, a huge family of such measures.

Examples

(1) If $X = \widetilde{M}$, then the *Liouville measure* m_{Liou} on $T^1 M = \Gamma \backslash (T^1 \widetilde{M})$ is the measure on $T^1 M$ which disintegrates, with respect to the canonical footpoint projection $T^1 M \to M$, over the Riemannian measure vol_M of the Riemannian orbifold $M = \Gamma \backslash \widetilde{M}$, with conditional measures on the fibers the spherical measures $\mathrm{vol}_{T_x^1 M}$ on the (orbifold) unit tangent spheres at the points x in M:

$$dm_{\mathrm{Liou}}(v) = \int_{x \in M} d\,\mathrm{vol}_{T_x^1 M}(v)\ d\,\mathrm{vol}_M(x).$$

---

[4]This assumption on the derivatives was forgotten in the statements of [20, 22], but is used in the proofs.

(2) For every periodic orbit $\mathcal{O}$ of the geodesic flow $(g^t)_t$ on $\Gamma\backslash\mathscr{G}X$, we denote by $\mathscr{L}_\mathcal{O}$ the Lebesgue measure[5] (when $X = \widetilde{M}$) or counting measure (when X is a tree) of $\mathcal{O}$. This is a $(g^t)_t$-invariant measure on $\Gamma\backslash\mathscr{G}X$ with support $\mathcal{O}$.

The main class of invariant measures we will study is the following one, and the terminology has been mostly introduced by Sinai, Ruelle, Bowen, see for instance [24]. A $(g^t)_t$-invariant probability measure μ on the phase space $\Gamma\backslash\mathscr{G}X$ is an *equilibrium state* for the potential F if it realizes the upper bound defining the pressure of F, that is, if

$$h_\mu + \int_{\Gamma\backslash\mathscr{G}X} F \, d\mu = P_F \ .$$

The remainder of this section is devoted to the problems of **existence, unique-ness and explicit construction** of equilibrium states.

Gibbs Cocycles As for instance defined by Hamenstädt, the (normalised) *Gibbs cocycle* of the potential F is the function $C : \partial_\infty X \times \widetilde{M} \times \widetilde{M} \to \mathbb{R}$ when $X = \widetilde{M}$ or the function $C : \partial_\infty X \times V\mathbb{X} \times V\mathbb{X} \to \mathbb{R}$ when X is a tree, defined by the following limit of difference of amplitudes for the renormalised potential

$$C_\xi(x, y) = \lim_{t \to +\infty} \int_y^{\xi_t} (\widetilde{F} - \delta_F) - \int_x^{\xi_t} (\widetilde{F} - \delta_F),$$

where $t \mapsto \xi_t$ is any geodesic ray converging to ξ. The limit does exist. The Gibbs cocycle is Γ-invariant (for the diagonal action) and locally Hölder-continuous. It does satisfy the cocycle property $C_\xi(x, z) = C_\xi(x, y) + C_\xi(y, z)$ for all x, y, z. Furthermore, there exist constants $c_1, c_2 > 0$ (depending only on the bounds on the potential $\widetilde{F}$ and on the pinching of the sectional curvature, when $X = \widetilde{M}$) such that if $d(x, y) \leq 1$, then $C_\xi(x, y) \leq c_1 d(x, y)^{c_2}$. See [3, §3.4].

---

[5]If the length of $\mathcal{O}$ is T and if $v \in T^1\widetilde{M}$ maps into $\mathcal{O}$ by the canonical projection $T^1\widetilde{M} \to T^1M$, the Lebesgue measure $\mathscr{L}_\mathcal{O}$ of $\mathcal{O}$ is the pushforward by $t \mapsto \Gamma g^t v$ of the Lebesgue measure on $[0, T]$.

Patterson Densities A (normalised) *Patterson density* of the potential F is a Γ-equivariant family $(\mu_x)_{x\in X}$ of pairwise absolutely continuous (positive, Borel) measures on $\partial_\infty X$, whose support is $\Lambda\Gamma$, such that

$$\gamma_*\mu_x = \mu_{\gamma x} \quad \text{and} \quad \frac{d\mu_x}{d\mu_y}(\xi) = e^{-C_\xi(x,\,y)} \tag{9.1.1}$$

for every $\gamma \in \Gamma$, for all $x, y \in X$, and for (almost) every $\xi \in \partial_\infty X$.

Patterson densities do exist and they satisfy the following Mohsen's shadow lemma (see for instance [3, §4.1]):

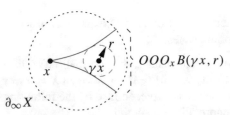

Define the *shadow* $\mathscr{O}_x E$ seen from x of a subset E of X as the set of points at infinity of the geodesic rays from x through E. Then for every $x \in X$, if $r > 0$ is large enough, there exists $\kappa > 0$ such that for every $\gamma \in \Gamma$, we have

$$\frac{1}{\kappa} \exp\left(\int_x^{\gamma x} (\widetilde{F} - \delta_F)\right) \le \mu_x\big(\mathscr{O}_x B(\gamma x, r)\big) \le \kappa \, \exp\left(\int_x^{\gamma x} (\widetilde{F} - \delta_F)\right). \tag{9.1.2}$$

Gibbs Measures The *Hopf parametrisation* of X at x_* is the map from $\mathscr{G}X$ to $(\partial_\infty X \times \partial_\infty X - \text{Diag}) \times R$, where $R = \mathbb{R}$ if $X = \widetilde{M}$ and $R = \mathbb{Z}$ if X is a tree, defined by

$$\ell \mapsto (\ell_-, \ell_+, t)$$

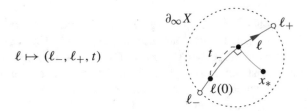

where ℓ_-, ℓ_+ are the original and terminal points at infinity of the geodesic line ℓ, and t is the algebraic distance along ℓ between the footpoint $\ell(0)$ and the closest point to x_* on the geodesic line. It is a Hölder-continuous homeomorphism (for the previously defined distances). Up to translations on the third factor, it does not depend on the basepoint x_* and is Γ-invariant, see for instance [3, §2.3 and §3.1]. The geodesic flow acts by translations on the third factor.

Let $(\mu_x)_{x\in X}$ and $(\mu_x^\iota)_{x\in X}$ be Patterson densities for the potentials F and $F \circ \iota$ respectively, where $\iota : \Gamma\ell \mapsto \Gamma\{t \mapsto \ell(-t)\}$ is the time reversal on the phase space $\Gamma\backslash\mathscr{G}X$. We denote by C^ι the Gibbs cocycle of the potential $F \circ \iota$. We denote by

dt the Lebesgue or counting measure on R. The measure on $\mathscr{G}X$ defined using the Hopf parametrisation at x_* by

$$d\widetilde{m}_F(\ell) = \frac{d\mu^\iota_{x_*}(\ell_-)\, d\mu_{x_*}(\ell_+)\, dt}{\exp\left(C^\iota_{\ell_-}(x_*,\, \ell(0)) + C_{\ell_+}(x_*,\, \ell(0))\right)}$$

is a σ-finite nonzero measure on $\mathscr{G}X$. By Eq. (9.1.1) and by the invariance of the measure dt under translations, it is independent of the choice of basepoint x_*, hence is Γ-invariant and $(\mathfrak{g}^t)_t$-invariant. Therefore it induces a σ-finite nonzero $(\mathfrak{g}^t)_t$-invariant measure on $\Gamma \backslash \mathscr{G}X$, called the *Gibbs measure* on the phase space and denoted by m_F.

Examples

(1) When $F = 0$, then the Gibbs measure is called the Bowen-Margulis measure (see for instance [23]).
(2) When $X = \widetilde{M}$ and $\widetilde{F}$ is the *unstable Jacobian*, that is, for every $v \in T^1\widetilde{M}$,

$$\widetilde{F}^{\mathrm{su}}(v) = -\frac{d}{dt}_{|t=0} \ln\left(\begin{array}{c}\text{Jacobian of restriction of } \mathfrak{g}^t \text{ to}\\ \text{strong unstable leaf } W^{su}(v)\end{array}\right),$$

we have the following result (see [22, §7], in particular for weaker assumptions). When M has variable sectional curvature, the Liouville measure and the Bowen-Margulis measure might be quite different. The following result in particular says that the huge family of Gibbs measures interpolates between the Liouville measure and the Bowen-Margulis measure. This sometimes provides common proofs of properties satisfied by both the Liouville measure and the Bowen-Margulis measure.

Theorem 9.3 (Paulin-Pollicott-Schapira) *If $X = \widetilde{M}$ has pinched sectional curvatures with uniformly bounded derivatives, then $\widetilde{F}^{\mathrm{su}}$ is Hölder-continuous and bounded. If $\widetilde{M}$ has a cocompact lattice and if $(\mathfrak{g}^t)_t$ is completely conservative[6] for the Liouville measure, then*

$$m_{F_{\mathrm{su}}} = m_{\mathrm{Liou}}. \qquad \square$$

---

[6]That is, every wandering set has measure zero.

The following result, due to Bowen and Ruelle when M is compact and to Otal-Peigné [20] when $F = 0$, completely solves the problems of existence, uniqueness and explicit construction of equilibrium states, see [22, §6].

Theorem 9.4 (Paulin-Pollicott-Schapira) *Assume that* $X = \widetilde{M}$ *has pinched sectional curvatures with uniformly bounded derivatives.[7] If the Gibbs measure* m_F *is finite, then* $\overline{m_F} = \frac{m_F}{\|m_F\|}$ *is the unique equilibrium state. Otherwise, there is no equilibrium state.* □

We refer to Sect. 9.3.2 for an analogous statement when X is a tree, whose proof uses completely different techniques.

9.2 Basic Ergodic Properties of Gibbs Measures

We refer to [22, Chap. 3, 5, 8] and [3, Chap. 4] for details and complements on this section.

9.2.1 The Gibbs Property

In this section, we justify the terminology of Gibbs measures used above.

For every $\ell \in \Gamma \backslash \mathscr{G} X$, say $\ell = \Gamma \widetilde{\ell}$, for every $r > 0$ and for all $t, t' \geq 0$, the (Bowen or) *dynamical ball* $B(\ell; t, t', r)$ in the phase space $\Gamma \backslash \mathscr{G} X$ centered at ℓ with parameters t, t', r is the image in $\Gamma \backslash \mathscr{G} X$ of the set of geodesic lines in $\mathscr{G} X$ following the lift $\widetilde{\ell}$ at distance less than r in the time interval $[-t', t]$, that is, the image in $\Gamma \backslash \mathscr{G} X$ of

$$B(\widetilde{\ell}; t, t', r) = \left\{ \ell' \in \mathscr{G} X \ : \ \sup_{s \in [-t', t]} d_X(\widetilde{\ell}(s), \ell'(s)) < r \right\}.$$

The following definition of the Gibbs property is well adapted to the possible noncompactness of the phase space $\Gamma \backslash \mathscr{G} X$. A $(\mathfrak{g}^t)_t$-invariant measure m' on $\Gamma \backslash \mathscr{G} X$ satisfies the *Gibbs property* for the potential F with *Gibbs constant* $c(F) \in \mathbb{R}$ if for every compact subset K of $\Gamma \backslash \mathscr{G} X$, there exists $r > 0$ and $c_{K,r} \geq 1$ such that for all $t, t' \geq 0$ large enough, for every ℓ in $\Gamma \backslash \mathscr{G} X$ with $\mathfrak{g}^{-t'} \ell, \mathfrak{g}^t \ell \in K$, we have

$$\frac{1}{c_{K,r}} \leq \frac{m'\big(B(\ell; t, t', r)\big)}{e^{\int_{-t'}^{t} (F(\mathfrak{g}^t \ell) - c(F)) \, dt}} \leq c_{K,r}.$$

---

[7]This assumption on the derivatives was forgotten in the statements of [20, 22].

The following result is due to [22, §3.8] when $X = \tilde{M}$ and [3, §4.2] in general.

Proposition 9.5 *The Gibbs measure m_F satisfies the Gibbs property for F with Gibbs constant $c(F)$ equal to the critical exponent δ_F.* □

Let us give a sketch of its proof, which explains the decorrelation of the influence of the two points at infinity of the geodesic lines, using the fact that the Gibbs measure is absolutely continuous with respect to a product measure in the Hopf parametrisation. The key geometric lemma is the following one.

Lemma 9.6 *For every $r > 0$, there exists $t_r > 0$ such that for all $t, t' \geq t_r$ and $\ell \in \mathscr{G}X$, we have, using the Hopf parametrisation at the footpoint $\ell(0)$,*

$$\mathscr{O}_{\ell(0)} B(\ell(-t'), r) \times \mathscr{O}_{\ell(0)} B(\ell(t), r) \times\ ]-1, 1[\ \subset\ B(\ell; t, t', 2r+2)$$

$$B(\ell; t, t', r) \subset\ \mathscr{O}_{\ell(0)} B(\ell(-t'), 2r) \times \mathscr{O}_{\ell(0)} B(\ell(t), 2r) \times\ ]-r, r[\ .$$

Let us give a proof-by-picture of the first claim, the second one being similar. See the following picture. If a geodesic line ℓ' has its points at infinity ℓ'_- and ℓ'_+ in the shadows seen from $\ell(0)$ of $B(\ell(-t'), r)$ and $B(\ell(-t'), r)$ respectively, then by the properties of triangles in negatively curved spaces, if t and t' are large, then the image of ℓ' is close to the union of the images of the geodesic rays from $\ell(0)$ to ℓ_- and ℓ_+. The control on the time parameter in Hopf parametrisation then says that ℓ' is staying at bounded distance from ℓ in the time interval $[-t', t]$.

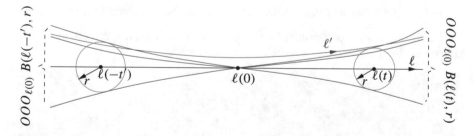

We now conclude the proof of Proposition 9.5 by using the boundedness of the Gibbs cocycles C and C' on a given compact subset K in order to control the denominator in the formula giving $\tilde{m}_F$, and by using Mohsen's shadow lemma (see Eq. (9.1.2)) which estimates the Patterson measures of shadows of balls.

9.2.2 Ergodicity

In this section, we study the ergodicity property of the Gibbs measures under the geodesic flow in the phase space.

The *Poincaré series* of the potential F is

$$Q_F(s) = \sum_{\gamma \in \Gamma} \exp \left(\int_{x_*}^{\gamma x_*} (\widetilde{F} - s) \right).$$

It depends on the basepoint x_*, but its convergence or divergence does not. It converges if $s > \delta_F$ and diverges for $s < \delta_F$, by the definition of the critical exponent δ_F.

The following result has a long history, and we refer for instance to [22, §5] and [3, §4.2] for proofs, and proofs of its following two corollaries.

Theorem 9.7 (Hopf-Tsuji-Sullivan-Roblin) *The following assertions are equivalent.*

(1) *The Poincaré series of F diverges at the critical exponent of F :* $Q_F(\delta_F) = +\infty$.
(2) *The group action* $(\partial_\infty X \times \partial_\infty X - \mathrm{Diag}, \mu_{x_*}^\iota \otimes \mu_{x_*}, \Gamma)$ *is ergodic and completely conservative.*
(3) *The geodesic flow on the phase space endowed with the Gibbs measure* $(\Gamma \backslash \mathscr{G} X, m_F, (\mathfrak{g}^t)_t))$ *is ergodic and completely conservative.* □

Corollary 9.8 *If* $Q_F(\delta_F) = +\infty$, *then there exists a Patterson density for F, unique up to a positive scalar. It is atomless, and the diagonal in* $\partial_\infty X \times \partial_\infty X$ *has measure 0 for the product measure* $\mu_{x_*}^\iota \otimes \mu_{x_*}$. □

Let us give a sketch of the very classical proof of the first claim of this corollary.

Existence Using the properties of negatively curved spaces, one can prove, denoting by $\mathscr{D}_x$ the Dirac mass at a point x, that one can take

$$\mu_x = \lim_{s_i \to \delta_F^+} \frac{1}{Q_F(s_i)} \sum_{\gamma \in \Gamma} \exp \left(\int_x^{\gamma x_*} (\widetilde{F} - s_i) \right) \mathscr{D}_{\gamma x_*},$$

where the atomic measure before taking the limit is, when $x = x_*$, a probability measure, hence has, for some sequence $(s_i)_{i \in \mathbb{N}}$ in $]\delta_F, +\infty[$ converging to δ_F, a weakstar converging subsequence in the compact space of probability measures on the compact space $X \cup \partial_\infty X$.

Uniqueness Let $(\mu_x')_x$ be another Patterson density. Up to positive scalars, we may assume that μ_{x_*} and μ_{x_*}' are probability measures. Then $(\omega_x = \frac{1}{2}(\mu_x + \mu_x'))_x$ is a Patterson density, μ_{x_*} is absolutely continuous with respect to ω_{x_*}, and by ergodicity, the Radon-Nikodym derivative $\frac{d\mu_{x_*}}{d\omega_{x_*}}$ is almost everywhere constant, hence the probability measures μ_{x_*} and ω_{x_*} are equal, hence $\mu_{x_*} = \mu_{x_*}'$.

Corollary 9.9 *If m_F is finite, then $Q_F(\delta_F) = +\infty$ (hence $(\mathfrak{g}^t)_t$) is ergodic) and the normalised Gibbs measure $\overline{m_F} = \frac{m_F}{\|m_F\|}$ is a cohomological invariant of the potential F.*   □

9.2.3   Mixing

In this section, we study the mixing property of the Gibbs measures under the geodesic flow in the phase space. Recall that the *length spectrum* for the action of Γ on X is the subgroup of $\mathbb{R}$ (hence of $\mathbb{Z}$ when X is a tree) generated by the set of lengths of the closed geodesic in $\Gamma\backslash X$ (or, in dynamical terms, of the set of lengths of periodic orbits of the geodesic flow on the phase space). See for instance [22, §8.1] when $X = \widetilde{M}$ and [3, §4.4] when X is a tree for a proof of the following result, which crucially uses the fact that the Gibbs measure is absolutely continuous with respect to a product measure in the Hopf parametrisation.

Theorem 9.10 (Babillot) *If the Gibbs measure m_F is finite, then the following assertions are equivalent.*

(1) *The Gibbs measure m_F is mixing under the geodesic flow $(\mathfrak{g}^t)_t$.*
(2) *The geodesic flow $(\mathfrak{g}^t)_t$ is topologically mixing on its nonwandering set in the phase space.*
(3) *The length spectrum of Γ is dense in $\mathbb{R}$ if $X = \widetilde{M}$ or equal to $\mathbb{Z}$ if X is a tree.*   □

We summarise in the following result the known properties of the rate of mixing of the geodesic flow in the manifold case when $X = \widetilde{M}$ (see [3, §9.1]), referring to Sect. 9.3 for the tree case, whose proof turns out to be quite different.

Let $\alpha \in]0, 1]$ and let $\mathscr{C}_b^\alpha(Z)$ be the Banach space[8] of bounded α-Hölder-continuous functions on a metric space Z. When $X = \widetilde{M}$, we will say that the (continuous time) geodesic flow on the phase space $T^1M = \Gamma\backslash T^1\widetilde{M}$ is *exponentially mixing for the α-Hölder regularity* or that it has *exponential decay of α-Hölder correlations* for the potential F if there exist two constants $c', \kappa > 0$ such that for all $\phi, \psi \in \mathscr{C}_b^\alpha(T^1M)$ and $t \in \mathbb{R}$, we have

$$\left| \int_{T^1M} \phi \circ \mathfrak{g}^{-t} \, \psi \, d\overline{m_F} - \int_{T^1M} \phi \, d\overline{m_F} \int_{T^1M} \psi \, d\overline{m_F} \right| \leq c' \, e^{-\kappa|t|} \, \|\phi\|_\alpha \, \|\psi\|_\alpha .$$

---

[8]Recall that its norm (taking into account the possible noncompactness of Z) is given by

$$\|f\|_\alpha = \|f\|_\infty + \sup_{\substack{x, y \in Z \\ 0 < d(x, y) \leq 1}} \frac{|f(x) - f(y)|}{d(x, y)^\alpha} .$$

Theorem 9.11 *Assume that* $X = \tilde{M}$ *and that* $M = \Gamma \backslash \tilde{M}$ *is compact. Then the geodesic flow on the phase space* $T^1 M$ *has exponential decay of Hölder correlations if*

- M *is two-dimensional, by [6],*
- M *is* $1/9$-*pinched and* $F = 0$, *by [7, Coro. 2.7],*
- *the potential* F *is the unstable Jacobian* F^{su}, *so that, up to a positive scalar,* m_F *is the Liouville measure* m_{Liou}, *by [15], see also [27], [19, Coro. 5] who give more precise estimates,*
- M *is locally symmetric by [26], see also [14, 18] for some noncompact cases.* □

Note that this gives only a very partial picture of the rate of mixing of the geodesic flow in negative curvature, and it would be interesting to have a complete result. Stronger results exist for the Sobolev regularity when $\tilde{M}$ is a symmetric space, $F = 0$ and Γ is an arithmetic lattice (the Gibbs measure then coincides, up to a multiplicative constant, with the Liouville measure): see for instance [12, Theorem 2.4.5], using spectral gap properties given by [5, Theorem 3.1]. But this still does not give a complete answer.

9.3 Coding and Rate of Mixing for Geodesic Flows on Trees

We refer to [3, Chap. 5 and 9.2] for details and complements on this section.

From now on, we assume that X is (the geometric realisation of) a simplicial tree $\mathbb{X}$, and we write $\mathscr{G}\mathbb{X}$ instead of $\mathscr{G}X$. We consider the discrete group Γ, the system of conductances $\tilde{c}$ and the associated potential F on the phase space $\Gamma \backslash \mathscr{G}\mathbb{X}$ as introduced in Sect. 9.1.

The study of the rate of mixing of the (discrete time) geodesic flow on the phase space uses coding theory. But since, as explained, we make no assumption of compactness on the phase space, and no hypothesis of being without torsion on the group Γ in the huge class of examples described in Sect. 9.1, the coding theory requires more sophisticated tools than subshifts of finite type.

9.3.1 Coding

Let $\mathscr{A}$ be a countable discrete set, called an *alphabet*, and let $A = (A_{i,j})_{i,j \in \mathscr{A}}$ be an element in $\{0, 1\}^{\mathscr{A} \times \mathscr{A}}$, called a *transition matrix*. The (two-sided, countable state) *topological shift*[9] with alphabet $\mathscr{A}$ and transition matrix A is the topological

---

[9]We prefer not to use the frequent terminology of *topological Markov shift* as it could be misleading, many probability measures invariant under general topological shifts do not satisfy

dynamical system (Σ, σ), where Σ, called the *shift space*, is the closed subset of the topological product space $\mathscr{A}^{\mathbb{Z}}$ of A-*admissible* two-sided infinite sequences, defined by

$$\Sigma = \left\{ x = (x_n)_{n \in \mathbb{Z}} \in \mathscr{A}^{\mathbb{Z}} \; : \; \forall\, n \in \mathbb{Z}, \quad A_{x_n, x_{n+1}} = 1 \right\},$$

and $\sigma : \Sigma \to \Sigma$ is the (two-sided) *shift* defined by

$$\forall\, x \in \Sigma, \; \forall\, n \in \mathbb{Z}, \quad (\sigma(x))_n = x_{n+1} \,.$$

We endow Σ with the distance

$$d(x, x') = \exp\left(- \sup \left\{ n \in \mathbb{N} \; : \; \forall\, i \in \{-n, \dots, n\}, \quad x_i = x_i' \right\} \right).$$

Let us denote by $\mathbb{Y}$ the (countable) quotient graph[10] $\Gamma \backslash \mathbb{X}$. For every vertex or edge $x \in V\mathbb{Y} \cup E\mathbb{Y}$, we fix a lift $\widetilde{x}$ in $V\mathbb{X} \cup E\mathbb{X}$, and we define $G_x = \Gamma_{\widetilde{x}}$ to be the stabiliser of $\widetilde{x}$ in Γ.

For every $e \in E\mathbb{Y}$, we assume that $\widetilde{\overline{e}} = \overline{\widetilde{e}}$. But there is no reason in general for the equality $t\widetilde{(e)} = t(\widetilde{e})$ to hold. We fix $g_e \in \Gamma$ mapping $t(\widetilde{e})$ to $t(\widetilde{e})$ (which does exist), and we denote by $\rho_e : G_e = \Gamma_{\widetilde{e}} \to \Gamma_{t\widetilde{(e)}} = G_{t(e)}$ the conjugation $g \mapsto g_e^{-1} g\, g_e$ by g_e on G_e (noticing that the stabiliser $\Gamma_{\widetilde{e}}$ is contained in the stabiliser $\Gamma_{t(\widetilde{e})}$).

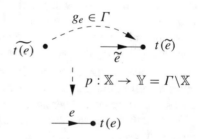

Let us try to code a geodesic line in the phase space $\Gamma \backslash \mathscr{G}\mathbb{X}$. The natural starting point is to write it as $\Gamma \ell$ for some $\ell \in \mathscr{G}X$, that is, to choose one of its lifts. We then have to construct a coding which is independent of the choice of this lift. For every $i \in \mathbb{Z}$, let us denote by $f_i = \ell([i, i+1])$ the i-th edge followed by ℓ, and by e_i (also denoted by $e_{i+1}^-(\ell)$ for later use) its image by the canonical $p : \mathbb{X} \to \mathbb{Y} = \Gamma \backslash \mathbb{X}$, which seems fit to be a natural part of the coding of ℓ. Since we will need to translate through our coding the fact that ℓ is geodesic, hence has no backtracking, the edge e_{i+1} (also denoted by $e_{i+1}^+(\ell)$ for later use) following e_i seems to have a role to play.

---

the Markov chain property that the probability to pass from one state to another depends only on the previous state, not of all past states.

[10]The fact that the canonical projection is a morphism of graphs is the reason why we assumed Γ to be acting without mapping an edge to its inverse.

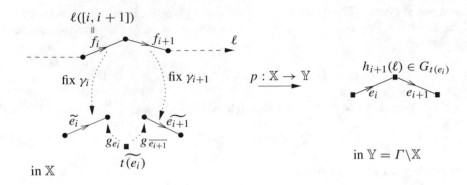

Since the terminal point of f_i is the original point of f_{i+1}, the terminal point of e_i is naturally also the original point of e_{i+1}. But there is no reason for the terminal point of the choosen lift $\widetilde{e_i}$ to also be the original point of the choosen lift $\widetilde{e_{i+1}}$. Since f_i and $\widetilde{e_i}$ both map by p to e_i, we may fix $\gamma_i \in \Gamma$ such that $\gamma_i f_i = \widetilde{e_i}$, for every $i \in \mathbb{Z}$.

Now, note that the vertex stabilizers in Γ of vertices of $\mathbb{X}$ are in general nontrivial (and we explained in Sect. 9.1 that it is important to allow them to become very large in order to have numerous dynamically interesting noncompact quotients of simplicial trees). The construction (see the above diagram) provides a natural element $g_{e_i}^{-1} \gamma_i \gamma_{i+1}^{-1} g_{\overline{e_{i+1}}}$ which stabilises the lifted vertex $t(\widetilde{e_i})$, hence belongs to $G_{t(e_i)}$. Since we made choices for the elements γ_i, the element $g_{e_i}^{-1} \gamma_i \gamma_{i+1}^{-1} g_{\overline{e_{i+1}}}$ gives a well-defined double class $h_{i+1}(\ell)$ in $\rho_{e_i}(G_{e_i})\backslash G_{t(e_i)}/\rho_{\overline{e_{i+1}}}(G_{e_{i+1}})$, which also seems fit to be another natural piece of the coding of ℓ.

It turns out that this construction is indeed working. We take as alphabet the (countable) set

$$\mathscr{A} = \left\{ (e^-, h, e^+) \ : \ \begin{array}{l} e^\pm \in E\mathbb{Y} \text{ with } t(e^-) = o(e^+) \\ h \in \rho_{e^-}(G_{e^-})\backslash G_{o(e^+)}/\rho_{\overline{e^+}}(G_{e^+}) \text{ with } h \neq [1] \text{ if } \overline{e^+} = e^- \end{array} \right\}.$$

This last assumption of conditional nontriviality of the double class codes the fact that ℓ being a geodesic line, the edge f_{i+1} is not the opposite edge of f_i, though e_{i+1} might be the opposite edge of e_i. And since in the tree $\mathbb{X}$, being locally geodesic implies being geodesic, it is very reasonable that we have captured through our coding all the geodesic properties of the geodesic lines and translated them into symbolic terms. We take as transition matrix A over the alphabet $\mathscr{A}$ the matrix with entries

$$A_{(e^-, h, e^+),\, (e'^-, h', e'^+)} = \begin{cases} 1 \text{ if } e^+ = e'^- \\ 0 \text{ otherwise,} \end{cases}$$

which just says that we are glueing together the coding of pairs of consecutive edges of the geodesic line. Note that since the tree is locally finite, the transition matrix A has finitely many nonzero entries on each row and column, hence the associated shift space Σ is locally compact.

We then refer to [3, §5.2] for a proof of the following result, though almost everything is in the above picture! We denote by $F_{\text{symb}} : \Sigma \to \mathbb{R}$ the locally constant map which associates to $((e_i^-, h_i, e_i^+))_{i \in \mathbb{Z}}$ the image $\tilde{c}(\tilde{e_0^+})$ by the system of conductances of the lift of its first edge.

Theorem 9.12 *The map*

$$\Theta : \begin{cases} \Gamma \backslash \mathscr{G} \mathbb{X} \longrightarrow \Sigma \\ \Gamma \ell \mapsto \left((e_i^-(\ell), h_i(\ell), e_i^+(\ell)) \right)_{i \in \mathbb{Z}} \end{cases}$$

is a bilipschitz homeomorphism, conjugating the time 1 map of the (discrete time) geodesic flow $(\mathfrak{g}^t)_{t \in \mathbb{Z}}$ to the shift σ. Furthermore,

(1) *(Σ, σ) is topologically transitive,*[11]
(2) *if the Gibbs measure m_F is finite and if the length spectrum of Γ is equal to $\mathbb{Z}$, then the probability measure $\mathbb{P} = \Theta_* \overline{m_F}$ is mixing for the shift σ on Σ,*
(3) *the measure $\mathbb{P}$ satisfies the Gibbs property on (Σ, σ) with Gibbs constant δ_F for the potential F_{symb},*[12]
(4) *if $(Z_n : x \mapsto x_n)_{n \in \mathbb{Z}}$ is the canonical random process in symbolic dynamics, then the pair $((Z_n)_{n \in \mathbb{Z}}, \mathbb{P})$ is not always a Markov chain.* □

This last claim has lead to an erratum in the paper [13]. The pair $((Z_n)_{n \in \mathbb{Z}}, \mathbb{P})$ is not a Markov chain for instance in Example (2) at the beginning of Sect. 9.1, when $\mathbb{X} = \mathbb{X}_q$ and $\Gamma = \text{PGL}_2(\mathbb{F}_q[Y])$.[13]

---

[11] This comes from the assumption that there is no nontrivial proper Γ-invariant subtree in $\mathbb{X}$, since then $\partial_\infty X = \Lambda \Gamma$, implying that the nonwandering set of the geodesic flow $(\mathfrak{g}^t)_{t \in \mathbb{Z}}$ is the full phase space $\Gamma \backslash \mathscr{G} \mathbb{X}$.

[12] That is, with a formulation adapted to the possibility that the alphabet $\mathscr{A}$ may be infinite, for every finite subset E of the alphabet $\mathscr{A}$, there exists $C_E \geq 1$ such that for all $p \leq q$ in $\mathbb{Z}$ and for every $x = (x_n)_{n \in \mathbb{Z}} \in \Sigma$ such that $x_p, x_q \in E$, we have

$$\frac{1}{C_E} \leq \frac{\mathbb{P}([x_p, x_{p+1}, \ldots, x_{q-1}, x_q])}{e^{-\delta_F(q-p+1) + \sum_{n=p}^{q} F_{\text{symb}}(\sigma^n x)}} \leq C_E .$$

where $[x_p, x_{p+1}, \ldots, x_{q-1}, x_q]$ is the cylinder $\{(y_n)_{n \in \mathbb{Z}} \in \Sigma : \text{if } p \leq n \leq q \text{ then } y_n = x_n\}$.

[13] As noticed by J.-P. Serre [25], the image of almost every geodesic line of $\mathbb{X}$ in the quotient ray $\Gamma \backslash X$ is a broken line which makes infinitely many back-and-forths from the origin of the quotient ray.

9.3.2 Variational Principle for Simplicial Trees

The first corollary of the coding results in the previous section is the following existence and uniqueness result of equilibrium states for the geodesic flow on the phase space $\Gamma \backslash \mathscr{G} \mathbb{X}$ for the potential F.

Corollary 9.13 If m_F is finite, then $\overline{m_F} = \frac{m_F}{\|m_F\|}$ is the unique equilibrium state for F under the geodesic flow $(\mathfrak{g}^t)_{t \in \mathbb{Z}}$ on $\Gamma \backslash \mathscr{G} \mathbb{X}$, and furthermore

$$P_F = \delta_F. \qquad \Box$$

We only give a sketch of a proof, refering to [3, §5.4] for a complete one. We use the coding given in Theorem 9.12 with its properties (in particular the fact that it satisfies the Gibbs property for a symbolic potential related to the potential F).

Let (Σ, σ) be a topological shift, with countable alphabet $\mathscr{A}$. A σ-invariant probability measure m on Σ is a *weak*[14] *Gibbs measure* for a map $\phi : \Sigma \to \mathbb{R}$ with Gibbs constant $c(m) \in \mathbb{R}$ if for every $a \in \mathscr{A}$, there exists a constant $c_a \geq 1$ such that for all $n \in \mathbb{N} - \{0\}$ and x in the cylinder $[a] = \{y = (y_n)_{n \in \mathbb{Z}} \in \Sigma : y_0 = a\}$ such that $\sigma^n(x) = x$, we have

$$\frac{1}{c_a} \leq \frac{m([x_0, x_1, \ldots, x_{n-1}])}{e^{\sum_{i=0}^{n-1} (\phi(\sigma^i x) - c(m))}} \leq c_a .$$

The following result of Buzzi is proved in [3, Appendix], with a much weaker regularity assumption on ϕ, and it concludes the proof of Corollary 9.13.

Theorem 9.14 (Buzzi) Let (Σ, σ) be a topological shift and $\phi : \Sigma \to \mathbb{R}$ a bounded Hölder-continuous function. If m is a weak Gibbs measure for ϕ with Gibbs constant $c(m)$, then $P_\phi = c(m)$ and m is the unique equilibrium state for the potential ϕ. $\qquad \Box$

9.3.3 Rate of Mixing for Simplicial Trees

Let us first recall the definition of an exponential mixing rate for discrete time dynamical systems.

There is absolutely no way to predict the probability of behaviour of the geodesic line image at a given time in terms of its recent past probabilities (except that when it starts to go down, it has to go down all the way to the origin).

[14]The terminology comes from the fact that the assumptions bear only on the periodic points of σ.

Let (Z, m, T) be a dynamical system with (Z, m) a metric probability space and $T : Z \mapsto Z$ a (not necessarily invertible) measure preserving map. For all $n \in \mathbb{N}$ and $\phi, \psi \in \mathbb{L}^2(m)$, the (well-defined) n-th *correlation coefficient* of ϕ, ψ is

$$\mathrm{cov}_{m, n}(\phi, \psi) = \int_Z (\phi \circ T^n) \, \psi \, dm - \int_Z \phi \, dm \int_Z \psi \, dm \,.$$

Let $\alpha \in \,]0, 1]$. As for the case of flows in Sect. 9.2.3, we will say that the dynamical system (Z, m, T) is *exponentially mixing for the α-Hölder regularity* or that it has *exponential decay of α-Hölder correlations* if there exist $c', \kappa > 0$ such that for all $\phi, \psi \in \mathscr{C}_b^\alpha(Z)$ and $n \in \mathbb{N}$, we have

$$|\mathrm{cov}_{m, n}(\phi, \psi)| \leq c' \, e^{-\kappa n} \, \|\phi\|_\alpha \, \|\psi\|_\alpha \,.$$

Note that this property is invariant under measure preserving conjugations of dynamical systems by bilipschitz homeomorphisms. In our case, T will be either the time 1 map of the geodesic flow $(\mathfrak{g}^t)_{t \in \mathbb{Z}}$ on the phase space $Z = \Gamma \backslash \mathscr{G}\mathbb{X}$ or the two-sided shift σ on a two-sided topological shift space Σ or (see below) the one-sided shift σ_+ on a one-sided topological shift space Σ_+.

The following result is one of the new results contained in the book [3]. For every finite subset E in $\Gamma \backslash V\mathbb{X}$, let $\tau_F : \Gamma \backslash \mathscr{G}\mathbb{X} \to \mathbb{N} \cup \{+\infty\}$ be the first positive passage time of geodesic lines in E, that is, the map

$$\ell \mapsto \inf\{n \in \mathbb{N} - \{0\} \ : \ \mathfrak{g}^n \ell(0) \in E\} \,.$$

The following result says that if the tree quotient contains a finite subset in which the geodesic lines with large return times have an exponentially decreasing mass, then the (discrete time) geodesic flow on the phase space has exponential decay of correlations. This condition turns out to be quite easy to check on practical examples, see for instance [3, §9.2].

Theorem 9.15 *If m_F is finite and mixing for $(\mathfrak{g}^t)_{t \in \mathbb{Z}}$, if there exist a finite subset E in $\Gamma \backslash V\mathbb{X}$ and $c'', \kappa' > 0$ such that*

$$\forall \, n \in \mathbb{N}, \quad m_F(\{\ell \in \Gamma \backslash \mathscr{G}\mathbb{X} \ : \ \ell(0) \in E, \tau_E(\ell) \geq n\}) \leq c'' e^{-\kappa' n} \,,$$

then for every $\alpha \in \,]0, 1]$, the (discrete time) dynamical system $(\Gamma \backslash \mathscr{G}\mathbb{X}, m_F, (\mathfrak{g}^t)_{t \in \mathbb{Z}})$ is exponentially mixing for the α-Hölder regularity.   □

The hypothesis of Theorem 9.15 is for instance satisfied for Example (2) at the beginning of Sect. 9.1 with $\mathbb{X} = \mathbb{X}_q$ and $\Gamma = \mathrm{PGL}_2(\mathbb{F}_q[Y])$, taking E consisting of the origin of the modular ray $\Gamma \backslash \mathbb{X}_q$, and using the exponential decay of the stabilisers orders along a lift of the modular ray in $\mathbb{X}_p$. In this case, the quotient graph $\Gamma \backslash \mathbb{X}$ has linear growth. We gave in [3, page 193] examples where the quotient graph $\Gamma \backslash \mathbb{X}$ has exponential growth.

Here is an example where the quotient graph has quadratic growth, for every even $q \geq 2$. The tree $\mathbb{X}$ is the regular tree of degrees $q + 2$. The vertex group of the top-left vertex x_* of the quotient graph is $\mathbb{Z}/(\frac{q}{2} + 1)\mathbb{Z}$. A set E as in Theorem 9.15 consists of the three vertices at distance at most 1 from x_*. The vertex group of a vertex at distance at least 1 from x_*, on the $(m + 1)$-th horizontal and $(n + 1)$-th vertical is $(\mathbb{Z}/q^n\mathbb{Z}) \times (\mathbb{Z}/(q + 1)^m\mathbb{Z})$. The number at the beginning of each edge represents the index of the edge group inside the vertex group of its origin.

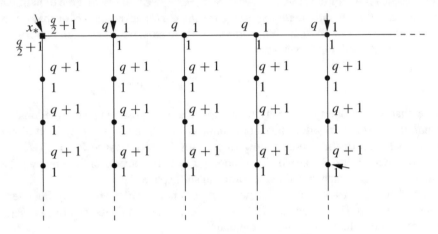

Recall that two *growth functions* f and f', that is, two increasing maps from $\mathbb{N}$ to $\mathbb{N} - \{0\}$, are *equivalent* if there exist two integers $c \geq 1$ and $c' \geq 0$ such that for every $n \in \mathbb{N}$ large enough, we have $f(\lfloor \frac{1}{c} n - c' \rfloor) \leq f'(n) \leq f(c\, n + c')$. The *type of growth* of an infinite, connected, locally finite graph Y is the equivalence class of the map $n \mapsto \operatorname{Card} B_{VY}(v_0, n)$, which does not depend on the choice of a base point $v_0 \in VY$, nor on the quasi-isometry type of Y.

It is well known (see for instance [4, 9] or [8, §6.2]) that every totally disconnected compact metric space is homeomorphic to the boundary at infinity of a simplicial tree with uniformly bounded degrees (possibly equal to 1 or 2), and that any increasing positive integer sequence $(a_n)_{n \in \mathbb{N}}$ with at most exponential speed (that is, there exists $k \in \mathbb{N}$ such that $a_{n+1} \leq k a_n$ for every $n \in \mathbb{N}$) is, up to the above equivalence, the sequence of orders of the balls of an infinite rooted simplicial tree with uniformly bounded degrees. Hence the following result (not contained in [3]) says that we can realize any space of ends, or any at most exponential type of growth, in the quotient graph of an action of a group on a tree satisfying the hypothesis of Theorem 9.15.

Proposition 9.16 *For every rooted tree $(\mathscr{T}, *)$ with uniformly bounded degrees such that $\mathscr{T} \neq \{*\}$, there exists a simplicial tree $\mathbb{X}$ and a discrete group Γ of automorphisms of $\mathbb{X}$ as in the beginning of Sect. 9.1 such that Γ is a lattice, $\Gamma \backslash \mathbb{X}$ is the union of $\mathscr{T}$ with a loop at $*$, and the geodesic flow $(\mathfrak{g}^t)_{t \in \mathbb{Z}}$ is exponentially mixing for the α-Hölder regularity on $\Gamma \backslash \mathscr{G}\mathbb{X}$ for the zero potential.* □

Proof We refer for instance to [25, §I.5] for background on graphs of groups.

Let us fix $q \in \mathbb{N}$ large enough compared with the maximum degree d of $\mathscr{T}$. We define a graph of groups $(\mathscr{T} \cup \{e_*, \overline{e_*}\}, G_\bullet)$ with underlying graph the union of $\mathscr{T}$ with a loop glued at the root $*$ as follows. Let $G_{e_*} = \{1\}$. For every vertex v of $\mathscr{T}$ at distance n of the root $*$, we define $G_v = \mathbb{Z}/q^n\mathbb{Z}$. For every edge $e \neq e_*, \overline{e_*}$, whose closest vertex to the root $*$ is at distance n from $*$, we define $G_e = \mathbb{Z}/q^n\mathbb{Z}$. For every edge $e \neq e_*, \overline{e_*}$, pointing away from the root, we define the monomorphism $G_e \to G_{o(e)}$ to be the identity, and the monomorphism $G_e \to G_{t(e)}$ to be the multiplication by q map, so that the index of G_e in $G_{o(e)}$ is 1 and the index of G_e in $G_{t(e)}$ is q.

Let Γ and $\mathbb{X}$ be respectively the fundamental group (using the root as the basepoint) and the Bass-Serre tree of the graph of groups $(\mathscr{T}, G_\bullet)$. Then the degrees of the vertices of $\mathbb{X}$ are at least 3 and at most $q + d - 1$, and for every n, we have

$$\sum_{x \in V\mathscr{T} \,:\, d(x,*)=n} \frac{1}{|G_x|} \leq d^n/q^n . \qquad (9.3.1)$$

Since q is large compared to d, this implies that the volume of $(\mathscr{T}, G_\bullet)$ is finite, hence Γ is a lattice.

Since the potential is the zero potential, the Gibbs measure m_0 is the Bowen-Margulis measure. Note that m_0 is finite since Γ is a lattice, by [3, Prop. 4.16]. Since we glued a loop at the root, there exists an element in Γ whose translation length is equal to 1, hence the length spectrum of Γ is equal to $\mathbb{Z}$. By Theorem 9.12, this implies that m_0 is mixing for the geodesic flow $(\mathfrak{g}t)_t$. If $E = \{*\}$ is the singleton in $V\mathscr{T}$ consisting of the root, since q is large compared to d, Eq. (9.3.1) then shows that the hypothesis of Theorem 9.15 is satisfied, and this concludes the proof of Proposition 9.16. □

We conclude this survey with a sketch of proof of Theorem 9.15, sending to [3, §9.2] for a complete proof. We thank Omri Sarig for a key idea in the proof of this theorem.

Step 1. The first step consists in passing from the geometric dynamical system to a two-sided symbolic dynamical system, using Sect. 9.3.1.

Let $\mathscr{A}, A, \Sigma, \sigma, \Theta, \mathbb{P}$ be as given in Theorem 9.12 for the coding of the (discrete time) geodesic flow on the phase space $\Gamma \backslash \mathscr{G}\mathbb{X}$. Let $\pi_+ : \Sigma \to \mathscr{A}^\mathbb{N}$ be the natural projection defined by $(x_n)_{n \in \mathbb{Z}} \mapsto (x_n)_{n \in \mathbb{N}}$. Let

$$\mathscr{E} = \{(e^-, h, e^+) \in \mathscr{A} \,:\, t(e^-) = o(e^+) \in E\}$$

which is a finite subset of the alphabet, and let $\tau_\mathscr{E} : \Sigma \to \mathbb{N}$ be the first positive passage time in $\mathscr{E}$ of the two-sided shift orbits, that is, the map

$$x = (x_n)_{n \in \mathbb{N}} \mapsto \inf\{n \in \mathbb{Z} - \{0\} \,:\, x_n \in \mathscr{E}\} .$$

The rate of mixing statement for two-sided symbolic dynamical systems, that we will prove in Step 2, is the following one.

Theorem 9.17 *Let* $(\mathscr{A}, A, \Sigma, \sigma)$ *be a locally compact transitive two-sided topological shift, and let* $\mathbb{P}$ *be a mixing* σ-*invariant probability measure with full support on* Σ. *Assume that*

(1) for every $n \in \mathbb{N}$ *and for every* A-*admissible finite sequence* $w = (w_0, \ldots, w_n)$ *in* $\mathscr{A}$, *the (measure theoretic) Jacobian of the map*

$$f_w : \{(x_k)_{k \in \mathbb{N}} \in \pi_+(\Sigma) \; : \; x_0 = w_n\}$$

$$\rightarrow \{(y_k)_{k \in \mathbb{N}} \in \pi_+(\Sigma) \; : \; y_0 = w_0, \ldots, y_n = w_n\}$$

defined by $(x_0, x_1, x_2, \ldots) \mapsto (w_0, \ldots, w_n, x_1, x_2, \ldots)$, *with respect to the restrictions of the pushforward measure* $(\pi_+)_* \mathbb{P}$, *is constant;*

(2) there exist a finite subset $\mathscr{E}$ *of* $\mathscr{A}$ *and* $c'', \kappa' > 0$ *such that for every* $n \in \mathbb{N}$, *we have*

$$\mathbb{P}\big(\{x \in \Sigma \; : \; x_0 \in \mathscr{E} \text{ and } \tau_\mathscr{E}(x) \geq n\}\big) \leq c'' \, e^{-\kappa' n} \; .$$

Then $(\Sigma, \sigma, \mathbb{P})$ *has exponential decay of* α-*Hölder correlations.* $\square$

Theorem 9.15 follows from Theorem 9.17 by using the coding given in Theorem 9.12. The verification of Assertion (2) is immediate as it corresponds to the assumption of Theorem 9.15. The one of Assertion (1) is a bit technical, using a strengthened version of Mohsen's shadow lemma for trees.

Step 2. The second step consists in passing from the two-sided symbolic dynamical system to a one-sided symbolic dynamical system.

Let (Σ_+, σ_+) be the one-sided topological shift with the same alphabet $\mathscr{A}$ and same transition matrix A as the two-sided one in the statement of Theorem 9.17, with $\Sigma_+ = \pi_+(\Sigma)$ where π_+ is the natural projection, and let $\mathbb{P}_+ = (\pi_+)_* \mathbb{P}$. Let $\tau_{\mathscr{E}, +} : \Sigma_+ \rightarrow \mathbb{N}$ be the first positive passage time in $\mathscr{E}$ of the one-sided shift orbits, that is, the map $(x_n)_{n \in \mathbb{N}} \mapsto \inf\{n \in \mathbb{N} - \{0\} \; : \; x_n \in \mathscr{E}\}$. . Recall that the *cylinders* in Σ_+ are the subsets defined for $k \in \mathbb{N}$ and $w_0, \ldots, w_k \in \mathscr{A}$ by

$$[w_0, \ldots, w_k] = \{x = (x_n)_{n \in \mathbb{N}} \in \Sigma_+ \; : \; x_0 = w_0, \ldots, x_k = w_k\} \; .$$

The rate of mixing statement for one-sided symbolic dynamical system, that we will prove in Step 3, is the following one.

Theorem 9.18 *Let* $(\mathscr{A}, A, \Sigma_+, \sigma_+)$ *be a locally compact transitive one-sided topological shift, and let* $\mathbb{P}_+$ *be a mixing* σ-*invariant probability measure with full support on* Σ_+. *Assume that*

(1) for every $n \in \mathbb{N}$ *and for every* A-*admissible finite sequence* $w = (w_0, \ldots, w_n)$ *in* $\mathscr{A}$, *the Jacobian of the map between cylinders*

$$f_w : [w_n] \rightarrow [w_0, \ldots, w_n]$$

defined by $(x_0, x_1, x_2, \dots) \mapsto (w_0, \dots, w_n, x_1, x_2, \dots)$, *with respect to the restrictions of* $\mathbb{P}_+$, *is constant;*

(2) *there exist a finite subset* $\mathscr{E}$ *of* $\mathscr{A}$ *and* $c'', \kappa' > 0$ *such that for every* $n \in \mathbb{N}$, *we have*

$$\mathbb{P}_+\big(\{x \in \Sigma_+ \; : \; x_0 \in \mathscr{E} \text{ and } \tau_{\mathscr{E},+}(x) \geq n\}\big) \leq c'' \, e^{-\kappa' n} \; .$$

Then $(\Sigma_+, \sigma_+, \mathbb{P}_+)$ *has exponential decay of* α-*Hölder correlations.*   □

Theorem 9.17 follows from Theorem 9.18 by a classical argument due to Sinai and Bowen (and explained to the authors by Buzzi), saying that if the one-sided symbolic dynamical system $(\Sigma_+, \sigma_+, (\pi_+)_* \mathbb{P})$ is exponentially mixing, then so is the two-sided symbolic dynamical system $(\Sigma, \sigma, \mathbb{P})$.

Step 3. The third and final step that we sketch is a proof of Theorem 9.18, using as main tool a Young's tower argument.

We implicitly throw away from Σ_+ the measure zero subset of points $x \in \Sigma_+$ whose orbit under the shift σ_+ does not pass infinitely many times in the open nonempty finite union of fundamental cylinders

$$\Delta_0 = \bigcup_{a \in \mathscr{E}} [a] \; .$$

We denote by $\Phi : \Sigma_+ \to \Delta_0$ the first positive time passage map, which is defined by $x \mapsto \sigma_+^{\tau_{\mathscr{E},+}(x)}(x)$. We denote by W the set of excursions outside $\mathscr{E}$, that is, the set of A-admissible finite sequences $(w_0, \dots, w_n)$ in $\mathscr{A}$ such that $w_0, w_n \in \mathscr{E}$ and $w_i \notin \mathscr{E}$ for $1 \leq i \leq n - 1$.

We have the following properties.

(1) The set $\{[a] \; : \; a \in \mathscr{E}\}$ is a finite measurable partition of Δ_0. For every $a \in \mathscr{E}$, the set $\{[w] \; : \; w \in W, w_0 = a\}$ is a countable measurable partition of $[a]$.
(2) For every $w \in W$, the first positive passage time $\tau_{\mathscr{E},+}$ is positive on every excursion cylinder $[w]$, and if w_n is the last letter of w, then the restriction $\Phi \mid_{[w]}: [w] \to [w_n]$ is a bijection with constant Jacobian with respect to $\mathbb{P}_+$ (actually much less is needed in order to apply Young's arguments).
(3) The first positive time passage map Φ satisfies strong dilation properties on the excursion cylinders. More precisely, for every excursion $w = (w_0, \dots, w_n) \in W$, for every $k \leq n - 1$, for all $x, y \in [w]$, we have $d(\Phi(x), \Phi(y)) \geq e \, d(x, y)$ and $d(\sigma_+^k x, \sigma_+^k y) < d(\Phi(x), \Phi(y))$.   □

Let us fix $\alpha \in \,]0, 1]$. Then an adaptation of [28, Theo. 3] implies that there exists $\kappa > 0$ such that for all $\phi, \psi \in \mathscr{C}_b^\alpha(\Sigma_+)$, there exists $c_{\phi, \psi} > 0$ such that for every $n \in \mathbb{N}$, we have

$$|\text{cov}_{\mathbb{P}_+, n}(\phi, \psi)| \leq c_{\phi, \psi} \, e^{-\kappa n} \; .$$

An argument using the Principle of Uniform Boundedness due to Chazotte then allows us to take $c_{\phi,\psi} = c' \, \|\phi\|_\alpha \, \|\psi\|_\alpha$ for some constant $c' > 0$.

References

1. H. Bass, A. Lubotzky, *Tree Lattices*. Progress in Mathematics, vol. 176 (Birkhäuser, Basel, 2001)
2. M.R. Bridson, A. Haefliger, *Metric Spaces of Non-positive Curvature*. Grundlehren der mathematischen Wissenschaften, vol. 319 (Springer Verlag, Berlin, Heidelberg, 1999)
3. A. Broise-Alamichel, J. Parkkonen, F. Paulin, *Equidistribution and Counting Under Equilibrium States in Negative Curvature and Trees. Applications to Non-Archimedean Diophantine Approximation.* With an Appendix by J. Buzzi. Progress in Mathematics, vol. 329 (Birkhäuser, Basel, 2019)
4. F.M. Choucroun, Arbres, espaces ultramétriques et bases de structure uniforme. Geom. Dedicata **53**, 69–74 (1994)
5. L. Clozel, *Démonstration de la conjecture τ*. Invent. Math. **151**, 297–328 (2003)
6. D. Dolgopyat, On decay of correlation in Anosov flows. Ann. Math. **147**, 357–390 (1998)
7. P. Giulietti, C. Liverani, M. Pollicott, Anosov flows and dynamical zeta functions. Ann. Math. **178**, 687–773 (2013)
8. R. Grigorchuk, V. Nekrashevich, V. Sushchanskii, Automata, dynamical systems, and groups. Proc. Steklov Inst. Math. **231**, 128–203 (2000)
9. B. Hughes, Trees and ultrametric spaces: a categorical equivalence. Adv. Math. **189**, 148–191 (2004)
10. S. Katok, *Fuchsian Groups* (University of Chicago Press, Chicago, 1992)
11. A. Katok, B. Hasselblatt, *Introduction to the Modern Theory of Dynamical Systems*. Encyclopedia of Mathematics and Its Applications, vol. 54, (Cambridge University Press, Cambridge, 1995)
12. D. Kleinbock, G. Margulis, Bounded orbits of nonquasiunipotent flows on homogeneous spaces, in *Sinai's Moscow Seminar on Dynamical Systems*. American Mathematical Society Translations Series, vol. 171 (American Mathematical Society, Providence, 1996), pp. 141–172
13. S. Kwon, Effective mixing and counting in Bruhat-Tits trees. Erg. Theo. Dyn. Sys. **38**, 257–283 (2018). Erratum **38**, 284 (2018)
14. J. Li, W. Pan, Exponential mixing of geodesic flows for geometrically finite hyperbolic manifolds with cusps. Preprint. arXiv:2009.12886
15. C. Liverani, On contact Anosov flows. Ann. Math. **159**, 1275–1312 (2004)
16. A. Livšic, Cohomology of dynamical systems. Math. USSR-Izv. **6**, 1278–1301 (1972)
17. G. Margulis, *Discrete Subgroups of Semi-simple Groupes*. Ergebnisse der Mathematik und ihrer Grenzgebiete, vol. 17 (Springer, Berlin, Heidelberg, 1991)
18. A. Mohammadi, H. Oh, Matrix coefficients, counting and primes for orbits of geometrically finite groups. J. Eur. Math. Soc. **17**, 837–897 (2015)
19. S. Nonnenmacher, M. Zworski, Decay of correlations for normally hyperbolic trapping. Invent. Math. **200**, 345–438 (2015)
20. J.-P. Otal, M. Peigné, Principe variationnel et groupes kleiniens. Duke Math. J. **125**, 15–44 (2004)
21. F. Paulin, On the critical exponent of discrete group of hyperbolic isometries. Differ. Geom. Appl. **7**, 231–236 (1997)
22. F. Paulin, M. Pollicott, B. Schapira, Equilibrium states in negative curvature, in Astérisque, vol. 373 (Société mathématique de France, Paris, 2015)
23. T. Roblin, *Ergodicité et équidistribution en courbure négative*. Mémoire Soc. Math. France, vol. 95 (Société mathématique de France, Paris, 2003)

24. D. Ruelle, *Thermodynamic Formalism: The Mathematical Structure of Equilibrium Statistical Mechanics*. Cambridge Mathematical Library, 2nd edn. (Cambridge University Press, Cambridge, 2004)
25. J.-P. Serre, Arbres, amalgames, SL_2. 3ème éd. corr. *Astérisque*, vol. 46 (Société Mathématique de France, Paris, 1983)
26. L. Stoyanov, Spectra of Ruelle transfer operators for axiom A flows. Nonlinearity **24**, 1089–1120 (2011)
27. M. Tsujii, Quasi-compactness of transfer operators for contact Anosov flows. Nonlinearity **23**, 1495–1545 (2010)
28. L.-S. Young, Recurrence times and rates of mixing. Israel J. Math. **110**, 153–188 (1999)
29. A. Zemanian, *Infinite Electrical Networks*. Cambridge Tracts in Mathematics, vol. 101 (Cambridge University Press, Cambridge, 1991)

Chapter 10
Statistical Properties
of the Rauzy-Veech-Zorich Map

Romain Aimino and Mark Pollicott

Abstract In this note we survey some very basic statistical properties of the Rauzy-Veech map and the Zorich acceleration. Our aim is to give a particularly thermodynamic perspective of well known results.

10.1 Introduction

In this note we will consider the Rauzy-Veech-Zorich renormalization map for interval exchange maps. The special case of interval exchange transformations on two intervals simply corresponds to rotations on the unit circle, and in this case the corresponding renormalization map reduces to the usual Farey map, and its acceleration to the continued fraction transformation. Thus, one might naturally view interval exchange maps on $m \geq 3$ intervals as generalizations of circle rotations; and the renormalization map as a generalization of the classical continued fraction transformation. It was shown by Masur and Veech that their original renormalization map $\mathscr{T}_0$ possesses an absolutely continuous ergodic invariant measure, and Zorich showed that for the accelerated version $\mathscr{T}_1$ there is a finite invariant measure.

A number of interesting statistical results already have already been established for the renormalization map, and related transformations (e.g., Central Limit Theorems and other Limit Theorems cf. [2, 4, 20]). The first aim of this paper is to present an alternative approach to some of these results, and to give some simple generalizations. Indeed, for dynamical systems in general there is a potential

R. Aimino
Centro de Matemática da Universidade do Porto, Porto, Portugal
e-mail: romain.aimino@fc.up.pt

M. Pollicott (✉)
Department of Mathematics, Warwick University, Coventry, UK
e-mail: masdbl@warwick.ac.uk

© The Author(s), under exclusive license to Springer Nature Switzerland AG 2021
M. Pollicott, S. Vaienti (eds.), *Thermodynamic Formalism*, Lecture Notes
in Mathematics 2290, https://doi.org/10.1007/978-3-030-74863-0_10

hierarchy of statistical properties that one may establish for such maps, beginning
with ergodicity; central limits theorems; functional central limit theorems, and
finally almost sure invariance principles. In this paper we will re-derive the central
limit theorem, the stronger functional central limit theorem, and establish the almost
sure invariance principle, from which the others then follow. A basic technique,
familiar from other non-uniformly hyperbolic settings, is to induce a hyperbolic
map $\mathcal{T}_2$ on a smaller set B in the domain of $\mathcal{T}_1$. In particular, statistical properties
are typically easier to establish for $\mathcal{T}_2$, and these can then be lifted to the map $\mathcal{T}_1^2$.
There is a well known application of related results to Teichmüller flows for abelian
differentials, which can be modeled in terms of suspended flows over these maps
(and their natural extensions).

One of the interesting applications of the (accelerated) Rauzy-Veech-Zorich map
is to the theory of Teichmüller flows. In particular, a suspension semi-flow for the
(accelerated) Rauzy-Veech-Zorich map corresponds to a well known model for the
Teichmüller flow.

Theorem 10.1.1 *The transformations $\mathcal{T}_1$ and $\mathcal{T}_2$ satisfy the functional central limit
theorem with respect to the natural absolutely continuous invariant probability
measure for Hölder continuous observables. In particular, they satisfy the law of
the iterated logarithm and the arcsine law for Hölder continuous observables.* $\square$

The second aim of this paper is to describe a "zeta function" associated to
$\mathcal{T}_2$. This is defined by analogy with the Ruelle zeta function for Axiom A
diffeomorphisms. The poles of these zeta functions (and the residues of associated
complex functions) encapsulate dynamical information about the maps. Moreover,
when these invariants vanish then the zeta function takes a particularly trivial form.

We will initially follow Morita in studying a transfer operator associated to $\mathcal{T}_2$
acting on Lipschitz (or, more generally, Hölder) continuous functions [20]. This
allows us to apply the method of Mackey and Tyran Kamiński [13, 14], to give a
simple and direct proof of the (Functional) Central Limit Theorem, and the method
of Philipp-Stout [22], as developed in the dynamical context by Melbourne and
Nicol [17], to show the almost everywhere invariance principles. Subsequently, we
will consider a transfer operator associated to $\mathcal{T}_2$ on a smaller space of analytic
functions and study the complex function $d(z, s)$ of two variables formally defined
by

$$d(z, s) = \exp\left(-\sum_{n=1}^{\infty} \frac{z^n}{n} \sum_{\mathcal{T}_2^n x = x} |\det(D\mathcal{T}_2^n)(x)|^{-s} \right), \quad z, s \in \mathbb{C},$$

in terms of the periodic points $\mathcal{T}_2^n x = x$ and the weights $|\det(D\mathcal{T}_2^n)(x)|$.

In particular, we can apply a powerful approach of Ruelle [24] (cf also Mayer
[15, 16] for particularly readable account in specific cases related to continued fractions) based on Fredholm determinants to show such functions have a meromorphic

extension, and we can give an alternative expression for (the sum of the Lyapunov exponents):

$$\Lambda = \int \log |\det(D\mathscr{T}_2)(x)| d\mu_2(x)$$

for the Kontsevich-Zorich cocycle, where μ_2 is the unique absolutely continuous invariant probability measure for $\mathscr{T}_2$.

Theorem 10.1.2 *The function $d(z, s)$ is analytic on $\mathbb{C}^2$. We can write*

$$\Lambda = \frac{\frac{\partial d(1,s)}{\partial s}|_{s=1}}{\frac{\partial d(z,1)}{\partial z}|_{z=1}}.$$

The methods in this note will work for other multidimensional continued fraction type algorithms, for which the (accelerated) Rauzy-Veech-Zorich algorithm forms a topical example.

In Sect. 10.2, we recall results on interval exchanges and their renormalizations. In Sect. 10.3, we introduce the transfer operator on Hölder continuous functions and recall the results of Morita on its spectra. In Sect. 10.4, we prove the statistical properties for the induced map $\mathscr{T}_2$. In Sect. 10.5, we derive the statistical properties for the Zorich map $\mathscr{T}_1$. In Sect. 10.6, we study the transfer operator on the smaller space of analytic functions, and in Sect. 10.7, we use these results to study Lyapunov exponents and $d(z, s)$. Finally, in Sect. 10.8, we describe the connection to Teichmüller flows and in the last section we speculate on the connection to pressure.

10.2 Interval Exchange Transformation

In this section we recall some of the basic constructions. We refer the reader to the excellent surveys [31] and [33] for further details.

Interval exchange transformations $T : [0, 1] \to [0, 1]$ are orientation preserving piecewise isometries of the unit interval. In the case of two intervals, this corresponds to a rotation of the circle, i.e., a translation of the interval (modulo one). More generally, assume that I is partitioned into m intervals $I_1, \cdots, I_m$ of lengths $\lambda_1, \cdots, \lambda_m$, respectively, upon each of which T acts isometrically. We can represent this partition as a vector λ in the standard $(m - 1)$-dimensional simplex

$$\Delta = \{\lambda = (\lambda_1, \cdots, \lambda_m) : 0 < \lambda_1, \cdots, \lambda_m < 1 \text{ and } \lambda_1 + \cdots + \lambda_m = 1\}$$

(Fig. 10.1).

Thus the transformation T is completely determined by these lengths, and by order of the images of the original intervals. This latter information is encapsulated

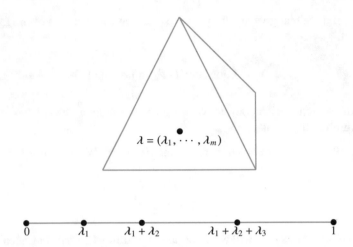

$$\lambda = (\lambda_1, \cdots, \lambda_m)$$

0   λ_1 $\lambda_1 + \lambda_2$ $\lambda_1 + \lambda_2 + \lambda_3$ 1

Fig. 10.1 A partition of the unit interval corresponds to a point in a simplex

by a permutation π on $\{1, \cdots, m\}$. In particular, every interval exchange transformation corresponds to a pair (λ, π), where $\lambda \in \Delta$ and π is a permutation. Moreover, corresponding to the natural assumption that T doesn't contain an invariant subsystem, we say that π is irreducible if there is no $1 \leq l < m$ such that $\pi(\{1, \cdots, l\}) = \{1, \cdots, l\}$. We will always assume from now on that π is irreducible.

The classical Keane Conjecture (proved by Masur and Veech, independently) states that the transformation T is uniquely ergodic for almost all $\lambda \in \Delta$. The method of proof lead to the development of an important renormalization scheme on such transformations, which we will briefly describe.

10.2.1 The Rauzy Class of Permutations

Given a permutation π, let us denote by $k = \pi^{-1}(n)$ (i.e., $\pi(k) = n$). A key idea of Rauzy was to replace the permutation π by one of two new permutations: either

$$a\pi(j) := \begin{cases} \pi(j) & \text{if } 1 \leq j \leq k \\ \pi(m) & \text{if } j = k+1 \\ \pi(j-1) & \text{if } k+2 \leq j \leq m \end{cases} \quad \text{or } b\pi(j) := \begin{cases} \pi(j) & \text{if } 1 \leq \pi(j) \leq \pi(m) \\ \pi(j)+1 & \text{if } \pi(m) < \pi(j) < n \\ \pi(m)+1 & \text{if } j = k \end{cases}$$

If we start from a given permutation we do not necessarily get all permutations by these two operations. This leads to the following definition.

Definition 10.2.1 Given a permutation π the Rauzy class $\mathscr{R}$ consists of all permutations that can be derived from π by repeatedly applying these two operations. $\square$

It can be shown that belonging to the same Rauzy class is an equivalence relation. The irreducible permutations are a union of a finite number of Rauzy classes.

Example 10.2.2 $(n = 4)$ The irreducible permutation $\pi_0 = \left(\begin{smallmatrix} 1 & 2 & 3 & 4 \\ 4 & 3 & 2 & 1 \end{smallmatrix}\right)$ lies in a Rauzy class of 7 permutations. These are illustrated in the following diagram, where an arrow labeled by a goes from π to $a\pi$ (and an arrow labeled by b goes from π to $b\pi$).

$$a \circlearrowleft \left(\begin{smallmatrix} 1 & 2 & 3 & 4 \\ 3 & 2 & 4 & 1 \end{smallmatrix}\right) \qquad\qquad\qquad \left(\begin{smallmatrix} 1 & 2 & 3 & 4 \\ 4 & 2 & 1 & 3 \end{smallmatrix}\right) \circlearrowleft b$$

$$\uparrow b \quad \searrow b \qquad\qquad a \swarrow \quad \uparrow a$$

$$\left(\begin{smallmatrix} 1 & 2 & 3 & 4 \\ 2 & 4 & 3 & 1 \end{smallmatrix}\right) \leftarrow_b \left(\begin{smallmatrix} 1 & 2 & 3 & 4 \\ 4 & 3 & 2 & 1 \end{smallmatrix}\right) \rightarrow_a \left(\begin{smallmatrix} 1 & 2 & 3 & 4 \\ 4 & 1 & 3 & 2 \end{smallmatrix}\right)$$

$$\updownarrow a \qquad\qquad\qquad\qquad \updownarrow b$$

$$\left(\begin{smallmatrix} 1 & 2 & 3 & 4 \\ 2 & 4 & 1 & 3 \end{smallmatrix}\right) \qquad\qquad\qquad \left(\begin{smallmatrix} 1 & 2 & 3 & 4 \\ 3 & 1 & 4 & 2 \end{smallmatrix}\right)$$

$$\circlearrowleft b \qquad\qquad\qquad\qquad \circlearrowleft a$$

Similarly, one can look at the Rauzy class of $\pi_0 = \left(\begin{smallmatrix} 1 & 2 & 3 & 4 \\ 2 & 3 & 4 & 1 \end{smallmatrix}\right)$ described by the following diagram.

$$a \circlearrowleft \left(\begin{smallmatrix} 1 & 2 & 3 & 4 \\ 2 & 3 & 4 & 1 \end{smallmatrix}\right) \rightarrow_b \left(\begin{smallmatrix} 1 & 2 & 3 & 4 \\ 3 & 4 & 2 & 1 \end{smallmatrix}\right) \leftrightarrow_a \left(\begin{smallmatrix} 1 & 2 & 3 & 4 \\ 3 & 4 & 1 & 2 \end{smallmatrix}\right) \leftrightarrow_b \left(\begin{smallmatrix} 1 & 2 & 3 & 4 \\ 4 & 3 & 1 & 2 \end{smallmatrix}\right) \leftarrow_a \left(\begin{smallmatrix} 1 & 2 & 3 & 4 \\ 4 & 1 & 2 & 3 \end{smallmatrix}\right) \circlearrowleft b$$

$$b \searrow \quad \swarrow b \qquad\qquad\qquad\qquad a \searrow \quad \nearrow a$$

$$\left(\begin{smallmatrix} 1 & 2 & 3 & 4 \\ 4 & 2 & 3 & 1 \end{smallmatrix}\right) \qquad\qquad\qquad\qquad \left(\begin{smallmatrix} 1 & 2 & 3 & 4 \\ 4 & 2 & 3 & 1 \end{smallmatrix}\right)$$

$$a \swarrow \quad \nwarrow a \qquad\qquad\qquad\qquad b \nearrow \quad \searrow b$$

$$b \circlearrowleft \left(\begin{smallmatrix} 1 & 2 & 3 & 4 \\ 4 & 1 & 2 & 3 \end{smallmatrix}\right) \rightarrow_a \left(\begin{smallmatrix} 1 & 2 & 3 & 4 \\ 4 & 3 & 1 & 2 \end{smallmatrix}\right) \leftrightarrow_b \left(\begin{smallmatrix} 1 & 2 & 3 & 4 \\ 3 & 4 & 1 & 2 \end{smallmatrix}\right) \leftrightarrow_a \left(\begin{smallmatrix} 1 & 2 & 3 & 4 \\ 3 & 4 & 2 & 1 \end{smallmatrix}\right) \leftarrow_b \left(\begin{smallmatrix} 1 & 2 & 3 & 4 \\ 2 & 3 & 4 & 1 \end{smallmatrix}\right) \circlearrowleft a$$

We notice a symmetry with respect to the centre of the diagram.

There are excellent descriptions of this procedure in [31] to which we refer the interested reader.

10.2.2 The Rauzy-Veech Renormalization $\mathscr{T}_0$

Consider some given $1 \leq k \leq m$. We can then apply one of the following two operations on the vector $\lambda = (\lambda_1, \cdots, \lambda_m)$, to produce a new vector $\lambda' = (\lambda'_1, \cdots, \lambda'_m)$: Either

Case I $(\lambda_m > \lambda_k)$: Let $\lambda \mapsto \lambda' = (\lambda_1, \cdots, \lambda_{m-1}, \lambda_m - \lambda_k)$; or
Case II $(\lambda_k > \lambda_m)$: Let $\lambda \mapsto \lambda' = (\lambda_1, \cdots, \lambda_{k-1}, \lambda_k - \lambda_m, \lambda_m, \lambda_{k+1}, \cdots, \lambda_{m-1})$.

Firstly, we would like to make a particular choice of case such that vector λ' is strictly positive. The case $\lambda_k = \lambda_m$ is therefore ambiguous, but atypical, and shall be ignored. Secondly, we observe that the definition of λ' is such that it does not lie in the simplex Δ. However, this will soon be corrected by rescaling.

We can define a map $\mathscr{T}_0$ from $\Delta \times \mathscr{R}$ to itself (modulo some codimension one planes, as described above, on which it is ambiguously defined). This will be a renormalization map, in the sense that it associates a new interval exchange map to an old one (with the same number of intervals, m). To be more precise, given $\pi \in \mathscr{R}$ we denote

$$\Delta_\pi^+ = \{(\lambda, \pi) \in \Delta \times \{\pi\} : \lambda_m > \lambda_{\pi^{-1}m}\} \text{ and}$$

$$\Delta_\pi^- = \{(\lambda, \pi) \in \Delta \times \{\pi\} : \lambda_m < \lambda_{\pi^{-1}m}\}.$$

We can define a transformation $\mathscr{T}_0 : \Delta \times \mathscr{R} \to \Delta \times \mathscr{R}$ a.e. by

$$\mathscr{T}_0(\lambda, \pi) = \left(\frac{\lambda'}{\|\lambda'\|_1}, \pi'\right) = \begin{cases} \left(\frac{(\lambda_1, \cdots, \lambda_{m-1}, \lambda_m - \lambda_k)}{1 - \lambda_k}, a\pi\right) & \text{if } \lambda \in \Delta_\pi^+ \\ \left(\frac{(\lambda_1, \cdots, \lambda_{k-1}, \lambda_k - \lambda_m, \lambda_m, \lambda_{k+1}, \cdots, \lambda_{m-1})}{1 - \lambda_m}, b\pi\right) & \text{if } \lambda \in \Delta_\pi^- \end{cases}$$

with $k = \pi^{-1}(m)$, where we divide by $\|\lambda'\|_1 = \sum_i \lambda_i'$ so as to rescale the image vectors to lie on the simplex Δ (Fig. 10.2).

Example 10.2.3 (Example 10.2.2 Revisited) Let $\underline{\lambda} = (\lambda_1, \lambda_2, \lambda_3, \lambda_4)$. We can again consider the Rauzy class $\mathscr{R}$ of $\pi = \left(\begin{smallmatrix} 1 & 2 & 3 & 4 \\ 4 & 3 & 2 & 1 \end{smallmatrix}\right)$ as described above. We can then consider, say, the restriction of the map to the simplex labelled by $\left(\begin{smallmatrix} 1 & 2 & 3 & 4 \\ 3 & 1 & 4 & 2 \end{smallmatrix}\right)$. Since $k = \pi^{-1}(4) = 3$ we have that

$$\mathscr{T}_0\left((\lambda_1, \lambda_2, \lambda_3, \lambda_4), \left(\begin{smallmatrix} 1 & 2 & 3 & 4 \\ 3 & 1 & 4 & 2 \end{smallmatrix}\right)\right)$$

$$= \begin{cases} \left((\frac{\lambda_1}{1-\lambda_3}, \frac{\lambda_2}{1-\lambda_3}, \frac{\lambda_3}{1-\lambda_3}, \frac{\lambda_4 - \lambda_3}{1-\lambda_3}), \left(\begin{smallmatrix} 1 & 2 & 3 & 4 \\ 3 & 1 & 4 & 2 \end{smallmatrix}\right)\right) & \text{if } \lambda_4 > \lambda_3. \\ \left((\frac{\lambda_1}{1-\lambda_4}, \frac{\lambda_2}{1-\lambda_4}, \frac{\lambda_3 - \lambda_4}{1-\lambda_4}, \frac{\lambda_4}{1-\lambda_4}), \left(\begin{smallmatrix} 1 & 2 & 3 & 4 \\ 4 & 1 & 3 & 2 \end{smallmatrix}\right)\right) & \text{if } \lambda_4 < \lambda_3. \end{cases}$$

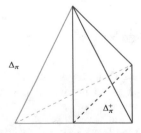
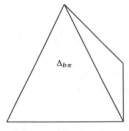

Fig. 10.2 The image of half of each copy of the simplex gets mapped to a copy of the simplex

Unfortunately, these transformations aren't uniformly hyperbolic, as one can readily see since some of the boundaries of the simplicies remain fixed (e.g., the side $\lambda_3 = 0$ in the simplex). This will be partly remedied by replacing $\mathscr{T}_0$ by maps which are "more hyperbolic".

10.2.3   The Zorich Accelerated Renormalization $\mathscr{T}_1$

Following Zorich, one can consider a map $\mathscr{T}_1 : \Delta \times \mathscr{R} \to \Delta \times \mathscr{R}$ defined a.e. by $\mathscr{T}_1(\lambda, \pi) = \mathscr{T}_0^{n(\lambda,\pi)}(\lambda, \pi)$ where

$$n(\lambda, \pi) = \inf\{k > 0 : \mathscr{T}_0^k(\lambda, \pi) \in \Delta^{\pm} \times \mathscr{R} \text{ where } \lambda \in \Delta^{\mp}\}$$

and where we denote $\Delta^+ = \bigcup_{\pi \in \mathscr{R}} \Delta_\pi^+$ and $\Delta^- = \bigcup_{\pi \in \mathscr{R}} \Delta_\pi^-$.

The following elegant result was proved by Zorich.

Proposition 10.2.4 (Zorich) *The transformation $\mathscr{T}_1$ preserves a finite absolutely continuous invariant measure μ_1 (i.e., $\mu_1(\Delta \times \mathscr{R}) < +\infty$). Moreover, the restriction $\mathscr{T}_1^2 : \Delta^+ \to \Delta^+$ is ergodic (and $\mathscr{T}_1^2 : \Delta^- \to \Delta^-$ is ergodic).*   □

Previously, Masur and Veech had shown the existence of a sigma finite $\mathscr{T}_0$-invariant measure μ_0, which can be easily recovered from μ_1.

However, to gain more control over the distortion properties of the transformations one can induce on a smaller set, so as to get a transformation which has even stronger properties.

10.2.4   The Induced Map $\mathscr{T}_2$ on a Smaller Set

Let $\mathscr{P} = \{\Delta_\pi^+, \Delta_\pi^- : \pi \in \mathscr{R}\}$ be the natural finite partition of $\Delta \times \mathscr{R}$ then we can define the refinements

$$\mathscr{P}_n := \vee_{k=0}^{n-1} \mathscr{T}_1^{-k} \mathscr{P} = \{P_{i_1} \cap \mathscr{T}_1^{-1} P_{i_2} \cap \cdots \cap \mathscr{T}_1^{-(n-1)} P_{i_{n-1}} : P_j \in \mathscr{P}\}$$

for any $n \geq 1$. Following a now standard approach we can choose $n_0 > 1$ and $B \in \mathscr{P}_{n_0}$, say, to be any image of an inverse branch of $\mathscr{T}^{n_0}$ which is a contraction.[1]

---

[1] All of these transformations are projective, i.e., matrices act linearly on vectors, followed by normalizing. Such a transformation is contracting in the projective metric when the simplex is mapped strictly inside itself, which happens when the matrix is strictly positive.

Finally, we can then consider the induced map $\mathcal{T}_2 : B \to B$ defined by $\mathcal{T}_2(\lambda, \pi) = \mathcal{T}_1^{\widehat{n}(\lambda, \pi)}(\lambda, \pi)$ where

$$\widehat{n}(\lambda, \pi) = \inf\{k > 0 : \mathcal{T}_1^k(\lambda, \pi) \in B\}$$

is the first return time to B. The following is immediate from the observation that the composition of projective transformation remains projective, see Morita [20, Lemma 3.1].

Lemma 10.2.5 *The induced map $\mathcal{T}_2 : B \to B$ is a piecewise projective expanding map of the general form*

$$(\lambda_1, \cdots, \lambda_n) \mapsto \left(\frac{\sum_{j=1}^d a_{1j}\lambda_j}{\sum_{i,j=1}^d a_{ij}\lambda_j}, \cdots, \frac{\sum_{j=1}^d a_{dj}\lambda_j}{\sum_{i,j=1}^d a_{ij}\lambda_{ij}} \right)$$

on each piece of the partition of smoothness of $\mathcal{T}_2$.   □

We are now in a position to use familiar techniques for the study of hyperbolic maps.

10.3 Transfer Operators

Let ω denote the natural volume form on B. We can formally define a linear map $\mathscr{L} : L^1(B, \omega) \to L^1(B, \omega)$ associated to $\mathcal{T}_2 : B \to B$ by the identity

$$\int_B \mathscr{L}f(x)g(x)d\omega(x) = \int_B f(x)g(\mathcal{T}_2 x)d\omega(x), \text{ where } f \in L^1(B), g \in L^\infty(B)$$

and we denote $x = (\lambda, \pi) \in B$. (The existence of such a $\mathscr{L}f \in L^1(B)$ follows immediately from the Riesz representation theorem.) Moreover, we can use the change of variables formula to formally write:

$$\mathscr{L}^k f(x) = \sum_{y \in \mathcal{T}_2^{-k} x} \frac{f(y)}{|\text{Jac}(\mathcal{T}_2^k)(y)|} \text{ a.e..}$$

In fact, a simple calculation, see Veech [28, Proposition 5.2], shows:

Lemma 10.3.1 *Let A be the matrix such that $y = \frac{Ax}{\|Ax\|_1}$. We can write the Jacobian as $Jac(\mathcal{T}_2^k)(y) = \|Ax\|_1^m$.*   □

From this explicit formula for the Jacobian one easily sees that $\mathscr{L}(C^0(B)) \subset C^0(B)$. In order to get stronger results on $\mathcal{T}_2$, we need to consider the operator acting on smaller Banach spaces than $C^0(B)$. In Sect. 10.6, we will consider the operator acting on analytic functions. However, for the present we shall follow the more classical approach of studying the operator acting on Hölder continuous functions.

Given $\beta > 0$ and a function $w : B \to \mathbb{C}$, we define $\|w\|_\beta = \|w\|_\infty + |w|_\beta$ where

$$|w|_\beta = \sup_{x \neq y} \frac{|w(x) - w(y)|}{\|x - y\|^\beta}$$

and let $C^\beta(B) = \{w : B \to \mathbb{C} : \|w\| < \infty\}$. When $\beta = 1$ these are simply the Lipschitz functions. The next result can be used to show that $\mathcal{L}$ preserves Hölder functions. Let $\mathcal{Q}$ be the partition of smoothness of $\mathcal{T}_2$, and let $\mathcal{Q}_k = \vee_{i=0}^{k-1} \mathcal{T}_2^{-i} \mathcal{Q}$. The following result is basically due to Morita [20]:

Lemma 10.3.2

(1) *There exists $C > 0$ and $\Theta > 1$ such that for any $n \geq 1$ and x, y in the same element of $\mathcal{Q}_n$ we have*

$$\|\mathcal{T}_2^n x - \mathcal{T}_2^n y\| \geq C\Theta^n \|x - y\|.$$

(2) *There exists $C > 0$ such that for any $n \geq 1$ and x, y lie in the same element of $\mathcal{Q}_n$ we have*

$$\left| \log \left(\frac{Jac(\mathcal{T}_2^n)(x)}{Jac(\mathcal{T}_2^n)(y)} \right) \right| \leq C \|\mathcal{T}_2^n x - \mathcal{T}_2^n y\|.$$

(3) *There exists $D > 1$ such that for any $A \in \mathcal{Q}_n$ and any $x \in A$ we can estimate*

$$\frac{1}{D} \leq \omega(A) \left| Jac(\mathcal{T}_2^n)(x) \right| \leq D.$$

Proof These results are based on the basic observation that the first return map $\mathcal{T}_2 : B \to B$ must be of the form $\mathcal{T}_2(x) = \mathcal{T}_1^{\widehat{n}(\lambda,\pi)}(x) = \mathcal{T}_1^{\widehat{n}(\lambda,\pi)-n_0} \circ \mathcal{T}_1^{n_0}(x)$, where $\mathcal{T}_1^{\widehat{n}(\lambda,\pi)-n}$ does not contract distances and $\mathcal{T}_1^{n_0}$ definitely expands them. Full details can be found in [20, Lemma 3.4]. □

Corollary 10.3.3 *The operator $\mathcal{L}$ preserves the space of Hölder functions, i.e., $\mathcal{L} : C^\beta(B) \to C^\beta(B)$ is well defined.* □

Many of the statistical results for $\mathcal{T}_2$ are related to the existence of a spectral gap for $\mathcal{L}$. In the case of the operator acting on analytic functions is essentially automatic since the operator is compact (as we will see later). However, in the present context of Hölder continuous functions it remains true.

Lemma 10.3.4 *The value 1 is a simple eigenvalue with a positive eigenfunction $\rho > 0$. The rest of the spectrum is contained in a disk of radius τ strictly smaller than 1.* □

Proof The proof follows a classical approach [21]. Given $g \in C^\beta(B)$, we can estimate for each $x \in B$ that

$$|(\mathscr{L}^n g)(x)| \le \|g\|_\infty \left(\sum_{\mathscr{T}_2^n y = x} \frac{1}{\mathrm{Jac}(\mathscr{T}_2^n)(y)} \right) \le D\|g\|_\infty$$

by part (3) of Lemma 10.3.2. Thus $\|\mathscr{L}^n g\|_\infty \le D\|g\|_\infty$. Similarly, in the special case $g = 1$ we can see that $D^{-1} \le \mathscr{L}^n(1)(x) \le D$, for all $x \in B$.

Given $x_1, x_2 \in B$, assume that $y_i \in (\mathscr{T}_2^n)^{-1} x_i$ ($i = 1, 2$) are chosen in the same inverse branch. With this convention, we write that

$$(\mathscr{L}^n g)(x_1) - (\mathscr{L}^n g)(x_2)$$

$$= \sum_{\mathscr{T}_2^n y_i = x_i} \left(\frac{1}{\mathrm{Jac}(\mathscr{T}_2^n)(y_1)} - \frac{1}{\mathrm{Jac}(\mathscr{T}_2^n)(y_2)} \right) g(y_1) + \sum_{\mathscr{T}_2^n y_2 = x_2} \frac{(g(y_1) - g(y_2))}{\mathrm{Jac}(\mathscr{T}_2^n)(y_2)}.$$

Note that by part (3) of Lemma 10.3.2, we have

$$D^{-2} \le \frac{\mathrm{Jac}(\mathscr{T}_2^n)(y_1)}{\mathrm{Jac}(\mathscr{T}_2^n)(y_2)} \le D^2,$$

and hence, we can write

$$\left| \frac{1}{\mathrm{Jac}(\mathscr{T}_2^n)(y_1)} - \frac{1}{\mathrm{Jac}(\mathscr{T}_2^n)(y_2)} \right| \le \frac{D^2}{\mathrm{Jac}(\mathscr{T}_2^n)(y_2)} \left| \log \left(\frac{\mathrm{Jac}(\mathscr{T}_2^n)(y_2)}{\mathrm{Jac}(\mathscr{T}_2^n)(y_1)} \right) \right|$$

$$\le \frac{D^2 C}{\mathrm{Jac}(\mathscr{T}_2^n)(y_2)} \|x_1 - x_2\|.$$

Thus we can bound

$$|(\mathscr{L}^n g)(x_1) - (\mathscr{L}^n g)(x_2)|$$

$$\le \sum_{\mathscr{T}_2^n y_2 = x_2} \frac{1}{|\mathrm{Jac}(\mathscr{T}_2^n)(y_2)|} \left(D^2 C \|g\|_\infty + \Theta^{-n} \|g\|_\beta \right) \|x_1 - x_2\|$$

$$\le D \left(D^2 C \|g\|_\infty + \frac{|g|_\beta}{\Theta^n} \right) \|x_1 - x_2\|.$$

(This gives the well known Doeblin-Fortet, Marinescu-Tulcea or Lasota-Yorke inequality for $\mathscr{L}$: there exists $C > 0$ such that $\|\mathscr{L}^n g\|_\beta \le C \left(\|g\|_\infty + \Theta^{-n} \|g\|_\beta \right)$ for all $n \ge 0$ and all $g \in C^\beta(B)$.)

In particular, the family $\{\frac{1}{N} \sum_{n=0}^{N-1} \mathscr{L}^n 1\}_{N=1}^\infty$ is equicontinuous and bounded, and thus has a uniform accumulation point $\rho \in C^\beta(B)$, say, where $D^{-1} \le \rho(x) \le$

D, for all $x \in B$. Clearly, $\mathscr{L}\rho = \rho$ is a positive eigenfunction for the eigenvalue 1. Let $d\mu_2(x) = \rho(x)d\omega(x)$ be the corresponding invariant probability measure. To see that 1 is a simple eigenvalue, assume that $\mathscr{L}\rho' = \rho'$, and then choose the largest $\epsilon > 0$ that the eigenfunction $\rho_\epsilon := \rho + \epsilon\rho' \geq 0$. Since we can find $x \in B$ with $\rho_\epsilon(x) = 0$, it then follows from $\mathscr{L}\rho_\epsilon = \rho_\epsilon$ that $\rho_\epsilon(y) = 0$, for all $y \in \mathscr{T}_2^{-1}x$. Proceeding inductively, we see that $\rho_\epsilon(y)$ vanishes on the dense set $y \in \cup_{n=0}^{\infty} \mathscr{T}_2^{-n}x$, and thus $\rho' = \epsilon\rho$, i.e., 1 is a simple eigenvalue. We can define $\widehat{\mathscr{L}} : C^\beta(B) \to C^\beta(B)$ by

$$\widehat{\mathscr{L}}w(x) = \frac{1}{\rho(x)}\mathscr{L}(w\rho)(x).$$

Then $\widehat{\mathscr{L}}1 = 1$ (and $\mathscr{L}^*\mu_2 = \mu_2$) and again the Doeblin-Fortet inequality holds for $\widehat{\mathscr{L}}$, i.e., $\|\widehat{\mathscr{L}}^n w\|_\beta \leq C\|w\|_\infty + \Theta^{-n}\|w\|_\beta$. Moreover, since for any positive $w \in C^\beta(B)$ we have $\sup w \geq \sup \widehat{\mathscr{L}}w \geq \sup \widehat{\mathscr{L}}^2w \geq \cdots$ we can deduce from the equicontinuity that there is a unique limit in the uniform norm which, using that $\widehat{\mathscr{L}}1 = 1$, we conclude must be the constant $\int w d\mu_2$, i.e., $\widehat{\mathscr{L}}^n w \to \int w d\mu_2$ as $n \to +\infty$, see [21, Theorem 2.2].

Finally, to show that the rest of the spectrum of $\mathscr{L}$ is contained strictly within the unit disc it suffices to show the same for $\widehat{\mathscr{L}}$ and, more particularly, $\widehat{\mathscr{L}} : C^\beta(B)/\mathbb{C} \to C^\beta(B)/\mathbb{C}$ has spectral radius strictly smaller than 1. However, the convergence of $\widehat{\mathscr{L}}^n w$ implies that $\|\widehat{\mathscr{L}}^n w + \mathbb{C}\|_\infty \to 0$ as $n \to +\infty$ and thus two applications of the Marinescu-Tulcea inequality gives

$$\|\widehat{\mathscr{L}}^{2n}w\|_\beta \leq C\left(\|\widehat{\mathscr{L}}^n w + \mathbb{C}\|_\infty + \Theta^{-n}\|\widehat{\mathscr{L}}^n w\|_\beta\right) + \Theta^{-n}\|\widehat{\mathscr{L}}^n w\|_\beta$$

$$\leq C\left(\|\widehat{\mathscr{L}}^n w + \mathbb{C}\|_\infty + \Theta^{-n}(C+1)\left(C\|w\|_\infty + \|w\|_\beta\Theta^{-n}\right)\right)$$

$$< 1$$

for large enough $n \geq 0$, uniformly on the unit ball of $C^\beta(B)/\mathbb{C}$. The result follows from the spectral radius theorem.   □

As usual, the probability measure μ_2 which is the eigenprojection associated to 1 (i.e., $\widehat{\mathscr{L}}\mu_2 = \mu_2$) is the unique absolutely continuous $\mathscr{T}_2$-invariant probability measure on B. In particular, μ_2 is the renormalized restriction of μ_1 to B.

Corollary 10.3.5

(1) *The transformation $\mathscr{T}_2 : B \to B$ is exponentially mixing on Hölder functions, i.e., there exists $0 < \tau < 1$ and $C > 0$ such that for all $F \in L^\infty(B)$ and $G \in C^\beta(B)$ with $\int F d\mu_2 = \int G d\mu_2 = 0$,*

$$\left|\int F \circ \mathscr{T}_2^n . G d\mu_2 - \int F d\mu_2 \int G d\mu_2\right| \leq C\tau^n \|F\|_{L^1(\mu_2)}\|G\|_\beta \text{ for all } n \geq 0.$$

(2) *For μ_2-almost all $x = (\lambda, \pi) \in B$ we have that*

$$\frac{1}{N} \sum_{n=0}^{N-1} F(\mathscr{T}_2^n(x, \lambda)) = \int F d\mu_2 + O\left(\frac{\log N}{\sqrt{N}}\right).$$

Proof For the first part, we can write

$$\int F \circ \mathscr{T}_2^n . G d\mu_2 - \int F d\mu_2 \int G d\mu_2 = \int \left(\mathscr{L}^n(G\rho) - \left(\int G d\mu_2\right) \rho\right) F d\omega.$$

Thus, $\left|\int F \circ \mathscr{T}_2^n . G d\mu_2 \int F d\mu_2 \int G d\mu_2\right| \leq \|\mathscr{L}^n(G\rho) - \left(\int G d\mu_2\right) \rho\|_\infty \|F\|_{L^1(\omega)}$.
By Lemma 10.3.4, $\|\mathscr{L}^n(G\rho) - \left(\int G d\mu_2\right) \rho\|_\infty \leq C\tau^n \|G\|_\beta$, since $C^\beta(B)$ embeds
into $L^\infty(B)$, is a Banach algebra and $\rho \in C^\beta(B)$. On the other hand, $\|F\|_{L^1(\omega)} \leq$
$c^{-1} \|F\|_{L^1(\mu_2)}$, where $c = \inf \rho$ is strictly positive.

The second part follows immediately from the first part by a standard spectral
result [10]. □

10.4   Statistical Properties for $\mathscr{T}_2$

Let $d\mu_2(x) = \rho(x) d\omega(x)$ be the unique absolutely $\mathscr{T}_2$-invariant probability
measure on B given by Proposition 10.2.4. This measure μ_2 is ergodic (cf. [4] or,
alternatively, by part (1) of Corollary 10.3.5) and so we can apply the Birkhoff
ergodic theorem which gives that for any $f \in L^1(X, \mu_2)$ and for μ_2-a.e. $x \in B$ we
have that

$$\frac{1}{n} \sum_{j=0}^{n-1} f(\mathscr{T}_2^j x) \to \int f d\mu_2, \text{ as } n \to +\infty,$$

pointwise and in L^1. In this section we want to discuss various generalizations of
this basic property.

10.4.1   The Central Limit Theorem and Functional Central
Limit Theorem

A classical result for expanding dynamical systems is the Central Limit Theorem,
and the stronger Functional Central Limit Theorem.

Definition 10.4.1 We say that $\mathscr{T}_2$ satisfies the Functional Central Limit Theorem
whenever for a Hölder continuous function $h \in C^\beta(B, \mathbb{R})$ with $\int h d\mu_2 = 0$ (not

equal to a coboundary) there exists $\sigma > 0$ such that for $0 \leq t \leq 1$,

$$w_n(t) = \frac{1}{\sigma\sqrt{n}} \left(\sum_{j=0}^{[nt]-1} h \circ \mathscr{T}_2^j + (nt - [nt])h \circ \mathscr{T}_2^{[nt]} \right)$$

converges weakly to the Wiener measure on $C([0, 1], \mathbb{R})$. □

This is sometimes called a weak invariance principle, in reference to the topology of convergence.

The Central Limit Theorem could be deduced directly from the spectral results on $\widehat{\mathscr{L}}$ in the previous section, but, with no additional work we can deduce the stronger Functional Central Limit Theorem.

Proposition 10.4.2 *The Functional Central Limit Theorem holds for $\mathscr{T}_2$.* □

Proof By a quite general result of Mackey and Tyran-Kamińska [13, 14] (cf. also [27]) if $h_0 \in L^2(B, \mu_2)$ satisfies $\int h_0 d\mu_2 = 0$ and $\widehat{\mathscr{L}}h_0 = 0$, and

$$\sum_{n=1}^{\infty} \frac{1}{n^{3/2}} \sqrt{\int \left(\sum_{k=0}^{n-1} \widehat{\mathscr{L}}^k h_0 \right)^2 d\mu_2} < \infty,$$

then setting $\sigma^2 = \int |h_0|^2 d\mu_2$ gives

$$w_n^0(t) = \frac{1}{\sqrt{n}} \sum_{j=0}^{[nt]-1} h_0 \circ \mathscr{T}_2^j \to \sigma w(t), \text{ for } t \in [0, 1].$$

(i.e., the Functional Central Limit Theorem for h_0). More generally, given a Hölder continuous function h with $\int h d\mu_2 = 0$, we recall from Lemma 10.3.4 that there exists $0 < \tau < 1$ such that $\|\widehat{\mathscr{L}}^n h\|_\beta = O(\theta^n)$, and therefore $u = \sum_{n=1}^{\infty} \widehat{\mathscr{L}}^n h$ converges in $C^\beta(B)$. Let $u = \sum_{n=1}^{\infty} \widehat{\mathscr{L}}^n h$ and set $h_0 := h - u \circ T + u$ then $\widehat{\mathscr{L}}(h_0) = \widehat{\mathscr{L}}h - u + \widehat{\mathscr{L}}u = 0$. Since h and h_0 are cohomologous we can bound $|w_n(t) - w_n^0(t)| \leq 2\|u\|_\infty/\sqrt{n}$ and thus deduce the Functional Central Limit Theorem for h. If $\sigma^2 = 0$, then we would have $h_0 \equiv 0$, and so h would be equal to a coboundary, which is not the case by assumption.

The following are standard corollaries for Hölder continuous functions f using the Continuous Mapping Theorem [8, 9] beginning with the central limit theorem.

Corollary 10.4.3 (Central Limit Theorem) *For $y \in \mathbb{R}$ we have that*

$$\lim_{n \to +\infty} \mu_2 \left\{ x \in B : \frac{1}{\sqrt{n}} \sum_{j=1}^{n} f(\mathscr{T}_2^j x) \leq y \right\} = \frac{1}{\sqrt{2\pi}\sigma} \int_{-\infty}^{y} e^{-t^2/2\sigma^2} dt$$

The Central Limit Theorem (and much more besides) has already been proved by Butetov [4] and Morita [20]. The approach of Bufetov involved studying the rate of mixing of $\mathcal{T}_2$; and the method of Morita involved perturbation theory of the transfer operator.

The following are other standard corollaries [8, 9].

Corollary 10.4.4 *For $y \geq 0$ we have that*

$$\lim_{n \to +\infty} \mu_2 \left\{ x \in B : \frac{1}{\sqrt{n}} \max_{1 \leq k \leq n} \sum_{j=1}^{k} f(\mathcal{T}_2 x) \leq y \right\} = \frac{\sqrt{2}}{\sqrt{\pi}\sigma} \int_{-\infty}^{y} e^{-t^2/2\sigma^2} dt - 1.$$

Corollary 10.4.5 (Arcsine Law) *For $0 \leq y \leq 1$ we have that*

$$\lim_{n \to +\infty} \mu_2 \left\{ x \in B : \frac{N_n(x)}{n} \leq y \right\} = \frac{2}{\sqrt{\pi}} \sin^{-1} \sqrt{y}$$

where $N_n(x) = Card \left\{ 1 \leq k \leq n : \sum_{j=1}^{k} f(\mathcal{T}_2^j x) > 0 \right\}$.   □

Corollary 10.4.6 (Law of the Iterated Logarithm) *For μ_2-a.e. $x \in B$ we have*

$$\limsup_{n \to +\infty} \frac{\sum_{j=1}^{n} f(\mathcal{T}_2^j x)}{\sigma \sqrt{2n \log \log n}} = 1.$$

Remark 10.4.7 There are a number of other statistical results which could be considered. For example, Morita has shown that there is a local limit theorem and Berry-Esseen estimates for $\mathcal{T}_2$. We could also consider Edgeworth expansions, following Fernando and Liverani [6].   □

10.4.2 Almost Sure Invariance Principles

With only a little further work, we next establish a class of stronger results, from which the preceding (and several others) can easily be deduced.

Given a Hölder continuous function $f : B \to \mathbb{R}$ with $\int f d\mu_2 = 0$ we can associate the summation $f^n(x) := \sum_{i=0}^{n-1} f(\mathcal{T}^i x)$, for each $n \geq 1$.

Definition 10.4.8 We say that $\mathcal{T}_2 : B \to B$ satisfies the *Almost Sure Invariance Principle* relative to Hölder continuous functions and the measure μ_2 if for any such function $f : B \to \mathbb{R}$ with $\int f d\mu_2 = 0$ not equal to a coboundary, there exists a sequence of random variables $\{S_n\}$, possibly on a larger probability space, equal in distribution under μ_2 with $\{f^n\}$ and there exists $\epsilon > 0$ such that $S_n = W_n + O(n^{\frac{1}{2}-\epsilon})$ as $n \to +\infty$, where $\{W_t\}_{t \geq 0}$ is a Brownian motion with variance $\sigma^2 > 0$.   □

The following result is a strengthening of Proposition 10.4.2.

Theorem 10.4.9 (Almost Sure Invariance Principle for $\mathscr{T}_2$) *The transformation $\mathscr{T}_2 : B \to B$ satisfies the Almost Sure Invariance Principle.* □

Proof The standard approach is to deduce this from an application of a result of Philipp and Stout [22] (cf. [17] for a dynamical reformulation). In particular, we only need to establish that the hypotheses there hold. More precisely, given a β-Hölder function $f : B \to \mathbb{R}$ with $\int f d\mu_2 = 0$ we observe that:

(1) $f \in L^{2+\delta}(B)$, for any $\delta > 0$ (since v is automatically bounded);
(2) for any $n \geq 1$,

$$\int |f^n|^2 d\mu_2 = n\sigma^2 + O(1)$$

   (by expanding the Left Hand Side and bounding the cross terms using Part (1) of Corollary 10.3.5), see [17, Proof of Corollary 2.3] for more details;
(3) for any $k \geq 0$,

$$E\left(|f - E(f| \vee_{i=0}^{k-1} \mathscr{T}_2^{-i} \mathscr{Q})|^{2+\delta})| \vee_{i=0}^{k-1} \mathscr{T}_2^{-i} \mathscr{Q}\right)$$

$$\leq \|f - E(f| \vee_{i=0}^{k-1} \mathscr{T}_2^{-i} \mathscr{Q}))\|_\infty^{2+\delta}$$

$$\leq (\|f\|_\beta \sup_{a \in \mathscr{Q}_k} \operatorname{diam}(a))^{2+\delta}$$

$$\leq (\|f\|_\beta \Theta^{-k})^{2+\delta},$$

   (where, as usual, $E(\cdot| \vee_{i=0}^{k-1} \mathscr{T}_2^{-i} \mathscr{Q}) = \sum_{a \in \mathscr{T}_2^{-i} \mathscr{Q}} \frac{1}{\mu(a)} \int_a (\cdot) d\mu$); and, finally,
(4) given any $A_1 \in \vee_{i=0}^{k-1} \mathscr{T}_2^{-i} \mathscr{Q}$ and any Borel measurable set $A_2 \subset B$, and for any $n, k \geq 0$, we can bound

$$\left|\mu_2(A_1 \cap \mathscr{T}_2^{-(k+n)} A_2) - \mu_2(A_1)\mu_2(A_2)\right|$$

$$= \left|\int \chi_{A_1}(\chi_{A_2} \circ \mathscr{T}_2^{k+n}) d\mu_2 - \int \chi_{A_1} d\mu_2 . \int \chi_{A_2} d\mu_2\right|$$

$$= \left|\int (\mathscr{L}^n \chi_{A_1})(\chi_{A_2} \circ \mathscr{T}_2^k) d\mu_2 - \int \mathscr{L}^n \chi_{A_1} d\mu_2 \int \chi_{A_2} \circ \mathscr{T}_2^k d\mu_2\right|$$

$$= \left|\int \left[\mathscr{L}^k \chi_{A_1} - \int \mathscr{L}^k \chi_{A_1} d\mu_2\right](\chi_{A_2} \circ \mathscr{T}_2^n) d\mu_2\right|$$

$$= \left|\int \mathscr{L}^n \left[\mathscr{L}^k \chi_{A_1} - \int \mathscr{L}^k \chi_{A_1} d\mu_2\right] \chi_{A_2} d\mu_2\right|$$

$$
\leq \left(\int \left| \mathscr{L}^n \left[\mathscr{L}^k \chi_{A_1} - \int \mathscr{L}^k \chi_{A_1} d\mu_2 \right] \right|^2 d\mu_2 \right)^{\frac{1}{2}} \left(\int \chi_{A_2}^2 d\mu_2 \right)^{\frac{1}{2}}
$$

$$
\leq C\tau^n \| \mathscr{L}^k \chi_{A_1} \|_\beta \mu_2(A_2)^{\frac{1}{2}},
$$

for some $C > 0$, using the Cauchy-Schwartz inequality, that $\widehat{\mathscr{L}}^* \mu_2 = \mu_2$ and (again) that $0 < \tau < 1$ is a bound on the modulus of the second eigenvalue of $\widehat{\mathscr{L}}$. Finally, we can observe that $\| \widehat{\mathscr{L}}^k \chi_{A_1} \|_\beta \leq D\mu(A_1)$, as in the proof of [17, Lemma 2.4], and so the bound can be taken to be $C\tau^n$.

We can then apply Theorem 7.1 in [22] (cf. Theorem A.1 in [17]) to deduce that the Almost Sure Invariance Principle holds for $\mathscr{T}_2$. $\qquad\square$

There is an immediate application of the preceding analysis to return times for $\mathscr{T}_2$. Given any Borel set A we denote by $r_A : A \to \mathbb{N}$ the first return time to A, i.e., $r_A(x) = \inf\{n \geq 1 : \mathscr{T}_1^n x \in A\}$. In particular, the value defined inductively by $r_A^{(n)}(x) = r_A^{(n-1)}(x) + r_A(\mathscr{T}_2^{r_A^{(n-1)}(x)})$ is the nth return time. Using Birkhoff's theorem and Kac's theorem on return times we have that

$$
\lim_{n \to +\infty} \frac{r_A^{(n)}(x)}{n} = \frac{1}{\mu_1(A)} \text{ for } \mu\text{-a.e. } x \in B.
$$

For the particular choice $A = B$ we can consider the function $r_B(x) = \widehat{n}(x)$ and by Kac's theorem $\int r_B d\mu_2 = 1/\mu_1(B)$. It is easy to see that the variance is non-zero and thus this leads, for example, to the following corollary:

Corollary 10.4.10 *There exists $\sigma > 0$ such that*

$$
\lim_{N \to +\infty} \mu_2 \left\{ x : \frac{1}{N} r_B^{(N)}(x) - \frac{1}{\mu_1(B)} \leq y \right\} = \frac{1}{\sqrt{2\pi}\sigma} \int_{-\infty}^y e^{-t^2/2\sigma^2} dt
$$

for $y \in \mathbb{R}$. $\qquad\square$

Remark 10.4.11 Finer results about recurrence properties and the statistical behavior of return times for $\mathscr{T}_1$ and $\mathscr{T}_2$ can also be deduced from the spectral gap (Lemma 10.3.4), see Aimino et al. [1]. $\qquad\square$

10.5   Statistical Properties for $\mathscr{T}_1$

The statistical properties of $\mathscr{T}_2$ described above can be used to establish analogous results for the original Zorich map $\mathscr{T}_1 : \Delta \times \mathscr{R} \to \Delta \times \mathscr{R}$, with respect to μ_1, by viewing it as a suspension. More precisely, we can associate to the map $\mathscr{T}_2 : B \to B$

and the return time $\widehat{n} : B \to \mathbb{Z}^+$ a suspension space

$$B^{\widehat{n}} := \{(x, k) \in B \times \mathbb{Z} : 0 \le k \le \widehat{n}(x) - 1\}/ \sim$$

where we identify $(\lambda, \pi; \widehat{n}(x))$ and $(\mathscr{T}_2(\lambda, \pi); 0)$. We can also define the natural map $\mathscr{T}_2^{\widehat{n}} : B^{\widehat{n}} \to B^{\widehat{n}}$ on this suspension space by

$$\mathscr{T}_2^{\widehat{n}}(x, k) = \begin{cases} (x, k + 1) & \text{if } 0 \le k \le \widehat{n}(x) - 2 \\ (\mathscr{T}_2 x, 0) & \text{if } k = \widehat{n}(x) - 1. \end{cases}$$

There is a natural $\mathscr{T}_2^{\widehat{n}}$-invariant measure $d\mu_2 \times d\mathbb{N}/ \int \widehat{n} d\mu_2$, where $d\mathbb{N}$ corresponds to the usual counting measure. The following result is standard.

Lemma 10.5.1 *The map* $\Psi : B^{\widehat{n}} \to \Delta \times \mathscr{R}$ *defined by* $\Psi(x, k) = \mathscr{T}_1^k(x)$ *is:*

(1) *a semi-conjugacy, i.e.,* $\mathscr{T}_1 \circ \Psi = \Psi \circ \mathscr{T}_2^{\widehat{n}}$, *and*
(2) *an isomorphism (with respect to* $d\mu_2 \times d\mathbb{N}/ \int \widehat{n} d\mu_2$ *and* $d\mu_1$*).* □

We can deduce the almost sure invariance principle for the Zorich map $\mathscr{T}_1$: $\Delta \times \mathscr{R} \to \Delta \times \mathscr{R}$, by applying a result given in a paper of Melbourne and Nicol [17] (which is formulated from the results of Melbourne and Török [19]), and whose proof is made precise by Korepanov [11]. The other statistical properties follow as a direct consequence.

The main technical condition we require is the following:

Lemma 10.5.2 *For any* $\delta > 0$ *we have that*

$$\sum_{k=1}^{\infty} \mu_2 \{x = (\lambda, \pi) \in B : \widehat{n}(x) = k\} k^{2+\delta} < +\infty.$$

Proof By an estimate of Avila-Bufetov [2, Lemma 1], there exists $C > 0$ and $0 < \theta < 1$ such

$$\mu_2 \{x \in B : \widehat{n}(\underline{\lambda}, \pi) \ge k\} \le C\theta^k, \text{ for all } k \ge 1.$$

Thus $\sum_{k=1}^{\infty} \mu_2 \{x \in B : \widehat{n}(\underline{\lambda}, \pi) = k\} k^{2+\delta} \le C \sum_{k=1}^{\infty} \theta^k k^{2+\delta} < +\infty.$ □

We now describe a general class of function for which the results will be established. Let $f : \Delta \times \mathscr{R} \to \mathbb{R}$ be Hölder continuous and satisfy $\int f d\mu_1 = 0$. We can associate to f a function $\overline{f} : B \to \mathbb{R}$ defined μ_2-a.e. by

$$\overline{f}(x) = \sum_{l=0}^{\widehat{n}(x)-1} f(\mathscr{T}_1^l x).$$

In particular, we have that $\int \overline{f} d\mu_2 = 0$. A key property is that Birkhoff sums of $\overline{f}$ with respect to $\mathcal{T}_2$ constitute a subsequence of Birkhoff sums of f with respect to $\mathcal{T}_1$. Thus, to obtain statistical properties for the latter, it is enough to prove them for the former, and to have some control on the gaps between two consecutive terms of the subsequence. This is the approach followed in [11, 19]. If, in the interests of expediency, we make the hypothesis that the function $\overline{f} : B \to \mathbb{R}$ is Hölder continuous, then we can lift the results for $\mathcal{T}_2$ in Theorem 10.4.9 (with respect to $\overline{f}$) to those for $\mathcal{T}_1$ (with respect to f). More generally, we can assume that f is Hölder continuous and the associated function $\overline{f}$ satisfies a weaker "local Hölder" condition that if x and y belong to the same element of $\mathcal{Q}$ with $\widehat{n}(x) = \widehat{n}(y) = n$, say, then $|\overline{f}(x) - \overline{f}(y)| \lesssim n\|\overline{f}\|_\beta \|x - y\|^\beta$. However, following [17] we can then consider the slightly larger Banach space $\mathcal{B}$ with respect to the norm

$$\|h\|_{\mathcal{B}} = \sup_{A \in \mathcal{Q}} \sup_{x \in A} \frac{|f(x)|}{\widehat{n}(A)} + \sup_{A \in Q} \sup_{\substack{x,y \in A \\ x \neq y}} \frac{1}{\widehat{n}(A)} \frac{|h(x) - h(y)|}{\|x - y\|^\beta},$$

for which the proofs of Lemma 10.3.4 and Theorem 10.4.9 readily generalize.

To extend the almost sure invariance principle from $\mathcal{T}_2$ to $\mathcal{T}_1$ we need first to check the hypotheses of the theorem of Melbourne and Török [19]. This will prove the almost sure invariance principle for $\mathcal{T}_1$ and the renormalized restriction of μ_1 to B, and we can then use the result of Korepanov [11, Theorem 3.7] to conclude the results for $(\mathcal{T}_1, \mu_1)$. In particular,

(1) by the Lemma 10.5.2, we can choose $\delta > 0$ so that $\widehat{n} \in L^{2+\delta}(B, \mu_2)$, and
(2) by the analogue of part (2) of Corollary 10.3.5 we have that

$$\frac{1}{N} \sum_{i=0}^{N-1} \widehat{n}(\mathcal{T}_2^i x) = \int \widehat{n} d\mu + O\left(\frac{1}{N^{1-\epsilon}}\right), \quad \mu_2\text{-a.e. } x \in B.$$

In particular, we can now conclude that the almost sure invariance principle holds for $\mathcal{T}_1$ with variance $\widehat{\sigma}^2 = \sigma^2 / \int \widehat{n} d\mu_2$.

Theorem 10.5.3 (Almost Sure Invariance Principle for $\mathcal{T}_1$) *The almost sure invariance principle holds for $\mathcal{T}_1$ and μ_1.*   □

Remark 10.5.4 It can be interesting to precise the error rates in the almost sure invariance principle above. Even if the result of Philipp and Stout [22] used to prove Theorem 10.4.9 does not provide very insightful bounds, it is possible, using different methods, to prove that, for $\mathcal{T}_2$ and $\mathcal{T}_1$, we have $S_n = W_n + o(n^\lambda)$ for every $\lambda > 0$, see Korepanov [12].   □

This theorem has several consequences for Hölder continuous functions f, including the analogues of Proposition 10.4.2 and Corollaries 10.4.3–10.4.6. for $\mathcal{T}_1$.

More precisely, we have the following results.

Proposition 10.5.5 *The Functional Central Limit Theorem holds for* $\mathcal{T}_1$. ☐

This completes the proof of Theorem 10.1.1.

Corollary 10.5.6 (Central Limit Theorem) *For* $y \in \mathbb{R}$ *we have that*

$$\lim_{n \to +\infty} \mu_1 \left\{ x \in B : \frac{1}{\sqrt{n}} \sum_{j=1}^{n} f(\mathcal{T}_1^j x) \le y \right\} = \frac{1}{\sqrt{2\pi}\sigma} \int_{-\infty}^{y} e^{-t^2/2\sigma^2} dt$$

Corollary 10.5.7 *For* $y \ge 0$ *we have that*

$$\lim_{n \to +\infty} \mu_1 \left\{ x \in B : \frac{1}{\sqrt{n}} \max_{1 \le k \le n} \sum_{j=1}^{k} f(\mathcal{T}_1^j x) \le y \right\} = \frac{\sqrt{2}}{\sqrt{\pi}\sigma} \int_{-\infty}^{y} e^{-t^2/2\sigma^2} dt - 1$$

Corollary 10.5.8 (Arcsine Law) *For* $0 \le y \le 1$ *we have that*

$$\lim_{n \to +\infty} \mu_1 \left\{ x \in B : \frac{N_n(x)}{n} \le y \right\} = \frac{2}{\sqrt{\pi}} \sin^{-1} \sqrt{y}$$

where $N_n(x) = Card \left\{ 1 \le k \le n : \sum_{j=1}^{k} f(\mathcal{T}_1^j x) > 0 \right\}$. ☐

Corollary 10.5.9 (Law of the Iterated Logarithm) *For* μ*-a.e.* $x \in B$ *we have*

$$\limsup_{n \to +\infty} \frac{\sum_{j=1}^{n} f(\mathcal{T}_1^j x)}{\sigma\sqrt{2n \log \log n}} = 1.$$

From the structure of the map $\mathcal{T}_1$, one can deduce many other interesting statistical properties. For instance, using Lemmata 10.3.2 and 10.5.2, we can obtain a local large deviations principle, thanks to Melbourne and Nicol [18, Theorem 2.1] (see also Rey-Bellet and Young [23, Theorem B]):

Theorem 10.5.10 (Local Large Deviations Principle for $\mathcal{T}_1$**)** *For any Hölder continuous function* $f : \Delta \times \mathcal{R} \to \mathbb{R}$ *not equal to a coboundary such that* $\int f d\mu_1 = 0$, *there exists* $\epsilon_0 > 0$ *and a rate function* $c : (-\epsilon_0, \epsilon_0) \to \mathbb{R}$ *continuous, strictly convex, vanishing only at 0, such that for every* $0 < \epsilon < \epsilon_0$,

$$\lim_{n \to \infty} \frac{1}{n} \log \mu_1(f^n > n\epsilon) = -c(\epsilon).$$

10.6   Transfer Operators and Analytic Functions

To take advantage of the transformation $\mathscr{T}_2$ being piecewise analytic, we can also consider the transfer operator acting on a space of analytic functions. This will prove useful in the proof of Theorem 10.1.2. Let us denote $\underline{\lambda} = (\lambda_1, \cdots, \lambda_m), \xi = (\xi_1, \cdots, \xi_m) \in \mathbb{R}^m$. For sufficiently small $\epsilon > 0$ we denote by

$$B_\epsilon^{\mathbb{R}} = \left\{ \underline{\lambda} \in \mathbb{R}^m : \sum_{j=1}^m \lambda_j = 1 \text{ and } |\lambda - B| < \epsilon \right\}$$

an ϵ-neighbourhood of B in the (hyperplane containing the) simplex and consider a simple complexification of the form

$$B_\epsilon^{\mathbb{C}} = \left\{ \underline{\lambda} + i\underline{\xi} \in \mathbb{C}^m : |\underline{\lambda} - B| < \epsilon, \sum_{j=1}^m \lambda_j = 1, \sum_{j=1}^m \xi_j = 0 \text{ and } |\xi_j| \le \epsilon \right\}.$$

Let $\mathscr{T}_2 : B_\epsilon^{\mathbb{C}} \to \mathbb{C}^n$ also denote the analytic extension from B to $B_{\mathbb{C}}$ provided $\epsilon > 0$ is sufficiently small.

In order to show that $\mathscr{L}$ preserves a space of analytic functions on this space we can use the following simple lemma.

Lemma 10.6.1 *Providing $\epsilon > 0$ is sufficiently small we have that $\overline{\mathscr{T}_2^{-1} B_\epsilon^{\mathbb{C}}} \subset int(B_\epsilon^{\mathbb{C}})$. Moreover, for $x = \underline{\lambda} + i\underline{\xi} \in B_\epsilon^{\mathbb{C}}$ we have that*

$$\sup_{x \in B_\epsilon^{\mathbb{C}}} \left| \sum_{\mathscr{T}_2^{-1} y = x} \frac{1}{\left(\sum_i (Ay)_i \right)^m} \right| < +\infty$$

Proof Since the inverse branches of $\mathscr{T}_2 : B \to B$ are uniformly contracting, we can choose $\epsilon > 0$ sufficiently small and $0 < \theta < 1$ such that $\mathscr{T}_2^{-1} B_\epsilon^{\mathbb{R}} \subset B_{\theta\epsilon}^{\mathbb{R}}$. We can show that their complexifications have a similar property with respect to $B_{\mathbb{C}}$. To begin, observe that the linear action of any of the positive matrices A corresponding to an inverse branch of $\mathscr{T}_2$ act on both the real and imaginary coordinates independently, and the complexification of the linear action is again a linear action:

$$(\lambda_1, \cdots, \lambda_m) + i(\xi_1, \cdots, \xi_m) \mapsto A(\lambda_1, \cdots, \lambda_m) + i A(\xi_1, \cdots, \xi_m).$$

The image under the projective action comes from dividing by $\sum_j (A\underline{\lambda})_j + i \sum_j (A\underline{\xi})_j$ (i.e., the complexification of $\|A\underline{\lambda}\|$) to get:

$$\frac{A\underline{\lambda} + iA\underline{\xi}}{\sum_j (A\underline{\lambda})_j + i \sum_j (A\underline{\xi})_j} = \frac{A\underline{\lambda}}{\sum_j (A\underline{\lambda})_j}$$

$$- \left(\frac{A\underline{\lambda} \frac{(\sum_j (A\underline{\xi})_j)^2}{(\sum_j (A\underline{\lambda})_j)} - A\underline{\xi}(\sum_j (A\underline{\xi})_j)}{\left(\sum_j (A\underline{\lambda})_j\right)\left(\sum_j (A\underline{\lambda})_j + \frac{(\sum_j (A\underline{\xi})_j)^2}{(\sum_j (A\underline{\lambda})_j)}\right)} \right)$$

$$+ i \frac{\left(A\underline{\xi} - A\underline{\lambda} \frac{\sum_j (A\underline{\xi})_j}{\sum_j (A\underline{\lambda})_j} \right)}{\sum_j (A\underline{\lambda})_j + \frac{(\sum_j (A\underline{\xi})_j)^2}{(\sum_j (A\underline{\lambda})_j)}} .$$

In particular, for $\theta' = (1 + \theta)/2$ and $\epsilon > 0$ sufficiently small we can deduce that $\mathscr{T}_2^{-1} B_\epsilon^{\mathbb{C}} \subset B_{\theta'\epsilon}^{\mathbb{C}}$. This completes the proof of the first part of the lemma.

For the second part of the lemma, we first observe that uniformly in $\underline{\lambda} + i\underline{\xi} \in B_\epsilon^{\mathbb{C}}$ we have

$$\frac{1}{(\sum_j (A\underline{\lambda})_j + i \sum_j (A\underline{\xi})_j)^m} = \frac{1}{(\sum_j (A\underline{\lambda})_j)^m} \frac{1}{\left(1 + i \frac{(\sum_j (A\underline{\xi})_j)^m}{(\sum_j (A\underline{\lambda})_j)^m}\right)}$$

$$= \left(\frac{1}{(\sum_j (A\underline{\lambda})_j)^m} \right)(1 + O(\epsilon)). \qquad (10.1)$$

However, from the formula of the transfer operator, we know that, as in the proof of Lemma 10.3.4, for $x \in B$,

$$\sup_{x \in B} \left| \sum_{\mathscr{T}_2^{-1} y = x} \frac{1}{(\sum_i (Ay)_i)^m} \right| < +\infty. \qquad (10.2)$$

Comparing (10.1) and (10.2) completes the proof. □

We can consider the Banach space $H(B_\epsilon^{\mathbb{C}})$ of analytic functions $f : B_\epsilon^{\mathbb{C}} \to \mathbb{C}$ with a continuous extension to the closure of $B_\epsilon^{\mathbb{C}}$ endowed with supremum norm $\|f\| = \sup_{B_\epsilon^{\mathbb{C}}} |f(z)|$. We can apply Lemma 10.6.1 to deduce that the operator $\mathscr{L} : H(B_\epsilon^{\mathbb{C}}) \to H(B_\epsilon^{\mathbb{C}})$ is well defined. In particular, that the series expression for $\mathscr{L}w(x)$ converges to an analytic function for $x \in B_\epsilon^{\mathbb{C}}$ merely follows by complex differentiation under summation sign.

This leads to the following definition and result.

Definition 10.6.2 Any bounded linear operator $L : B \to B$ on a Banach space B with norm $\| \cdot \|$ is called *nuclear* (of order α) if there exist:

(i)   vectors $u_n \in B$ (with $\|u_n\| = 1$);
(ii) bounded linear functionals $l_n \in B^*$ (with $\|l_n\| = 1$); and
(iii) a sequence (ρ_n) of complex numbers such that $\sum_{n=0}^{\infty} |\rho_n|^{\alpha} < +\infty$, with

$$L(v) = \sum_{n=0}^{\infty} \rho_n l_n(v) u_n, \quad \text{for all } v \in B.$$

We say that L has order zero, if property holds for any $\alpha > 0$.   □

In particular, a nuclear operator is automatically a compact operator, for which the non-zero eigenvalues are of finite multiplicity (and the eigenspaces and dual spaces are of finite multiplicity).

Proposition 10.6.3 *The operator $\mathcal{L} : H(B_\epsilon^{\mathbb{C}}) \to H(B_\epsilon^{\mathbb{C}})$ is nuclear (of order zero).*   □

Proof The proof follows the same lines as that in [15, 16], see also [24]. We denote by $C^{\omega}(B_\epsilon^{\mathbb{C}})$ the Fréchet space of analytic functions on $B_\epsilon^{\mathbb{C}}$, endowed with the compact-open topology. We observe that $\mathcal{L} : H(B_\epsilon^{\mathbb{C}}) \to C^{\omega}(B_\epsilon^{\mathbb{C}})$ is a bounded linear operator and recall that the space $C^{\omega}(B_\epsilon^{\mathbb{C}})$ is nuclear [7]. In particular, if we compose $\mathcal{L}$ with the continuous inclusion $H(B_\epsilon^{\mathbb{C}}) \hookrightarrow C^{\omega}(B_\epsilon^{\mathbb{C}})$, we conclude that the operator $\mathcal{L}$ is nuclear (or order zero) [7] (cf. [16], proof of Lemma 3).   □

Many of the statistical results for $\mathcal{T}_2$ described in the previous sections are related to the existence of a spectral gap for $\mathcal{L}$. In the present analytic context this is essentially automatic since the operator is compact. Moreover, one can apply an approach of Mayer [16, p. 12] to recover that the value 1 is a simple eigenvalue of maximal modulus, and that eigenfunction ρ is real analytic.

We can recover the following:

Corollary 10.6.4 *The invariant density of $\mathcal{T}_2$ (and thus $\mathcal{T}_1$) is real analytic.*   □

Remark 10.6.5 Zorich [32, Theorem 1] actually proved that the invariant density is, when restricted to a subset of the form Δ_π^+ or Δ_π^-, a function which is rational, positive and homogeneous of degree $-m$ on $\mathbb{R}^m$.   □

We can again define $\widehat{\mathcal{L}} : C^{\omega}(B) \to C^{\omega}(B)$ by

$$\widehat{\mathcal{L}}w(x) = \frac{1}{\rho(x)} \mathcal{L}(w\rho)(x).$$

then $\widehat{\mathcal{L}}1 = 1$ and $\widehat{\mathcal{L}}^* \mu = \mu$.

10.7   Zeta Functions and Lyapunov Exponents

We now turn to the proof of Theorem 10.1.2. Recall that we can write the sum Λ of the Lyapunov exponents of $\mathscr{T}_2$ as

$$\Lambda = \int_B \log |\det D\mathscr{T}_2(x)| d\mu_2(x).$$

We shall describe an approach to the Lyapunov exponents using complex functions. The connection between zeta functions and both the standard and multidimensional continued fraction transformations was explored by Mayer in [15] (cf. also [16]). We also refer the reader to the monograph of Baladi [3] for an account of the theory of dynamical zeta functions and determinants for hyperbolic maps.

Definition 10.7.1 We can associate to $\mathscr{T}_2$ a complex function $d(z, s)$ in two variables defined by

$$d(z, s) = \exp\left(-\sum_{n=1}^{\infty} \frac{z^n}{n} \sum_{\mathscr{T}_2^n x = x} |\det(D\mathscr{T}_2^n)(x)|^{-s}\right)$$

where we interpret the periodic points as points in the disjoint union. This converges for $|z|$ and $Re(s)$ sufficiently small.   □

The function $d(z, s)$ can be viewed as the reciprocal of a zeta function (in the sense of Ruelle).

The main technical result on such functions is the following.

Proposition 10.7.2

(1) If $|s|$ is sufficiently small, then $d(z, s)$ is an entire function in z;
(2) Moreover, if we expand $d(z, s) = 1 + \sum_{n=1}^{\infty} a_n(s)z^n$, then there exists $c > 0$ such that $|a_n| = O(e^{-cn^{1+1/(m-1)}})$;
(3) The zeros z_0 for $d(z, 1)$ correspond to eigenvalues $\lambda = 1/z_0$. In particular, 1 is the zero of smallest modulus; and
(4) We can write

$$\frac{\frac{\partial d(1,s)}{\partial s}|_{s=1}}{\frac{\partial d(z,1)}{\partial z}|_{z=1}} = \int \log |\det(D\mathscr{T}_2)(x)| d\mu_2(x).$$

Proof This follows from the method of Ruelle [24] and Grothendieck [7]. The only additional feature is that the operator has infinitely many inverse branches but, as in [15, 16], this presents no additional complications to the proof.   □

This gives an alternative expression for Lyapunov exponent in terms of the fixed points of powers of $\mathscr{T}_2$.

Corollary 10.7.3 *We can write* Λ *in terms of rapidly convergent series*

$$\Lambda = \frac{\sum_{n=1}^{\infty} c_n}{\sum_{n=1}^{\infty} b_n}$$

where

(1) b_n *and* c_n *are explicit values (given below) using fixed points of powers of* $\mathcal{T}_2$; *and*

(2) $|b_n| = O(e^{-cn^{1+1/(m-1)}})$ *and* $|c_n| = O(e^{-cn^{1+1/(m-1)}})$.

□

Proof By Proposition 10.7.2 we can write

$$\Lambda = \frac{\frac{\partial d(1,s)}{\partial s}|_{s=1}}{\frac{\partial d(z,1)}{\partial z}|_{z=1}} = \frac{\sum_{n=1}^{\infty} a_n'(1)}{\sum_{n=1}^{\infty} n a_n(1)}.$$

Using the expansion $\exp(z) = 1 + \sum_{l=1}^{\infty} z^l/l!$ we can write that for $Re(s)$ sufficiently large and $|z|$ sufficiently small

$$d(z,s) = 1 + \sum_{l=1}^{\infty} \frac{1}{l!} \left(-\sum_{k=1}^{\infty} \frac{z^k}{k} \sum_{\mathcal{T}_2^k x = x} |\det(D\mathcal{T}_2^k)(x)|^{-s} \right)^l$$

$$= 1 + \sum_{n=1}^{\infty} z^n \left(\sum_{k_1 + \cdots + k_l = n} \frac{(-1)^l}{l!} \prod_{i=1}^{l} \left(\frac{1}{k_i} \sum_{\mathcal{T}_2^{k_i} x = x} |\det(D\mathcal{T}_2^{k_i})(x)|^{-s} \right) \right)$$

by grouping together terms with the same power of z. Thus by

$$a_n(s) = \left(\sum_{k_1 + \cdots + k_l = n} \frac{(-1)^l}{l!} \prod_{i=1}^{l} \left(\frac{1}{k_i} \sum_{\mathcal{T}_2^{k_i} x = x} |\det(D\mathcal{T}_2^{k_i})(x)|^{-s} \right) \right)$$

and thus by part (2) of Proposition 10.7.2

$$b_n = n a_n(1) = n \left(\sum_{k_1 + \cdots + k_l = n} \frac{(-1)^l}{l!} \prod_{i=1}^{l} \left(\frac{1}{k_i} \sum_{\mathcal{T}_2^{k_i} x = x} |\det(D\mathcal{T}_2^{k_i})(x)|^{-1} \right) \right)$$

and

$$c_n = a_n'(1) = \frac{d}{ds}|_{s=1}\left(\sum_{k_1+\cdots+k_l=n}\frac{(-1)^l}{l!}\prod_{i=1}^{l}\left(\frac{1}{k_i}\sum_{\mathscr{T}_2^{k_i}x=x}|\det(D\mathscr{T}_2^{k_i})(x)|^{-s}\right)\right).$$

The bounds on b_n come directly from the bounds on $a_n(1)$ in part (2) of Proposition 10.7.2.

Using the bounds on $a_n(s)$ in part (2) of Proposition 10.7.2 applied to s small neighbourhood of $s = 1$ we get bounds on $c_n = a_n'(1)$ using Cauchy's theorem, i.e., for small enough $\epsilon > 0$ we let

$$|a_n'(0)| \leq \frac{1}{2\pi}\left|\int_{|\xi|=\epsilon} a_n(\xi)\xi^{-2}d\xi\right| = O(e^{-cn^{1+1/(m-1)}})$$

and so the bounds on $|a_n(\cdot)|$ also serve to bound c_n.   □

By the estimate in Part (2) of Proposition 10.7.2 we see that for each fixed t the function $d(z, t)$ is an entire function of order 1 in z. In particular, if $\{z_n(t)\}$ are poles of $d(z, t)$ then by the Hadamard Weierstrauss theorem the function $d(z, t)$ takes the form

$$d(z, t) = e^{A(t)z+B(t)}\prod_n\left(1 - \frac{z}{z_n(t)}\right)e^{\frac{z}{z_n(t)}}$$

where $A(t), B(t) \in \mathbb{C}$ and each $z_n(t)$ depend analytically on t by the Implicit Function Theorem.

Remark 10.7.4 Following Zorich [32], we can also consider the largest Lyapunov exponent θ_1 for these transformations. Let E_{ij} ($1 \leq i, j \leq m$) denote the $m \times m$ matrix with entries 1 on the diagonal and in the (i, j)th place and 0 otherwise, and let P_π denote the permutation matrix associated to π. Consider the matrices

$$A(\pi, a) = (I + I_{\pi^{-1}m,m}).P(\tau^{\pi^{-1}(m)}) \text{ and } A(\pi, a) = E + I_{m,\pi^{-1}m}.$$

We then define a matrix valued function $B(\lambda, \mu)$ on $\cup_{\pi\in\mathscr{R}}\Delta_\pi^+ \cup \Delta_\pi^-$ by

$$B(\lambda, \pi) = A(\lambda, \pi)(A\mathscr{T}_0(\lambda, \pi))\cdots\left(A\mathscr{T}_0^{\hat{n}(\lambda,\pi)-1}(\lambda, \pi)\right).$$

The general definition for the (leading) Lyapunov exponent for this matrix is

$$\theta_1 = \inf_{n\geq 1}\left\{\frac{1}{n}\int \log\|B(\lambda, \pi)B\mathscr{T}_1(\lambda, \pi)\cdots B\mathscr{T}_1^n(\lambda, \pi)\|d\mu_1\right\}.$$

Zorich [32, Theorem 4] proved the following elegant result: The Lyapunov exponent can be written

$$\theta_1 = - \sum_{\pi \in \mathcal{R}} \int_{\Delta_\pi^\pm} \left| \log(1 - \lambda_m) - \log(1 - \lambda_{\pi^{-1}m}) \right| d\mu_1(\lambda)$$

$$= \frac{1}{m} \sum_{\pi \in \mathcal{R}} \int_{\Delta_\pi^\pm} \log | \det D\mathcal{T}_1 | d\mu_1.$$

To complete this section, we briefly consider a related complex function. We can formally define

$$\eta(z) = - \sum_{n=1}^{\infty} \frac{z^n}{n} \sum_{\mathcal{T}_2^n x = x} \frac{\log | \det(D\mathcal{T}_2^n)(x) |}{| \det(D\mathcal{T}_2^n)(x) |}, \quad z \in \mathbb{C}.$$

In particular, we observe that since $\eta(z) = \frac{\partial \log d(z,t)}{\partial t} |_{t=1}$ then by part (1) of Proposition 10.7.2 we see that $\eta(z)$ is meromorphic in the entire complex plane and we can write

$$\eta(z) = B'(1) + \sum_n \frac{\frac{zz_n'(1)}{z_n(1)}}{(z_n(1) - z)} + z \left(A'(1) + \frac{z_n'(1)}{[z_n(1)]^2} \right),$$

for which the poles are $\{z_n\}$ and the residues are $\mu_n := \frac{z_n'(1)}{z_n(1)}$ ($n \geq 1$). Moreover, by part (3) of Proposition 10.7.2 the poles also correspond to derivatives of the eigenvalues of the associated transfer operator. This gives a connection to the approach to resonances considered by Ruelle in the context of Axiom A diffeomorphisms and is suggestive of an analogous interpretation.

Finally, we conclude with the following curiosity.

Proposition 10.7.5 *Assume that* $\mu_n = 0$ *for every* $n \geq 1$ *then* $\eta(z) = 0$ *for all* $z \in \mathbb{C}$. □

Of course, the conclusion of the Proposition is equivalent to $\sum_{\mathcal{T}_2^n x = x} \frac{\log | \det(D\mathcal{T}_2^n)(x) |}{| \det(D\mathcal{T}_2^n)(x) |} = 0$ for each $n \geq 1$.

10.8 A Glimpse into Teichmüller Flows

Thus far we have only considered the case of discrete transformations ($\mathcal{T}_1$, $\mathcal{T}_2$, etc.), but not the case for continuous flows. For completeness, we briefly describe in this section a small piece of the relationship with Teichmüller flows and suggest a connection with the preceding statistical results for $\mathcal{T}_1$ and $\mathcal{T}_2$. We begin by recalling a well known connection between flat surfaces (or translation surfaces) and interval exchange transformations, although we will keep our description brief and informal and the refer the reader to one of the several excellent surveys in this area, such as Veech [29], Viana [30, Chapter 2], Bufetov [4, Section 1.6] or Zorich [32, Section 5] to name a few.

There is a close connection between interval exchange maps and flat metrics on surfaces. A particularly convenient presentation of a flat surface is as a union of m rectangles in the plane based on the intervals I_i and of height l_i, for $i = 1, \cdots, m$. Thus the information we need to reconstruct the flat torus begins with

(a) The lengths λ_i of the intervals I_i ($i = 1, \cdots, m$);
(b) The heights h_i of the rectangles ($i = 1, \cdots, m$).

Since we will assume that the surface has unit area we can write that $\lambda_1 h_1 + \cdots + \lambda_m h_m = 1$. In addition in order to attach the tops of the rectangles back to their bottoms in the correct order we need:

(c) The permutation π on $\{1, \cdots, m\}$ which tells the change in order in which we reattach the tops of the rectangles.

In addition, to define the flow and invariant measure it is convenient to introduce two other coordinates (which obviously depend on those above):

(d) $a_0, \cdots, a_m$, which are actually dependent on the other variables by $h_i - a_i = h_{\pi^{-1}(\pi(i)+1)} - a_{\pi^{-1}(\pi(i)+1)-1}$ for $i = 1, \cdots, m - 1$, with the convention $a_0 = a_{m+1} = 0$; and
(e) $\delta_i = a_{i-1} - a_i$, for $i = 1, \cdots, m$

and the heights of other singularities (which lie in the sides of the rectangles) (Fig. 10.3).

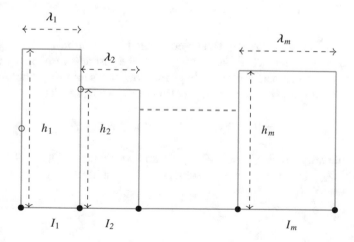

Fig. 10.3 A zippered rectangle

This construction is usually called a zippered rectangle. Let $\Omega_{\mathscr{R}}$ denote the space of all unit area (zippered) rectangles. There is a natural volume $d\lambda_1 \cdots d\lambda_m d\delta_1 \cdots d\delta_m$ on $\Omega_{\mathscr{R}}$. Let μ denote the normalized measure. A version of the Teichmüller flow $T_t : \Omega_{\mathscr{R}} \to \Omega_{\mathscr{R}}$ is defined locally by $T_t(\lambda, h, a, \pi) = (e^t\lambda, e^{-t}h, e^{-t}a, \pi)$ (i.e., flattening the rectangles from above) and this preserves the volume. There is a natural projection from $\Omega_{\mathscr{R}}$ to the moduli space of flat metrics $\mathscr{M}$ and the corresponding semi-conjugate flow $S_t : \mathscr{M} \to \mathscr{M}$ is the *Teichmüller flow*. However, to emphasize the connection to our previous discussion we will persist with the model flow T_t. We can consider the cross section

$$\mathscr{Y} = \left\{ (\lambda, h, a, \pi) \in \Omega_{\mathscr{R}} : \sum_{i=1}^{m} \lambda_i = 1 \right\}$$

to the flow T_t. Under the natural identification on $\Omega_{\mathscr{R}}$ corresponding to different presentations of surfaces as rectangles: the return time function to $\mathscr{Y}$ corresponds to the natural extension of the map $\mathscr{T}_0$ and the return time function is simply $r(\lambda, \pi) = \log\left(1 - \min\{\lambda_m, \lambda_{\pi^{-1}m}\}\right)$. This shows that the properties of the flow T_t are closely related to those of the maps related to the Rauzy-Veech map.

In particular, the Teichmüller flow T_t is a finite-to-one factor of the natural extension of the suspended semi-flow associated to the map $\mathscr{T}_0$ and the function r, i.e., let

$$(\Delta \times \mathscr{R})^r = \left\{ \underbrace{(\lambda, \pi, u)}_{=:x} \in \Delta \times \mathscr{R} \times \mathbb{R} : 0 \leq u \leq r(\lambda, \pi) \right\}$$

where we identify $(x, r(x)) = (\mathscr{T}_0(x), 0)$ and we define the semi-flow

$$(\mathscr{T}_0)_t^r : (\Delta \times \mathscr{R})^r \to (\Delta \times \mathscr{R})^r$$

locally by $(\mathscr{T}_0)_t^r(x, u) = (x, u + t)$, subject to the identifications.

Since inducing on $B \subset \Delta$ (as described in the discrete case) gives the map $\mathscr{T}_2 : B \to B$, we can also represent this semi flow as a suspension semiflow over $\mathscr{T}_2 : B \to B$ with respect to a related function $r_2 : B \to \mathbb{R}$, i.e., let

$$B^{r_2} = \{(x, u) \in B \times \mathbb{R} : 0 \leq u \leq r_2(x)\} / \sim$$

where we identify $(x, r_2(x)) = (\mathscr{T}_2(x), 0)$ and we define $(\mathscr{T}_2)_t^{r_2} : B^{r_2} \to B^{r_2}$ locally by $(\mathscr{T}_2)_t^{r_2}(x, u) = (x, u + t)$, subject to the identifications.

The following lemma was established by Bufetov [4].

Lemma 10.8.1

(1) $r_2 \in L^\gamma(B, \mu_2)$, for every $\gamma > 1$; and
(2) if $F : \Omega_{\mathscr{R}} \to \mathbb{R}$ is Hölder and $f : B \to \mathbb{R}$ is defined by $f(x) :=$
$\int_0^{r_2(x)} F(S_t x) dt$ then there exists $\delta > 0$ such that $f \in L^{2+\delta}(B, \mu_2)$. □

Since r_2 is integrable, the Teichmüller flow preserves the probability measure μ_{r_2}
defined by $d\mu_{r_2} = \left(\int_B r_2 d\mu_2\right)^{-1} d\mu_2 \times ds$.

We now recall the continuous analogue of the Almost Sure Invariance Principle.

Definition 10.8.2 A flow $\psi_t : X \to X$ with invariant probability measure μ is
said to satisfy the *Almost Sure Invariance Principle* with respect to a probability
measure ν if for a Hölder function $\Phi : X \to \mathbb{R}$ not equal to a coboundary such
that $\int \Phi d\mu = 0$ there is a $\epsilon > 0$ and a random variable $\{S_t\}_{t \geq 0}$ and a Brownian
motion B with variance σ^2 such that $\left\{\int_0^t \Phi(\psi_s) ds\right\}_{t \geq 0}$, seen as a random process
defined on (X, ν), is equal in distribution to random variables $\{S_t\}_{t \geq 0}$ and $S_t =$
$B_t + O(t^{1/2 - \epsilon})$. □

The result for Teichmüller flows corresponding to Theorem 10.4.9 is the
following.

Theorem 10.8.3 *The Teichmüller flow satisfies the almost sure invariance principle*
with respect to the probability measure μ_2 seen as a measure on B^{r_2} supported on
$B \times \{0\}$. □

Proof It suffices to show the result for the associated semi-flow (the result for the
natural extension requiring a standard argument involving changing functions by a
coboundary, see for instance [17, Lemma 3.2]). Let $\Phi : B^{r_2} \to B^{r_2}$ be a Hölder
function with $\phi(x) = \int_0^{r(x)} \Phi((\mathscr{T}_2)_t^{r_2} x)) dt$.

(1) $\bar{r}_2 \in L^{2+\beta}(B, \mu_2)$, for some $\beta > 1$ (by part (1) of Lemma 10.8.1)
(2) $\phi \in L^{2+\delta}(B, \mu_2)$, for some $\delta > 0$ (by part (2) of Lemma 10.8.1); and
(3) $\mathscr{T}_2 : B \to B$ satisfies the Almost Sure Invariance Principle (by Theorem 10.4.9)

The Teichmüller flow then satisfies the Almost Sure Invariance Principle by the
results of [19]. □

Even though the Almost Sure Invariance Principle has been proven only for the
measure μ_2, and not for the measure μ_{r_2}, this is enough to deduce the following
corollaries for Hölder continuous functions Φ, see Denker and Philipp [5].

Corollary 10.8.4 (Central Limit Theorem) *For $y \in \mathbb{R}$ we have that*

$$\lim_{T \to +\infty} \mu_{r_2} \left\{ (x, s) : \frac{1}{\sqrt{T}} \int_0^T \Phi((\mathcal{F}_2)_t^{r_2}(x, s)) dt \le y \right\} = \frac{1}{\sqrt{2\pi}\sigma} \int_{-\infty}^y e^{-t^2/2\sigma^2} dt$$

The central limit theorem for Teichmüller flows was proved by Bufetov [4].

Corollary 10.8.5 *For $y \ge 0$ we have that*

$$\lim_{T \to +\infty} \mu_{r_2} \left\{ (x, s) : \frac{1}{\sqrt{T}} \max_{1 \le t \le T} \int_0^t \Phi((\mathcal{F}_2)_t^{r_2}(x, s)) dt \le y \right\} = \frac{\sqrt{2}}{\sqrt{\pi}\sigma} \int_{-\infty}^y e^{-t^2/2\sigma^2} dt - 1$$

Corollary 10.8.6 (Arcsine Law) *For $0 \le y \le 1$ we have that*

$$\lim_{T \to +\infty} \mu_{r_2} \left\{ (x, s) : \frac{N_T(x)}{T} \le y \right\} = \frac{2}{\sqrt{\pi}} \sin^{-1} \sqrt{y}$$

where $N_T(x, s) = Leb \left\{ 0 \le t \le T : \int_0^t \Phi((\mathcal{F}_2)_t^{r_2}(x, s)) dt > 0 \right\}$.   □

Corollary 10.8.7 (Law of the Iterated Logarithm) *For μ_{r_2}-a.e. (x, s) we have*

$$\limsup_{T \to +\infty} \frac{\int_0^T \Phi((\mathcal{F}_2)_t^{r_2}(x, s)) dt}{\sigma \sqrt{2T \log \log T}} = 1.$$

Remark 10.8.8 If Korepanov's results [11] can be extended to suspended flows, then it would be possible to prove Theorem 10.8.3 for the invariant measure μ_{r_2}, and then, using arguments from Melbourne and Nicol [17], to pass to the natural extension, thus obtaining the Almost Sure Invariance Principle and all its corollaries for the original (invertible) Teichmüller flow defined on $\Omega_{\mathscr{R}}$.   □

10.9 Comments on Pressure

It is natural to relate these statistical properties to classical ideas on pressure. To accommodate the complication of having a countable-to-one map $\mathcal{F}_2 : B \to B$ (and also an unbounded return time $\widehat{n} : B \to \mathbb{Z}^+$ when we look at the tower to reconstruct $\mathcal{F}_1$) it is convenient to work with the Gurevich pressure (as developed by Sarig [25], by analogy with the more familiar Gurevich entropy for countable subshifts of finite type). Let us consider a fairly general formulation of these results.

Recall that $\mathcal{Q}$ is the partition of smoothness of $\mathcal{F}_2$, and that $\mathcal{Q}_n = \vee_{i=0}^{n-1} \mathcal{F}_2^{-i} \mathcal{Q}$ is its nth level refinement. For $x, y \in B$, we denote by

$$s(x, y) = \inf \{ n \ge 0 : x \text{ and } y \text{ belong to two different elements of } \mathcal{Q}_n \}$$

their separation time with respect to $\mathscr{T}_2$. Assume that $\phi : B \to \mathbb{R}$ is (locally) Hölder continuous, in the sense that there exists $0 < \theta < 1$ and $A > 0$ such that $V_n(\theta) \leq A\theta^n$ for $n \geq 0$, where

$$V_n(\phi) := \sup\{|\phi(x) - \phi(y)| : x, y \in B, s(x, y) \geq n\}$$

(i.e., the variation of the function over elements of the nth level refinement of the partition associated with $\mathscr{T}_2$). In particular, Lipschitz functions satisfy these conditions. On the other hand, more generally this condition doesn't require ϕ to be bounded, say.

Definition 10.9.1 To define the (Gurevich) pressure can fix any element $A \in \mathscr{Q}_{n_0}$, for chosen value n_0. We then define

$$P(\phi) := \lim_{n \to \infty} \frac{1}{n} \log \left(\sum_{\mathscr{T}_2^n x = x \in A} e^{\phi^n(x)} \right),$$

where $\phi^n(x) := \phi(x) + \phi(\mathscr{T}_2 x) + \cdots + \phi(\mathscr{T}_2^{n-1} x)$. □

Under very modest mixing conditions (i.e., the "Big Images Property" which applies in the case of $\mathscr{T}_2$) we can see that the definition is independent of the choice of n_0 and A. However, in general some additional assumptions are required to ensure that the pressure is finite.

One would anticipate that the properties of $P(\phi)$ would be useful in further studies of the properties of these maps and flows.

Remark 10.9.2 Sarig [26] has results which suggest that the map $t \mapsto P(\phi + t\psi)$ is analytic for suitable Hölder continuous functions ϕ, ψ whenever the pressure is finite and $t \in (-\epsilon, \epsilon)$. This is based on spectral properties of a suitable transfer operator. □

Acknowledgments The first author was partially supported by FCT fellowship SFRH/BPD/1236 30/2016, by FCT project PTDC/MAT-PUR/28177/2017 and by CMUP, which is financed by national funds through FCT - Fundação para a Ciência e a Tecnologia, I.P., under the project with reference UIDB/00144/2020. The second author was partially supported by ERC-Advanced Grant 833802-Resonances and EPSRC grant EP/T001674/1. □

References

1. R. Aimino, M. Nicol, M. Todd, Recurrence statistics for the space of Interval Exchange maps and the Teichmüller flow on the space of translation surfaces. Ann. Inst. H. Poincaré Probab. Stat. **53**, 1371–1401 (2017)
2. A. Avila, A. Bufetov, Exponential decay of correlations for the Rauzy-Veech-Zorich induction map, in *Partially Hyperbolic Dynamics, Laminations, and Teichmüller Flow*, Fields Inst. Commun., vol. 51 (Amer. Math. Soc., Providence, RI, 2007), pp. 203–211

3. V. Baladi, *Dynamical Zeta Functions and Dynamical Determinants for Hyperbolic Maps - A Functional Approach.* Ergebnisse der Mathematik und ihrer Grenzgebiete. 3. Folge/A Series of Modern Surveys in Mathematics, vol. 68 (Springer International Publishing, Cham, 2018)
4. A. Bufetov, Decay of correlations for the Rauzy-Veech-Zorich induction map on the space of interval exchange transformations and the central limit theorem for the Teichmüller flow on the moduli space of Abelian differentials. J. Am. Math. Soc. **19**, 579–623 (2006)
5. M. Denker, W. Philipp, Approximation by Brownian motion for Gibbs measures and flows under a function. Ergodic Theory Dynam. Syst. **4**, 541–552 (1984)
6. K. Fernando, C. Liverani, Edgeworth expansions for weakly dependent random variables. Ann. Inst. H. Poincaré Probab. Stat. **57**, 469–505 (2021)
7. A. Grothendieck, *Produits tensoriels topologiques et espaces nucléaires.* Mem. Amer. Math. Soc., vol. 16 (Amer. Math. Soc., Providence, 1955)
8. P. Hall, C. Heyde, *Martingale Limit Theory and Its Application* (Academic, New York, 1980)
9. C. Heyde, Invariance principles in statistics. Int. Stat. Rev. **49**, 143–152 (1981)
10. A.G. Kachurovskiĭ, Rates of convergence in ergodic theorems. Russ. Math. Surv. **51**, 653–703 (1996)
11. A. Korepanov, Equidistribution for nonuniformly expanding dynamical systems, and application to the almost sure invariance principle. Commun. Math. Phys. **359**, 1123–1138 (2018)
12. A. Korepanov, Rates in almost sure invariance principle for dynamical systems with some hyperbolicity. Commun. Math. Phys. **363**, 173–190 (2018)
13. M. Mackey, M. Tyran Kamińska, Deterministic Brownian motion: the effects of perturbing a dynamical system by a chaotic semi-dynamical system. Phys. Rep. **422**, 167–222 (2006)
14. M. Mackey, M. Tyran Kamińska, Central limit theorems for non-invertible measure preserving maps. Colloq. Math. **110**, 167–191 (2008)
15. D. Mayer, On a ζ function related to the continued fraction transformations. Bull. Soc. Math. France. **104**, 195–203 (1976)
16. D. Mayer, Approach to equilibrium for locally expanding maps in $\mathbb{R}^k$. Commun. Math. Phys. **95**, 1–15 (1984)
17. I. Melbourne, M. Nicol, Almost sure invariance principle for nonuniforlmy hyperbolic systems. Commun. Math. Phys. **260**, 131–145 (2005)
18. I. Melbourne, M. Nicol, Large deviations for nonuniformly hyperbolic systems. Trans. Am. Math. Soc., **360**, 6661–6676 (2008)
19. I. Melbourne, A. Török, Statistical limit theorems for suspension flows. Israel J. Math. **144**, 191–209 (2004)
20. T. Morita, Renormalized Rauzy inductions. Adv. Stud. Pure Math. **43**, 1–25 (2005)
21. W. Parry, M. Pollicott, Zeta functions and the periodic orbit structure of hyperbolic systems. Astérisque **187–188**, 1–268 (1990)
22. W. Philipp, W. Stout, *Almost Sure Invariance Principles for Partial Sums of Weakly Dependent Random Variables.* Mem. Amer. Math. Soc., vol. 161 (Amer. Math. Soc, Providence, 1975)
23. L. Rey-Bellet, L.-S. Young, Large deviations in non-uniformly hyperbolic dynamical systems. Ergodic Theory Dynam. Syst. **28**, 587–612 (2008)
24. D. Ruelle, Zeta-functions for expanding maps and Anosov flows. Invent. Math. **34**, 231–242 (1976)
25. O. Sarig, Thermodynamic formalism for countable Markov shifts. Ergodic Theory Dynam. Syst. **19**, 1565–1593 (1999)
26. O. Sarig, Phase transitions for countable Markov shifts. Commun. Math. Phys. **217**, 555–577 (2001)
27. M. Tyran Kamińska, An invariance principle for maps with polynomial decay of correlations. Commun. Math. Phys. **260**, 1–15 (2005)
28. W. Veech, Interval exchange transformations. J. Anal. Math. **33**, 222–272 (1978)
29. W. Veech, Gauss measures for transformations on the space of interval exchange maps. Ann. Math. **115**, 201–242 (1982)
30. M. Viana, *Dynamics of Interval Exchange Maps and Teichmüller Flows.* IMPA (2008). http://w3.impa.br/~viana/out/ietf.pdf

31. J.-C. Yoccoz, Continued fraction algorithms for interval exchange maps: an introduction, in *Frontiers in Number Theory, Physics, and Geometry I* (Springer, Berlin, 2006), pp. 401–435
32. A. Zorich, Finite Gauss measure on the space of interval exchange transformations. Lyapunov exponents. Ann. Inst. Fourier **46**, 325–370 (1996)
33. A. Zorich, Flat surfaces, in *Frontiers in Number Theory, Physics, and Geometry I* (Springer, Berlin, 2006), pp. 437–583

Chapter 11
Entropy Rigidity, Pressure Metric, and Immersed Surfaces in Hyperbolic 3-Manifolds

Lien-Yung Kao

Abstract In this paper, we show an entropy rigidity result for immersed surfaces in hyperbolic 3-manifolds that relates dynamic and geometric quantities including entropy, critical exponent, and geodesic stretch. We then apply this result to $\mathscr{H}$ the minimal hyperbolic germs (a deformation space corresponding to the quasifuchsian space $\mathscr{QF}$ proposed by Taubes). As a consequence, we recover the famous Bowen rigidity theorem for quasifuchsian representations. Moreover, we construct a Riemannian metric, i.e., the pressure metric, on the Fuchsian space $\mathscr{F} \subset \mathscr{H}$. We also discuss relations between the pressure metric, Sander's metric, and Weil-Petersson metric on $\mathscr{F}$.

11.1 Introduction

Entropy rigidity problems have drawn a lot of attention since the late twentieth century. It shows that entropy, a dynamics quantity, can characterize the geometry of the ambient space. This phenomenon occurs in many different geometric or dynamic settings, such as the seminal work of Bowen [1], Katok [2], Burger [3], Knieper [4], etc. In this paper, we investigate a version of the entropy rigidity phenomenon arising in immersed surfaces in hyperbolic 3-manifolds.

In the view of entropy rigidity results, it is natural to ask if one can use these dynamics quantities to gauge the ambient geometric structures, such as a metric on deformation spaces. The geometric object or the deformation space tightly related to the immersed surfaces in hyperbolic 3-manifolds setting is the minimal hyperbolic germ $\mathscr{H}$ of a given closed surface. This deformation space is, inspired by minimal surfaces in hyperbolic 3-manifolds, introduced in Taubes [5]. $\mathscr{H}$ shares many features with $\mathscr{QF}$ the quasifuchsian space. Inspired by the work of McMullen [6] and Bridgeman [7], we construct a Riemannian metric, the pressure

L.-Y. Kao (✉)
Department of Mathematics, The George Washington University, Washington, DC, USA
e-mail: lkao@gwu.edu

© The Author(s), under exclusive license to Springer Nature Switzerland AG 2021
M. Pollicott, S. Vaienti (eds.), *Thermodynamic Formalism*, Lecture Notes in Mathematics 2290, https://doi.org/10.1007/978-3-030-74863-0_11

metric, on the Fuchsian space $\mathscr{F} \subset \mathscr{H}$. Furthermore, the pressure metric can also be constructed through the relationship to Manhattan curves. Relations between Riemannian metrics on $\mathscr{F}$ are also discussed. (See Sect. 11.3 for precise definitions of terminology used above.)

To put our results in context, we shall introduce our notation. Throughout this paper, S denotes a closed surface and M denotes a hyperbolic 3-manifold with the hyperbolic metric h. Let $f : S \to M$ be a π_1-injective immersion. We denote the induced Riemannian metric on S by g, that is, $g = f^*h$.

We recall that when (N, g) is a negatively curved Riemannian manifold, each conjugacy class $[\gamma] \in [\pi_1 N]$ corresponds to a unique closed geodesic on (N, g). In this case, $l_g[\gamma]$ denotes the length of a closed geodesic corresponding to $[\gamma]$ with respect to the Riemannian metric g.

Definition 11.1 (Asymptotic Geodesic Distortions) $C_g(f)$ the *asymptotic geodesic distortion* of f with respect to g is defined as

$$C_g(f) = \limsup_{T \to \infty} \frac{\displaystyle\sum_{[\gamma] \in R_T(g)} l_h[\gamma]}{\displaystyle\sum_{[\gamma] \in R_T(g)} l_g[\gamma]}$$

where $R_T(g) := \{[\gamma] \in [\pi_1 S] : l_g[\gamma] < T\}$. Similarly, $C_h(f)$ the *asymptotic geodesic distortion* of f with respect to h is given by

$$C_h(f) = \limsup_{T \to \infty} \frac{\displaystyle\sum_{[\gamma] \in R_T(h)} l_h[\gamma]}{\displaystyle\sum_{[\gamma] \in R_T(h)} l_g[\gamma]}$$

where $R_T(h) := \{[\gamma] \in [\pi_1 S] : l_h[\gamma] < T\}$. $\qquad\qquad\qquad\qquad\qquad\qquad\qquad\square$

The following result links critical exponent, entropy and asymptotic geodesic distortions through inequalities with rigidity features. Please see Sect. 11.3 for precise definitions.

Theorem A *Let $f : S \to M$ be a π_1-injective immersion from a closed surface S to a hyperbolic 3-manifold M, and let Γ be the copy of $\pi_1 S$ in $\mathrm{Isom}(\mathbb{H}^3)$ induced by the immersion f. Suppose Γ is convex cocompact and (S, g) is negatively curved. Then*

(1) *The limit-sups in $C_h(f)$ and $C_g(f)$ are limits.*
(2) $0 < C_h(f) \leq C_g(f) \leq 1$.
(3) *Let $h_{top}(S)$ be the topological entropy of the geodesic flow on $T^1 S$ and δ_Γ be the critical exponent, then*

$$C_h(f) \cdot \delta_\Gamma \leq h_{top}(S) \leq C_g(f) \cdot \delta_\Gamma. \qquad\qquad (11.1.1)$$

(4) *The first (resp. second) equality in (11.1.1) holds if and only if the marked length spectrum of (S, g) is proportional to the marked length spectrum of (M, h), and the proportion is the ratio $\frac{\delta_\Gamma}{h_{top}(S)}$ (resp. $\frac{h_{top}(S)}{\delta_\Gamma}$).*
(5) *If $C_h(f) = 1$ or $C_g(f) = 1$, then S is a totally geodesic submanifold in M.*

Remark 11.2 Theorem A extends several results in [8] by relaxing the embedding condition to immersion.

In particular, when $f : S \to M$ is an *embedding*, inspired by Glorieux [8], we can strengthen Theorem A and relating asymptotic geodesic distortions to *geodesic stretches*. Let $\phi : T^1 S \to T^1 S$ be the geodesic flow on the unit tangent bundle of (S, g). Geodesic stretches are dynamics quantities that introduced by Knieper [4] which characterize how geodesics are stretched from one metric to the other with respect to a given ϕ–invariant probability measure. The precise definition of geodesic stretch can be found in Sect. 11.4.2.

Theorem 11.3 *Under the same assumptions in Theorem A and assuming $f : S \to M$ is an embedding, then the asymptotic geodesic distortions $C_h(f)$ and $C_g(f)$ match the geodesic stretches relative to ϕ–invariant measures. More precisely,*

$$C_h(f) = I_\mu(S, M) \text{ and } C_g(f) = I_{\mu_{BM}}(S, M),$$

where μ is some ϕ–invariant measure and μ_{BM} is the Bowen-Margulis measure of ϕ. $\square$

The Manhattan curve can be regarded as the 2-dimensional generalization of the critical exponent. Motivated by Burger [3] and Sharp [9], we adapt their argument to our setting. Precisely, the Manhattan curve χ_f corresponding to the immersion $f : S \to M$ is defined as the boundary of the convex set

$$\{(a, b) \in \mathbb{R}^2 : \sum_{\gamma \in \pi_1 S} e^{-(a \cdot l_g[\gamma] + b \cdot l_h[\gamma])} < \infty\}.$$

We have the following result.

Theorem B *Under the same assumption as Theorem A, then*

(1) $(0, 1)$ *and* $(h_{top}(g), 0) \in \chi_f$.
(2) χ_f *is a real analytic curve.*
(3) χ_f *is strictly convex unless the marked length spectrum of (S, g) is proportional to the marked length spectrum of (M, h).*
(4) χ_f *is a straight line if and only if the marked length spectrum of (S, g) is proportional to the marked length spectrum of (M, h).*

Using the convexity of the Manhattan curve χ_f, we derive an entropy rigidity result similar to that which Bishop and Steger discovered for Fuchsian representations [10].

Corollary 11.4 (Bishop-Steger Rigidity) *Under the same assumption as Theorem A, then*

$$\lim_{T \to \infty} \frac{1}{T} \log \#\{\gamma \in \pi_1 S : l_g[\gamma] + l_h[\gamma] < T\} \leq \frac{h_{top}(g)}{h_{top}(g) + 1}$$

and the equality holds if and only if the marked length spectrum of (S, g) is proportional to the marked length spectrum of M.

We now change gear to the applications of the above rigidity results. Roughly speaking, the space of minimal hyperbolic germs $\mathcal{H}$ for a closed surface S is a set of pairs (g, B) consisting of a Riemannian metric g and a symmetric two tensor B on S. Each pair $(g, B) \in \mathcal{H}$ can be thought of as an induced metric and a second fundamental form of a minimal immersion of S into some hyperbolic 3-manifold. (See precise definition in Sect. 11.3.6.)

Pointed out in Uhlenbeck [11], one can relate $\mathcal{H}$ with the character variety $\mathcal{R}(\pi_1(S), \mathsf{PSL}(2, \mathbb{C}))$. $\mathcal{R}(\pi_1(S), \mathsf{PSL}(2, \mathbb{C}))$ is the space of conjugacy classes of representations of $\pi_1(S)$ into $\mathsf{PSL}(2, \mathbb{C})$. Using this relation, we are interested in the quasifuchsian spaces $\mathcal{QF}$ drawn from $\mathcal{R}(\pi_1(S), \mathsf{PSL}(2, \mathbb{C}))$ into $\mathcal{H}$. (See the precise definitions in Sect. 11.3.6.)

With Theorem A, we recover the famous Bowen rigidity theorem for quasifuchsian representations [1].

Corollary 11.5 (Bowen's Rigidity, [1]) *A quasifuchsian representation $\rho \in \mathcal{QF}$ is Fuchsian if and only if $\dim_H \Lambda(\Gamma) = 1$ where $\dim_H \Lambda(\Gamma)$ is the Hausdorff dimension of the limit set $\Lambda(\Gamma)$ of Γ.* $\qquad\square$

By the definition of $\mathcal{H}$, for $(g, 0) \in \mathcal{H}$ we know g is a hyperbolic metric on S. In other words, the Teichmüller space of S is a subspace of $\mathcal{H}$. This copy of Teichmüller space in $\mathcal{H}$ is called the *Fuchsian space* $\mathcal{F}$.

To study the geometry of the Fuchsian space $\mathcal{F}$, we will need to investigate a bigger space $\mathcal{AF}$ the almost-Fuchsian space. A pair $(g, B) \in \mathcal{H}$ is called *almost-Fuchsian* if (g, B) is close to a Fuchsian pair, in the sense that $\|B\|_g^2 < 2$. In particular, Uhlenbeck [11] showed that for a hyperbolic metric $(m, 0) \in \mathcal{F}$ and a holomorphic quadratic differential $\alpha \in Q([m])$ there exists a smooth path $r_B(t) = (g_t, tB) \in \mathcal{AF}$ where $g_0 = m$ and $B = \mathrm{Re}(\alpha)$. Therefore, the study on this particular path $r_B(t) = (g_t, tB) \in \mathcal{AF}$ will help us to see the geometry of $\mathcal{F}$ and derive the pressure metric for $\mathcal{F}$.

The path $r_B \subset \mathcal{AF}$ defines a smooth family of Riemann metrics $\{g_t\}$ over S. There are two related dynamics objects to the path, namely, $h_{top}(g_t)$ the topological entropy of the geodesic flow over (S, g_t) and the Manhattan curve $\chi_t = \chi(g_0, g_t)$. By the structural stability of Anosov flows, we know $h_{top}(g_t)$ and χ_t vary smoothly along the path $r_B(t)$ when t is small.

Given $(m, 0) \in \mathscr{F}$ and $\alpha \in Q([m])$, we define the *normalized intersection number* $J_{m,\alpha}(t)$ with respect to m and α by

$$J_{m,\alpha}(t) = h_{top}(g_t) \cdot \limsup_{T \to \infty} \frac{\sum_{\tau \in R_T(m)} l_{g_t}[\tau]}{\sum_{\tau \in R_T(m)} l_m[\tau]}$$

where g_t is given by the path $\gamma_\alpha(t) = (g_t, t\mathrm{Re}\alpha)$ with $g_0 = m$.

Theorem C (Intersection Number and the Pressure Metric) *For* $(m, 0) \in \mathscr{F}$ *and* $\alpha \in Q([m])$, *we know*

(1) $J_{m,\alpha}(t) \leq 1$ *and the equality holds if and only* $t = 0$.
(2) $\frac{d^2}{dt^2} J_{m,\alpha}(t)\big|_{t=0} \geq 0$ *and is equal to zero if and only* $\alpha = 0$.
(3) $\|\alpha\|_P^2 := \frac{d^2}{dt^2} J_{m,\alpha}(t)\big|_{t=0}$ *defines a Riemannian metric for* $\mathscr{F}$, *and which is called the pressure metric.*

Notice that Sanders [12, Theorem 3.8] showed one can use $h_{top}(g_t)$ to construct a Riemannian metric for $\mathscr{F}$, and which is bounded below by the Weil-Petersson metric. The following result describes a relation between Sanders' metric $\| \cdot \|_S$, the pressure metric $\| \cdot \|_P$ and the Weil-Petersson metric $\| \cdot \|_{WP}$.

Corollary 11.6 *Let* $(m, 0)$ *be a Fuchsian pair and* $\alpha \in Q([m])$ *be a holomorphic quadratic differential. Then*

$$\|\alpha\|_S^2 := \frac{d^2}{dt^2} h_{top}(g_t)\bigg|_{t=0} \geq \|\alpha\|_P^2 + 2\pi \|\alpha\|_{WP}^2.$$

We remark that [12, Theorem 3.8] is an immediate consequence of Corollary 11.6.

Lastly, inspired by the work of Pollicott and Sharp [13], we relate the pressure metric and the Manhattan curve χ_t.

Corollary 11.7 *One can define a family of metrics on the Fuchsian space* $\mathscr{F}$ *by using the second derivatives of* $\chi_t(s)$ *for* $s \in [0, \varepsilon)$ *for some* $\varepsilon > 0$. *More precisely, for* $(m, 0)$ *a Fuchsian pair and* $\alpha \in Q([m])$, *we have*

$$\|\alpha\|_{\chi(s)}^2 := \frac{d^2}{dt^2} \chi_t(s)\bigg|_{t=0} = \ddot{h}_0(1-s) - (s-s^2)\|\alpha\|_P^2.$$

□

11.1.1 Outline of the Paper

In Sects. 11.2 and 11.3, we will give a fair amount of related background knowledge of dynamics and geometry. In Sect. 11.4, we will present the main dynamics rigidity results in Proposition A and Theorem 11.3 as well as Manhattan curve results in Theorem B. In Sect. 11.5, we will investigate the space minimal hyperbolic germs and the construction of the pressure metrics and present the proof of Theorem C.

11.2 Background from the Thermodynamic Formalism

11.2.1 Flows and Reparametrization

Let X be a compact metric space with a continuous flow $\phi = \{\phi_t\}_{t \in \mathbb{R}}$ on X without any fixed point and μ a ϕ–invariant probability measure on X. Consider a positive continuous function $F : X \to \mathbb{R}_{>0}$ and define, for $t > 0$

$$\kappa(x, t) := \int_0^t F(\phi_s(x)) ds,$$

and $\kappa(x, t) := -\kappa(\phi_t(x), -t)$ for $t < 0$. The function κ satisfies the cocycle property

$$\kappa(x, t + s) = \kappa(x, t) + \kappa(\phi_t x, s)$$

for all $x, t \in \mathbb{R}$ and $x \in X$.

Since $F > 0$ and X is compact, F has a positive minimum and $\kappa(x, \cdot)$ is an increasing homeomorphism of $\mathbb{R}$. We then have a map $\alpha : X \times \mathbb{R} \to \mathbb{R}$ such that

$$\alpha(x, \kappa(x, t)) = \kappa(x, \alpha(x, t)) = t.$$

for all $(x, t) \in X \times \mathbb{R}$.

Definition 11.8 Let $F : X \to \mathbb{R}$ be a positive continuous function. The *reparametrization* of the flow ϕ by F is the flow $\phi^F = \{\phi_t^F\}_{t \in \mathbb{R}}$ defined as $\phi_t^F(x) = \phi_{\alpha(x,t)}(x)$. □

11.2.2 Periods and Measures

Let O be the set of closed orbits of ϕ. For $\tau \in O$, let $l(\tau)$ be the period of τ with respect to ϕ. Then the period of τ with respect to the reparametrized flow ϕ^F is

$$\int_\tau F := \langle \delta_\tau, F \rangle = \int_0^{l(\tau)} F(\phi_s(x)) ds,$$

where x is any point on τ and δ_τ is the Lebesgue measure supported by the orbit τ.

Let μ be a ϕ−invariant probability measure on X, $F : X \rightarrow \mathbb{R}$ be a continuous function, and ϕ^F be the reparametrization of ϕ by F. We define $\widehat{F \cdot \mu}$ to be the probability measure: for any continuous function G on X

$$\widehat{F \cdot \mu}(G) = \frac{\int_X G \cdot F d\mu}{\int_X F d\mu}.$$

Then $\widehat{F \cdot \mu}$ is a ϕ^F−invariant probability measure.

11.2.3 Entropy, Pressure, and Equilibrium States

We denote by $h_\phi(\mu)$ the *measure-theoretic entropy* of ϕ with respect to a ϕ−invariant probability measure μ. Let $\mathscr{M}^\phi$ denote the set of ϕ−invariant probability measures, and $C(X)$ denote the set of continuous functions on X.

Definition 11.9 The *pressure* of a function $F : X \rightarrow \mathbb{R}$ is defined as

$$P_\phi(F) := \sup_{m \in \mathscr{M}^\phi} \left(h_\phi(m) + \int_X F dm \right)$$

A measure $m \in \mathscr{M}^\phi$ is called an *equilibrium state* if it realizes the equality.

We define the *topological entropy* of the flow ϕ by $h_\phi = P_\phi(0)$, and an equilibrium state for the function $F \equiv 0$ is called a *measure of maximum entropy*.

If there is no ambiguity as to which flow we refer to, such as ϕ, then we might drop the subscript ϕ and use h to denote the topological entropy, and $h(\mu)$ to denote the measure theoretic entropy of ϕ with respect to μ.   □

The following Abramov formula relates the measure theoretic entropies of the flow ϕ and its reparametrization ϕ^F.

Theorem 11.10 (Abramov Formula [14]) *Suppose ϕ is a continuous flow on X and ϕ^F is the reparametrization of ϕ by a positive continuous function F, then for all $\mu \in \mathscr{M}^\phi$*

$$h_{\phi^F}(\widehat{F.\mu}) = \frac{h_\phi(\mu)}{\int_X F d\mu}.$$

11.2.4 Anosov Flows

A $C^{1+\alpha}$ flow $\phi_t : X \rightarrow X$ on a compact manifold X is called *Anosov* if there is a continuous splitting of the unit tangent bundle $T^1 X = E^0 \oplus E^s \oplus E^u$, where E^0 is the one-dimensional bundle tangent to the flow direction, and there exists $C, \lambda > 0$

such that $\|D\phi_t|E^s\| \leq Ce^{-\lambda t}$ and $\|D\phi_{-t}|E^u\| \leq Ce^{-\lambda t}$ for $t \geq 0$. We say that the flow is *transitive* if there is a dense orbit.

We know that if M is a compact negatively curved Riemannian manifold, then the geodesic flow $\phi : T^1M \to T^1M$ is a transitive Anosov flow.

Recall that a function $F : X \to \mathbb{R}$ is called $\alpha-$*Hölder continuous* if there exists $C > 0$ and $\alpha \in (0, 1]$ such that for all $x, y \in X$ we have $|F(x) - F(y)| \leq C \cdot d_X(x, y)^\alpha$. In most cases, we will abbreviate $\alpha-$*Hölder continuous* to *Hölder continuous*.

Let O be the set of periodic orbits of ϕ. For a continuous function $F : X \to \mathbb{R}_{>0}$ and $T \in \mathbb{R}$, we define

$$R_T(F) = \{\tau \in O : \langle \delta_\tau, F \rangle < T\}.$$

Proposition 11.11 (Bowen [15]) *The topological entropy h_ϕ of a transitive Anosov flow ϕ is finite and positive. Moreover,*

$$h_\phi = \lim_{T \to \infty} \frac{1}{T} \log \#\{\tau \in O : l(\tau) < T\}.$$

If $F : X \to \mathbb{R}$ is a positive Hölder continuous function, then

$$h_F := h_{\phi^F} = \lim_{T \to \infty} \frac{1}{T} \log \#R_T(F),$$

is finite and positive. □

Recall that two Hölder continuous $F, G : X \to \mathbb{R}$ are called *Livšic cohomologous* if there exists a Hölder continuous $V : X \to \mathbb{R}$ which is C^1 in the flow direction such that

$$F(x) - G(x) = \left.\frac{\partial}{\partial t}\right|_{t=0} V(\phi_t(x)).$$

The following theorem shows that equilibrium states can distinguish Hölder continuous functions up to a coboundary in the Livšic cohomology.

Theorem 11.12 (Bowen-Ruelle [16], cf. [17] Proposition 6.1) *If ϕ_t is a transitive Anosov flow on a compact manifold X, then for each $F : X \to \mathbb{R}$ Hölder continuous function, there exists a unique equilibrium state m_F of F. Moreover, if F and G are Hölder continuous functions such that $m_F = m_G$, then $F - G$ is Livšic cohomologous to a constant.* □

In particular, the we also call the unique measure of maximum entropy m_0 of a transitive Anosov flow ϕ the *Bowen-Margulis measure* and denote it by m_{BM}.

Theorem 11.13 (Equidistribution, Bowen [15], Cf. [17] Theorem 9.4) *Suppose ϕ is a transitive Anosov flow on a compact manifold X. Then m_{BM}, the Bowen-*

Margulis measure, satisfies that for any continuous function G on X

$$\int_X G \, dm_{BM} = m_{BM}(G) = \lim_{T \to \infty} \frac{1}{\#R_T(1)} \sum_{\tau \in R_T(1)} \frac{\langle \delta_\tau, G \rangle}{\langle \delta_\tau, 1 \rangle} = \lim_{T \to \infty} \frac{\displaystyle\sum_{\tau \in R_T(1)} \langle \delta_\tau, G \rangle}{\displaystyle\sum_{\tau \in R_T(1)} \langle \delta_\tau, 1 \rangle}.$$

The following Bowen's formula links the topological entropy of the reparametrized flow ϕ^F and the reparametrization function F.

Theorem 11.14 (Bowen's Formula, Cf. [17] Proposition 6.1) *If ϕ is a transitive Anosov flow on a compact metric space X and $F : X \to \mathbb{R}$ is a positive Hölder continuous function, then*

$$P_\phi(-hF) = 0$$

if and only if $h = h_{\phi^F}$. Moreover, if $h = h_{\phi^F}$ and m is an equilibrium state of $-hF$, then $\widehat{F.m}$ is a measure of maximal entropy of the reparametrized flow ϕ^F.

11.2.5   A Livšic Type Theorem

By the definition of the Livšic cohomology, the following properties are immediate:

1. If F and G are Livšic cohomologous then they have the same integral over any $\phi-$invariant measure.
2. The pressure $P_\phi(F)$ only depends on the Livšic cohomology class of F.
3. $R_T(F)$ only depends on the Livšic cohomology class of F.

Theorem 11.15 (Nonnegative Livšic Theorem, Lopes-Thieullen [18]) *Suppose X is a compact Riemannian manifold with negative sectional curvature. Let $\phi_t : T^1X \to T^1X$ be the geodesic flow on T^1X. Let $F : T^1X \to \mathbb{R}$ be a Hölder continuous function such that $\langle \delta_\tau, F \rangle \geq 0$ for each $\phi-$closed orbit τ. Then F is cohomologous to a Hölder continuous function $G(x)$ such that $G(x) \geq 0$, $\forall x \in T^1X$.*

11.2.6   Variance and Derivatives of the Pressure

Let $\phi_t : X \to X$ be a transitive Anosov flow on a compact metric space X, and $C^\alpha(X)$ be the set of α-Hölder continuous function on X.

Definition 11.16 Suppose $F \in C^{\alpha}(X)$ and m_F is the equilibrium state of F. For any $H, G \in C^{\alpha}(X)$ with mean zero (i.e., $\int G \, dm_F = \int H \, dm_F = 0$), the *covariance* of G and H with respect to m_F is given by

$$\mathrm{Cov}(G, H, m_F) := \lim_{T \to \infty} \frac{1}{T} \int_X \left(\int_0^T H(\phi_t(x)) dt \right) \left(\int_0^T G(\phi_t(x)) dt \right) dm_F(x).$$

Similarly, for $G \in C^{\alpha}(X)$ with mean zero we define the *variance* of G with respect to m_F as

$$\mathrm{Var}(G, m_F) := \mathrm{Cov}(G, G, m_F).$$

The following properties are some handy formulas of the derivatives of the pressure.

Proposition 11.17 (Proposition 4.10, 4.11 [17]) *Suppose that $\phi_t : X \to X$ is a transitive Anosov flow on a compact metric space X, and $F, G \in C^{\alpha}(X)$. If m_F is the equilibrium state of F, then*

(1) *The function $t \mapsto P_{\phi}(F + tG)$ is analytic.*
(2) *The first derivative is given by*

$$\left. \frac{d P_{\phi}(F + tG)}{dt} \right|_{t=0} = \int_X G \, dm_F.$$

(3) *If $\int G \, dm_F = 0$, then the second derivative can be formulated as*

$$\left. \frac{d^2 P_{\phi}(F + tG)}{dt^2} \right|_{t=0} = \mathrm{Var}(G, m_F).$$

(4) $\mathrm{Var}(G, m_F) = 0$ *if and only if G is Livšic cohomologous to a constant.*

Corollary 11.18 (Theorem 2.2 [6]) *Let ψ_t be an smooth path in $C^{\alpha}(X)$, $m_0 = m_{\psi_0}$ be the equilibrium state of ψ_0, and $\dot{\psi}_0 := \left. \frac{d\psi_t}{dt} \right|_{t=0}$. Then*

$$\left. \frac{d P(\psi_t)}{dt} \right|_{t=0} = \int_{\Sigma_A^+} \dot{\psi}_0 \, dm_0. \tag{11.2.1}$$

Moreover, if the first derivative is zero (i.e., $\int_{\Sigma_A^+} \dot{\psi}_0 m_0 = 0$) then

$$\left. \frac{d^2 P(\psi_t)}{dt^2} \right|_{t=0} = \mathrm{Var}(\dot{\psi}_0, m_0) + \int_{\Sigma_A^+} \ddot{\psi}_0 \, dm_0. \tag{11.2.2}$$

11.2.7 The Pressure Metric

Let $\phi_t : X \to X$ be a transitive Anosov flow on a compact metric space X. We consider the space $\mathscr{P}(X)$ of Livšic cohomology classes of pressure zero Hölder continuous functions on X, i.e.,

$$\mathscr{P}(X) := \left\{ F : F \in C^\alpha(X) \text{ for some } \alpha \text{ and } P_\phi(F) = 0 \right\} / \sim .$$

The tangent space of $\mathscr{P}(X)$ at F is

$$T_F \mathscr{P}(X) = \ker\left((\mathbf{D}P_\phi)(F)\right) = \left\{ G : G \in C^\alpha(X) \text{ for some } \alpha \text{ and } \int G \, dm_F = 0 \right\} / \sim .$$

For $G \in T_F \mathscr{P}(X)$, we define the *pressure metric* as

$$\|G\|_P^2 := \frac{\mathrm{Var}(G, m_F)}{-\int F \, dm_F}.$$

Proposition 11.19 *If $\{c_t\}_{t \in (-1,1)}$ is an analytic one parameter family contained in $\mathscr{P}(X)$, then*

$$\|\dot{c}_0\|_P^2 = \frac{\int \ddot{c}_0 \, dm_{c_0}}{\int c_0 \, dm_{c_0}},$$

where $\dot{c}_0 = \frac{d}{dt} c_t \big|_{t=0}$ and $\ddot{c}_0 = \frac{d^2}{dt^2} c_t \big|_{t=0}$. □

Proof This follows from the direct computation of the (Gâteaux) second derivative of $P(c_t)$:

$$\frac{d^2}{dt^2} P(c_t) \bigg|_{t=0} = (\mathbf{D}^2 P)(c_0)(\dot{c}_0, \dot{c}_0) + (\mathbf{D}P)(c_0)(\ddot{c}_0)$$

$$= \mathrm{Var}(\dot{c}_0, m_{c_0}) + \int \ddot{c}_0 \, dm_{c_0}.$$

Since $P(c_t) = 0$, we have

$$\|\dot{c}_0\|_P^2 := \frac{\mathrm{Var}(\dot{c}_0, m_{c_0})}{-\int c_0 \, dm_{c_0}} = \frac{\int \ddot{c}_0 \, dm_{c_0}}{\int c_0 \, dm_{c_0}}.$$

□

11.3 Background from Geometry

In this section, we survey several facts of $\delta-$hyperbolic spaces and their group of isometries. A good reference is the book [19] edited by Ghys and de la Harpe and Kapovich's book [20].

11.3.1 δ–Hyperbolic Spaces

A metric space (X, d) is said to be *geodesic* if any two points $x, y \in X$ can be joined be a *geodesic segment* $[x, y]$ that is a naturally parametrized path from x to y whose length is equal to $d(x, y)$, and is called *proper* if all closed balls are compact.

Definition 11.20 A geodesic metric space (X, d) is called $\delta-$*hyperbolic* (where $\delta \geq 0$ is some real number) if for any geodesic triangle in X each side of the triangle is contained in the $\delta-$neighborhood of the union of two other sides. A metric space (X, d) is called *hyperbolic* if it is $\delta-$hyperbolic for some $\delta \geq 0$.

It is well-known that a *Pinched Hadamard manifold* $\widetilde{M}$ is $\delta-$hyperbolic space where $(\widetilde{M}, \widetilde{g})$ is a complete and simply connected Riemannian manifold with bounded negative sectional curvature. Recall that a group G is *hyperbolic* if for one (and for all) finite generating set the Cayley graph is hyperbolic. For example, finitely generated free groups and surface groups for surfaces with genus > 1 are hyperbolic groups. □

We say that two geodesic rays $\tau_1 : [0, \infty) \to X$ and $\tau_2 : [0, \infty) \to X$ are *equivalent* and write $\tau_1 \sim \tau_2$ if there is a $K > 0$ such that for all $t > 0$

$$d(\tau_1(t), \tau_2(t)) < K.$$

It is easy to see that $\sim$ is indeed an equivalence relation on the set of geodesic rays. We then define the *geometric boundary* $\partial_\infty X$ of X by

$$\partial_\infty X := \{[\tau] : \tau \text{ is a geodesic ray in } X\}.$$

Moreover, we know that when X is proper, $\partial_\infty X$ is metrizable by the *visual metric* (see [21, Theorem 1.5.2]).

11.3.2 Quasi-Isometries

Definition 11.21 A function $q : X \to Y$ from a metric space (X, d_X) to a metric space (Y, d_Y) is called a (C, L)-*quasi-isometry embedding* if there is $C, L > 0$ such that:

For any $x, x' \in X$, we have

$$\frac{1}{C} d_X(x, x') - L \le d_Y(q(x), q(x')) \le C \cdot d_X(x, x') + L.$$

If, in addition, there exists an *approximate inverse* map $\bar{q} : Y \to X$ that is a (C, L)-quasi-isometric embedding such that for all $x \in X$ and $y \in Y$

$$d_X(\bar{q}q(x), x) \le L, \qquad d_Y(q\bar{q}(y), y) \le L,$$

then we call q a (C, L)-*quasi-isometry*. In this case, (X, d_X) and (Y, d_Y) are called *quasi-isometric*.   □

In most cases, the quasi-isometry constants C and L do not matter, so we shall use the words quasi-isometry and quasi-isometry embedding without specifying constants.

Theorem 11.22 ([21], Theorem 1.6.4) *Let (X, d_X) and (Y, d_Y) be hyperbolic spaces. Suppose the boundaries are equipped with visual metrics. Then*

(1) Any quasi-isometry embedding $q : X \to X'$ extends to a bi-Hölder embedding $q : \partial_\infty X \to \partial_\infty Y$ with respect to the corresponding visual metrics.
(2) Any quasi-isometry $q : X \to X'$ extends to a bi-Hölder homeomorphism $q : \partial_\infty X \to \partial_\infty Y$ with respect to the corresponding visual metrics.

Definition 11.23 A (C, L)-*quasi-geodesic* is a (C, L)-quasi-isometry embedding $q : \mathbb{R} \to X$.   □

Theorem 11.24 (Morse Lemma, cf. Ch.5, Theorem 6 [19]) *Suppose X and Y are hyperbolic spaces, and $q : X \to Y$ is a (C, L)-quasi-isometry. Then, for every geodesic $\gamma \subset X$, its image $q(\gamma)$ is a quasi-geodesic on Y and is within a bounded distance R from a geodesic on Y. Moreover, this constant R is only depends on X, Y, and the quasi-isometry constants C and L.*   □

Remark 11.25 When the space Y is a pinched Hadamard manifold, we have a stronger result of the above theorem. Namely, every geodesic $\gamma \subset X$ its image $q(\gamma)$ is a quasi-geodesic on Y and is within a bounded distance R from a *unique* geodesic on Y.   □

Let X be a hyperbolic space. We denote its group of isometries by $\mathrm{Isom}(X)$. The following lemma connects some subgroups of $\mathrm{Isom}(X)$ and the hyperbolic space X.

Theorem 11.26 (Švarc-Milnor lemma, cf. Lemma 3.37 [20]) *Let X be a proper geodesic metric space. Let G be a subgroup of $\mathrm{Isom}(X)$ acting properly discontinuously and cocompactly on X. Pick a point $o \in X$. Then the group G is finitely generated; for any choice of finitely generating set S of G, the map $q : G \to X$, given by $q(\gamma) = \gamma(o)$, is a quasi-isometry. Here G is given the word metric induced from $C(G, S)$.*   □

11.3.3   Negatively Curved Manifolds and the Group of Isometries

Let (X, g) be a negatively curved compact Riemannian manifold. The universal covering $(\widetilde{X}, \widetilde{g})$ of (X, g) is a pinched Hadamard manifold. Let Γ denote the group of deck transformations of the covering $\widetilde{X}$.

Since (X, g) is negatively curved, every $\gamma \in \Gamma$ corresponds to a unique geodesic τ_{γ}^{X} on X. Furthermore, each conjugacy class $[\gamma] \in [\Gamma]$ corresponds to a unique closed geodesic τ_{γ}^{X} on X and vice versa. Moreover, the length of the closed geodesic τ_{γ}^{X} is exactly the *translation distance* of $\gamma \in \pi_1 X$ (i.e., $l_g(\tau_{\gamma}^{X}) = l_g[\gamma] := \inf_{x \in X} d_g(x, \gamma \cdot x)$).

Recall that the *marked length spectrum* is a function $l : [\tau] \mapsto l[\tau] \in \mathbb{R}^+$ which assigns to a homotopy class $[\tau]$ the length $l[\tau]$.

By a famous result of Margulis [22], we know that $h_{top}(X)$ the geodesic flow for (X, g) can be characterized by the (exponential) growth rate of closed geodesics, that is,

$$h_{top}(X) = \lim_{T \to \infty} \frac{1}{T} \log \# \left\{ [\gamma] \in [\pi_1 X] : l_g[\gamma] < T \right\}.$$

Now let us consider a compact 3–manifold M equipped with a hyperbolic metric h. Then there exists a discrete and faithful representation $\rho : \pi_1 M \to \mathrm{Isom}(\mathbb{H}^3)$ such that $M \cong \rho(\pi_1 M) \backslash \mathbb{H}^3$ where $(\mathbb{H}^3, \widetilde{h})$ is the universal covering of (M, h). For the sake of brevity, in what follows we will denote the lifted metric of $\widetilde{h}$ on $\mathbb{H}^3$ by h.

Let Γ be a discrete subgroup of $\mathrm{Isom}(\mathbb{H}^3)$. Recall that The *limit set* $\Lambda(\Gamma)$ is the set of limit points Γx for any $x \in \mathbb{H}^3$, and Γ is called *convex cocompact* if Γ acts cocompactly on the convex hull $\mathrm{Conv}(\Lambda(\Gamma))$ of the limit set of Γ.

Definition 11.27 The *critical exponent* δ_Γ is defined as

$$\delta_\Gamma := \inf\{s > 0 : \sum_{\gamma \in \Gamma} e^{-s d_h(x, \gamma x)} < \infty\},$$

for any point $x \in \mathbb{H}^3$ and d_h is the hyperbolic distance on $\mathbb{H}^3$.   □

The following result of Sullivan links critical exponent, Hausdorff dimension, and entropy.

Theorem 11.28 (Sullivan [23, 24]) *Suppose Γ is a non-elementary, convex cocompact, and discrete subgroup of* $\mathrm{Isom}(\mathbb{H}^3)$*, then*

$$\delta_\Gamma = \dim_H \Lambda(\Gamma) = h_{top}(M) = \lim_{T \to \infty} \frac{1}{T} \log \#\{[\gamma] \in [\Gamma] : l_h[\gamma] < T\},$$

where $\dim_H \Lambda(\Gamma)$ is the Hausdorff dimension of $\Lambda(\Gamma)$, $M = \Gamma \backslash \mathbb{H}^3$ and $l_h(\gamma) = d_h(o, \gamma o)$, o is the origin of $\mathbb{H}^3$. □

11.3.4 Hölder Cocycles

Let (X, g) be a compact negatively curved manifold, $\widetilde{X}$ be its universal covering, and Γ be the group of deck transformations of the covering $\widetilde{X}$. Recall that the $\pi_1 X$-action on $\widetilde{X}$ is defined by $\gamma \cdot x = i_X(\gamma)(x)$, where i is the isomorphism $i_X : \pi_1 S \to \Gamma$.

Definition 11.29 A *Hölder cocycle* is a function $c : \pi_1 X \times \partial_\infty \widetilde{X} \to \mathbb{R}$ such that

$$c(\gamma_0 \gamma_1, x) = c(\gamma_0, \gamma_1 \cdot x) + c(\gamma_1, x)$$

for any $\gamma_0, \gamma_1 \in \pi_1 X$ and $x \in \partial_\infty \widetilde{X}$, and $c(\gamma, \cdot)$ is Hölder continuous for every $\gamma \in \pi_1 X$. □

Given a Hölder cocycle c we define the *periods* of c to be the number

$$l_c[\gamma] := c(\gamma, \gamma_X^+)$$

where γ_X^+ is the attracting fixed point of $\gamma \in \pi_1 X \backslash \{e\}$ on $\partial_\infty \widetilde{X}$.

Two cocycles c and c' are said to be cohomologous if there exists a Hölder continuous function $U : \partial_\infty \widetilde{X} \to \mathbb{R}$ such that, for all $\gamma \in \pi_1 X$, one has

$$c(\gamma, x) - c'(\gamma, x) = U(\gamma \cdot x) - U(x).$$

One easily deduces from the definition that the set of periods of a cocycle is a cohomological invariant.

Definition 11.30 The *exponential growth rate* for a Hölder cocycle c is defined as

$$h_c := \limsup_{T \to \infty} \frac{1}{T} \log \#\{[\gamma] \in [\pi_1 X] : l_c[\gamma] < T\}.$$

The following theorem of Ledrappier gives a precise method to construct a Hölder function F_c from a Hölder cocycle c. The main point of this construction is that the exponential growth rate h_c of the Hölder cocycle is exactly the topological entropy for the reparametrized flow ϕ^{F_c}.

Theorem 11.31 (Ledrappier [25]) *For each Hölder cocycle $c : \pi_1 X \times \partial_\infty \widetilde{X} \to \mathbb{R}$, there exists a Hölder continuous function $F_c : T^1 X \to \mathbb{R}$, such that, for all $\gamma \in \pi_1 X - \{e\}$, one has*

$$l_c[\gamma] = \int_{[\gamma]} F_c.$$

The map $c \mapsto F_c$ induces a bijection between the set of cohomology classes of $\mathbb{R}$−valued Hölder cocycles, and the set of Livšic cohomology classes of Hölder continuous functions from $T^1 X$ to $\mathbb{R}$. Moreover,

$$h_c = h_{F_c} = \lim_{T \to \infty} \frac{1}{T} \log \#\{[\gamma] : \langle \delta_{[\gamma]}, F_c \rangle < T\}.$$

11.3.5   Immersed Surfaces in Hyperbolic 3–Manifolds

Let S be a differentiable surface and M be a 3-manifold, we say a differentiable mapping $f : S \to M$ is an *immersion* if $df_p : T_p S \to T_{f(p)} M$ is injective for all $p \in S$. If, in addition, f is a homemorphism onto $f(S) \subset M$, where $f(S)$ has the subspace topology induced from M, we say that f is an *embedding*. Moreover, if the induced homomorphism $f_* : \pi_1 S \to \pi_1 M$ is injective, then we say f is π_1–*injective*.

Throughout, we consider that M is a 3–manifold equipped with a hyperbolic metric h and S is a closed surface with negative Euler characteristic. Before moving further, we recall some terminology from differential geometry. Given an immersion $f : S \to M$, let $g = f^* h$ be the induced Riemannian metric on S, ∇ the Levi-Civita connection on (M, h), N be the unit outward normal vector field to the surface $f(S) \subset M$, and ∂_1 and ∂_2 be the coordinate fields of TS.

The *second fundamental form* $B : TS \times TS \to \mathbb{R}$ of $f(S)$ is the symmetric 2-tensor on S defined by, locally,

$$B(\partial_i, \partial_j) = \langle \partial_i, -\nabla_{\partial_j} N \rangle_h,$$

where $\langle, \rangle_h$ is the hyperbolic metric h on M.

The *shape operator* $S_g : TS \to TS$ is the symmetric self-adjoint endomorphism defined by raising one index of the second fundamental form B with respect to the metric g. The *mean curvature* H of the immersion $f : S \to M$ (or, of the immersed surface (S, g)) is the trace of the shape operator. We call an immersion $f : S \to M$ *minimal* if the mean curvature H vanishes identically.

Moreover, we can relate the induced Riemannian metric g and shape operator S_g by Gauss-Codazzi equations:

$$K_g = -1 + \det S_g, \qquad\qquad \text{(Gauss eq.)} \qquad (11.3.1)$$

$$\nabla_{df(X)}(S_g(Y)) - \nabla_{df(Y)} S_g(X) = S_g([X, Y]). \qquad \text{(Codazzi eq.)} \qquad (11.3.2)$$

where $X, Y \in TS$ and $[\cdot, \cdot]$ is the Lie bracket on TS.

Remark 11.32 If f is a minimal immersion, then Gauss-Codazzi equations can be expressed in terms of B by

$$K_g = -1 - \frac{1}{2} \|B\|_g^2,$$

$$(\nabla_{\partial_i} B)_{jk} = (\nabla_{\partial_j} B)_{ik},$$

where $\| \cdot \|_g$ is the tensor norm w.r.t. metric g and ∂_1 and ∂_2 are coordinate fields of TM. Moreover, in this case the Gauss equation implies $K_g \leq -1$, i.e., (S, g) is a negatively curved surface.   □

11.3.6 Minimal Hyperbolic Germs

In this subsection, we continue the discussion under the same assumption as in the previous subsection. Let (g, B) be a pair consisting of a Riemannian metric g and a symmetric 2-tensor B on S.

Definition 11.33 (Minimal Hyperbolic Germ) A pair (g, B) is called a *minimal hyperbolic germ* if it satisfies the following equations

$$\begin{cases} K_g = -1 - \frac{1}{2} \|B\|_g^2, \\ (\nabla_{\partial_i} B)_{jk} = (\nabla_{\partial_j} B)_{ik}, \\ B \text{ is traceless w.r.t. } g. \end{cases} \qquad (11.3.3)$$

Recall that $\mathrm{Diff}_0(S)$ is the space of orientation preserving diffeomorphisms of S isotopic to the identity. There is a natural $\mathrm{Diff}_0(S)$ action (i.e., by pullback) on the space of minimal hyperbolic germs, and we are mostly interested in the following quotient space.

Definition 11.34 The space $\mathscr{H}$ of minimal hyperbolic germs is the quotient:

$$\mathscr{H} = \{\text{minimal hyperbolic germs}\}/\mathrm{Diff}_0(S).$$

Taubes [5] showed that $\mathscr{H}$ is a smooth manifold of dimension $12g - 12$ where g is the genus of S. The fundamental theorem of surface theory ensures that each $(g, B) \in \mathscr{H}$ can be integrated into an immersed minimal surface in a hyperbolic 3-manifold with the Riemannian metric g and the second fundamental form B.

Moreover, $\mathscr{H}$ is closely related with Teichmüller space. Recall that the *Teichmüller space* $\mathscr{T}$ of S is the space of conformal classes of Riemannian metrics with curvature -1. It is clear that we can identify $\mathscr{T}$ with a subspace $\mathscr{F}$ of $\mathscr{H}$. Namely, the *Fuchsian space* $\mathscr{F}$ is the set

$$\mathscr{F} = \{(m, 0) \in \mathscr{H} : m \text{ is a Reimannian metric of constant curvature } -1\}.$$

Let $[g]$ be the conformal class of a Riemannian metric g on S and $X = (S, [g])$ be the Riemann surface associated with g. It is well-known that $T_X^* \mathcal{T}$ the fiber of the holomorphic cotangent bundle over X can be identified with $Q(X)$ the space of holomorphic quadratic differentials on X.

The following theorem of Hopf [26] helps us see the relation between $\mathcal{H}$ and $Q(X)$.

Theorem 11.35 (Hopf [26]) *If $(g, B) \in \mathcal{H}$, then B is the real part of a (unique) holomorphic quadratic differential $\alpha \in Q(X)$. More precisely, if $(x_1, x_2) = x_1 + ix_2 = z$ is a local isothermal coordinate of X and $B = B_{11}dx_1^2 + B_{22}dx_2^2 + 2B_{12}dx_1dx_2$, then*

$$\alpha(g, B) = (B_{11} - iB_{12})(x_1, x_2)dz^2. \tag{11.3.4}$$

Remark 11.36 In fact, $B_{11} = -B_{22}$ because (S, g) is minimal. It is not hard to see that, given a holomorphic quadratic differential $\alpha \in Q(X)$, one can derivative a symmetric 2-tensor B on S by Eq. (11.3.4) and $2||\alpha||_g^2 = ||B||_g^2$.   □

Moreover, the space $\mathcal{H}$ admits a smooth map to $T^* \mathcal{T}$ given by

$$\Psi : \mathcal{H} \to \quad T^* \mathcal{T}$$

$$(g, B) \mapsto ([g], \alpha(g, B)).$$

For any two holomorphic quadratic differentials α and β in $Q(X)$, the *Weil-Petersson pairing* is given by

$$\langle \alpha, \beta \rangle_{WP} = \int_S \frac{\alpha \bar{\beta}}{m},$$

where m is the hyperbolic metric on S conformal to g. It's also well-known that this pairing defines a Kähler metric, *the Weil-Petersson metric*, on the Teichmüller space whose geometry has been intensely studied. In the last section, we will discuss several applications of our results related with the Weil-Petersson metric on $\mathcal{F}$.

11.4 Immersed and Embedded Surfaces in Hyperbolic 3-Manifolds

Let $f : S \to M$ be a π_1−injective immersion from S to a hyperbolic 3-manifold M and Γ be the copy of $\pi_1 S$ in $\text{Isom}(\mathbb{H}^3)$ induced by the immersion f. More precisely, let $\rho : \pi_1 M \to \text{Isom}(\mathbb{H}^3)$ be the discrete and faithful representation, up to conjugacy, corresponding to M, i.e., $M = \rho(\pi_1 M)\backslash \mathbb{H}^3$. Then $\Gamma = \rho(f_*(\pi_1 S))$ where f_* is the induced homomorphism of $f : S \to M$.

The standing hypotheses throughout here are: Γ is a convex cocompact and (S, g) is negatively curved where $g = f^*h$ and h is the given hyperbolic metric on M.

Notice that because (S, g) is a closed negatively curved surface, its universal covering $(\widetilde{S}, \widetilde{g})$ is a pinched Hadamard manifold. Let Γ_S denote the group of deck transformations of the covering $\widetilde{S}$. Then we know $\Gamma_S \cong \pi_1 S$ and $\Gamma_S \subset \text{Isom}(\widetilde{S})$.

The following lemma is an immediate consequence of Theorems 11.22 and 11.26.

Lemma 11.37 *There exists a quasi-isometry* $q : \widetilde{S} \to \text{Conv}(\Lambda(\Gamma))$*, where* $\text{Conv}(\Lambda(\Gamma))$ *is the convex hull of* $\Lambda(\Gamma)$ *in* $\mathbb{H}^3$. *Moreover,* q *extends to a bi-Hölder and* Γ*–equivariant map between boundaries, and* q *sends the attracting limit point* γ_S^+ *of the hyperbolic element* $\gamma_S \in \Gamma_S \subset \text{Isom}(\widetilde{S})$ *to the attracting limit point* γ_M^+ *of* $\gamma_M \in \Gamma \subset \text{Isom}(\mathbb{H}^3)$. $\qquad\square$

Now we are ready to state and prove Theorem A. The proof of Theorem A consists of several lemmas. In the following we indicate how one should read the Proof of Theorem A from these lemmas.

Proof of Theorem A Assertion 1 follows Lemma 11.40. Assertions 2, 3, 4 are consequences of Lemmas 11.38 and 11.40. Assertion 5 follows Lemma 11.39. $\qquad\square$

Lemma 11.38 *Under the same assumptions as Theorem A, the following holds.*

(1) There exists a Hölder continuous function $F : T^1S \to \mathbb{R}$ *such that* $0 < F \le 1$ *and* $\int_\tau F = l_h[\tau]$ *for all closed orbits* τ *on* T^1S *where* $l_h[\tau]$ *is the length of the closed geodesic in the free homotopy class containing* $f(\tau) \subset T^1M$ *with respect to the hyperbolic metric* h.

(2) Let $\mu_{-h_F F}$ *be the equilibrium for* $-h_F F$ *and* μ_{BM} *be the Bowen-Margulis measure for the geodesic flow on* T^1S *where*

$$h_F := \lim_{T \to \infty} \frac{1}{T} \log \#\{\tau \text{ is a closed orbit on } T^1S : \int_\tau F < T\}.$$

We have $C_1 := \int F d\mu_{-h_F F}$ *and* $C_2 := \int F d\mu_{BM}$ *satisfy*

$$C_1 \delta_\Gamma \le h_{top}(S) \le C_2 \delta_\Gamma. \tag{11.4.1}$$

(3) Each equality in (11.4.1) holds if and only if the marked length spectrum of S *is proportional to the marked length spectrum of* M.

Proof Let ϕ denote the geodesic flow on the unit tangent bundle of (S, g).

The first step is to construct a Hölder reparametrization function $F : T^1S \to \mathbb{R}_{>0}$ such that the topological entropy h_F of the reparametrized flow ϕ^F is the critical exponent δ_Γ of Γ in $\mathbb{H}^3$.

Recall the Busemann function $B_\eta^h(x, y) : \partial_\infty \mathbb{H}^3 \times \mathbb{H}^3 \times \mathbb{H}^3 \to \mathbb{R}$, for $\eta \in \partial_\infty \mathbb{H}^3$ and $x, y \in \mathbb{H}^3$ is given by

$$B_\eta^h(x, y) := \lim_{z \to \eta} d_h(x, z) - d_h(y, z).$$

Using the quasi-isometry q given in Lemma 11.37, we define a map $c : \pi_1 S \times \partial_\infty \widetilde{S} \to \mathbb{R}$ by

$$c : \pi_1 S \times \partial_\infty \widetilde{S} \to \mathbb{R}$$

$$(\gamma, \xi) \mapsto B_{q(\xi)}^h(f(o), \gamma^{-1} \cdot f(o)),$$

for $o \in \widetilde{S}$.

Claim c is a Hölder cocycle.

Proof of Claim: By Lemma 11.37

$$\begin{aligned}
c(\gamma_1 \gamma_2, \xi) &= B_{q(\xi)}^h(f(o), (\gamma_1 \gamma_2)^{-1} \cdot f(o)) = B_{q(\xi)}^h(f(o), (\gamma_2^{-1} \gamma_1^{-1}) \cdot f(o)) \\
&= B_{q(\xi)}^h(f(o), \gamma_2^{-1} \cdot f(o)) + B_{q(\xi)}^h(\gamma_2^{-1} \cdot f(o), (\gamma_2^{-1} \gamma_1^{-1}) \cdot f(o)) \\
&= c(\gamma_2, \xi) + B_{\gamma_2 q(\xi)}^h(f(o), \gamma_1^{-1} \cdot f(o)) \\
&= c(\gamma_2, \xi) + B_{q(\gamma_2 \xi)}^h(f(o), \gamma_1^{-1} \cdot f(o)) \\
&= c(\gamma_2, \xi) + c(\gamma_1, \gamma_2 \xi).
\end{aligned}$$

Therefore, c is a cocycle. To see c is Hölder, we first notice that the boundary map $q : \partial_\infty \widetilde{S} \to \Lambda(\Gamma) \subset \partial_\infty \mathbb{H}^3$ is bi-Hölder by Lemma 11.37. Moreover, we know that $\Lambda(\Gamma)$ embeds in $\partial_\infty \mathbb{H}^3$ and $B_\eta^h(x, y)$ is smooth on $\partial_\infty \mathbb{H}^3$. Therefore, $c(\gamma, \cdot)$ is Hölder continuous on $\partial_\infty \widetilde{S}$, and we have finished the proof of this claim.

Notice that the period $c(\gamma, \gamma_S^+) = B_{q(\gamma_S^+)}^h(f(o), \gamma^{-1} f(o)) = l_h[\gamma] > 0$ for all $[\gamma] \in [\pi_1 S]$. Thus, $l_c[\gamma] = l_h[\gamma]$ for all $[\gamma] \in [\pi_1 S]$, and we can easily see that

$$h_c = \delta_\Gamma = \lim_{T \to \infty} \frac{1}{T} \log \#\{[\gamma] \in [\pi_1 S] : l_h[\gamma] < T\} < \infty.$$

Thus, by Theorem 11.31, there exists a positive Hölder continuous maps F_c on $T^1 S$ such that for all $[\gamma] \in [\pi_1 S]$

$$c(\gamma, \gamma_S^+) = \int_{[\gamma]} F_c = l_h[\gamma],$$

and the topological entropy of the flow ϕ^{F_c} is exactly the exponential growth rate of c, i.e., $h_{F_c} = h_c$.

Notice that for the constant function 1 on $T^1 S$, we have $l_g[\gamma] = \int_{[\gamma]} 1$ for all $[\gamma] \in [\pi_1 S]$. Therefore, we have the pressure of the function $-h_1 \cdot 1$ is zero where

$$h_1 = \lim_{T \to \infty} \frac{1}{T} \log \#\{[\gamma] \in [\pi_1 S] : l_g[\gamma] < T\}$$

is the topological entropy of the geodesic flow ϕ on $T^1 S$.

From now on we denote F_c by F.

The second step is to show that

$$h_{top}(S) \leq \underbrace{\int F d\mu_{BM}}_{C_2} \cdot h_F,$$

where μ_{BM} is the Bowen-Margulis measure of the geodesic flow $\phi : T^1 S \to T^1 S$.

Note that

$$P(-h_F \cdot F) = 0 = h(\mu_{-h_F F}) - h_F \int F d\mu_{-h_F F}$$

$$P(-h_{top}(S) \cdot 1) = 0 = h(\mu_{BM}) - h_{top}(S) \cdot \int 1 d\mu_{BM} = h(\mu_{BM}) - h_{top}(S).$$

where $\mu_{-h_F F}$ is the equilibrium state of $-h_F F$. Since $\mu_{BM} \in \mathcal{M}^\phi$, by the variational principle we have

$$P(-h_F \cdot F) = 0 \geq h(\mu_{BM}) - h_F \int F d\mu_{BM}.$$

Furthermore,

$$h_F \int F d\mu_{BM} \geq h(\mu_{BM}) = h_{top}(S).$$

The third step is to show the inequality

$$\underbrace{\int F d\mu_{-Fh_F}}_{C_1} \cdot h_F \leq h_{top}(S).$$

We know

$$h_{top}(S) \geq h(\mu_{-Fh_F})$$

$$\iff h_{top}(S) - h_F \int F \mathrm{d}\mu_{-Fh_F} \geq \underbrace{h(\mu_{-Fh_F}) - h_F \int F \mathrm{d}\mu_{-Fh_F}}_{=0}$$

$$\iff h_{top}(S) \geq h_F \cdot \int F \mathrm{d}\mu_{-Fh_F}.$$

The fourth step is to show that $0 \leq C_1 \leq 1$ and $0 \leq C_2 \leq 1$.

Because $C_1 = \int F \mathrm{d}\mu_{-Fh_F}$, $C_2 = \int F \mathrm{d}\mu_{BM}$ and F is positive, it is enough to show that F can be chosen to be smaller than or equal to 1.

Claim $F \leq 1$.

Proof of Claim: This is a consequence of Theorem 11.15. For each conjugacy class $[\gamma] \in [\pi_1 S]$ there exists a unique closed geodesic τ_γ^S on S such that $l_g[\gamma] = l_g(\tau_\gamma^S)$. Because f is π_1–injective, f maps τ_γ^S to a closed curve $f(\tau_\gamma^S)$ on M which is in the same free homotopy class generated by $[\gamma]$. More precisely, let τ_γ^M denote the closed geodesic on M in the conjugacy class $[\gamma]$. Then we know that $f(\tau_\gamma^S)$ and τ_γ^M are in the same free homotopy class. Moreover, because g is the induced metric f^*h, we know that (S, g) is Riemannian isometric to $(f(S), h)$. Thus, $l_g(\tau_\gamma^S) = l_h(f(\tau_\gamma^S))$. Therefore, $\forall [\gamma] \in [\pi_1 S]$,

$$l_g[\gamma] = l_g(\tau_\gamma^S) = l_h(f(\tau_\gamma^S)) \geq l_h(\tau_\gamma^M) = l_h[\gamma].$$

Therefore, for all $[\gamma] \in [\pi_1 S]$

$$\int_{[\gamma]} 1 = l_g[\gamma] \geq l_h[\gamma] = \int_{[\gamma]} F.$$

By Theorem 11.15, we have $1 - F$ is cohomologous to a nonnegative Hölder continuous function H, and H is unique up to cohomology. Thus, we have that $F \sim 1 - H$ and $1 - H \leq 1$. By choosing F to be $1 - H$, we have finished the proof of this claim.

The fifth step is to examine the equality cases.

If $h_{top}(S) = h_F \int F \mathrm{d}\mu_{-Fh_F}$, then $h_{top}(S) = h(\mu_{-Fh_F})$, i.e., μ_{-Fh_F} is the equilibrium state of the constant function $-h_{top}(S) \cdot 1$. By the uniqueness part of Theorem 11.12, we have that Fh_F is cohomologous to the constant $h_{top}(S)$, i.e., $F \sim \frac{h_{top}(S)}{h_F}$. Similarly, if $h_{top}(S) = h_F \cdot \int F \mathrm{d}\mu_{BM}$, then $\mu_{BM} = \mu_{-h_F F}$. Hence, again, $h_{top}(S) \sim F \cdot h_F$. $\square$

Lemma 11.39 *If $h_{top}(S) = \delta_\Gamma$, then S is a totally geodesic submanifold in M.* $\square$

Proof Notice $h_{top}(S) = \delta_\Gamma$ implies $F = 1$. This means that the length of each closed geodesic on S has the same length with the corresponding closed geodesic on M. Furthermore, we know that the closed geodesics in S are dense; that is, for any point $p \in S$, the set of tangent vectors $v \in T_p S$ such that the exponential map $\exp_p tv$ gives a closed geodesic is dense in $T_p S$. Therefore, the shape operator S_g is zero when evaluating on this dense subset of vectors on $T_p S$. By the continuity of the shape operator S_g, we have $S_g \equiv 0$. Therefore S is totally geodesic in M. $\square$

Lemma 11.40 *Let μ_{BM} be the Bowen-Margulis measure of the geodesic flow $\phi : T^1 S \to T^1 S$ and $\mu_{-h_F F}$ be the equilibrium state for $-h_F F$ defined in Lemma 11.38. Then*

$$C_2 := \int F d\mu_{BM} = \lim_{T \to \infty} \frac{1}{\#R_T(g)} \sum_{[\gamma] \in R_T(g)} \frac{l_h[\gamma]}{l_g[\gamma]}$$

$$= \lim_{T \to \infty} \frac{\displaystyle\sum_{[\gamma] \in R_T(g)} l_h[\gamma]}{\displaystyle\sum_{[\gamma] \in R_T(g)} l_g[\gamma]} = C_g(f)$$

and

$$C_1 := \int F d\mu_{-h_F F} = \left(\lim_{T \to \infty} \frac{1}{\#R_T(h)} \sum_{[\gamma] \in R_T(h)} \frac{l_g[\gamma]}{l_h[\gamma]} \right)^{-1}$$

$$= \lim_{T \to \infty} \frac{\displaystyle\sum_{[\gamma] \in R_T(h)} l_h[\gamma]}{\displaystyle\sum_{[\gamma] \in R_T(h)} l_g[\gamma]} = C_h(f)$$

where

$$R_T(g) := \{[\gamma] \in [\pi_1 S] : l_g[\gamma] < T\} \text{ and } R_T(h) := \{[\gamma] \in [\pi_1 S] : l_h[\gamma] < T\}.$$

Proof This is a consequence of the equidistribution theorem (Theorem 11.13).
By Theorem 11.13, we have

$$C_2 := \int F d\mu_{BM} = \lim_{T \to \infty} \frac{1}{\#R_T(1)} \sum_{\tau \in R_T(1)} \frac{\langle \delta_\tau, F \rangle}{\langle \delta_\tau, 1 \rangle} = \lim_{T \to \infty} \frac{\displaystyle\sum_{\tau \in R_T(1)} \langle \delta_\tau, F \rangle}{\displaystyle\sum_{\tau \in R_T(1)} \langle \delta_\tau, 1 \rangle}.$$

Notice that every closed orbit τ of the geodesic flow ϕ on $T^1 S$ corresponds to a unique conjugacy class $[\gamma^\tau]$ of $\pi_1 \Sigma$, and vice versa. Moreover, the period of τ is the length of γ^τ on S, i.e.,

$$l_h(\gamma^\tau) = \langle \delta_\tau, F \rangle, \quad l_g(\gamma^\tau) = \langle \delta_\tau, 1 \rangle.$$

Since there is a one-to-one correspondence between $R_T(1)$ and $R_T(g)$, we can rewrite the equation above by

$$C_2 := \int F d\mu_{BM} = \lim_{T \to \infty} \frac{1}{\# R_T(g)} \sum_{[\gamma] \in R_T(g)} \frac{l_h[\gamma]}{l_g[\gamma]}$$

$$= \lim_{T \to \infty} \frac{\displaystyle\sum_{[\gamma] \in R_T(g)} l_h[\gamma]}{\displaystyle\sum_{[\gamma] \in R_T(g)} l_g[\gamma]} =: C_g(f).$$

For the other equation, by Theorem 11.14, we know that $\mu_{\phi^F} = \widehat{F \cdot \mu_{-h_F F}}$. Therefore

$$\mu_{\phi^F}\left(\frac{1}{F}\right) = \widehat{F \cdot \mu_{-h_F F}}\left(\frac{1}{F}\right) = \frac{\int \left(\frac{1}{F}\right) \cdot F d\mu_{-Fh_F}}{\int F d\mu_{-Fh_F}} = \frac{1}{\int F d\mu_{-Fh_F}}.$$

By Theorem 11.13, we have

$$\mu_{\phi^F}\left(\frac{1}{F}\right) = \lim_{T \to \infty} \frac{1}{\# R_T(F)} \sum_{\tau' \in R_T(F)} \frac{\langle \delta_{\tau'}^F, \frac{1}{F} \rangle}{\langle \delta_{\tau'}^F, 1 \rangle} = \lim_{T \to \infty} \frac{\displaystyle\sum_{\tau' \in R_T(F)} \langle \delta_{\tau'}^F, \frac{1}{F} \rangle}{\displaystyle\sum_{\tau' \in R_T(F)} \langle \delta_{\tau'}^F, 1 \rangle}.$$

Notice that for a closed geodesic τ' of the geodesic flow $\phi : T^1 S \to T^1 S$, $\langle \delta_{\tau'}^F, \frac{1}{F} \rangle = \int_0^{l_g(\tau')} \frac{1}{F(\phi_t)} \cdot F(\phi_t) dt = l_g(\tau')$ and similarly $\langle \delta_{\tau'}^F, F \rangle = \int_0^{l_g(\tau')} F(\phi_t) dt = l_h(\tau')$. By the one-to-one correspondence between closed orbit τ' and conjugacy class $[\gamma^{\tau'}]$, we have a one-to-one correspondence between $R_T(F)$ and $R_T(h)$.

Hence, we have the following equation

$$C_1 = \int F \mathrm{d}\mu_{-h_F F} = \left(\mu_{\phi^F} \left(\frac{1}{F} \right) \right)^{-1}$$

$$= \left(\lim_{T \to \infty} \frac{1}{\#R_T(h)} \sum_{[\gamma] \in R_T(h)} \frac{l_g[\gamma]}{l_h[\gamma]} \right)^{-1} = \lim_{T \to \infty} \frac{\sum\limits_{[\gamma] \in R_T(h)} l_h[\gamma]}{\sum\limits_{[\gamma] \in R_T(h)} l_g[\gamma]} =: C_h(f).$$

$\square$

11.4.1 Immersed Minimal Surfaces

Recall that $f : S \to M$ is called a minimal immersion if f is an immersion and the mean curvature H vanishes identically. Let $g = f^*h$ denote the induced metric on S via the immersion f. By the Gauss equation, when $f : S \to M$ is a minimal immersion, the Gaussian curvature $K_g \leq -1$.

Applying the Theorem A to this case, we have the following corollary.

Corollary 11.41 *Let* $f : S \to M$ *be a* π_1*–injective minimal immersion from a closed surface* S *to a hyperbolic 3–manifold* M, *and* Γ *be the copy of* $\pi_1 S$ *in* $\mathrm{Isom}(\mathbb{H}^3)$ *induced by the immersion. Suppose* Γ *is convex cocompact. Then assertions* $(1) - (5)$ *in Theorem A are true.* $\square$

11.4.2 Embedded Surfaces in Hyperbolic 3-Manifolds

In this subsection, we assume that $f : S \to M$ is an embedding. To state our results more precisely and to put it in context, we first introduce the geodesic stretch and discuss the relation between the geodesic stretch and $C_h(f)$, $C_g(f)$.

Notice that we can lift $f : S \to M$ to an embedding between their universal coverings, i.e., $\tilde{f} : \tilde{S} \to \tilde{M} = \mathbb{H}^3$. Moreover, one can easily check this lifting is $\pi_1 S$-equivariant. Specifically, for each $\gamma \in \pi_1 S$, let $\gamma_S \in \Gamma_S$ and $\gamma_M \in \Gamma$ be the corresponding element of γ in the deck transformation groups $\Gamma_S \subset \mathrm{Isom} \tilde{S}$ and $\Gamma \subset \mathrm{Isom}(\mathbb{H}^3)$, respectively. Then for each $\tilde{x} \in \tilde{S}$ we have

$$\tilde{f}(\gamma \cdot \tilde{x}) := \tilde{f}(\gamma_S(\tilde{x})) = \gamma_M(\tilde{f}(\tilde{x})) =: \gamma \cdot \tilde{f}(\tilde{x}).$$

Using this embedding $\widetilde{f} : \widetilde{S} \to \mathbb{H}^3$ we can define a tangent map $\mathbf{f} : T^1\widetilde{S} \to T^1\mathbb{H}^3$ by

$$\mathbf{f} : (\widetilde{x}_0, w) \mapsto (\widetilde{f}(\widetilde{x}_0), d\widetilde{f}_{\widetilde{x}_0}(w))$$

where $\widetilde{x}_0 \in \widetilde{S}$ and w is a unit vector on the tangent plane $T_{\widetilde{x}_0}\widetilde{S}$. Notice that $\pi_1 S$ acts on $T^1\widetilde{S}$ and $T^1\mathbb{H}^3$ in an obvious way. Thus $\mathbf{f}$ is also $\pi_1 S$–equivariant. More precisely, $\gamma \cdot \mathbf{f}(\widetilde{x}_0, w) = (\gamma \cdot \widetilde{f}(\widetilde{x}_0), d\widetilde{f}_{\widetilde{x}_0}(w)) = (\widetilde{f}(\gamma \cdot x_0), d\widetilde{f}_{\widetilde{x}_0}(w)) = \mathbf{f}(\gamma \cdot (\widetilde{x}_0, w))$.

The following lemma depicts a key feature of the embedding $f : S \to M$. By Theorem 11.26, we have the following result.

Lemma 11.42 $(\widetilde{S}, d_g)$ *is quasi-isometric to* $(\widetilde{f}(\widetilde{S}), d_h) \subset (\mathbb{H}^3, d_h)$ *where* d_g *is the distance on* $\widetilde{S}$ *induced by g and* d_h *is the hyperbolic distance on* $\mathbb{H}^3$. ☐

Definition 11.43 For all $v \in T^1\widetilde{S}$ and $t > 0$, we define

$$a(v, t) := d_h(\pi \circ \mathbf{f}(v), \pi \circ \mathbf{f} \circ \widetilde{\phi}_t(v))$$

where $\pi : T^1\widetilde{S} \to \widetilde{S}$ is the natural projection and $\widetilde{\phi}$ is the lift of ϕ.

The following corollary is a consequence of Kingman's sub-additive ergodic theorem [27].

Corollary 11.44 *Let* μ *be a* ϕ_t–*invariant probability measure on* $T^1 S$. *Then for* $\mu - a.e.\ v \in T^1 S$

$$I_\mu(S, M, v) := \lim_{t \to \infty} \frac{a(v, t)}{t},$$

exists and defines a μ–*integrable function on* $T^1 S$, *invariant under the geodesic flow* ϕ_t. ☐

Proof To adapt Kingman's sub-additive ergodic theorem [27] to flow case, it is sufficient to check:

$$\sup\{a(v, t); v \in T^1\widetilde{S},\ 0 \le t \le 1\} \in L^1(\mu).$$

We notice that $(T^1\widetilde{S}, d_g)$ and $(\mathbf{f}(T^1\widetilde{S}), d_h)$ are quasi-isometric because (S, d_g) and $(f(S), d_h)$ are. Therefore we have

$$a(v, 1) = d_h(\pi \circ \mathbf{f}(v), \pi \circ \mathbf{f} \circ \widetilde{\phi}_1(v)) \le Cd_g(v, \widetilde{\phi}_1(v)) + L < C + L$$

where C, L are the quasi-isometry constants. Hence, $a(v, 1)$ is bounded. ☐

From the above corollary, we can define the geodesic stretch as the following. Recall that $\mathcal{M}^\phi$ is the set of ϕ_t–invariant probability measures.

Definition 11.45 The *geodesic stretch* $I_\mu(S, M)$ of S relative to M and $\mu \in \mathcal{M}^\phi$ is defined as

$$I_\mu(S, M) := \int_{T^1\Sigma} I_\mu(S, M, v)\mathrm{d}\mu.$$

Since $f : (\widetilde{S}, d_g) \to (f(\widetilde{S}), d_h)$ is a quasi-isometry, by Theorem 11.22 we know that f extends to a bi-Hölder map between $\partial_\infty \widetilde{S}$ and $\partial_\infty f(\widetilde{S}) = \Lambda(\Gamma)$. By the same discussion as in Lemma 11.37, we know that f maps the attracting (resp. repelling) fixed point γ_S^+ (resp. γ_S^-) of $\gamma_S \in \Gamma_S$ to the corresponding attracting (resp. repelling) fixed point γ_M^+ (resp. γ_M^-) of $\gamma_M \in \Gamma$.

Moreover, each conjugacy class $[\gamma] \in [\pi_1 S]$ corresponds to a unique closed geodesic τ_γ^S on S and τ_γ^M on M, and τ_γ^S also corresponds to the unique geodesic $\widetilde{\tau_\gamma^S}$ connecting γ_S^- and γ_S^+ on $\partial_\infty \widetilde{S}$. Notice that $\widetilde{f}(\gamma_S^-) = \gamma_M^-$ and $\widetilde{f}(\gamma_S^+) = \gamma_S^+$ on $\partial_\infty \widetilde{f}(\widetilde{S}) = \Lambda(\Gamma) \subset \partial_\infty \mathbb{H}^3$, so $f(\widetilde{\tau_\gamma^S})$ is a quasi-geodesic on $\mathbb{H}^3$ within a bounded Hausdorff distance from the geodesic $\widetilde{\tau_\gamma^M}$ on $\mathbb{H}^3$, where $\widetilde{\tau_\gamma^M}$ is the geodesic on $\mathrm{Conv}(\Lambda(\Gamma)) \subset \mathbb{H}^3$ connecting γ_M^- and γ_M^+ on $\Lambda(\Gamma)$.

Now we are ready to state the main result of this section. However, because its proof consists of several lemmas, we postpone the proof until the end of this section.

Theorem 11.46 (Theorem 11.3) *Assume the same assumptions as in Theorem A, and, additionally, assume that $f : S \to M$ is an embedding. Then the geometric constants $C_h(f)$ and $C_g(f)$ in Theorem A are geodesic stretches relative to invariant measures. More precisely,*

$$C_h(f) = I_\mu(S, M),$$

$$C_g(f) = I_{\mu_{BM}}(S, M),$$

where μ is a ϕ−invariant measure and μ_{BM} is the Bowen-Margulis measure of the geodesic flow ϕ_t on $T^1 S$. □

Remark 11.47 The invariant measure μ in $C_h(f) = I_\mu(S, M)$ is indeed the equilibrium state $\mu_{-h_F F}$ derived in the proof of Theorem A.

Before we start proving Theorem 11.3, we shall introduce two useful lemmas.

Lemma 11.48 *Suppose $\mu \in \mathcal{M}^\phi$ and ergodic. Then there exists a sequence of conjugacy classes $\{[\gamma_n]\} \subset [\pi_1 S]$, i.e., closed geodesics, such that*

$$\int F\mathrm{d}\mu = \lim_{n\to\infty} \frac{l_h[\gamma_n]}{l_g[\gamma_n]},$$

where F is the reparametrization function defined in Theorem A. □

Proof First, by the sub-additive ergodic theorem we know that for $\mu - a.e.\ v \in T^1 S$

$$\lim_{t \to \infty} \frac{a(v, t)}{t} = I_\mu(S, M). \tag{11.4.2}$$

By the Birkhoff ergodic theorem we have for $\mu - a.e.\ v \in T^1 S$

$$\lim_{t \to \infty} \frac{1}{t} \int_0^t F(\phi_s v) \mathrm{d}s = \int F \mathrm{d}\mu. \tag{11.4.3}$$

We define two sets

$$A := \{v \in T^1 S : v \text{ satisfies } (11.4.2)\}$$

$$B := \{v \in T^1 S : v \text{ satisfies } (11.4.3)\}.$$

Since A and B are both full μ-measure, we have $A \cap B \neq \emptyset$.

Pick $v \in A \cap B$, and $\varepsilon_n \searrow 0$ as $n \to \infty$. By the Anosov Closing Lemma [28], for each ε_n, there exists $\delta_n = \delta_n(\varepsilon_n)$ such that for $v \in T^1 S$ and $T_n = T_n(\delta_n) > 0$ satisfying $D_g(\phi_{T_n}(v), v) < \varepsilon_n$, then there exists $w_n \in T^1 S$ which generates a periodic orbit τ_n^S on S of period $l_g(\tau_n^S) = T_n'$ such that $|T_n - T_n'| < \varepsilon_n$ and $D_g(\phi_s(v), \phi_s(w_n)) < \varepsilon_n$ for all $s \in [0, T_n]$.

Furthermore, because the geodesic flow ϕ_t on $T^1 S$ is a transitive Anosov flow and $T^1 S$ is compact, by the Poincaré Recurrent Theorem, for each δ_n given as above, we can pick T_n to be the n-th return time of the flow ϕ_t to the set $B_{\delta_n}(v)$, i.e., $D_g(\phi_{T_n}(v), v) < \delta_n$ for each n.

Suppose τ_n^S corresponds to $[\gamma_n] \in [\pi_1 S]$, then since μ is ergodic, by the Birkhoff ergodic theorem we have

$$\int_{T^1 S} F \mathrm{d}\mu = \lim_{T \to \infty} \frac{1}{T} \int_0^T F(\phi_t v) \mathrm{d}t.$$

Claim $\displaystyle \int F \mathrm{d}\mu = \lim_{n \to \infty} \frac{\int_{\gamma_n} F}{l_g[\gamma_n]}.$

Proof of Claim: Notice that

$$\frac{1}{l_g[\gamma_n] + \varepsilon_n} \int_0^{l_g(\gamma_n) - \varepsilon_n} F(\phi_t v) \leq \frac{1}{t_n} \int_0^{t_n} F(\phi_t v) \leq \frac{1}{l_g[\gamma_n] - \varepsilon_n} \int_0^{l_g(\gamma_n) + \varepsilon_n} F(\phi_t v).$$

Because F is Hölder, we know that $|F(\phi_t v) - F(\phi_t w_n)| \leq C \cdot D_g(\phi_t v, \phi_t w_n)^\alpha \leq C \cdot \varepsilon_n^\alpha$.

When n is big enough such that $l_g[\gamma_n] > 2\varepsilon_n$ (notice that $\varepsilon_n \searrow 0$ and $l_g(\gamma_n) \nearrow \infty$), we have

$$
\left| \frac{1}{t_n} \int_0^{t_n} F(\phi_t v) - \frac{1}{l_g[\gamma_n]} \int_0^{l_g(\gamma_n)} F(\phi_t w_n) \right|
$$

$$
\leq \frac{l_g[\gamma_n] \int_0^{l_g(\gamma)} |F(\phi_t v) - F(\phi_t w_n)| \, dt}{l_g[\gamma_n] \cdot (l_g(\gamma_n) - \varepsilon_n)} + \frac{2 \cdot l_g[\gamma_n] \cdot \varepsilon_n \cdot \|F\|_\infty}{l_g[\gamma_n] \cdot (l_g[\gamma_n] - \varepsilon_n)}
$$

$$
\leq \frac{1}{l_g[\gamma_n] - \varepsilon_n} \left(l_g[\gamma_n] \cdot C \cdot \varepsilon_n^\alpha + 2\varepsilon_n \cdot \|F\|_\infty \right)
$$

$$
\leq 2C \cdot \varepsilon_n^\alpha + \frac{2\varepsilon_n}{l_g[\gamma_n] - \varepsilon_n} \cdot \|F\|_\infty .
$$

So, we can now finish the proof of this claim.

Moreover, from the construction of F, $\forall [\gamma_n] \in [\pi_1 S]$ we have

$$
\int_{[\gamma_n]} F = l_h[\gamma_n].
$$

Therefore,

$$
\int F d\mu = \lim_{n \to \infty} \frac{\int_{\gamma_n} F}{l_g[\gamma_n]} = \lim_{n \to \infty} \frac{l_h[\gamma_n]}{l_g[\gamma_n]}.
$$

$\square$

Lemma 11.49 *Let* $\{[\gamma_n]\} \subset [\pi_1 S]$ *be the sequence constructed in the proof of Lemma 11.48. Then*

$$
\lim_{n \to \infty} \frac{l_h[\gamma_n]}{l_g[\gamma_n]} = I_\mu(S, M).
$$

Proof We claim that

$$
\lim_{n \to \infty} \frac{a(w_n, l_g[\gamma_n])}{l_g[\gamma_n]} = \lim_{n \to \infty} \frac{l_h[\gamma_n]}{l_g[\gamma_n]}.
$$

To see this, by definition,

$$
a(w_n, l_g[\gamma_n]) := d_h(\pi \circ \mathbf{f} \circ w_n, \pi \circ \mathbf{f} \circ \widetilde{\phi}_{l_g[\gamma_n]} w_n).
$$

For such $[\gamma_n] \in [\pi_1 S]$, let τ_n^S, τ_n^M denote the corresponding closed geodesics on S and M, and $\widetilde{\tau_n^S}$ and $\widetilde{\tau_n^M}$ denote their lifting on $\widetilde{S}$ and $\mathrm{Conv}(\Lambda(\Gamma))$, respectively. Then we know that $\widetilde{f}(\tau_n^S)$ and $\widetilde{\tau_n^M}$ are at most Hausdorff distance R from each

other. Therefore we can choose $x_n \in \widetilde{\tau_n^M}$ such that $d_h(\pi w_n, x_n) < R$. Because d_h is
Γ−invariant, $\widetilde{f} : \widetilde{S} \to \mathbb{H}^3$ is an embedding, and $\pi \circ \mathbf{f} \circ w_n$ and $\pi \circ \mathbf{f} \circ \widetilde{\phi}_{l_g(\gamma_n)} w_n$ project
to the same point on S, we have $d_h(\gamma_n \cdot x_n, \pi \circ \mathbf{f} \circ \widetilde{\phi}_{l_g(\tau_n)} w_n) = d_h(\pi \circ \mathbf{f} \circ w_n, x_n) < R$.
Hence, by the triangle inequality

$$
\left| d_h(\pi \circ \mathbf{f} \circ w_n, \pi \circ \mathbf{f} \circ \widetilde{\phi}_{l_g(\tau_n)} w_n) - \underbrace{d_h(x_n, \gamma_n \cdot x_n)}_{=l_h(\tau_n)} \right| \leq \underbrace{d_h(\pi \circ \mathbf{f} \circ w_n, x_n)}_{\leq R}
$$

$$
+ \underbrace{d_h(\gamma_n \cdot x_n, \pi \circ \mathbf{f} \circ \widetilde{\phi}_{l_g(\tau_n)} w_n)}_{\leq R}
$$

$$
= 2R.
$$

Therefore,

$$
\lim_{n \to \infty} \frac{l_h[\gamma_n]}{l_g[\gamma_n]} = \lim_{n \to \infty} \frac{l_h[\gamma_n] - 2R}{l_g[\gamma_n]} \leq \lim_{n \to \infty} \frac{a(w_n, l_g[\gamma_n])}{l_g[\gamma_n]}
$$

$$
\leq \lim_{n \to \infty} \frac{l_h[\gamma_n] + 2R}{l_g[\gamma_n]} = \lim_{n \to \infty} \frac{l_h[\gamma_n]}{l_g[\gamma_n]},
$$

and we can now finish the proof of this claim.

We claim

$$
I_\mu(S, M) = \lim_{t \to \infty} \frac{a(v, t)}{t} = \lim_{n \to \infty} \frac{l_h[\gamma_n]}{l_g[\gamma_n]}.
$$

To see this pick the t_n as we mentioned in Lemma 11.48. Then

$$
\left| a(v, t_n) - a(w_n, l_g[\gamma_n]) \right|
$$

$$
\leq \left| d_h(\pi \circ \mathbf{f} \circ v, \pi \circ \mathbf{f} \circ \widetilde{\phi}_{t_n} v) - d_h(\pi \circ \mathbf{f} \circ w_n, \pi \circ \mathbf{f} \circ \widetilde{\phi}_{t_n} w_n) \right|
$$

$$
\leq d_h(\pi \circ \mathbf{f} \circ v, \pi \circ \mathbf{f} \circ w_n) + d_h(\pi \circ \mathbf{f} \circ \widetilde{\phi}_{l_g(\gamma_n)} w_n, \pi \circ \mathbf{f} \circ \widetilde{\phi}_{t_n} v)
$$

(quas-isometry Lemma 11.42) $\leq C \cdot \big(d_g(\pi \circ v, \pi \circ w_n)$

$$
+ d_g(\pi \circ \widetilde{\phi}_{l_g(\gamma_n)} w_n, \pi \circ \widetilde{\phi}_{t_n} v)\big) + 2L
$$

(Anosov closing lemma) $\leq C \cdot (\delta_2 + \varepsilon) + 2L$,

where C and L are the quasi-isometry constants only depending on the embedding
$f : S \to M$.

Therefore,

$$\lim_{t_n \to \infty} \frac{a(v, t_n)}{t_n} = \lim_{n \to \infty} \frac{a(w_n, l_g[\gamma_n])}{l_g[\gamma_n]} = \lim_{n \to \infty} \frac{l_h[\gamma_n]}{l_g[\gamma_n]}.$$

□

Proof of Theorem 11.3 It is not hard to see the result follows Lemmas 11.48 and 11.49.

□

11.4.3 The Manhattan Curve for Immersed Surfaces

In this subsection we prove Theorem B.

Proof of Theorem B We first recall that by Lemma 11.38, there exists a Hölder continuous function $F : T^1 S \to T^1 S$ such that $\int_\tau F = l_h[\tau]$ for all closed orbit τ on $T^1 S$. Therefore, we have $-a l_g[\tau] - b l_h[\tau] = \int_\tau -a - bF$.

Moreover, the pressure for a Hölder continuous function G over the geodesic flow $\phi_t : T^1 S \to T^1 S$ can be written as

$$P_\phi(G) = \lim_{T \to \infty} \frac{1}{T} \log \sum_{l_g[\tau] \leq T} e^{\int_\tau G}.$$

Thus we deduce that $\sum_\tau e^{-a l_g[\tau] - b l_h[\tau]}$ is convergent if $P_\phi(-a - bF) < 0$ and is divergent if $P_\phi(-a - bF) > 0$. Hence we can identify the Manhattan curve χ_f with

$$\{(a, b) \in \mathbb{R}^2 : P_\phi(-a - bF) = 0\} = \{(a, b) \in \mathbb{R}^2 : P_\phi(-bF) = a\}.$$

Therefore, we have $(0, 1)$, $(h_{top}(g), 0) \in \chi_f$.

Recall by Proposition 11.17 that we know $P_\phi(-sF)$ is analytic. Moreover,

$$\frac{d}{ds} P_\phi(-sF) = \int -F d\mu_{-F} \neq 0.$$

Thus, by the Implicit Function Theorem, we know the solution of $P_\phi(-bF) = t$ is an analytic curve, i.e., $b = b(t)$ is analytic. In other words, the Manhattan curve χ_f and be parametrized and written as $(t, b(t))$ for $t \in \mathbb{R}$.

This implies, again by Proposition 11.17,

$$1 = \frac{d}{dt} P_\phi(-b(t)F) = -b'(t) \int F d\mu_{-b(t)F},$$

and

$$0 = \frac{d^2}{dt^2} P_\phi(-b(t)F) = \text{Var}(-b'(t)F, \mu_{-b(t)F}) - b''(t) \int F d\mu_{-b(t)F}.$$

Therefore, we have

$$b'(t) = \frac{1}{-\int F d\mu_{-b(t)F}} < 0$$

and

$$b''(t) = \frac{\text{Var}(-b'(t)F, \mu_{-b(t)F})}{\int F d\mu_{-b(t)F}}.$$

Hence, χ_f is strictly convex unless $\text{Var}(-b'(t)F, m_{-b(t)F}) = 0$, that is, F is cohomologus to a constant.

Lastly, it is clear that χ_f is a straight line when F is cohomologous to a constant.   □

Using the above theorem, we immediately have the following entropy rigidity result, which generalizes the Bishop-Steger entropy rigidity given in [10].

Corollary 11.50 *Under the same assumption as Theorem A, then*

$$h^{(1,1)} := \lim_{T \to \infty} \frac{1}{T} \log \#\{\gamma \in \pi_1 S : l_g[\gamma] + l_h[\gamma] < T\} \le \frac{h_{top}(g)}{h_{top}(g) + 1}$$

and the equality holds iff the marked length spectrum of (S, g) is proportional to the marked length spectrum of M.   □

Proof First we notice that since $1 + F$ is Hölder, there exists a unique s_0 such that

$$P_\phi(-s_0(1 + F)) = 0.$$

Moreover, by Lemma 1 and the remark after Lemma 1 in [25], we know $s_0 = h^{(1,1)}$ and thus

$$h^{(1,1)} \cdot (1, 1) \in \chi_f.$$

Since χ_f is strictly convex and the point $\frac{h_{top}(g)}{h_{top}(g)+1}(1, 1)$ is the intersection of the line connecting $(0, 1)$ and $(h_{top}(g), 0)$ and the line connecting $(0, 0)$ and $(1, 1)$, we know that $\frac{h_{top}(g)}{h_{top}(g)+1}(1, 1)$ sits above $h^{(1,1)} \cdot (1, 1)$. See Fig. 11.1. Hence we have

$$h^{(1,1)} \le \frac{h_{top}(g)}{h_{top}(g) + 1}.$$

Fig. 11.1 Rigidity for $h^{(1,1)}$

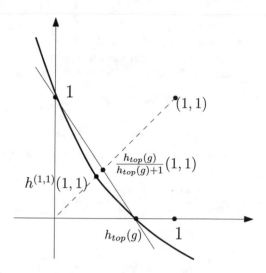

The second assertion is because the convexity of the Manhattan curve. More precisely, the convexity implies that the equality holds if and only if χ_f is a straight line.
□

Remark 11.51 Notice that since the ambient metric h may not be a Riemannian metric on the immersed surface $f(S)$, Otal's marked length spectrum theorem is not applicable here.
□

11.5 Minimal Hyperbolic Germs

Recall that $\mathscr{H}$ is the set of the isotopy classes of pairs consisting of a Riemann metric g and a symmetric 2-tensor B on S such that the trace of B with respect to g is zero and (g, B) satisfies the Gauss-Codazzi equations, i.e. Eq. (11.3.3).

The following corollary is an obvious consequence of Theorem A. Recall that $h_{top}(g, B)$ denotes the topological entropy of the geodesic flow for the immersed surface (S, g) with second fundamental form B.

Corollary 11.52 *Let $\rho \in \mathscr{R}(\pi_1(S), \mathrm{PSL}(2, \mathbb{C}))$ be a discrete, convex cocompact representation and suppose $(g, B) \in \Phi^{-1}(\rho) \neq \emptyset$. Then there are explicit geometric constants $C_1(g, B)$ and $C_2(g, B)$ with $0 \leq C_1(g, B) \leq C_2(g, B) \leq 1$ such that*

$$C_1(g, B) \cdot \delta_{\rho(\pi_1 S)} \leq h_{top}(g, B) \leq C_2(g, B) \cdot \delta_{\rho(\pi_1 S)} \leq \delta_{\rho(\pi_1 S)}$$

with the last inequality being a equality if and only if B is identically zero which holds if and only if ρ is Fuchsian.
□

Proof Notice that $(g, B) \in \Phi^{-1}(\rho)$ means that there exists a π_1–injective immersion $f : S \to \rho(\pi_1 S) \backslash \mathbb{H}^3 = M$ such that the induced metric is g and the second fundamental form is B, where (M, h) is a convex cocompact hyperbolic 3–manifold. Therefore, by Theorem A we have

$$C_h(f) \cdot \delta_{\rho(\pi_1 S)} \leq h_{top}(S) \leq C_g(f) \cdot \delta_{\rho(\pi_1 S)}.$$

Then we pick $C_1(g, B) = C_h(f)$ and $C_2(g, B) = C_g(f)$. The rightmost inequality is because $C_2(g, B) = C_g(f) \leq 1$, and the rigidity is the consequence of Corollary 11.39. □

Remark 11.53 By Sullivan's results (i.e., Theorem 11.28), we can replace the critical exponent by the Hausdorff dimension in the above corollary. □

11.5.1 Quasifuchsian Spaces

We call a discrete faithful representation $\rho : \pi_1(S) \to \text{Isom}(\mathbb{H}^3)$ *quasifuchsian* if and only if the limit set $\Lambda(\rho(\pi_1 S))$ of $\rho(\pi_1 S)$ is a Jordan curve and the domain of discontinuity $\partial_\infty \mathbb{H}^3 \backslash \Lambda(\rho(\pi_1 S))$ is composed by two invariant, connected, simply-connected components. $\mathscr{QF}$ denotes the set of quasifuchsian representations.

Uhlenbeck [11] pointed out that we can relate the space of minimal hyperbolic germs $\mathscr{H}$ with the character variety $\mathscr{R}(\pi_1(S), \text{PSL}(2, \mathbb{C}))$, that is the space of conjugacy classes of representations of $\pi_1(S)$ into $\text{PSL}(2, \mathbb{C})$. More precisely, Uhlenbeck [11] proved that for each data $(g, B) \in \mathscr{H}$ there exists a representation $\rho : \pi_1(S) \to \text{Isom}(\mathbb{H}^3) \cong \text{PSL}(2, \mathbb{C})$ leaving this minimal immersion invariant, i.e., there is a map

$$\Phi : \mathscr{H} \to \mathscr{R}(\pi_1(S), \text{PSL}(2, \mathbb{C})). \tag{11.5.1}$$

We notice that if $\rho \in \mathscr{QF}$, then elements in $\Phi^{-1}(\rho)$ are $\pi_1(S)$–injective minimal immersions from S to $\rho(\pi_1(S)) \backslash \mathbb{H}^3$. Moreover, Uhlenbeck in [11] showed that for $\rho \in \mathscr{QF}$, $\Phi^{-1}(\rho)$ is always a non-empty set.

Corollary 11.54 *Let $\rho \in \mathscr{QF}$ be a quasifuchsian representation and $(g, B) \in \Phi^{-1}(\rho)$. Then there are explicit geometric constants $C_1(g, B)$ and $C_2(g, B)$ with $0 \leq C_1(g, B) \leq C_2(g, B)$ such that*

$$C_1(g, B) \cdot \delta_{\rho(\pi_1 S)} \leq h_{top}(g, B) \leq C_2(g, B) \cdot \delta_{\rho(\pi_1 S)} \leq \delta_{\rho(\pi_1 S)}$$

with the last equality if and only if B is identically zero which holds if and only if ρ is Fuchsian. □

Using the above corollary, we can give another proof the famous Bowen's rigidity theorem.

Corollary 11.55 (Bowen's Rigidity [1]) *A quasifuchsian representation $\rho \in \mathcal{QF}$ is Fuchsian if and only if $\dim_H \Lambda_\Gamma = 1$.*   □

Proof For any $(g, B) \in \Phi^{-1}(\rho)$, we have S is an immersed minimal surface in a quasifuchsian manifold $M = \rho(\pi_1 S) \backslash \mathbb{H}^3$ with the induced metric g and the second fundamental form B. Let $K(S)$ denote the Gaussian curvature of S in M, then by the Gauss-Codazzi equation $K(S) \leq -1$. Therefore using the Theorem B in [2], we have

$$h_{top}(g, B) \geq \left(\frac{-\int K(S)\mathrm{d}A}{\mathrm{Area}(S)} \right)^{\frac{1}{2}} \geq 1.$$

Hence the result is derived by the above lower bound of $h_{top}(S)$ plus the above corollary.   □

11.5.2 Manhattan Curve for Almost-Fuchsian Space

Recall that the *almost-Fuchsian* space $\mathcal{AF}$ is the space of minimal hyperbolic germs $(g, B) \in \mathcal{H}$ such that $\|B\|_g < 2$. Given a hyperbolic metric $m \in \mathcal{F}$ and a holomorphic quadratic differential $\alpha \in Q([m])$, we discuss an informative smooth path

$$r_\alpha(t) = (g_t, tB) \subset \mathcal{AF},$$

where $g_t = e^{2u_t} m$ and $B = \mathrm{Re}(\alpha)$ satisfying $\|tB\|_{g_t}^2 < 2$. Notice that $u_t : S \to \mathbb{R}$ is well-defined and smooth on t (cf. [11, Theorem 4.4]). Moreover, Sanders [12] pointed out that the entropy $h_{top}(g_t)$ varies smoothly along $r_\alpha(t)$.

Before we start proving our results, we recall several important concepts of Anosov flows. We first notice that by the structural stability of the Anosov flow (cf. Prop. 1 in [29] or [30]), when t is small, for the geodesic flow $\psi : T^1(S, g_t) \to T^1(S, g_t)$, there exists is a Hölder continuous function such that ψ is conjugated to the reparametrized flow $\phi^{F_t} : T^1(S, m) \to T^1(S, m)$ where $\phi : T^1(S, m) \to T^1(S, m)$ is the geodesic flow. Moreover, when $\{g_t\}$ is a smooth one parameter family, then the structure stability theorem also indicates that $\{F_t\}$ form a smooth one parameter family of Hölder continuous functions on $T^1(S, m)$.

Since we shall only be interested in metrics g_t close to $g_0 (= m)$, it suffices to consider one parameter families given by perturbation series: for t small,

$$g_t = g_0 + t \cdot \dot{g}_0 + \frac{t^2}{2} \ddot{g}_0 + \dots, \text{ and } F_t = F_0 + t \cdot \dot{F}_0 + \frac{t^2}{2} \ddot{F}_0 + \dots,$$

where $\dot{g}_0, \ddot{g}_0, \dots$ are symmetric 2-tensors on $T^1(S, m)$ and $\dot{F}_0, \ddot{F}_0, \dots$ are Hölder continuous functions on $T^1(S, m)$.

Definition 11.56 *The Manhattan curve* χ_t *associated with* m *and* g_t *is defined as* the boundary of the convex set

$$\{(a, b) \in \mathbb{R}^2 : \sum_{\gamma \in \pi_1 S} e^{-(a \cdot l_m[\gamma] + b \cdot l_{g_t}[\gamma])} < \infty\}.$$

We can reparametrize the $\chi_t : \mathbb{R} \to \mathbb{R}$ by writing it as

$$\chi_t(a) := \inf\{b \geq 0 : \sum_{\gamma \in \pi_1 S} e^{-(a \cdot l_m[\gamma] + b \cdot l_{g_t}[\gamma])} < \infty\}$$

Using the structural stability of Anosov flows, we can adapt Sharp's method to our (non-constant curvature) setting and derive the following result. The same proof of Theorem B gives us the following result.

Theorem 11.57 *We have*

(1) $(0, h_t)$, $(1, 0) \in \chi_t$.
(2) $\chi_t(s)$ *is real analytic.*
(3) $\chi_t(s)$ *is strictly convex unless* $t = 0$ *and* $\chi_0(s)$ *is a straight line.*

In the following, we discuss the variation of Manhattan curves χ_t when varying t. For convenience, we first consider the *renormalized Manhattan curve* $\widetilde{\chi}_t(s)$ defined as

$$\widetilde{\chi}_t(s) := \frac{\chi_t(s)}{h_{top}(g_t)}.$$

Moreover, let $c_t := -h_{top}(g_t)F_t$ denote the smooth path of zero pressure Hölder continuous functions. Also, we use dots to denote derivatives with respect to t.

Lemma 11.58 *We have*

$$\frac{d^2}{dt^2}\widetilde{\chi}_t(s)\bigg|_{t=0} = (s - s^2)\text{Var}(\dot{c}_0, \mu_{c_0}).$$

Proof First, we know $\chi_t(s)$ satisfies $P_\phi(-sF_0 - \chi_t(s)F_t) = 0$, therefore $\widetilde{\chi}_t$ satisfies

$$P_\phi(sc_0 + \widetilde{\chi}_t(s)c_t) = 0.$$

For a fixed s, let $\psi_t := sc_0 + \widetilde{\chi}_t(s)c_t$. Notice that when $t = 0$, $\widetilde{\chi}_0$ is a straight line satisfying $s + \widetilde{\chi}_t(s) = 1$. So we have $\psi_0 = c_0$.

Moreover, we have $c_0 = -h_{top}(m) \cdot 1 = -1$, $\dot{\psi}_0 = \dot{\widetilde{\chi}}_0 c_0 + \widetilde{\chi}_0 \dot{c}_0$ and $\ddot{\psi}_0 = \ddot{\widetilde{\chi}}_t c_0 + 2\dot{\widetilde{\chi}}_0 \dot{c}_0 + \widetilde{\chi}_0 \ddot{c}_0$.

By Proposition 11.17, we have

$$0 = \frac{d}{dt} P_\phi(\psi_t)\Big|_{t=0} = \int \dot{\psi}_0 d\mu_{\psi_0} = \int \tilde{\chi}_0 c_0 + \tilde{\chi}_0 \dot{c}_0 d\mu_{c_0}$$

$$= \tilde{\chi}_0(s) \int c_0 d\mu_{c_0} + \tilde{\chi}_0(s) \int \dot{c}_0 d\mu_{c_0}.$$

Since $\int \dot{c}_0 d\mu_{c_0} = 0$ (because $P_\phi(c_t) = 0$) and $\int c_0 d\mu_{c_0} < 0$, we have

$$\tilde{\chi}_0(s) = 0 \ \forall s \in \mathbb{R}.$$

Furthermore,

$$0 = \frac{d^2}{dt^2} P_\phi(\psi_t)\Big|_{t=0} = \text{Var}(\dot{\psi}_0, \mu_{\psi_0}) + \int \ddot{\psi}_0 d\mu_{\psi_0}$$

$$= \text{Var}(\underbrace{\tilde{\chi}_0}_{0} c_0 + \tilde{\chi}_0 \dot{c}_0, \mu_{c_0}) + \int \left(\tilde{\chi}_t c_0 + 2 \underbrace{\tilde{\chi}_0}_{0} \dot{c}_0 + \tilde{\chi}_0 \ddot{c}_0 d\mu_{c_0} \right)$$

$$= (\tilde{\chi}_0(s))^2 \cdot \text{Var}(\dot{c}_0, \mu_{c_0}) - \tilde{\chi}_t + \tilde{\chi}_0 \int \ddot{c}_0 d\mu_{c_0}.$$

Notice that $P_\phi(c_t) = 0$, then by taking the derivative twice we know,

$$0 = \text{Var}(\dot{c}_0, \mu_{c_0}) + \int \ddot{c}_0 d\mu_{c_0},$$

Therefore, we have

$$\tilde{\chi}_0(s) = (\tilde{\chi}_0(s)^2 - \tilde{\chi}_0(s)) \text{Var}(\dot{c}_0, \mu_{c_0})$$

$$= \left((1-s)^2 - (1-s) \right) \text{Var}(\dot{c}_0, \mu_{c_0}) = (s^2 - s) \text{Var}(\dot{c}_0, \mu_{c_0}).$$

$$\square$$

Theorem 11.59 *Let $h_t := h_{top}(g_t)$ be the topological entropy for (S, g_t). Then*

$$\frac{d^2}{dt^2} \chi_t(s)\Big|_{t=0} = \ddot{h}_0(1-s) - (s - s^2) \text{Var}(\dot{c}_0, \mu_{c_0}).$$

Proof It is a direct computation that follows $\chi_t = h_t \cdot \tilde{\chi}_t$ where $h_t := h_{top}(g_t)$.
More precisely, we have

$$\left. \frac{d^2}{dt^2} \chi_t(s) \right|_{t=0} = \ddot{h}_0 \tilde{\chi}_0(s) + 2 \underbrace{\dot{\tilde{\chi}}_0 (s)\dot{h}_0}_0 + \tilde{\chi}_0 \underbrace{\ddot{h}_0}_1 .$$

$\square$

11.5.3 Metrics on $\mathscr{F}$

We see from previous sections that the space of minimal hyperbolic germs $\mathscr{H}$ for a closed surface S shares many similar structures as the quasifuchsian space $\mathscr{QF}$ for S. In [7] Bridgeman proved one can construct a nonnegative two (pressure) form on $\mathscr{QF}$ such that it is the Weil-Petersson form on $\mathscr{F}$. In this section, inspired by Bridgeman, we discuss a similar construction in our setting.

We shall note that Sanders constructs a metric on the Fuchsian space $\mathscr{F} \subset \mathscr{H}$ by taking the Hessian of the topological entropy along the path $r(t) = (e^{2u_t}m, tB)$. Moreover, this metric is bounded below by the Weil-Petersson metric on $\mathscr{F}$. At the end of this section, we will provide many other metrics on $\mathscr{F}$ using Manhattan curves, renormalized Manhattan curves, and some other geometric quantities. Through studying them we can construct a pressure metric as well as other metrics on $\mathscr{F}$. Moreover, Sanders' metric is a special case including in our construction. We will also compare the Weil-Petersson metric, Sanders' metric and the pressure metric.

Recall that the fiber of the cotangent bundle of $m \in \mathscr{F}$ is identified with the space of holomorphic quadratic differentials $Q([m])$ on the Riemann surface (S, m). Moreover, for each $\alpha \in Q([m])$ and t is small the path $r_\alpha(t) = (g_t, t \cdot \operatorname{Re}\alpha)$ induces a path c_t in pressure zero Hölder continuous functions over $T^1(S, m)$, namely, $c_t = -h_{top}(g_t)F_t$ where F_t is the reparametrized function given by the structure stability of Anosov flow. Therefore, for each $\alpha \in Q([m])$ can be identified with $\dot{c}_0$, that is, the derivative of c_t at $t = 0$.

Definition 11.60 For the path $r_\alpha(t) \subset \mathscr{AF}$ and t is small, *the renormalized intersection number $J_{m,\alpha}(t)$* is defined as

$$J_{m,\alpha}(t) := h_{top}(g_t) \cdot \limsup_{T \to \infty} \frac{\displaystyle\sum_{\tau \in R_T(m)} l_{g_t}[\tau]}{\displaystyle\sum_{\tau \in R_T(m)} l_m[\tau]}$$

where $g_t = e^{2u_t}m$.

$\square$

Theorem 11.61 *We have*

$$J_{m,\alpha}(t) = h_{top}(g_t) \int F_t d\mu_0 = h_{top}(g_t) \cdot \lim_{T \to \infty} \frac{\sum_{\tau \in R_T(m)} l_{g_t}[\tau]}{\sum_{\tau \in R_T(m)} l_m[\tau]}$$

where μ_0 is the Bowen-Margulis measure for the geodesic flow $\phi : T^1(S, m) \to T^1(S, m)$. □

Proof It follows immediately from the equidistribution property of μ_0 (cf. Theorem 11.13). □

Corollary 11.62 (Theorem C (1)) *We have*

$$J_{m,\alpha}(t) \leq 1$$

and equal to 1 if and only if $t = 0$. □

Proof By (the proof of) [12, Theorem 3.5] we know $h_{top}(g_t)$ and g_t are decreasing for $t > 0$. Moreover, g_t is decreasing implies that F_t is decreasing. □

Lemma 11.63 (Theorem C (2))

$$\frac{d^2}{dt^2} J_{m,\alpha}(t) \bigg|_{t=0} = \mathrm{Var}(\dot{c}_0, \mu_0)$$

where $c_t := -h_{top}(g_t) F_t$ and μ_0 is the Bowen-Margulis measure of the geodesic flow $T^1(S, m) \to T^1(S, m)$.

Proof

$$\frac{d^2}{dt^2} J_{m,\alpha}(t) = \frac{d^2}{dt^2} \int h_{top}(g_t) F_t d\mu_0 = -\frac{d^2}{dt^2} \int c_t d\mu_0 = -\int \ddot{c}_t d\mu_0.$$

Moreover, $P_\phi(c_t) = 0$ implies

$$0 = \mathrm{Var}(\dot{c}_0, \mu_0) + \int \ddot{c}_0 d\mu_0.$$

□

Let us recall a computational lemma from Pollicott [29, Lemma 5] and Sanders [12, P. 12].

Lemma 11.64 *Let* $r(t) = (e^{2u_t}m, t \cdot \text{Re}\alpha)$ *and for* t *small* $g_t = m + t\dot{g}_0 + \frac{t^2}{2}\ddot{g}_0 \dots$.
We have

$$\int_{T^1S} \frac{\ddot{g}_0(v, v)}{2} d\mu_0 = \int_{T^1S} \ddot{u}_0 d\mu_0 = -2\pi \int_S \|\alpha\|_m^2 \, dV_m = -2\pi \|\alpha\|_{WP}^2,$$

$$\int_{T^1S} \dot{F}_0 d\mu_0 = \int_{T^1S} \dot{g}_0(v, v) d\mu_0 \quad and \quad \int_{T^1S} \ddot{F}_0 d\mu_0 \leq \int_{T^1S} \frac{\ddot{g}_0(v, v)}{2} d\mu_0.$$
$$(11.5.2)$$

Lemma 11.65 (Theorem C (2)) *For* t *is small, let* $r_B(t) = (g_t, tB) \in \mathscr{AF}$
and F_t *be the Hölder reparametrization function corresponding to* g_t. *Then* $\dot{c}_0$ *is
cohomologous to 0 if and only if* $B = 0$.   □

Proof The only if part is clear. Suppose $\dot{c}_0 = 0$. Let's denote $h_{top}(g_t)$ by h_t, and
notice that $h_0 = 1$ and $F_0 = 1$. Therefore,

$$\dot{c}_0 \sim 0 \iff h_0 \dot{F}_0 \sim -\dot{h}_0 F_0$$

$$\iff \dot{F}_0 \sim -\dot{h}_0$$

Then by Lemma 11.64, we have $-\dot{h}_0 = 0$, and which implies $h_t \equiv 1$ and $F_t \sim 1$ for
t is small. In other words, (S, g_t) and (S, m) share the same marked length spectrum.
Thus, $g_t = m$ for t is small; however, it is impossible unless $B = 0$.   □

Theorem 11.66 (Theorem C (3)) *The second derivative of the normalized asymptotic geodesic distortion defines a metric on* $\mathscr{F}$, *and which is called the pressure
metric and denoted by* $\| \cdot \|_P$.   □

Proof By Lemma 11.63, we know $\frac{d^2}{dt^2} J_{m,\alpha}(t)\big|_{t=0} = \text{Var}(\dot{c}_0, \mu_0)$ which is the
pressure metric we defined in Sect. 11.2.7 (since $c_0 = -1$).

Lastly, notice that Lemma 11.65 guarantees that

$$\|\alpha\|_P^2 := \frac{d^2}{dt^2} J_{m,\alpha}(t)\big|_{t=0} = \text{Var}(\dot{c}_0, \mu_0)$$

is non-degenerate, i.e., $\text{Var}(\dot{c}_0, \mu_0) = 0$ implies $\alpha = 0$.   □

Corollary 11.67 *The second derivatives of renormalized Manhattan curves defines
a family of metrics on* $\mathscr{F}$.   □

Proof This is because for $s \in (0, 1)$

$$-\frac{d^2}{dt^2} \tilde{\chi}_t(s)\big|_{t=0} = (s - s^2)\text{Var}(\dot{c}_0, \mu_0) = (s - s^2) \cdot \|\alpha\|_P^2.$$

□

In Sander's paper, he showed a relation between the Sanders' metric $|| \cdot ||_S^2$ (defined by the second derivative of entropy) and the Weil-Petersson metric $|| \cdot ||_{WP}$.

Theorem 11.68 (Sanders, Theorem 3.8 [12]) *For $\alpha \in Q([m])$ and $r_\alpha(t) = (g_t, t\mathrm{Re}\alpha) \in \mathscr{AF}$ we have*

$$||\alpha||_S^2 := \frac{d^2}{dt^2} h_{top}(g_t)\Big|_{t=0} \geq 2\pi ||\alpha||_{WP}^2$$

where $|| \cdot ||_{WP}$ is the Weil-Petersson metric on $\mathscr{F}$. □

We strengthen the above result and add the pressure metric $|| \cdot ||_P$ into the comparison. Before we write the statement we give a computational lemma.

Corollary 11.69 (Corollary 11.6) *Let $(m, 0)$ be a Fuchsian pair and $\alpha \in Q([m])$ be holomorphic quadratic differential. Then*

$$||\alpha||_S^2 := \frac{d}{dt^2} h_{top}(g_t)\Big|_{t=0} \geq ||\alpha||_P^2 + 2\pi ||\alpha||_{WP}^2.$$

Proof Let us consider $c_t := -h_{top}(g_t, tB) \cdot F_t = -h_t F_t$ and thus $c_0 = -1$. Since $P_\phi(c_t) = 0$, we know that $\int \dot{c}_0 d\mu_0 = 0$ where μ_0 is the Bowen-Margulis measure of the flow $\phi : T^1 S_{r_\alpha(0)} \to T^1 S_{r_\alpha(0)}$, i.e., $m_{c_0} = \mu_0$ and $\dot{c}_0 \in T_{c_0}\mathscr{P}(T^1(S, m))$. Therefore, by Proposition 11.19, the pressure metric of $\dot{c}_0$ is

$$||\alpha||_P^2 := ||\dot{c}_0||_P^2 = -\frac{\mathrm{Var}(\dot{c}_0, \mu_0)}{\int c_0 d\mu_0} = \frac{\int \ddot{c}_0 d\mu_0}{\int c_0 d\mu_0} = -\int \ddot{c}_0 d\mu_0.$$

Notice that $h_0 = 1$, $F_0 = 1$, and $\dot{u}_0 = 0$, so by Lemma 11.64

$$\int \dot{F} d\mu_0 = \int \dot{g}_0 d\mu_0 = \int 2\dot{u}_0 m(v, v) d\mu_0 = 0,$$

and hence

$$0 \leq ||\dot{c}_0||_P^2 = \ddot{h}_0 + 2\dot{h}_0 \int \dot{F}_0 d\mu_0 + h_0 \int \ddot{F}_0 d\mu_0$$

$$= \ddot{h}_0 + \int \ddot{F}_0 d\mu_0.$$

Therefore, by Lemma 11.64, we know

$$\ddot{h}_0 = ||\alpha||_P^2 - \int_{T^1 S} \ddot{F}_0 d\mu_0 \geq ||\alpha||_P^2 - \int_{T^1 S} \frac{\ddot{g}_0(v, v)}{2} d\mu_0$$

$$= ||\alpha||_P^2 + 2\pi ||\alpha||_{WP}^2 .$$

□

Corollary 11.70 (Corollary 11.7) *One can define a family of metrics on the Fuchsian space $\mathscr{F}$ by using the Hessian of $\chi_t(s)$ for $s \in [0, \varepsilon)$ for some $\varepsilon > 0$.* $\square$

Proof By Theorem 11.59, we get

$$||\alpha||^2_{\chi_s} := \frac{d^2}{dt^2}\chi_t(s)\Big|_{t=0} = \ddot{h}_0(1-s) - (s-s^2)||\alpha||^2_P.$$

Furthermore, since when $s = 0$ by the above theorem we know $||\alpha||^2_{\chi_0} > 0$ and by the continuity of $||\alpha||^2_{\chi_s}$, there exists $\varepsilon > 0$ such that for $s \in [0, \varepsilon)$

$$||\alpha||^2_{\chi_s} > 0.$$

$\square$

Acknowledgments The author is grateful to François Ledrappier. This work would have never been possible without François' support, guidance, patience, and sharing of his insightful ideas. The author also would like to thank Olivier Glorieux, Andy Sanders, and Zeno Huang for useful discussions on their works.

References

1. R. Bowen, Hausdorff dimension of quasicircles. Inst. Hautes Études Sci. Publ. Math. **50**, 11–25 (1979)
2. A. Katok, Entropy and closed geodesics. Ergodic Theory Dyn. Syst. **2**, 339–365 (1982)
3. M. Burger, Intersection, the Manhattan curve, and Patterson-Sullivan theory in rank 2. Internat. Math. Res. Notices **7** 217–225 (1993)
4. G. Knieper, Volume growth, entropy and the geodesic stretch. Math. Res. Lett. **2**, 39–58 (1995)
5. C.H. Taubes, Minimal surfaces in germs of hyperbolic 3-manifolds, in Proceedings of the Casson Fest. Geom. Topol. Monogr. **7**, 69–100 (2004)
6. C. McMullen, , Thermodynamics, dimension and the Weil-Petersson metric. Invent. Math. **173**, 365–425 (2008)
7. M. Bridgeman, Hausdorff dimension and the Weil-Petersson extension to quasifuchsian space. Geomet. Topol. **14**, 799–831 (2010)
8. O. Glorieux, Entropy of embedded surfaces in quasi-Fuchsian manifolds. Pacific J. Math. **294**, 375–400 (2018)
9. R. Sharp, The Manhattan curve and the correlation of length spectra on hyperbolic surfaces. Math. Zeitschr. **228**, 745–750 (1998)
10. C. Bishop, T. Steger, Representation-theoretic rigidity in PSL(2, R). Acta Math. **170**, 121–149 (1993)
11. K. Uhlenbeck, Closed minimal surfaces in hyperbolic 3-manifolds in *Seminar on Minimal Submanifolds* (Princeton University Press, Princeton, 1983), pp. 147–168
12. A. Sanders, Entropy, minimal surfaces and negatively curved manifolds. Ergodic Theory Dyn. Syst. **38**, 336–370 (2018)
13. M. Pollicott, R. Sharp, Weil-Petersson metrics, Manhattan curves and Hausdorff dimension. Math. Z. **282**, 1007–1016 (2016)
14. L.M. Abramov, On the entropy of a flow. Dokl. Akad. Nauk SSSR **128**, 873–875 (1959)

15. R. Bowen, Periodic orbits for hyperbolic flows. Amer. J. Math. **94**, 1–30 (1972)
16. R. Bowen, D. Ruelle, The ergodic theory of Axiom A flows Invent. Math. **29**, 181–202 (1975)
17. W. Parry, M. Pollicott, Zeta functions and the periodic orbit structure of hyperbolic dynamics. Astérisque **187**(188), 1–268 (1990)
18. A.O. Lopes, Ph. Thieullen, Sub-actions for Anosov flows. Ergodic Theory Dyn. Syst. **25**, 605–628 (2005)
19. É. Ghys, P. de la Harpe, *Sur les groupes hyperboliques d'après Mikhael Gromov*. Progress in Mathematics, vol. 83 (Birkhäuser Boston, Boston, 1990)
20. M. Kapovich, *Hyperbolic Manifolds and Discrete Groups*. Modern Birkhäuser Classics (Birkhäuser Boston, Boston, 2009)
21. Bourdon, M., Structure conforme au bord et flot géodésique d'un CAT(−1)-espace. Enseign. Math. **41**, 63–102 (1995)
22. G. Margulis, Applications of ergodic theory to the investigation of manifolds of negative curvature. Funct. Analy. Appl. **3**, 335–336 (1969)
23. D. Sullivan, Entropy, Hausdorff measures old and new, and limit sets of geometrically finite Kleinian groups. Acta Math. **153**, 259–277 (1984)
24. D. Sullivan, The density at infinity of a discrete group of hyperbolic motions, Institut des Hautes Études Scientifiques. Publ. Math. **50**, 171–202 (1979)
25. F. Ledrappier, Structure au bord des variétés à courbure négative. Séminaire de Théorie Spectrale et Géométrie **13**, 97–122 1994–1995
26. H. Hopf, Über Flächen mit einer Relation zwischen den Hauptkrümmungen. Mathematische Nachr. **4**, 232–249 (1951)
27. J.F.C. Kingman, Subadditive ergodic theory. Ann. Probab. **1**, 883–909 (1973)
28. D. Anosov, Geodesic flows on closed Riemannian manifolds of negative curvature. Akad. Nauk SSSR **90**, 209 (1967). Trudy Matematicheskogo Instituta imeni V. A. Steklova
29. M. Pollicott, Derivatives of topological entropy for Anosov and geodesic flows. J. Differential Geom. **39**, 457–489 (1994)
30. R. de la Llave, J.M. Marco, R. Moriyon, Canonical Perturbation theory of Anosov systems and regularity results for the livsic cohomology equation, Ann. Math. **123**, 537–611 (1986)

Chapter 12
Higher Teichmüller Theory for Surface Groups and Shifts of Finite Type

Mark Pollicott and Richard Sharp

Abstract The Teichmüller space of Riemann metrics on a compact oriented surface V without boundary comes equipped with a natural Riemannian metric called the Weil–Petersson metric. Bridgeman, Canary, Labourie and Sambarino generalised this to Higher Teichmüller Theory, i.e. representations of $\pi_1(V)$ in $\mathrm{SL}(d, \mathbb{R})$, and showed that their metric is analytic. In this note we will present a new equivalent definition of the Weil–Petersson metric for Higher Teichmüller Theory and also give a short proof of analyticity. Our approach involves coding $\pi_1(V)$ in terms of a symbolic dynamical system and the associated thermodynamic formalism.

12.1 Introduction

Given a compact oriented surface V without boundary of genus $g \geq 2$, the classical Teichmüller space $\mathcal{T}(V)$ is the space of hyperbolic structures on V, i.e. Riemannian metrics of constant curvature -1. Then $\mathcal{T}(V)$ is diffeomorphic to $\mathbb{R}^{6g-6}$ and it supports a number of natural metrics. One of the best known of these is the Weil–Petersson metric, which is negatively curved but incomplete. Let Γ denote the fundamental group $\pi_1(V)$ of V. By the uniformisation theorem, each element of $\mathcal{T}(V)$ can be realised as $\mathbb{H}^2/\rho(\Gamma)$, where $\rho : \Gamma \to \mathrm{PSL}(2, \mathbb{R})$ is a discrete co-compact representation of Γ into $\mathrm{PSL}(2, \mathbb{R}) = \mathrm{Isom}^+(\mathbb{H}^2)$ and where the action on $\mathbb{H}^2$ is by Möbius transformations. In fact, Goldman [8] and Hitchin [9] showed that $\mathcal{T}(V)$ can be identified with a connected component in the representation space

$$\mathrm{Rep}(\Gamma, \mathrm{PSL}(2, \mathbb{R})) = \mathrm{Hom}(\Gamma, \mathrm{PSL}(2, \mathbb{R}))/\mathrm{PSL}(2, \mathbb{R}),$$

where $\mathrm{PSL}(2, \mathbb{R})$ acts by conjugation. (Some modification is needed to obtain a Hausdorff quotient space, see [2] or [7] for details. However this does not affect

M. Pollicott (✉) and R. Sharp
Mathematics Institute, University of Warwick, Coventry, UK
e-mail: M.Pollicott@warwick.ac.uk; R.J.Sharp@warwick.ac.uk

© The Author(s), under exclusive license to Springer Nature Switzerland AG 2021
M. Pollicott, S. Vaienti (eds.), *Thermodynamic Formalism*, Lecture Notes
in Mathematics 2290, https://doi.org/10.1007/978-3-030-74863-0_12

the Teichmüller component or the Hitchin components introduced below.) Another (homeomorphic) connected component $\mathscr{T}'(V)$ is obtained through the action of an outer automorphism of $\mathrm{PSL}(2, \mathbb{R})$ (corresponding to reversing the orientation of V).

In 1958, Weil defined the Weil–Petersson metric on Teichmüller space, using the Petersson inner product on modular forms. An alternative definition of the metric was introduced by Thurston, the equivalence of which to the Weil–Petersson metric was shown by Wolpert in 1986 [26]. In 2008, McMullen gave a more thermodynamic formulation, using the pressure metric [16].

In recent years, particularly through the work of Bridgeman, Canary, Labourie and Sambarino, there have been significant advances in generalizing the definition of Weil–Petersson metric to more general classes of representation spaces. In this broader context this metric was also called the pressure metric. In particular, in [3], they considered representations of hyperbolic groups into the higher rank groups $\mathrm{SL}(d, \mathbb{R})$, $d \geq 3$, and showed real analyticity of their metric on the natural analogue of Teichmüller space in this setting. One of the aims of the present paper is to provide an alternative definition of Weil–Petersson metric and a simpler proof of the analyticity result, albeit in the more restrictive setting of surface groups.

We begin by recalling another equivalent definition of the classical Weil–Petersson metric from [20], based on [22]. Given an analytic family of representations

$$(-\epsilon, \epsilon) \to \mathrm{Rep}(\Gamma, \mathrm{PSL}(2, \mathbb{R})) = \mathrm{Hom}(\Gamma, \mathrm{PSL}(2, \mathbb{R}))/\mathrm{PSL}(2, \mathbb{R}) : \lambda \mapsto \rho_\lambda,$$

with expansion

$$\rho_\lambda = \rho_0 + \rho^{(1)}\lambda + o(\lambda),$$

where $\mathrm{tr}(\rho^{(1)}) = 0$, it suffices to define the norm of the tangent $\rho^{(1)}$. (We assume that the familiy is non-trivial and thus $\epsilon > 0$ can be chosen sufficiently small that $\rho_\lambda \neq \rho_0$ for $\lambda \neq 0$.) One approach to doing this is given in Proposition 12.1 below.

For $g \in \Gamma$, let $[g]$ denote its conjugacy class and let $\mathscr{C}(\Gamma)$ denote the set of non-trivial conjugacy classes in Γ. To each conjugacy classes $[g] \in \mathscr{C}(\Gamma)$ and $\lambda \in (-\epsilon, \epsilon)$, we associate the length $l_{\rho_\lambda}([g])$ of corresponding closed geodesic in $\mathbb{H}^2/\rho_\lambda(\Gamma)$. We recall that

$$\lim_{T \to +\infty} \frac{1}{T} \log \#\{[g] \in \mathscr{C}(\Gamma) : l_{\rho_0}([g]) \leq T\} = 1.$$

The next result describes how the growth rate changes if, for a given $\lambda \neq 0$, we restrict to conjugacy classes $[g]$ for which $l_{\rho_\lambda}([g])$ is close to $l_{\rho_0}([g])$.

Proposition 12.1 ([20, 22]) *For each* $\lambda \in (-\epsilon, \epsilon) \setminus \{0\}$, *there exists* $0 < \alpha(\lambda) < 1$ *such that*

$\alpha(\lambda)$

$$= \lim_{\delta \to 0} \lim_{T \to +\infty} \frac{1}{T} \log \# \left\{ [g] \in \mathscr{C}(\Gamma) : l_{\rho_0}([g]) \le T \text{ and } \frac{l_{\rho_\lambda}([g])}{l_{\rho_0}([g])} \in (1 - \delta, 1 + \delta) \right\}.$$

(12.1.1)

Furthermore, if we define $\alpha : (-\epsilon, \epsilon) \to [0, 1]$ by setting $\alpha(0) = 1$ then the Weil–Petersson norm is given by

$$\|\rho^{(1)}\| = -\frac{1}{12\pi(g-1)} \frac{\partial^2 \alpha(\lambda)}{\partial \lambda^2} \bigg|_{\lambda=0}.$$

Indeed, [22] contains a stronger asymptotic result than (1.1), but the above statement suffices for our purpose of studying the Weil–Petersson metric.

It is natural to consider the generalisation of this approach to representations in the higher rank group $\mathrm{PSL}(d, \mathbb{R})$ (for $d \ge 3$). As we discuss in Sect. 12.2, there is a natural representation from $\mathrm{PSL}(2, \mathbb{R})$ to $\mathrm{PSL}(d, \mathbb{R})$ (induced by the action on homogeneous polynomials in two variables of degree $d-1$) and a representation $R : \Gamma \to \mathrm{PSL}(d, \mathbb{R})$ is called *Fuchsian* if it is obtained from a representation in $\mathscr{T}(V)$ or $\mathscr{T}'(V)$ by composition. Unlike the $d = 2$ case, the Fuchsian representations do not fill out a whole connected component of the representation space

$$\mathrm{Rep}(\Gamma, \mathrm{PSL}(d, \mathbb{R})) = \mathrm{Hom}(\Gamma, \mathrm{PSL}(d, \mathbb{R}))/\mathrm{PSL}(d, \mathbb{R})$$

but a component containing a Fuchsian representation is called a *Hitchin component*. Such a component is an analytic manifold diffeomorphic to an open ball of dimension $(2g - 2) \dim(\mathrm{PSL}(d, \mathbb{R}))$ [10].

Let $\mathscr{H}$ be a Hitchin component. The natural problem of defining an analogue of the Weil–Petersson metric on $\mathscr{H}$ has already been considered by Bridgeman, Canary, Labourie and Sambarino (in the even more general setting of Gromov hyperbolic groups) in [3]. We start by defining a numerical characteristic called the *entropy* of a representation. Representations in the Hitchin component have the key proximality property that for $g \in \Gamma \setminus \{1_\Gamma\}$, the matrix $R(g)$ (which we can think of as lifted to $\mathrm{SL}(d, \mathbb{R})$) has a unique simple eigenvalue $\lambda(g)$ which is strictly maximal in modulus, satisfies $|\lambda(g)| > 1$, and which only depends of the conjugacy class $[g]$. This then allows us to define the entropy, $h(R)$, of a representation $R \in \mathscr{H}$ by

$$h(R) = \lim_{T \to +\infty} \frac{1}{T} \log \left(\#\{[g] \in \mathscr{C}(\Gamma) : d_R([g]) \le T\} \right),$$

where $d_R([g]) = \log |\lambda(g)|$. Bridgeman, Canary, Labourie and Sambarino have shown that the entropy is analytic on $\mathscr{H}$:

Theorem 12.2 (Bridgeman et al. [3]) *The map $h : \mathscr{H} \to \mathbb{R}$ is real analytic.* $\square$

In the particular case $d = 2$ then, as noted above, we always have that $h(R) = 2$ and the result is trivial. (This is because, in this case, we have $\lambda(g) = \exp(l([g])/2)$, where $l([g])$ is the length of the unique closed geodesic in the free homotopy class determined by the conjugacy class $[g]$, the claim then following from the Prime Geodesic Theorem of Huber [11]. This is closely related to the geodesic flow which has entropy one, the factor of two coming from the normalization.) In [3], Bridgeman, Canary, Labourie and Sambarino introduced a generalised intersection form and a generalised Weil–Petersson norm on $\mathcal{H}$.

Definition 12.1.1 The *intersection* is defined on the Hitchin component by

$$I(R_0, R_1) = \lim_{T \to +\infty} \frac{\sum_{d_{R_0}([g]) \leq T} \frac{d_{R_1}([g])}{d_{R_0}([g])}}{\#\{[g] \in \mathcal{C}(\Gamma) : d_{R_0}([g]) \leq T\}}.$$

The *normalised* intersection is then defined by

$$J(R_0, R_1) = \frac{h(R_1)}{h(R_0)} I(R_0, R_1).$$

Given an analytic family of representations $R_\lambda \in \mathcal{H}$, $\lambda \in (-\epsilon, \epsilon)$, with expansion

$$R_\lambda = R_0 + \lambda R^{(1)} + o(\lambda),$$

one can define the *Weil–Petersson norm* of the tangent $R^{(1)}$ by

$$\|R^{(1)}\|^2 = \frac{\partial^2 J(R_0, R_\lambda)}{\partial \lambda^2}\bigg|_{\lambda=0}.$$

A key property of the Weil–Petersson norm is the following.

Theorem 12.3 (Bridgeman et al. [3]) *The normalised intersection J and the norm $\|\cdot\|$ are real analytic.*   □

We will present short proofs of Theorem 12.2 and Theorem 12.3 in Sect. 12.5.

Our main result is the following new equivalent definition of the Weil–Petersson norm, which is inspired by Proposition 12.1.

Theorem 12.4 *Let $R_\lambda \in \mathcal{H}$, $\lambda \in (-\epsilon, \epsilon)$ be a (non-constant) analytic family of representations. Then for each $\lambda \in (-\epsilon, \epsilon) \setminus \{0\}$, there exists $0 < \alpha(\lambda) < h(R_0)$ such that*

$$\alpha(\lambda) = \lim_{\delta \to 0} \lim_{T \to +\infty} \frac{1}{T} \log \# \Big\{[g] : d_{R_0}([g]) \leq T \text{ and }$$

$$\frac{d_{R_\lambda}([g])}{d_{R_0}([g])} \in \left(\frac{h(R_0)}{h(R_\lambda)} - \delta, \frac{h(R_0)}{h(R_\lambda)} + \delta\right)\Big\}.$$

Furthermore, if we define $\alpha : (-\epsilon, \epsilon) \to [0, h(R_0)]$ *by setting* $\alpha(0) = h(R_0)$ *then the Weil–Petersson metric is given by*

$$\| R^{(1)} \|^2 = -\frac{4}{h(R_0)} \left. \frac{\partial^2 \alpha(\lambda)}{\partial \lambda^2} \right|_{\lambda=0}.$$

The approach of Bridgeman et al. in [3] is to use the thermodynamic approach of McMullen [16] (involving the *pressure metric*). In the present note, we will also use the thermodynamic approach, but we introduce two new ingredients which help simplify the analysis. Firstly, we introduce the thermodynamics directly via the strongly Markov structure of Γ and an associated one-sided subshift of finite type, rather than more indirectly via the construction of a flow and the associated symbolic dynamics for that flow. Secondly, we will bypass many of the complications associated with studying the analyticity properties of pressure using Banach manifolds by the introduction of a suitable family of complex functions.

12.2 Representations and Proximality

In this section we discuss the generalisation of the classical Teichmüller theory of representations into $PSL(2, \mathbb{R})$ to $PSL(d, \mathbb{R})$ (for $d \geq 3$). In particular, we discuss the Hitchin components and the associate proximality property introduced in the introduction. We then describe the key ideas that link the geometry of the representation space to a readily analysed dynamical system.

There is an irreducible representation of $\iota : PSL(2, \mathbb{R}) \to PSL(d, \mathbb{R})$, induced by the natural action on the space of homogeneous polynomials of degree $d - 1$,

$$\begin{pmatrix} a & b \\ c & d \end{pmatrix} \cdot P(X, Y) = P(aX + bY, cX + dY),$$

and representations of the form $R = \iota \circ \rho$, with $\rho : \Gamma \to PSL(2, \mathbb{R})$ in $\mathscr{T}(V)$ or $\mathscr{T}'(V)$, are called *Fuchsian representations*. More generally, a representation of $R : \Gamma \to PSL(d, \mathbb{R})$ is said to be in a *Hitchin component* $\mathscr{H}$ if it is in the same connected component of the representation space

$$\mathrm{Rep}(\Gamma, PSL(d, \mathbb{R})) = \mathrm{Hom}(\Gamma, PSL(d, \mathbb{R}))/PSL(d, \mathbb{R})$$

as a Fuchsian representation. (If d is odd there is a single Hitchin component but if d is even there are two Hitchin components.)

For future use, we note that a representation in the Hitchin component can be lifted to a representation over Γ in $SL(d, \mathbb{R})$. To see this, note first that since Γ is torsion free, a discrete faithful representation $\rho : \Gamma \to PSL(2, \mathbb{R})$ can be lifted to a representation $\tilde{\rho} : \Gamma \to SL(2, \mathbb{R})$ [4]. Furthermore, ι is actually obtained from a representation $\tilde{\iota} : \mathrm{SL}(2,\mathbb{R}) \to \mathrm{SL}(d,\mathrm{R})$ and $\tilde{\iota} \circ \tilde{\rho}$ is a lift of $\iota \circ \rho$. Finally, Theorem 4.1 of [4] tells us that any representation in a Hitchin component also has a lift to $\mathrm{SL}(d,\mathrm{R})$. We will use the same symbol to denote both the original and lifted representations.

We next discuss the notion of a hyperconvex representation and describe how it relates the boundary of the group Γ to something akin to a limit set in $\mathbb{R}P^{d-1}$. The boundary of Γ, denoted $\partial\Gamma$, is the well-defined topological space obtained from the set of (one-sided) infinite geodesic paths in the Cayley graph of Γ by declaring that two paths are equivalent if they remain a bounded distance apart. In the case where Γ is the fundamental group of a compact surface without boundary, $\partial\Gamma$ is homeomorphic to S^1.

We recall that a flag space $\mathscr{F}$ for $\mathbb{R}^d$ is a collection of subspaces $V_1 \subset V_2 \subset \cdots \subset V_d$ of $\mathbb{R}^d$ with $\dim(V_i) = i$. There is a natural linear action of each $R(g) \in \mathrm{SL}(d,\mathbb{R})$ on $\mathbb{R}^d$ which induces a corresponding action on the vector subspaces, and thus on the flags.

Definition 12.2.1 A representation of $R : \Gamma \to \mathrm{SL}(d,\mathbb{R})$ is hyperconvex if there exist Γ-equivariant (Hölder) continuous maps $(\xi,\theta) : \partial\Gamma \to \mathscr{F} \times \mathscr{F}$ such that for distinct $x, y \in \mathscr{F}$ the images $\xi(x) = (V_i(x))_{i=0}^d$ and $\theta(x) = (W_i(x))_{i=0}^d$ satisfy $V_i(x) \oplus W_{d-i}(x) = \mathbb{R}^d$, for $i = 0, \cdots, d$. □

By Γ-equivariance we mean that $R(g)\xi(x) = \xi(gx)$, where $R(g)\xi(x)$ is the image under the linear action of $R(g)$ for $g \in \Gamma$.

The following fundamental result of Labourie tells us that the representations in a Hitchin component are hyperconvex.

Proposition 12.5 (Labourie [15]) *If $R \in \mathscr{H}$ then R hyperconvex.* □

For our purposes it suffices for us to focus on one component of $\xi : \partial\Gamma \to \mathscr{F}$, say, and furthermore take the one dimensional subspace $V_1(x)$ in the flag given by $\xi_0(x) = V_1(x)$, say. This corresponds to a point in projective space and thus we have a Hölder continuous Γ-equivariant map from $\partial\Gamma$ to $\mathbb{R}P^{d-1}$.

Let $R \in \mathscr{H}$. An important consequence of the hyperconvexity of R is that, for each $g \in \Gamma \setminus \{1_\Gamma\}$, the matrix $R(g) \in \mathrm{SL}(d,\mathbb{R})$ is proximal, i.e. it has a unique simple eigenvalue $\lambda(g)$ which is strictly maximal in modulus (and which only depends on the conjugacy class $[g]$) [15, 21]. Since $\det R(g) = 1$, we have $|\lambda(g)| > 1$. As above, we will write $d_R([g]) = \log|\lambda(g)| > 0$.

It will prove important to characterise $d_R([g])$ in terms of the action that $R(g)$ induces on projective space. We can consider the projective action $\widehat{R}(g) : \mathbb{R}P^{d-1} \to \mathbb{R}P^{d-1}$ of the representation $R(g) \in \mathrm{SL}(d,\mathbb{R})$ defined by $\widehat{R}(g)[v] = R(g)v/\|R(g)v\|_2$ (where $v \in \mathbb{R}^d \setminus \{0\}$ is a representative element).

The proximality of $R(g)$ ensures that $\widehat{R}(g) : \mathbb{R}P^{d-1} \to \mathbb{R}P^{d-1}$ has a unique attracting fixed point $\xi_g \in \mathbb{R}P^{d-1}$. We can use the following simple lemma to relate the weight $d_R([g])$ to the action of $R(g)$ on $\mathbb{R}P^{d-1}$.

Lemma 12.2.1 *If $g \in \Gamma \setminus \{1_\Gamma\}$ and $\xi_g \in \mathbb{R}P^{d-1}$ is the attracting fixed point for $\widehat{R}(g) : \mathbb{R}P^{d-1} \to \mathbb{R}P^{d-1}$ then*

$$d_R([g]) = -\frac{1}{d} \log \det(D_{\xi_g} \widehat{R}(g)).$$

Proof We can consider the linear action of $R(g)$ on $\mathbb{R}^d$, then the fixed point corresponds to an eigenvector v and the result follows from a simple calculation using that the linear action of $R(g) \in SL(d, \mathbb{R})$ preserves area in $\mathbb{R}^d$. More precisely, ξ_g corresponds to an eigenvector v for the maximal eigenvalue $\lambda(g)$, with $|\lambda(g)| > 1$, for the matrix $R(g)$. We can assume without loss of generality that $\|v\| = 1$ and then for arbitrarily small $\delta > 0$ we can consider a δ-neighbourhood of v which is the product of a $(d - 1)$-dimensional neighbourhood in $\mathbb{R}P^{d-1}$ and a δ-neighbourhood in the radial direction. The effect of the linear action of $R(g)$ is to replace v by $\lambda(g)v$, and thus stretch the neighbourhood in the radial direction by a factor of $|\lambda(g)|$. Since $R(g)$ has determinant one, the volume of the $(d - 1)$-dimensional neighbourhood contracts by $|\lambda(g)|^{-1}$. To calculate the effect of the projective action $\widehat{R}(g)$, we need to rescale $\lambda(g)v$ to have norm one, which corresponds to multiplication by the diagonal matrix $\mathrm{diag}(|\lambda(g)|^{-1}, \ldots, |\lambda(g)|^{-1})$. In particular, the $(d - 1)$-dimensional neighbourhood in $\mathbb{R}P^{d-1}$ shrinks by a factor of approximately $|\lambda(g)|^{-d}$, giving the result. □

12.3   Symbolic Dynamics

The structure of the group Γ allows us to code it in terms of a symbolic dynamical system, namely a subshift of finite type. We will describe this and then discuss how the geometric information given by the numbers $d_R([g])$ may also be encoded. This in turn enables use to use the machinery of thermodynamic formalism to define a form of pressure function and hence an associated metric on spaces of representations.

As the fundamental group of a compact orientable surface without boundary of genus $g \geq 2$, Γ has the standard presentation

$$\Gamma = \left\langle a_1, \ldots, a_g, b_1, \ldots, b_g \mid \prod_{i=1}^{g} [a_i, b_i] = 1 \right\rangle.$$

We write $\Gamma_0 = \{a_1^{\pm 1}, \cdots, a_g^{\pm 1}, b_1^{\pm 1}, \cdots, b_g^{\pm 1}\}$ for the symmetrised generating set.

The surface group Γ is a particular example of a Gromov hyperbolic group and as such it is a strongly Markov group in the sense of Ghys and de la Harpe [6], i.e., they can be encoded using a directed graph and an edge labelling by elements in Γ_0. Lemma 12.3.1 below deals with the particular case of surface groups, where the more precise statements, including the relationship between closed paths and conjugacy classes, follow from the work of Adler and Flatto [1] and Series [23] on coding the action on the boundary and the associated shift of finite type is mixing.

Lemma 12.3.1 *We can associate to* (Γ, Γ_0) *a directed graph* $G = (V, E)$, *with a distinguished vertex* $* \in V$, *and an edge labelling* $\rho : E \to \Gamma_0$, *such that:*

(1) *no edge terminates at* $*$;
(2) *there is at most one directed edge between each ordered pair of vertices;*
(3) *the map from the set finite paths in the graph starting at* $*$ *to* $\Gamma \setminus \{e\}$ *defined by*

$$(e_1, \ldots, e_n) \mapsto \rho(e_1) \cdots \rho(e_n)$$

is a bijection and $|\rho(e_1) \cdots \rho(e_n)| = n$;
(4) *after removing finitely many closed paths, the map from closed paths in* G *(modulo cyclic permutation) to* $\mathscr{C}(\Gamma)$ *induced by* ρ *is a bijection and for such a closed path* $(e_1, \ldots, e_n, e_1)$, n *is the minimum word length in the conjugacy class of* $\rho(e_1) \cdots \rho(e_n)$; *and*
(5) *a conjugacy class in* $\mathscr{C}(\Gamma)$ *is primitive (i.e. it does not contain an element of the form* g^n *with* $g \in \Gamma$ *and* $n \in \mathbb{Z} \setminus \{-1, 1\}$) *if and only if the corresponding closed path is not a power of a shorter path.*

Furthermore, the subgraph obtained be deleting the vertex $*$ *has the aperiodicity property that there exists* $N \geq 1$ *such that, given any two* $v, v' \in V \setminus \{*\}$, *there is a directed path of length* N *from* v *to* v'. □

We now introduce a dynamical system. We can associate to the directed graph G a subshift of finite type where the states are labelled by the edges in the graph after deleting the edges that originate in the vertex $*$. In particular, if there are k such edges then we can define a $k \times k$ matrix A by $A(e, e') = 1$ if e' follows e in the directed graph and then define a space

$$\Sigma = \{x = (x_n)_{n=0}^{\infty} \in \{1, \ldots, k\}^{\mathbb{Z}^+} : A(x_n, x_{n+1}) = 1, n \geq 0\},$$

where for convenience we have labelled the edges $1, \ldots, k$. This is a compact space with respect to the metric

$$d(x, y) = \sum_{n=0}^{\infty} \frac{1 - \delta(x_n, y_n)}{2^n}.$$

The shift map is the local homeomorphism $\sigma : \Sigma \to \Sigma$ defined by $(\sigma x)_n = x_{n+1}$. By Lemma 12.3.1, A is aperiodic (i.e. there exists $N \geq 1$ such that A^N has all entries positive) and, equivalently, the shift $\sigma : \Sigma \to \Sigma$ is mixing (i.e. for all open non-empty $U, V \subset \Sigma$, there exists $N \geq 1$ such that $\sigma^{-n}(U) \cap V \neq \emptyset$ for all $n \geq N$). The periodic orbits for σ correspond exactly to the conjugacy classes in $\mathscr{C}(\Gamma)$ and they are prime if and only if the corresponding conjugacy class is primitive.

There is a natural surjective Hölder continuous map $\pi : \Sigma \to \partial\Gamma$ defined by setting $\pi((x_n)_{n=0}^\infty)$ to be the equivalence class of the infinite geodesic path $(\rho(x_n))_{n=0}^\infty$ in $\partial\Gamma$.

However, the shift $\sigma : \Sigma \to \Sigma$ only encodes information about Γ as an abstract group. In order to keep track of the additional information given by the representation of Γ in $\mathrm{PSL}(d, \mathbb{R})$ we need to introduce a Hölder continuous function $r : \Sigma \to \mathbb{R}$.

Definition 12.3.1 We can associate a map $r : \Sigma \to \mathbb{R}$ defined by

$$ r(x) = -\frac{1}{d} \log \det(D_{\Xi(x)} \widehat{R}(g_{x_0})), $$

(i.e., the Jacobian of the derivative of the projective action) where $\Xi = \xi_0 \circ \pi$ and where $g_{x_0} = \rho(x_0)$ is the generator corresponding to the first term in $x = (x_n)_{n=0}^\infty \in \Sigma$.   □

Given $r : \Sigma \to \mathbb{R}$ and $x \in \Sigma$ we denote $r^n(x) := r(x) + r(\sigma x) + \cdots + r(\sigma^{n-1}x)$ for $n \geq 1$. We now have the following simple but key result.

Lemma 12.3.2 *The function $r : \Sigma \to \mathbb{R}$ is Hölder continuous, and if $\sigma^n x = x$ is a periodic point corresponding to an element $g \in \Gamma$ then $r^n(x) = d_R([g])$.*   □

Proof The Hölder continuity of r follows immediately from the Hölder continuity of ξ_0, which in turn comes from Proposition 12.5. The second part of the lemma follows from the equivariance and the observation $\Xi(\sigma x) = R(g_{x_0})\Xi(x)$. Moreover, that the periodic point x has an image $\Xi(x)(= \xi_g)$ which is fixed by $\widehat{R}(g)$ and the result follows from Lemma 12.2.1.   □

The next lemma shows how the analytic dependence of the representations translates into analytic dependence of the associated function r.

Lemma 12.3.3 *For a C^ω family $(-\epsilon, \epsilon) \ni \lambda \mapsto R_\lambda$ of representations, the associated maps r_λ have a C^ω dependence.*   □

Proof The proof is very similar to Proposition 2.2 in [12], which is in turn based on the classical approach of Mather, and the refinement of de la Llave-Marco-Moriyón [5], to showing the existence of, and analytic dependence of, a conjugating (Hölder) homeomorphism between nearby expanding maps on a manifold (i.e., structural stability). Given this similarity, it suffices to only outline the main steps in the proof. The main objective is to construct a natural family of (Hölder) continuous equivariant maps $\Xi_\lambda : \Sigma \to \mathbb{R}P^{d-1}$, that is a family of (Hölder) continuous maps satisfying $R_\lambda(g_{x_0})\Xi_\lambda(x) = \Xi_\lambda(\sigma x)$, for $x \in \Sigma$. Given any $0 < \alpha < 1$, we let $C^\alpha(\Sigma, \mathbb{R}P^{d-1})$ denote the Banach manifold of α-Hölder continuous functions on Σ taking values in the projective space $\mathbb{R}P^{d-1}$. We can now consider the family of maps $H_\lambda : C^\alpha(\Sigma, \mathbb{R}P^{d-1}) \to C^\alpha(\Sigma, \mathbb{R}P^{d-1})$ defined by $H_\lambda(\Xi)(x) = R_\lambda(g_{x_0}^{-1})\Xi(\sigma x)$, for $x \in \Sigma$ with first symbol x_0, and $\Xi \in C^\alpha(\Sigma, \mathbb{R}P^{d-1})$. In particular, providing $0 < \alpha < 1$ is sufficiently small then one can show that for each

$\lambda \in (-\epsilon, \epsilon)$ there exists a unique continuous family Ξ_λ which is a fixed point (i.e., $H_\lambda(\Xi_\lambda) = \Xi_\lambda$) and, moreover, the maps $(-\epsilon, \epsilon) \in \lambda \mapsto \Xi_\lambda \in C^\alpha(\Sigma, \mathbb{R}P^{d-1})$ are analytic. This follows from an application of the Implicit Function Theorem. More precisely, in order to apply the Implicit Function Theorem we first observe that we can identify the tangent space $T_v \mathbb{R}P^{d-1}$ at $v \in \mathbb{R}P^{d-1}$ with $\mathbb{R}^{d-1}$. We can then consider the derivative $DH_\lambda : C^\alpha(\Sigma, \mathbb{R}^{d-1}) \to C^\alpha(\Sigma, \mathbb{R}^{d-1})$ which can be defined by $DH_\lambda(\Pi)(x) = Dg_{x_0}^{-1} \Pi(x)$, for $\Pi \in C^\alpha(\Sigma, \mathbb{R}^{d-1})$ and $x \in \Sigma$. For $0 < \alpha < 1$ sufficiently small the hyperbolic nature of $\widehat{R}_\lambda(g_{x_0}^{-1})$ ensures that the operator $(DH_\lambda - I) : C^\alpha(\Sigma, \mathbb{R}^{d-1}) \to C^\alpha(\Sigma, \mathbb{R}^{d-1})$ is invertible. (This is more readily seen in the case of $(DH_\lambda - I) : C(\Sigma, \mathbb{R}^{d-1}) \to C(\Sigma, \mathbb{R}^{d-1})$ on continuous functions, the setting of Mather's original proof, but then the result extends to Hölder functions providing α is sufficiently small, as in the article of de la Llave-Marco-Moriyón [5]). It then follows from the Implicit Function Theorem that there is a unique fixed point Ξ_λ and also that this depends analytically on $\lambda \in (-\epsilon, \epsilon)$. Finally, writing $r_\lambda(x) = \log \det(R_\lambda(g_{x_0}))(\Xi_\lambda(x))$ we see that this too depends analytically on $\lambda \in (-\epsilon, \epsilon)$. $\qquad \square$

12.4 Thermodynamic Formalism

In this section we discuss the thermodynamic formalism associated to the map $\sigma : \Sigma \to \Sigma$ and, subsequently, to an associated suspended semiflow. (We refer the reader to [18] for a more detailed account.) We say that two Hölder continuous functions $f_1, f_2 : \Sigma \to \mathbb{R}$ are cohomologous if $f_1 - f_2 = u \circ \sigma - u$, for some continuous function $u : \Sigma \to \mathbb{R}$. Then f_1 and f_2 are cohomologous of and only if $f_1^n(x) = f_2^n(x)$ whenever $\sigma^n x = x, n \geq 1$.

Let $\mathscr{M}_\sigma$ denote the set of σ-invariant probability measures on Σ. For a Hölder continuous function $f : \Sigma \to \mathbb{R}$, its pressure $P(f)$ is defined by

$$P(f) := \sup_{\mu \in \mathscr{M}_\sigma} \left\{ h_\sigma(\mu) + \int f \, d\mu \right\},$$

where $h_\sigma(\mu)$ denotes the measure-theoretic entropy, and its equilibrium state μ_f is the unique σ-invariant probability measure for which the supremum is attained. If f is not cohomologous to a constant then

$$\mathscr{I}_\sigma(f) := \left\{ \int f \, d\mu : \mu \in \mathscr{M}_\sigma \right\}$$

is a non-trivial closed interval and, for $\xi \in \text{int}(\mathscr{I}(f))$,

$$\sup \left\{ h(\mu) : \mu \in \mathscr{M}_\sigma \text{ and } \int f \, d\mu = \xi \right\} > 0.$$

The following result is standard (see [18]).

Lemma 12.4.1 *The map $t \mapsto P(tf_1 + f_2)$ is real analytic on $\mathbb{R}$ and satisfies*

$$\left. \frac{dP(tf_1 + f_2)}{dt} \right|_{t=0} = \int f_1 \, d\mu_{f_2}.$$

We will also need some material about suspended semi-flows over $\sigma : \Sigma \to \Sigma$. Let $f : \Sigma \to \mathbb{R}$ be strictly positive and Hölder continuous.

Definition 12.4.1 We define

$$\Sigma^f = \{(x, s) : x \in \Sigma, \ 0 \leq s \leq f(x)\}/\sim,$$

where we have quotiented by the relation $(x, f(x)) \sim (\sigma x, 0)$. The associated suspended semiflow $\sigma_t^f : \Sigma^f \to \Sigma^f$, $t \geq 0$, is defined by $\sigma_t^f(x, s) = (x, s+t)$, modulo the identifications.   □

Let $\mathcal{M}_{\sigma^f}$ denote the set of σ^f-invariant probability measures on Σ^f. Each $m \in \mathcal{M}_{\sigma^f}$ takes the form $dm = (d\mu \times dt)/\int f \, d\mu$, where $\mu \in \mathcal{M}_\sigma$ and their entropies are related by

$$h_{\sigma^f}(m) = \frac{h_\sigma(\mu)}{\int f \, d\mu}.$$

For a Hölder continuous function $G : \Sigma^f \to \mathbb{R}$, its equilibrium state m_G is the unique σ^f-invariant probability measure for which

$$h_{\sigma^f}(m_G) + \int G \, dm_G = P(G) := \sup_{m \in \mathcal{M}_{\sigma^f}} \left\{ h_{\sigma^f}(m) + \int G \, dm \right\}.$$

Then $dm_G = (d\mu_{g-P(G)f} \times dt)/\int f \, d\mu_{g-P(G)f}$, where $g : \Sigma \to \mathbb{R}$ is defined by

$$g(x) = \int_0^{f(x)} G(x, s) \, ds.$$

In particular, σ^f has a unique measure of measure of maximal entropy m_0 for σ^f, i.e. a unique measure m_0 such that

$$h_{\sigma^f}(m_0) = \sup_{m \in \mathcal{M}_{\sigma^f}} h_{\sigma^f}(m).$$

Furthermore, $h_{\sigma^f}(m_0)$ is equal to the topological entropy

$$h(\sigma^f) := \lim_{T \to \infty} \frac{1}{T} \log \left(\sum_{n=1}^{\infty} \# \left\{ \sigma^n x = x : f^n(x) \leq T \right\} \right).$$

This measure is given by $dm_0 = (d\mu_{-h(\sigma^f)f} \times dt) / \int f \, d\mu_{-h(\sigma^f)f}$ and we have

$$h(\sigma^f) = h_{\sigma^f}(m_0) = \frac{h_{\sigma}(\mu_{-h(\sigma^f)f})}{\int f \, d\mu_{-h(\sigma^f)f}}.$$

The topological entropy is also characterised by the equation $P(-h(\sigma^f)f) = 0$. We have the following analogue of Lemma 12.4.1 (see Lemma 1 of [24]).

Lemma 12.4.2 *The map* $t \mapsto P(tG_1 + G_2)$ *is real analytic on* $\mathbb{R}$ *and satisfies*

$$\left. \frac{dP(tG_1 + G_2)}{dt} \right|_{t=0} = \int G_1 \, dm_{G_2}.$$

If G is not cohomologous to a constant then

$$\mathscr{I}_{\sigma^f}(G) := \left\{ \int G \, dm : m \in \mathscr{M}_{\sigma}^f \right\}$$

is a non-trivial closed interval. Furthermore,

$$\left\{ \int G \, dm_{tG} : t \in \mathbb{R} \right\} = \mathrm{int}(\mathscr{I}_{\sigma^f}(G)).$$

We use the following large deviation type result.

Lemma 12.4.3 *Let* $f_1, f_2 : \Sigma \to \mathbb{R}$ *be strictly positive Hölder continuous functions such that* $0 \in \mathrm{int}(\mathscr{I}_{\sigma}(f_1 - f_2))$. *Then*

$$\beta(f_1, f_2) :=$$

$$\lim_{\delta \to 0} \limsup_{T \to \infty} \frac{1}{T} \log \left(\sum_{n=1}^{\infty} \# \left\{ \sigma^n x = x : f_1^n(x) \leq T \text{ and } \frac{f_2^n(x)}{f_1^n(x)} \in (1 - \delta, 1 + \delta) \right\} \right)$$

satisfies

$$\beta(f_1, f_2) = \sup \left\{ \frac{h(\mu)}{\int f_1 \, d\mu} : \mu \in \mathscr{M}_{\sigma}, \int f_1 \, d\mu = \int f_2 \, d\mu \right\}.$$

In particular, $0 < \beta(f_1, f_2) \leq h := h(\sigma^{f_1})$ *and* $\beta(f_1, f_2) = h$ *if and only if* $\int f_1 \, d\mu_{-hf_1} = \int f_2 \, d\mu_{-hf_1}$, *where* μ_{-hf_1} *is the equilibrium state for* $-hf_1$. $\qquad \square$

Proof We apply results about periodic orbits for hyperbolic flows, which also apply to suspended semiflows over subshifts of finite type. We have that

$$\sum_{n=1}^{\infty} \# \left\{ \sigma^n x = x : f_1^n(x) \le T \text{ and } \frac{f_2^n(x)}{f_1^n(x)} \in (1 - \delta, 1 + \delta) \right\}$$

$$= \# \left\{ \tau : l(\tau) \le T \text{ and } \int F \, dm_\tau \in (1 - \delta, 1 + \delta) \right\},$$

where τ denotes a periodic orbit of the suspended semi-flow σ^{f_1} with least period $l(\tau)$, m_τ is the corresponding orbital measure (of total mass $l(\tau)$) and $F : \Sigma^{f_1} \to \mathbb{R}$ satisfies $\int F \, dm_\tau = f_2^n(x)$. (The function F may be constructed as follows. Choose a smooth function $\kappa : [0, 1] \to \mathbb{R}^+$ such that $\kappa(0) = \kappa(1) = 0$ and $\int_0^1 \kappa(s) \, ds = 1$. Then set $F(x, s) = (f_2(x)/f_1(x))\kappa(s/f_1(x))$.) Using Kifer's large deviations results for hyperbolic flows [13], we have

$$\lim_{T \to \infty} \frac{1}{T} \log \# \left\{ \tau : l(\tau) \le T \text{ and } \frac{1}{l(\tau)} \int F \, dm_\tau \in (1 - \delta, 1 + \delta) \right\}$$

$$= \sup \left\{ h(m) : m \in \mathscr{M}_{\sigma^{f_1}} \text{ and } \int F \, dm \in (1 - \delta, 1 + \delta) \right\}$$

$$= \sup \left\{ \frac{h(\mu)}{\int f_1 \, d\mu} : \mu \in \mathscr{M}_\sigma \text{ and } \frac{\int f_2 \, d\mu}{\int f_1 \, d\mu} \in (1 - \delta, 1 + \delta) \right\}$$

$$= \sup_{\xi \in (1 - \delta, 1 + \delta)} H(\xi),$$

where

$$H(\xi) = \sup \left\{ \frac{h(\mu)}{\int f_1 \, d\mu} : \mu \in \mathscr{M}_\sigma \text{ and } \frac{\int f_2 \, d\mu}{\int f_1 \, d\mu} = \xi \right\}.$$

Since $H(\xi)$ is analytic, letting $\delta \to 0$ gives the required formula for $\beta(f_1, f_2)$. (The analyticity of H follows from the fact that $-H$ is the Legendre transform of the pressure function $t \mapsto P(tF)$.) That $\beta(f_1, f_2) \le h$ is immediate and $\beta(f_1, f_2) > 0$ follows from $0 \in \text{int}(\mathscr{I}_\sigma(f_1 - f_2))$, since $\int f_2 \, d\mu / \int f_1 \, d\mu = 1$ is equivalent to $\int f_1 - f_2 \, d\mu = 0$. If $\int f_1 \, d\mu_{-hf_1} = \int f_2 \, d\mu_{-hf_1}$ then it is clear that $\beta(f_1, f_2) = h$. On the other hand, if

$$h = \beta(f_1, f_2) = \frac{h_\sigma(\mu)}{\int f_1 \, d\mu},$$

for some $\mu \in \mathcal{M}_\sigma$, then

$$h_\sigma(\mu) - h \int f_1 \, d\mu = 0 = P(-hf_1)$$

so uniqueness of equilibrium states gives $\mu = \mu_{-hf_1}$. This completes the proof. $\square$

12.5 Analyticity of the Metric and the Entropy

In this section we will establish the analyticity of the metric and the entropy. We will do this by considering certain complex functions, which provides a fairly direct proof avoiding the use of Lemma 12.3.3. We want to establish analyticity of the intersection form, normalised intersection form and metric by using the analytic function $\eta(s, R_0, R_\lambda)$ defined below, where R_λ depends analytically on λ.

We begin by establishing the analyticity of individual weights $d_{R_\lambda}([g])$ as functions of λ.

Lemma 12.5.1 *For each* $[g] \in \mathscr{C}(\Gamma)$, *the weight* $d_{R_\lambda}([g]) \in \mathbb{R}$ *has a real analytic dependence on* $\lambda \in (-\epsilon, \epsilon)$. *Moreover, we can choose an open neighbourhood* $(-\epsilon, \epsilon) \subset U \subset \mathbb{C}$ *so that we have an analytic extension* $U \ni \lambda \mapsto d_{R_\lambda}([g]) \in \mathbb{C}$ *for each* $[g] \in \mathscr{C}(\Gamma)$. $\square$

Proof We need only modify the approach in Proposition 1.1 of [12]. For each generator $g_0 \in \Gamma_0$ we can consider the image $X_{g_0} \subset \mathbb{R}P^{d-1}$ of the corresponding 1-cylinder $[x_0] \subset \Sigma$, say. In particular X_{g_0} is a compact set in $\mathbb{R}P^{d-1}$. Since $\mathbb{R}P^{d-1}$ is a real analytic manifold it has a (local) complexification and we can then choose a (small) neighbourhood $U_{g_0} \supset X_{g_0}$ in this complexification of $\mathbb{R}P^{d-1}$. We will still denote by $R_\lambda(g_0)^{-1}$ the unique extension of the action of $R(g_0)^{-1}$ to the neighbourhood $U_{g_0} \supset X_{g_0}$. Providing the neighbourhoods U_{g_0} are sufficiently small we have by continuity of the extension $R_\lambda(g_0)^{-1}$ that $R_\lambda(g_0)^{-1}U_{g_0} \supset \overline{U}_{g_1}$, for $\lambda \in (-\epsilon, \epsilon)$, where $g_1 \in \Gamma_0$ satisfies $R_\lambda(g_0)^{-1}X_{g_0} \supset X_{g_1}$, since we know that the restriction $R_\lambda(g_0)^{-1}|X_{g_0}$ is a contraction. Moreover, by continuity and by choosing U_{g_0} smaller, if necessary, we can assume that the inclusion $R_\lambda(g_0)^{-1}U_{g_0} \supset \overline{U}_{g_1}$ also holds for each g_0 for the complexification of R_λ for λ lying in a suitably small open subset $\mathbb{C} \supset V \supset (-\epsilon, \epsilon)$, say.

The key observation now is that when we extend these inclusions to conjugacy classes of more general elements $g \in \Gamma \setminus \{1_\Gamma\}$ without further reducing the neighbourhood $(-\epsilon, \epsilon) \subset V \subset \mathbb{C}$. More precisely, for each reduced word $g = g_{i_0} \cdots g_{i_{n-1}}$ (where $g_{i_0}, \ldots, g_{i_{n-1}} \in \Gamma_0$) we have from the above construction that $R_\lambda(g)^{-1}U_{g_{i_0}} \supset \overline{U}_{g_{i_{n-1}}}$ for $\lambda \in V$. Moreover, writing $\xi_g^\lambda \in \mathbb{R}P^{d-1}$ for the fixed point for $R_\lambda(g_\lambda)^{-1}$, we see that $V \ni \lambda \mapsto \xi_g^\lambda$ is analytic and $V \ni \lambda \mapsto d_{R_\lambda}([g]) = -\frac{1}{2}\log\det(D_{\xi_g^\lambda}\widehat{R}_\lambda(g)) \in \mathbb{C}$ is analytic as the sum of analytic terms. In particular, these functions are analytic on the region V. $\square$

We now define a complex function using these weights.

Definition 12.5.1 We can associate to the two representations $R_0, R_\lambda \in \mathcal{H}$ a complex function

$$\eta(s, R_0, R_\lambda) = \sum_{[g]} d_{R_\lambda}([g]) e^{-s d_{R_0}([g])}$$

which converges for $\mathrm{Re}(s)$ sufficiently large. □

From now on, we shall write $h(R_0) = h$.

Lemma 12.5.2 *The function $\eta(s, R_0, R_\lambda)$ is analytic for $\mathrm{Re}(s) > h$. Moreover, $s = h$ is a simple pole with residue equal to*

$$\frac{\int r_\lambda \, d\mu_{-hr_0}}{\int r_0 \, d\mu_{-hr_0}},$$

where r_0, r_λ correspond to R_0, R_λ using Lemma 12.3.2. In particular, $\eta(s, R_\lambda, R_0)$ has a simple pole at $h(R_\lambda)$. □

Proof We will write $\mathcal{C}'(\Gamma) \subset \mathcal{C}(\Gamma)$ for the set of primitive conjugacy classes in Γ. We can associate to R_0 and R_λ a zeta function formally defined by

$$\zeta(s, z, R_0, R_\lambda) = \prod_{[g] \in \mathcal{C}'(\Gamma)} \left(1 - e^{-s d_{R_0}([g]) + z d_{R_\lambda}([g])}\right)^{-1}, \quad \text{for } s \in \mathbb{C}, z \in \mathbb{R},$$

which converges for $\mathrm{Re}(s)$ sufficiently large and $|z|$ sufficiently small (depending on s). We can rewrite this in terms of the shift $\sigma : \Sigma \to \Sigma$ and the functions r_0, r_λ as

$$\zeta(s, z, R_0, R_\lambda) = \exp\left(\sum_{n=1}^{\infty} \frac{1}{n} \sum_{\sigma^n x = x} e^{-s r_0^n(x) - z r_\lambda^n(x)}\right).$$

(Here we use the fact that primitive conjugacy classes correspond to prime periodic orbits for the shift map and then there is convergence to an analytic function $P(-\mathrm{Re}(s)r_0 - zr_\lambda) < 0$ [18].) Using the analysis of [18], we see that $\zeta(s, z, R_0, R_\lambda)$ converges for $\mathrm{Re}(s) > h$. Furthermore, for s close to h and z close to zero,

$$\zeta(s, z, R_0, R_\lambda) = \frac{A(s, z)}{1 - e^{P(-sr_0 + zr_\lambda)}},$$

where $A(s, z)$ is non-zero and analytic and $e^{P(-sr_0 + zr_\lambda)}$ is the standard analytic extension of the exponential of the pressure function to complex arguments (obtained via perturbation theory applied to the maximal eigenvalue of the associated transfer operator cf. [18]).

It is easy to show that

$$\eta(s, R_0, R_\lambda) = \left. \frac{\partial}{\partial z} \log \zeta(s, z, R_0, R_\lambda) \right|_{z=0} + \phi(s),$$

where $\phi(s)$ is analytic for $\mathrm{Re}(s) > h/2$, while, for s close to h,

$$\left. \frac{\partial}{\partial z} \log \zeta(s, z, R_0, R_\lambda) \right|_{z=0} = \frac{\partial A(s, z)/\partial z|_{z=0}}{A(s, 0)} + \frac{\left. \frac{\partial P(-sr_0 + zr_\lambda)}{\partial z} \right|_{z=0}}{1 - e^{P(-sr_0)}}$$

$$= \frac{\int r_\lambda \, d\mu_{-hr_0}}{\int r_0 \, d\mu_{-hr_0}} \frac{1}{s - h} + B(s)$$

where $B(s)$ is analytic in a neighbourhood of $s = h$. The final statement follows by reversing the roles of R_0 and R_λ. $\square$

We have the following result (which implies Theorem 12.2)

Lemma 12.5.3 *The function* $(-\epsilon, \epsilon) \ni \lambda \mapsto h(R_\lambda)$ *is real analytic.* $\square$

Proof We note that $\zeta(h(R_\lambda), 0, R_0, R_\lambda) = 0$. By Lemma 5.1, the function $1/\zeta(s, \lambda)$, where $\zeta(s, \lambda) := \zeta(s, 0, R_0, R_\lambda)$, has an analytic dependence on $\lambda \in (-\epsilon, \epsilon)$ for $\mathrm{Re}(s)$ sufficiently large. It follows from [19] that, for each $\lambda \in (-\epsilon, \epsilon)$, $1/\zeta(s, \lambda)$ has an analytic extension to a half plane $\mathrm{Re}(s) > \nu(\lambda)$, where $\nu(\lambda) < h(R_\lambda)$ depends continuously on λ. We can therefore find a common domain $\mathscr{D}$, containing $\bigcup_{-\epsilon < \lambda < \epsilon} \{s \in \mathbb{C} : \mathrm{Re}(s) \geq h(R_\lambda)\}$, such that $1/\zeta(s, \lambda)$ is separately analytic for $s \in \mathscr{D}$ and $\lambda \in (-\epsilon, \epsilon)$. We may then apply Theorem 1 of [25] to conclude that $(s, \lambda) \mapsto 1/\zeta(s, \lambda)$ is real analytic on $\mathscr{D} \times (-\epsilon, \epsilon)$. Finally, we can use the Implicit Function Theorem to show that $\lambda \mapsto h(R_\lambda)$ is real analytic. $\square$

In order to establish further analyticity results, we need to show that the intersection $I(R_0, R_\lambda)$ is equal to the residue of $\eta(s, R_0, R_\lambda)$ at $s = h$. To do this, it will be convenient to use the following technical result.

Lemma 12.5.4 *Let* $R \in \mathscr{H}$. *Then there does not exist* $\alpha > 0$ *such that* $\{d_R([g]) : g \in \Gamma \setminus \{1_\Gamma\}\} \subset \alpha \mathbb{Z}$. $\square$

Proof Let $g, h \in \Gamma \setminus \{1_\Gamma\}$ be two distinct elements of the group. For any $N > 0$ we can consider $g^N, h^N \in \Gamma$. The linear maps on $\mathbb{R}^d$ for the associated matrices $R(g^N), R(h^N) \in \mathrm{SL}(d, \mathbb{R})$ can be written in the form $\lambda(g)^N \pi_g + U_{g^N}$ and $\lambda(h)^N \pi_h + U_{h^N}$, respectively, where $\lambda(g), \lambda(h)$ are the largest simple eigenvalues, $\pi_g, \pi_h : \mathbb{R}^d \to \mathbb{R}^d$ are the eigenprojections onto their one dimensional eigenspaces, $\limsup_{N \to +\infty} \|U_{g^N}\|^{1/N} < \lambda(g)$ and $\limsup_{N \to +\infty} \|U_{h^N}\|^{1/N} < \lambda(h)$.

Let us now consider $g^N h^N \in \Gamma$ and associated matrix $R(g^N h^N)$. The associated linear map will be of the form $\lambda(g^N h^N) \pi_{g^N h^N} + U_{g^N h^N}$ where $\lambda(g^N h^N)$ is the largest simple eigenvalue, $\pi_{g^N h^N} : \mathbb{R}^d \to \mathbb{R}^d$ is the eigenprojection onto their one dimensional eigenspaces, and $\lim \sup_{N \to +\infty} \|U_{g^N h^N}\|^{1/N} < \lambda(g^N h^N)$. However, since we have the identity $R(g^N h^N) = R(g^N) R(g^N)$ for the matrix representations we can also write the corresponding relationship for the linear maps:

$$\lambda(g^N h^N)\pi_{g^N h^N} + U_{g^N h^N} = \left(\lambda(g^N)\pi_{g^N} + U_{g^N}\right)\left(\lambda(h^N)\pi_{h^N} + U_{h^N}\right).$$

$$(12.5.1)$$

In particular, we see that as N becomes larger

$$\lim_{N \to +\infty} \exp\left((d_R([g^N h^N]) - d_R([g^N]) - d_R([h^N]))\right) = \lim_{N \to +\infty} \frac{\lambda(g^N h^N)}{\lambda(g^N)\lambda(h^N)}$$

$$= \langle \pi_h, \pi_g \rangle$$

where $\langle \pi_h, \pi_g \rangle$ is simply the cosine of the angle between the eigenvectors associated to $\lambda(g)$ and $\lambda(h)$, respectively. However, if we assume for a contradiction that the conclusion of the lemma does not hold, then the right hand side of (5.1) must be of the form $e^{n\alpha}$, for some $n \in \mathbb{Z}$. However, the directions for the associated eigenprojections form an infinite set in $\mathbb{R}P^{d-1}$ and have an accumulation point. Thus for suitable choices of g, h we can arrange that $0 < \langle \pi_h, \pi_g \rangle < e^\alpha$, leading to a contradiction. This completes the proof of the lemma.   □

Corollary 12.6 *Apart from the simple pole at $s = h$, $\eta(s, R_0, R_\lambda)$ has an analytic extension to a neighbourhood of* $\mathrm{Re}(s) \geq h$.   □

Proof Given Lemma 12.5.4, it follows from the analysis of [18] that $\zeta(s, z, R_0, R_\lambda)$ has an analytic and non-zero extension to a neighbourhood of each point $s = h + it$, $t \neq 0$, for $|z|$ sufficiently small depending on s. Using again that

$$\eta(s, R_0, R_\lambda) = \frac{\partial}{\partial z} \log \zeta(s, z, R_0, R_\lambda)\Big|_{z=0} + \phi(s),$$

where $\phi(s)$ is analytic for $\mathrm{Re}(s) > h/2$, we obtain the result.   □

We now have the following result which characterises the intersection number of $I(R_0, R_\lambda)$.

Lemma 12.5.5 *We can write*

$$I(R_0, R_\lambda) = \frac{\int r_\lambda \, d\mu_{-hr_0}}{\int r_0 \, d\mu_{-hr_0}}$$

Proof Recall from Lemma 12.5.2 that the right hand side in the statement is the residue of $\eta(s, R_0, R_\lambda)$ at $s = h$. In view of Corollary 12.6, we can apply the Ikehara

Tauberian theorem to $\eta(s, R_0, R_\lambda)$ to deduce that

$$\sum_{d_{R_0}([g]) \leq T} d_{R_\lambda}([g]) \sim \frac{\int r_\lambda \, d\mu_{-hr_0}}{\int r_0 \, d\mu_{-hr_0}} \, e^{hT}, \text{ as } T \to +\infty.$$

Moreover, upon taking $R_\lambda = R_0$, we deduce that

$$\sum_{d_{R_0}([g]) \leq T} d_{R_0}([g]) \sim e^{hT}, \text{ as } T \to +\infty.$$

An elementary argument given in [17] shows that

$$\lim_{T \to +\infty} \frac{\sum_{d_{R_0}([g]) \leq T} \frac{d_{R_\lambda}([g])}{d_{R_0}([g])}}{\sum_{d_{R_0}([g]) \leq T} 1} = \lim_{T \to +\infty} \frac{\sum_{d_{R_0}([g]) \leq T} d_{R_\lambda}([g])}{\sum_{d_{R_0}([g]) \leq T} d_{R_0}([g])},$$

so that

$$I(R_0, R_\lambda) = \frac{\int r_\lambda \, d\mu_{-hr_0}}{\int r_0 \, d\mu_{-hr_0}},$$

as required. $\square$

We can now use the characterisation of $I(R_0, R_\lambda)$ in terms of a complex function to deduce the following.

Lemma 12.5.6 *The function* $(-\epsilon, \epsilon) \to \mathbb{R} : \lambda \mapsto I(R_0, R_\lambda)$ *is real analytic.* $\square$

Proof By Lemma 12.5.1, $\eta(s, R_0, R_\lambda)$ has an analytic dependence on $\lambda \in U$. More precisely, it is a uniformly convergent series with individually analytic terms in $\lambda \in U$ for $\text{Re}(s) > h$ and thus bi-analytic for $\lambda \in U$. Moreover, by Hartogs' Theorem for functions of several complex variables [14], $1/\eta(s, R_0, R_\lambda)$ is bi-analytic for s in a neighbourhood of h and $\lambda \in U$. Thus the residue of $\eta(s, R_0, R_\lambda)$ at $s = h$ is analytic. Thus, using the residue theorem, $I(R_0, R_\lambda)$, which is the residue of $\eta(s, R_0, R_\lambda)$, depends analytically on λ. $\square$

Since $h(R_\lambda)$ and $I(R_0, R_\lambda)$ both depend analytically on λ, we have the following.

Corollary 12.7 *The function* $(-\epsilon, \epsilon) \to \mathbb{R} : \lambda \mapsto J(R_0, R_\lambda)$ *is real analytic.* $\square$

By differentiating twice and using that $\|R^{(1)}\|^2 = \left.\frac{\partial^2 J(R_0, R_\lambda)}{\partial \lambda^2}\right|_{\lambda=0}$ we have the following result.

Corollary 12.8 *The function* $(-\epsilon, \epsilon) \to \mathbb{R} : \lambda \mapsto \|R^{(1)}\|$ *is real analytic.* $\square$

12.6 Proof of Theorem 12.4

The first part of Theorem 12.4 will follow from Lemma 12.4.3 once we formulate things appropriately. Given an analytic family of representations $\lambda \mapsto R_\lambda$, we define strictly positive Hölder continuous functions $r_\lambda : \Sigma \to \mathbb{R}$ as in Sect. 12.3 so that if $\sigma^n x = x$ corresponds to a conjugacy class $[g]$ then $d_{R_\lambda}([g]) = r_\lambda^n(x)$, using Lemma 12.3.2. By Lemma 12.3.3, r_λ depends analytically on λ. We then have $h(\sigma^{r_0}) = h(R_0)$ and $h(\sigma^{r_\lambda}) = h(R_\lambda)$. We now define $f_0 = h(R_0)r_0$ and $f_\lambda = h(R_\lambda)r_\lambda$, so that, in particular, $P(-f_0) = P(-f_\lambda) = 0$. Since periodic point measures are dense in $\mathcal{M}_\sigma$, it is clear that $0 \in \text{int}(\mathcal{I}_\sigma(f_0 - f_\lambda))$ if and only if there exist two conjugacy classes $[g]$ and $[g']$ such that $h(R_0)d_{R_0}([g]) < h(R_\lambda)d_{R_\lambda}([g])$ and $h(R_0)d_{R_0}([g']) > h(R_\lambda)d_{R_\lambda}([g'])$ (since these correspond to measure μ_0, μ_1 supported on two periodic orbits for which $\int f_0 - f_\lambda d\mu_0 < 0 < \int f_0 - f_\lambda d\mu_1$ and $\text{int}(\mathcal{I}_\sigma(f_0 - f_\lambda))$ is convex). We will show that this latter condition holds provided the representations R_0 and R_λ are not equal up to conjugacy.

Lemma 12.6.1 *If R_0 and R_λ are not conjugate then there exist two conjugacy classes $[g]$ and $[g']$ such that we have $h(R_0)d_{R_0}([g]) < h(R_\lambda)d_{R_\lambda}([g])$ and $h(R_0)d_{R_0}([g']) > h(R_\lambda)d_{R_\lambda}([g'])$.* □

Proof We will prove the contrapositive. Without loss of generality, suppose that $h(R_0)d_{R_0}([g]) \leq h(R_1)d_{R_\lambda}([g])$ for all $[g] \in \mathscr{C}(\Gamma)$, i.e. that $(f_0 - f_\lambda)^n(x) \leq 0$ whenever $\sigma^n x = x$. Then $\int (f_0 - f_\lambda) \, d\mu \leq 0$ for every $\mu \in \mathcal{M}_\sigma$.

Now consider the real analytic map $Q : [0, 1] \to \mathbb{R}$ defined by $Q(t) = P(-f_0 + t(f_0 - f_\lambda))$. This has derivative $Q'(t) = \int (f_0 - f_\lambda) \, d\mu_t \leq 0$, where μ_t is the equilibrium state for $-f_0 + t(f_0 - f_\lambda)$. Since $Q(0) = Q(1) = 0$ we deduce that $Q(t) = 0$ for all $t \in [0, 1]$ and then the strict convexity of pressure implies that $f_0 - f_\lambda$ is cohomologous to a constant. Since $P(f_0) = P(-f_\lambda)$, the constant must be zero and so $f_0^n(x) = f_\lambda^n(x)$, whenever $\sigma^n x = x$. This implies that $h(R_0)d_{R_0}([g]) = h(R_\lambda)d_{R_\lambda}([g])$ for all g and hence that $J(R_0, R_\lambda) = 1$. It then follows by Corollary 1.5 of [3] that the representations are equal up to conjugacy. □

Write $h = h(R_0)$. We may now apply Lemma 12.4.3 to show that, for each $\lambda \in (-\epsilon, \epsilon)$, the limit

$$\alpha(\lambda)$$

$$= \lim_{\delta \to 0} \lim_{T \to +\infty} \frac{1}{T} \log \# \left\{ [g] : d_{R_0}([g]) \leq T \text{ and} \right.$$

$$\left. \frac{d_{R_\lambda}([g])}{d_{R_0}([g])} \in \left(\frac{h(R_0)}{h(R_\lambda)} - \delta, \frac{h(R_0)}{h(R_\lambda)} + \delta \right) \right\}$$

$$= \lim_{\delta \to 0} \limsup_{T \to \infty} \frac{1}{T} \log \left(\sum_{n=1}^{\infty} \# \left\{ \sigma^n x = x : \frac{f_0^n(x)}{h} \leq T \text{ and} \right. \right.$$

$$\left. \left. \frac{f_\lambda^n(x)}{f_0^n(x)} \in (1 - \delta, 1 + \delta) \right\} \right)$$

$$= h\beta(f_0, f_\lambda)$$

exists and satisfies $0 < \alpha(\lambda) \le h$. (Here we have used that $h(\sigma^{f_0}) = 1$.) The next result shows that we have a strict inequality when $\lambda \ne 0$.

Lemma 12.6.2 *For $\lambda \in (-\epsilon, \epsilon) \setminus \{0\}$, $\alpha(\lambda) < h$.*   □

Proof By Proposition 12.4.3, we will have $\alpha(\lambda) < h$ unless $\int f_0 \, d\mu_{-f_0} = \int f_\lambda \, d\mu_{-f_0}$. The latter condition may be rewritten as

$$\frac{\int f_\lambda \, d\mu_{-f_0}}{\int f_0 \, d\mu_{-f_0}} = 1 = \frac{h(\sigma^{f_0})}{h(\sigma^{f_\lambda})} = \frac{h_\sigma(\mu_{-f_0})}{\int f_0 \, d\mu_{-f_0}} \frac{\int f_\lambda \, d\mu_{-f_\lambda}}{h_\sigma(\mu_{-f_\lambda})}.$$

Rearranging, this becomes

$$\frac{h_\sigma(\mu_{-f_\lambda})}{\int f_\lambda \, d\mu_{-f_\lambda}} = \frac{h_\sigma(\mu_{-f_0})}{\int f_\lambda \, d\mu_{-f_0}},$$

which, by uniqueness of the measure of maximal entropy for σ^{f_λ}, forces $\mu_{-f_0} = \mu_{-f_\lambda}$. The latter equality implies that $f_0 - f_\lambda$ is cohomologous to a constant and, since $P(-f_0) = P(-f_\lambda)$, the constant is necessarily zero. This means that $h(R_0)d_{R_0}([g]) = h(R_\lambda)d_{R_\lambda}([g])$ for all $[g] \in \mathscr{C}(\Gamma)$, contradicting Lemma 12.6.1.□

We now complete the proof of Theorem 12.4 by establishing the characterisation of the Weil–Petersson metric in terms of the growth rate $\alpha(\lambda)$. It is more convenient to work with $\beta(\lambda) := \beta(f_0, f_\lambda) = \alpha(\lambda)/h$. For $t \in \mathbb{R}$, consider the pressure $P(-tf_0 - f_\lambda)$ and define $\chi_\lambda(t)$ by the equation $P(-tf_0 - \chi_\lambda(t) f_\lambda) = 0$. We trivially have $\chi_0(t) = 1 - t$ but we are interested in the function when $\lambda \ne 0$.

Lemma 12.6.3 *For each $\lambda \in (-\epsilon, \epsilon) \setminus \{0\}$, the function $\chi_\lambda(t)$ is well-defined and real analytic. Furthermore,*

$$\lim_{t \to \pm\infty} \chi_\lambda(t) = \mp\infty.$$

Proof That $\chi_\lambda(t)$ is well-defined and real analytic follows from the Implicit Function Theorem. Suppose $\lim_{t \to +\infty} \chi_\lambda(t) \ne -\infty$. Then there exists a sequence $t_n \to +\infty$ and a constant $A \ge 0$ such that $\chi_\lambda(t_n) \ge -A$ for all n. We have

$$-t_n f_0 - \chi_\lambda(t_n) f_\lambda \le -t_n f_0 + A\|f_\lambda\|_\infty$$

and so

$$0 = P(-t_n f_0 - \chi_\lambda(t_n) f_\lambda) \le P(-t_n f_0 + A\|f_\lambda\|_\infty) = P(-t_n f_0) + A\|f_\lambda\| \to -\infty,$$

as $n \to \infty$, a contradiction. A similar argument show that $\lim_{t \to -\infty} \chi_\lambda(t) = +\infty$.□

We want to show that there is a unique number $0 < t_\lambda < 1$ for which $\chi'_\lambda(t_\lambda) = -1$. To do this, it is convenient to use the following alternative characterisation of $\chi_\lambda(t)$ in terms of the semiflow σ^{f_0}. Let $F_\lambda : \Sigma^{f_0} \to \mathbb{R}$ be a Hölder continuous function such that, for a periodic σ^{f_0}-orbit τ corresponding to a periodic σ-orbit $\sigma^n x = x$, $\int F_\lambda \, dm_\tau = f^n_\lambda(x)$. (We can define F_λ by the same procedure we used to define F in the proof of Lemma 12.4.3.) Here, as above, m_τ is the associated periodic orbit measure of total mass $l(\tau) = f^n_0(x)$. The function F_λ also satisfies $\int F_\lambda \, dm > 0$, for every $m \in \mathcal{M}_{\sigma^{f_0}}$. We then have that $\chi_\lambda(t)$ is defined by

$$P(-\chi_\lambda(t) F_\lambda) = t.$$

It is then easy to calculate that

$$\chi'_\lambda(t) = \frac{-1}{\int F_\lambda \, dm_{-\chi_\lambda(t) F_\lambda}}.$$

In particular, $\chi_\lambda(t)$ is strictly decreasing. By Lemma 12.6.3, χ_λ takes all real values and so

$$\left\{ \int F_\lambda \, dm_{-\chi_\lambda(t) F_\lambda} : t \in \mathbb{R} \right\} = \left\{ \int F_\lambda \, dm_{t F_\lambda} : t \in \mathbb{R} \right\} = \text{int}(\mathcal{I}_{\sigma^{f_0}}(F_\lambda)).$$

However, by Lemma 12.6.1, we can find periodic σ^{f_0}-orbits τ and τ' (corresponding to conjugacy classes $[g]$ and $[g']$) such that

$$\frac{1}{l(\tau)} \int F_\lambda \, dm_\tau > 1 \quad \text{and} \quad \frac{1}{l(\tau')} \int F_\lambda \, dm_{\tau'} < 1.$$

Hence, in particular, for $\lambda \neq 0$, there exists a unique t_λ such that $\chi'_\lambda(t_\lambda) = -1$.

Lemma 12.6.4 *We have* $\beta(\lambda) = t_\lambda + \chi_\lambda(t_\lambda)$. $\qquad\qquad\square$

Proof By Proposition 12.4.3, we have

$$\beta(\lambda) = \sup \left\{ \frac{h_\sigma(\mu)}{\int f_0 \, d\mu} : \mu \in \mathcal{M}_\sigma \text{ and } \int f_0 \, d\mu = \int f_\lambda \, d\mu \right\}.$$

Let ν denote the equilibrium state of $-t_\lambda f_0 - \chi_\lambda(t_\lambda) f_\lambda$. By the definition of t_λ,

$$\frac{\int f_\lambda \, d\nu}{\int f_0 \, d\nu} = \int F_\lambda \, dm_{-\chi_\lambda(t_\lambda) F_\lambda} = 1.$$

Thus

$$0 = P(-t_\lambda f_0 - \chi_\lambda(t_\lambda) f_\lambda) = h_\sigma(\nu) - t_\lambda \int f_0 \, d\nu - \chi_\lambda(t_\lambda) \int f_\lambda \, d\nu$$

$$= h_\sigma(\nu) - (t_\lambda + \chi_\lambda(t_\lambda)) \int f_0 \, d\nu,$$

so that

$$t_\lambda + \chi_\lambda(t_\lambda) = \frac{h_\sigma(\nu)}{\int f_0 \, d\nu}$$

and $\int f_0 \, d\nu = \int f_\lambda \, d\nu$. On the other hand, if $\mu \in \mathcal{M}_\sigma$, $\mu \neq \nu$ satisfies $\int f_0 \, d\mu = \int f_\lambda \, d\mu$ then

$$0 = P(-t_\lambda f_0 - \chi_\lambda(t_\lambda) f_\lambda) > h_\sigma(\mu) - t_\lambda \int f_0 \, d\mu - \chi_\lambda(t_\lambda) \int f_\lambda \, d\mu$$

$$= h_\sigma(\mu) - (t_\lambda + \chi_\lambda(t_\lambda)) \int f_0 \, d\mu,$$

so that

$$t_\lambda + \chi_\lambda(t_\lambda) > \frac{h_\sigma(\nu)}{\int f_0 \, d\nu}.$$

Combining these two observations shows that $t_\lambda + \chi_\lambda(t_\lambda) = \beta(\lambda)$.   □

Since $\lambda \mapsto r_\lambda$ and $\lambda \mapsto h(R_\lambda)$ are analytic, we can write $f_\lambda = f_0 + f_0^{(1)} \lambda + f_0^{(2)} \lambda^2/2 + o(\lambda^2)$. It follows from the definition of the Weil–Petersson metric in terms of $J(R_0, R_\lambda)$ and Lemma 12.5.5 that

$$\|R^{(1)}\|^2 = \frac{\int f_0^{(2)} \, d\mu_{-f_0}}{\int f_0 \, d\mu_{-f_0}}.$$

We may then use the calculation in the proof of Lemma 4.2 of [20] to show that

$$\frac{\partial^2 \chi_\lambda}{\partial \lambda^2}(t) \bigg|_{\lambda=0} = t(t-1)\|R^{(1)}\|^2.$$

The next lemma establishes the final part of Theorem 12.4.

Lemma 12.6.5 *The function* $\alpha : (-\epsilon, \epsilon) \to (0, h(R_0)]$ *satisfies*

$$\|R^{(1)}\|^2 = -4 \frac{\partial^2 \beta(\lambda)}{\partial \lambda^2} \bigg|_{\lambda=0} = -\frac{4}{h(R_0)} \frac{\partial^2 \alpha(\lambda)}{\partial \lambda^2} \bigg|_{\lambda=0}.$$

Proof This follows from the calculations in the proof of Theorem 4.3 of [20], once one replaces the function D_λ there with $\beta(\lambda)$, combined with Lemma 12.6.4.   □

References

1. R. Adler, L. Flatto, Geodesic flows, interval maps, and symbolic dynamics. Bull. Amer. Math. Soc. **25**, 229–334 (1991)
2. S. Bradlow, O. García-Prada, W. Goldman, A. Weinhard, Representations of surface groups: background material for AIM workshop, American Institute of Mathematics, Palo Alto (2007), Available at http://www.math.u-psud.fr/~repsurf/ANR/aim.pdf. Accessed 27 Dec 2015
3. M. Bridgeman, R. Canary, F. Labourie, A. Sambarino, The pressure metric for Anosov representations. Geom. Funct. Anal. **25**, 1089–1179 (2015)
4. M. Culler, Lifting representations to covering groups. Adv. Math. **59**, 64–70 (1986)
5. R. de la Llave, J. Marco, R. Moriyón, Canonical perturbation of Anosov systems and regularity results for the Livšic cohomology equation. Ann. Math. **123**, 537–611 (1986)
6. E. Ghys, P. de la Harpe, *Sur les Groupes Hyperboliques d'après Mikhael Gromov*. Progress in Mathematics (Birkhäuser, Boston, 1990)
7. W. Goldman, Representations of fundamental groups of surfaces, in *Geometry and Topology, Proceedings*, ed. by J. Alexander, J. Harer. Lecture Notes in Mathematics, vol. 1167, University of Maryland 1983–84 (Springer, New York, 1985)
8. W. Goldman, Topological components of spaces of representations. Invent. Math. **93**, 557–607 (1988)
9. N. Hitchin, The self-duality equations on a Riemann surface. Proc. London Math. Soc. **55**, 59–126 (1987)
10. N. Hitchin, Lie groups and Teichmüller space. Topology **31**, 449–473 (1992)
11. H. Huber, Zur analytischen Theorie hyperbolischer Raumformen und Bewegungsgruppen, II. Math. Ann. **142**, 385–398 (1961)
12. A. Katok, G. Knieper, M. Pollicott, H. Weiss, Differentiability and analyticity of topological entropy for Anosov and geodesic flows. Invent. Math. **98**, 581–597 (1989)
13. Y. Kifer, Large deviations, averaging and periodic orbits of dynamical systems. Comm. Math. Phys. **162**, 33–46 (1994)
14. S. Krantz, *Function Theory of Several Complex Variables*, Reprint of the 1992 edition (AMS Chelsea Publishing, Providence, 2001)
15. F. Labourie, Anosov flows, surface groups and curves in projective space. Invent. Math. **165**, 51–114 (2006)
16. C. McMullen, Thermodynamics, dimension and the Weil–Petersson metric. Invent. Math. **173**, 365–425 (2008)
17. W. Parry, Bowen's equidistribution theory and Dirichlet's density theorem. Ergodic Theory Dyn. Syst. **4**, 117–134 (1984)
18. W. Parry, M. Pollicott, Zeta functions and the periodic orbit structure of hyperbolic dynamics. Astérisque **187**(180), 1–268 (1990)
19. M. Pollicott, Meromorphic extensions of generalised zeta functions. Invent. Math. **85**, 147–164 (1986)
20. M. Pollicott, R. Sharp, Weil–Petersson metrics, Manhattan curves and Hausdorff dimension. Math. Z. **282**, 1007–1016 (2016)
21. A. Sambarino, Quantitative properties of convex representations. Comment. Math. Helv. **89**, 443–488 (2014)
22. R. Schwartz, R. Sharp, The correlation of length spectra of two hyperbolic surfaces. Comm. Math. Phys. **153**, 423–430 (1993)

23. C. Series, Geometrical methods of symbolic coding, in *Ergodic Theory, Symbolic Dynamics, and Hyperbolic Spaces (Trieste, 1989)*, ed. by T. Bedford, M. Keane, C. Series, vol. 125–151 (Oxford University Press, Oxford, 1991)
24. R. Sharp, Prime orbit theorems with multi-dimensional constraints for Axiom A flows. Monat. Math. **114**, 261–304 (1992)
25. B. Shiffman, Separate analyticity and Hartogs theorems. Indiana Univ. Math. J. **38**, 943–957 (1989)
26. S. Wolpert, Thurston's Riemannian metric for Teichmüller space. J. Diff. Geom. **23**, 146–174 (1986)

Part V
Fractal Geometry

Chapter 13
Dimension Estimates for C^1 Iterated Function Systems and C^1 Repellers, a Survey

De-Jun Feng and Károly Simon

Abstract In this note we give a survey about some of the results related to fractal dimensions of attractors and ergodic measures of non-linear and non-conformal Iterated Function Systems (IFS) and the repellers of expanding maps on $\mathbb{R}^d$. The only new result in this note is the proof of the fact that Theorem 13.1.1 implies Theorem 13.1.2.

13.1 Introduction

We consider attractors of Iterated Function Systems (IFS) and repellers of expanding maps and we estimate their various fractal dimensions. The most important results this note is focused on are as follows:

Theorem 13.1.1 ([29]) *Let $\mathcal{F} = \{f_i\}_{i=1}^{\ell}$ be a C^1 IFS on a compact subset of $\mathbb{R}^d$. Let $\Sigma := \{1, \ldots, \ell\}^{\mathbb{N}}$ and let $\Pi : \Sigma \to \Lambda$ be the natural projection from the symbolic space to the attractor Λ. We write $\dim_S X$ and $\dim_L \mu$ for the singularity dimension of X and Lyapunov dimension of μ (defined in Sect. 13.7.2.1). Then*

(1) $\overline{\dim}_B \Lambda \leq \dim_S \Sigma$. *More generally, if $X \subset \Sigma$ is a subshift then*

$$\overline{\dim}_B \Pi(X) \leq \dim_S X. \tag{13.1.1}$$

(2) Let $\mu \in \mathcal{E}(\Sigma, \sigma)$. Then

$$\overline{\dim}_P(\mu \circ \Pi^{-1}) \leq \dim_L \mu. \tag{13.1.2}$$

D.-J. Feng
Department of Mathematics, The Chinese University of Hong Kong, Shatin, Hong Kong
e-mail: djfeng@math.cuhk.edu.hk

K. Simon (✉)
Department of Stochastics, Institute of Mathematics and MTA-BME Stochastics Research Group,
Budapest University of Technology and Economics, Budapest, Hungary
e-mail: simonk@math.bme.hu

© The Author(s), under exclusive license to Springer Nature Switzerland AG 2021
M. Pollicott, S. Vaienti (eds.), *Thermodynamic Formalism*, Lecture Notes
in Mathematics 2290, https://doi.org/10.1007/978-3-030-74863-0_13

Theorem 13.1.2 ([29]) *Let* $\Lambda \subset \mathbb{R}^d$ *be a repeller of the expanding* C^1 *mapping* ψ *and let* μ *be an ergodic invariant measure. Let* $\dim_{S*} \Lambda$ *and* $\dim_{L*} \mu$ *be the singularity dimension of* Λ *and the Lyapunov dimension of* μ *respectively (defined Sect. 13.7.2.2). Then*

(1)

$$\overline{\dim}_B \Lambda \leq \dim_{S*} \Lambda. \tag{13.1.3}$$

(2)

$$\overline{\dim}_P \mu \leq \dim_{L*} \mu. \tag{13.1.4}$$

In Part I we give a detailed review of the classical results. In Part II we introduce some tools to handle the neither conformal nor linear attractors. We review earlier results in this field. Finally we point out how part (a) of Theorem 13.1.2 follows from part (a) of Theorem 13.1.1.

Review of Classical Results

13.2   Notation

13.2.1   Definitions of Fractal Dimensions of Sets and Measures

First we recall the definitions of the Hausdorff- box- and packing dimensions. For a detailed discussion about their properties see Falconer's book [11].

Definition 13.2.1

(1) **Hausdorff measure and dimension**. For $t \geq 0$ we define the t-dimensional Hausdorff measure:

$$\mathcal{H}^t(\Lambda) = \lim_{\delta \to 0} \left\{ \inf \left\{ \sum_{k=1}^{\infty} |A_k|^t : \Lambda \subset \bigcup_{i=1}^{\infty} A_k; |A_i| < \delta \right\} \right\}, \tag{13.2.1}$$

where $|A|$ denotes the diameter of the set A. Then

$$\dim_H \Lambda := \inf \left\{ t : \mathcal{H}^t(\Lambda) = 0 \right\} = \sup \left\{ t : \mathcal{H}^t(\Lambda) = \infty \right\}.$$

(2) **Box dimension**. Let $E \subset \mathbb{R}^d$, $E \neq \emptyset$, bounded and let $N_\delta(E)$ be the number of δ-mesh cubes that intersect E. Then the lower and upper box dimensions of E are

$$\underline{\dim}_B(E) := \liminf_{r \to 0} \frac{\log N_\delta(E)}{-\log \delta}, \quad \overline{\dim}_B(E) := \limsup_{r \to 0} \frac{\log N_\delta(E)}{-\log \delta}.$$

$$(13.2.2)$$

If the limit exists then we call it the box dimension of E and we denote it by $\dim_B E$.

(3) **Packing measure and dimension**. For $\delta > 0$ and $E \subset \mathbb{R}^d$ we say that a finite or countable collection of disjoint balls $\{B_i\}_i$ of radii at most δ and with centers in E is a δ-packing of $E \subset \mathbb{R}^d$. Then for any $s \geq 0$ and $\delta > 0$ we define

$$\mathcal{P}_\delta^s(E) := \sup \left\{ \sum_{i=1}^\infty |B_i|^s : \{B_i\} \text{ is a } \delta\text{-packing of E} \right\}.$$

Since $\mathcal{P}_0^s(E) := \lim_{\delta \to 0} \mathcal{P}_\delta^s(E)$ is NOT countably sub-additive, we need one more step to get the s-dimensional packing measure:

$$\mathcal{P}^s(E) := \inf \left\{ \sum_{i=1}^\infty \mathcal{P}_0^s(E_i) : E \subset \bigcup_{i=1}^\infty E_i \right\}.$$

Feng, Hua and Wen [17] proved that for a compact set $E \subset \mathbb{R}^d$, if $\mathcal{P}_0^s(E) < \infty$ then $\mathcal{P}_0^s(E) = \mathcal{P}^s(E)$. The packing dimension of the set E is

$$\dim_P(E) := \inf \left\{ s : \mathcal{P}^s(E) = 0 \right\}$$
$$= \sup \left\{ s : \mathcal{P}^s(E) = \infty \right\}.$$

Alternatively, we can define the packing dimension (see [11]) as

$$\dim_P(E) = \inf \left\{ \sup_i \overline{\dim}_B E_i : E \subset \bigcup_{i=1}^\infty E_i \right\}.$$

It is well known (see [11]) that

$$\dim_H(E) \leq \dim_P(E) \leq \overline{\dim}_B E.$$

Definition 13.2.2 (Lower and Upper Hausdorff and Packing Dimensions of a Measure) Let μ be a Borel probability measure on $\mathbb{R}^d$.

$$\underline{\dim}_H\mu := \inf\{\dim_H A : \mu(A) > 0\}, \quad \overline{\dim}_H\mu := \inf\{\dim_H A : \mu(A^c) = 0\},$$

$$\underline{\dim}_P\mu := \inf\{\dim_P A : \mu(A) > 0\}, \quad \overline{\dim}_P\mu := \inf\{\dim_P A : \mu(A^c) = 0\}.$$

If we want to estimate these dimensions, we often use their equivalent definitions in terms of the local densities of the measure.

Definition 13.2.3 Let μ be a Borel measure on a metric space X. Then the lower and upper local dimensions of μ at $x \in X$ are:

$$\underline{d}(\mu, x) := \liminf_{n \to \infty} \frac{\log \mu(B(x, r))}{\log r} \tag{13.2.3}$$

and

$$\overline{d}(\mu, x) := \limsup_{n \to \infty} \frac{\log \mu(B(x, r))}{\log r}. \tag{13.2.4}$$

We say that the measure μ is exact dimensional if for μ-almost all x the limit $\lim_{r \downarrow 0} \frac{\log \mu(B(x,r))}{\log r}$ exists and equals to a constant. This constant is denoted by $d(\mu)$.

$\square$

Let $A \subset \mathbb{R}^d$ for an integer $d \geq 1$. We write $\mathcal{M}(A)$ for the collection of Borel measures μ

- whose support $\mathrm{spt}(\mu) \subset A$ and
- $\mathrm{spt}(\mu)$ is compact and
- $0 < \mu(A) < \infty$.

For a proof of the following lemma see [20, p. 234]. $\square$

Lemma 13.2.1 *Let $\mu \in \mathcal{M}(\mathbb{R}^d)$. Then*

$$\underline{\dim}_H\mu = \mathrm{essinf}_{x \sim \mu}\underline{d}(\mu, x), \quad \overline{\dim}_H\mu = \mathrm{esssup}_{x \sim \mu}\underline{d}(\mu, x) \tag{13.2.5}$$

$$\underline{\dim}_P\mu = \mathrm{essinf}_{x \sim \mu}\overline{d}(\mu, x), \quad \overline{\dim}_P\mu = \mathrm{esssup}_{x \sim \mu}\overline{d}(\mu, x). \tag{13.2.6}$$

13.2.2 Singular Value Function

To study the dimension theory of non-conformal Iterated Function Systems, Falconer [10] introduced the singular value functions. Let A be a $d \times d$ non-singular matrix. The positive square roots of the eigenvalues of $A^T A$ are the singular values of the matrix A. We number the singular values in decreasing order: $\alpha_1(A) \geq \alpha_2(A) \geq \cdots \geq \alpha_d(A) > 0$. Clearly, $\alpha_1(A) = \|A\|$ and $\alpha_d(A) = \|A^{-1}\|^{-1}$. The singular value function $\phi^t(A)$ is defined for all $t \geq 0$ by

$$\phi^t(A) := \begin{cases} \alpha_1(A) \cdots \alpha_{[t]}(A)\alpha_{[t]+1}^{t-[t]}(A), & t \leq d; \\ \det(A)^{t/d}, & t \geq d. \end{cases} \qquad (13.2.7)$$

Assume that $\|A\| < 1$. Then it is easy to see that $t \mapsto \phi^t(A)$ is a continuous and strictly decreasing function. Moreover, it is sub-multiplicative (see [10, Lemma 2.1]). Namely, for any $d \times d$ matrices A, B we have

$$\phi^s(A \cdot B) \leq \phi^s(A) \cdot \phi^s(B). \qquad (13.2.8)$$

13.3 Iterated Function Systems

First we introduce the most frequently used notation related to the iterated function systems then we mention some tools and results of their dimension theory.

13.3.1 The Basic Notations Related to the IFSs

In general, a finite family $\mathcal{F} = \{f_i : Z \to Z\}_{i=1}^{\ell}$ of strict contractions of a complete metric space Z is called an Iterated Function System (IFS). Hutchinson [21] showed that there exists a unique non-empty compact set Λ satisfying

$$\Lambda = \bigcup_{i=1}^{\ell} f_i(\Lambda) = \bigcup_{\mathbf{i} \in [\ell]^n} f_{\mathbf{i}}(\Lambda). \qquad (13.3.1)$$

We say that Λ is the attractor of the IFS $\mathcal{F}$ since for every $z \in Z$ the limit $\lim_{n \to \infty} f_{i_1 \ldots i_n}(z)$ exists and is contained in Λ for every infinite sequence $(i_1, i_2, \ldots)$ such that $i_k \in [\ell]$ for all k, where we used the shorthand notation

$$[\ell] := \{1, \ldots, \ell\} \quad \text{and} \quad f_{i_1 \ldots i_n} := f_{i_1} \circ \cdots \circ f_{i_n}.$$

The sets $\Lambda_i := \{f_i(\Lambda)\}_{i=1}^{\ell}$ are called cylinder or one-cylinders. In general the n-fold iterations of the elements of $\mathcal{F}$ applied on Λ are the n-cylinders $\{f_{\mathbf{i}}(\Lambda)\}_{\mathbf{i} \in [\ell]^n}$. We say that the cylinders are well separated if either

- the cylinders are disjoint (then we say that the Strong Separation Property (SSP) holds) or
- there exists a non-empty bounded open set V such that

(1) $f_i(V) \subset V$ for all $i \in [\ell]$ and
(2) $f_i(V) \cap f_j(V) = \emptyset$ for all distinct $i, j \in [\ell]$.

In this case we say that the Open Set Condition (OSC) holds.

The points of the attractor Λ are coded by elements of the symbolic space $\mathbf{i} = (i_1, i_2, \dots) \in \Sigma := [\ell]^{\mathbb{N}}$. Namely we frequently use the natural projection

$$\Pi : \Sigma \to \Lambda, \qquad \Pi(\mathbf{i}) := \lim_{n \to \infty} f_{i_1 \dots i_n}(z), \qquad (13.3.2)$$

for an arbitrary $z \in Z$. The natural projection is continuous (actually Hölder continuous) in the usual topology on Σ. This topology is generated by the metric

$$\operatorname{dist}(\mathbf{i}, \mathbf{j}) := \ell^{-|\mathbf{i} \wedge \mathbf{j}|}, \qquad (13.3.3)$$

where $|\mathbf{i} \wedge \mathbf{j}|$ is the length of the commons prefix $\mathbf{i} \wedge \mathbf{j}$ of the distinct $\mathbf{i}, \mathbf{j} \in \Sigma$. We write σ for the left-shift on Σ. For an element $\mathbf{i} = (i_1, i_2, \dots) \in \Sigma$ we write $\mathbf{i}|_n := (i_1, \dots, i_n)$ and for $\omega \in \Sigma_n := [\ell]^n$ we set $[\omega] := \{\mathbf{i} \in \Sigma : \mathbf{i}|_n = \omega\}$. In this note we mostly consider the case when the complete metric space mentioned above is $\mathbb{R}^d$.

Definition 13.3.1 Let $\mathcal{F} = \{f_i\}_{i=1}^{\ell}$ be an IFS. We say that

(1) $\mathcal{F}$ is a self-similar IFS (self-affine IFS) if for all $i \in [\ell]$ the mapping f_i is a contracting similitude (affine mapping), respectively.
(2) $\mathcal{F}$ is a self-conformal IFS on a compact set $Z \subset \mathbb{R}^d$ if there exists a bounded open convex set $V \supset Z$ such that for all $i \in [\ell]$

    a. $f_i(Z) \subset Z$,
    b. f_i extends to an injective conformal mapping $f_i : V \to V$. This means that the differential $f_i'(z) : \mathbb{R}^d \to \mathbb{R}^d$ is a similarity mapping for all $z \in V$,
    c. $\|f_i'\| := \sup_{x \in V} |f_i'(x)| < 1$.
    d. The differentials are Hölder continuous. That is, there exist L, β such that

$$|f_i'(x) - f_i'(y)| \leq L \cdot |x - y|^{\beta}, \quad \text{for all } x, y \in V. \qquad (13.3.4)$$

We remark that (iv) follows from conformality and injectivity if $d \geq 2$.   □

(3) Let $\gamma \geq 1$. We say that $\mathcal{F}$ is a C^{γ} IFS on a compact set $Z \subset \mathbb{R}^d$ if there is an open set $U \supset Z$ such that for every $i \in [\ell] := \{1, \dots, \ell\}$, the mapping f_i extends to a contracting C^{γ} diffeomorphism $f_i : U \to f_i(U)$. In this case we write $\xi(\mathcal{F})$ and $\zeta(\mathcal{F})$ for the minimal and maximal contraction rates on Z. That is

$$0 < \xi(\mathcal{F}) := \min_{z \in Z, i \in [\ell]} \alpha_d(D_z f_i) \leq \max_{w \in Z, j \in [\ell]} \alpha_1(D_w f_j) =: \zeta(\mathcal{F}) < 1. \qquad (13.3.5)$$

13.3.2   The Basics of the Dimension Theory of Self-Conformal IFSs

For a very detailed treatment of the dimension theory of conformal IFS see [20]. First we would like to guess what should be the Hausdorff dimension of the attractor Λ of a self-conformal IFS $\mathcal{F}$ which satisfies the OSC.

The most natural cover of the attractor Λ is the cover by n-cylinders $\{f_\omega(\Lambda)\}_{\omega \in [\ell]^n}$. For this cover the sum that appears in the definition (13.2.1) of the t-dimensional Hausdorff measure (for any $t \geq 0$) is $\sum_{\omega \in [\ell]^n} |f_\omega(\Lambda)|^t$. Since we would like to obtain a heuristic and sensible guess for the Hausdorff dimension of Λ, we assume that this cover is not only the most natural but also the most economic covering system in the sense of minimizing the sum that appears in the definition of the t-dimensional Hausdorff measure. Then we should understand the exponential growth rate (in n, for a fixed t) of the sum $\sum_{\omega \in [\ell]^n} |f_\omega(\Lambda)|^t$. To do so, we recall that for a self-conformal IFS the so-called Bounded Distortion Property (BDP) (see [16, Proposition 20.1]) holds. That is, there exists $C_1 > 1$ such that

$$C_1^{-1} < \frac{|f_\omega(\Lambda)|}{\|f_\omega'\|} < C_1 \quad \text{for all } n \geq 1 \text{ and } \omega \in [\ell]^n. \tag{13.3.6}$$

This implies that the sum of the t-th power of the diameter of the elements of the most natural cover satisfies:

$$\left(C_1^{-1}\right)^t < \frac{\sum_{\omega \in [\ell]^n} |f_\omega(\Lambda)|^t}{\sum_{\omega \in [\ell]^n} \|f_\omega'\|^t} < C_1^t. \tag{13.3.7}$$

So, for a fixed $t \geq 0$ the exponential growth rates (in n) of the sums $\sum_{\omega \in [\ell]^n} |f_\omega(\Lambda)|^t$ and $\sum_{\omega \in [\ell]^n} \|f_\omega'\|^t$ are the same. We call this common exponential growth rate the pressure function $P : [0, \infty) \to \mathbb{R}$,

$$P(t) := \lim_{n \to \infty} \frac{1}{n} \log \sum_{\omega \in [\ell]^n} \|f_\omega'\|^t. \tag{13.3.8}$$

It can be proved that the pressure function is convex and strictly decreasing, $P(0) = \log \ell > 0$ and $P(d) \leq 0$ (since we assumed that the OSC holds). In this way the pressure function $P(\cdot)$ has a unique zero

$$t_0 := P^{-1}(0).$$

It is easy to see that for $t < t_0$ and $t \geq t_0$ the sum $\sum\limits_{\omega \in [\ell]^n} |f_\omega(\Lambda)|^t$ tends to infinity and zero respectively. That is, by the definition of the Hausdorff dimension, the zero of the pressure function t_0, is the best heuristic guess for $\dim_H \Lambda$. In fact it follows form the argument above that $\dim_H \Lambda \leq t_0$ and even $\mathcal{H}^{t_0}(\Lambda) < \infty$ always holds.

Theorem 13.3.1 ([4, 15, 18, 19]) *Let Λ be the attractor of a self-conformal IFS $\mathcal{F}$ and let $P(\cdot)$ be the pressure function defined in (13.3.8). As above we write t_0 for the zero of the pressure function. That is, $t_0 = P^{-1}(0)$.*

(1) *If the OSC holds then $\dim_H \Lambda = t_0$.*
(2) *We have $\mathcal{H}^{t_0}(\Lambda) > 0$ if and only if the OSC holds.*
(3) *$\dim_H \Lambda = \dim_B \Lambda = \dim_P \Lambda$.*

$\square$

Part (a) of the theorem follows from the work of Bowen [4] and Ruelle [15] and part (b) is due to Peres, Rams, Simon and Solomyak [18] and a second proof was given by Lau, Rao and Ye [34]. Part (c) was proved by Falconer [19].

Now we consider the special case when the self-conformal IFS is even self-similar. That is, $\mathcal{F} = \{f_i\}_{i=1}^\ell$ and f_i are similitudes with contraction ratio $r_i < 1$. In this case the sum of the definition of the pressure function is

$$\sum_{\omega \in [\ell]^n} \|f'_\omega\|^t = \left(\sum_{i=1}^\ell r_i^t\right)^n.$$

So, the pressure function in the self-similar case is $P(t) = \log\left(\sum\limits_{i=1}^\ell r_i^t\right)$. Hence the zero of the pressure function is the solution s of the equation

$$\sum_{i=1}^\ell r_i^s = 1. \tag{13.3.9}$$

This s is called the similarity dimension of the self-similar IFS $\mathcal{F}$.

13.3.3 Self-Affine IFSs

Recently there has been a very intense development in the theory of self-affine IFSs. Here we mention only the most basic classical method (which is called Falconer's cutting up ellipses method). This method yields a natural upper bound on the Hausdorff dimension of the self-affine attractor. As a result of Bárány, Hochman, Rapaport and Hochman, Rapaport, it turned out that at least on the plane, this upper bound is actually the Hausdorff dimension of the self-affine set under some mild conditions. See [31] and [32]. Let A_i be $d \times d$ non-singular matrices with $\|A_i\| < 1$ and $t_i \in \mathbb{R}^d$ for $i = 1, \ldots, m$. Then the following IFS is self-affine:

$$\mathcal{F} := \{f_i(x) = A_i \cdot x + t_i\}_{i=1}^m . \tag{13.3.10}$$

We want to estimate the dimension of the attractor Λ of the IFS $\mathcal{F}$.

For simplicity we assume here that $f_i([0, 1]^d) \subset [0, 1]^d$ for all i. After n iterations we have m^n (not necessarily different) cylinders:

$$\left\{ f_{i_1} \circ \cdots \circ f_{i_n} ([0, 1]^d) \right\}_{i_1 \ldots i_n \in \{1 \ldots m\}^n} .$$

It is difficult to understand their relative positions, in the general case. So, in general, we cover each cylinders individually. In our case a cylinder is a parallelepiped $f_{i_1} \circ \cdots \circ f_{i_n} ([0, 1]^d)$. Falconer [10] introduced the most natural covering system for these cylinders. For simplicity assume that $d = 3$ and for a moment we also assume that $f_{i_1} \circ \cdots \circ f_{i_n} ([0, 1]^d)$ are boxes like in Fig. 13.1.

In this special case the figure shows that there are potentially three different natural ways to cover the cylinder with cubes: we can cover by one cube of the longest side (α_1) or by $\frac{\alpha_1}{\alpha_2}$ cubes of side α_2 or by $\frac{\alpha_1}{\alpha_3} \cdot \frac{\alpha_1}{\alpha_3}$ cubes of side α_3.

The contribution of the cylinder $f_{i_1} \circ \cdots \circ f_{i_n} ([0, 1]^d)$ to the covering sum in the definition of the Hausdorff dimension is

$$\phi^t(i_1, \ldots, i_n) := \min_i \alpha_1 \cdots \alpha_i \alpha_{i+1}^{t-i} = \alpha_1 \cdots \alpha_{[t]} \alpha_{[t]+1}^{t-[t]} = \frac{\alpha_1 \cdots \alpha_{[t]}}{\alpha_{[t]+1}^{[t]}} \alpha_{[t]+1}^t ,$$

$$\tag{13.3.11}$$

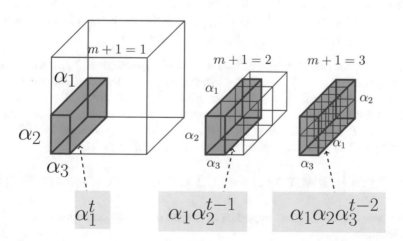

Fig. 13.1 The most economic covering systems

where $\alpha_k = \alpha_k(A_{i_1} \cdots A_{i_n})$ is the k-th largest singular value of the matrix $A_{i_1} \cdots A_{i_n}$. This is why Falconer defined the singular value function by the formula (13.2.7).

Let $n \geq 1$ and $\boldsymbol{\omega} = (\omega_1, \ldots, \omega_n) \in [\ell]^n$. We consider the matrix $A_{\boldsymbol{\omega}} := A_{\omega_1} \cdots A_{\omega_n}$. Like in the conformal case the best guess for the Hausdorff dimension of the attractor is the zero of the sub-additive pressure function $P : [0, \infty) \to \mathbb{R}$

$$P(t) := \lim_{n \to \infty} \frac{1}{n} \log \left(\sum_{|\boldsymbol{\omega}|=n} \phi^t(A_{\boldsymbol{\omega}}) \right). \tag{13.3.12}$$

As in the conformal case, the function $P(t)$ is strictly decreasing and continuous (see [22]). It has a unique zero $P^{-1}(0)$ which is called the affinity dimension of the attractor. Similarly, as in the conformal case we have $\dim_H \Lambda \leq P^{-1}(0)$. However, in this non-conformal situation the box and Hausdorff dimensions of the attractor can be different.

13.4 Some Elements of Thermodynamical Formalism

Assume that either both $\mathcal{U} = \{U_i\}_{i \in I}$ and $\mathcal{V} = \{V_j\}_{j \in J}$ are covers of a set X (that is $X = \bigcup_{i \in I} U_i = \bigcup_{j \in J} V_j$) or both $\mathcal{U}$ and $\mathcal{V}$ are partitions of X.

- We say that $\mathcal{V}$ is finer than $\mathcal{U}$, $(\mathcal{U} \prec \mathcal{V})$ if every element of $\mathcal{V}$ is contained in an element of $\mathcal{U}$.
- The joint refinement $\mathcal{U} \vee \mathcal{V}$: If $\mathcal{U}$ and $\mathcal{V}$ are both covers of X then $\mathcal{U} \vee \mathcal{V}$ is the cover of X the sets by $\{U_i \cap V_j\}_{i \in I, j \in J}$. If $\mathcal{U}$ and $\mathcal{V}$ are both partitions of X then $\mathcal{U} \vee \mathcal{V}$ is a partition of X with classes $\{U_i \cap V_j\}_{i \in I, j \in J}$. $\qquad\square$

In this and in the following subsections we always assume that (X, ρ) is a compact metric space and $T : X \to X$ is a continuous transformation and we say that (X, T) is a topological dynamical system. We write $\mathcal{B}(X)$ for the Borel σ-algebra of X and $\mathcal{M}(X, T)$ for the set of all T-invariant Borel probability measures. That is $\mu(H) = \mu(T^{-1}H)$ for all $H \in \mathcal{B}(X)$ if μ is an invariant measure. Moreover, we denote by $\mathcal{E}(X, T)$ the set of invariant and ergodic measures. That is

$$\mathcal{E}(X, T) := \left\{ \mu \in \mathcal{M}(X, T) : (A \in \mathcal{B}(X) \& T^{-1}(A) = A) \Longrightarrow \mu(A) \in \{0, 1\} \right\}.$$

In this note one of the most important examples of topological dynamical systems is as follows:

Example 13.1 (Subshift) Let $\ell \geq 2$ and $\Sigma := [\ell]^{\mathbb{N}}$ endowed with the metric $\text{dist}(\mathbf{i}, \mathbf{j}) := \ell^{-|\mathbf{i} \wedge \mathbf{j}|}$ and $\sigma : \Sigma \to \Sigma$, $\sigma(i_1, i_2, \ldots) := (i_2, i_3, \ldots)$. Let $X \subset \Sigma$

be compact and $\sigma X \subset X$. Clearly, (X, σ) is a topological dynamical system which is called subshift. In this case for $n \geq 1$ we set

$$X_n^* := \{ \mathbf{i}|_n : \mathbf{i} \in X \}. \tag{13.4.1}$$

In particular, if there is a matrix $A = (a(i, j))_{i,j=1}^{\ell}$ such that for every $\mathbf{i} = (i_1, i_2, \dots) \in \Sigma$ we have

$$\mathbf{i} \in X \text{ if and anly if } a(i_k, i_{k+1}) = 1 \text{ holds for all } k \geq 1,$$

then we say that X is a subshift of finite type or topological Markov chain and we write $X = \Sigma_A$ and

$$\Sigma_{A,n} := \left\{ \omega \in \{1, \dots, \ell\}^n : \exists \mathbf{i} \in \Sigma_A, \ \mathbf{i}|_n = \omega \right\}. \tag{13.4.2}$$

13.4.1 Measure Theoretical and Topological Entropy

Definition 13.4.1 Let (X, T) be a topological dynamical system.

(1) The measure theoretic entropy of T with respect to $\mu \in \mathcal{M}(X, T)$ is defined by

$$h_\mu(T) := \sup_{\mathcal{A}} h_\mu(T, \mathcal{A}) \quad \text{where} \quad h_\mu(T, \mathcal{A}) = \lim_{n \to \infty} \frac{1}{n} H_\mu \left(\bigvee_{i=0}^{n-1} T^{-i} \mathcal{A} \right),$$

and the supremum above is taken over all finite partitions $\mathcal{A}$ (which consists of Borel sets) of X, and the entropy of a partition $\mathcal{U} = \{U_1, \dots, U_n\}$ is defined by

$$H_\mu(\mathcal{U}) := - \sum_{k=1}^{n} \mu(U_k) \log \mu(U_k).$$

(2) Let β be an open cover of X. By compactness of X we can select a finite subcover of β. The number of sets in such a minimal subcover is denoted by $N(\beta)$. We define the entropy of β by $H(\beta) := \log N(\beta)$. The topological entropy of T is

$$h_{\text{top}}(T) := \sup_{\alpha} \lim_{n \to \infty} \frac{1}{n} H \left(\bigvee_{i=0}^{n-1} T^{-i} \alpha \right),$$

where α ranges over all finite open covers of X. $\qquad\Box$

One can find very nice and detailed treatments of the measure theoretical and the topological entropies in the books [13, 20] and [25]. We just mention here four important properties. Their proofs can be found in [20, Theorem 3.4.1, Theorem 3.5.6] and [35] respectively.

Theorem 13.4.1 *Let (X, T) be a topological dynamical system. Then we have*

(1) *for $k \geq 1$ and $\mu \in M(X, T)$ we have $h_\mu(T^k) = kh_\mu(T)$ and $h_{\text{top}}(T^k) = kh_{\text{top}}(T)$.*

(2) *The Variational Principle holds:*

$$h_{\text{top}}(T) = \sup \left\{ h_\mu(T) : \mu \in M(X, T) \right\}.$$

(3) *If T is expansive (like in the case of the subshifts, see Definition 13.5.1) then the function $M(X, T) \ni \mu \mapsto h_\mu(T)$ is upper semi-continuous.*

(4) *Let (X_i, T_i) be topological dynamical systems for $i = 1, 2$. Suppose $\pi : X_1 \to X_2$ is a continuous surjection such that the following diagram commutes:*

$$
\begin{array}{ccc}
X_1 & \xrightarrow{T_1} & X_1 \\
\pi \downarrow & & \downarrow \pi \\
X_2 & \xrightarrow{T_2} & X_2
\end{array}
$$

   Then

   a. $\pi_* : M(X_1, T_1) \to M(X_2, T_2)$ *(defined by $\mu \mapsto \mu \circ \pi^{-1}$) is surjective.*
   b. If $\sup\limits_{y \in X_2} \#\pi^{-1}(y) < \infty$ then

$$h_\mu(T_1) = h_{\mu \circ \pi^{-1}}(T_2)$$

   for each $\mu \in M(X_1, T_1)$.   □

13.4.2 Topological Pressure

The topological pressure was introduced by Ruelle [23] and studied in the general case by Walters [24]. However, below we follow Przytycki, Urbanski's book [20, Section 3]. Let $C(X, \mathbb{R})$ be the space of the real valued continuous functions on the compact metric space (X, ρ) and let $T : X \to X$ be a continuous transformation. For a $\phi \in C(X, \mathbb{R})$ we define the topological pressure $P(T, \phi)$ below. To do so, first we consider covers of the compact metric space (X, ρ).

Definition 13.4.2 Let $\mathcal{U}$ be a finite open cover of the compact metric space (X, ρ) and let $\phi \in C(X, \mathbb{R})$. For every $n \geq 1$, $x \in X$ and for a set $Y \subset X$ we write

$$S_n\phi(x) := \sum_{k=0}^{n-1} \phi \circ T^k(x) \quad \text{and} \quad S_n\phi(Y) := \sup \left\{ \sum_{k=0}^{n-1} \phi \circ T^k(x) : x \in Y \right\}.$$

$$\tag{13.4.3}$$

Then for every $n \geq 1$ we write

$$\mathcal{U}^n := \mathcal{U} \vee T^{-1}\mathcal{U} \vee \cdots \vee T^{1-n}\mathcal{U}. \tag{13.4.4}$$

Finally, we define the partition function

$$Z_n(T, \phi, \mathcal{U}) := \inf \left\{ \sum_{V \in \mathcal{V}} \exp S_n\phi(V) : \mathcal{V} \text{ is a subcover of } \mathcal{U}^n \right\}. \tag{13.4.5}$$

Then the following limit exists (see [20, Lemma 3.2.1])

$$P(T, \phi, \mathcal{U}) := \lim_{n \to \infty} \frac{1}{n} \log Z_n(T, \phi, \mathcal{U}) \tag{13.4.6}$$

and $P(T, \phi, \mathcal{U}) \geq -\|\phi\|_\infty$.

Definition 13.4.3 (Topological Pressure) Let $\{\mathcal{U}_n\}_{n=1}^\infty$ be a sequence of open finite covers of the compact metric space (X, ρ) satisfying $\lim_{n \to \infty} \mathrm{diam}(\mathcal{U}_n) = 0$, where $\mathrm{diam}(\mathcal{U}_n) := \max\{|U| : U \in \mathcal{U}_n\}$. Then the following limit exists (see [20, Lemma 3.2.4]) and we call it `topological pressure`

$$P(T, \phi) := \lim_{n \to \infty} P(T, \phi, \mathcal{U}_n). \tag{13.4.7}$$

The topological pressure does not depend on which equivalent metric we choose. An alternative definition of the topological pressure is as follows:

Definition 13.4.4 We say that $E \subset X$ is an (n, ε)-`separated set` if for every distinct $x, y \in E$ we have

$$\rho_n(x, y) := \max_{0 \leq i \leq n-1} \rho\left(T^i(x), T^i(y)\right) \geq \varepsilon. \tag{13.4.8}$$

Then [20, Theorem 3.3.2] asserts that

Theorem 13.4.2 Let (X, T) be a topological dynamical system and let $\phi \in C(X, \mathbb{R})$. Let $E_n(\varepsilon)$ be an arbitrary (n, ε)-separated set in X for every $\varepsilon > 0$ and $n \geq 1$. Then

$$P(T, \phi) = \lim_{\varepsilon \to 0} \limsup_{n \to \infty} \frac{1}{n} \log \sum_{x \in E_n(\varepsilon)} \exp S_n\phi(x)$$

$$= \lim_{\varepsilon \to 0} \liminf_{n \to \infty} \frac{1}{n} \log \sum_{x \in E_n(\varepsilon)} \exp S_n\phi(x) \tag{13.4.9}$$

For various important properties of the topological pressure see [20, Section 3] and [13, Section 9].

Lemma 13.4.1 ([13, Theorem 9.7]) *Let (X, T) be a topological dynamical system, $f, g \in C(X, \mathbb{R})$. Then we have*

(1) $P(T, 0) = h_{\text{top}}(T)$.
(2) $f \leq g$ *implies* $P(T, f) \leq P(T, g)$.
(3) $P(T, \cdot)$ *is either finite valued or constantly* ∞.
(4) *If* $P(T, \cdot) < \infty$ *then* $|P(T, f) - P(T, g)| \leq \|f - g\|$.
(5) *If* $P(T, \cdot) < \infty$ *then* $P(T, \cdot)$ *is convex.*
(6) $P(T, f + g) \leq P(T, f) + P(T, g)$.
(7) $P(T, c \cdot f) \leq c P(T, f)$ *if* $c \geq 1$ *and* $P(T, c \cdot f) \geq c P(T, f)$ *if* $c \leq 1$.   □

Finally, we state the very important Variational Principle for the topological pressure:

Theorem 13.4.3 *Let (X, T) be a topological dynamical system and $\phi \in C(X, \mathbb{R})$. Then*

$$P(T, \phi) = \sup \left\{ h_\mu(T) + \int \phi d\mu : \mu \in M(X, T) \right\}. \tag{13.4.10}$$

The measures for which the supremum in (13.4.10) is attained are called `equilibrium states for the transformation` T `and the function` ϕ.

Theorem 13.4.4 ([20, Theorem 3.5.6]) *Let (X, T) be a topological dynamical system and $\phi \in C(X, \mathbb{R})$. If the function $M(X, T) \ni \mu \mapsto h_\mu(T)$ is upper semi-continuous (this holds for expansive transformations, in particular for subshifts) then there exists an equilibrium state.*   □

13.5   General Distance-Expanding Open Mappings on a Compact Metric Space

Expanding mappings on a repeller are special cases of distance-expanding open mappings on a compact metric space. So, we start with the more general theory first. Here we follow the book [20, Sections 4,5].

Definition 13.5.1 Let (X, ρ) be a compact metric space and let $T : X \to X$ be a continuous transformation. We say that (X, T) is a `topological dynamical system`. Moreover, we say that (X, T) is

(1) `distance-expanding` if there exists $\eta > 0$ such that

$$\rho(x, y) < 2\eta \implies \rho(T(x), T(y)) \geq \widetilde{\lambda}\rho(x, y), \forall x, y \in \Lambda, \tag{13.5.1}$$

for some $\widetilde{\lambda} > 1$.

(2) `expansive` (see [20, Section 3.5]) if

$$\exists \delta > 0 \text{ such that } \left(\rho \left(T^n(x), T^n(y) \right) \leq \delta, \ \forall n \geq 0 \right) \Longrightarrow x = y \qquad (13.5.2)$$

Such a $\delta > 0$ is called the `expansive constant` of T. $\qquad\qquad\square$

If X is a subshift then the left-shift σ is an expansive transformation with expansive constant $\delta = \ell^{-1}$.

All distance-expanding transformations are expansive (see [20, Theorem 4.1.1]). On the other hand, expansive mappings are distance-expanding in some compatible metric (see [20, Section 4.6]). According to [20, Proposition 3.5.8], in the expansive case we have

Lemma 13.5.1 *If (X, T) is expansive with expansive constant δ and* $\mathrm{diam}(\mathcal{U}) \leq \delta$ *then*

$$P(T, \phi) = P(T, \phi, \mathcal{U}). \qquad (13.5.3)$$

Using this, the topological pressure in the case of subshifts can be presented in the following simpler form:

Example 13.2 Let $X \subset \Sigma := [\ell]^{\mathbb{N}}$ be a subshift (see Example 13.1) and $\phi \in C(X, \mathbb{R})$. Then we have

$$P(\sigma, \phi) = \lim_{n \to \infty} \frac{1}{n} \log \sum_{\omega \in X_n^*} \exp \sup_{\mathbf{i} \in [\omega] \cap X} S_n \phi(\mathbf{i}), \qquad (13.5.4)$$

where X_n^* was defined in (13.4.1). This implies that the pressure function $P(t)$ in (13.3.8) satisfies

$$P(t) = P(\sigma, t\varphi(\mathbf{i})) \quad \text{for} \quad \varphi(\mathbf{i}) := \log \| f_{i_1}' \left(\Pi(\sigma \mathbf{i}) \right) \|.$$

Let (X, T) be a distance-expanding topological dynamical system as in Definition 13.5.1. Then according to [20, Lemma 4.1.2] there exists $\xi > 0$ such that

$$T(B(x, \eta)) \supset B(T(x), \xi), \quad \forall x \in X, \qquad (13.5.5)$$

where η was defined in (13.5.1). Hence we obtain that the definitions below make sense.

Definition 13.5.2 (Local Inverses) Let (X, T) be a distance-expanding, open topological dynamical system. For every $x \in X$, we can define the local inverse $T_x^{-1} :$ $B(T(x), \xi) \to B(x, \eta)$ of T as $T^{-1}|_{B(T(x), \xi)}$. Moreover, for every $x \in X$, $n \geq 1$ and $j \in \{0, 1, \ldots, n - 1\}$ we write $x_j := T^j(x)$. Then by Przytycki and Urbanski

[20, Lemma 4.1.4] the composition $T_{x_0}^{-1} \circ T_{x_1}^{-1} \circ \ldots \circ T_{x_{n-1}}^{-1} : B\left(T^n(x), \xi\right) \to X$ is well defined and we set

$$T_x^{-n} := T_{x_0}^{-1} \circ T_{x_1}^{-1} \circ \ldots \circ T_{x_{n-1}}^{-1}. \tag{13.5.6}$$

We have

$$T^{-n}(A) = \bigcup_{x \in T^{-n}(y)} T_x^{-n}(A), \quad \forall y \in X \text{ and } A \subset B(y, \xi). \tag{13.5.7}$$

13.5.1 Markov Partition for Distance Expanding Maps

As in [20, Definition 4.5.1] we set

Definition 13.5.3 (Markov Partition) Let (X, T) be a distance-expanding open mapping on a compact space X as in part (i) of Definition 13.5.1. A finite cover $\mathcal{R} = \{R_1, \ldots, R_M\}$ of X is said to be a Markov partition of the space X for the mapping T if $\mathrm{diam}(\mathcal{R}) < \min\{\eta, \xi\}$ and the following conditions are satisfied.

(1) $R_i = \overline{\mathrm{Int}\, R_i}$ for all $i = 1, 2, \ldots, M$.
(2) $\mathrm{Int}\, R_i \cap \mathrm{Int}\, R_j = \emptyset$ for all $i \neq j$.
(3) $\mathrm{Int}\, R_j \cap T(\mathrm{Int}\, R_i) \neq \emptyset \Longrightarrow R_j \subset T(R_i)$ for all $i, j = 1, 2, \ldots, M$. $\qquad \square$

The following theorem will be essential for us:

Theorem 13.5.1 ([20, Theorem 4.5.2]) *Let (X, T) be as in Definition 13.5.3. Then there is a Markov partition for X of arbitrarily small diameters.* $\qquad \square$

Every Markov partition $\mathcal{R}$ generates a natural coding of the elements of X. Namely,

Definition 13.5.4 Let (X, T) and the Markov partition $\mathcal{R}$ be as in Definition 13.5.3.

(1) We say that the $M \times M$, 0, 1-matrix $A^{\mathcal{R}} = \left(a_{i,j}^{\mathcal{R}}\right)_{i,j=1}^{M}$ is the transition matrix associated to the Markov partition $\mathcal{R}$ if

$$a_{ij}^{\mathcal{R}} = \begin{cases} 1 & \text{if } \mathrm{Int}\, T(R_i) \cap \mathrm{Int}\, R_j \neq \emptyset \\ 0 & \text{if } \mathrm{Int}\, T(R_i) \cap \mathrm{Int}\, R_j = \emptyset \end{cases}. \tag{13.5.8}$$

We consider the topological Markov chain $(\Sigma_{A^{\mathcal{R}}}, \sigma)$, where σ is the left shift on

$$\Sigma_{A^{\mathcal{R}}} := \left\{ \mathbf{i} = (i_0, i_1, i_2, \ldots) \in \{1, \ldots, M\}^{\mathbb{N}} : a_{i_k, i_{k+1}}^{\mathcal{R}} = 1, \ \forall k \geq 0 \right\}.$$

(2) A sequence $\boldsymbol{\omega} = (\omega_0, \ldots, \omega_{n-1}) \in \{1, \ldots, M\}^n$ is an admissible sequence if $a^{\mathcal{R}}_{\omega_k, \omega_{k+1}} = 1$ for all $0 \leq k < n$. The collection of all such sequences is denoted by $\Sigma^n_{A^{\mathcal{R}}}$.

(3) Let $\boldsymbol{\omega} \in \Sigma^n_{A^{\mathcal{R}}}$. Then the corresponding n-cylinder is defined by

$$R_{\boldsymbol{\omega}} := \left\{ x \in X : T^k(x) \in R_{\omega_k}, \text{ for all } 0 \leq k < n \right\}. \tag{13.5.9}$$

(4) We define the natural projection $\Pi : \Sigma_{A^{\mathcal{R}}} \to X$ by

$$\Pi(\mathbf{i}) := \bigcap_{k=0}^{\infty} T^{-k}(R_{i_k}) = \bigcap_{k=1}^{\infty} R_{\mathbf{i}|_k}. \tag{13.5.10}$$

Recall that a function $\phi : X \to \mathbb{R}$ is called Hölder continuous with exponent $\alpha \in (0, 1]$ if there exists a constant $C > 0$ such that $|\phi(x) - \phi(y)| \leq C\rho(x, y)^{\alpha}$.

Then it was proved in [20, Section 4] that

Lemma 13.5.2 *Let (X, T) and the Markov partition $\mathcal{R}$ be as in Definition 13.5.3. Then $\Pi : \Sigma_{A^{\mathcal{R}}} \to X$ is well defined and onto. Moreover,*

(1) *the following diagram is commutative:*

$$
\begin{array}{ccc}
\Sigma_{A^{\mathcal{R}}} & \xrightarrow{\;\;\sigma\;\;} & \Sigma_{A^{\mathcal{R}}} \\
{\scriptstyle \Pi}\downarrow & & \downarrow{\scriptstyle \Pi} \\
X & \xrightarrow[\;\;T\;\;]{} & X
\end{array}
$$

$$\tag{13.5.11}$$

(2) $\Pi : \Sigma_{A^{\mathcal{R}}} \to X$ *is a Hölder continuous mapping which is injective on the set* $\Pi|_{\Pi^{-1}(X \setminus \bigcup_{n=0}^{\infty} T^{-n}(\cup_i \partial R_i))}$.

(3) *If $\phi : X \to \mathbb{R}$ is Hölder continuous then $\phi \circ \Pi : \Sigma_{A^{\mathcal{R}}} \to \mathbb{R}$ is also Hölder continuous and the pressures coincide:*

$$P(T, \phi) = P(\sigma, \phi \circ \Pi). \tag{13.5.12}$$

(4) *Let $\mu \in \mathcal{E}(\Sigma_{A^{\mathcal{R}}}, \sigma)$ which is positive on the non-empty open sets. Then Π is an isomorphism between the probability spaces: $(\Sigma_{A^{\mathcal{R}}}, \mathcal{B}(\Sigma_{A^{\mathcal{R}}}), \mu)$ and $(X, \mathcal{B}(X), \mu \circ \Pi^{-1})$*

Now we consider a family of very important measures, the so-called Gibbs measures.

13.5.2   Gibbs Measures

Definition 13.5.5 (Gibbs Measures) Let (X, ρ) be a compact metric space and we assume that (X, T) is a distance-expanding, open, continuous and topologically transitive (there is a point whose orbit is dense) topological dynamical system and we also assume that $\phi \in C(X, \mathbb{R})$ is a Hölder continuous potential. We say that a measure μ is a Gibbs measure for the potential ϕ if there exists a constant $C \geq 1$ such that

$$C^{-1} \leq \frac{\mu\left(T_x^{-n}\left(B\left(T^n(x), \xi\right)\right)\right)}{\exp\left(S_n\phi(x) - nP(T, \phi)\right)} \leq C, \quad \text{for all } n \geq 0, \tag{13.5.13}$$

where T_x^{-n} was defined in Definition 13.5.2. If $\mu \in M(X, T)$ then we say that μ is an invariant Gibbs measure for the potential ϕ.   □

We remark that the corresponding statement holds for the n-cylinders:

Corollary 13.5.1 ([20, Remark 5.1.3]) *Let (X, T) and ϕ be as in Definition (13.5.5). Moreover, let μ be an invariant Gibbs measure for the potential ϕ. Let $\mathcal{R}$ be a Markov partition of diameter smaller than ξ. By Theorem 13.5.1 we can choose such Markov partitions. Then we can find $\widetilde{C} > 1$ which depends on $\mathcal{R}$ such that for all $n \geq 1$ and $\omega \in \Sigma_{A^{\mathcal{R}}}^n$*

$$\widetilde{C}^{-1} \leq \frac{\mu\left(R_\omega\right)}{\exp\left(S_n\phi(x) - n \cdot P(T, \phi)\right)} \leq \widetilde{C}, \quad \text{for all } x \in R_\omega. \tag{13.5.14}$$

It is proved in [20, Theorem 5.3.2, Corollary 5.2.13, Proposition 1.5.1 and Lemma 5.4.12] that

Theorem 13.5.2 *Let (X, T) and ϕ be as in Definition 13.5.5. Then we have:*

(1) *There exist a unique invariant Gibbs measure for ϕ. Let us denote it by μ_ϕ.*
(2) *μ_ϕ is ergodic and μ_ϕ is the unique equilibrium state for T and ϕ.*
(3) *$\mu_\phi = \mu_{\phi \circ \Pi} \circ \Pi^{-1}$.*   □

13.6 C^r Repellers

We define C^r expanding maps for an $r \geq 1$ and their repellers. This definition is a special case of the one in [16, p. 197].

Definition 13.6.1 Let $r \geq 1$, U be an open subset of $\mathbb{R}^d$ and $\Lambda \subset U$ be compact. Finally, let $\psi : U \to U$ be a C^r mapping such that $\psi : \Lambda \to \Lambda$. We say that ψ is a

C^r expanding mapping on Λ and Λ is a C^r-repeller of ψ if conditions (a) and (b) below hold:

(a) there exists $\lambda > 1$ such that $\|(D_z\psi)v\| \geq \lambda\|v\|$ for all $z \in \Lambda$, $v \in \mathbb{R}^d$;
(b) there exists an open neighborhood $V \subset U$ of Λ such that

$$\Lambda = \{z \in V : \psi^n(z) \in V \text{ for all } n \geq 0\}.$$

If in addition condition (c) also holds then we say that Λ is a topologically mixing repeller of ψ:

(c) If W is an open set that intersects Λ then $\Lambda \subset \psi^n(W)$ for some $n \geq 0$. □

Remark 13.1 We remark (see [16, p. 197]) that if (a) and (b) above hold then ψ is a local homeomorphism. That is, there exists r_0 such that for every $x \in \Lambda$ the mapping $\psi|_{B(x,r_0)}$ is a homeomorphism onto its image. Hence there exist two constants $b \geq a > 1$ such that

$$B(\psi(x), ar) \subset \psi(B(x, r)) \subset B(\psi(x), br), \quad \forall x \in \Lambda, \text{ and } 0 < r < r_0.$$
$$(13.6.1)$$

In particular (Λ, ψ) is an open and distance-expanding mapping with a constant $1 < \tilde{\lambda} < \lambda$ and expansive with the constant $\delta = 2\eta$. Hence the corresponding results of Sect. 13.5 apply. □

13.6.1 Markov Partitions and the Corresponding Symbolic Dynamics

Definition 13.6.2 Let ψ, Λ and the Markov partition $\mathcal{R}$ as in Definition 13.5.3. Now we use the notation of Definition 13.5.4. Let $\tilde{R}_i$ be a sufficiently small open neighborhood of R_i as detailed in [15, Appendix 1] and [6, Section 3]. In particular $\psi|_{\tilde{R}_i}$ is injective and

$$\tilde{R}_i \cap \tilde{R}_j \neq \emptyset \iff R_i \cap R_j \neq \emptyset \text{ and } \psi(\tilde{R}_i) \supset \tilde{R}_j \iff \psi(R_i) \supset R_j \overset{\text{by def.}}{\iff} a_{ij}^{\mathcal{R}} = 1.$$
$$(13.6.2)$$

Now we write

$$R_{i_0,\ldots,i_n} := \bigcap_{j=0}^{n} \psi^{-j}(R_{i_j}) \text{ and } \tilde{R}_{i_0,\ldots,i_n} := \bigcap_{j=0}^{n} \psi^{-j}(\tilde{R}_{i_j}). \quad (13.6.3)$$

Then we define the local inverses of ψ. Namely, for $i, j \in \{1, \ldots, M\}$ with $a_{ij}^{\mathcal{R}} = 1$ we define the local inverse $f_{i,j} : \widetilde{R}_j \to \widetilde{R}_{i,j}$ of ψ by

$$f_{i,j} := \left(\psi|_{\widetilde{R}_i} \right)^{-1} |_{\widetilde{R}_j}. \tag{13.6.4}$$

Like in [6], we can define $f_{i_0 \ldots i_n} : \widetilde{R}_{i_n} \to \widetilde{R}_{i_0, \ldots, i_n}$ by

$$\left(\psi^n|_{\widetilde{R}_{i_1, \ldots, i_n}} \right)^{-1} = f_{i_0, i_1} \circ f_{i_1, i_2} \circ \cdots \circ f_{i_{n-1}, i_n} =: f_{i_0, \ldots, i_n}. \tag{13.6.5}$$

For $\mathbf{i} \in \Sigma_{A^{\mathcal{R}}}$ the set $\bigcap_{n=0}^{\infty} f_{i_0, i_1, \ldots, i_n}(\widetilde{R}_{i_n})$ consists of exactly one element. This element is denoted by $\Pi_{\mathcal{R}}(\mathbf{i})$. Then

$$\{\Pi_{\mathcal{R}}(\mathbf{i})\} := \bigcap_{n=1}^{\infty} f_{i_0 i_1 \ldots i_n}(\widetilde{R}_{i_n}) = \bigcap_{n=1}^{\infty} \widetilde{R}_{i_0, \ldots, i_n}, \tag{13.6.6}$$

Using that $\psi(\Lambda) = \Lambda$ we have

$$\Lambda = \left\{ x : \psi^n x \in \bigcup_{i=1}^{M} \widetilde{R}_i, \ \forall n \geq 0 \right\} = \bigcap_{n=1}^{\infty} \bigcup_{\omega \in \Sigma_{A^{\mathcal{R}}}^n} f_\omega \left(\widetilde{R}_{\omega_n} \right)$$

$$= \bigcup_{u=1}^{M} \underbrace{\bigcap_{n=1}^{\infty} \bigcup_{\omega \in \Sigma_{A^{\mathcal{R}}}^n, \omega_0 = u} f_\omega \left(\widetilde{R}_{\omega_n} \right)}_{\Lambda_u} = \bigcup_{u=1}^{M} \Lambda_u, \tag{13.6.7}$$

$\square$

where $\Lambda_1, \ldots, \Lambda_M$ are non-empty compact set satisfying:

$$\Lambda_u = \bigcup_{v: a_{u,v}^{\mathcal{R}} = 1} f_{u,v}(\Lambda_v). \tag{13.6.8}$$

Hence, $\Pi_{\mathcal{R}} : \Sigma_{A^{\mathcal{R}}} \to \Lambda$ is onto and the following diagram is commutative:

$$\begin{array}{ccc}
\Sigma_{A^{\mathcal{R}}} & \xrightarrow{\sigma} & \Sigma_{A^{\mathcal{R}}} \\
\Pi_{\mathcal{R}} \downarrow & & \downarrow \Pi_{\mathcal{R}} \\
\Lambda & \xrightarrow{\psi} & \Lambda
\end{array} \tag{13.6.9}$$

Remark 13.2 We remark that

(1) it follows from [26, Proposition 2.2] that there is an integer q such that

$$\text{card}\left\{\Pi_{\mathcal{R}}^{-1}\{x\}\right\} \leq q, \quad \text{for all } x \in \Lambda. \tag{13.6.10}$$

(2) The mapping $\Pi \mid_{\Pi^{-1}\left(\Lambda \setminus \bigcup_{n=0}^{\infty} \psi^{-n}\left(\bigcup_{i=1}^{M} \partial R_i\right)\right)}$ is injective as we noted in Part (b) of
Remark 13.2.

(3) By the definition of the Markov partition:

$$z \in \widetilde{R}_{i_0 \ldots i_n} \implies \psi^k(z) \in \widetilde{R}_{i_k \ldots i_n} \quad \text{for all } k = 0, \ldots, n. \tag{13.6.11}$$

Combining this with (13.6.4) and with the Inverse Function Theorem we get

$$z \in \widetilde{R}_{i_0 \ldots i_n} \implies \left(D_{\psi^k z}\psi\right)^{-1} = D_{\psi^{k+1} z} f_{i_k i_{k+1}} \quad \text{for all } k = 0, \ldots, n. \tag{13.6.12}$$

From this and the chain rule we get

$$z \in \widetilde{R}_{i_0 \ldots i_n} \implies \left(D_z \psi^n\right)^{-1} = D_{\psi^n z} f_{i_0 \ldots i_{n-1} i_n} \quad \text{for all } n \geq 1. \tag{13.6.13}$$

13.6.2 Dimension of Conformal Repellers

In this subsection we always assume (as in Definition 13.6.1) that ψ is a topologically mixing expanding mapping on the repeller Λ in $\mathbb{R}^d$. Moreover, we also always assume here that ψ is a conformal mapping (see Part (ii) of Definition 13.3.1). We express this as (Λ, ψ) is a mixing CER.

Definition 13.6.3 Let (Λ, ψ) be a mixing CER.

(1) For $t \geq 0$ we define the Hölder continuous function $\varphi_t : \Lambda \to \Lambda$ by

$$\varphi_t(x) := -t \cdot \log \|\psi'\|(x).$$

(2) Moreover, we define the geometric pressure function for $t \geq 0$ by

$$P(t) := P(\Lambda, \varphi_t).$$

The function $P(t)$ is strictly decreasing from ∞ to $-\infty$. So, it has a unique zero $t_0 = t_0(\Lambda, \psi)$. That is, $P(t_0) = 0$.

Theorem 13.6.1 ([20, Theorems 9.1.6, 9.14, Corollaries 9.1.11, 9.17]) *Let* (Λ, ψ) *be a mixing CER. As in Theorem 13.5.2 we write* $\mu_{\varphi_{t_0}}$ *for the unique invariant Gibbs measure for the potential* φ_{t_0}.

(1) *Then* $\mu_{\varphi_{t_0}}$ *is a geometric measure. This means that there exists a constant* $C \geq 1$ *such that*

$$C^{-1} \leq \frac{\mu_{\varphi_{t_0}}(B(x,r))}{r^{t_0}} \leq C, \qquad \forall x \in \Lambda, \quad \forall r \in (0,1]. \tag{13.6.14}$$

Consequently,

$$\lim_{r \to 0} \frac{\log \mu_{\varphi_{t_0}}(B(x,r))}{\log r} = t_0. \tag{13.6.15}$$

(2) *All dimensions are equal to* t_0:

$$\dim_H \mu_{\varphi_{t_0}} = \dim_P \mu_{\varphi_{t_0}} = \dim_H \Lambda = \dim_B \Lambda = \dim_P \Lambda = t_0. \tag{13.6.16}$$

(3) *The measures* $\mu_{\varphi_{t_0}}$, $\mathcal{H}^{t_0}$ *and* $\mathcal{P}^{t_0}$ *are mutually equivalent with bounded Radon-Nikodym derivatives.*

(4) *For a general* $\mathfrak{m} \in \mathcal{E}(\Lambda, \psi)$ *we have*

$$\dim_H \mathfrak{m} = \frac{h_{\mathfrak{m}}(\psi)}{\lambda_{\mathfrak{m}}(\psi)}, \tag{13.6.17}$$

where $\lambda_{\mathfrak{m}}(\psi) := \lim_{n \to \infty} \frac{1}{n} \log \|(f^n)'(x)\|$ *for* $\mathfrak{m}$*-almost all* $x \in X$. $\qquad \square$

Remark 13.3 We remark that the combination of part (b) of the previous theorem and Theorem 13.5.2 yields that for a mixing CER the Hausdorff dimension of the repeller is the supremum (actually the maximum) of the Hausdorff dimensions of ergodic measures. $\qquad \square$

The Neither Conformal Nor Affine Attractors and Repellers

13.7 History of Neither Conformal Nor Affine Attractors and Repellers

Here we give a brief account about some of the developments of the field.

(1) In 1994 Falconer [6] introduced a generalization of the usual pressure. He called it sub-additive pressure and proved that the zero of the corresponding

sub-additive pressure formula is an upper bound on the box-dimension of an expanding C^2 repeller which satisfies the so-called 1-bunched property. This condition means that if the expansion in a certain direction is $a > 1$ then the expansion in all directions are not stronger than a^2.

(2) In 1996 Barreira [2] introduced a version of non-additive pressure (which is equivalent to Falconer's sub-additive pressure under some conditions (see [1, 5])). Using this, he gave upper bounds on various Cantor sets of very general geometric constructions. Moreover, he proved that the variational principle holds for his pressure. Barreira gave conditions under which the box and the Hausdorff dimensions are equal.

(3) In 1997 Hu [8] extended the scope of Falconer's theorem. Namely, he considered expanding C^2 maps on the plane that leave invariant the strong unstable foliation. Under this condition he gave effective upper bound on the box dimension of the repeller.

(4) In 1997 Zhang [14] extended Falconer's result to C^1 expanding maps and dropped the 1-bunched property but gave upper bound only for the Hausdorff dimension of the repeller. Even used a different notion of pressure.

(5) In 2003 Barreira [3] claimed a generalization of Falconer's theorem (above) but the proof was incorrect.

(6) In 2007 Manning and Simon [9] gave counter examples to the previously mentioned Barreira's paper and proved that if the so called one bunched property does not hold then it can happen that the bounded distortion does not hold either.

(7) In 2008 Cao et al. [5] proved a variational principle result for the sub-additive pressure.

(8) In 2009 Ban et al. [1] proved the equivalence of some seemingly different definitions of singularity dimension.

(9) In 2017 Das and Simmons [28] proved that the supremum of the dimensions of ergodic measures may be smaller than the Hausdorff dimension of the attractor for a self-affine carpet in three dimension. This means that the assertion of Remark 13.3 does not hold in the non-conformal case.

(10) In 2020 Falconer and Fraser [27] investigated the L^q-dimensions of measures on the plane for certain non-conformal attractors.   □

13.7.1 The Sub-additive Topological Pressure and Lyapunov Exponents

In the non-conformal case the most important tool is the sub-additive pressure introduced by Falconer in [6]. It was reformulated by Zhang [14], Barreira [2]. In Cao, Feng and Huang [5] it was proved that these different formulations yield the same non-additive pressure. First recall that the "additive" topological pressure was defined in Definition 13.4.3 and an equivalent definition was given in Theorem 13.4.2. This second definition is the one along which the topological pressure is extended to the sub-additive case. Namely, in the definition of topological pressure in formula (13.4.9), the role of the sequence of functions $\{S_n\phi(x)\}_{n=1}^\infty$ is taken by a more general sequence called subadditive valuation.

Definition 13.7.1 (Sub-additive Valuation) Let (X, T) be a topological dynamical system. A sub-additive valuation on X is a sequence of continuous functions $\mathcal{G} = \{g_n\}_{n=1}^\infty$ satisfying

$$g_{m+n}(x) \leq g_n(x) + g_m(T^n x), \quad \forall n, m \geq 1 \text{ and } x \in X. \tag{13.7.1}$$

Clearly the sequence $\{S_n\phi(x)\}_{n=1}^\infty$ satisfies (13.7.1) with equality. That is, if $g_n(x) = S_n\phi(x)$ then

$$g_{m+n}(x) = g_n(x) + g_m(T^n x), \quad \forall n, m \geq 1 \text{ and } x \in X.$$

Now by replacing $\{S_n\phi(x)\}_{n=1}^\infty$ by a sub-additive valuation in the second definition (13.4.9) of the pressure we obtain the sub-additive pressure:

Definition 13.7.2 (Sub-additive Topological Pressure) Let (X, T) be a topological dynamical system.

(1) For an $n \in \mathbb{N}$, $\varepsilon > 0$ and sub-additive valuation $\mathcal{G} = \{g_n\}_{n=1}^\infty$ we define

$$P_n(T, X, \mathcal{G}, \varepsilon) := \sup\left\{\sum_{x \in E} \exp g_n(x) : E \text{ is an } (n, \varepsilon)\text{-separated set}\right\},$$
$$\tag{13.7.2}$$

where the notation of (n, ε)-separated set was introduced in Definition 13.4.4.
(2) Then the sub-additive topological pressure of $\mathcal{G}$ with respect to T is

$$P(T, X, \mathcal{G}) := \lim_{\varepsilon \to 0} \limsup_{n \to \infty} \frac{1}{n} \log P_n(T, X, \mathcal{G}, \varepsilon). \tag{13.7.3}$$

As we mentioned above, in the special case when there is a continuous function $\phi : X \to \mathbb{R}$ such that $g_n(x) = S_n\phi(x)$ we get back the classical or traditional topological pressure. This is why from now on we call the traditional topological pressure additive topological pressure as opposed to the more general sub-additive topological pressure defined above.

Example 13.3 It was proved in [5, p. 649] that in the special case when we consider a subshift (X, σ) the sub-additive pressure of the sub-additive valuation $\mathcal{G}$ can be presented in the form (cf. Example 13.2):

$$P(X, \sigma, \mathcal{G}) = \lim_{n \to \infty} \frac{1}{n} \log \left(\sum_{\omega \in X_n^*} \exp \left(\sup_{\mathbf{i} \in [\omega] \cap X} g_n(\mathbf{i}) \right) \right). \tag{13.7.4}$$

Definition 13.7.3 (Lyapunov Exponent of a Sub-additive Valuation) Let (X, T) be a topological dynamical system and let $\mu \in \mathcal{M}(X, T)$ then the Lyapunov exponent   of the sub-additive valuation $\mathcal{G}$ with respect to μ is

$$\mathcal{G}_*(\mu) := \inf_n \frac{1}{n} \int g_n d\mu = \lim_{n \to \infty} \frac{1}{n} \int g_n d\mu, \tag{13.7.5}$$

where the second equality follows from subadditivity. □

We remark that the inequality $\mathcal{G}_*(\mu) < \infty$ always holds, although $\mathcal{G}_*(\mu) = -\infty$ can happen. The following variational principle was proved in [5, Theorem 1.1]:

Theorem 13.7.1 (Cao, Feng and Huang) *Let (X, T) be a topological dynamical system such that $h_{\text{top}}(T) < \infty$ and let $\mathcal{G}$ be a sub-additive valuation. Then*

$$P(T, X, \mathcal{G}) = \sup \left\{ h_\mu(T) + \mathcal{G}_*(\mu) : \mu \in \mathcal{M}(X, T) \right\}. \tag{13.7.6}$$

If $\mu \in \mathcal{M}(X, T)$ is a measure that achieves the supremum in (13.7.6) then we say that μ is an equilibrium measure for the valuation $\mathcal{G}$. It follows from [7, Proposition 3.5] that

Proposition 13.1 *If (X, T) is a subshift then there exists at least one ergodic equilibrium measure.* □

Now we consider the two most important examples where we use the Lyapunov exponents in this note.

Example 13.4 Let $s \geq 0$ and $\mathcal{F} = \{f_i\}_{i=1}^{\ell}$ be a C^1 IFS with attractor $\Lambda \subset \mathbb{R}^d$ (recall the definitions from Sect. 13.3.1). Moreover, let $\Sigma := [\ell]^{\mathbb{N}}$. Then for every $s \in [0, d]$ the sub-additive valuation $\widehat{\mathcal{G}}^s := \{\widehat{g}_n^s\}_{n=1}^{\infty}$ corresponding to s and $\mathcal{F}$ is

$$\widehat{g}_n^s(x) := \log \phi^s \left(D_{\Pi \sigma^n x} f_{x|n} \right), \quad x \in \Sigma, \tag{13.7.7}$$

where $\Pi : \Sigma \to \Lambda$ is the natural projection as defined in (13.3.2) and ϕ^s is the singular value function defined in (13.2.7). It follows from the definition (13.2.7) of the singular value function that for an ergodic measure $\mathfrak{m} \in \mathcal{E}(X, T)$, the Lyapunov exponent of $\widehat{G}$ is

$$\widehat{G}_*^s(\mathfrak{m}) = \begin{cases} \widehat{\lambda}_1(\mathfrak{m}) + \cdots + \widehat{\lambda}_{[s]}(\mathfrak{m}) + (s - [s])\widehat{\lambda}_{[s]+1}(\mathfrak{m}), & \text{if } s < d, \\ \frac{s}{d}(\widehat{\lambda}_1(\mathfrak{m}) + \cdots + \widehat{\lambda}_d(\mathfrak{m})), & \text{if } s \geq d, \end{cases} \tag{13.7.8}$$

where

$$\widehat{\lambda}_i(\mathfrak{m}) := \lim_{n \to \infty} \frac{1}{n} \int \log\left(\alpha_i(D_{\Pi \sigma^n x} f_{x|n})\right) \, d\mathfrak{m}(x) \tag{13.7.9}$$

is the i-th Lyapunov exponent of the measure $\mathfrak{m}$ for $1 \leq i \leq d$. We remind the reader that $\alpha_i(A)$ was defined in Sect. 13.2.2 as the i-th singular value of the matrix A. $\qquad\square$

Example 13.5 Let $\Lambda \subset \mathbb{R}^d$ be the C^1-repeller of an expanding map $\psi : U \to U$, where $U \supset \Lambda$ is an open subset of $\mathbb{R}^d$ like in Definition 13.6.1. Then for every $s \in [0, d]$ the sub-additive valuation $\mathcal{G}^s := \{g_n^s\}_{n=1}^\infty$ corresponding to s and (Λ, ψ) is

$$g_n^s(z) := -\log\left(\prod_{k=d-[s]+1}^{d} \alpha_k(D_z\psi) \cdot \alpha_{d-[s]}^{s-[s]}(D_z\psi)\right) \tag{13.7.10}$$

$$= \log \phi^s((D_z\psi^n)^{-1}), \quad z \in \Lambda.$$

Then for every $s \in [0, d]$ and ergodic measure $\mathfrak{m} \in \mathcal{E}(X, T)$ we have

$$G_*^s(\mathfrak{m}) = \lambda_1(\mathfrak{m}) + \cdots + \lambda_{[s]}(\mathfrak{m}) + (s - [s])\lambda_{[s]+1}(\mathfrak{m}),$$

where for every $i = 1, \ldots, d$,

$$\lambda_i(\mathfrak{m}) = \lim_{n \to \infty} \frac{1}{n} \int \log\left(\alpha_i((D_z\psi^n)^{-1})\right) \, d\mathfrak{m}(z) \tag{13.7.11}$$

$$= -\lim_{n \to \infty} \frac{1}{n} \int \log \alpha_{d-i+1}\left(D_z\psi^n\right) d\mathfrak{m}(z).$$

Lemma 13.7.1 *Let (Λ, ψ) be as in Example 13.5 and let $\mathcal{R}$ be an arbitrary Markov partition. Using the notation of Sect. 13.6.1, for every $s \geq 0$ we introduce*

$$\widehat{\mathcal{G}}^s = \{\widehat{g}_n^s\}_{n=1}^\infty, \quad \text{where } \widehat{g}_n^s(\mathbf{i}) := \log \phi^s\left(D_{\Pi_{\mathcal{R}}(\sigma^n \mathbf{i})} f_{i_0 \ldots i_n}\right). \tag{13.7.12}$$

Then

$$P\left(\Sigma_{A^{\mathcal{R}}}, \sigma, \widehat{\mathcal{G}}^s\right) = P(\psi, \Lambda, \mathcal{G}^s). \tag{13.7.13}$$

Proof Using (13.6.13) and the fact that for an arbitrary $z = \Pi_{\mathcal{R}}(\mathbf{i})$ we have $\psi^n z = \Pi_{\mathcal{R}}(\sigma^n \mathbf{i})$, we get that for all $\mathbf{i} \in \Sigma_{A\mathcal{R}}$ and $n \geq 1$,

$$\widehat{g}_n^s(\mathbf{i}) = \log \phi^s \left(D_{\Pi_{\mathcal{R}}(\sigma^n \mathbf{i})} f_{i_0 \ldots i_{n-1} i_n} \right) = \log \phi^s \left((D_{\Pi(\mathbf{i})} \psi^n)^{-1} \right) = g_n^s(\Pi(\mathbf{i})).$$

(13.7.14)

Hence for all $\mathrm{m} \in \mathcal{M}(X, \sigma)$

$$\int_{\Sigma_{A\mathcal{R}}} \widehat{g}_n^s(\mathbf{i}) d\mathrm{m}(\mathbf{i}) = \int_{\Sigma_{A\mathcal{R}}} g_n^s(\Pi(\mathbf{i})) d\mathrm{m}(\mathbf{i}) = \int_{\Lambda} g_n^s(z) d(\Pi_* \mathrm{m})(z).$$

This yields that by definition

$$(\widehat{\mathcal{G}}^s)_*(\mathrm{m}) = \left(\mathcal{G}^s \right)_* (\Pi_* \mathrm{m}), \quad \mathrm{m} \in \mathcal{M}\left(\Sigma_{A\mathcal{R}}, \sigma \right).$$

(13.7.15)

By Part (iv) of Theorem 13.4.1, $\mathrm{m} \mapsto \Pi_* \mathrm{m}$ is a surjective map from $\mathcal{M}(X, \sigma)$ to $\mathcal{M}(\Lambda, f)$. Moreover, the combination of Part (a) of Remark 13.2 and Part (iv) of Theorem 13.4.1 yields that

$$h_{\mathrm{m}}(\sigma) = h_{\Pi_* \mathrm{m}}(\psi), \quad \mathrm{m} \in \mathcal{M}\left(\Sigma_{A\mathcal{R}}, \sigma \right).$$

(13.7.16)

Now the assertion of the lemma follows directly from the combination of (13.7.15), (13.7.16) and the variational principle for sub-additive pressure (see Theorem 13.7.1). Namely,

$$\begin{aligned}
P\left(\psi, \Lambda, \mathcal{G}^s \right) &= \sup \left\{ h_\mu(\psi) + \left(\mathcal{G}^s \right)_* (\mu) : \mu \in \mathcal{M}(\Lambda, \psi) \right\} \\
&= \sup \left\{ h_{\Pi_* \mathrm{m}}(\psi) + \left(\mathcal{G}^s \right)_* (\Pi_* \mathrm{m}) : \mathrm{m} \in \mathcal{M}\left(\Sigma_{A\mathcal{R}}, \sigma \right) \right\} \\
&= \sup \left\{ h_{\mathrm{m}}(\sigma) + (\widehat{\mathcal{G}}^s)_*(\mathrm{m}) : \mathrm{m} \in \mathcal{M}\left(\Sigma_{A\mathcal{R}}, \sigma \right) \right\} \\
&= P(\Sigma_{A\mathcal{R}}, \sigma, \widehat{\mathcal{G}}^s).
\end{aligned}$$

13.7.1.1 Zhang's Approach to the Sub-additive Topological Pressure

Let (Λ, ψ) be as in Example 13.5. Zhang [14, p.743] defined $P_n : [0, d] \to \mathbb{R}$ by

$$P_n(s) := P\left(\psi, \frac{1}{n} g_n^s(x) \right),$$

(13.7.17)

where P on the right hand-side is the (additive) pressure defined in (13.4.7). Zhang proved [14, Lemma 2] that the following limit exists

$$\lim_{n \to \infty} P_n(s) = \inf_{n \in \mathbb{Z}^+} P_n(s) =: P^*(s).$$

(13.7.18)

It was proved in [1, Propositions 2.1 and 2.2] that for all $s \in [0, d]$,

$$P(\psi, \Lambda, \mathcal{G}^s) = P^*(s) = \lim_{n \to \infty} P\left(\psi, \frac{1}{n} g_n^s(x)\right) = \lim_{n \to \infty} \frac{1}{n} P\left(\psi^n, g_n^s\right),$$

$$(13.7.19)$$

where on the left-hand side we have the sub-additive pressure defined in (13.7.3) and the last two P stand for the additive pressure.

13.7.2 The Singularity and the Lyapunov Dimensions

First we define the singularity dimension and Lyapunov dimension for IFSs.

13.7.2.1 The Singularity and the Lyapunov Dimensions for C^1 IFSs

Here we always assume that $\mathcal{F} = \{f_i\}_{i=1}^{\ell}$ is a C^1 IFS and $X \subset \Sigma := [\ell]^{\mathbb{N}}$ is a subshift. For every $s \geq 0$ we define the sub-additive valuation $\widehat{\mathcal{G}}^s$ as in (13.7.7). We consider the pressure function corresponding to $(X, \mathcal{F})$ by

$$P_{X,\mathcal{F}}(s) := P(X, \sigma, \widehat{\mathcal{G}}^s).   \tag{13.7.20}$$

By the definition of the topological entropy we have

$$P_{X,\mathcal{F}}(0) = h_{\text{top}}(X) \geq 0.   \tag{13.7.21}$$

Moreover, using (13.7.4) and (13.3.5) we obtain

$$\log \xi(\mathcal{F}) \leq \frac{P_{X,\mathcal{F}}(s_2) - P_{X,\mathcal{F}}(s_1)}{s_2 - s_1} \leq \log \zeta(\mathcal{F}) < 0, \quad \text{for all } 0 \leq s_1 \leq s_2.$$

$$(13.7.22)$$

In this way the function $P_{X,\mathcal{F}}(s)$ is strictly decreasing, continuous, non-negative at zero and tends to negative infinity when s tends to infinity. Hence, $P_{X,\mathcal{F}}(s)$ has a unique non-negative zero.

Definition 13.7.4 The singularity dimension of X (we denote it by $\dim_S X$) is the unique $s \geq 0$ for which

$$P_{X,\mathcal{F}}(s) = P(X, \sigma, \widehat{\mathcal{G}}^s) = 0.   \tag{13.7.23}$$

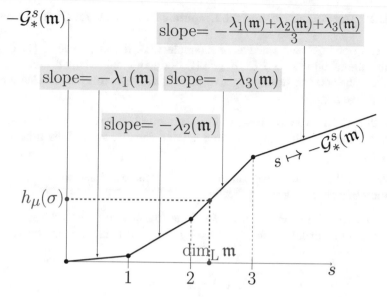

Fig. 13.2 The connection between Lyapunov dimension, entropy and the function $s \mapsto -\mathcal{G}_*^s(m)$ when $d = 3$

Definition 13.7.5 Let $\mathcal{F} = \{f_i\}_{i=1}^{\ell}$ be a C^1 IFS. For every $s \geq 0$ we define the sub-additive valuation $\widehat{G}^s$ as in (13.7.7). Moreover, let $m \in \mathcal{E}(\Sigma, \sigma)$. The Lyapunov dimension of m with respect to $\mathcal{F}$ is denoted by $\dim_L m$ and is defined (see Fig. 13.2) as the unique non-negative s satisfying

$$h_m(\sigma) + \widehat{\mathcal{G}}_*^s(m) = 0. \tag{13.7.24}$$

Such an s clearly exists since by definition, the function $s \mapsto -\widehat{\mathcal{G}}_*^s(m)$ is continuous and increases from zero to infinity. If $- (\widehat{\lambda}_1(m) + \cdots + \widehat{\lambda}_d(m)) > h_m(\sigma)$ then we can present the Lyapunov dimension in a form which may be more familiar for some of the readers:

$$\dim_L m = k + \frac{h_m(\sigma) + \widehat{\lambda}_1(m) + \cdots + \widehat{\lambda}_k(m)}{-\widehat{\lambda}_{k+1}(m)}, \tag{13.7.25}$$

where

$$k := \max \left\{ i : -(\widehat{\lambda}_1(m) + \cdots + \widehat{\lambda}_i(m)) < h_m(\sigma) \right\}.$$

By assumption, $k < d$.

13.7.2.2   The Singularity and the Lyapunov Dimensions for C^1 Repellers

Let (Λ, ψ) and $\mathcal{G}^s$ be defined as in Example 13.5. It was proved in [1, Theorem 2.4] that the function $s \mapsto P(X, \sigma, \mathcal{G}^s)$ is continuous and strictly decreasing, takes positive value at 0 and strictly negative value at d. Hence this function has a unique non-negative zero.

Definition 13.7.6 The singularity dimension of Λ with respect to ψ (denoted by $\dim_{S*} \Lambda$) is the unique non-negative zero of the function $s \mapsto P(X, \sigma, \mathcal{G}^s)$.

Similarly to the IFS case we can define the Lyapunov dimension for the ergodic measures of a C^1-repeller:

Definition 13.7.7 Let $m \in \mathcal{E}(\Lambda, \psi)$. Then the Lyapunov dimension of m with respect to ψ (denoted by $\dim_{L*} m$) is the unique non-negative s satisfying

$$h_m(\sigma) + \mathcal{G}^s_*(m) = 0. \tag{13.7.26}$$

13.8   Falconer's Bounded Distortion Result and Box-Dimension Estimates

Falconer [6] considered repellers satisfying the following assumptions:

Assumption 13.8.1

(1) Let (Λ, ψ) be a mixing C^2 repeller on $\mathbb{R}^d$.
(2) The one-bunched property holds:

$$\left\| (D_x f)^{-1} \right\|^2 \cdot \| D_x f \| < 1, \quad \forall x \in \Lambda. \tag{13.8.1}$$

Then Falconer proved that

Theorem 13.8.2 ([6, Theorem 5.3 (a)]) *Assume that (Λ, ψ) is a repeller satisfying Assumption 13.8.1. Then*

(1) $\overline{\dim}_B \Lambda \leq \dim_{S*} \Lambda$.
(2) *Moreover, the equality holds if Λ contains a non-differentiable arc.* □

The proof of part (a) of this theorem contains many essential properties of the repeller which will play an important role later. Therefore we elaborate on the steps of this proof. The heart of the proof of Theorem 13.8.2 is the following assertion:

Proposition 13.2 ([6, Proposition 4.3]) *Assume that (Λ, ψ) is a repeller as in Assumption 13.8.1. Then there exist $C_1 > 0$ such that for all admissible sequences $i_0, \ldots, i_n$ and $x, y \in R_{i_n}$ we have*

$$\|(D_x f_{i_0 \ldots i_n})^{-1} \cdot (D_y f_{i_0 \ldots i_n}) - I\| \leq C_1 \|x - y\|, \tag{13.8.2}$$

where the local inverses $f_{i,j}$ were defined in (13.6.4). □

This implies that the so called bounded distortion property holds:

Corollary 13.8.1 *There exist two constants $C_2, C_3 > 0$ such that for every $1 \leq j \leq d$,*

$$C_2 < \frac{\alpha_j(D_x f_{\mathbf{i}})}{\alpha_j(D_y f_{\mathbf{i}})} < C_3, \quad \forall n, \forall \mathbf{i} \in \Sigma_{A,n}, \forall x, y \in \widetilde{R}_{i_n}. \tag{13.8.3}$$

We remark that this is a very important property! Using this, Falconer proved [6, p. 330] that:

Proposition 13.3 *There exists a constant $C_4 > 0$ such that for all admissible $\mathbf{i} \in \Sigma_A^*$ the cylinder $\widetilde{R}_{\mathbf{i}}$ can be covered by a rectangular parallelepiped of sides*

$$C_4 \alpha_1(D_x f_{\mathbf{i}}), \ldots, C_4 \alpha_d(D_x f_{\mathbf{i}}),$$

where $x \in \widetilde{R}_{i_n}$ is arbitrary. □

Using Falconer's cutting up ellipses method (reviewed in Sect. 13.8.1 for self-affine sets) we obtain from the assertion of Proposition 13.3 that the Hausdorff dimension of the repeller Λ is less than or equal to the singularity dimension. To see that the same holds for the upper box dimension, first observe that Proposition 13.3 immediately implies (see Fig. 13.1) that

Corollary 13.8.2 *There exist constants $C_5, C_6 > 0$ such that for all admissible sequence $\omega = (\omega_0, \ldots, \omega_{n-1})$ and for all $y \in \widetilde{R}_{\omega_{n-1}}$ the cylinder $R_\omega := f_\omega(R_{\omega_{n-1}})$ can be covered by $C_5 \cdot N_{m,y}(\omega)$ balls of radius $C_6 \cdot \alpha_{m+1}(D_y f_\omega)$ for all $1 \leq m \leq d - 1$, where*

$$N_{m,y}(\omega) := \frac{\alpha_1(D_y f_\omega)}{\alpha_{m+1}(D_y f_\omega)} \cdots \frac{\alpha_m(D_y f_\omega)}{\alpha_{m+1}(D_y f_\omega)}. \tag{13.8.4}$$

Now we present the proof of how this corollary implies that part (a) of Theorem 13.8.2 holds. For a more detailed proof see [6, Section 5].

Proof of Part (a) of Theorem 13.8.2 The collection of admissible words $\omega = (\omega_0, \ldots, \omega_n)$ of length $n + 1$ is denoted by S_n. That is

$$S_n := \{\omega = (\omega_0, \ldots, \omega_n) : a_R(\omega_k, \omega_{k+1}) = 1, \ k = 0, \ldots, n - 1\}. \tag{13.8.5}$$

Let $s := \dim_{S^*} \Lambda$. We may assume that $s < d$. Choose $t \in (s, d)$. Then by the definition of s we have

$$P\left(\Sigma_A, \sigma, \mathcal{G}^t\right) < 0. \tag{13.8.6}$$

For $\omega \in S_n$ we write

$$\overline{\phi}^t(\omega) := \sup_{\mathbf{i} \in [\omega] \cap \Sigma_A} \phi^t\left(D_{\Pi_{\mathcal{R}}(\sigma^n \mathbf{i})} f_\omega\right) \text{ and } \overline{\alpha}_k(\omega) := \sup_{\mathbf{i} \in [\omega] \cap \Sigma_A} \alpha_k\left(D_{\Pi_{\mathcal{R}}(\sigma^n \mathbf{i})} f_\omega\right). \tag{13.8.7}$$

Then by (13.7.4),

$$P\left(\Sigma_A, \sigma, \mathcal{G}^t\right) = \lim_{n \to \infty} \frac{1}{n} \log \left(\sum_{\omega \in S_n} \overline{\phi}^t(\omega)\right) < 0. \tag{13.8.8}$$

This yields that we can find $q \in \mathbb{N}$ such that

$$\sum_{\omega \in S_q} \overline{\phi}_k^t(\omega) < 1. \tag{13.8.9}$$

Set $m := [t]$ the integer part of t and let

$$Q_r := \left\{\omega \in \Sigma_A^* : \quad |\omega| \text{ is a multiple of } q, \quad \overline{\alpha}_m(\omega) \leq r, \quad \overline{\alpha}_m(\sigma^q \omega) > r\right\}.$$

Repeated applications of (13.8.9) and (13.2.8) imply that for every $r > 0$ small enough we have

$$\sum_{\omega \in Q_r} \overline{\phi}_k^t(\omega) < 1. \tag{13.8.10}$$

For every $\mathbf{i} \in \Sigma_A$ there exists n such that $\mathbf{i}|_n \in Q_r$. That is

$$\Lambda \subset \bigcup_{\omega \in Q_r} R_\omega. \tag{13.8.11}$$

By the definition of Q_r and the Bounded Distortion Property (Corollary 13.8.1) we have

$$\alpha_{m+1}(D_y f_\omega) \approx r, \quad \text{for all } y \in \widetilde{R}_{\omega_n}, \tag{13.8.12}$$

where $\approx$ means that the ratio of the two sides is in between two positive constants. Using this and the Bounded Distortion Property again, we get that there exists a constant $C_7 > 0$ such that for every $r > 0$ small enough:

$$\widetilde{R}_\omega \text{ can be covered by } C_7 \cdot N_{m,y}(\omega) \text{ balls of radius } r \text{ for all } \omega \in Q_r, \ y \in \widetilde{R}_{\omega_n}.$$
(13.8.13)

On the other hand, by (13.8.12) and the definitions of ϕ^t and $N_{m,y}$ we obtain that there exists a constant $C_8 > 0$ such that for admissible $\omega \in \Sigma_A^*$,

$$N_{m,y}(\omega) = \phi^t\left(D_y f_\omega\right) \cdot \left(\alpha_{m+1}\left(D_y f_\omega\right)\right)^{-t} \leq C_8 \phi^t\left(D_y f_\omega\right) \cdot r^{-t} \text{ for all } y \in \widetilde{R}_{\omega_n}.$$
(13.8.14)

In this way, we have proved that for every $\omega \in Q_r$, we can cover R_ω by $C_8 \cdot \overline{\phi}^t(\omega) r^{-t}$ balls of radius r. Putting together this, (13.8.10) and (13.8.11) we get that Λ can be covered by $C_8 \cdot r^{-t}$ balls of radius r. This means that $\overline{\dim}_B \Lambda \leq t$. But this holds for all $s \leq t$. Hence $\overline{\dim}_B \Lambda \leq s$. $\qquad \square$

All of these properties rely on the assumption that the one-bunched property holds, which implies for example that the cylinder set $R_\mathbf{i}$ for every admissible $\mathbf{i} = (l_0, \ldots, l_n)$ is a convex set [6, p. 317].

13.9   No One-Bunched Property, Possibly No Bounded Distortion

Informally speaking the example given by Manning and Simon [9] shows that if we drop the one-bunched condition assumption then not only that the cylinders are not necessarily convex sets but also the bounded distortion property no longer holds. More precisely, in the example, the one-bunched condition does not hold, but by changing the parameter τ appropriately, we can get as close to one-bunched as we wish and the Bounded Distortion Property (13.8.3) does not hold.

Example 13.6 ([9]) Let $Q := [0, 1]^2$. Let

$$\lambda_1 = \frac{1}{2}, \quad \tilde{\lambda}_1 = \frac{1}{2^{2+2\tau}}, \quad \lambda_2 = \frac{1}{2^{2+4\tau}}, \quad \tilde{\lambda}_2 = \frac{1}{2^{1+\tau}}$$

and $f_1(x) := \lambda_1 x$, $f_2(x) := \lambda_2 x + 1 - \lambda_2$. Let $C \subset [0, 1]$ be the attractor of the IFS $\{f_1, f_2\}$ (Fig. 13.3). There exists $h \in C^2[0, 1]$ satisfying:

- $h(x)$ is strictly increasing, $h(1) = 1 - \tilde{\lambda}_2$.
- $\forall x \in C$ we have $h'(x) = 0$.

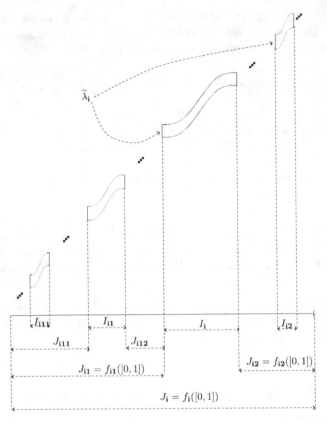

Fig. 13.3 The cylinder $F_{\mathbf{i}}([0, 1]^2)$ contains infinitely many non-convex strips of height $\widetilde{\lambda}_{\mathbf{i}}$. The figure is from [9]

- For an arbitrary n and $\mathbf{i} = (i_1, \ldots, i_n) \in \{1, 2\}^n$, let $G_{\mathbf{i}} = f_{\mathbf{i}}([\lambda_1, 1 - \lambda_2])$ be the biggest gap in $f_{\mathbf{i}}[0, 1]$. Then we have

$$\max_{x \in G_{\mathbf{i}}} h'(x) \geq 4c_7 \cdot \frac{|G_{\mathbf{i}}|}{\log(n + 2)}.$$

We consider the IFS $\mathcal{F} = \{F_1, F_2\}$ on $Q = [0, 1]^2$:

$$F_1(x_1, x_2) := \left(f_1(x_1), h(x_1) + \widetilde{\lambda}_1 x_2\right),$$
$$F_2(x_1, x_2) := \left(f_2(x_1), h(x_1) + \widetilde{\lambda}_2 x_2 + 1 - h(1) - \widetilde{\lambda}_2\right), \tag{13.9.1}$$

$$D_{(x_1, x_2)} F_1 = \begin{bmatrix} \lambda_1 & 0 \\ h'(x_1) & \widetilde{\lambda}_1 \end{bmatrix}, \quad D_{(x_1, x_2)} F_2 = \begin{bmatrix} \lambda_2 & 0 \\ h'(x_1) & \widetilde{\lambda}_2 \end{bmatrix}. \tag{13.9.2}$$

Theorem 13.9.1 (Manning and Simon [9]) *For all $c > 0$, there exists $L > 0$ such that for all $x \in [0, 1]^2$, for all $n > L$, and*

$$\mathbf{i} = (\underbrace{2, \ldots, 2}_{n}, \underbrace{1, \ldots, 1}_{n+1}),$$

the cylinder $F_{\mathbf{i}}[0, 1]^2$ cannot be covered by any rectangles with sides $c \cdot \alpha_1(D_x F_{\mathbf{i}})$ and $c \cdot \alpha_2(D_x F_{\mathbf{i}})$. $\square$

László Mikolás observed in his Master of Sciences Thesis [33] that essentially the same proof leads to the following stronger statement:

Theorem 13.9.2 (Mikolás) *There exists $\varepsilon_0 > 0$ such that for all $c > 0$, there exists L such that for all $x \in [0, 1]^2$, for all $n > L$, and*

$$\mathbf{i} = (\underbrace{2, \ldots, 2}_{n}, \underbrace{1, \ldots, 1}_{n+1}),$$

the cylinder $F_{\mathbf{i}}[0, 1]^2$ cannot be covered by any rectangles with sides $c \cdot e^{n\varepsilon_0}\alpha_1(D_x F_{\mathbf{i}})$ and $c \cdot e^{n\varepsilon_0}\alpha_2(D_x F_{\mathbf{i}})$. $\square$

Theorem 13.9.3 (Manning and Simon [9]) *The Bounded Distortion Property does not hold. That is,*

$$\sup_{\mathbf{i} \in \Sigma_A^*} \frac{\displaystyle\sup_{x \in \widetilde{R}_{i_{|\mathbf{i}|}}} \|D_x F_{\mathbf{i}}\|}{\displaystyle\inf_{y \in \widetilde{R}_{i_{|\mathbf{i}|}}} \|D_y F_{\mathbf{i}}\|} = \infty.$$

13.10 The Proof of Part (a) of Theorem 13.1.2 Assuming Part (a) of Theorem 13.1.1

In this section we always assume that the compact set $\Lambda \subset U \subset \mathbb{R}^d$ is the attractor of the C^1 expanding map $\psi : U \to U$ as in Definition 13.6.1. We frequently use the notation of Sects. 13.6.1 and 13.5.1. For every $s \geq 0$ we consider the sub-additive valuation $\mathcal{G}^s$ introduced in (13.7.10):

$$\mathcal{G}^s = \{g_n^s\}_{n=1}^{\infty}, \quad \text{where} \quad g_n^s(z) = \log \phi^s\left((D_z\psi)^{-1}\right). \quad (13.10.1)$$

Proposition 13.4 *There exist an integer $\ell \geq 2$, a constant $\tau > 1$, ℓ mappings $H_1, \ldots, H_\ell : \mathbb{R}^d \to \mathbb{R}^d$ and a subshift $X \subset \Sigma_\ell := \{1, \ldots, \ell\}^{\mathbb{N}}$ such that*

(1) $H_i \in \mathrm{Diff}^1(\mathbb{R}^d, \mathbb{R}^d)$, $i = 1, \ldots, \ell$.

(2) *For all $y \in \mathbb{R}^d$ and for all $v \in \mathbb{R}^d$ with $\|v\| = 1$ we have $0 < \|D_y H_i v\| < \frac{1}{\tau}$,*
    $i = 1, \ldots, \ell$.
(3) *Let Π be the natural projection for the IFS $\mathcal{H} := \{H_1, \ldots, H_\ell\}$ as defined*
    in (13.3.2). Then

$$\Lambda = \Pi(X). \tag{13.10.2}$$

(4) *There exists $q \in \mathbb{N}$ such that*

$$\operatorname{card}\left\{(\Pi|_X)^{-1}(z)\right\} < q \quad \text{for all } z \in \Lambda. \tag{13.10.3}$$

(5)

$$(D_{\Pi(\mathbf{i})}\psi^n)^{-1} = D_{\Pi\sigma^n\mathbf{i}}H_{i_0\ldots i_{n-1}} \text{ for all } \mathbf{i} \in X \text{ and } n \geq 1. \tag{13.10.4}$$

Proof We present the proof of Part (a) of Theorem 13.1.2 assuming Part (a) of Theorem 13.1.1 and Proposition 13.4.
    Write

$$\mathfrak{h}_n^s(\mathbf{i}) := \log \phi^s \left(D_{\Pi\sigma^n\mathbf{i}}H_{i_0\ldots i_{n-1}}\right) \text{ and } \mathfrak{H}^s := \left\{\mathfrak{h}_n^s\right\}_{n=1}^{\infty}.$$

Recall that $\mathcal{G}^s$ was defined as (13.7.10). Using the same reasoning as in the proof of Lemma 13.7.1 we obtain that

$$P\left(X, \sigma, \mathfrak{H}^s\right) = P(\psi, \Lambda, \mathcal{G}^s). \tag{13.10.5}$$

Let s_0 be the unique zero of the function $s \mapsto P(\psi, \Lambda, \mathcal{G}^s)$. That is, $s_0 = \dim_{S*} \Lambda$ is the singularity dimension of Λ. We know from Part (a) of Theorem 13.1.1 that $\overline{\dim}_B(\Pi(X))$ is less than or equal to the zero of the function $s \mapsto P(\Sigma_A, \sigma, \mathfrak{H}^s)$, which is equal to s_0 by (13.10.5). Moreover, we know from (13.10.2) that $\Pi(X) = \Lambda$. Putting these together we get that $\overline{\dim}_B(\Lambda) \leq \dim_{S*} \Lambda$. □

In the rest of this section we prove Proposition 13.4.
    For a set $E \subset \mathbb{R}^d$ we write E^ε for the closed ε-neighborhood of E. It is easy to check that we can choose such a small $\delta > 0$ which satisfies the following conditions:

Definition 13.10.1 Let $\delta > 0$ be so small that the following conditions hold:

(i) $\Lambda^{4\delta} \subset V$, where V was defined in Definition 13.6.1.
(ii) For every $z \in \Lambda^{3\delta}$, $\psi : B(z, \delta) \to \psi(B(z, \delta))$ is a C^1 diffeomorphism.
(iii) Put $\widetilde{\lambda} := \frac{2\lambda+1}{3}$, where λ was defined in Definition 13.6.1. Then

$$\|(D_z\psi)v\| > \widetilde{\lambda}\|v\|, \quad \forall z \in \Lambda^{2\delta}, \quad v \in \mathbb{R}^d. \tag{13.10.6}$$

For every $z \in \Lambda$ we define a diffeomorphism which is equal to ψ in a very small neighborhood of z and this diffeomorphism to be defined is almost like the linear approximation of ψ out of a little bit bigger but still small neighborhood of z.

Lemma 13.10.1 *Let $\delta > 0$ be as in Definition 13.10.1. For every $z \in \Lambda^{2\delta}$, we can find $t(z) \in (0, \delta)$ and $\delta_0 > 0$ (independent of z) such that for an arbitrary $d \times d$ matrix A with $\|A - D_z \psi\| < \delta_0$ there exists $F_z^A \in \mathrm{Diff}^1(\mathbb{R}^d, \mathbb{R}^d)$ satisfying*

(a) $F_z^A|_{B(z,t(z))} = \psi|_{B(z,t(z))}$,
(b) $F_z^A(y) = \psi(z) + A \cdot (y - z)$ if $y \notin B(z, 2t(z))$,
(c) $\|D_y F_z^A v\| > \widehat{\lambda}$ for all $y, v \in \mathbb{R}^d$ with $\|v\| = 1$, where $\widehat{\lambda} = \frac{1+\widetilde{\lambda}}{2} > 1$.   □

We prove Lemma 13.10.1 at the end of this section.

Consider the open cover $\{B(z, t(z)/2)\}_{z \in \Lambda}$ of Λ. We choose a minimal finite sub-cover

$$C := \{B(z_1, t_1), \dots, B(z_N, t_N)\},$$

where $t_i := t(z_i)/2$, $i = 1, \dots, N$.

Fix $\alpha \in (0, \delta)$ such that 2α is smaller than the Lebesgue number of the open cover C. Choose a Markov partition $\mathcal{R} = \{R_1, \dots, R_\ell\}$ of Λ such that

$$\mathrm{diam}(\widetilde{R}_u) < \alpha, \qquad \text{for all } u \in \{1, \dots, \ell\}, \tag{13.10.7}$$

where $\widetilde{R}_u$ is a small neighborhood of R_u introduced in Sect. 13.6.1. We can define a mapping $\kappa : \{1, \dots, \ell\} \to \{1, \dots, N\}$ such that

$$\widetilde{R}_u \subset B(z_{\kappa(u)}, t_{\kappa(u)}). \tag{13.10.8}$$

For this Markov partition we define the matrix $A = A^{\mathcal{R}}$ as in Definition 13.5.4. Let

$$\Sigma_\ell := \{1, \dots, \ell\}^{\mathbb{N}}, \text{ and } S_n := \Sigma_A^n := \{(i_0, \dots, i_n) : a_{i_k, i_{k+1}} = 1, k = 0, \dots, n-1\}. \tag{13.10.9}$$

The elements of S_n are called admissible words of length $n + 1$. Moreover, recall that we also introduced the local inverses of ψ in Sect. 13.6.1 as

$$f_{i,j} : \widetilde{R}_j \to \widetilde{R}_{i,j} \subset \widetilde{R}_i, \quad f_{i,j} := \left(\psi|_{\widetilde{R}_i}\right)^{-1}|_{\widetilde{R}_j} \quad \text{if} \quad a_{i,j} = 1. \tag{13.10.10}$$

Definition 13.10.2 For every $u \in \{1, \dots, \ell\}$ we choose a $d \times d$ non-singular matrix A_u such that

(i) The matrices $\{A_u\}_{u=1}^\ell$ are different and
(ii) $\|A_u - D_{z_{\kappa(u)}}\| < \delta_0$, where δ_0 was defined in Lemma 13.10.1.   □

Now we define the diffeomorphisms $h_u \in \text{Diff}^1(\mathbb{R}^d, \mathbb{R}^d)$ by

$$h_u(x) := F_{z_\kappa(u)}^{A_u}(x), \qquad u \in \{1, \ldots, \ell\}.$$

It follows from Lemma 13.10.1 and (13.10.8) that

Lemma 13.10.2 *The diffeomorphisms* $\{h_u\}_{u=1}^\ell$ *have the following properties:*

(a) $h_u|_{\widetilde{R}_u} = \psi|_{\widetilde{R}_u}$ *for all* $u \in \{1, \ldots, \ell\}$.
(b) $\|D_y h_u v\| > \widetilde{\lambda} > 1$ *for all* $y \in \mathbb{R}^d$, $u \in \{1, \ldots, \ell\}$ *and* $v \in \mathbb{R}^d$ *with* $\|v\| = 1$.
(c) *For all distinct* $u, v \in \{1, \ldots, \ell\}$ *the diffeomorphisms* h_u *and* h_v *are different.*

$\square$

We define

$$H_i := h_i^{-1}, \qquad \text{for all } i \in \{1, \ldots, \ell\}. \tag{13.10.11}$$

Similar to Part (c) of Remark 13.2 we get some important properties of H_i:

Fact 13.1 Let $(i, j) \in S_1$, $(i_0, \ldots, i_n) \in S_n$. By the definition of H_i we have:

$$\left(\psi|_{\widetilde{R}_i}\right)^{-1}|_{\widetilde{R}_j} = H_i|_{\widetilde{R}_j} = f_{i,j}. \tag{13.10.12}$$

By the definition of the Markov partition:

$$z \in \widetilde{R}_{i_0 \ldots i_n} \implies \psi^k(z) \in \widetilde{R}_{i_k \ldots i_n} \qquad \text{for all } k = 0, \ldots, n. \tag{13.10.13}$$

Combining the previous two identities with the Inverse Function Theorem we get

$$z \in \widetilde{R}_{i_0 \ldots i_n} \implies \left(D_{\psi^k z}\psi\right)^{-1} = D_{\psi^{k+1}z} H_{i_k} \qquad \text{for all } k = 0, \ldots, n. \tag{13.10.14}$$

From this and the chain rule we get

$$z \in \widetilde{R}_{i_0 \ldots i_n} \implies \left(D_z \psi^n\right)^{-1} = D_{\psi^n z} H_{i_0, \ldots i_{n-1}} = D_{\Pi \sigma^n \mathbf{i}} f_{i_0 \ldots i_{n-1} i_n} \qquad \text{for all } n \geq 1. \tag{13.10.15}$$

Proof of Proposition 13.4 We consider the C^1 IFS $\mathcal{H} := \{H_i\}_{i=1}^\ell$. Clearly $H_i \in \text{Diff}^1(\mathbb{R}^d, \mathbb{R}^d)$ and by part (b) of Lemma 13.10.2 $0 < \|D_y H_i\| < 1/\widetilde{\lambda} < 1$ for all $i \in \{1, \ldots, \ell\}$ and $y \in \mathbb{R}^d$. These yield parts (a) and (b) of Proposition 13.4. Let $\Sigma_\ell := \{1, \ldots, \ell\}^{\mathbb{N}}$. Then $X := \Sigma_A \subset \Sigma_\ell$ is a subshift (of finite type). We write $\Pi : \Sigma_\ell \to \mathbb{R}^d$ for the natural projection corresponding to the IFS $\mathcal{H}$ defined in (13.3.2). In this settings, the subshift whose existence is claimed in Proposition 13.4 is $X := \Sigma_A$. Now we prove that

$$\Lambda = \Pi(\Sigma_A). \tag{13.10.16}$$

To do so, first recall that we introduced a natural projection $\Pi_{\mathcal{R}} : \Sigma_A \to \Lambda$ in (13.6.6) as

$$\{\Pi_{\mathcal{R}}(\mathbf{i})\} := \bigcap_{n=1}^{\infty} f_{i_0 i_1 \dots i_n}(\widetilde{R}_{i_n}) = \bigcap_{n=1}^{\infty} \widetilde{R}_{i_0 \dots i_n} = \bigcap_{n=1}^{\infty} R_{i_0 \dots i_n}, \tag{13.10.17}$$

It is obvious that

$$\Pi|_{\Sigma_A} = \Pi_{\mathcal{R}}. \tag{13.10.18}$$

Namely, let $z \in \Lambda$. Then by (13.3.2), for every $\mathbf{i} \in \Sigma_{\ell}$ we have $\Pi(\mathbf{i}) = \lim_{n \to \infty} H_{i_0 \dots i_n}(z)$. Assume that $\mathbf{i} \in \Sigma_A$ and fix an $n \geq 1$. Then $R_{i_0 \dots i_n}$ is defined and $\mathrm{dist}(R_{i_0, \dots i_n}, H_{i_0 \dots i_{n-1}}(z)) < (1/\widehat{\lambda})^n \cdot |\Lambda|$. This implies that (13.10.18) holds. However, we pointed out in Sect. 13.6.1 that $\Pi_{\mathcal{R}}(\Sigma_A) = \Lambda$. Hence part (c) of Proposition 13.4 holds. Part (d) of Proposition 13.4 follows from Part (a) of Remark 13.2 and (13.10.18). Finally, Part (e) of Proposition 13.4 is immediate from (13.10.15) since $\psi^n(\Pi(\mathbf{i})) = \Pi(\sigma^n \mathbf{i})$ for an $\mathbf{i} \in \Sigma_A$. $\square$

We are left only to prove Lemma 13.10.1. This is standard but for the convenience of the reader we enclose a proof which uses the idea of https://math.stackexchange.com/questions/148808/the-extension-of-diffeomorphism.

Proof of Lemma 13.10.1 It is well known that for every $\tau > 0$ we can find a C^{∞} cutoff function $\varphi_{\tau} : \mathbb{R} \to [0, 1]$ so that

 (i) $\varphi_{\tau}(x) \equiv 0$ if $|x| \geq \tau$,
 (ii) $\varphi_{\tau}(x) \equiv 1$ if $|x| \leq \tau/2$,
 (iii) $|\varphi'_{\tau}(x)| \leq \frac{4}{\tau}$ for every $x \in \mathbb{R}$.

Fix an arbitrary $z \in \mathbb{R}^d$. We define $\zeta_{\tau,z} : \mathbb{R}^d \to \mathbb{R}^+$ such that $\zeta_{\tau,z}(x) := \varphi_{\tau}(\|x - z\|)$. Clearly,

$$\zeta'_{\tau,z}(x) = \varphi'_{\tau}(\|x - z\|) \cdot \frac{1}{\|x - z\|}(x - z) \text{ for } x \neq z. \tag{13.10.19}$$

From here we get

$$\|x - z\| \leq \tau \implies \zeta'_{\tau,z}(x) \cdot \|x - z\| \leq 4. \tag{13.10.20}$$

Let

$$F^A_{\tau,z}(x) := \left(1 - \zeta_{\tau,z}(x)\right)\left(\psi(z) + A(x - z)\right) + \zeta_{\tau,z}(x)\psi(x). \tag{13.10.21}$$

Clearly,

$$F_{\tau,z}^{A}(x) = \begin{cases} \psi(x), & \text{if } \|x - z\| \leq \frac{\tau}{2}, \\ \psi(z) + A \cdot (x - z), & \text{if } \|x - z\| \geq \tau. \end{cases} \tag{13.10.22}$$

We are only interested in the derivative of $F_{\tau,z}^{A}(x)$ when

$$z \in \Lambda^{3\delta}, \quad \|x - z\| \leq \tau \text{ and } x \neq z. \tag{13.10.23}$$

For such $x's$ we have

$$\|D_x F_{\tau,z}^{A} - A\| \leq \underbrace{|\zeta_{\tau,z}(x)|}_{\leq 1} \cdot \|D_x \psi - A\|$$

$$+ \underbrace{|\zeta_{\tau,z}'(x)| \cdot \|x - z\|}_{\leq 4} \cdot \frac{\|\psi(x) - \psi(z) - A(x - z)\|}{\|x - z\|}$$

$$\leq \|D_x \psi - A\| + 4 \cdot \frac{\|\psi(x) - \psi(z) - D_z \psi \cdot (x - z)\|}{\|x - z\|}$$

$$+ 4 \cdot \|A - D_z \psi\|$$

$$\leq 5 \cdot \|A - D_z \psi\| + 4 \cdot \frac{\|\psi(x) - \psi(z) - D_z \psi \cdot (x - z)\|}{\|x - z\|}.$$

If we choose a very small $\tau > 0$ (which may depend on z and denote it by $t(z)$) then the second summand above is as small as we wish. Moreover, we choose also very small $\delta_0 > 0$ and require that $\|D_z \psi - A\| < \delta_0$. Then also the first summand above is as small as we wish. Putting together this and (13.10.6) yields part (c) of Lemma 13.10.1. Parts (a) and (b) follow from (13.10.22). □

13.11   Almost All Type Results in the Lower Triangular Case

In this section we compute the Hausdorff dimension of the attractors of some one-parameter families of non-conformal C^2 IFSs on the plane of the form

$$\mathcal{F}^{t} := \left\{ F_i^{t}(x, y) := \left(f_{i,1}^{t}(x), f_{i,2}^{t}(x, y) \right) \right\}_{i=1}^{\ell}. \tag{13.11.1}$$

13.11.1 Notation

More precisely, let $\ell \geq 2$ and for every $i \in [\ell] := \{1, \ldots, \ell\}$ we are given the C^2 functions such that for some $\tau > 0$,

$$f_{i,1} : [-\tau, 1+\tau] \to (0, 1) \text{ and } f_{i,2} : [-\tau, 1+\tau] \to (0, 1). \qquad (13.11.2)$$

For each $i \in [\ell]$ and $\mathbf{t}_i = (t_{i,1}, t_{i,2}) \in \mathbb{R}^2$ let

$$F_i^{\mathbf{t}}(x, y) := F_i(x, y) + \mathbf{t}_i := \left(f_{i,1}(x) + t_{i,1}, f_{i,2}(x, y) + t_{i_2} \right) \text{ and } \mathcal{F}^{\mathbf{t}} := \left\{ F_i^{\mathbf{t}} \right\}_{i \in [\ell]} \qquad (13.11.3)$$

We consider the set of admissible translations:

$$U := \left\{ \mathbf{t} = (\mathbf{t}_1, \ldots \mathbf{t}_\ell) \in \underbrace{\mathbb{R}^2 \times \cdots \times \mathbb{R}^2}_{\ell} : F_i^{\mathbf{t}}[0, 1]^2 \subset (0, 1)^2 \right\}. \qquad (13.11.4)$$

According to (13.11.2), $F_i[0, 1]^2$ are compact subsets of $(0, 1)^2$, hence the origin of $\mathbb{R}^{2\ell}$ is in U:

$$\underline{\mathbf{0}} := \underbrace{(\mathbf{0}, \ldots, \mathbf{0})}_{\ell} \in U \subset \mathbb{R}^{2\ell}. \qquad (13.11.5)$$

Set $\mathcal{F} := \mathcal{F}^{\mathbf{t}}$ when $\mathbf{t} = \underline{\mathbf{0}}$. We assume about the derivative of $F_i^{\mathbf{t}}$

$$D_{(x,y)} F_i^{\mathbf{t}} = D_{(x,y)} F_i = \begin{pmatrix} a_i(x) & 0 \\ c_i(x, y) & d_i(x, y) \end{pmatrix} \qquad (13.11.6)$$

that there exist $\beta_1, \beta_2 \in (0, 1)$ such that

$$\beta_1 \leq \|(D_{(x,y)} F_i)\| \leq \beta_2, \qquad \text{for all } i \in [\ell] \text{ and } (x, y) \in [0, 1]^2. \qquad (13.11.7)$$

Definition 13.11.1 If all of the assumptions above hold then we say that $\left\{ \mathcal{F}^{\mathbf{t}} \right\}_{\mathbf{t} \in U}$ is the translation family of the non-conformal triangular IFS $\mathcal{F}$. We denote the attractors of $\mathcal{F}^{\mathbf{t}}$ and $\mathcal{F}$ by $\Lambda^{\mathbf{t}}$ and Λ respectively. As always we write $\Sigma := [\ell]^{\mathbb{N}}$ and the natural projection corresponding to $\mathcal{F}^{\mathbf{t}}$ and $\mathcal{F}$ are denoted by $\Pi^{\mathbf{t}}$ and Π respectively. The singularity dimension of Σ (cf. Definition 13.7.23) corresponding to the IFS $\mathcal{F}^{\mathbf{t}}$ is denoted by $d(\mathbf{t})$. Similarly, the Lyapunov dimension of an ergodic measure μ corresponding to the IFS $\mathcal{F}^{\mathbf{t}}$ is denoted by $d_\mu(\mathbf{t})$. $\qquad \square$

We set the following conditions:

(C1) The diagonal case: $F_i(x, y) = (f_{i,1}(x), f_{i,2}(y))$ for all $i \in [\ell]$ and $(x, y) \in [0, 1]^2$.

(C2) The triangular case:

    a. The derivative $D_{(x,y)} F_i$ is of the form as in (13.11.6), $(x, y) \in (0, 1)^2$, $i \in [\ell]$.

    b. For each $i \in [\ell]$ there exists $\bar{a}_i \in (0, 1)$ such that

$$0 < |d_i(x, y)| < |a_i(x)| < \bar{a}_i < 1, \quad \forall (x, y) \in [0, 1]^2. \tag{13.11.8}$$

    c. Moreover,

$$\max_{k \neq \ell} \{\bar{a}_k + \bar{a}_\ell\} < 1. \tag{13.11.9}$$

Theorem 13.11.1 *Let $\left\{\mathcal{F}^t\right\}_{t \in U}$ be the translation family of the non-conformal triangular IFS $\mathcal{F}$ satisfying either **(C1)** or **(C2)**. Let μ be an ergodic measure on Σ. Then*

(1) *for $\mathcal{L}_{2\ell}$-a.e. $t \in U$,*

$$\dim_H \left(\Lambda^t\right) = \min\{2, d(t)\}, \tag{13.11.10}$$

where $d(t)$ stands for the singularity dimension of Λ^t.

(2) *For $\mathcal{L}_{2\ell}$-a.e. $t \in U$ we have*

$$\dim_H \Pi^t_* \mu = \min\{2, d_\mu(t)\}, \tag{13.11.11}$$

where $d_\mu(t)$ is the Lyapunov dimension of μ for $\left\{\mathcal{F}^t\right\}_{t \in U}$. Combining this with Theorem 13.1.1, we obtain that the measure $\Pi^t_* \mu$ is exact dimensional for $\mathcal{L}_{2\ell}$-a.e. $t \in U$.

(3) *Let $U^{\mathrm{big}}_\mu := \left\{t \in U : d_\mu(t) > 2\right\}$. Then*

$$\Pi^t_* \mu \ll \mathcal{L}_{2\ell}, \quad \text{for } \mathcal{L}_{2\ell}\text{-a.e. } t \in U^{\mathrm{big}}_\mu. \tag{13.11.12}$$

(4) *Let $U^{\mathrm{big}} := \{t \in U : d(t) > 2\}$. Then*

$$\mathcal{L}_2 \left(K^t\right) > 0 \quad \text{for } \mathcal{L}_{2\ell}\text{-a.e. } t \in U^{\mathrm{big}}. \tag{13.11.13}$$

In what follows we always assume that $\left\{\mathcal{F}^t\right\}_{t \in U}$ is the translation family of the non-conformal triangular IFS $\mathcal{F}$ and at least one of the conditions (C1) or (C2) hold.

13.11.2 The Idea of the Proof

The proof follows the main idea of Jordan, Pollicott and Simon's paper [12]. There the authors introduced the so-called self-affine transversality condition. Analogously, the proof of Theorem 13.11.1 relies on the Non-Conformal Transversality Condition (NCTC) introduced in [30, Section 3.3]. We describe this condition below. The heart of the matter is that if NCTC holds then it implies all of the assertions of Theorem 13.11.1 hold. So the steps of the proof are as follows:

(Step 1) We prove that either of the conditions (C1) or (C2) implies that the following four assumprtions hold: There exist constants $b, c_1 \geq 1, c_2, c_3, \delta_* > 0$ such that for any $t, t_0 \in U, k \in \{1, 2\}, n \in \mathbb{N}$ and $\omega \in \Sigma_n$, any $\mathbf{x}, \mathbf{y}, \mathbf{z} \in [0, 1]^2$ and $s \in [0, 2]$ we have

(A1) Bounded distortion assumption:

$$ e^{-b} \leq \frac{\alpha_k(D_\mathbf{x} F_\omega^\mathbf{t})}{\alpha_k(D_\mathbf{y} F_\omega^\mathbf{t})} \leq e^b, \quad e^{-b} \leq \frac{\phi^s(D_\mathbf{x} F_\omega^\mathbf{t})}{\phi^s(D_\mathbf{y} F_\omega^\mathbf{t})} \leq e^b. $$

(A2)

$$ \left\| D_\mathbf{z} F_\omega^{\mathbf{t_0}} \cdot (D_\mathbf{z} F_\omega^\mathbf{t})^{-1} \right\| < c_1 e^{nc_2 \|\mathbf{t} - \mathbf{t_0}\|} \quad \text{if} \quad \|\mathbf{t} - \mathbf{t_0}\| < \delta_*. $$

(A3)

$$ \left\| D_\mathbf{z} F_\omega^\mathbf{t} \cdot (D_{\mathbf{x,y}}^* F_\omega^\mathbf{t})^{-1} \right\| < c_3, $$

where for a differentiable function $H(\mathbf{x}) = (h_1(\mathbf{x}), h_2(\mathbf{x}))$ and $\mathbf{a}, \mathbf{b} \in \mathbb{R}^2$ we write

$$ D_{\mathbf{a,b}}^* H = \begin{pmatrix} \text{grad}(h_1(\mathbf{a})) \\ \text{grad}(h_2(\mathbf{b})) \end{pmatrix}. $$

(A4) Let $r_i := \max_{x \in [0,1]^2} \|D_x F_i\|$ for $i \in [\ell]$. We assume that

$$ \max_{i \neq j} \{r_i + r_j\} := r < 1. $$

We remark that Assumption (A1) has recently been proved by Falconer and Fraser [27] independently.

(Step 2) Assumptions (A1)–(A4) imply that NCTC holds.

(Step 3) NCTC implies the so-called Almost Lower Semi-Continuous (ALSC) property holds. Namely, there exists a function $\xi : \mathbb{R}^+ \to \mathbb{R}^+$ such that $\lim_{\delta \to 0} \xi(\delta) = 0$ and for every ergodic measure μ on Σ, $\hat{t} \in U$, every sufficiently small $\delta > 0$ we have:

$$\dim_H \Pi_*^t \mu \geq \min\left\{2, d_\mu(\hat{t})\right\} - \xi(\delta) \quad \text{for } \mathcal{L}_{2\ell}\text{-a.a. } t \in B^{(2\ell)}(\hat{t}, \delta),$$
$$(13.11.14)$$

where $B^{(2\ell)}(\hat{t}, \delta) := \left\{\mathbf{y} \in \mathbb{R}^{2\ell} : \|\hat{t} - \mathbf{y}\| < \delta\right\}$. Moreover,

$$d_\mu(\hat{t}) > 2 \implies \Pi_*^t \mu \ll \mathcal{L}_2 \quad \text{for } \mathcal{L}_{2\ell}\text{-a.a. } t \in B^{(2\ell)}(\hat{t}, \delta). \qquad (13.11.15)$$

(Step 4) ALSC implies the assertions of Theorem 13.11.1.   □

The heart of the matter is the usage of NCTC (introduced in [30]) which we present now.

13.11.3 Non-conformal Transversality Condition (NCTC)

In order to introduce the NCTC we need some notation that we use throughout in this subsection. For $\omega \in \Sigma^*$, $t \in U$ and $s \in [0, 2]$ we define $z = z_{t,\omega}^s \in [0, 1]^2$ for which

$$\phi^s\left(D_{z_{t,\omega}^s} F_\omega^t\right) = \max_{y \in [0,1]^2} \phi^s\left(D_y F_\omega^t\right). \qquad (13.11.16)$$

The following function that we call Z-function, plays a crucial role in our argument. It is defined similarly (but not identically) as a corresponding function on [12, p.527].

Definition 13.11.2 (The Z function) Fix an arbitrary $k \in \{1, 2\}$, $\omega \in \Sigma^*$, $\hat{t} \in U$, $s \in [0, 2]$ and a number $\delta > 0$ so small that $B^{(2\ell)}(\hat{t}, \delta) \subset U$. We write

$$\widetilde{\alpha}_k := \widetilde{\alpha}_k^{\hat{t},\delta}(\omega) := e^{-nc_2\delta}\alpha_k\left(D_{z_{t,\omega}^s} F_\omega^{\hat{t}}\right), \qquad (13.11.17)$$

where the constant c_2 was introduced in Assumption (A2). Now we can define the function $\widetilde{Z}_\omega^{\hat{t},\delta} : [0, \infty) \to [0, 1]$ (see Fig. 13.4) by

$$\widetilde{Z}(r) := \widetilde{Z}_\omega^{\hat{t},\delta}(r) := \prod_{k=1}^{2} \frac{\min\{r, \widetilde{\alpha}_k\}}{\widetilde{\alpha}_k}$$

$$= \mathbb{1}_{[0,\widetilde{\alpha}_2]}(r) \cdot \frac{r^2}{\widetilde{\alpha}_1 \cdot \widetilde{\alpha}_2} + \mathbb{1}_{[\widetilde{\alpha}_2,\widetilde{\alpha}_1]}(r) \cdot \frac{r}{\widetilde{\alpha}_1} + \mathbb{1}_{[\widetilde{\alpha}_1,\infty]}(r). \qquad (13.11.18)$$

□

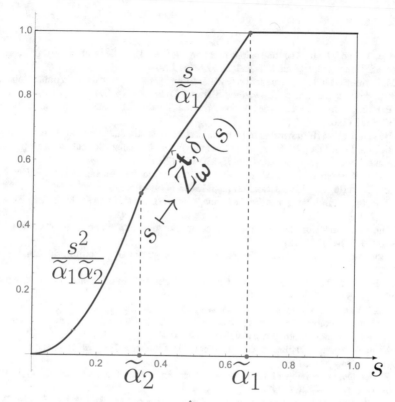

Fig. 13.4 For any fixed $\widehat{t}$, δ, ω set $\tilde{\alpha}_\ell(\omega) := \hat{\alpha}_\ell^{\widehat{t},\delta_1}(\omega)$, which is defined in (13.11.17)

Definition 13.11.3 (Non-conformal Transversality Condition (NCTC)) We say that the Non-Conformal Transversality Condition (NCTC) holds if there exists a constant $C > 0$ such that for every $\widehat{t} \in U$ and for every $\delta > 0$ satisfying $B^{(2m)}(\widehat{t}, \delta) \subset U$, for all distinct $\mathbf{i}, \mathbf{j} \in \Sigma$,

$$\mathcal{L}_{2m} \left\{ t \in B(\widehat{t}, \delta) : \left| \Pi^t(\mathbf{i}) - \Pi^t(\mathbf{j}) \right| < r \right\} < C \cdot \tilde{Z}_{\mathbf{i} \wedge \mathbf{j}}^{\widehat{t},\delta}(r). \tag{13.11.19}$$

Below we roughly indicate the methods of the steps of the proof:

(1) Step 1 of the proof, namely that ((C1) or (C2)) $\Longrightarrow$ ((A1)–(A4) hold), requires some technical calculations.
(2) Steps 2 and 3 ((A1)–(A4)$\Longrightarrow$ NCTC and NTCT $\Longrightarrow$ ALSC use the technique introduced in [12].
(3) Step 4 (ALSC $\Longrightarrow$ Theorem 13.11.1) is the really original part of the proof.

Acknowledgments The research of Feng was partially supported by a HKRGC GRF grant and the Direct Grant for Research in CUHK. The research of Simon was partially supported by the grant OTKA K104745.

References

1. J. Ban, Y. Cao, H. Hu, The dimensions of a non-conformal repeller and an average conformal repeller. Trans. Amer. Math. Soc. **362**(2), 727–751 (2010)
2. L.M. Barreira, A non-additive thermodynamic formalism and applications to dimension theory of hyperbolic dynamical systems. Ergodic Theory Dyn. Syst. **16**, 871–927 (1996)
3. L. Barreira, Dimension estimates in nonconformal hyperbolic dynamics. Nonlinearity **16**(5), 1657–1672 (2003)
4. R. Bowen, Hausdorff dimension of quasi-circles. Publ. Math. de l'IHÉS **50**, 11–25 (1979)
5. Y.-L. Cao, D.-J. Feng, W. Huang, The thermodynamic formalism for sub-additive potentials. Discrete Contin. Dyn. Syst. Ser. A **20**(3), 639–657 (2008)
6. Falconer, Kenneth J., Bounded distortion and dimension for nonconformal repellers. Math. Proce. Cambridge Philosoph. Soc. **115**(2), 315–334 (1994)
7. D.-J. Feng, Equilibrium states for factor maps between subshifts. Adv. Math. **226**(3), 2470–2502 (2011)
8. H. Hu, Box dimensions and topological pressure for some expanding maps. Commun. Math. Phys. **191** 397–407 (1998)
9. A. Manning, K. Simon, Subadditive pressure for triangular maps. Nonlinearity **201**, 133–149 (2007)
10. K.J. Falconer, The Hausdorff dimension of self-affine fractals. Math. Proce. Cambridge Philosoph. Soc. **103**, 339–350 (1988)
11. K.J. Falconer, *Techniques in Fractal Geometry* (Wiley, Chichester, 1997)
12. T. Jordan, M. Pollicott, K. Simon, Hausdorff dimension for randomly perturbed self affine attractors. Commun. Math. Phys. **270**, 519–544 (2007)
13. P. Walters, An introduction to ergodic theory, in *Graduate Texts in Mathematics*, vol. 79 (Springer, New York, 1982)
14. Y. Zhang, Dynamical upper bounds for Hausdorff dimension of invariant sets. Ergodic Theory Dyn. Syst. **17**, 739–756 (1997)
15. D. Ruelle, Repellers for real analytic maps. Ergodic Theory Dyn. Syst. **2**, 99–107 (1982)
16. Y.B. Pesin, *Dimension Theory in Dynamical Systems*. Chicago Lectures in Mathematics (University of Chicago Press, Chicago, 1997)
17. D.-J. Feng, S. Hua, Z.-Y. Wen, Some relations between packing premeasure and packing measure. Bull. London Math. Soc. **31**, 665–670 (1999)
18. Y. Peres, M. Rams, K. Simon, B. Solomyak, Equivalence of positive Hausdorff measure and the open set condition for self-conformal sets. Proc. Amer. Math. Soc. **129**, 2689–2699 (2001)
19. K.J. Falconer, Dimensions and measures of quasi self-similar sets. Proc. Amer. Math. Soc. **106**, 543–554 (1989)
20. F. Przytycki, M. Urbanski, *Conformal Fractals: Ergodic Theory Methods*, vol. 371 (Cambridge University Press, Cambridge, 2010)
21. J.E. Hutchinson, Fractals and self similarity. Ind. Univer. Math. J. **30**, 713–747 (1981)
22. D.-J. Feng, P. Shmerkin, Non-conformal repellers and the continuity of pressure for matrix cocycles. Geometr. Funct. Analy. **24** 1101–1128 (2014)
23. D. Ruelle, Statistical mechanics on a compact set with Z^v action satisfying expansiveness and specification. Bull. Amer. Math. Soc. **78**, 988–991 (1972)
24. P. Walters, A variational principle for the pressure of continuous transformations. Amer. J. Math. **97**, 937–971 (1975)
25. W. Parry, Ergodic theory, in *Time Series and Statistics* (Springer, Berlin, 1990), pp. 73–81
26. D. Ruelle, The thermodynamic formalism for expanding maps. Commun. Math. Phys. **125**, 239–262 (1989)
27. K.J. Falconer, J.M. Fraser, L.D. Lee, L^q-spectra of measures on planar non-conformal attractors (2020). Preprint arXiv:2005.09361
28. T. Das, D. Simmons, The Hausdorff and dynamical dimensions of self-affine sponges: a dimension gap result, Invent. Math. **210**, 85–134 (2017)

29. D.-J. Feng, K. Simon, Dimension estimates for C^1 iterated function systems and repellers. Part I (2020). arXiv preprint arXiv:2007.15320
30. D.-J. Feng, K. Simon, Dimension Estimates for C^1 Iterated Function Systems and Repellers. Part II (2021). arXiv preprint arXiv:2106.14393
31. B. Bárány, M. Hochman, A. Rapaport, Hausdorff dimension of planar self-affine sets and measures. Invent. Math. **216**, 601–659 (2019)
32. M. Hochman, A. Rapaport, Hausdorff dimension of planar self-affine sets and measures with overlaps (2019). Preprint arXiv:1904.09812
33. L. Mikolás, Dimension of non-conformal iteratedfunction systems on the plane, Master of Science Thesis, Budapest University of Technology and Economics, 2020
34. K.-S. Lau, H. Rao, Y.-L. Ye, Corrigendum: "Iterated function system and Ruelle operator" [Journal of Mathematical Analysis and Applications **231** (1999), no. 2, 319–344; MR1669203 (2001a:37013)] by Lau and A. H. Fan. J. Math. Anal. Appl. **262**, 446–451 (2001)
35. T. Bogenschütz, H. Crauel, The Abramov-Rokhlin formula, in *Ergodic Theory and Related Topics, III (Güstrow, 1990)*. Lecture Notes in Mathematics, vol. 1514, (Springer, Berlin, 1992), pp. 32–35

Chapter 14
Intermediate Dimensions: A Survey

Kenneth J. Falconer

Abstract This article surveys the θ-intermediate dimensions that were introduced recently which provide a parameterised continuum of dimensions that run from Hausdorff dimension when $\theta = 0$ to box-counting dimensions when $\theta = 1$. We bring together diverse properties of intermediate dimensions which we illustrate by examples.

14.1 Introduction

Many interesting fractals, for example many self-affine carpets, have differing box-counting and Hausdorff dimensions. A smaller value for Hausdorff dimension can result because covering sets of widely ranging scales are permitted in the definition, whereas box-counting dimensions essentially come from counting covering sets that are all of the same size. Intermediate dimensions were introduced in [12] in 2019 to provide a continuum of dimensions between Hausdorff and box-counting; this is achieved by restricting the families of allowable covers in the definition of Hausdorff dimension by requiring that $|U| \leq |V|^{\theta}$ for all sets U, V in an admissible cover, where $\theta \in [0, 1]$ is a parameter. When $\theta = 1$ only covers using sets of the same size are allowable and we recover box-counting dimension, and when $\theta = 0$ there are no restrictions giving Hausdorff dimension.

This article brings together what is currently known about intermediate dimensions from a number of sources, especially [1, 3, 4, 12, 21]; in particular Banaji [1] has very recently obtained many detailed results. We first consider basic properties of θ-intermediate dimensions, notably continuity when $\theta \in (0, 1]$, and discuss some tools that are useful when working with intermediate dimensions. We look at some examples to show the sort of behaviour that occurs, before moving onto the more challenging case of Bedford-McMullen carpets. Finally we consider a potential-

K. J. Falconer (✉)
Mathematical Institute, University of St Andrews, St Andrews, Fife, UK
e-mail: kjf@st-andrews.ac.uk

© The Author(s), under exclusive license to Springer Nature Switzerland AG 2021
M. Pollicott, S. Vaienti (eds.), *Thermodynamic Formalism*, Lecture Notes
in Mathematics 2290, https://doi.org/10.1007/978-3-030-74863-0_14

theoretic characterisation of intermediate dimensions which turns out to be useful for studying the dimensions of projections and other images of sets. Proofs for most of the results can be found elsewhere and are referenced, though some are sketched to provide a feeling for the subject.

We work with subsets of $\mathbb{R}^n$ throughout, although much of the theory easily extends to more general metric spaces, see [1]. To avoid problems of definition, we assume throughout this account that all the sets $F \subset \mathbb{R}^n$ whose dimensions are considered are non-empty and bounded.

Whilst Hausdorff dimension $\dim_H$ is usually defined via Hausdorff measure, it may also be defined directly, see [7, Section 3.2]. For $F \subset \mathbb{R}^n$ we write $|F|$ for the *diameter* of F and say that a finite or countable collection of subsets $\{U_i\}$ of $\mathbb{R}^n$ is a *cover* of F if $F \subset \bigcup_i U_i$. Then the Hausdorff dimension of F is given by:

$$\dim_H F = \inf \Big\{ s \geq 0 : \text{for all } \varepsilon > 0 \text{ there exists a cover } \{U_i\} \text{ of } F \text{ such that}$$

$$\sum_i |U_i|^s \leq \varepsilon \Big\}.$$

(Lower) box-counting dimension $\underline{\dim}_B$ may be expressed in a similar manner except that here we require the covering sets all to be of equal diameter. For bounded $F \subset \mathbb{R}^n$,

$$\underline{\dim}_B F = \inf \Big\{ s \geq 0 : \text{for all } \varepsilon > 0 \text{ there exists a cover } \{U_i\} \text{ of } F$$

$$\text{such that } |U_i| = |U_j| \text{ for all } i, j \text{ and } \sum_i |U_i|^s \leq \varepsilon \Big\}.$$

From this viewpoint, Hausdorff and box-counting dimensions may be regarded as extreme cases of the same definition, one with no restriction on the size of covering sets, and the other requiring them all to have equal diameters; one might regard these two definitions as the extremes of a continuum of dimensions with increasing restrictions on the relative sizes of covering sets. This motivates the definition of *intermediate dimensions* where the coverings are restricted by requiring the diameters of the covering sets to lie in a geometric range $\delta^{1/\theta} \leq |U_i| \leq \delta$ where $0 \leq \theta \leq 1$ is a parameter.

Definition 14.1 Let $F \subset \mathbb{R}^n$. For $0 \leq \theta \leq 1$ the *lower θ-intermediate dimension* of F is defined by

$$\underline{\dim}_\theta F = \inf \Big\{ s \geq 0 : \text{for all } \varepsilon > 0 \text{ and all } \delta_0 > 0, \text{ there exists } 0 < \delta \leq \delta_0$$

$$\text{and a cover } \{U_i\} \text{ of } F \text{ such that } \delta^{1/\theta} \leq |U_i| \leq \delta \text{ and } \sum |U_i|^s \leq \varepsilon \Big\}.$$

Analogously the *upper θ-intermediate dimension* of F is defined by

$$\overline{\dim}_\theta F = \inf\Big\{s \geq 0 : \text{for all } \varepsilon > 0 \text{ there exists } \delta_0 > 0 \text{ such that for all } 0 < \delta \leq \delta_0,$$

$$\text{there is a cover } \{U_i\} \text{ of } F \text{ such that } \delta^{1/\theta} \leq |U_i| \leq \delta \text{ and } \sum |U_i|^s \leq \varepsilon\Big\}.$$

Note that, except when $\theta = 0$, these definitions are unchanged if $\delta^{1/\theta} \leq |U_i| \leq \delta$ is replaced by $\delta \leq |U_i| \leq \delta^\theta$.

It is immediate that

$$\dim_{\mathrm{H}} F = \underline{\dim}_0 F = \overline{\dim}_0 F, \quad \underline{\dim}_{\mathrm{B}} F = \underline{\dim}_1 F \quad \text{and} \quad \overline{\dim}_{\mathrm{B}} F = \overline{\dim}_1 F,$$

where $\overline{\dim}_{\mathrm{B}}$ is upper box-counting dimension. Furthermore, for a bounded $F \subset \mathbb{R}^n$ and $\theta \in [0, 1]$,

$$0 \leq \dim_{\mathrm{H}} F \leq \underline{\dim}_\theta F \leq \overline{\dim}_\theta F \leq \overline{\dim}_{\mathrm{B}} F \leq n \text{ and } 0 \leq \underline{\dim}_\theta F \leq \underline{\dim}_{\mathrm{B}} F \leq n.$$

As with box-counting dimensions we often have $\underline{\dim}_\theta F = \overline{\dim}_\theta F$ in which case we just write $\dim_\theta F = \underline{\dim}_\theta F = \overline{\dim}_\theta F$ for the *θ-intermediate dimension* of F.

We remark that a continuum of dimensions of a different form, known as the *Assouad spectrum*, has also been investigated recently, see [14, 16, 17]; this provides a parameterised family of dimensions which interpolate between upper box-counting dimension and quasi-Assouad dimension, but we do not pursue this here.

14.2 Properties of Intermediate Dimensions

14.2.1 Basic Properties

We start by reviewing some basic properties of intermediate dimensions of a type that are familiar in many definitions of dimension.

1. *Monotonicity.* For all $\theta \in [0, 1]$ if $E \subset F$ then $\underline{\dim}_\theta E \leq \underline{\dim}_\theta F$ and $\overline{\dim}_\theta E \leq \overline{\dim}_\theta F$.
2. *Finite stability.* For all $\theta \in [0, 1]$ if $E, F \subset \mathbb{R}^n$ then $\overline{\dim}_\theta(E \cup F) = \max\{\overline{\dim}_\theta E, \overline{\dim}_\theta F\}$. Note that, analogously with box-counting dimensions, $\underline{\dim}_\theta$ is not finitely stable, and neither $\underline{\dim}_\theta$ or $\overline{\dim}_\theta$ are countably stable (i.e. it is not in general the case that $\overline{\dim}_\theta \cup_{i=1}^\infty F_i = \sup_{1 \leq i < \infty} \overline{\dim}_\theta F_i$).
3. *Monotonicity in θ.* For all bounded F, $\underline{\dim}_\theta F$ and $\overline{\dim}_\theta F$ are monotonically increasing in $\theta \in [0, 1]$.
4. *Closure.* For all $\theta \in (0, 1]$, $\underline{\dim}_\theta F = \underline{\dim}_\theta \overline{F}$ and $\overline{\dim}_\theta F = \overline{\dim}_\theta \overline{F}$ where $\overline{F}$ is the closure of F. (This follows since for $\theta \in (0, 1]$ it is enough to consider finite covers of closed sets in the definitions of intermediate dimensions.)

5. *Lipschitz and Hölder properties.* Let $f : F \to \mathbb{R}^m$ be an α-Hölder map, i.e.
$|f(x) - f(y)| \le c|x - y|^\alpha$ for $\alpha \in (0, 1]$ and $c > 0$. Then for all $\theta \in [0, 1]$,

$$\underline{\dim}_\theta f(F) \le \frac{1}{\alpha} \underline{\dim}_\theta F \quad \text{and} \quad \overline{\dim}_\theta f(F) \le \frac{1}{\alpha} \overline{\dim}_\theta F. \tag{14.2.1}$$

(To see this, if $\{U_i\}$ is a cover of F with $\delta \le |U_i| \le \delta^\theta$ consider the cover of $f(F)$
by the sets $\{f(U_i)\}$ if $c\delta^\alpha \le |f(U_i)|$ and by sets $V_i \supset f(U_i)$ with $|V_i| = c\delta^\alpha$
otherwise.)

In particular, if $f : F \to f(F) \subset \mathbb{R}^m$ is bi-Lipschitz then $\underline{\dim}_\theta f(F) = \underline{\dim}_\theta F$ and $\overline{\dim}_\theta f(F) = \overline{\dim}_\theta F$, i.e. $\underline{\dim}_\theta$ and $\overline{\dim}_\theta$ are bi-Lipschitz invariants.
For further Lipschitz and Hölder estimates see Banaji [1, Section 4].

14.2.2 Continuity

A natural question is whether, for a fixed bounded set F, $\underline{\dim}_\theta F$ and $\overline{\dim}_\theta F$ vary
continuously for $\theta \in [0, 1]$. It turns out that this is the case except possibly at
$\theta = 0$ where the intermediate dimensions may or may not be continuous, see the
examples in Sect. 14.4. Continuity on $(0, 1]$ follows immediately from the following
inequalities which relate $\underline{\dim}_\theta F$, respectively $\overline{\dim}_\theta F$, for different values of θ.

Proposition 14.2 *Let F be a bounded subset of $\mathbb{R}^n$ and let $0 < \theta < \phi \le 1$. Then*

$$\overline{\dim}_\theta F \le \overline{\dim}_\phi F \le \frac{\phi}{\theta} \overline{\dim}_\theta F \tag{14.2.2}$$

and

$$\overline{\dim}_\theta F \le \overline{\dim}_\phi F \le \overline{\dim}_\theta F + \left(1 - \frac{\theta}{\phi}\right)(n - \overline{\dim}_\theta F), \tag{14.2.3}$$

with corresponding inequalities where $\overline{\dim}_\theta$ and $\overline{\dim}_\phi$ are replaced by $\underline{\dim}_\theta$ and $\underline{\dim}_\phi$. □

Proof We include the proof of (14.2.2) to give a feel for this type of argument. The
left-hand inequality is just monotonicity of $\overline{\dim}_\theta F$.

With $0 < \theta < \phi \le 1$ let $t > \frac{\phi}{\theta} \overline{\dim}_\theta F$ and choose s such that $\overline{\dim}_\theta F < s < \frac{\theta}{\phi} t$.
Given $\varepsilon > 0$, for all sufficiently small $0 < \delta < 1$ we may find countable or finite
covers $\{U_i\}_{i \in I}$ of F such that

$$\sum_{i \in I} |U_i|^s < \varepsilon \quad \text{and} \quad \delta \le |U_i| \le \delta^\theta \quad \text{for all } i \in I. \tag{14.2.4}$$

Let

$$I_0 = \{i \in I : \delta \le |U_i| < \delta^{\theta/\phi}\} \quad \text{and} \quad I_1 = \{i \in I : \delta^{\theta/\phi} \le |U_i| \le \delta^{\theta}\}.$$

For each $i \in I_0$ let V_i be a set with $V_i \supset U_i$ and $|V_i| = \delta^{\theta/\phi}$. Let $0 < s < t\theta/\phi \le n$. Then $\{W_i\}_{i \in I} := \{V_i\}_{i \in I_0} \cup \{U_i\}_{i \in I_1}$ is a cover of F by sets with diameters in the range $[\delta^{\theta/\phi}, \delta^{\theta}]$. Taking sums with respect to this cover:

$$\sum_{i \in I} |W_i|^t = \sum_{i \in I_0} |V_i|^t + \sum_{i \in I_1} |U_i|^t = \sum_{i \in I_0} \delta^{t\,\theta/\phi} + \sum_{i \in I_1} |U_i|^t$$

$$\le \sum_{i \in I_0} |U_i|^{t\,\theta/\phi} + \sum_{i \in I_1} |U_i|^{t\,\theta/\phi} = \sum_{i \in I} |U_i|^{t\,\theta/\phi} \le \sum_{i \in I} |U_i|^s < \varepsilon.$$

$$(14.2.5)$$

Thus for all $t > \dfrac{\phi}{\theta} \overline{\dim}_\theta F$, for all $\varepsilon > 0$, for all sufficiently small δ (equivalently, for all sufficiently small δ^θ) there is a cover $\{W_i\}_i$ of F by sets with $(\delta^\theta)^{1/\phi} \le |W_i| \le \delta^\theta$ satisfying (14.2.5), so $\overline{\dim}_\phi F \le \dfrac{\phi}{\theta} \overline{\dim}_\theta F$.

The analogue of (14.2.2) for $\underline{\dim}_\theta$ follows by exactly the same argument by choosing covers of F with $\delta \le |U_i| \le \delta^\theta$ for arbitrarily small δ.

The proof of (14.2.3) is given in [12]: essentially, given a cover of F by sets $\{U_i\}$ with $\delta \le |U_i| \le \delta^\theta$ one breaks up those U_i with $\delta^\phi \le |U_i| \le \delta^\theta$ into smaller pieces to get a cover of F by sets with diameters in the range $[\delta, \delta^\phi]$. □

Note that the right hand inequality of (14.2.2) is stronger than that in (14.2.3) precisely when $\dfrac{\theta}{\phi} \le \dfrac{n}{\overline{\dim}_\phi F} - 1$, which is the case for all $0 < \theta < \phi \le 1$ if $\overline{\dim}_\phi F \le \tfrac{1}{2} n$; similarly for lower dimensions.

Inequality (14.2.2) implies that $\dfrac{\overline{\dim}_\theta F}{\theta}$ and $\dfrac{\underline{\dim}_\theta F}{\theta}$ are monotonic decreasing in $\theta \in (0, 1]$; Banaji [1, Proposition 3.9] points out that they are strictly decreasing if $\overline{\dim}_B F > 0$, respectively $\underline{\dim}_B F > 0$. Thus the graphs of $\theta \mapsto \overline{\dim}_\theta F$ and $\theta \mapsto \underline{\dim}_\theta F$ $(0 < \theta \le 1)$ are starshaped with respect to the origin (i.e. each half-line from the origin in the first quadrant cuts the graphs in a single point).

The following corollary is immediate.

Corollary 14.3 *The maps $\theta \mapsto \underline{\dim}_\theta F$ and $\theta \mapsto \overline{\dim}_\theta F$ are continuous for $\theta \in$ (0, 1].*   □

By setting $\phi = 1$ in Proposition 14.2 and rearranging we get useful comparisons with box-counting dimensions.

Corollary 14.4 *Let F be a bounded subset of* $\mathbb{R}^n$. *Then*

$$\overline{\dim}_\theta F \ \geq \ n - \frac{\left(n - \overline{\dim}_B F\right)}{\theta} \qquad\qquad (14.2.6)$$

and

$$\overline{\dim}_\theta F \ \geq \ \theta\, \overline{\dim}_B F, \qquad\qquad (14.2.7)$$

with corresponding inequalities where $\overline{\dim}_\theta$ *and* $\overline{\dim}_B$ *are replaced by* $\underline{\dim}_\theta$ *and* $\underline{\dim}_B$. $\qquad\qquad\qquad\qquad\qquad\qquad\qquad\qquad\qquad\qquad\qquad\qquad\square$

Again (14.2.7) gives a better lower bound than (14.2.6) if and only if $\theta \leq \dfrac{n}{\overline{\dim}_B F} - 1$ which is the case for all $\theta \in (0, 1]$ if $\overline{\dim}_B F \leq \frac{1}{2}n$, and similarly for lower dimensions.

Intermediate dimensions may or may not be continuous when $\theta = 0$, see Sect. 14.4.2 for examples. Indeed, determining whether a given set has intermediate dimensions that are continuous at $\theta = 0$, which relates to the distribution of scales of covering sets for Hausdorff and box dimensions, is one of the key questions in this subject.

Banaji [1] introduced a generalisation of intermediate dimensions by replacing the condition $\delta^{1/\theta} \leq |U_i| \leq \delta$ in Definition 14.1 by $\Phi(\delta) \leq |U_i| \leq \delta$, where $\Phi : (0, Y) \to \mathbb{R}$ is monotonic and satisfies $\lim_{\delta \searrow 0} \Phi(\delta)/\delta = 0$ for some $Y > 0$, to obtain families of dimensions $\underline{\dim}^\Phi F$ and $\overline{\dim}^\Phi F$; clearly when $\Phi(x) = x^{1/\theta}$ we recover $\underline{\dim}_\theta F$ and $\overline{\dim}_\theta F$. He provides an extensive analysis of these Φ-*intermediate dimensions*. In particular they interpolate all the way between Hausdorff and box-dimensions, that is there exist such functions Φ^s for $s \in [\dim_H F, \underline{\dim}_B F]$ that are increasing with s with respect to a natural ordering and are such that $\overline{\dim}^\Phi F = s$ and $\overline{\dim}^\Phi F = \min\{s, \underline{\dim}_B F\}$, see [1, Theorem 6.1].

14.3   Some Tools for Intermediate Dimension

As with other notions of dimension, there are some basic techniques that are useful for studying intermediate dimensions and calculating them in specific cases.

14.3.1   A Mass Distribution Principle

The *mass distribution principle* is frequently used for finding lower bounds for Hausdorff dimension by considering local behaviour of measures supported on the set, see [7, Principle 4.2]. Here are the natural analogues for $\underline{\dim}_\theta$ and $\overline{\dim}_\theta$

which are proved using an easy modification of the standard proof for Hausdorff dimensions.

Proposition 14.5 ([12, Proposition 2.2]) *Let F be a Borel subset of $\mathbb{R}^n$ and let $0 \leq \theta \leq 1$ and $s \geq 0$. Suppose that there are numbers $a, c > 0$ such that for arbitrarily small $\delta > 0$ we can find a Borel measure μ_δ supported on F such that $\mu_\delta(F) \geq a$, and with*

$$\mu_\delta(U) \leq c|U|^s \quad \text{for all Borel sets } U \subset \mathbb{R}^n \text{ with } \delta \leq |U| \leq \delta^\theta. \tag{14.3.1}$$

Then $\overline{\dim}_\theta F \geq s$. Alternatively, if measures μ_δ with the above properties can be found for all sufficiently small δ, then $\underline{\dim}_\theta F \geq s$. $\qquad\square$

Note that in Proposition 14.5 a different measure μ_δ is used for each δ, but it is essential that they all assign mass at least $a > 0$ to F. In practice μ_δ is often a finite sum of point masses.

14.3.2 A Frostman Type Lemma

Frostman's lemma is another powerful tool in fractal geometry which is a sort of dual to Proposition 14.5. We state here a version for intermediate dimensions. As usual $B(x, r)$ denotes the closed ball of centre x and radius r.

Proposition 14.6 ([12, Proposition 2.3]) *Let F be a compact subset of $\mathbb{R}^n$, let $0 < \theta \leq 1$, and let $0 < s < \underline{\dim}_\theta F$. Then there exists $c > 0$ such that for all $\delta \in (0, 1)$ there is a Borel probability measure μ_δ supported on F such that for all $x \in \mathbb{R}^n$ and $\delta^{1/\theta} \leq r \leq \delta$,*

$$\mu_\delta(B(x, r)) \leq c r^s. \tag{14.3.2}$$

Fraser has pointed out a nice alternative proof of (14.2.2) using the Frostman's lemma and the mass distribution principle. Briefly, let $0 < \theta < \phi \leq 1$. if $s < \underline{\dim}_\phi F$, Proposition 14.6 gives probability measures μ_δ on F (which we may take to be compact) such that $\mu_\delta(B(x, r)) \leq c r^s$ for $\delta^{1/\phi} \leq r \leq \delta$. If $\delta^{1/\theta} \leq r \leq \delta^{1/\phi}$ then

$$\mu_\delta(B(x, r)) \leq \mu_\delta(B(x, \delta^{1/\phi})) \leq c \, \delta^{s/\phi} \leq c \, r^{s\theta/\phi},$$

so $\mu_\delta(B(x, r)) \leq c \, r^{s\theta/\phi}$ for all $\delta^{1/\theta} \leq r \leq \delta$. Using Proposition 14.5 $\underline{\dim}_\theta F \geq s\phi/\theta$. This is true for all $s < \underline{\dim}_\phi F$ so $\underline{\dim}_\theta F \geq \frac{\theta}{\phi}\underline{\dim}_\phi F$.

14.3.3 Relationship with Assouad Dimension

Assouad dimension has been studied intensively in recent years, see the books [14, 26] and paper [13]. Although Assouad dimension does not *a priori* seem closely related to intermediate dimensions, it turns out that information about the Assouad dimension of a set can refine estimates of intermediate dimensions and under certain conditions imply discontinuity at $\theta = 0$.

The *Assouad dimension* of $F \subset \mathbb{R}^n$ is defined by

$$\dim_A F = \inf \left\{ s \geq 0 \; : \; \text{there exists } C > 0 \text{ such that } N_r(F \cap B(x, R)) \leq C \left(\frac{R}{r} \right)^s \right.$$

$$\left. \text{for all } x \in F \text{ and all } 0 < r < R \right\},$$

where $N_r(A)$ denotes the smallest number of sets of diameter at most r that can cover a set A. In general $\underline{\dim}_B F \leq \overline{\dim}_B F \leq \dim_A F \leq n$, but equality of these three dimensions often occurs, even if the Hausdorff dimension and box-counting dimension differ, for example if the box-counting dimension is equal to the ambient spatial dimension.

The following proposition due to Banaji, which extends an earlier estimate in [12, Proposition 2.4], gives lower bounds for intermediate dimensions in terms of Assouad and box dimensions. This lower bound is sharp, taking F to be the F_p of Sect. 14.4.1, and can be particular useful near $\theta = 1$ where the estimate approaches the box dimension.

Proposition 14.7 ([1, Proposition 3.10]) *For a bounded set $F \subset \mathbb{R}^n$ and $\theta \in (0, 1]$,*

$$\underline{\dim}_\theta F \geq \frac{\theta \dim_A F \, \underline{\dim}_B F}{\dim_A F - (1 - \theta)\underline{\dim}_B F},$$

with a similar inequality for upper dimensions. In particular, if $\underline{\dim}_B F = \dim_A F$ (which is always the case if $\underline{\dim}_B F = n$), then $\underline{\dim}_\theta F = \overline{\dim}_\theta F = \underline{\dim}_B F = \dim_A F$ for all $\theta \in (0, 1]$. □

One consequence of Proposition 14.7 is that if $\dim_H F < \underline{\dim}_B F = \dim_A F$, then the intermediate dimensions $\underline{\dim}_\theta F$ and $\overline{\dim}_\theta F$ are constant on $(0, 1]$ and discontinuous at $\theta = 0$. This will help us analyse examples that exhibit a range of behaviours in Sect. 14.4.2.

Banaji also shows [1, Proposition 3.8] that (14.2.2), (14.2.3) and (14.2.6) may be strengthened by incorporating the Assouad dimension of F into the right-hand estimates.

14.3.4 Product Formulae

It is natural to relate dimensions of products of sets to those of the sets themselves. The following product formulae for intermediate dimensions are of interest in their own right and are also useful in constructing examples.

Proposition 14.8 ([12, Proposition 2.5]) *Let $E \subset \mathbb{R}^n$ and $F \subset \mathbb{R}^m$ be bounded and let $\theta \in [0, 1]$. Then*

$$\underline{\dim}_\theta E + \underline{\dim}_\theta F \leq \underline{\dim}_\theta(E \times F) \leq \overline{\dim}_\theta(E \times F) \leq \overline{\dim}_\theta E + \overline{\dim}_B F.$$

(14.3.3)

Sketch Proof The cases $\theta = 0, 1$ are well-known, see [7, Chapter 7]. For other θ the left hand inequality follows by using Proposition 14.6 to put measures on E and F satisfying inequalities of the form (14.3.2) and then applying Proposition 14.5 to the product of these two measures.

The middle inequality is trivial. For the right hand inequality let $s > \overline{\dim}_\theta E$ and $d > \overline{\dim}_B F$. We can find a cover of E by sets $\{U_i\}$ with $\delta^{1/\theta} \leq |U_i| \leq \delta$ for all i and with $\sum_i |U_i|^s \leq \varepsilon$. Then, for each i, we find a cover $\{U_{i,j}\}_j$ of F by at most $|U_i|^{-d}$ sets with diameters $|U_{i,j}| = |U_i|$ for all j. Thus $E \times F \subset \bigcup_i \bigcup_j (U_i \times U_{i,j})$ where $\delta^{1/\theta} \leq |U_i \times U_{i,j}| \leq \sqrt{2}\delta$ for all i, j. A simple estimate gives $\sum_i \sum_j |U_i \times U_{i,j}|^{s+d} \leq 2^{(s+d)/2}\varepsilon$, leading to the right hand inequality. □

Banaji [1, Theorem 5.5] extends such product inequalities to Φ-intermediate dimensions.

14.4 Some Examples

The following basic examples in $\mathbb{R}$ or $\mathbb{R}^2$ serve to give a feel for intermediate dimensions and indicate some possible behaviours of $\underline{\dim}_\theta$ and $\overline{\dim}_\theta$ as θ varies.

14.4.1 Convergent Sequences

The pth *power sequence* for $p > 0$ is given by

$$F_p = \left\{0, \frac{1}{1^p}, \frac{1}{2^p}, \frac{1}{3^p}, \dots\right\}.$$

(14.4.1)

Since F_p is countable $\dim_H F_p = 0$ and a standard exercise shows that $\dim_B F_p = 1/(p+1)$, see [7, Chapter 2]. We obtain the intermediate dimensions of F_p.

Proposition 14.9 ([12, Proposition 3.1]) *For $p > 0$ and $0 \leq \theta \leq 1$,*

$$\underline{\dim}_\theta F_p = \overline{\dim}_\theta F_p = \frac{\theta}{p + \theta}. \tag{14.4.2}$$

Sketch Proof This is clearly valid when $\theta = 0$. Otherwise, to bound $\overline{\dim}_\theta F_p$ from above, let $0 < \delta < 1$ and let $M = \lceil \delta^{-(s+\theta(1-s))/(p+1)} \rceil$. Take a covering $\mathcal{U}$ of F_p consisting of the M intervals $B(k^{-p}, \delta/2)$ of length δ for $1 \leq k \leq M$ together with $\lceil M^{-p}/\delta^\theta \rceil \leq M^{-p}/\delta^\theta + 1$ intervals of length δ^θ that cover the left hand interval $[0, M^{-p}]$. Then

$$\sum_{U \in \mathcal{U}} |U|^s \leq M\delta^s + \delta^{\theta s}\left(\frac{1}{M^p \delta^\theta} + 1\right) \tag{14.4.3}$$

$$\leq 2\delta^{(\theta(s-1)+sp)/(p+1)} + \delta^s + \delta^{\theta s} \to 0$$

as $\delta \to 0$ if $s(\theta + p) > \theta$. Thus $\overline{\dim}_\theta F_p \leq \theta/(p + \theta)$. [Note that M was chosen essentially to minimise the expression (14.4.3) for given δ.]

For the lower bound we put a suitable measure on F_p and apply Proposition 14.5. Let $s = \theta/(p + \theta)$ and $0 < \delta < 1$ and, as with the upper bound, let $M = \lceil \delta^{-(s+\theta(1-s))/(p+1)} \rceil$. Define μ_δ as the sum of point masses on the points $1/k^p$ $(1 \leq k < \infty)$ with

$$\mu_\delta\left(\left\{\frac{1}{k^p}\right\}\right) = \begin{cases} \delta^s & \text{if } 1 \leq k \leq M \\ 0 & \text{if } M+1 \leq k < \infty \end{cases}. \tag{14.4.4}$$

Then

$$\mu_\delta(F_p) = M\delta^s \geq \delta^{-(s+\theta(1-s))/(p+1)}\delta^s = 1$$

by the choice of s. To check (14.3.1), note that the gap between any two points of F_p carrying mass is at least p/M^{p+1}. A set U such that $\delta \leq |U| \leq \delta^\theta$, intersects at most $1 + |U|/(p/M^{p+1}) = 1 + |U|M^{p+1}/p$ of the points of F_p which have mass δ^s. Hence

$$\mu_\delta(U) \leq \delta^s + \frac{1}{p}|U|\delta^s \delta^{-(s+\theta(1-s))} \leq \left(1 + \frac{1}{p}\right)|U|^s,$$

Proposition 14.5 gives $\underline{\dim}_\theta F_p \geq s = \theta/(p + \theta)$.   □

Here is a generalisation of Proposition 14.9 to sequences with 'decreasing gaps'. Let $a \in \mathbb{R}$ and let $f : [a, \infty) \to (0, 1]$ be continuously differentiable with $f'(x)$ negative and increasing and $f(x) \to 0$ as $x \to \infty$. Considering integer values, the mean value theorem gives that $f(n) - f(n+1)$ is decreasing, so the sequence $\{f(n)\}_n$ is a 'decreasing sequence with decreasing gaps'.

Proposition 14.10 *With f as above, let*

$$F = \{0, f(1), f(2), \dots\}.$$

Suppose that $\dfrac{xf'(x)}{f(x)} \to -p$ *as* $x \to \infty$, *where* $0 \le p \le \infty$. *Then for all* $0 < \theta \le 1$,

$$\underline{\dim}_\theta F = \overline{\dim}_\theta F = \frac{\theta}{p+\theta},$$

taking this expression to be 0 *when* $p = \infty$.   □

This may be proved in a similar way to Proposition 14.9 using that $xf'(x)/f(x)$ is close to, rather than equal to, $-p$ when x is large.

For example, taking $f(x) = 1/\log(x+1)$, the sequence

$$F_{\log} = \left\{0, \frac{1}{\log 2}, \frac{1}{\log 3}, \frac{1}{\log 4}, \dots\right\} \tag{14.4.5}$$

has $\dim_\theta F_{\log} = 1$ if $\theta \in (0, 1]$ and $\dim_0 F_{\log} = 0$, so there is a discontinuity at 0. On the other hand, with $f(x) = e^{-x}$,

$$F_{\exp} = \left\{0, e^{-1}, e^{-2}, e^{-3}, \dots\right\}$$

has $\dim_\theta F_{\exp} = 0$ for all $\theta \in [0, 1]$.

14.4.2 Simple Examples Illustrating Different Behaviours

Using the examples above together with tools from Sect. 14.3 we can build up simple examples of sets exhibiting various behaviours as θ ranges over $[0, 1]$, shown in Fig. 14.1.

Example 14.11 (Continuous at 0, Part Constant, Then Strictly Increasing) Let $F = F_1 \cup E$ where F_1 is as in (14.4.1) and let $E \subset \mathbb{R}$ be any compact set with $\dim_H E = \overline{\dim}_B E = 1/4$ (for example a suitable self-similar set). Then

$$\dim_\theta F = \max\left\{\frac{\theta}{1+\theta}, 1/4\right\} \qquad (\theta \in [0, 1]).$$

This follows using (14.4.2) and the finite stability of upper intermediate dimensions.

Example 14.12 (Discontinuous at 0, Part Constant, Then Strictly Increasing) Let $F = F_1 \cup E$ where this time $E \subset \mathbb{R}$ is any closed countable set with $\underline{\dim}_B E =$

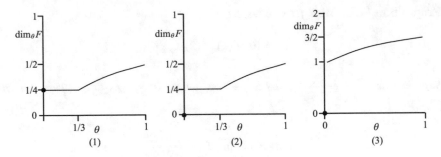

Fig. 14.1 Graphs of $\underline{\dim}_\theta F$ for the three examples in Sect. 14.4.2

$\dim_A E = 1/4$. Using Proposition 14.7 and finite stability of upper intermediate dimensions,

$$\dim_\theta F = \max\left\{\frac{\theta}{1+\theta}, 1/4\right\} \qquad (\theta \in (0, 1].$$

Note that the intermediate dimensions are exactly as in Example 14.11 except when $\theta = 0$ and a discontinuity occurs.

Example 14.13 (Discontinuous at 0, Smooth and Strictly Increasing) Consider the countable set

$$F = F_1 \times F_{\log} \subset \mathbb{R}^2.$$

Then $\dim_0 F = \dim_H F = 0$ and

$$\dim_\theta F = \frac{\theta}{1+\theta} + 1 \qquad (\theta \in (0, 1]),$$

noting that $\dim_\theta F_{\log} = \dim_B F_{\log} = \dim_A F_{\log} = 1$ for $\theta \in (0, 1]$ using (14.4.5) and Propositions 14.7 and 14.8.

14.4.3 Circles, Spheres and Spirals

Infinite sequences of concentric circles and spheres with radii tending to 0 might be thought of as higher dimensional analogues of the sets F_p defined in (14.4.1). A countable union of concentric circles will have Hausdorff dimension 1, but the box and intermediate dimensions may be greater as a result of the accumulation of circles at the centre. For $p > 0$ define the family of circles

$$C_p = \left\{x \in \mathbb{R}^2 : |x| \in F_p\right\}.$$

Tan [27] showed, using the mass distribution principle and the Frostman lemma, Proposition 14.6, that

$$\underline{\dim}_\theta C_p = \overline{\dim}_\theta C_p = \begin{cases} \frac{2p+2\theta(1-p)}{2p+\theta(1-p)} & \text{if } 0 < p \le 1 \\ 1 & \text{if } 1 \le p \end{cases}$$

with analogous formulae for concentric spheres in $\mathbb{R}^n$ and also for families of circles or spheres with radii given by other monotonic sequences converging to 0. He also considers families of points evenly distributed across such sequences of circles or spheres for which the intermediate dimension may be discontinuous at 0.

Closely related to circles are spirals. For $0 < p \le q$ define

$$S_{p,q} = \left\{ \left(\frac{1}{t^p} \sin \pi t, \frac{1}{t^q} \cos \pi t \right) : t \ge 1 \right\} \subset \mathbb{R}^2.$$

Then $S_{p,q}$ is a spiral winding into the origin, if $p = q$ it is a circular polynomial spiral, otherwise it is an elliptical polynomial spiral. Burrell, Falconer and Fraser [5] calculated that

$$\dim_\theta S_{p,q} = \overline{\dim}_\theta S_{p,q} = \begin{cases} \frac{p+q+2\theta(1-p)}{p+q+\theta(1-p)} & \text{if } 0 < p \le 1 \\ 1 & \text{if } 1 \le p \end{cases}.$$

Not unexpectedly, when $p = q$ these circular polynomial spirals have the same intermediate dimensions as the concentric circles C_p.

Another variant is the 'topologist's sine curve' given, for $p > 0$ by

$$T_p = \left\{ \left(\frac{1}{t^p}, \sin \pi t \right) : t \ge 1 \right\} \subset \mathbb{R}^2,$$

that is the graph of the function $f : (0, 1] \to \mathbb{R}$ given by $f(x) = \sin(\pi x^{-1/p})$. Tan [27] used related methods show that

$$\underline{\dim}_\theta T_p = \overline{\dim}_\theta T_p = \frac{p + 2\theta}{p + \theta},$$

as well as finding the intermediate dimensions of various generalisations of this curve.

14.5   Bedford-McMullen Carpets

Self affine carpets are a well-studied class of fractals where the Hausdorff and box-counting dimensions generally differ; this is a consequence of the alignment of the component rectangles in the iterated construction. The dimensions of planar self-affine carpets were first investigated by Bedford [2] and McMullen [24] independently, see also [25], and these carpets have been widely studied and generalised, see [6, 15] and references therein. Finding the intermediate dimensions of these carpets gives information about the range of scales of covering sets needed to realise their Hausdorff and box-counting dimensions. Deriving exact formulae seems a major challenge, but some lower and upper bounds have been obtained, in particular enough to demonstrate continuity of the intermediate dimensions at $\theta = 0$ and that they attain a strict minimum when $\theta = 0$.

Bedford-McMullen carpets are attractors of iterated function systems of a set of affine contractions, all translates of each other which preserve horizontal and vertical directions. More precisely, for integers $n > m \geq 2$, an $m \times n$-carpet is defined in the following way. Let $I = \{0, \ldots, m - 1\}$ and $J = \{0, \ldots, n - 1\}$ and let $D \subset I \times J$ be a *digit set* with at least two elements. For each $(p, q) \in D$ we define the affine contraction $S_{(p,q)} \colon [0, 1]^2 \to [0, 1]^2$ by

$$S_{(p,q)}(x, y) = \left(\frac{x + p}{m}, \frac{y + q}{n} \right).$$

Then $\{S_{(p,q)}\}_{(p,q) \in D}$ is an iterated function system so there exists a unique non-empty compact set $F \subset [0, 1]^2$ satisfying

$$F = \bigcup_{(p,q) \in D} S_{(p,q)}(F)$$

called a *Bedford-McMullen self-affine carpet*, see Fig. 14.2 for examples. The carpet can also be thought of as the set constructed using a 'template' consisting of the selected rectangles $\{S_{(p,q)}([0, 1]^2)\}_{(p,q) \in D}$ by repeatedly substituting affine copies of the template in each of the selected rectangles.

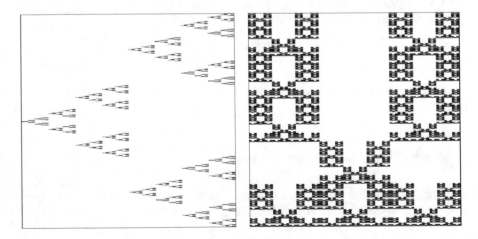

Fig. 14.2 A 2×3 and a 3×5 Bedford-McMullen carpet

Bedford [2] and McMullen [24] showed that the box-counting dimension of F exists with

$$\dim_B F = \frac{\log M}{\log m} + \frac{\log N - \log M}{\log n} \qquad (14.5.1)$$

where N is the total number of selected rectangles and M is the number of p such that there is a q with $(p, q) \in D$, that is the number of columns of the template containing at least one rectangle. They also showed that

$$\dim_H F = \frac{\log\left(\sum_{p=1}^m N_p^{\log_n m}\right)}{\log m}, \qquad (14.5.2)$$

where N_p $(1 \le p \le m)$ is the number of q such that $(p, q) \in D$, that is the number of rectangles in the pth column of the template. The Hausdorff and box-counting dimensions of F are equal if and only if the number of selected rectangles in every non-empty column is constant.

Virtually all work on these carpets depends on dividing the iterated rectangles into 'approximate squares'. The box-counting dimension result (14.5.1) is then a straightforward counting argument. The Hausdorff dimension (14.5.1) argument is more involved; McMullen's approach defined a Bernoulli-type measure μ on F via the iterated rectangles and obtained an upper bound for the local upper density of μ that is valid everywhere and a lower bound valid μ-almost everywhere. These ideas have been adapted and extended for estimating intermediate dimensions, but with the considerable complication that one seeks good density estimates that are valid over a restricted range of scales, but even getting close estimates for the intermediate dimensions seems a considerable challenge.

The best upper bounds known at the time of writing are:

$$\overline{\dim}_\theta F \le \dim_H F + \left(\frac{2\log(\log_m n)\log a}{\log n}\right)\frac{1}{-\log\theta} \quad \left(0 < \theta < \tfrac{1}{4}(\log_n m)^2\right),$$
$$(14.5.3)$$

proved in [12]. The $-1/\log\theta$ term makes this a very poor upper bound as θ increases away from 0, but at least it implies that $\underline{\dim}_\theta F$ and $\overline{\dim}_\theta F$ are continuous at $\theta = 0$ and so are continuous on [0, 1]. An upper bound for θ that is better except close to 0 was given in [21]:

$$\overline{\dim}_\theta F \le \dim_B F - \frac{\Delta_0(\theta)}{\log n}(1 - \theta) < \dim_B F \quad (\log_n m \le \theta < 1), \qquad (14.5.4)$$

where $\Delta_0(\theta)$ is the solution an equation involving a large deviation rate term which can be found numerically in particular cases. This upper bound is strictly increasing near 1 and by monotonicity also gives a constant upper bound if $0 < \theta < \log_n m$.

A reasonable lower bound that is linear in θ is

$$\underline{\dim}_\theta F \geq \dim_H F + \theta \frac{\log |D| - H(\mu)}{\log n} \quad (0 \leq \theta \leq 1), \tag{14.5.5}$$

where $H(\mu)$ is the entropy of McMullen's measure μ; this was essentially proved in [12], but see [21] for a note on the constant. In particular this implies that there is a strict minimum for the intermediate dimensions at $\theta = 0$. An alternative lower bound depending on optimising a certain function was given by [21]:

$$\underline{\dim}_\theta F \geq \sup_{t>0} \psi(t, \theta) \quad (0 \leq \theta \leq 1) \tag{14.5.6}$$

Here $\psi(t, \theta)$ depends on entropies of linear interpolants of probability measures of the form $\theta^t \widetilde{\mathbf{p}} + (1 - \theta^t)\widehat{\mathbf{p}}$ and $\theta^t \widetilde{\mathbf{q}} + (1 - \theta^t)\widehat{\mathbf{q}}$ where $\widetilde{\mathbf{p}}, \widetilde{\mathbf{q}}$ and $\widehat{\mathbf{p}}, \widehat{\mathbf{q}}$ are measures that occur naturally in the calculations for, respectively, the box-counting and Hausdorff dimensions of the carpets. Of course, the lower bounds given by Corollary 14.4 for a general F in terms of box-counting dimensions also apply here. In particular, Banaji's general lower bound [1, Proposition 3.10] in terms of the box and Assouad dimensions of F gives the best-known lower bound for θ close to 1 for some, though not all, Bedford-McMullen carpets.

Many questions on the intermediate dimensions of these carpets remain, most notably finding the exact forms of $\underline{\dim}_\theta F$ and $\overline{\dim}_\theta F$. Towards that we would at least conjecture that the lower and upper intermediate dimensions are equal and strictly monotonic.

14.6 Potential-Theoretic Formulation

The potential-theoretic approach for estimating Hausdorff dimensions goes back to Kaufman [20]. More recently box-counting dimensions have been defined in terms of energies and potentials with respect to suitable kernels and these have been used to obtain results on the box-counting dimensions of projections of sets in terms of 'dimension profiles', see [8, 9]. In particular the box-counting dimension of the projection of a Borel set $F \subset \mathbb{R}^n$ onto m-dimensional subspaces is constant for almost all subspaces (with respect to the natural invariant measure) generalising the long-standing results of Marstrand [22] and Mattila [23] for Hausdorff dimensions.

As with Hausdorff and box-counting dimensions, it turns out that θ-intermediate dimensions can be characterised in terms of capacities with respect to certain kernels, and this can be extremely useful as will be seen in Sect. 14.7. Let $\theta \in (0, 1]$

and $0 < m \leq n$ (m is often an integer, though it need not be so). For $0 \leq s \leq m$ and $0 < r < 1$, define the kernels

$$\phi_{r,\theta}^{s,m}(x) = \begin{cases} 1 & 0 \leq |x| < r \\ \left(\frac{r}{|x|}\right)^s & r \leq |x| < r^\theta \\ \frac{r^{\theta(m-s)+s}}{|x|^m} & r^\theta \leq |x| \end{cases} \qquad (x \in \mathbb{R}^n). \qquad (14.6.1)$$

If $s = m$ this reduces to

$$\phi_{r,\theta}^{m,m}(x) = \begin{cases} 1 & 0 \leq |x| < r \\ \left(\frac{r}{|x|}\right)^m & r \leq |x| \end{cases} \qquad (x \in \mathbb{R}^n), \qquad (14.6.2)$$

which are the kernels $\phi_r^m(x)$ used in the context of box-counting dimensions [8, 9]. Note that $\phi_{r,\theta}^{s,m}(x)$ is continuous in x and monotonically decreasing in $|x|$. Let $\mathcal{M}(F)$ denote the set of Borel probability measures supported on a compact $F \subset \mathbb{R}^n$. The *energy* of $\mu \in \mathcal{M}(F)$ with respect to $\phi_{r,\theta}^{s,m}$ is

$$\int \int \phi_{r,\theta}^{s,m}(x - y) \, d\mu(x) d\mu(y) \qquad (14.6.3)$$

and the *potential* of μ at $x \in \mathbb{R}^n$ is

$$\int \phi_{r,\theta}^{s,m}(x - y) \, d\mu(y). \qquad (14.6.4)$$

The *capacity* $C_{r,\theta}^{s,m}(F)$ of F is the reciprocal of the minimum energy achieved by probability measures on F, that is

$$C_{r,\theta}^{s,m}(F) = \left(\inf_{\mu \in \mathcal{M}(E)} \int \int \phi_{r,\theta}^{s,m}(x - y) \, d\mu(x) d\mu(y) \right)^{-1}. \qquad (14.6.5)$$

Since $\phi_{r,\theta}^{s,m}(x)$ is continuous in x and strictly positive and F is compact, $C_{r,\theta}^{s,m}(F)$ is positive and finite. For general bounded sets we take the capacity of a set to be that of its closure.

The existence of energy minimising measures and the relationship between the minimal energy and the corresponding potentials is standard in classical potential theory, see [8, Lemma 2.1] and [4] in this setting. In particular, there exists an *equilibrium measure* $\mu \in \mathcal{M}(E)$ for which the energy (14.6.3) attains a minimum value, say γ. Moreover, the potential (14.6.4) of this equilibrium measure is at least γ for all $x \in F$ (otherwise perturbing μ by a point mass where the potential is less than γ reduces the energy) with equality for μ-almost all $x \in F$. These properties turn out to be key in expressing these dimensions in terms of capacities.

Let $F \subset \mathbb{R}^n$ be compact, $m \in (0, n]$, $\theta \in (0, 1]$ and $r \in (0, 1)$. It may be shown that

$$\frac{\log C_{r,\theta}^{s,m}(F)}{-\log r} - s \qquad (14.6.6)$$

is continuous in s and decreases monotonically from positive when $s = 0$ to negative or 0 when $s = m$. Thus there is a unique s for which (14.6.6) equals 0. Moreover, the rate of decrease of (14.6.6) is bounded away from 0 and from $-\infty$ uniformly for $r \in (0, 1)$. This means we can pass to the limit as $r \to 0$ and for each $m \in (0, n]$ define the *lower θ-intermediate dimension profile* of $F \subset \mathbb{R}^n$ as

$$\underline{\dim}_\theta^m F = \text{ the unique } s \in [0, m] \text{ such that } \liminf_{r \to 0} \frac{\log C_{r,\theta}^{s,m}(F)}{-\log r} = s \qquad (14.6.7)$$

and the *upper θ-intermediate dimension profile* as

$$\overline{\dim}_\theta^m F = \text{ the unique } s \in [0, m] \text{ such that } \limsup_{r \to 0} \frac{\log C_{r,\theta}^{s,m}(F)}{-\log r} = s. \qquad (14.6.8)$$

Since the kernels $\phi_{r,\theta}^{t,m}(x)$ are decreasing in m the intermediate dimension profiles (14.6.7) and (14.6.8) are increasing in m.

The reason for introducing (14.6.7) and (14.6.8) is that they not only permit an equivalent definition of θ-intermediate dimensions but also give the intermediate dimensions of the images of sets under certain mappings, as we will see in Sect. 14.7. The following theorem states the equivalence between intermediate dimensions when defined by sums of powers of diameters as in Definition 14.1 and using this capacity formulation.

Theorem 14.14 *Let $F \subset \mathbb{R}^n$ be bounded and $\theta \in (0, 1]$. Then*

$$\underline{\dim}_\theta F = \underline{\dim}_\theta^n F$$

and

$$\overline{\dim}_\theta F = \overline{\dim}_\theta^n F.$$

The proof of these identities involve relating the potentials to s-power sums of diameters of covering balls of F with diameters in the required range, using a decomposition into annuli to relate this to the kernels, see [4, Section 4].

We defined the intermediate dimension profiles $\underline{\dim}_\theta^m F$ and $\overline{\dim}_\theta^m F$ for $F \subset \mathbb{R}^n$ but Theorem 14.14 refers just to the case when $m = n$. The significance of these dimension profiles when $0 < m < n$ will become clear in the next section.

14.7 Projections and Other Images

The relationship between the dimensions of a set $F \subset \mathbb{R}^n$ and its orthogonal projections $\pi_V(F)$ onto subspaces $V \in G(n, m)$, where $G(n, m)$ is the Grassmannian of m-dimensional subspaces of $\mathbb{R}^n$ and $\pi_V : \mathbb{R}^n \to V$ denotes orthogonal projection, goes back to the foundational work on Hausdorff dimension by Marstrand [22] for $G(2, 1)$ and Mattila [23] for general $G(n, m)$. They showed that for a Borel set $F \subset \mathbb{R}^n$

$$\dim_H \pi_V(F) = \min\{\dim_H F, m\} \tag{14.7.1}$$

for almost all m-dimensional subspaces V with respect to the natural invariant probability measure $\gamma_{n,m}$ on $G(n, m)$, where $\dim_H$ denotes Hausdorff dimension. Later Kaufman [20] gave a potential-theoretic proof of these results. See, for example, [11] for a survey of the many generalisations, specialisations and consequences of these projection results. In particular, there are theorems that guarantee that the lower and upper box-counting dimensions and the packing dimensions of the projections $\pi_V(F)$ are constant for almost all $V \in G(n, m)$, see [8–10, 18]. This constant value is *not* the direct analogue of (14.7.1) but rather it is given by a dimension profile of F.

Thus a natural question is whether there is a Marstrand-Mattila-type theorem for intermediate dimensions, and it turns out that this is the case with the θ-intermediate dimension profiles $\underline{\dim}_\theta^m F$ and $\overline{\dim}_\theta^m F$ defined in (14.6.7) and (14.6.8) providing the almost sure values for orthogonal projections from $\mathbb{R}^n$ onto m-dimensional subspaces. Intuitively, we think of $\underline{\dim}_\theta^m F$ and $\overline{\dim}_\theta^m F$ as the intermediate dimensions of F when regarded from an m-dimensional viewpoint.

Theorem 14.15 *Let $F \subset \mathbb{R}^n$ be bounded. Then, for all $V \in G(n, m)$*

$$\underline{\dim}_\theta \pi_V F \leq \underline{\dim}_\theta^m F \quad \text{and} \quad \overline{\dim}_\theta \pi_V F \leq \overline{\dim}_\theta^m F \tag{14.7.2}$$

for all $\theta \in (0, 1]$. Moreover, for $\gamma_{n,m}$-almost all $V \in G(n, m)$,

$$\underline{\dim}_\theta \pi_V F = \underline{\dim}_\theta^m F \quad \text{and} \quad \overline{\dim}_\theta \pi_V F = \overline{\dim}_\theta^m F \tag{14.7.3}$$

for all $\theta \in (0, 1]$. □

The upper bounds in (14.7.2) utilise the fact that orthogonal projection does not increase distances, so does not increase the values taken by the kernels, that is

$$\phi_{r,\theta}^{s,m}(\pi_V x - \pi_V y) \geq \phi_{r,\theta}^{s,m}(x - y) \quad (x, y \in \mathbb{R}^n).$$

By comparing the energy of the equilibrium measure on F with its projections onto each $\pi_V F$ it follows that $C_{r,\theta}^{s,m}(\pi_V F) \geq C_{r,\theta}^{s,m}(F)$ and using (14.6.7) or (14.6.8) gives the θ-intermediate dimensions of $\pi_V F$ as a subset of the m-dimensional space V.

The almost sure lower bounds in (14.7.3) essentially depend on the relationship between the kernels on $\mathbb{R}^n$ and on their averages over $V \in G(n, m)$. More specifically, for $m \in \{1, \ldots, n-1\}$ and $0 \le s < m$ there is a constant $a > 0$, depending only on n, m and s, such that for all $x \in \mathbb{R}^n$, $\theta \in (0, 1)$ and $0 < r < \frac{1}{2}$,

$$\int \phi_{r,\theta}^{s,m}(\pi_V x - \pi_V y) d\gamma_{n,m}(V) \le a\, \phi_{r,\theta}^{s,m}(x - y) \log \frac{r}{|x - y|}.$$

Using this for a sequence $r = 2^{-k}$ with a Borel-Cantelli argument gives (14.7.3). Full details may be found in [4, Section 5].

Theorem 14.15 has various consequences, firstly concerning continuity at $\theta = 0$.

Corollary 14.16 *Let $F \subset \mathbb{R}^n$ be such that $\underline{\dim}_\theta F$ is continuous at $\theta = 0$. Then $\underline{\dim}_\theta \pi_V F$ is continuous at $\theta = 0$ for almost all V. A similar result holds for the upper intermediate dimensions.* □

Proof If $\dim_H F \ge m$ then for almost all V, $\dim_H \pi_V(F) = m = \underline{\dim}_\theta \pi_V F$ for all $\theta \in [0, 1]$ by (14.7.1). Otherwise, for almost all V and all $\theta \in [0, 1]$,

$$\dim_H F = \dim_H \pi_V F \le \underline{\dim}_\theta \pi_V F \le \underline{\dim}_\theta^m F \le \underline{\dim}_\theta F \to \dim_H F$$

as $\theta \to 0$, where we have used (14.7.1) and (14.7.2). □

For example, taking $F \subset \mathbb{R}^2$ to be an $m \times n$ Bedford-McMullen carpet (see Sect. 14.5), it follows from (14.5.3) and Corollary 14.16 that the intermediate dimensions of projections of F onto almost all lines are continuous at 0. In fact more is true: if $\log m / \log n \notin \mathbb{Q}$ then $\underline{\dim}_\theta \pi_V F$ and $\overline{\dim}_\theta \pi_V F$ are continuous at 0 for projections onto *all* lines V, see [4, Corollaries 6.1 and 6.2] for more details.

The following surprising corollary shows that continuity of intermediate dimensions of a set at 0 is enough to imply a relationship between the *Hausdorff dimension* of a set and the *box-counting* dimensions of its projections.

Corollary 14.17 *Let $F \subset \mathbb{R}^n$ be a bounded set such that $\underline{\dim}_\theta F$ is continuous at $\theta = 0$. Then*

$$\underline{\dim}_B \pi_V F = m$$

for almost all $V \in G(n, m)$ if and only if

$$\dim_H F \ge m.$$

A similar result holds on replacing lower by upper dimensions. □

Proof The 'if' direction is clear even without the continuity assumption, since if $\dim_H F \geq m$, then

$$m \geq \underline{\dim}_B \pi_V F \geq \dim_H \pi_V F \geq m$$

for all V using (14.7.1).

On the other hand, suppose that $\underline{\dim}_B \pi_V F = m$ for almost all V. The final statement of Proposition 14.7 gives that $\underline{\dim}_\theta \pi_V F = m$ for all $\theta \in (0, 1]$ for almost all V. As $\underline{\dim}_\theta F$ is assumed continuous at $\theta = 0$, Corollary 14.16 implies that $\underline{\dim}_\theta \pi_V F$ is continuous at 0 for almost all V and so $\dim_H F = \dim_H \pi_V F = \underline{\dim}_0 \pi_V F = m$ for almost all V, using (14.7.1). $\square$

An striking example of this is given by products of the sequence sets F_p of (14.4.1) for $p > 0$. By Proposition 14.9 $\dim_B F_p = \theta/(\theta + p)$ so by Proposition 14.8

$$\dim_\theta (F_p \times F_p) = \frac{2\theta}{\theta + p} \quad (\theta \in [0, 1]),$$

which is continuous at $\theta = 0$. Since $\dim_H (F_p \times F_p) = 0$, Corollary 14.17 implies that

$$\overline{\dim}_B \pi_V (F_p \times F_p) < 1$$

for almost all V. This is particularly striking if p is close to 0, as $\dim_B (F_p \times F_p) = 2/(1 + p)$ is close to 2 but still the box-counting dimensions of its projections never reach 1. In fact, a calculation not unlike that in Proposition 14.9 shows that for all projections onto lines V, apart from the horizontal and vertical projections,

$$\overline{\dim}_B \pi_V (F_p \times F_p) = 1 - \left(\frac{p}{p+1}\right)^2.$$

Analogous ideas using dimension profiles can be used to find dimensions of images of a given set F under other parameterised families of mappings. These include images under certain stochastic processes (which are parameterised by points in the probability space). For example, let $B_\alpha : \mathbb{R} \to \mathbb{R}^m$ be index-α fractional Brownian motion where $0 < \alpha < 1$, see for example [7, Section 16.3]. The following theorem generalises the result of Kahane [19] on the Hausdorff dimension of fractional Brownian images and that of Xiao [28] for box-counting and packing dimensions of fractional Brownian images.

Theorem 14.18 *Let $F \subset \mathbb{R}^n$ be compact. Then, almost surely, for all $0 \leq \theta \leq 1$,*

$$\underline{\dim}_\theta B_\alpha(F) = \frac{1}{\alpha}\underline{\dim}_\theta^{m\alpha} F \quad \text{and} \quad \underline{\dim}_\theta B_\alpha(F) = \frac{1}{\alpha}\underline{\dim}_\theta^{m\alpha} F. \tag{14.7.4}$$

The proof of this is along the same lines as for projections, see [3] for details. The upper bound uses that for all $\varepsilon > 0$ fractional Brownian motion satisfies an almost sure Hölder condition $|B_\alpha(x) - B_\alpha(y)| \le M|x - y|^{1/2-\varepsilon}$ for $x, y \in F$, where M is a random constant. The almost sure lower bound uses that

$$\mathbb{E}\big(\phi_{r,\theta}^{sm}(B_\alpha(x) - B_\alpha(y))\big) \le c\,\phi_{r,\theta}^{sm}(x - y)$$

where c depends only on m and s.

We can get an explicit form of the intermediate dimensions of these Brownian images taking $F = F_p$ of (14.4.1).

Proposition 14.19 *For index-α Brownian motion $B_\alpha : \mathbb{R} \to \mathbb{R}$, almost surely, for all $0 \le \theta \le 1$ and $p > 0$,*

$$\underline{\dim}_\theta B_\alpha(F_p) = \overline{\dim}_\theta B_\alpha(F_p) = \frac{\theta}{p\alpha + \theta}. \tag{14.7.5}$$

In particular (14.7.5) is less than the upper bound $\theta/\alpha(p + \theta)$ that comes from directly applying the almost sure Hölder condition (14.2.1) for B_α to the intermediate dimensions of F_p.

14.8   Open Problems

Finally here are a few open questions relating to intermediate dimensions. A general problem is to find the possible forms of intermediate dimension functions. At the very least they are constrained by the inequalities of Proposition 14.2.

> **Question**
> Characterise the possible functions $\theta \mapsto \underline{\dim}_\theta F$ and $\theta \mapsto \overline{\dim}_\theta F$ that may be realised by some set $F \subset \mathbb{R}$ or $F \subset \mathbb{R}^n$.

It may be easier to answer more specific questions about the form of the dimension functions. I am not aware of any counter-example to the following suggestion.

> **Question**
> Is it true that if $\overline{\dim}_\theta F$, respectively $\underline{\dim}_\theta F$, is constant for $\theta \in [a, b]$ where $0 < a < b \le 1$ then it must be constant for $\theta \in (0, b]$?

Similarly, the following question suggested by Banaji seems open.

Question

Can $\overline{\dim}_\theta F$ or $\underline{\dim}_\theta F$ be convex functions of θ, or even (non-constant) linear functions?

As far as I know, in all cases where explicit values have been found, the intermediate dimensions equal upper bounds obtained using coverings by sets of just the two diameters $\delta^{1/\theta}$ and δ (or constant multiples thereof). It seems unlikely that this is enough for every set, indeed Kolossváry [21, Section 5] suggests that three or more diameters of covering sets may be needed to get close upper bounds for the intermediate dimensions of Bedford-McMullen carpets.

Question

Are there (preferably fairly simple) examples of sets F for which the intermediate dimensions $\overline{\dim}_\theta F$ or $\underline{\dim}_\theta F$ cannot be approximated from above using coverings by sets just of two diameters? Are there even sets where the number of different scales of covering sets needed to get arbitrary close approximations to the intermediate dimensions is unbounded?

Coming to more particular examples, the Bedford-McMullen carpets are a class of sets where current knowledge of the intermediate dimensions is limited.

Question

Find the exact form of the intermediate dimensions $\underline{\dim}_\theta F$ and $\overline{\dim}_\theta F$ for the Bedford McMullen carpets F discussed in Sect. 14.5, or at least improve the existing bounds.

Getting exact formulae for these dimensions is likely to be challenging, but better bounds, in particular the asymptotic form near $\theta = 0$ and $\theta = 1$, would be of interest. It would also be useful to know more about the behaviour of the intermediate dimensions of these carpets as functions of θ.

Question

Are the intermediate dimensions $\underline{\dim}_\theta F$ and $\overline{\dim}_\theta F$ of Bedford McMullen carpets F equal? Are they strictly increasing in θ? Are they differentiable, or even analytic, as functions of θ or can they exhibit phase transitions?

Acknowledgments The author thanks Amlan Banaji, Stuart Burrell, Jonathan Fraser, Tom Kempton and István Kolossváry for many discussions around this topic. The work was supported in part by an EPSRC Standard Grant EP/R015104/1.

References

1. A. Banaji, Generalised intermediate dimensions. arxiv: 2011.08613
2. T. Bedford, Crinkly curves, Markov partitions and box dimensions in self-similar sets. PhD dissertation, University of Warwick, 1984
3. S. Burrell, Dimensions of fractional Brownian images. arxiv: 2002.03659
4. S. Burrell, K.J. Falconer, J. Fraser, Projection theorems for intermediate dimensions. J. Fractal Geom. arxiv: 1907.07632. Online First, 1 May 2021, https://doi.org/10.4171/JFG/99
5. S. Burrell, K.J. Falconer, J. Fraser, The fractal structure of elliptical polynomial spirals. arxiv: 2008.08539
6. K.J. Falconer, Dimensions of self-affine sets: a survey, in *Further Developments in Fractals and Related Fields*, ed. by J. Barrel, S. Seuret (Birkhauser, Basel, 2013), pp. 115–134
7. K.J. Falconer, *Fractal Geometry - Mathematical Foundations and Applications*, 3rd edn. (Wiley, New York, 2014)
8. K.J. Falconer, A capacity approach to box and packing dimensions of projections and other images, in *Analysis, Probability and Mathematical Physics on Fractals*, ed. by P. Ruiz, J. Chen, L. Rogers, R. Strichartz, A. Teplyaev (World Scientific, Singapore, 2020), pp. 1–19
9. K.J. Falconer, A capacity approach to box and packing dimensions of projections of sets and exceptional directions. J. Fractal Geom. **8**, 1–26 (2021)
10. K.J. Falconer, J.D. Howroyd, Packing dimensions of projections and dimension profiles. Math. Proc. Camb. Philos. Soc. **121**, 269–286 (1997)
11. K. Falconer, J. Fraser, X. Jin, Sixty years of fractal projections, in *Fractal Geometry and Stochastics V*, ed. by C. Bandt, K. Falconer, M. Zähle. Progress in Probability, vol. 70 (Birkhäuser, Basel, 2015), pp. 3–25
12. K.J. Falconer, J.M. Fraser, T. Kempton, Intermediate dimensions. Math. Zeit. **296**, 813–830 (2020)
13. J.M. Fraser, Assouad type dimensions and homogeneity of fractals. Trans. Am. Math. Soc. **366**, 6687–6733 (2014)
14. J.M. Fraser, *Assouad Dimension and Fractal Geometry* (Cambridge University Press, Cambridge, 2020). arxiv: 2005.03763
15. J.M. Fraser, Fractal geometry of Bedford-McMullen carpets, in *These Proceedings* (2021), pp. 495–517
16. J.M. Fraser, Interpolating between dimensions, in *Fractal Geometry and Stochastics VI*, ed. by U. Freiberg, B. Hambly, M. Hinz, S. Winter. Progress in Probability, vol. 76 (Birkhäuser, Basel, 2021)
17. J.M. Fraser, H. Yu, New dimension spectra: finer information on scaling and homogeneity. Adv. Math. **329**, 273–328 (2018)
18. J.D. Howroyd, Box and packing dimensions of projections and dimension profiles. Math. Proc. Camb. Philos. Soc. **130**, 135–160 (2001)
19. J.-P. Kahane, *Some Random Series of Functions* (Cambridge University Press, Cambridge, 1985)
20. R. Kaufman, On Hausdorff dimension of projections. Mathematika **15**, 153–155 (1968)
21. I. Kolossváry, On the intermediate dimensions of Bedford-McMullen carpets. arxiv: 2006.14366
22. J.M. Marstrand, Some fundamental geometrical properties of plane sets of fractional dimensions. Proc. Lond. Math. Soc. **4**, 257–302 (1954)

23. P. Mattila, Hausdorff dimension, orthogonal projections and intersections with planes. Ann. Acad. Sci. Fenn. Ser. A I Math. **1**, 227–244 (1975)
24. C. McMullen, The Hausdorff dimension of general Sierpiński carpets. Nagoya Math. J. **96**, 1–9 (1984)
25. Y. Peres, The self-affine carpets of McMullen and Bedford have infinite Hausdorff measure. Math. Proc. Camb. Philos. Soc. **116**, 513–526 (1994)
26. J.C. Robinson, *Dimensions, Embeddings, and Attractors* (Cambridge University Press, Cambridge, 2011)
27. J.T. Tan, On the intermediate dimensions of concentric spheres and related sets. Carpets. arxiv: 2008.10564
28. Y. Xiao, Packing dimension of the image of fractional Brownian motion. Stat. Probab. Lett. **33**, 379–387 (1997)

Chapter 15
Fractal Geometry of Bedford-McMullen Carpets

Jonathan M. Fraser

Abstract In 1984 Bedford and McMullen independently introduced a family of self-affine sets now known as *Bedford-McMullen carpets*. Their work stimulated a lot of research in the areas of fractal geometry and non-conformal dynamics. In this survey article we discuss some aspects of Bedford-McMullen carpets, focusing mostly on dimension theory.

15.1   Bedford-McMullen Carpets

One of the most important and well-studied methods for generating interesting fractal sets is via iterated function systems (IFSs). Roughly speaking, an *IFS* is a finite collection of contraction mappings acting on a common compact domain, and the associated *attractor* is the unique non-empty compact set which may be expressed as the union of scaled down copies of itself under the maps in the IFS. *Self-similar sets* are attractors of IFSs where the contractions are similarities, and *self-affine sets* are attractors of IFSs where the maps act on a Euclidean domain and are affine contractions (the composition of a linear contraction and a translation). See [16] for more background on IFSs and the survey [15] for a detailed history of self-affine sets and measures as well as [7] for a recent breakthrough in the dimension theory of general self-affine sets.

Affine maps may scale by different amounts in different directions (as well as skewing and shearing) and this leads to self-affine sets being rather more complicated than self-similar sets. Bedford-McMullen carpets are the simplest possible family of (genuinely) self-affine sets. They preserve the key feature of self-affinity: different scaling in different directions, but everything else about the construction is as simple as possible. The simplicity of the model, combined with the ability to capture a key aspect of the theory, has contributed greatly to its popularity.

J. M. Fraser (✉)
University of St Andrews, St Andrews, Scotland
e-mail: jmf32@st-andrews.ac.uk

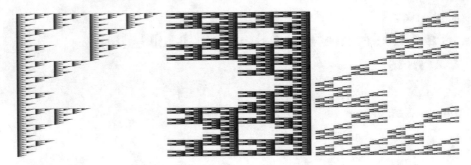

Fig. 15.1 Three examples of Bedford-McMullen carpets based on the 2×3 grid

In fact, Bedford-McMullen carpets provide an excellent example to aspiring mathematicians: a good model should reveal and capture a new phenomenon, but remain as simple as possible.

Let us first recall the Bedford-McMullen construction following [11, 49]. We work in the Euclidean plane, and begin with the unit square, $[0, 1]^2$. Fix integers $n > m > 1$, and divide the unit square into an $m \times n$ grid. Select a subset of the rectangles formed by the grid and consider the IFS consisting of the affine maps which map $[0, 1]^2$ onto each chosen rectangle preserving orientation (that is, the affine part of each map is the diagonal matrix with diagonal entries $1/m$ and $1/n$). The attractor of this IFS is a self-affine set, and such self-affine sets are known as *Bedford-McMullen carpets*, see Fig. 15.1.

Bedford-McMullen carpets also have an important role in the theory of expanding dynamical systems. Viewed as subsets of the 2-torus $[0, 1)^2$, Bedford-McMullen carpets are invariant under the toral endomorphism $(x, y) \mapsto (mx \bmod 1, ny \bmod 1)$, which provides a simple model of a non-conformal dynamical system. Since the work of Bedford and McMullen, the study of self-affine carpets has received sustained interest in the literature. Generally speaking a 'carpet' is an attractor of an IFS acting on the plane consisting of affine maps whose linear parts are given by diagonal matrices (or possibly anti-diagonal matrices). There are now many popular families of carpet, generalising the Bedford-McMullen model in various ways. Lalley-Gatzouras carpets [32] maintain the column structure but allow the diagonal matrix to vary, Barański carpets [5] maintain the grid structure but allow the matrices to vary. A crucial difference between the Barański and Lalley-Gatzouras models is that Barański allows the strongest contraction to be in either direction, whereas Lalley-Gatzouras insists that the strongest contraction be in the vertical direction. Generalising both Lalley-Gatzouras and Barański carpets is the family introduced by Feng-Wang [21] which allows arbitrary non-negative diagonal matrices and generalising the Feng-Wang family is a family we introduced which allows arbitrary diagonal and anti-diagonal matrices [25]. There are also models which step out of the carpet programme whilst maintaining several of the key features, such as excessive alignment of cylinders, for example, [42]. In order to keep this survey concise, we make no further mention of carpets outside the

Bedford-McMullen family. There is a vast literature on Bedford-McMullen carpets and as such there are a lot of interesting research directions which we will not discuss here. This survey is mostly focused on the fractal geometry and dimension theory of Bedford-McMullen carpets and associated self-affine measures.

15.2   Dimension Theory

A central aspect of fractal geometry is dimension theory. In fractal settings, fine structure makes the task of simply *defining* dimension already an interesting problem. Roughly speaking, a 'dimension' should describe how an object fills up space on small scales. There are many ways to describe this, however, and an important aspect of the subject is in understanding the relationships and differences between the many different notions of dimension; each of interest in its own right. In this survey we focus on Hausdorff, packing, box, Assouad and lower dimension, which we denote by $\dim_H$, $\dim_P$, $\dim_B$, $\dim_A$, $\dim_L$, respectively. Often one needs to consider upper and lower box dimension separately, but for the sets we discuss these coincide and so we brush over this detail. We will not define these notions here, but refer the reader to [16, 47] for the definitions and an in depth discussion of the Hausdorff, packing, and box dimensions and [28] for the Assouad and lower dimensions. It is useful to keep in mind that for all non-empty compact sets $E \subseteq \mathbb{R}^d$ (with equal upper and lower box dimension),

$$0 \leq \dim_L E \leq \dim_H E \leq \dim_P E \leq \dim_B E \leq \dim_A E \leq d.$$

Bedford [11] and McMullen [49] independently obtained explicit formulae for the Hausdorff, packing, and box dimensions of Bedford-McMullen carpets. More recently, in 2011, Mackay [45] computed the Assouad dimension, and, in 2014, Fraser computed the lower dimension [26]. We need more notation in order to state these results.

Let N be the number of maps in the defining IFS (that is the number of chosen rectangles), and let M be the number of columns containing at least one chosen rectangle. Finally, let $N_i > 0$ be the number of rectangles chosen from the ith non-empty column.

Theorem 15.2.1 *Let F be a Bedford-McMullen carpet. Then*

$$\dim_A F = \frac{\log M}{\log m} + \max_i \frac{\log N_i}{\log n},$$

$$\dim_P F = \dim_B F = \frac{\log M}{\log m} + \frac{\log(N/M)}{\log n},$$

$$\dim_H F = \frac{\log \sum_{i=1}^M N_i^{\log m / \log n}}{\log m}$$

and $$\dim_L F = \frac{\log M}{\log m} + \min_i \frac{\log N_i}{\log n}.$$

Sketch Proof We sketch the argument giving the box dimension and only discuss the rough ideas for the others. The box dimension of a bounded set E captures the polynomial growth rate of $N_r(E)$ as $r \to 0$, where $N_r(E)$ denotes the smallest number of open sets of small diameter $r \in (0, 1)$ required to cover E. That is, the box dimension can loosely be defined by $N_r(E) \approx r^{-\dim_B E}$.

Let $r > 0$ be very small and k be an integer such that $r \approx n^{-k}$. The kth level cylinders in the construction of the carpet F are rectangular sets of height $\approx r$ and length m^{-k} (which is rather longer). Therefore, when looking for optimal r-covers of F we may treat the kth level cylinders separately. Let l be an integer such that $r \approx m^{-l}$ and consider the lth level cylinders inside a given kth level cylinder. This forms a grid and cylinders in the same column may be covered efficiently by ≈ 1 set of diameter r and therefore we only need to count the non-empty columns, of which there are M^{l-k}. The total number of kth level cylinders is N^k and therefore

$$N_r(F) \approx N^k M^{l-k} \approx r^{-(\log N / \log n + \log M / \log m - \log M / \log n)}$$

as required.

The Hausdorff dimension is more awkward to compute. The lower bound is usually handled via measures, either by the mass distribution principle (see [16, Chapter 4]) or by direct computation of the Hausdorff dimension of a suitable measure. In fact we sketch this part of the proof in Sect. 15.4. The upper bound is proved by a delicate covering argument. The key difference between this argument and the covering argument we sketched for box dimension above is that Hausdorff dimension allows different sized sets in the cover and so the difficulty is in deciding which cylinders to cover together and which to break up into smaller pieces. A more direct approach to proving the upper bound, which ultimately boils down to constructing a delicate cover, is to show that the lower local dimension of a suitable measure is at most h for all points $x \in F$, where $h = \dim_H F$ is the intended Hausdorff dimension. In contrast to the mass distribution principle, which asks for the measure of a ball never to be too big, this approach asks for balls around all points to have large mass infinitely often. See [16, Chapter 4] for more on this approach to finding upper bounds for Hausdorff dimension in general. The McMullen measure, defined later by (15.4.2), can be used for both the upper and lower bounds.

The Assouad and lower dimensions are in some sense dual to each other and so we only discuss Assouad dimension. The lower bound is most efficiently proved via *weak tangents*. See [28, Chapter 5] for more on weak tangents in the context of dimension theory. Mackay [45] constructed a weak tangent which is the product of two self-similar sets, one of dimension $\log M / \log m$ (the projection of F onto the first coordinate) and the other of dimension $\max_i \log N_i / \log n$ (the maximal vertical slice of F). Then the lower bound follows since the Hausdorff dimension of a weak tangent is a lower bound for the Assouad dimension. Mackay proved the upper bound by a direct covering argument, similar to that sketched above for box dimension. An alternative argument giving the upper bound using the Assouad dimension of measures is given in [29]. □

The tangent structure of self-affine sets is particularly interesting since the small scale structure typically differs greatly from the large scale structure. This is very different from self-similar sets, for example. As mentioned in the above proof, Mackay [45] used tangent sets with a product structure to study the Assouad dimension. This product structure is seen much more generally. Bandt and Käenmäki [4] gave a general description of the tangent structure of Bedford-McMullen carpets in the case where $M = m$. This result has been generalised in various ways, see [2, 10, 36, 37].

Returning to Theorem 15.2.1, the following amusing commonality was pointed out to me by Kenneth Falconer. For $p \in [0, \infty] \cup \{-\infty\}$, write

$$\|\underline{N}\|_p = \left(\frac{1}{M} \sum_{i=1}^{M} N_i^p \right)^{1/p}$$

for the "pth average" of the vector $\underline{N} := (N_1, \ldots, N_M)$ describing the number of rectangles in each of the non-empty columns. We adopt the natural interpretation of $\|\underline{N}\|_\infty = \max_i N_i$ and $\|\underline{N}\|_{-\infty} = \min_i N_i$. Then, the expression

$$\frac{\log M}{\log m} + \frac{\log \|\underline{N}\|_p}{\log n} \tag{15.2.1}$$

gives the Assouad dimension when $p = \infty$, the box and packing dimensions when $p = 1$, the Hausdorff dimension when $p = \log m / \log n$, and the lower dimension when $p = -\infty$. This observation warrants the question of whether there are sensible, perhaps yet undiscovered, notions of fractal dimension corresponding to other values of p. Moreover, the expression (15.2.1) has a useful interpretation as the dimension of the projection of F onto the first coordinate plus the 'average' column dimension.

It is no surprise that the box and packing dimensions coincide in Theorem 15.2.1. Indeed, the packing dimension and upper box dimensions coincide much more generally: see [16, Corollary 3.10], which applies to very general IFS attractors. This identity aside, we see that Bedford-McMullen carpets provide an excellent model for understanding the differences between the different notions of dimension. If all non-empty columns contain the same number of rectangles (that is, $N_i = N/M$ for all i), then we say the carpet has *uniform fibres* and otherwise it has *non-uniform fibres*. There is a simple dichotomy: in the uniform fibres case

$$\dim_L F = \dim_H F = \dim_B F = \dim_A F$$

and, in the non-uniform fibres case,

$$\dim_L F < \dim_H F < \dim_B F < \dim_A F.$$

The question of Hausdorff and packing *measure* for Bedford-McMullen carpets is subtle. Peres [52, 53] proved that in the non-uniform fibres case both the Hausdorff and packing measures are infinite in their respective dimensions. It is instructive to compare this with the situation for self-similar sets where the open set condition is enough to ensure that the Hausdorff and packing measures are positive and finite in their dimension. We write $\mathscr{P}^h$ and $\mathscr{H}^h$ for the packing and Hausdorff measures with respect to a gauge function h, see [16]. In [52] it is shown that $\mathscr{P}^h(F) = \infty$ for

$$h(x) = \frac{x^{\dim_P F}}{|\log x|}$$

but $\mathscr{P}^h(F) = 0$ for

$$h(x) = \frac{x^{\dim_P F}}{|\log x|^{1+\varepsilon}}$$

for all $\varepsilon \in (0, 1)$. In [53] it is shown that $\mathscr{H}^h(F) = \infty$ for

$$h(x) = x^{\dim_H F} \exp\left(\frac{-c|\log x|}{(\log|\log x|)^2}\right)$$

with $c > 0$ small enough, but $\mathscr{H}^h(F) = 0$ for

$$h(x) = x^{\dim_H F} \exp\left(\frac{-|\log x|}{(\log|\log x|)^{2-\varepsilon}}\right)$$

for all $\varepsilon \in (0, 1)$. In particular these results show that

$$\mathscr{H}^{\dim_H F}(F) = \mathscr{P}^{\dim_P F}(F) = \infty.$$

15.3 Interpolating Between Dimensions

A new perspective in dimension theory is that of 'dimension interpolation', see [27]. Roughly speaking, the idea is to consider two distinct notions of dimension dim and Dim, which satisfy $\dim E \leq \operatorname{Dim} E$ for all sets $E \subseteq \mathbb{R}^d$, and introduce a continuously parametrised family of dimensions $\dim_\theta$ for $\theta \in (0, 1)$ which satisfy $\dim E \leq \dim_\theta E \leq \operatorname{Dim} E$. Crucially, the dimensions $\dim_\theta$ should capture some key features of both dim and Dim in a geometrically interesting way and the hope is to provide more nuanced information than that provided by dim and Dim when considered in isolation. Since the most commonly studied dimensions are typically distinct in this case, Bedford-McMullen carpets provide the ideal testing ground for this approach.

15.3.1 The Assouad Spectrum

The Assouad spectrum, introduced in [30] and denoted by $\dim_A^\theta$, interpolates between the (upper) box dimension and the (quasi-)Assouad dimension. The Assouad dimension of a bounded set $E \subseteq \mathbb{R}^d$ is defined by considering $N_r(B(x, R) \cap E)$, that is, the size of an optimal r-cover of an R-ball where $0 < r < R$ are two independent scales. The Assouad spectrum fixes the relationship between these two scales by forcing $R = r^\theta$, where $\theta \in (0, 1)$ is the interpolation parameter. The result is a family of dimensions $\dim_A^\theta E$ which is continuous in θ and satisfies $\dim_B E \leq \dim_A^\theta E \leq \dim_A E$ for all $\theta \in (0, 1)$. The analogous lower spectrum $\dim_L^\theta$ is defined by a similar modification of the definition of lower dimension. The Assouad and lower spectra of Bedford-McMullen carpets were computed in [31]. We write $N_{\max} = \max_i N_i$ and $N_{\min} = \min_i N_i$.

Theorem 15.3.1 *Let F be a Bedford-McMullen carpet. Then, for all $0 < \theta \leq \log m / \log n$,*

$$\dim_A^\theta F = \frac{\log M - \theta \log(N/N_{\max})}{(1 - \theta) \log m} + \frac{\log(N/M) - \theta \log N_{\max}}{(1 - \theta) \log n}$$

and

$$\dim_L^\theta F = \frac{\log M - \theta \log(N/N_{\min})}{(1 - \theta) \log m} + \frac{\log(N/M) - \theta \log N_{\min}}{(1 - \theta) \log n}$$

and, for all $\log m / \log n \leq \theta < 1$,

$$\dim_A^\theta F = \frac{\log M}{\log m} + \frac{\log N_{\max}}{\log n}$$

and

$$\dim_L^\theta F = \frac{\log M}{\log m} + \frac{\log N_{\min}}{\log n}.$$

In both the Assouad and lower spectrum there is a single phase transition at $\theta = \frac{\log m}{\log n}$ (see Fig. 15.2). In many other examples where the Assouad spectrum is known there is a similar phase transition occurring at a value of θ with a particular geometric significance, see [28]. In this case it is the ratio of the Lyapunov exponents

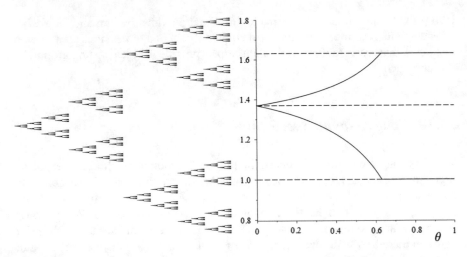

Fig. 15.2 Left: a Bedford McMullen carpet with $m = 2$, $n = 3$ and $N_1 = 1 < N_2 = 2$. Right: plots of the Assouad and lower spectra. Dotted lines at the lower, box, and Assouad dimensions are shown for comparison

(for any ergodic measure) or the 'logarithmic eccentricity' of the cylinders in the construction.

The curious reader may at this point wonder if the Assouad spectrum is a partial solution to the problem of finding dimensions satisfying (15.2.1). By continuity and monotonicity, for each $0 < \theta < \frac{\log m}{\log n}$ there is a unique $p \in (1, \infty)$ such that

$$\dim_A^\theta F = \frac{\log M}{\log m} + \frac{\log \|\underline{N}\|_p}{\log n},$$

however, the function mapping θ to p is not satisfying: it depends too strongly on F. Ideally we would want a function depending only on θ and possibly m and n.

15.3.2 Intermediate Dimensions

The intermediate dimensions, denoted by $\dim_\theta$ and introduced in [19], interpolate between the (upper or lower) box dimension and the Hausdorff dimension. The Hausdorff and box dimension are both defined by considering efficient covers of the set. The Hausdorff dimension places no restriction on the relative sizes of the sets used in the cover and weights their contribution to the dimension according to their size (see definition of Hausdorff measure [16]) and box dimension considers covers by sets of the same size. The intermediate dimensions impose partial restrictions on the relative sizes of the covering sets by insisting that $|U| \leq |V|^\theta$ for all covering sets U, V with diameters $|U|, |V| \leq 1$. The intermediate dimensions are continuous in $\theta \in (0, 1]$ with $\dim_1 E = \dim_B E$ and satisfy

$$\dim_H E \leq \dim_\theta E \leq \dim_B E$$

for all bounded $E \subseteq \mathbb{R}^d$. The intermediate dimensions are not necessarily continuous at $\theta = 0$, that is, they do not necessarily approach the Hausdorff dimension as $\theta \to 0$. Establishing continuity at 0 for particular examples turns out to be a key problem. For example, if the intermediate dimensions are continuous at 0, then strong applications can be derived concerning the box dimensions of projections and images under stochastic processes, see [12, 13].

Computing an explicit formula for the intermediate dimensions of Bedford-McMullen carpets seems to be a difficult problem, investigated in [19] and [41]. See also the survey [17]. We summarise what is known so far, for F a Bedford-McMullen carpet with non-uniform fibres:

1. $\dim_\theta F$ is continuous at 0, and therefore interpolates between the Hausdorff and box dimensions [19];
2. for all $\theta \in (0, 1)$, $\dim_\theta F > \dim_H F$ [19];
3. for all $\theta \in (0, 1)$, $\dim_\theta F < \dim_B F$ [41];
4. there are upper and lower bounds for $\dim_\theta F$ with the best known given in [41];
5. $\dim_\theta F$ is not necessarily concave [41], which is noteworthy since most of the basic examples have turned out to have concave intermediate dimensions. □

15.4 Invariant Measures

The interplay between invariant sets and invariant measures is central in fractal geometry and ergodic theory. There are many natural $(\times m, \times n)$ invariant measures living on Bedford-McMullen carpets but perhaps the most natural are the self-affine measures. Given a Bedford-McMullen carpet with first level rectangles indexed by the set $\{1, \ldots, N\}$, associate a probability vector $\{p_1, \ldots, p_N\}$, that is, with $0 < p_d < 1$ for all d and $\sum_d p_d = 1$. Let μ be the measure formed by iteratively subdividing unit measure among the N rectangles at each stage in the construction of the carpet F according to the probability vector. More formally, let $\mathbb{P}$ denote the Bernoulli measure on the symbolic space $\{1, \ldots, N\}^{\mathbb{N}}$ consisting of infinite one-sided words over the alphabet $\{1, \ldots, N\}$. Then $\mu = \mathbb{P} \circ \Pi^{-1}$ where Π is the associated coding map which sends a word $(d_1, d_2, \ldots)$ to the point in F coded by the corresponding sequence of rectangles. In particular, μ is a Borel probability measure fully supported on F and invariant under the endomorphism $(\times m, \times n)$. The measure μ is a self-affine measure since it is the unique Borel probability measure invariant under the IFS weighted by the probability vector. There is a rich dimension theory of measures, which we will not dwell on here. Measures invariant under nice enough dynamical systems are often 'exact dimensional', which means that many of the familiar notions of dimension for measures coincide (e.g.

Hausdorff, packing, entropy dimensions). A Borel measure ν is *exact dimensional* (with dimension α) if for ν almost all x

$$\lim_{r \to 0} \frac{\log \nu(B(x, r))}{\log r} = \alpha. \qquad (15.4.1)$$

The expression on the left is the *local dimension* of ν at x (when the limit exists) and so exact dimensionality can be characterised as the local dimension being almost surely constant. Self-affine measures on Bedford-McMullen carpets are known to be exact dimensional and, moreover, the dimension satisfies the *Ledrappier-Young formula*. The Ledrappier-Young formula stems from influential papers [43, 44] which established a deep connection between dimension, entropy and Lyapunov exponents in the general context of measures invariant under C^1-diffeomorphisms. Recently there has been a lot of progress establishing Ledrappier-Young formulae for measures invariant under expanding maps, such as those associated with IFSs. In particular, it is now known that *all* self-affine measures are exact dimensional and satisfy the appropriate Ledrappier-Young formula, see [6, 8, 20]. The case of self-affine measures on Bedford-McMullen carpets was resolved in [38]. In this setting the Lyapunov exponents are $\log m < \log n$ and the entropy is given by

$$h(\mu) = -\sum_{d=1}^{N} p_d \log p_d.$$

The entropy of the projection $\pi\mu$ of μ onto the first coordinate also plays a role. Here we have to sum weights belonging to the same column and so the entropy is

$$h(\pi\mu) = -\sum_{i=1}^{M} \left(\sum_{d:\pi d=i} p_d \right) \log \left(\sum_{d:\pi d=i} p_d \right).$$

Theorem 15.4.1 *Self-affine measures μ on Bedford-McMullen carpets are exact dimensional with dimension given by*

$$\dim \mu = \frac{h(\pi\mu)}{\log m} + \frac{h(\mu) - h(\pi\mu)}{\log n}.$$

We give a sketch proof here, which is deliberately not at all rigorous but hopefully shows where the formula comes from and what the key ideas are. We learned this argument from Thomas Jordan and Natalia Jurga, although it is based on a proof from [54] which applies to more general constructions without an underlying grid structure.

Sketch Proof Let $r > 0$ be very small and $x \in F$ be a 'μ-typical point', which is to say we have selected it from a set of large μ measure guaranteeing it to behave as expected. This is made precise using the ergodic theorem and Egorov's theorem.

Let k be such that $r \approx n^{-k}$ and let $E \subseteq F$ be the kth-level cylinder (basic rectangle after k steps in the IFS construction) which contains x. Since x is typical, we can assume this is uniquely defined. We wish to estimate the measure of $B(x, r)$ which, since the height of E is $\approx r$, is roughly the measure of a vertical strip of width r passing through E 'centred' at x. Since x is typical, the measure of E is roughly $\exp(-kh(\mu))$ (recall the Shannon-McMillan-Breiman theorem). Moreover, using the self-affinity of μ, the proportion of the measure of E which lies inside the vertical strip we are interested in will be (roughly) the same as the proportion of an rm^k-ball centred at a $\pi\mu$-typical point x' in the projection πF. The m^k factor comes from scaling the base of E up to match the unit interval (where $\pi\mu$ lives). Since $\pi\mu$ is a self-similar measure satisfying the open set condition, it is well-known and easily shown that it is exact dimensional with dimension given by entropy divided by Lyapunov exponent, that is,

$$\dim \pi\mu = \frac{h(\pi\mu)}{\log m}.$$

Therefore,

$$\mu(B(x, r)) \approx \mu(E) \cdot \pi\mu(B(x', rm^k)) \approx \exp(-kh(\mu)) \left(rm^k\right)^{\frac{h(\pi\mu)}{\log m}}.$$

Then, using $k \approx -\log r / \log n$,

$$\frac{\log \mu(B(x, r))}{\log r} \approx \frac{h(\pi\mu)}{\log m} + \frac{h(\mu) - h(\pi\mu)}{\log n}$$

and the error vanishes as $r \to 0$. $\qquad\square$

The (Hausdorff) dimension of a measure cannot exceed the Hausdorff dimension of its support and so it is natural to ask if the Hausdorff dimension of a Bedford-McMullen carpet can be realised as the Hausdorff dimension of an invariant measure. Define μ by

$$p_d = N_i^{(\log m / \log n) - 1} / m^{\dim_H F} \tag{15.4.2}$$

for all d corresponding to rectangles in the ith non-empty column. Note that $\sum_{d=1}^{N} p_d = 1$, by Theorem 15.2.1. Then, applying Theorem 15.4.1, we have

$$\dim \mu = \dim_H F.$$

This self-affine measure is known as the *McMullen measure* and was first used in [49]. In fact, the McMullen measure is the unique invariant probability measure of maximal Hausdorff dimension, see [38].

15.5   Dimensions of Projections onto Lines

How fractal sets and measures behave under projection onto subspaces is a well-studied and central question in fractal geometry and geometric measure theory. We refer the reader to [18, 48] for an overview of the dimension theory of projections in general. Here we focus only on the planar case and projections of Bedford-McMullen carpets. Consider the set of lines L passing through the origin in $\mathbb{R}^2$ and write π_L for orthogonal projection from $\mathbb{R}^2$ onto the line L. In 1954 Marstrand [46] proved that, for Borel $E \subseteq \mathbb{R}^2$,

$$\dim_H \pi_L E = \min\{\dim_H E, 1\} \qquad\qquad (15.5.1)$$

for almost all L. Here 'almost all' is with respect to the natural length measure on the space of lines. This result stimulated much further work in the area and recently there has been a lot of activity concerning the exceptional set of lines in (15.5.1); especially when the exceptional set can be shown to be empty or small in some other sense, see [57]. Ferguson, Jordan and Shmerkin [24] proved the following projection theorem for Bedford-McMullen carpets.

Theorem 15.5.1 *Let F be a Bedford-McMullen carpet with $\log m / \log n \notin \mathbb{Q}$. Then*

$$\dim_H \pi_L F = \min\{\dim_H F, 1\}$$

*for **all** L apart from possibly when L is one of the two principle coordinate axes.*   □

The strategy for proving this result is to first establish it in the uniform fibres case and then upgrade to the general case by approximating the carpet from within by subsystems with uniform fibres. Proving the result in the uniform fibres setting is far from straightforward, but we will focus on the 'approximating from within' part of the proof. This trick seems to have useful applications elsewhere.

Lemma 15.5.2 *Let F be a Bedford-McMullen carpet and $\varepsilon > 0$. Then there exists a subset $E \subseteq F$ which is itself a Bedford-McMullen carpet with uniform fibres and, moreover, satisfies*

$$\dim_H E \geq \dim_H F - \varepsilon.$$

Sketch Proof Let

$$p_d = N_i^{(\log m / \log n)-1} / m^{\dim_H F}$$

be the probability weights defining the McMullen measure, see (15.4.2), that is the unique invariant probability measure of maximal dimension. For a large integer $k \geq 1$, let

$$l(k) = \sum_{d=1}^{N} \lfloor kp_d \rfloor$$

and

$$\mathscr{I}(k) = \left\{ (d_1, \ldots, d_{l(k)}) \in \{1, \ldots, N\}^{l(k)} \right.$$

$$\left. : \text{for all } 1 \le d \le N, \#\{t : d_t = d\} = \lfloor kp_d \rfloor \right\}.$$

Here, $\mathscr{I}(k)$ is a subset of the $l(k)$th iteration of the defining IFS for F chosen such that digits appear with the 'correct frequency', as determined by the McMullen measure.

Consider the IFS consisting of compositions of maps according to $\mathscr{I}(k)$ and denote its attractor by $E(k)$. It follows that $E(k)$ is a Bedford-McMullen carpet with uniform fibres. Then Theorem 15.2.1 gives

$$\dim_{\mathrm{H}} E(k) = \frac{\log M(k)}{\log m(k)} + \frac{\log(N(k)/M(k))}{\log n(k)}$$

where $m(k) = m^{l(k)}$ and $n(k) = n^{l(k)}$ are the integers defining the grid associated with $E(k)$ and $M(k)$ and $N(k)$ are the number of non-empty columns and the total number of rectangles in the construction of $E(k)$, respectively. Moreover,

$$N(k) = \frac{l(k)!}{\prod_{d=1}^{N} \lfloor kp_d \rfloor!}$$

and

$$M(k) = \frac{l(k)!}{\prod_{i=1}^{M} \left(\sum_{d:\pi d=i} \lfloor kp_d \rfloor \right)!}.$$

We can apply Stirling's approximation to estimate the dimension of $E(k)$ from below. A lengthy but straightforward calculation yields

$$\dim_{\mathrm{H}} E(k) = \frac{\log l(k)! - \sum_{i=1}^{M} \log \left(\sum_{d:\pi d=i} \lfloor kp_d \rfloor \right)!}{l(k) \log m}$$

$$+ \frac{\sum_{i=1}^{M} \log \left(\sum_{d:\pi d=i} \lfloor kp_d \rfloor \right)! - \sum_{d=1}^{N} \log \lfloor kp_d \rfloor!}{l(k) \log n}$$

$$\ge s - \varepsilon(k)$$

where $\varepsilon(k) \to 0$ as $k \to \infty$, proving the lemma. $\qquad\square$

We give one further example where Lemma 15.5.2 can be applied. The *modified lower dimension*, denoted by $\dim_{\mathrm{ML}}$, is a modification of the lower dimension to make it monotone, see [28]. Specifically, it is defined as $\dim_{\mathrm{ML}} F = \sup\{\dim_{\mathrm{L}} E : E \subseteq F\}$ and for compact sets F we have $\dim_{\mathrm{L}} F \leq \dim_{\mathrm{ML}} F \leq \dim_{\mathrm{H}} F$. Perhaps surprisingly it turns out to be equal to the *Hausdorff* dimension, not the lower dimension, in the case of Bedford-McMullen carpets. This was first observed in [31].

Corollary 15.5.3 *Let F be a Bedford-McMullen carpet. Then*

$$\dim_{\mathrm{ML}} F = \dim_{\mathrm{H}} F = \frac{\log \sum_{i=1}^{M} N_i^{\log m / \log n}}{\log m}.$$

Corollary 15.5.3 follows immediately from Lemma 15.5.2. This shows how to construct subsets of carpets which have lower dimension arbitrarily close to the Hausdorff dimension of the original carpet, using the fact that the carpets E provided by Lemma 15.5.2 have uniform fibres. This proves the lower bound and the upper bound always holds.

Ferguson et al. [22] considered the projections of measures supported on Bedford-McMullen carpets. Here the subsystem argument is not applicable and the proof relies on CP-chains and theory developed by Hochman and Shmerkin [33].

Theorem 15.5.4 *Let μ be a self-affine measure supported on a Bedford-McMullen carpet with $\log m / \log n \notin \mathbb{Q}$. Then*

$$\dim_{\mathrm{H}} \pi_L \mu = \min\{\dim_{\mathrm{H}} \mu, 1\}$$

*for **all** L apart from possibly when L is one of the two principle coordinate axes.* □

Theorem 15.5.4 was subsequently generalised by Almarza [3] to include Gibbs measures on transitive subshifts of finite type.

Related to the dimension theory of orthogonal projections, is the dimension theory of slices, see [47, Chapter 10]. In the plane, a *slice* is the intersection of a given set with a line. Write $L^{\perp}$ for the orthogonal complement of a line $L \subseteq \mathbb{R}^2$ and let $E \subseteq \mathbb{R}^2$ be a Borel set with $\mathscr{H}^s(E) > 0$. Then 'in typical directions, there are many big slices', that is,

$$\dim_{\mathrm{H}} E \cap (L^{\perp} + x) \geq s - 1$$

for almost all L and positively many $x \in L$. Here 'positively many' x means there is a set of x of positive length. Moreover, for an arbitrary Borel set $E \subseteq \mathbb{R}^2$ 'in all directions typical slices cannot be too big', that is,

$$\dim_{\mathrm{H}} E \cap (L^{\perp} + x) \leq \max\{\dim_{\mathrm{H}} E - 1, 0\}$$

for all L and almost all $x \in L$. Again, there has been a lot of interest in the dimension theory of slices in specific situations, such as for dynamically defined sets E. Slices of Bedford-McMullen carpets were considered by Algom [1] and the following was proved.

Theorem 15.5.5 *Let F be a Bedford-McMullen carpet with $\log m / \log n \notin \mathbb{Q}$. Then*

$$\dim_B F \cap (L^\perp + x) \leq \max\{\dim_A F - 1, 0\}$$

for all L and all $x \in L$ provided L is not one of the two principle coordinate axes.

□

15.6 Survivor Sets, Hitting Problems, and Diophantine Approximation

In this section we briefly touch upon a large and varied literature concerning the study of dynamically or number theoretically defined subsets of a given fractal. The literature goes far beyond the Bedford-McMullen setting but, as usual, Bedford-McMullen carpets provide an excellent testing ground for the theory. For example, the 'survivor set problem' studies points which do not fall into a given 'hole' under iteration of the $(\times m, \times n)$ dynamics. The 'hitting target problem' considers the complementary phenomenon where one focuses on points which hit a given target. Then there are various related problems in Diophantine approximation, where one is interested in how well points may be approximated by rationals. This may be interpreted as hitting a prescribed target.

Ferguson et al. [23] considered the survivor set problem as follows. Let T denote the $(\times m, \times n)$ endomorphism which leaves a given Bedford-McMullen carpet F invariant. Let $U \subseteq F$ be an open set (in the subspace topology) and define the *survivor set* by

$$F_U = \{x \in F : T^k(x) \notin U \text{ for all } k\}.$$

The dimensions of F_U were considered in [23] where U is a fixed finite collection of open cylinders, or a shrinking metric ball. They found that the box dimension is related to the escape rate of the measure of maximal entropy through the hole, and the Hausdorff dimension is related to the escape rate of the measure of maximal dimension.

Bárány and Rams [9] considered the shrinking target problem as follows. Let $B_k \subseteq F$ be a sequence of targets where each B_k is a 'dynamically defined rectangle'. Let

$$\Gamma = \{x \in F : T^k(x) \in B_k \text{ for infinitely many } k\}$$

that is, the set of points which, upon iteration of T, hit the (moving) target infinitely often. The Hausdorff dimension of Γ is given in [9] in terms of various complicated entropy functions.

A point $x \in \mathbb{R}^d$ is said to be *badly approximable* if there exists $c > 0$ such that, for all $\mathbf{p} = (p_1, \dots, p_d) \in \mathbb{Z}^d$ and all $q \in \mathbb{N}$,

$$\|x - \mathbf{p}/q\|_\infty \geq \frac{c}{q^{1+1/d}}.$$

That is, the badly approximable numbers are those for which Dirichlet's theorem can be improved by at most a constant factor. The badly approximable numbers $B(d)$ have zero Lebesgue measure but full Hausdorff dimension in $\mathbb{R}^d$ and a well-studied problem is to determine how $B(d)$ intersects a given fractal set. Das et al. [14] proved that

$$\dim_H B(2) \cap F = \dim_H F \qquad (15.6.1)$$

when F is a Bedford-McMullen carpet with at least two non-empty columns and at least two non-empty rows. Interestingly, there is a connection with the (modified) lower dimension here. It is proved in [14] that for an arbitrary closed set $E \subseteq \mathbb{R}^d$ satisfying a natural non-degeneracy condition called 'hyperplane diffuseness', we have

$$\dim_H B(d) \cap E \geq \dim_L E.$$

The result (15.6.1) may then be deduced from this and Corollary 15.5.3.

15.7 Multifractal Analysis

We saw in Theorem 15.4.1 that self-affine measures on Bedford-McMullen carpets are exact dimensional, meaning that the local dimension exists and takes a common value at almost every point. It is an interesting and difficult problem to study the exceptional set, that is, the set of points where the local dimension is not as expected. This is a μ-null set, but turns out to have full Hausdorff dimension, $\dim_H F$. This type of problem is common in multifractal analysis, see [16, Chapter 17]. Let $\alpha \geq 0$ and form multifractal decomposition sets

$$\Delta(\alpha) = \{x \in F : \dim_{\text{loc}}(\mu, x) = \alpha\},$$

where $\dim_{\text{loc}}(\mu, x)$ is the local dimension of μ at x, recall (15.4.1). In order to understand the fractal complexity of $\Delta(\alpha)$, and thus μ, define the *Hausdorff* and *packing multifractal spectra* as

$$f_H^\mu(\alpha) = \dim_H \Delta(\alpha)$$

and

$$f_P^\mu(\alpha) = \dim_P \Delta(\alpha),$$

respectively. It is not useful to define box or Assouad multifractal spectra here since the sets $\Delta(\alpha)$ tend to be dense in F and therefore we need a dimension which is not stable under taking closure to distinguish between different α.

Multifractal analysis has been considered in great detail in the context of self-affine measures on Bedford-McMullen carpets. These measures constitute one of the most complicated examples where the Hausdorff multifractal spectrum is known and given by an explicit formula. That said, many interesting questions remain.

Closely connected to multifractal analysis is the study of the L^q-spectrum. Given a Borel probability measure μ, the L^q-spectrum of μ is a function $\tau_\mu : \mathbb{R} \to \mathbb{R}$ which captures the coarse structure of the measure by considering qth-moment type expressions. Many interesting fractal features may be analysed via this function. For example, $\tau_\mu(0)$ coincides with the box dimension of the support of the measure and, provided τ_μ is differentiable at $q = 1$, $\dim_H \mu = -\tau_\mu'(1)$, see [50]. More importantly for us, one always has

$$f_H^\mu(\alpha) \le f_P^\mu(\alpha) \le \tau_\mu^*(\alpha) \tag{15.7.1}$$

where $\tau_\mu^*(\alpha)$ is the Legendre transform of τ_μ. In many cases of interest there is equality throughout in (15.7.1) in which case we say the multifractal formalism holds. For example, this holds for self-similar measures satisfying the open set condition. Self-affine measures on Bedford-McMullen carpets fail to satisfy the multifractal formalism in this sense in general but, nevertheless, the Hausdorff multifractal spectrum is known and is given by the Legendre transform of an auxiliary moment scaling function $\beta : \mathbb{R} \to \mathbb{R}$. This was proved by King [40] assuming an additional separation condition known as the *very strong separation condition*, and in full generality by Jordan and Rams [34]. Moreover, the L^q-spectrum is also known and given by an explicit formula. This is due to Olsen [51].

Theorem 15.7.1 *Let μ be a self-affine measure on a Bedford-McMullen carpet. There is an explicitly defined real analytic moment scaling function $\beta : \mathbb{R} \to \mathbb{R}$ such that $f_H^\mu(\alpha) = \beta^*(\alpha)$. Moreover, the L^q-spectrum is real analytic and given by an explicit formula. In general β and τ_μ do not coincide.* □

The above theorem provides explicit upper and lower bounds for the packing multifractal spectrum. The problem of computing the packing multifractal spectrum in general was considered in detail by Reeve [56] and Jordan and Rams [35] and it turns out to be a subtle problem. Reeve [56] considered multifractal analysis of Birkhoff averages, which is different to the multifractal analysis of local dimensions we consider here. However, in certain cases they can be related and it was shown in [56] that the upper bound given by the Legendre transform of the L^q-spectrum is generally not sharp. In [35] it was shown that usually the packing multifractal spectrum does not peak at the packing dimension of the Bedford-McMullen carpet.

This is in stark contrast to the Hausdorff case where the Hausdorff multifractal spectrum always peaks at the Hausdorff dimension of the carpet. In [35] the packing multifractal spectrum was computed for a special family of self-affine measures supported on Bedford-McMullen carpets. The carpet was allowed only two non-empty columns and the same Bernoulli weight was associated to each rectangle in the same column. Within this class they were able to show that the packing multifractal spectrum can be discontinuous as a function of the Bernoulli weights. Again, this is in stark contrast to the Hausdorff case.

Related to multifractal analysis is the study of the quantisation dimensions of a measure. These were computed in [39] for self-affine measures on Bedford-McMullen carpets. Roughly speaking the problem is to determine how well a measure can be approximated by a collection of point masses (quantised).

15.8 Open Problems

We conclude this survey article by collecting some open problems relating to Bedford-McMullen carpets. The following question was explicitly asked in [17, 19, 27, 41] and seems to be technically challenging.

Problem 15.8.1 Find a precise formula for the intermediate dimensions $\dim_\theta F$ for F a Bedford-McMullen carpet with non-uniform fibres. □

Theorem 15.5.1 completely describes the Hausdorff dimensions of the projections of Bedford-McMullen carpets onto lines provided $\log m / \log n \notin \mathbb{Q}$. The 'rational case' remains open, where it seems unlikely that the conclusion of Theorem 15.5.1 holds in general.

Problem 15.8.2 What can be said about the Hausdorff dimensions of the projections of a Bedford-McMullen carpet onto lines when $\log m / \log n \in \mathbb{Q}$? What about projections of associated self-affine measures? □

There are many interesting open problems in dimension theory and ergodic theory due to Furstenberg which ask about the independence of $\times 2$ and $\times 3$ actions. Many of these are formulated in terms of projections or slices of products of $\times 2$ and $\times 3$ invariant sets, see recent breakthroughs [58, 59]. Bedford-McMullen carpets (and more general $(\times m, \times n)$ invariant sets) therefore provide a natural extension of many of these conjectures since being $(\times m, \times n)$ invariant is more general than being the product of a $\times m$ invariant set and a $\times n$ invariant set. Theorems 15.5.1, 15.5.4 and 15.5.5 are all examples of this in action. Many questions and conjectures can be formulated and we highlight one example, implicit in [1].

Problem 15.8.3 Let F be a Bedford-McMullen carpet with $\log m / \log n \notin \mathbb{Q}$. Is it true that

$$\dim_H (F \cap L) \leq \max\{\dim_H F - 1, 0\}$$

for all lines L which are not parallel to the coordinate axes? $\square$

It remains an interesting and challenging open problem to fully describe the multifractal analysis of self-affine measures on Bedford-McMullen carpets in the setting of packing dimension. The following question was explicitly asked in [51, 56] and shown to be rather subtle in [35].

Problem 15.8.4 Find a precise formula for the packing multifractal spectrum $f_P^\mu (\alpha)$ for μ a self-affine measure on a Bedford-McMullen carpet with non-uniform fibres. $\square$

A compact set $E \subseteq \mathbb{R}^2$ is called *tube null* if it can be covered by a collection of tubes of arbitrarily small total area. A *tube* is an ε-neighbourhood of a line segment. If $\dim_H E < 1$, then E is immediately tube null since one can find a line L such that the $\mathcal{H}^1 (\pi_L (E)) = 0$ and then the collection of tubes can be taken transversal to L. In general, the tubes need not be all in the same direction. In [55] it was shown that Sierpiński carpets $E \subseteq \mathbb{R}^2$ are tube null provided $\dim_H E < 2$. Sierpiński carpets are constructed in the same way as Bedford-McMullen carpets but with $m = n$ and, as such, are self-similar rather than (strictly) self-affine.

The approach in [55] does not work for Bedford-McMullen carpets in general. The non-trivial case is when there are no empty columns and no empty rows. It seems especially difficult to prove tube-nullity in this case if $\log m / \log n \notin \mathbb{Q}$, since then there are no 'special projections', see Theorem 15.5.1.

Problem 15.8.5 Are Bedford-McMullen carpets tube null, provided they are not the whole unit square? $\square$

Acknowledgments The author was supported by an *EPSRC Standard Grant* (EP/R015104/1) and a *Leverhulme Trust Research Project Grant* (RPG-2019-034). He thanks Natalia Jurga and Istvan Kolossváry for making several helpful comments and suggestions. He is also grateful to Amir Algom and Meng Wu for interesting discussions.

References

1. A. Algom, Slicing theorems and rigidity phenomena for self affine carpets. Proc. Lond. Math. Soc. **121**, 312–353 (2020)
2. A. Algom, M. Hochman, Self embeddings of Bedford-McMullen carpets. Ergodic Theory Dynam. Syst. **39**, 577–603 (2019)
3. J.I. Almarza, CP-chains and dimension preservation for projections of $(\times m, \times n)$-invariant Gibbs measures. Adv. Math. **304**, 227–265 (2017)

4. C. Bandt, A. Käenmäki, Local structure of self-affine sets. Ergodic Theory Dynam. Syst. **33**, 1326–1337 (2013)
5. K. Barański, Hausdorff dimension of the limit sets of some planar geometric constructions. Adv. Math. **210**, 215–245 (2007)
6. B. Bárány, On the Ledrappier-Young formula for self-affine measures. Math. Proc. Camb. Philos. Soc. **159**, 405–432 (2015)
7. B. Bárány, M. Hochman, A. Rapaport, Hausdorff dimension of planar self-affine sets and measures. Invent. Math. **216**, 601–659 (2019)
8. B. Bárány, A. Käenmäki, Ledrappier-Young formula and exact dimensionality of self-affine measures. Adv. Math. **318**, 88–129 (2017)
9. B. Bárány, M. Rams, Shrinking targets on Bedford-McMullen carpets. Proc. Lond. Math. Soc. **117**, 951–995 (2018)
10. B. Bárány, A. Käenmäki, E. Rossi, Assouad dimension of planar self-affine sets. Trans. Am. Math. Soc. **374**(2), 1297–1326 (2021)
11. T. Bedford, Crinkly curves, Markov partitions and box dimensions in self-similar sets. PhD thesis, University of Warwick, 1984
12. S.A. Burrell, Dimensions of fractional Brownian images. Preprint, https://arxiv.org/abs/2002.03659
13. S.A. Burrell, K.J. Falconer, J.M. Fraser, Projection theorems for intermediate dimensions. J. Fractal Geom. https://arxiv.org/abs/1907.07632
14. T. Das, L. Fishman, D. Simmons, M. Urbański, Badly approximable points on self-affine sponges and the lower Assouad dimension. Ergodic Theory Dynam. Syst. **39**, 638–657 (2019)
15. K.J. Falconer, Dimensions of self-affine sets - a survey, in *Further Developments in Fractals and Related Fields* (Birkhäuser, Boston, 2013), pp. 115–134
16. K.J. Falconer, *Fractal Geometry: Mathematical Foundations and Applications*, 3rd edn. (Wiley, Hoboken, 2014)
17. K.J. Falconer, Intermediate dimensions - a survey (2020). Preprint
18. K.J. Falconer, J.M. Fraser, X. Jin, Sixty years of fractal projections, in *Fractal Geometry and Stochastics V*, ed. by C. Bandt, K.J. Falconer , M. Zähle. Progress in Probability (Birkhäuser, Basel, 2015)
19. K.J. Falconer, J.M. Fraser, T. Kempton, Intermediate dimensions. Math. Z. **296**, 813–830 (2020)
20. D.-J. Feng, Dimension of invariant measures for affine iterated function systems. Preprint
21. D.-J. Feng, Y. Wang, A class of self-affine sets and self-affine measures. J. Fourier Anal. Appl. **11**, 107–124 (2005)
22. A. Ferguson, J.M. Fraser, T. Sahlsten, Scaling scenery of $(\times m, \times n)$ invariant measures. Adv. Math. **268**, 564–602 (2015)
23. A. Ferguson, T. Jordan, M. Rams, Dimension of self-affine sets with holes. Ann. Acad. Sci. Fenn. Math. **40**, 63–88 (2015)
24. A. Ferguson, T. Jordan, P. Shmerkin, The Hausdorff dimension of the projections of self-affine carpets. Fundam. Math. **209**, 193–213 (2010)
25. J.M. Fraser, On the packing dimension of box-like self-affine sets in the plane. Nonlinearity, **25**, 2075–2092 (2012)
26. J.M. Fraser, Assouad type dimensions and homogeneity of fractals. Trans. Am. Math. Soc. **366**, 6687–6733 (2014)
27. J.M. Fraser, Interpolating between dimensions, in *Fractal Geometry and Stochastics VI*. Progress in Probability (Birkhäuser, Basel, 2019). https://doi.org/10.1007/978-3-030-59649-1
28. J.M. Fraser, *Assouad Dimension and Fractal Geometry*. Tracts in Mathematics Series, vol. 222 (Cambridge University Press, Cambridge, 2020)
29. J.M. Fraser, D.C. Howroyd, Assouad type dimensions for self-affine sponges. Ann. Acad. Sci. Fenn. Math. **42**, 149–174 (2017)
30. J.M. Fraser, H. Yu, New dimension spectra: finer information on scaling and homogeneity. Adv. Math. **329**, 273–328 (2018)

31. J.M. Fraser, H. Yu, Assouad type spectra for some fractal families. Indiana Univ. Math. J. **67**, 2005–2043 (2018)
32. D. Gatzouras, S.P. Lalley, Hausdorff and box dimensions of certain self-affine fractals. Indiana Univ. Math. J. **41**, 533–568 (1992)
33. M. Hochman, P. Shmerkin, Local entropy averages and projections of fractal measures. Ann. Math. **175**, 1001–1059 (2012)
34. T. Jordan, M. Rams, Multifractal analysis for Bedford-McMullen carpets. Math. Proc. Camb. Philos. Soc. **150**, 147–156 (2011)
35. T. Jordan, M. Rams, Packing spectra for Bernoulli measures supported on Bedford-McMullen carpets. Fund. Math. **229**, 171–196 (2015)
36. A. Käenmäki, H. Koivusalo, E. Rossi, Self-affine sets with fibred tangents. Ergodic Theory Dynam. Syst. **37**, 1915–1934 (2017)
37. A. Käenmäki, T. Ojala, E. Rossi, Rigidity of quasisymmetric mappings on self-affine carpets. Int. Math. Res. Not. IMRN, **12**, 3769–3799 (2018)
38. R. Kenyon, Y. Peres, Measures of full dimension on affine-invariant sets. Ergodic Theory Dynam. Syst. **16**, 307–323 (1996)
39. M. Kesseböhmer, S. Zhu, On the quantization for self-affine measures on Bedford-McMullen carpets. Math. Z. **283**, 39–58 (2016)
40. J.F. King, The singularity spectrum for general Sierpiński carpets. Adv. Math. **116**, 1–11 (1995)
41. I. Kolossváry, On the intermediate dimensions of Bedford-McMullen carpets. Preprint. https://arxiv.org/abs/2006.14366
42. I. Kolossváry, K. Simon, Triangular Gatzouras-Lalley-type planar carpets with overlaps. Nonlinearity **32**, 3294–3341 (2019)
43. F. Ledrappier, L.-S. Young, The metric entropy of diffeomorphisms. I. Characterization of measures satisfying Pesin's entropy formula. Ann. Math. **122**, 509–539 (1985)
44. F. Ledrappier, L.-S. Young, The metric entropy of diffeomorphisms. II. Relations between entropy, exponents and dimension. Ann. Math. **122**, 540–574 (1985)
45. J.M. Mackay, Assouad dimension of self-affine carpets. Conform. Geom. Dyn. **15**, 177–187 (2011)
46. J.M. Marstrand, Some fundamental geometrical properties of plane sets of fractional dimensions. Proc. Lond. Math. Soc. (3) **4**, 257–302 (1954)
47. P. Mattila, *Geometry of Sets and Measures in Euclidean Spaces*. Cambridge Studies in Advanced Mathematics, vol. 44 (Cambridge University Press, Cambridge, 1995)
48. P. Mattila, Recent progress on dimensions of projections, in *Geometry and Analysis of Fractals*, ed. by D.-J. Feng, K.-S. Lau. Springer Proceedings in Mathematics & Statistics, vol. 88 (Springer, Berlin, 2014), pp. 283–301
49. C. McMullen, The Hausdorff dimension of general Sierpiński carpets. Nagoya Math. J. **96**, 1–9 (1984)
50. S.-M. Ngai, A dimension result arising from the L^q-spectrum of a measure. Proc. Am. Math. Soc. **125**, 2943–2951 (1997)
51. L. Olsen, Self-affine multifractal Sierpiński sponges in $\mathbb{R}^d$. Pacific J. Math. **183**, 143–199 (1998)
52. Y. Peres, The packing measure of self-affine carpets. Math. Proc. Camb. Philos. Soc. **115**, 437–450 (1994)
53. Y. Peres, The self-affine carpets of McMullen and Bedford have infinite Hausdorff measure. Math. Proc. Camb. Philos. Soc. **116**, 513–526 (1994)
54. F. Przytycki, M. Urbański, On the Hausdorff dimension of some fractal sets. Studia Math. **93**, 155–186 (1989)
55. A. Pyörälä, P. Shmerkin, V. Suomala, M. Wu, Covering the Sierpiński carpet with tubes. Preprint. https://arxiv.org/abs/2006.00499
56. H.W.J. Reeve, The packing spectrum for Birkhoff averages on a self-affine repeller. Ergodic Theory Dynam. Syst. **32**, 1444–1470 (2012)

57. P. Shmerkin, Projections of self-similar and related fractals: a survey of recent developments, in *Fractal Geometry and Stochastics V*, ed. by C. Bandt, K.J. Falconer, M. Zähle. Progress in Probability (Birkhäuser, Basel, 2015)
58. P. Shmerkin, On Furstenberg's intersection conjecture, self-similar measures, and the L^q norms of convolutions. Ann. Math. **189**, 319–391 (2019)
59. M. Wu, A proof of Furstenberg's conjecture on the intersections of $\times p$ and $\times q$-invariant sets. Ann. Math. **189**, 707–751 (2019)

Chapter 16
Some Variants of Orponen's Theorem on Visible Parts of Fractal Sets

Carlos Matheus

Abstract It was recently established by T. Orponen that the visible parts from almost every direction of a compact subset of $\mathbb{R}^n$ have Hausdorff dimension at most $n - \frac{1}{50n}$.

In this note, we refine Orponen's argument in order to show that the visible parts from almost every direction of a compact subset of $\mathbb{R}^n$ have Hausdorff dimension at most $n - \min\{\frac{1}{5}, \frac{1}{n+2}\}$.

Moreover, we also show that some classes of dynamically defined Cantor sets $K \subset \mathbb{R}^n$ with Hausdorff dimension $d > \max\{\sqrt{3}, \frac{(n-1)+\sqrt{(n-1)(n+3)}}{2}\}$ have visible parts of Hausdorff dimension at most $\max\{\frac{3d+3}{d+3}, \frac{(n+1)d+(n-1)}{d+2}\}$ from almost every direction.

16.1 Introduction

Let K be a compact subset of the Euclidean space $\mathbb{R}^n$, $n \geq 2$. Intuitively, the *visible part* $\text{Vis}_e(K)$ of K in the direction $e \in S^{n-1}$ is the subset of K consisting of the points which are first hit by a light beam travelling in the direction e emanating from a certain affine hyperplane orthogonal to e.

More concretely, if $\pi_e : \mathbb{R}^n \to e^\perp$ denotes the orthogonal projection to the hyperplane $e^\perp$ orthogonal to e and $\langle ., . \rangle$ stands for the usual Euclidean inner product, then $\text{Vis}_e(K)$ is the collection of $\leq_e$-minimal points of K where $\leq_e$ is the partial order defined by $x \leq_e y$ if and only if $\pi_e(x) = \pi_e(y)$ and $\langle x, e \rangle \leq \langle y, e \rangle$.

In general, the visible parts $\text{Vis}_e(K)$ are Borel sets because they are the graphs of lower semi-continuous functions, cf. [4, Remark 2.2 (a)].

By definition, $\pi_e(\text{Vis}_e(K)) = \pi_e(K)$ for all $e \in S^{n-1}$. Therefore, Mattila's extension of Marstrand's theorem [7] provides the following *lower bound* on the

C. Matheus (✉)
CMLS, CNRS, École Polytechnique, Institut Polytechnique de Paris, Palaiseau, France
e-mail: carlos.matheus@math.cnrs.fr

© The Author(s), under exclusive license to Springer Nature Switzerland AG 2021
M. Pollicott, S. Vaienti (eds.), *Thermodynamic Formalism*, Lecture Notes
in Mathematics 2290, https://doi.org/10.1007/978-3-030-74863-0_16

Hausdorff dimension of typical visible parts:

$$\dim_H(\text{Vis}_e(K)) \geq \min\{\dim_H(K), n-1\}$$

for Lebesgue almost every $e \in S^{n-1}$.

The *visibility conjecture* asserts that the converse inequality is true, i.e., if $\dim_H(K) > n-1$, then $\dim_H(\text{Vis}_e(K)) = n-1$ for Lebesgue almost every $e \in S^{n-1}$ (see, e.g., [8, Problem 11]).

It is known that this conjecture admits a positive answer for several particular classes of compact subsets of $\mathbb{R}^n$ (cf. [2, 4] and [1]). Furthermore, we know that if $K \subset \mathbb{R}^n$ is a compact subset with d-Hausdorff measure $0 < \mathscr{H}^d(K) < \infty$ for $d > n-1$, then the d-Hausdorff measure of $\text{Vis}_e(K)$ is zero for Lebesgue almost every $e \in S^{n-1}$ (see [5, Theorem 1.1]).

More recently, T. Orponen [9] obtained an *unconditional* estimate on the Hausdorff dimension of typical visible parts of compact subsets K of $\mathbb{R}^n$: in a nutshell, he proved that $\dim_H(\text{Vis}_e(K)) \leq n - \frac{1}{50n}$ for Lebesgue almost every $e \in S^{n-1}$.

In this note, we refine Orponen's methods to establish the following two results:

Theorem 16.1.1 *Let* $K \subset \mathbb{R}^n$ *be a compact subset. Then, for Lebesgue almost every* $e \in S^{n-1}$, *the Hausdorff dimension of* $\text{Vis}_e(K)$ *is at most* $n - \min\{\frac{1}{5}, \frac{1}{n+2}\}$.

□

Theorem 16.1.2 *Let* $K \subset \mathbb{R}^n$ *be a product of* C^2-*dynamically defined Cantor sets of the real line or a self-similar set defined by a finite collection of Euclidean similarities verifying the strong open set condition. If the Hausdorff dimension of* K *is* $\dim_H(K) > \max\{\sqrt{3}, \frac{(n-1)+\sqrt{(n-1)(n+3)}}{2}\}$, *then, for Lebesgue almost every* $e \in S^{n-1}$, *the Hausdorff dimension of* $\text{Vis}_e(K)$ *is at most* $\max\{\frac{3d+3}{d+3}, \frac{(n+1)d+(n-1)}{d+2}\}$.

□

The remainder of this note is divided into two sections: its first half contains the proof of Theorem 16.1.1 and its second half is devoted to the proof of Theorem 16.1.2.

16.2 Visible Parts of General Compact Subsets

Let K be a compact subset of $\mathbb{R}^n$, $n \geq 2$. Up to rescaling, we can (and do) assume that $K \subset [0, 1]^n$. Since the conclusion of Theorem 16.1.1 always holds when K has Hausdorff dimension $\leq n - \min\{\frac{1}{5}, \frac{1}{n+2}\}$, we can (and do) also assume that

$$n - \min\left\{\frac{1}{5}, \frac{1}{n+2}\right\} < d := \dim_H(K) \leq n. \tag{16.2.1}$$

16.2.1   Some Preliminaries

Recall that the s-dimensional Hausdorff measure at scale $0 < \rho \leq \infty$ of a subset $E \subset \mathbb{R}^n$ is

$$\mathcal{H}_\rho^s(E) := \inf\left\{\sum_{i=1}^\infty \operatorname{diam}(U_i)^s : E \subset \bigcup_{i\geq 1} U_i \text{ and } \operatorname{diam}(U_i) < \delta \ \forall i \geq 1\right\}$$

and the s-dimensional Hausdorff measure of E is $\mathcal{H}^s(E) = \lim\limits_{\rho \to 0} \mathcal{H}_\rho^s(E)$, so that the Hausdorff dimension of E is

$$\dim_H(E) := \inf\{s : \mathcal{H}^s(E) = 0\} = \sup\{s : \mathcal{H}^s(E) = \infty\}.$$

Recall also that a dyadic cube $Q \subset [0, 1]^n$ is a cube of the form $Q = \prod_{j=1}^n [\frac{i_j}{2^N}, \frac{i_j+1}{2^N}]$ for some $N \in \mathbb{N}$ and $(i_1, \ldots, i_n) \in \{1, \ldots, 2^N - 1\}^n$. In the sequel, the collection of dyadic cubes with sides of fixed size 2^{-N} is denoted by $\mathcal{D}_{2^{-N}}$.

In [9, Lemma A.1], Orponen showed the following version of Frostman's lemma:

Lemma 16.2.1 (Orponen) *Let $E \subset [0, 1]^n$ be a compact subset and $n - 1 < s \leq n$. Then, there exists a Radon measure μ supported on E and a constant $0 < C = C(n) < \infty$ such that $\mu(B(x, r)) \leq Cr^s$ for all $x \in \mathbb{R}^n$, $r > 0$, and*

$$\mu(Q) \geq C^{-1} \min\{\mathcal{H}_\infty^s(E \cap Q), \mathcal{H}^n(Q)\}$$

for all dyadic cube $Q \subset [0, 1]^n$.   □

Similarly to Orponen [9], our long-term goal is to apply this lemma to estimate the Hausdorff dimension of visible parts in typical directions.

For this sake, we fix first some rational parameters

$$n - \min\left\{\frac{1}{5}, \frac{1}{n+2}\right\} < n - \varepsilon_0 < s_0'' < s_0' < s_0 < d \leq n, \tag{16.2.2}$$

$$\alpha := \min\left\{s_0'' - 1, 2 - \frac{s_0' - (n-1)}{2}\right\}, \tag{16.2.3}$$

$$\frac{\varepsilon_0 n}{2} < \frac{\varepsilon_1}{2} < \min\{s_0'' - 1, 1\} - \frac{\varepsilon_0 n}{2} - 2\varepsilon_0, \tag{16.2.4}$$

and

$$0 < \varepsilon_* < \min\left\{ s_0' + \varepsilon_0 - n, \frac{2}{3}\left(\min\{s_0'' - 1, 1\} - \frac{\varepsilon_0 n}{2} - 2\varepsilon_0 - \frac{\varepsilon_1}{2}\right)\right\}. \tag{16.2.5}$$

Note that these conditions are mutually compatible: indeed, our assumption (16.2.1) allows us to choose s_0, s_0', s_0'' and ε_0 in (16.2.2); since $\varepsilon_0 < \min\left\{\frac{1}{5}, \frac{1}{n+2}\right\}$ and $s_0' > n - \varepsilon_0$, we can select ε_1 in (16.2.4) and ε_* in (16.2.5).

Now, we use Lemma 16.2.1 to get μ supported on K such that

$$\mu(B(x, r)) \leq C r^{s_0} \tag{16.2.6}$$

for all $x \in \mathbb{R}^n$ and $r > 0$, and

$$\mu(Q) \geq C^{-1} \min\{\mathcal{H}_\infty^{s_0}(K \cap Q), \mathcal{H}^n(Q)\} \tag{16.2.7}$$

for all dyadic cube $Q \subset [0, 1]^n$ (where $0 < C = C(n) < \infty$ is a constant).

Recall that (16.2.6) implies that the s_0'-energy of μ is finite, i.e.,

$$I_{s_0'}(\mu) := \int \int \frac{d\mu(x)\, d\mu(y)}{|x - y|^{s_0'}} < \infty. \tag{16.2.8}$$

Remark 16.1 For later reference, let us remind that the s-energy of a measure θ can be expressed in terms of the Fourier transform as

$$I_s(\theta) = \int \int \frac{d\theta(x)\, d\theta(y)}{|x - y|^s} = c_1(s, n) \int |\widehat{\theta}(\xi)|^2 \cdot |\xi|^{s-n}\, d\mathcal{H}^n(\xi)$$

where $0 < c_1(s, n) < \infty$ is a constant. $\quad\square$

In the sequel, $\delta = 2^{-N}$, $N \in \mathbb{N}$, is an arbitrary (small) dyadic scale such that δ^{ε_0} is also a dyadic scale.

16.2.2 Contribution of Light Cubes

We say that a dyadic cube $Q \in \mathscr{D}_\delta$ is δ-*light* when $\mu(Q) \leq \delta^{n+\varepsilon_*}$. The portion of K contained in δ-light cubes is denoted by $K_{\delta,\text{light}}$. Since $\mathscr{D}_\delta$ has cardinality δ^{-n}, it follows from (16.2.7) that:

Lemma 16.2.2 $\mathcal{H}_\infty^{s_0}(K_{\delta,\text{light}}) \leq C(n) \cdot \delta^{\varepsilon_*}.$ $\quad\square$

In particular, this lemma says that we can safely focus on the δ-*heavy* portion $K_{\delta,\text{heavy}} := K \setminus K_{\delta,\text{light}}$ of K.

16.2.3   Exceptional Directions

Given a dyadic cube $Q \in \mathscr{D}_{\delta^{\varepsilon_0}}$, the restriction of μ to Q is denoted by μ_Q. The set of δ-*exceptional directions* associated to Q is

$$E_{\delta,Q} := \left\{ e \in S^{n-1} : \int_{e^\perp} |\widehat{\mu_Q}(\zeta)|^2 \cdot |\zeta|^{s_0'-(n-1)} \, d\mathscr{H}^{n-1}(\zeta) \geq \delta^{-\varepsilon_1} \right\}.$$

Since $I_{s_0'}(\mu_Q) \leq I_{s_0'}(\mu)$, it follows from (16.2.8) and a change of variables to polar coordinates in Remark 16.1 that:

Lemma 16.2.3 $\mathscr{H}^{n-1}(E_{\delta,Q}) \leq c_2(s_0', n) I_{s_0'}(\mu) \delta^{\varepsilon_1}$ *for all* $Q \in \mathscr{D}_{\delta^{\varepsilon_0}}$.   □

16.2.4   Good and Bad Lines

Denote by $\mathscr{L}_e$ the space of lines parallel to $e \in S^{n-1}$. Given a dyadic cube $Q \in \mathscr{D}_{\delta^{\varepsilon_0}}$ intersecting K, the set $\mathscr{L}_{e,\delta,\text{bad},Q}$ of δ-*bad lines* in direction e associated to Q consists of all lines $\ell \in \mathscr{L}_e$ disjoint from $K \cap Q$ whose 2δ-neighborhood $\ell(2\delta)$ satisfy

$$\#\{R \in \mathscr{D}_\delta : R \subset Q, R \cap K \neq \emptyset, R \text{ is not light}, R \cap \ell(2\delta) \neq \emptyset\} \geq \delta^{2\varepsilon_0-1}.$$

We say that $\ell \in \mathscr{L}_e$ is a δ-*good line* in the direction e whenever $\ell \notin \mathscr{L}_{e,\delta,\text{bad},Q}$ for all $Q \in \mathscr{D}_{\delta^{\varepsilon_0}}$ intersecting K. The collection of δ-good lines in the direction e is denoted by $\mathscr{L}_{e,\delta,\text{good}}$ and we define

$$L_{e,\delta,\text{good}} := \bigcup_{\ell \in \mathscr{L}_{e,\delta,\text{good}}} \ell.$$

Lemma 16.2.4 $\mathscr{H}_\delta^{s_0'}(\text{Vis}_e(K) \cap K_{\delta,\text{heavy}} \cap L_{e,\delta,\text{good}}) \leq \delta^{\varepsilon_*}$ *for all* δ *sufficiently small*.   □

Proof Let us use a collection $\mathscr{T}_{e,\delta}$ tubes of width δ whose bases are perpendicular to e in order to cover $[0, 1]^n$. Since $\#\mathscr{T}_{e,\delta} \leq c_3(n)\delta^{-(n-1)}$, our task is reduced to prove that, for each $T \in \mathscr{T}_{e,\delta}$, the minimal number $N(\text{Vis}_e(K) \cap K_{\delta,\text{heavy}} \cap L_{e,\delta,\text{good}} \cap T, \delta)$ of δ-balls needed to cover $\text{Vis}_e(K) \cap K_{\delta,\text{heavy}} \cap L_{e,\delta,\text{good}} \cap T$ is at most

$$N(\text{Vis}_e(K) \cap K_{\delta,\text{heavy}} \cap L_{e,\delta,\text{good}} \cap T, \delta) \leq c_5(n)\delta^{\varepsilon_0-1}.$$

Indeed, the estimates above imply that

$$\mathscr{H}_\delta^{s_0'}(\text{Vis}_e(K) \cap K_{\delta,\text{heavy}} \cap L_{e,\delta,\text{good}}) \leq c_3(n)c_5(n)\delta^{-(n-1)}\delta^{\varepsilon_0-1}\delta^{s_0'} \leq \delta^{\varepsilon_*}$$

for all δ sufficiently small thanks to the fact that $s_0' + \varepsilon_0 - n > \varepsilon_*$ (cf. (16.2.5)).

In order to estimate $N(\mathrm{Vis}_e(K) \cap K_{\delta,\mathrm{heavy}} \cap L_{e,\delta,\mathrm{good}} \cap T, \delta)$ for a given $T \in \mathscr{T}_{e,\delta}$, we consider two scenarios:

(i) for all $Q \in \mathscr{D}_{\delta^{\varepsilon_0}}$ intersecting K, one has

$$\#\{R \in \mathscr{D}_\delta : R \subset Q, R \cap K \neq \emptyset, R \text{ is not light}, R \cap T \neq \emptyset\} < \delta^{2\varepsilon_0 - 1};$$

(ii) there exists $Q_1 \in \mathscr{D}_{\delta^{\varepsilon_0}}$ intersecting K with

$$\#\{R \in \mathscr{D}_\delta : R \subset Q_1, R \cap K \neq \emptyset, R \text{ is not light}, R \cap T \neq \emptyset\} \geq \delta^{2\varepsilon_0 - 1}.$$

In the first scenario, we have that $N(\mathrm{Vis}_e(K) \cap K_{\delta,\mathrm{heavy}} \cap L_{e,\delta,\mathrm{good}} \cap T, \delta) \leq \delta^{\varepsilon_0 - 1}$ simply because T can meet at most $\delta^{-\varepsilon_0}$ dyadic cubes $Q \in \mathscr{D}_{\delta^{\varepsilon_0}}$.

In the second scenario, we take Q_1 to be a $\leq_e$-minimal dyadic cube with the property described in (ii) (in the sense that Q_1 minimizes $\inf\{\langle x, e\rangle : x \in Q_1\}$ among all dyadic cubes in (ii)). Since the 2δ-neighborhood of any line $\ell \subset T$ contains T, we also have

$$\#\{R \in \mathscr{D}_\delta : R \subset Q_1, R \cap K \neq \emptyset, R \text{ is not light}, R \cap \ell(2\delta) \neq \emptyset\} \geq \delta^{2\varepsilon_0 - 1}.$$

Therefore, it follows from the definition of δ-good line that any $\ell \in \mathscr{L}_{e,\delta,\mathrm{good}}$ included in T must intersect $K \cap Q_1$.

We affirm that

$$\mathrm{Vis}_e(K) \cap L_{e,\delta,\mathrm{good}} \cap T \cap Q = \emptyset$$

for any dyadic cube $Q \in \mathscr{D}_{\delta^{\varepsilon_0}}$ with $\inf\{\langle x, e\rangle : x \in Q\} > \sup\{\langle y, e\rangle : y \in Q_1\}$. In fact, if $x \in \mathrm{Vis}_e(K) \cap L_{e,\delta,\mathrm{good}} \cap T \cap Q$, then $\pi_e(x) = \pi_e(y)$ for some $y \in Q_1$. Since $\langle x, e\rangle > \langle y, e\rangle$, one would get $x \notin \mathrm{Vis}_e(K)$, a contradiction.

Hence, $\mathrm{Vis}_e(K) \cap L_{e,\delta,\mathrm{good}} \cap T$ is covered by the collection of dyadic cubes $Q \in \mathscr{D}_{\delta^{\varepsilon_0}}$ with $\inf\{\langle x, e\rangle : x \in Q\} \leq \sup\{\langle y, e\rangle : y \in Q_1\}$. Now, we observe that

- the number of dyadic cubes $Q \in \mathscr{D}_{\delta^{\varepsilon_0}}$ intersecting T with

$$\inf\{\langle z, e\rangle : z \in Q_1\} \leq \inf\{\langle x, e\rangle : x \in Q\} \leq \sup\{\langle y, e\rangle : y \in Q_1\}$$

is bounded by an absolute constant $c_4(n)$; for each of them, we will use the crude bound $N(\mathrm{Vis}_e(K) \cap L_{e,\delta,\mathrm{good}} \cap T, \delta) \leq \delta^{\varepsilon_0 - 1}$ coming from the fact that $Q \cap T$ can be covered using at most $\delta^{\varepsilon_0 - 1}$ balls of radius δ;
- any dyadic cube $Q \in \mathscr{D}_{\delta^{\varepsilon_0}}$ intersecting $T \cap K$ with

$$\inf\{\langle z, e\rangle : z \in Q_1\} > \inf\{\langle x, e\rangle : x \in Q\}$$

satisfies

$$\#\{R \in \mathscr{D}_\delta : R \subset Q_1, R \cap K \neq \emptyset, R \text{ is not light}, R \cap \ell(2\delta) \neq \emptyset\} \leq \delta^{2\varepsilon_0 - 1}$$

because of the $\leq_e$-minimality of Q_1; the number of such cubes Q is at most $\leq \delta^{-\varepsilon_0}$ because T meets at most $\delta^{-\varepsilon_0}$ dyadic cubes $Q \in \mathscr{D}_{\delta^{\varepsilon_0}}$.

By combining the estimates above, we conclude that

$$N(\text{Vis}_e(K) \cap K_{\delta,\text{heavy}} \cap L_{e,\delta,\text{good}} \cap T, \delta) \leq c_4(n)\delta^{\varepsilon_0 - 1} + \delta^{-\varepsilon_0}\delta^{2\varepsilon_0 - 1} = c_5(n)\delta^{\varepsilon_0 - 1}.$$

This completes the proof.   □

16.2.5   Typical Visible Parts in Bad Lines

The last step towards the proof of Theorem 16.1.1 is the following estimate:

Lemma 16.2.5 *Let $Q \in \mathscr{D}_{\delta^{\varepsilon_0}}$ be a dyadic cube intersecting K, consider a direction $e \notin E_{\delta,Q}$, and denote $L_{e,\delta,\text{bad},Q} := \bigcup\limits_{\ell \in \mathscr{L}_{e,\delta,\text{bad},Q}} \ell$. Then,*

$$\mathscr{H}_\infty^{s_0' - 1}\left(\pi_e(L_{e,\delta,\text{bad},Q})\right) \leq \delta^{\varepsilon_* + \varepsilon_0 n}$$

for all δ sufficiently small.   □

Proof By contradiction, suppose that $\mathscr{H}_\infty^{s_0' - 1}\left(\pi_e(L_{e,\delta,\text{bad},Q})\right) \geq \delta^{\varepsilon_* + \varepsilon_0 n}$. By Orponen's version of Frostman's lemma (cf. Lemma 16.2.1), we have a *probability* measure ν supported on $H_{e,\delta,Q} := \pi_e(L_{e,\delta,\text{bad},Q})$ such that

$$\nu(B(x,r)) \leq C(n-1)\delta^{-\varepsilon_* - \varepsilon_0 n} r^{s_0' - 1}$$

for all $x \in H$ and $r > 0$. Thus, our choice of $\alpha \leq s_0'' - 1 < s_0' - 1$ in (16.2.3) (and Remark 16.1) means that the α-energy of ν satisfies

$$c_1(\alpha, n-1) \int |\hat{\nu}(\xi)|^2 \cdot |\xi|^\alpha \, d\xi = I_\alpha(\nu) \leq c_6(s_0'', s_0', n)\delta^{-\varepsilon_* - \varepsilon_0 n}. \qquad (16.2.9)$$

Next, we observe that, by definition, any line $\ell \in \mathscr{L}_{e,\delta,\text{bad},Q}$ misses $K \cap Q$. Therefore, $\mu_{Q,e} := (\pi_e)_*(\mu_Q)$ and ν have disjoint supports. Hence, if we fix a non-negative smooth bump function φ on $e^\perp \simeq \mathbb{R}^{n-1}$ with total integral one and

$\varphi(0) = 1$, then

$$0 = \int \varphi_\eta * \mu_{Q,e}\, dv = \int \widehat{\varphi}(\eta\xi)\widehat{\mu_{Q,e}}(\xi)\overline{\widehat{v}(\xi)}\, d\xi$$

$$= \int (1 - \widehat{\varphi}(c_7(n)\delta\xi))\widehat{\varphi}(\eta\xi)\widehat{\mu_{Q,e}}(\xi)\overline{\widehat{v}(\xi)}\, d\xi$$

$$+ \int \widehat{\varphi}(c_7(n)\delta\xi)\widehat{\varphi}(\eta\xi)\widehat{\mu_{Q,e}}(\xi)\overline{\widehat{v}(\xi)}\, d\xi$$

$$:= A_2 - A_1$$

for all $0 < \eta \ll \delta$, where $\varphi_\eta(x) = \varphi(\eta x)/\eta^{n-1}$.

In the sequel, we will reach a contradiction with the identity in the previous paragraph by showing that $|A_2| < |A_1|$. For this sake, we observe that $\widehat{\varphi}$ is a bounded Lipschitz function with $\widehat{\varphi}(0) = 1$, so that $|1 - \widehat{\varphi}(c_7(n)\delta\xi)| \leq c_8(n)\delta|\xi|$ and, *a fortiori*,

$$|A_2| \leq c_8(n)\delta^{\frac{s_0'-(n-1)}{2}+\frac{\alpha}{2}} \left(\int |\widehat{\mu_{Q,e}}(\xi)|^2 \cdot |\xi|^{s_0'-(n-1)}\, d\xi \right)^{1/2} \left(\int |\widehat{v}(\xi)|^2 \cdot |\xi|^\alpha\, d\xi \right)^{1/2}$$

thanks to our choice of $\frac{s_0'-(n-1)}{2} + \frac{\alpha}{2} \leq 1$ in (16.2.3) and the Cauchy–Schwarz inequality. By plugging into the previous inequality the facts that our choices in (16.2.2) and (16.2.3) imply $\frac{s_0'-(n-1)}{2} + \frac{\alpha}{2} \geq \min\{s_0'' - 1, 1\}$, our assumption $e \notin E_{\delta,Q}$ allows (by definition) to control $|\widehat{\mu_Q}(\xi)|$ $(= |\widehat{\mu_{Q,e}}(\xi)|$ for $\xi \in e^\perp)$, and the α-energy of v is controlled by (16.2.9), we derive that

$$A_2 \leq c_9(s_0'', s_0', n)\delta^{\min\{s_0''-1,1\}}\delta^{-\varepsilon_1/2}\delta^{-(\varepsilon_*+\varepsilon_0 n)/2}.$$

On the other hand, if we write

$$A_1 = \int \varphi_{c_7(n)\delta} * \varphi_\eta * \mu_{Q,e}(r)\, dv(r),$$

and we recall that v is supported in $H_{e,\delta,Q} := \pi_e(L_{e,\delta,\text{bad},Q})$, then we can use the fact that $r \in H_{e,\delta,Q}$ means $\ell := \pi_e^{-1}(r) \in \mathscr{L}_{e,\delta,\text{bad},Q}$, i.e., $\ell(2\delta)$ meets at least $\delta^{2\varepsilon_0-1}$ dyadic cubes $R \in \mathscr{D}_\delta$ included in Q which are not light, to deduce that $\mu_Q(\ell(2\delta)) \geq \delta^{2\varepsilon_0-1+n+\varepsilon_*}$ and, *a fortiori*,

$$\varphi_{c_7(n)\delta} * \varphi_\eta * \mu_{Q,e}(r) \geq c_{10}(n)\delta^{2\varepsilon_0+\varepsilon_*}$$

for all $r \in H$ and $0 < \eta \ll \delta$. Therefore,

$$A_1 \geq c_{10}(n)\delta^{2\varepsilon_0+\varepsilon_*}$$

because ν is a probability measure on H.

At this point, we get the desired contradiction $A_1 > |A_2|$ for δ is sufficiently small because our choice (16.2.5) implies that $2\varepsilon_0 + \varepsilon_* < \min\{s_0'' - 1, 1\} - \frac{\varepsilon_1}{2} - \frac{\varepsilon_* + \varepsilon_0 n}{2}$. $\qquad\qquad\square$

16.2.6   End of the Proof of Theorem 16.1.1

Let us take a decreasing sequence of dyadic scales $\delta_j \to 0$ such that $\delta_j^{\varepsilon_0}$ also a dyadic scale. We define the set E_{δ_j} of δ_j-exceptional directions as

$$E_{\delta_j} := \bigcup_{Q \in \mathscr{D}_{\delta_j}^{\varepsilon_0}} E_{\delta_j, Q}.$$

Since $\#\mathscr{D}_\eta = \eta^{-n}$, it follows from Lemma 16.2.3 that

$$\mathscr{H}^{n-1}(E_{\delta_j}) \le c_2(s_0', n) I_{s_0'}(\mu) \delta_j^{\varepsilon_1 - \varepsilon_0 n}.$$

Therefore, our choice of $\varepsilon_1 > \varepsilon_0 n$ in (16.2.4) implies

$$\sum_{j=1}^{\infty} \mathscr{H}^{n-1}(E_{\delta_j}) < \infty,$$

so that the set

$$E = E(s_0, s_0', s_0'', \varepsilon_0, \varepsilon_1, \varepsilon_*) := \bigcap_{n=1}^{\infty} \bigcup_{j \ge n} E_{\delta_j}$$

has zero $\mathscr{H}^{n-1}$-measure.

We affirm that $\dim_H(\mathrm{Vis}_e(K)) \le s_0$ whenever $e \in S^{n-1} \setminus E$. In fact, an element $e \notin E$ belongs to finitely many E_{δ_j}'s, say $e \notin E_{\delta_j}$ for all $j \ge j_e$.

By Lemma 16.2.2, we have $\mathscr{H}_\infty^{s_0}(\mathrm{Vis}_e(K) \cap K_{\delta_j,\text{light}}) \le \mathscr{H}_\infty^{s_0}(K_{\delta_j,\text{light}}) \le C(n) \cdot \delta_j^{\varepsilon_*}$ for all j. Also, by Lemma 16.2.4, $\mathscr{H}_{\delta_j}^{s_0'}(\mathrm{Vis}_e(K) \cap K_{\delta_j,\text{heavy}} \cap L_{e,\delta_j,\text{good}}) \le \delta_j^{\varepsilon_*}$ for all j sufficiently large. Moreover, $\mathscr{H}_\infty^{s_0'}\left(\mathrm{Vis}_e(K) \cap K_{\delta_j,\text{heavy}} \cap \bigcup_{\substack{Q \in \mathscr{D}_{\delta_j}^{\varepsilon_0}, \\ Q \cap K \ne \emptyset}} L_{e,\delta_j,\text{bad},Q} \right) \le \delta_j^{\varepsilon_*}$ for all $j \ge j_e$ sufficiently large by Lemma 16.2.5 (and the fact that $\#\mathscr{D}_{\delta_j}^{\varepsilon_0} = \delta_j^{-\varepsilon_0 n}$).

By putting these three estimates together, we derive that if $e \notin E$, then

$$\mathscr{H}_{\infty}^{s_0}(\mathrm{Vis}_e(K)) \leq (C(n) + 2)\delta_j^{\varepsilon_*}$$

for all $j \geq j_e$ sufficiently large, and, consequently, $\dim_H(\mathrm{Vis}_e(K)) \leq s_0$ for all $e \notin E(s_0, s_0', s_0'', \varepsilon_0, \varepsilon_1, \varepsilon_*)$.

Since $s_0, s_0', s_0'', \varepsilon_0, \varepsilon_1, \varepsilon_*$ are arbitrary rational parameters satisfying (16.2.2)–(16.2.5), we conclude that

$$\dim_H(\mathrm{Vis}_e(K)) \leq n - \min\left\{\frac{1}{5}, \frac{1}{n+2}\right\}$$

for Lebesgue almost every $e \in S^{n-1}$.

16.3 Typical Visible Parts of Dynamical Cantor Sets

In this section, we revisit Orponen's method described above in order to establish Theorem 16.1.2.

16.3.1 Some Preliminaries

It is well-known (see, e.g., [6] and [3]) that the products of C^2-dynamically defined Cantor sets of the real line and the self-similar sets given by a finite collection of Euclidean similarities verifying the strong open set condition defined a class of compact subsets $K \subset \mathbb{R}^n$ with the following properties:

- K supports a measure μ equivalent to $\mathscr{H}^d|_K$, $d := \dim_H(K)$, such that $C^{-1}r^d \leq \mu(B(x,r)) \leq Cr^d$ for all $x \in K$, $r > 0$;
- there exists $\lambda > 1$ such that, for all $\rho > 0$, K can be covered by a collection $\mathscr{C}_\rho(K)$ of disjoint cubes with sizes belonging to the interval $[\rho, \lambda\rho]$ such that their mutual distances are at least $\lambda^{-1}\rho$ and each of them contain a ball of radius $\lambda^{-1}\rho$ about some point of K.

In the context of Theorem 16.1.2, recall that we are also assuming that

$$n \geq d > \max\{\sqrt{3}, \frac{(n-1) + \sqrt{(n-1)(n+3)}}{2}\}. \tag{16.3.1}$$

Furthermore, up to rescaling, we can suppose that $K \subset [0, 1]^n$.

Let us now fix some rational parameters

$$\max\{\frac{3d+3}{d+3}, \frac{(n+1)d+(n-1)}{d+2}\} < n - \varepsilon_0 < s_0'' < s_0' < s_0 < d \le n,$$

(16.3.2)

$$\alpha := \min\left\{s_0'' - 1, 2 - \frac{s_0' - (n-1)}{2}\right\},$$

(16.3.3)

$$\frac{\varepsilon_0 d}{2} < \frac{\varepsilon_1}{2} < \min\{s_0'' - 1, 1\} - \frac{\varepsilon_0 d}{2} - 2\varepsilon_0 - d + n,$$

(16.3.4)

and

$$0 < \varepsilon_* < \min\left\{s_0' + \varepsilon_0 - n, 2\left(\min\{s_0'' - 1, 1\} - \frac{\varepsilon_0 d}{2} - 2\varepsilon_0 - d + n - \frac{\varepsilon_1}{2}\right)\right\}.$$

(16.3.5)

Note that these conditions are mutually compatible: indeed, our assumption (16.3.1) allows us to choose s_0, s_0', s_0'' and ε_0 in (16.3.2); since $\varepsilon_0 < \min\left\{\frac{3-d}{d+3}, \frac{n-d+1}{d+2}\right\}$ and $s_0' > n - \varepsilon_0$, we can select ε_1 in (16.2.4) and s_* in (16.2.5).

In what follows, $\delta = 2^{-N}$, $N \in \mathbb{N}$, is an arbitrary (small) dyadic scale such that δ^{ε_0} is also a dyadic scale.

Our plan is to show Theorem 16.1.2 by following the same arguments from the previous section *after* some adjustments in the definitions and arguments.

16.3.2 Absence of Light Cubes

In comparison with the previous section, our current setting is technically easier because there are no δ-light cubes in the sense that any $Q \in \mathscr{C}_\delta(K)$ satisfies

$$\mu(Q) \ge C^{-1}\lambda^{-d}\delta^d =: c_{11}\delta^d.$$

(16.3.6)

16.3.3 Exceptional Directions

Given a cube $Q \in \mathscr{C}_{\delta^{\varepsilon_0}}(K)$, we define

$$E_{\delta,Q} := \left\{e \in S^{n-1} : \int_{e^\perp} |\widehat{\mu_Q}(\zeta)|^2 \cdot |\zeta|^{s_0'-(n-1)} \, d\mathscr{H}^{n-1}(\zeta) \ge \delta^{-\varepsilon_1}\right\}$$

where $\mu_Q = \mu|_Q$. Since $s'_0 < d$, we have that

$$\mathcal{H}^{n-1}(E_{\delta,Q}) \leq c_2(s'_0, n) I_{s'_0}(\mu) \delta^{\varepsilon_1} \tag{16.3.7}$$

for all $Q \in \mathscr{C}_{\delta^{\varepsilon_0}}(K)$.

16.3.4 Good and Bad Lines

Denote by $\mathscr{L}_e$ the space of lines parallel to $e \in S^{n-1}$. Given a cube $Q \in \mathscr{C}_{\delta^{\varepsilon_0}}(K)$, the set $\mathscr{L}_{e,\delta,\text{bad},Q}$ of δ-*bad lines* in direction e associated to Q consists of all lines $\ell \in \mathscr{L}_e$ disjoint from $K \cap Q$ whose 2δ-neighborhood $\ell(2\delta)$ satisfy

$$\#\{R \in \mathscr{C}_\delta(K) : R \cap Q \neq \emptyset, R \cap \ell(2\delta) \neq \emptyset\} \geq \delta^{2\varepsilon_0 - 1}.$$

We say that $\ell \in \mathscr{L}_e$ is a δ-*good line* in the direction e whenever $\ell \notin \mathscr{L}_{e,\delta,\text{bad},Q}$ for all $Q \in \mathscr{C}_{\delta^{\varepsilon_0}}(K)$. The collection of δ-good lines in the direction e is denoted by $\mathscr{L}_{e,\delta,\text{good}}$ and we define

$$L_{e,\delta,\text{good}} := \bigcup_{\ell \in \mathscr{L}_{e,\delta,\text{good}}} \ell.$$

Lemma 16.3.1 $\mathcal{H}^{s'_0}_{\lambda\delta}(\text{Vis}_e(K) \cap L_{e,\delta,\text{good}}) \leq \delta^{\varepsilon_*}$ *for all δ sufficiently small.* $\square$

Proof The argument below is parallel to the proof of Lemma 16.2.4 above. Once again, let $\mathscr{T}_{e,\delta}$ be a collection of tubes of width δ whose bases are perpendicular to e in order to cover $[0, 1]^n$, so that our task is reduced to prove that, for each $T \in \mathscr{T}_{e,\delta}$, the minimal number $N(\text{Vis}_e(K) \cap L_{e,\delta,\text{good}} \cap T, \delta)$ of balls of radii in the interval $[\delta, \lambda\delta]$ needed to cover $\text{Vis}_e(K) \cap L_{e,\delta,\text{good}} \cap T$ is at most

$$N(\text{Vis}_e(K) \cap L_{e,\delta,\text{good}} \cap T, \delta) \leq c_5(n) \delta^{\varepsilon_0 - 1}.$$

In order to estimate $N(\text{Vis}_e(K) \cap L_{e,\delta,\text{good}} \cap T, \delta)$ for a given $T \in \mathscr{T}_{e,\delta}$, we consider two scenarios:

(i) for all $Q \in \mathscr{C}_{\delta^{\varepsilon_0}}(K)$, one has

$$\#\{R \in \mathscr{C}_\delta(K) : R \cap Q \neq \emptyset, R \cap T \neq \emptyset\} < \delta^{2\varepsilon_0 - 1};$$

(ii) there exists $Q_1 \in \mathscr{C}_{\delta^{\varepsilon_0}}(K)$ with

$$\#\{R \in \mathscr{C}_\delta(K) : R \cap Q_1 \neq \emptyset, R \cap T \neq \emptyset\} \geq \delta^{2\varepsilon_0 - 1}.$$

In the first scenario, we have that $N(\text{Vis}_e(K) \cap L_{e,\delta,\text{good}} \cap T, \delta) \leq \delta^{\varepsilon_0-1}$ simply because T can meet at most $\delta^{-\varepsilon_0}$ cubes $Q \in \mathscr{C}_{\delta^{\varepsilon_0}}(K)$.

In the second scenario, we take Q_1 to be a $\leq_e$-minimal cube with the property described in (ii) (in the sense that Q_1 minimizes $\inf\{\langle x, e\rangle : x \in Q_1\}$ among all cubes in (ii)). Since the 2δ-neighborhood of any line $\ell \subset T$ contains T, we also have

$$\#\{R \in \mathscr{C}_\delta(K) : R \cap Q_1, R \cap \ell(2\delta) \neq \emptyset\} \geq \delta^{2\varepsilon_0-1}.$$

Therefore, it follows from the definition of δ-good line that any $\ell \in \mathscr{L}_{e,\delta,\text{good}}$ included in T must intersect $K \cap Q_1$.

We affirm that

$$\text{Vis}_e(K) \cap L_{e,\delta,\text{good}} \cap T \cap Q = \emptyset$$

for any cube $Q \in \mathscr{C}_{\delta^{\varepsilon_0}}(K)$ with $\inf\{\langle x, e\rangle : x \in Q\} > \sup\{\langle y, e\rangle : y \in Q_1\}$. In fact, if $x \in \text{Vis}_e(K) \cap L_{e,\delta,\text{good}} \cap T \cap Q$, then $\pi_e(x) = \pi_e(y)$ for some $y \in Q_1$. Since $\langle x, e\rangle > \langle y, e\rangle$, one would get $x \notin \text{Vis}_e(K)$, a contradiction.

Hence, $\text{Vis}_e(K) \cap L_{e,\delta,\text{good}} \cap T$ is covered by the collection of cubes $Q \in \mathscr{C}_{\delta^{\varepsilon_0}}(K)$ with $\inf\{\langle x, e\rangle : x \in Q\} \leq \sup\{\langle y, e\rangle : y \in Q_1\}$. Now, we observe that

- the number of cubes $Q \in \mathscr{C}_{\delta^{\varepsilon_0}}(K)$ intersecting T with

$$\inf\{\langle z, e\rangle : z \in Q_1\} \leq \inf\{\langle x, e\rangle : x \in Q\} \leq \sup\{\langle y, e\rangle : y \in Q_1\}$$

is bounded by an absolute constant $c_4(n)$; for each of them, we will use the crude bound $N(\text{Vis}_e(K) \cap L_{e,\delta,\text{good}} \cap T, \delta) \leq \delta^{\varepsilon_0-1}$ coming from the fact that $Q \cap T$ can be covered using at most δ^{ε_0-1} balls of radius δ;
- any cube $Q \in \mathscr{C}_{\delta^{\varepsilon_0}}(K)$ intersecting $T \cap K$ with

$$\inf\{\langle z, e\rangle : z \in Q_1\} > \inf\{\langle x, e\rangle : x \in Q\}$$

satisfies

$$\#\{R \in \mathscr{C}_\delta(K) : R \cap Q_1 \neq \emptyset, R \cap \ell(2\delta) \neq \emptyset\} \leq \delta^{2\varepsilon_0-1}$$

because of the $\leq_e$-minimality of Q_1; the number of such cubes Q is at most $\leq \delta^{-\varepsilon_0}$ because T meets at most $\delta^{-\varepsilon_0}$ cubes $Q \in \mathscr{C}_{\delta^{\varepsilon_0}}(K)$.

By combining the estimates above, we conclude that

$$N(\text{Vis}_e(K) \cap L_{e,\delta,\text{good}} \cap T, \delta) \leq c_4(n)\delta^{\varepsilon_0-1} + \delta^{-\varepsilon_0}\delta^{2\varepsilon_0-1} = c_5(n)\delta^{\varepsilon_0-1}.$$

This completes the proof.   □

16.3.5 Typical Visible Parts in Bad Lines

Similarly to the previous section, the last step towards the proof of Theorem 16.1.2 is the following estimate:

Lemma 16.3.2 *Let $Q \in \mathscr{C}_{\delta^{\varepsilon_0}}(K)$ be a cube, consider a direction $e \notin E_{\delta,Q}$, and denote $L_{e,\delta,\mathrm{bad},Q} := \bigcup_{\ell \in \mathscr{L}_{e,\delta,\mathrm{bad},Q}} \ell$. Then,*

$$\mathscr{H}_{\infty}^{s_0'-1}\left(\pi_e(L_{e,\delta,\mathrm{bad},Q})\right) \leq \delta^{\varepsilon_*+\varepsilon_0 d}$$

for all δ sufficiently small.   □

Proof By contradiction, suppose that $\mathscr{H}_{\infty}^{s_0'-1}\left(\pi_e(L_{e,\delta,\mathrm{bad},Q})\right) \geq \delta^{\varepsilon_*+\varepsilon_0 d}$. By Lemma 16.2.1, we have a probability measure ν supported on $H_{e,\delta,Q} := \pi_e(L_{e,\delta,\mathrm{bad},Q})$ with

$$\nu(B(x,r)) \leq C(n-1)\delta^{-\varepsilon_*-\varepsilon_0 d}r^{s_0'-1}$$

for all $x \in H$ and $r > 0$. Thus, our choice of $\alpha \leq s_0'' - 1 < s_0' - 1$ in (16.3.3) (and Remark 16.1) means that the α-energy of ν satisfies

$$c_1(\alpha, n-1)\int |\widehat{\nu}(\xi)|^2 \cdot |\xi|^{\alpha}\, d\xi = I_{\alpha}(\nu) \leq c_6(s_0'', s_0', n)\delta^{-\varepsilon_*-\varepsilon_0 d}. \qquad (16.3.8)$$

Next, we observe that, by definition, any line $\ell \in \mathscr{L}_{e,\delta,\mathrm{bad},Q}$ misses $K \cap Q$. Therefore, $\mu_{Q,e} := (\pi_e)_*(\mu_Q)$ and ν have disjoint supports. Hence, if we fix a non-negative smooth bump function φ on $e^{\perp} \simeq \mathbb{R}^{n-1}$ with total integral one and $\varphi(0) = 1$, then

$$0 = \int \varphi_{\eta} * \mu_{Q,e}\, d\nu = \int \widehat{\varphi}(\eta\xi)\widehat{\mu_{Q,e}}(\xi)\overline{\widehat{\nu}(\xi)}\, d\xi$$

$$= \int (1 - \widehat{\varphi}(c_7(n)\delta\xi))\widehat{\varphi}(\eta\xi)\widehat{\mu_{Q,e}}(\xi)\overline{\widehat{\nu}(\xi)}\, d\xi$$

$$+ \int \widehat{\varphi}(c_7(n)\delta\xi)\widehat{\varphi}(\eta\xi)\widehat{\mu_{Q,e}}(\xi)\overline{\widehat{\nu}(\xi)}\, d\xi$$

$$:= A_2 - A_1$$

for all $0 < \eta \ll \delta$, where $\varphi_{\eta}(x) = \varphi(\eta x)/\eta^{n-1}$.

Once more, we will reach a contradiction with the identity in the previous paragraph by showing that $|A_2| < |A_1|$. For this sake, we observe that $\widehat{\varphi}$ is a bounded Lipschitz function with $\widehat{\varphi}(0) = 1$, so that $|1 - \widehat{\varphi}(c_7(n)\delta\xi)| \leq c_8(n)\delta|\xi|$

and, *a fortiori*,

$$|A_2| \leq c_8(n)\delta^{\frac{s_0'-(n-1)}{2}+\frac{\alpha}{2}} \left(\int |\widehat{\mu_{Q,e}}(\xi)|^2 \cdot |\xi|^{s_0'-(n-1)} \, d\xi \right)^{1/2} \left(\int |\widehat{\nu}(\xi)|^2 \cdot |\xi|^\alpha \, d\xi \right)^{1/2}$$

thanks to our choice of $\frac{s_0'-(n-1)}{2} + \frac{\alpha}{2} \leq 1$ in (16.3.3) and the Cauchy–Schwarz inequality. By plugging into the previous inequality the facts that our choices in (16.3.2) and (16.3.3) imply $\frac{s_0'-(n-1)}{2} + \frac{\alpha}{2} \geq \min\{s_0'' - 1, 1\}$, our assumption $e \notin E_{\delta,Q}$ allows (by definition) to control $|\widehat{\mu_Q}(\xi)|$ $(= |\widehat{\mu_{Q,e}}(\xi)|$ for $\xi \in e^\perp)$, and the α-energy of ν is controlled by (16.3.8), we derive that

$$A_2 \leq c_9(s_0'', s_0', n)\delta^{\min\{s_0''-1,1\}}\delta^{-\varepsilon_1/2}\delta^{-(\varepsilon_*+\varepsilon_0 d)/2}.$$

On the other hand, if we write

$$A_1 = \int \varphi_{c_7(n)\delta} * \varphi_\eta * \mu_{Q,e}(r) \, d\nu(r),$$

and we recall that ν is supported in $H_{e,\delta,Q} := \pi_e(L_{e,\delta,\mathrm{bad},Q})$, then we can use the fact that $r \in H_{e,\delta,Q}$ means $\ell := \pi_e^{-1}(r) \in \mathcal{L}_{e,\delta,\mathrm{bad},Q}$, i.e., $\ell(2\delta)$ meets at least $\delta^{2\varepsilon_0-1}$ cubes $R \in \mathscr{C}_\delta(K)$ intersecting Q and verifying (16.3.6), to deduce that $\mu_Q(\ell(2\delta)) \geq c_{11}\delta^{2\varepsilon_0-1+d}$ and, *a fortiori*,

$$\varphi_{c_7(n)\delta} * \varphi_\eta * \mu_{Q,e}(r) \geq c_{11}c_{10}(n)\delta^{2\varepsilon_0-1+d-(n-1)}$$

for all $r \in H$ and $0 < \eta \ll \delta$. Therefore,

$$A_1 \geq c_{11}c_{10}(n)\delta^{2\varepsilon_0+d-n}$$

because ν is a probability measure on H.

At this point, we get the desired contradiction $A_1 > |A_2|$ for δ is sufficiently small because our choice (16.3.5) implies that $2\varepsilon_0 + d - n < \min\{s_0'' - 1, 1\} - \frac{\varepsilon_1}{2} - \frac{\varepsilon_*+\varepsilon_0 d}{2}$. □

16.3.6   End of the Proof of Theorem 16.1.2

Let us take a decreasing sequence of dyadic scales $\delta_j \to 0$ such that $\delta_j^{\varepsilon_0}$ also a dyadic scale. We define the set E_{δ_j} of δ_j-exceptional directions as

$$E_{\delta_j} := \bigcup_{Q \in \mathscr{C}_{\delta_j}(K)} E_{\delta_j,Q}.$$

Since $\#\mathscr{C}_\eta(K) \leq c_{12}\eta^{-d}$ (thanks to (16.3.6) and the finiteness of μ), it follows from (16.3.7) that

$$\mathscr{H}^{n-1}(E_{\delta_j}) \leq c_2(s_0', n)I_{s_0'}(\mu)\delta_j^{\varepsilon_1 - \varepsilon_0 d}.$$

Therefore, our choice of $\varepsilon_1 > \varepsilon_0 d$ in (16.3.4) implies

$$\sum_{j=1}^\infty \mathscr{H}^{n-1}(E_{\delta_j}) < \infty,$$

so that the set

$$E = E(s_0, s_0', s_0'', \varepsilon_0, \varepsilon_1, \varepsilon_*) := \bigcap_{n=1}^\infty \bigcup_{j \geq n} E_{\delta_j}$$

has zero $\mathscr{H}^{n-1}$-measure.

We affirm that $\dim_H(\mathrm{Vis}_e(K)) \leq s_0'$ whenever $e \in S^{n-1} \setminus E$. In fact, an element $e \notin E$ belongs to finitely many E_{δ_j}'s, say $e \notin E_{\delta_j}$ for all $j \geq j_e$.

By Lemma 16.3.1, $\mathscr{H}_{\lambda\delta_j}^{s_0'}(\mathrm{Vis}_e(K) \cap L_{e,\delta_j,\mathrm{good}}) \leq \delta_j^{\varepsilon_*}$ for all j sufficiently

large. Moreover, $\mathscr{H}_\infty^{s_0'}\left(\mathrm{Vis}_e(K) \cap \bigcup_{Q \in \mathscr{C}_{\delta_j^{\varepsilon_0}}(K)} L_{e,\delta_j,\mathrm{bad},Q}\right) \leq c_{12}\delta_j^{\varepsilon_*}$ for all $j \geq j_e$

sufficiently large by Lemma 16.3.2 (and the fact that $\#\mathscr{C}_{\delta_j^{\varepsilon_0}}(K) \leq c_{12}\delta_j^{-\varepsilon_0 d}$).

By putting these three estimates together, we derive that if $e \notin E$, then

$$\mathscr{H}_\infty^{s_0'}(\mathrm{Vis}_e(K)) \leq (c_{12} + 1)\delta_j^{\varepsilon_*}$$

for all $j \geq j_e$ sufficiently large, and, consequently, $\dim_H(\mathrm{Vis}_e(K)) \leq s_0'$ for all $e \notin E(s_0, s_0', s_0'', \varepsilon_0, \varepsilon_1, \varepsilon_*)$.

Since $s_0, s_0', s_0'', \varepsilon_0, \varepsilon_1, \varepsilon_*$ are arbitrary rational parameters satisfying (16.3.2)–(16.3.5), we conclude that

$$\dim_H(\mathrm{Vis}_e(K)) \leq \max\{\frac{3d+3}{d+3}, \frac{(n+1)d+(n-1)}{d+2}\}$$

for Lebesgue almost every $e \in S^{n-1}$.

References

1. I. Arhosalo, E. Järvenpää, M. Järvenpää, M. Rams, P. Shmerkin. Visible parts of fractal percolation. Proc. Edinb. Math. Soc. **55**, 311–331 (2012)
2. K. Falconer, J. Fraser, The visible part of plane self-similar sets. Proc. Amer. Math. Soc. **141**, 269–278 (2013)
3. J. Hutchinson, Fractals and self-similarity. Indiana Univ. Math. J. **30**, 713–747 (1981)
4. E. Järvenpää, M. Järvenpää, P. MacManus, T. O'Neil, Visible parts and dimensions. Nonlinearity **16**, 803–818 (2003)
5. E. Järvenpää, M. Järvenpää, J. Niemelä, Transversal mappings between manifolds and non-trivial measures on visible parts. Real Anal. Exchange **30**(2), 675–687 (2004/2005)
6. Y. Lima, C. Moreira, A combinatorial proof of Marstrand's theorem for products of regular Cantor sets. Expo. Math. **29**, 231–239 (2011)
7. J. Marstrand, Some fundamental geometrical properties of plane sets of fractional dimensions. Proc. London Math. Soc. **4**, 257–302 (1954)
8. P. Mattila, Hausdorff dimension, projections, and the Fourier transform. Publ. Mat. **48**, 3–48 (2004)
9. T. Orponen, On the dimension of visible parts (2019). Preprint available at arXiv:1912.10898

LECTURE NOTES IN MATHEMATICS   ⚘ Springer

Editors in Chief: J.-M. Morel, B. Teissier;

Editorial Policy

1. Lecture Notes aim to report new developments in all areas of mathematics and their applications – quickly, informally and at a high level. Mathematical texts analysing new developments in modelling and numerical simulation are welcome.

   Manuscripts should be reasonably self-contained and rounded off. Thus they may, and often will, present not only results of the author but also related work by other people. They may be based on specialised lecture courses. Furthermore, the manuscripts should provide sufficient motivation, examples and applications. This clearly distinguishes Lecture Notes from journal articles or technical reports which normally are very concise. Articles intended for a journal but too long to be accepted by most journals, usually do not have this "lecture notes" character. For similar reasons it is unusual for doctoral theses to be accepted for the Lecture Notes series, though habilitation theses may be appropriate.

2. Besides monographs, multi-author manuscripts resulting from SUMMER SCHOOLS or similar INTENSIVE COURSES are welcome, provided their objective was held to present an active mathematical topic to an audience at the beginning or intermediate graduate level (a list of participants should be provided).

   The resulting manuscript should not be just a collection of course notes, but should require advance planning and coordination among the main lecturers. The subject matter should dictate the structure of the book. This structure should be motivated and explained in a scientific introduction, and the notation, references, index and formulation of results should be, if possible, unified by the editors. Each contribution should have an abstract and an introduction referring to the other contributions. In other words, more preparatory work must go into a multi-authored volume than simply assembling a disparate collection of papers, communicated at the event.

3. Manuscripts should be submitted either online at www.editorialmanager.com/lnm to Springer's mathematics editorial in Heidelberg, or electronically to one of the series editors. Authors should be aware that incomplete or insufficiently close-to-final manuscripts almost always result in longer refereeing times and nevertheless unclear referees' recommendations, making further refereeing of a final draft necessary. The strict minimum amount of material that will be considered should include a detailed outline describing the planned contents of each chapter, a bibliography and several sample chapters. Parallel submission of a manuscript to another publisher while under consideration for LNM is not acceptable and can lead to rejection.

4. In general, **monographs** will be sent out to at least 2 external referees for evaluation.

   A final decision to publish can be made only on the basis of the complete manuscript, however a refereeing process leading to a preliminary decision can be based on a pre-final or incomplete manuscript.

   Volume Editors of **multi-author works** are expected to arrange for the refereeing, to the usual scientific standards, of the individual contributions. If the resulting reports can be forwarded to the LNM Editorial Board, this is very helpful. If no reports are forwarded or if other questions remain unclear in respect of homogeneity etc, the series editors may wish to consult external referees for an overall evaluation of the volume.

5. Manuscripts should in general be submitted in English. Final manuscripts should contain at least 100 pages of mathematical text and should always include

   – a table of contents;
   – an informative introduction, with adequate motivation and perhaps some historical remarks: it should be accessible to a reader not intimately familiar with the topic treated;
   – a subject index: as a rule this is genuinely helpful for the reader.
   – For evaluation purposes, manuscripts should be submitted as pdf files.

6. Careful preparation of the manuscripts will help keep production time short besides ensuring satisfactory appearance of the finished book in print and online. After acceptance of the manuscript authors will be asked to prepare the final LaTeX source files (see LaTeX templates online: https://www.springer.com/gb/authors-editors/book-authors-editors/manuscriptpreparation/5636) plus the corresponding pdf- or zipped ps-file. The LaTeX source files are essential for producing the full-text online version of the book, see http://link.springer.com/bookseries/304 for the existing online volumes of LNM). The technical production of a Lecture Notes volume takes approximately 12 weeks. Additional instructions, if necessary, are available on request from lnm@springer.com.

7. Authors receive a total of 30 free copies of their volume and free access to their book on SpringerLink, but no royalties. They are entitled to a discount of 33.3 % on the price of Springer books purchased for their personal use, if ordering directly from Springer.

8. Commitment to publish is made by a *Publishing Agreement*; contributing authors of multiauthor books are requested to sign a *Consent to Publish form*. Springer-Verlag registers the copyright for each volume. Authors are free to reuse material contained in their LNM volumes in later publications: a brief written (or e-mail) request for formal permission is sufficient.

Addresses:
Professor Jean-Michel Morel, CMLA, École Normale Supérieure de Cachan, France
E-mail: moreljeanmichel@gmail.com

Professor Bernard Teissier, Equipe Géométrie et Dynamique,
Institut de Mathématiques de Jussieu – Paris Rive Gauche, Paris, France
E-mail: bernard.teissier@imj-prg.fr

Springer: Ute McCrory, Mathematics, Heidelberg, Germany,
E-mail: lnm@springer.com

Printed in the United States
by Baker & Taylor Publisher Services